Evolutionary Relationships among Rodents

A Multidisciplinary Analysis

NATO ASI Series

Advanced Science Institutes Series

A series presenting the results of activities sponsored by the NATO Science Committee, which aims at the dissemination of advanced scientific and technological knowledge, with a view to strengthening links between scientific communities.

The series is published by an international board of publishers in conjunction with the NATO Scientific Affairs Division

A Life Sciences **B Physics**	Plenum Publishing Corporation New York and London
C Mathematical and Physical Sciences	D. Reidel Publishing Company Dordrecht, Boston, and Lancaster
D Behavioral and Social Sciences **E Engineering and Materials Sciences**	Martinus Nijhoff Publishers The Hague, Boston, and Lancaster
F Computer and Systems Sciences **G Ecological Sciences**	Springer-Verlag Berlin, Heidelberg, New York, and Tokyo

Recent Volumes in this Series

Volume 86—Wheat Growth and Modelling
edited by W. Day and R. K. Atkin

Volume 87—Industrial Aspects of Biochemistry and Genetics
edited by N. Gurdal Alaeddinoglu, Arnold L. Demain, and Giancarlo Lancini

Volume 88—Radiolabeled Cellular Blood Elements
edited by M. L. Thakur

Volume 89—Sensory Perception and Transduction in Aneural Organisms
edited by Giuliano Colombetti, Francesco Lenci, and Pill-Soon Song

Volume 90—Liver, Nutrition, and Bile Acids
edited by G. Galli and E. Bosisio

Volume 91—Recent Advances in Biological Membrane Studies: Structure and Biogenesis, Oxidation and Energetics
edited by Lester Packer

Volume 92—Evolutionary Relationships among Rodents: A Multidisciplinary Analysis
edited by W. Patrick Luckett and Jean-Louis Hartenberger

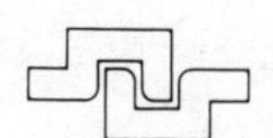

Series A: Life Sciences

Evolutionary Relationships among Rodents

A Multidisciplinary Analysis

Edited by

W. Patrick Luckett
School of Medicine
University of Puerto Rico
San Juan, Puerto Rico

and

Jean-Louis Hartenberger
Institute of Evolutionary Sciences
University of Montpellier II
Montpellier, France

Plenum Press
New York and London
Published in cooperation with NATO Scientific Affairs Division

Proceedings of a NATO Advanced Research Workshop on
Multidisciplinary Analysis of Evolutionary Relationships among Rodents,
held July 2-6, 1984, under the auspices of NATO and Centre
National de la Recherche Scientifique in Paris, France

Library of Congress Cataloging in Publication Data

NATO Advanced Research Workshop on Multidisciplinary Analysis of Evolutionary Relationships among Rodents (1984: Paris, France)
Evolutionary relationships among rodents.

(NATO ASI series. Series A, Life sciences; v. 92)
"Proceedings of a NATO Advanced Research Workshop on Multidisciplinary Analysis of Evolutionary Relationships among Rodents held July 2-6, 1984, in Paris, France"—T.p. verso.
"Published in cooperation with NATO Scientific Affairs Division."
Bibliography: p.
Includes index.
1. Rodents—Evolution—Congresses. 2. Mammals—Evolution—Congresses. I. Luckett, W. Patrick (Winter Patrick) II. Hartenberger, Jean-Louis. III. North Atlantic Treaty Organization. Scientific Affairs Division. IV. Title. V. Series.
QL737.R6N25 1984 599.32′30438 85-9480
ISBN 0-306-42061-9

A Division of Plenum Publishing Corporation
233 Spring Street, New York, N.Y. 10013

Printed in the United States of America

For more than 40 years, contributions to the knowledge of rodent evolution by Professors René Lavocat and Albert E. Wood have dominated and enriched this scientific field. All of the speakers and discussants at the Paris Rodent Evolution Symposium have had the opportunity to read papers by these investigators, and to exchange ideas with them. In addition, their friendship and warmth are well known to all of us. Thus, it is with great pleasure and respect that we dedicate this volume to these two outstanding scientists and friends.

René Lavocat

Albert E. Wood

PREFACE

The order Rodentia is the most abundant and successful group of mammals, and it has been a focal point of attention for comparative and evolutionary biologists for many years. In addition, rodents are the most commonly used experimental mammals for biomedical research, and they have played a central role in investigations of the genetic and molecular mechanisms of speciation in mammals. During recent decades, a tremendous amount of new data from various aspects of the biology of living and fossil rodents has been accumulated by specialists from different disciplines, ranging from molecular biology to paleontology. Paradoxically, our understanding of the possible evolutionary relationships among different rodent families, as well as the possible affinities of rodents with other eutherian mammals, has not kept pace with this information "explosion." This abundance of new biological data has not been incorporated into a broad synthesis of rodent phylogeny, in part because of the difficulty for any single student of rodent evolution to evaluate the phylogenetic significance of new findings from such diverse disciplines as paleontology, embryology, comparative anatomy, molecular biology, and cytogenetics.

The origin and subsequent radiation of the order Rodentia were based primarily on the acquisition of a key character complex: specializations of the incisors, cheek teeth, and associated musculoskeletal features of the jaws and skull for gnawing and chewing. Consequently, the most commonly used characters for studying rodent systematics have been components of a single functional system - the _masticatory apparatus_. Moreover, it is commonly believed that major changes in the teeth and skull for increased efficiency of the masticatory apparatus have evolved in parallel several times within unrelated higher taxonomic groups of rodents. This acceptance of widespread parallelism has inhibited assessments of phylogenetic relationships among rodent higher taxa during the past 30 years. Indeed, it is commonly believed that the traditionally defined major groups of rodents (Sciuromorpha, Myomorpha, Hystricomorpha, and Protrogomorpha) represent _grade_ levels of organization, rather than monophyletic clades, and that many of the shared similarities of each grade (such as hystricomorphy) have been attained independently by some of its members.

Several factors have contributed to the disagreements that surround analyses of rodent evolution. (1) The diversity of the group is a consequence of its extensive evolutionary radiation during Eocene-Oligocene times, and the fossil record for these epochs remains incomplete in many important aspects. (2) The extensive number of species and families of rodents makes it difficult to study and evaluate any character complex in most or all rodent taxa. (3) Much of the study of rodent systematics has focused on a search for key characters that can be used in classifications, rather than an assessment of the biological basis for shared similarities among different groups. (4) When evaluating phylogenetic relationships among rodents, many investigators have failed to distinguish among shared primitive retentions, shared derived traits, and shared similarities that may be due to parallel or convergent evolution. (5) Emphasis has been placed on the discovery of ancestor-descendant relationships during rodent phylogeny, and less attention has been devoted to the assessment of sister-group relationships among taxa. Hypotheses of ancestor-descendant relationship are restricted to studies of the incomplete and fragmentary fossil record, whereas assessments of sister-group relationships can utilize observations from both fossil and living forms.

In recent years, there has been a renewed interest in the evolutionary analysis of mammalian relationships, related in part to an increase in the kinds of data that are being evaluated (soft anatomical, embryological, molecular, and chromosomal, as well as the more traditional dental and cranioskeletal features). This resurgence has also profited by the increased emphasis and consideration given to problems of character analysis and the assessment of homology and convergence. As yet, most of this "new" (non-dental or cranial) data has been presented in isolation from other character analyses; consequently, we have evolutionary "trees" or classifications based on molecular, chromosomal, or dental traits, with little or no attempt to integrate these findings into a broader assessment of rodent (or mammalian) phylogeny.

As an initial step toward evaluating and integrating a vast amount of biological data into an analysis of evolutionary relationships among rodents, we organized a NATO Advanced Research Workshop that was held in Paris, France during July 2-6, 1984. This symposium provided a forum for a broad exchange of views among specialists from the different disciplines of paleontology, comparative anatomy, embryology, molecular biology, and cytogenetics, and this volume is the proceedings of that symposium. Participants were asked to discuss their methodology of phylogenetic reconstruction, and, in particular, their criteria for evaluating homology, convergence, and the relatively primitive or derived nature of shared similarities used in evolutionary analyses. The focal point of the symposium was on character analysis and phylogenetic reconstruction from a wide range of data, and the use of these data to test hypotheses of ancestor-descendant and cladistic relationships, both within the order Rodentia and between rodents and other eutherian orders. The initial portion of the volume is devoted to an assessment of the eutherian affinities of the Rodentia, and later sections focus on specific questions of intraordinal relationships at the family level or above.

We have concentrated on the multidisciplinary analyses of evolutionary relationships among rodents, and we have avoided the temptation of proposing new classifications of rodents, based on our preliminary analyses. Several hypotheses of higher taxon phylogeny were corroborated by analyses in this volume, including the monophyly of the Hystricognathi and Muroidea. Other problems of systematic relationships remain to be clarified, and it is our hope that future investigations will attempt to resolve these controversies by multidisciplinary analyses. If future research on rodent evolution is accomplished by collaborative efforts between paleontologists and neontologists, then one of the main goals of this symposium and volume will have been attained.

The editors are indebted to the following organizations for their financial support for the Paris Symposium, and for the subsequent preparation of this volume: North Atlantic Treaty Organization (Scientific Affairs Division); Centre National de la Recherche Scientifique (secteurs Sciences de la Vie et Terre, Océan, Atmosphère, Espace); Ministère de l'Industrie et de la Recherche (Fonds d'aide à la recherche No. 84.C.0431); and the National Science Foundation. We wish to offer our special thanks to Madame Christiane Denys for her administrative assistance during the organization of the Paris conference. We acknowledge and appreciate the scientific advice of our colleagues Drs. Jean Chaline, René Lavocat, and John Wahlert during the initial phases of organizing the symposium. Technical assistance for the symposium and volume was provided by Madame Geneviéve Jean, Monsieur Serge Legendre, and Mademoiselle Elisabeth Natale in Montpellier, and Señora Ivette Meléndez in San Juan. Various aspects of editorial assistance in preparing this volume were furnished by Ms. Pat Vann and Mr. John Matzka at Plenum Press, and we thank them for their efforts.

A special debt of gratitude is owed to Dr. Nancy Hong for her extensive help in all phases of the organization of the Paris Symposium, as well as for her aid in the preparation and editing of this volume. This project could not have been produced without her tireless and unselfish assistance.

W. P. Luckett
J.-L. Hartenberger

CONTENTS

ORIGIN AND EUTHERIAN AFFINITIES OF RODENTIA

PHYLETIC RELATIONSHIPS AMONG RODENT HIGHER CATEGORIES

SPECIFIC PROBLEMS OF INTRAORDINAL RELATIONSHIPS

CONCLUDING REMARKS

THE ORDER RODENTIA: MAJOR QUESTIONS ON THEIR EVOLUTIONARY ORIGIN, RELATIONSHIPS AND SUPRAFAMILIAL SYSTEMATICS

Jean-Louis Hartenberger

Institut des Sciences de l'Evolution
LA 327, U. S. T. L., Place Eugène Bataillon
34060 Montpellier, France

INTRODUCTION

During the last 55 million years of mammalian evolution, the order Rodentia has had an increasing success in terrestrial mammalian communities. No less than 330 to 400 genera, and 1800 to 2300 species, have been recorded in the Recent fauna, and the last number represents half of all species of mammals. All continents and all islands, including the smallest, are colonized by these small animals, sometimes with the aid of man himself. In deserts, mountains, rivers and cities, rodents are joining man in many of his activities, including exploitation of the earth.

Beginning this introductory presentation, I will discuss first some practical aspects of rodent studies. All authors who participated in the Symposium of Paris (July, 1984), and subsequently prepared a chapter for this book, are involved in fundamental scientific disciplines related to rodents, but we should not forget the implications and applications of these studies to applied sciences, and more generally, to everyday life. With an increasing human population, all countries, especially the less developed ones, need more food. In gigantic urban communities, health problems are more and more difficult to solve. In both cases, rodents can be considered as obstacles to human development. From another point of view, rodents play a central role as biological subjects for scientific studies, and these small animals have made major contributions to solving some of the problems of man and modern societies. A brief review of these two different aspects of rodent influence on human activities will serve to illustrate the point.

Five species of rodents live in close association with man

(Rattus rattus, R. norvegicus, Mus musculus, Bandicota bengalensis, and Mastomys natalensis), damaging stored foods and other materials and spreading diseases. These five overpopulated urban species are difficult to limit with low cost, and even if it would be possible to eradicate them from the earth's surface, others among the 2000 Recent species of rodents would soon replace our commensal "friends". One aspect about rodent populations known for a long time, but not well understood, relates to their demography; numerous species show periodic population explosions, with evident consequences on crops, vegetation, and spread of disease. With modern agriculture, risks of outbreak are increasing, and biocontrol of rodent populations is a necessity. While Northern countries have developed different strategies against these predators of human resources, Southern economies are often effected considerably by this problem, mainly because we are often ignorant about which particular species of rodents have direct influence on agriculture. Everyone is aware of the waste made by the introduction of the rabbit (Oryctolagus) into Australia. For a single rodent example, I will cite Prakash (1975), who estimated that existing densities of Meriones hurrianae (477 animals per hectare) in Rajastan utilized virtually all plant growth. When a determined effort is made, strategies against rodents have surprising results; in Malaysia, limitation of rodent populations showed an increase in rice crops of about 200% (Wood, 1971).

Clearly, rodents have had a direct influence on many human activities. This is evident in their negative influence on the growing and storage of crops, pastures, disruption of irrigation systems by burrowing, and in the diffusion of diseases in domestic animals and in wildlife. Hundreds of species of rodents are reservoirs of human disease. The origin of these diseases may be bacterial, viral, rickettsial, protozoan, fungal, or spirochetal. The list of diseases where rodents are implicated is very long (for an overview, see Arata, 1975), and I will cite here only the well known plague. If Rattus in Europe, Meriones in Iran, Tatera in India, and Rhambdomys in Africa are considered as the main reservoirs of plague, with fleas as vectors, no less than 200 species of rodents from all major families (Cricetidae, Muridae, Sciuridae, Dipodidae, Caviidae, Echimyidae, and Heteromyidae), as well as the lagomorphs Sylvilagus and Lepus, were recognized as naturally infected.

Another perspective allows us to consider rodents from a less negative viewpoint, because they have also served as collaborators in scientific studies that are of special interest for man. For a long time they have been the major laboratory mammals for biomedical research, and an extensive amount of knowledge on the anatomy and physiology of vertebrates was obtained from this group. Modern developments in genetics and molecular biology were recently emphasized in a French newspaper by this sentence: "Les biotechnologies vont nous permettre de fabriquer des souris grosses comme des autobus!" Reality is probably less exciting, but this brief introduction serves to illustrate in a few words some of the practical aspects of our systematic studies on rodents.

With this book our aim is to present an overview of rodent evolutionary relationships and the possible origins of the order, looking at it with our eyes as paleontologists, embryologists, anatomists, molecular biologists, or geneticists. First, we will take a brief glance back at the conclusions reached by our predecessors in the study of evolutionary relationships of these animals with each other, and with other mammals. My purpose will be to point out which characters have been used up to the present by systematists, and to evaluate the significance of the characters used. Taxonomy is a language spoken in many specialized fields of investigation; thus, it is important to appreciate how taxonomy and systematics have been built and with which materials. The first steps of Rodentia systematics were based on zoological or paleontological data, and my task will be to provide insight to other biological specialists about these methods, and also about the criticisms that we can direct toward them. To proceed as briefly as possible, I will consider the following major points: what are rodents; what is a rodent; the problem of the Glires concept; review of the different hypotheses proposed concerning the origin of rodents, different approaches used for assessing affinities, and I will conclude with some reflections about the choice of characters for evaluating the relationships among families of rodents, as viewed by a paleontologist.

WHAT ARE RODENTS?

Diversified, small, terrestrial mammals, specialized for gnawing, and with a high rate of reproduction, are the most commonly used terms to describe rodents. We note that, of about 32 extant rodent families, 12 are monogeneric (Fig. 1). The families Muridae, Cricetidae (including arvicolines and gerbillines), Sciuridae, and Echimyidae, the only families with more than 10 genera, comprise 75% of all genera and 80% of all species of rodents. The great diversity of rodent species has received different explanations. For instance, Beintema (this volume) has raised the possibility that some rapidly evolving proteins may have facilitated quick metabolic adaptations of murid species to diverse environments.

If there is good agreement among investigators concerning the familial grouping of rodent genera and species, their arrangement at higher levels and relationships among higher taxa have been subject to numerous discussions, since the time of Brandt (1855). Reviewing the different systematic positions, Simpson (1945) divided them into four schools: (1) the explosive, like Winge (1924), with little or no effort at higher groupings; (2) the wastebasket (Miller and Gidley, 1918); (3) the orthodox (Tullberg, 1899); and (4) the

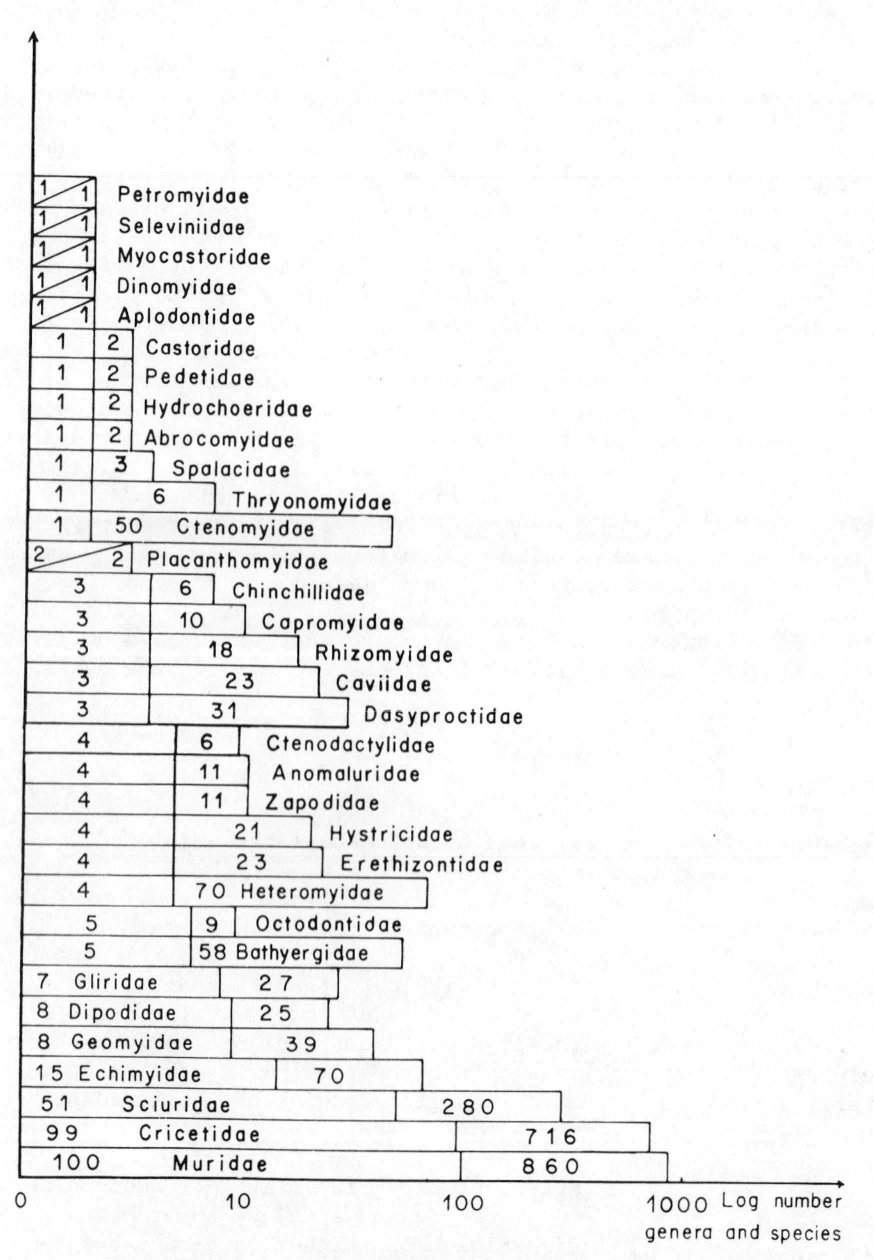

Fig. 1. Diversity of genera (first number) and species (second number) in Recent families of rodents.

eclectic (like Wood, 1937). Following Romer (1968), it is possible to add a fifth one: the *pessimist*, such as Thaler (1966), who proposed 10 suborders, but with six families *incertae sedis*; or McKenna (1975), who did not even attempt to subdivide the order! I must acknowledge that I should join this last group of workers, because my recent efforts at rodent classification, fortunately unpublished, led me to propose 16 suprafamilial taxa for the 32 Recent families!

Table 1 gives a brief review of some proposals made by several authors concerning rodent systematics. The main basis for these classifications was deduced from characters noted on the facial skull, and, more precisely, on the masticatory apparatus. Size and shape of the infraorbital foramen, attachments and development of the masseter muscle, and sciurognathy or hystricognathy of the jaw (see Wood, this volume) were the main characteristics chosen in the elaboration of rodent classifications.

The size of rodents is usually small (Fig. 2); most are between 80 to 350mm in length, with the large-bodied *Hydrochoerus* and *Castor* being exceptions. The fossil record shows the same general pattern, even though some Dinomyidae reached 400 to 600kg during the Pliocene in South America.

Table 2 illustrates the adaptations and specializations that are recognized in rodents. In fact, they are temperate to tropical terrestrial animals. If burrowing, gliding, or saltatorial adaptations have anatomical correlates, it can be observed that these adaptations are rare. Conversely, it should be noted that rodents have very diversified behaviors within the environment. Frequently, species of the same genus, although not very different from an anatomical viewpoint, may have an arboreal behavior on the one hand, while others are more terrestrial, or even semi-aquatic. This kind of eclecticism is almost the rule for rodents. In agreement with one of the participants of the Symposium (see Szalay, this volume), I must deplore the poor number of published studies on postcranial morphology for both Recent and fossil rodents. This is an important area in which a lot of work is needed.

Some data on reproduction and gestation for Recent families of rodents are provided in Table 3. In general, they can be divided into two major groups: (1) Sciurognaths with numerous young and a short duration of gestation, often with more than two litters per year; and (2) Hystricognaths with few young and a long gestation period. However, in the first group we note two important exceptions: Castoridae and Ctenodactylidae. If there is a correlation between body size and duration of gestation, it can not be proved solely by these data. We need more information about body size of the young in relation to maternal size, and also a more concise review of the problem (also see George, this volume).

Table 1. Several suprafamilial classifications of the order Rodentia. Only fossil families are named; numbers refer to extant families numbered in Tables 2-4.

Grimpeurs Fouisseurs Marcheurs	Sciuromorpha Myomorpha Hystricomorpha	Sciurognathi Sciuromorphi (1-5) Myomorphi (6-9,11-13,16,17) Hystricognathi Hystricomorphi (18,19,21-32) Bathyergomorphi (20)
BLAINVILLE, 1816	BRANDT, 1855	TULLBERG, 1899

Sciuromorpha Aplodontoidea (1, Eomyidae, Ischyromyidae, Mylagaulidae, Protoptychidae) Sciuroidea (2) Geomyoidea (3,4) Castoroidea (5, Eutypomyidae) Sciuromorpha *incertae sedis* Anomaluroidea (6) Anomaluroidea *incertae sedis* (7, Theridomyidae) Myomorpha Muroidea (8-11) Gliroidea (14-16) Dipodoidea (12,13) Hystricomorpha Hystricoidea (21) Erethizontoidea (22) Cavioidea (23-25,30, Cephalomyidae, Eocardiidae, Heptaxodontidae) Chinchilloidea (29) Octodontoidea (18,19,26-28,31, 32) Hystricomorpha *incertae sedis* Bathyergoidea (20) Hystricomorpha or Myomorpha *incertae sedis* Ctenodactyloidea (17)	Protrogomorpha Ischyromyoidea (Paramyidae, Sciuravidae, Cylindrodontidae, Protoptychidae, Ischyromyidae) Aplodontoidea (1, Mylagaulidae Caviomorpha Octodontoidea (26-28,32) Chinchilloidea (25,29,30,31, Heptaxodontidae) Cavioidea (23,24, Eocardiidae) Erethizontoidea (22) Myomorpha Muroidea (8,11) Geomyoidea (3,4, Eomyidae) Dipodoidea (12,13) Spalacoidea (9,10) Gliroidea (15,16) Clades not in suborders Castoroidea (5, Eutypomyidae) Theridomyoidea (Theridomyidae) 2,6,7,17,20,21 Thryonomyoidea (18,19, Phiomyidae)
SIMPSON, 1945	WOOD, 1965

Table 1. (Continued)

Sciurognathi	Myomorpha
Protrogomorpha	Gliroidea (16)
Ischyromyoidea (Paramyidae, Sciuravidae, Ischyromyidae, Cylindrodontidae)	Geomyoidea (3,4, Eomyidae)
Aplodontoidea (1, Mylagaulidae)	Dipodoidea (12,13)
Theridomorpha	Muroidea (8-11)
Theridomyoidea (Theridomyidae)	Hystricognathi
Anomaluroidea (6)	Franimorpha (Reithroparamyidae, Protoptychidae)
Sciuromorpha	Phiomorpha
Sciuroidea (2)	Thryonomyoidea (19, Phiomyidae, Diamantomyidae)
Castoroidea (5, Eutypomyidae, Rhyzospalacidae)	Bathyergoidea (20)
Ctenodactylomorpha	Hystricoidea (21)
Ctenodactyloidea (17, Chapattimyidae)	Caviomorpha
Incertae sedis Pedetoidea (7)	Octodontoidea (26-28)
	Erethizontoidea (22)
	Chinchilloidea (25,29-31)
	Cavioidea (23,24, Eocardiidae)

CHALINE AND MEIN, 1979

Table 4 gives an idea about the age of Recent families of rodents. Ctenodactylidae from the lower Eocene, about 55 million years ago (MYA), and Gliridae from the middle Eocene (45 MYA), are the oldest recorded. By 35 MYA, near the Eocene-Oligocene boundary, paleontologists have good ancestors or representatives for 20 of the 32 families reported here. Vianey-Liaud (this volume) provides an up-to-date synthesis of these findings.

WHAT IS A RODENT?

I must acknowledge that this is an hypocritical question, and perhaps I am the first to ask it. When studying other orders of mammals, taxonomists often have difficulties in proposing a good definition for these orders, especially one that is applicable for both fossil and extant forms, including the neontological and paleontological points of view. This problem reaches a high level of difficulty mainly when we are looking for ancestors of these groups. For example, it is difficult to provide a good definition for the order Primates, which includes early Paleocene supposed ancestors and Recent forms. Some authors (Szalay and Delson, 1979) consider *Purgatorius* as a primate, others as a nonprimate (but without proposing another relationship for *Purgatorius*). Specialists of rodents

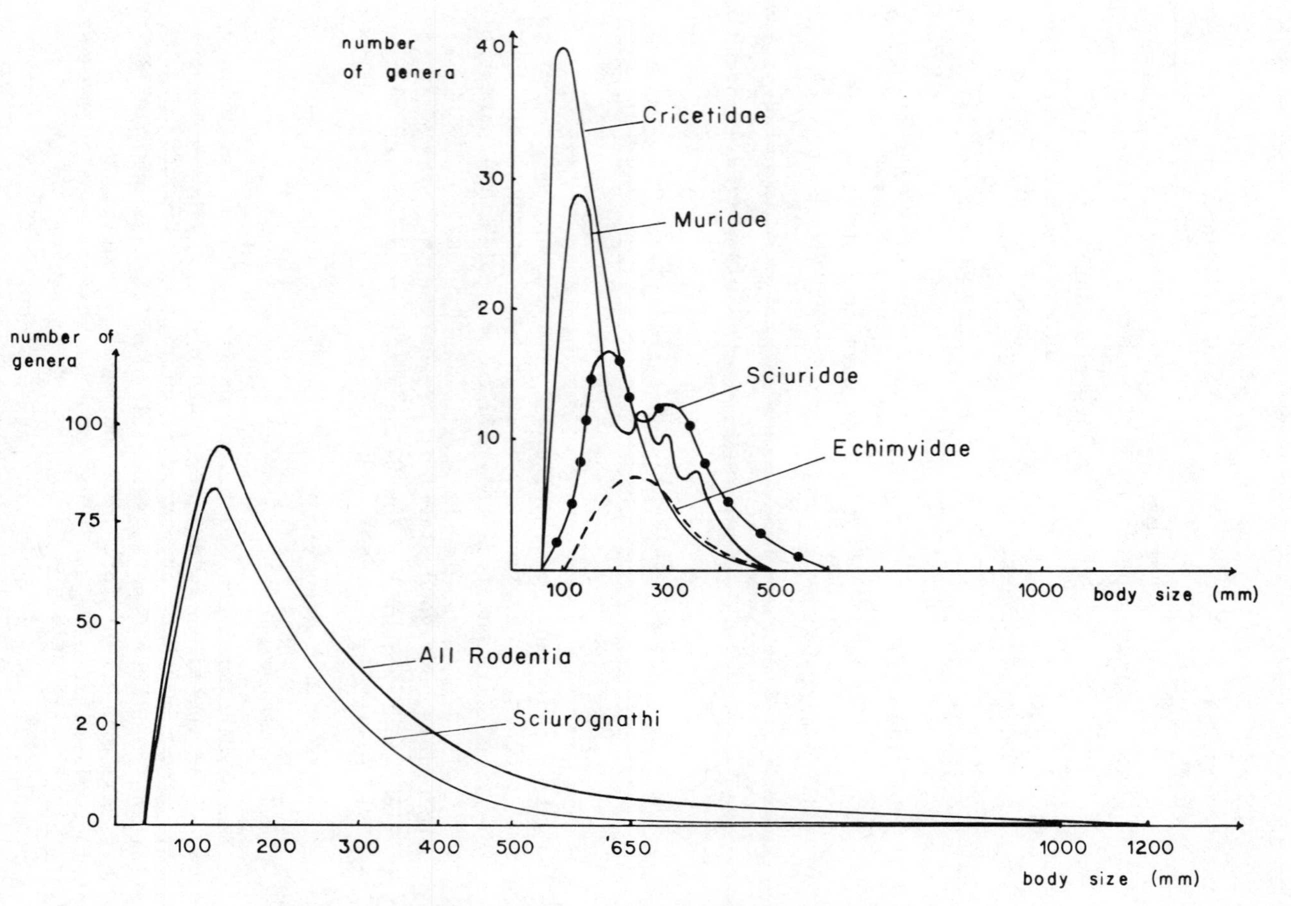

Fig. 2. Body size frequency curve for Rodentia.

Table 2. Occurrence of specialized adaptations in Recent families of rodents.

	Arboreal	Gliding	Fossorial	Saltatorial	Semiaquatic	Rabbit-rat-like	Spiny	Desertic	Arctic
1. Aplodontidae			X						
2. Sciuridae	X	X						X	
3. Geomyidae			X					X	
4. Heteromyidae				X				X	
5. Castoridae					X				
6. Anomaluridae	X	X							
7. Pedetidae				X					
8. Cricetidae	X		X	X	X		X	X	X
9. Spalacidae			X					X	
10. Rhizomyidae			X						
11. Muridae	X			X			X	X	
12. Zapodidae				X					
13. Dipodidae				X				X	
14. Platacanthomyidae	X								
15. Seleviniidae									
16. Gliridae	X								
17. Ctenodactylidae								X	
18. Petromyidae									
19. Thryonomyidae									
20. Bathyergidae			X						
21. Hystricidae							X		
22. Erethizontidae							X		
23. Caviidae									
24. Hydrochoeridae					X				
25. Dinomyidae									
26. Octodontidae									
27. Echimyidae									
28. Abrocomyidae									
29. Chinchillidae									
30. Dasyproctidae						X			
31. Myocastoridae					X				
32. Ctenomyidae			X						

do not have this kind of difficulty; when the oldest fossil rodents were found in late Paleocene (or early Eocene) beds, they were unquestionably Rodentia. The gap between rodent and nonrodent is greater than that between condylarths and artiodactyls, or between non-"insectivores" and true insectivores. Thus, if at first sight the systematic problem is simplified, the search for an ancestor is more complicated. This also has a direct influence on the manner in which rodents are studied; in fact, "Rodentology" is a natural research field for exclusive specialists, and other scientists enter this field only very cautiously!

For the purpose of this Symposium, we must consider that this taxonomic distinctness is a direct consequence of the great number of uniquely derived (autapomorphic) characters of Rodentia. These corroborate the monophyly of the group, and they can be listed as:

(1) one upper and one lower enlarged, evergrowing incisor;
(2) long diastema between incisors and premolars of both jaws, due to loss of canines and anterior premolars;
(3) incisor enamel restricted to buccal (= anterior) face;
(4) paraconid lost on lower cheek teeth;
(5) orbital cavity lying just dorsal to cheek teeth;
(6) ramus of zygoma lies anterior to first cheek teeth;
(7) glenoid fossa appears as an antero-posterior trough.

These seven characters are clearly advanced or derived eutherian features. Three more characters are often added to this list, although their advanced condition has not been demonstrated clearly:

(8) wall of orbital cavity formed mainly by frontal;
(9) optic foramina separated;
(10) orbitosphenoid small.

Specialists of cranial anatomy (see Novacek, this volume; Wahlert, this volume) have provided opinions and detailed analyses of these and other features. The fragmentary late Paleocene or early Eocene rodent fossil record does not allow us to identify all these characters on each specimen. However, rodent paleontologists, using a now qualified gradualistic approach, have for a long time recognized lineages from good middle Eocene records (often with skulls or even complete skeletons), which took root in taxa from late Paleocene beds. This deductive approach allows us to attribute rare jaws, isolated teeth, and the characteristic enlarged incisors to the order Rodentia. Such incisors, with pauciserial enamel only on their anterior face, are certainly the most frequently found and most characteristic elements for recognizing rodents in the Eocene fossil record. Does this mean that chisel-shaped incisors can be readily synonymized with rodents? The fossil record indicates that such modified incisors appeared several times independently in different orders or even subclasses of vertebrates.

The mammal-like reptiles _Bienotherium_, _Likkoelia_ and _Oligokyphus_

Table 3. Reproduction and gestation in rodents (data from Asdell, 1964; Nowak and Paradiso, 1983).

	Number of young/litter	Duration of gestation (days)	Number of litters/ year
1. Aplodontidae	2 - 6	42	1
2. Sciuridae	2 - 6	28 - 45	1 - 2
3. Geomyidae	2 - 10	18 - 19	1 - 2
4. Heteromyidae	1 - 7	24 - 33	1 - 3
5. Castoridae	2 - 4	90 - 110	1
6. Anomaluridae	1 - 4	?	2
7. Pedetidae	1	72 - 82	1 - 4
8. Cricetidae	1 - 12	16 - 40	1 - 6
9. Spalacidae	2 - 4	30	1
10. Rhizomyidae	1 - 5	21 - 45	1 - 2
11. Muridae	1 - 12	17 - 45	1 - 4
12. Zapodidae	1 - 9	17 - 23	1 - 3
13. Dipodidae	1 - 8	25 - 42	1 - 3
14. Platacanthomyidae	?	?	?
15. Seleviniidae	6 - 8	?	1 - 2
16. Gliridae	2 - 10	21 - 30	1 - 2
17. Ctenodactylidae	1 - 3	56 - 67	1
18. Petromyidae	1 - 2	?	?
19. Thryonomyidae	2 - 6	137 - 172	1 - 2
20. Bathyergidae	2 - 5	60 - 75?	1
21. Hystricidae	1 - 4	100 - 112	1 - 2
22. Erethizontidae	1 - 2	205 - 230	1
23. Caviidae	1 - 5	50 - 74	1 - 4
24. Hydrochoeridae	1 - 8	104 - 156	1
25. Dinomyidae	1 - 2	222 - 283	1
26. Octodontidae	1 - 10	90 - 109	1 - 2
27. Echimyidae	1 - 6	60 - 89	1 - 5
28. Abrocomyidae	1 - 2	115 - 118	1 - 2?
29. Chinchillidae	1 - 6	112 - 166	1 - 2
30. Dasyproctidae	1 - 4	104 - 120	1 - 2
31. Myocastoridae	1 - 9	110 - 140	2 - 3
32. Ctenomyidae	1 - 5	102 - 120	1

(Tritylodonta) from Triassic beds had large incisors separated by a long diastema from the postcanines. With a touch of irony I note that the enlarged incisors of *Bienotherium* are interpreted as upper I2 and lower I1. Because of these large incisors and other considerations, earlier authors (see Gregory, 1910) proposed to link the Triassic tritylodonts with Mesozoic multituberculates; this hypothesis was later abandoned, and tritylodonts are now considered to be without descendents.

Table 4. Oldest recognized genera of Recent families of rodents (Lower Eocene, 55 M.Y.A.; Upper Eocene, 35 M.Y.A.; Middle Miocene, 15 M.Y.A.).

Family	Genus	Age	Locality
1. Aplodontidae	Prosciurus	U. Eocene	N. America
2. Sciuridae	Palaeosciurus	L. Oligocene	France
3. Geomyidae	Diplolophus	L. Oligocene	N. America
	Nonomys		
4. Heteromyidae	Heliscomys	L. Oligocene	N. America
5. Castoridae	? Janimus	U. Eocene	N. America
	? Eutypomys	L. Oligocene	N. America
	Steneofiber	L. Oligocene	Belgium
6. Anomaluridae	Paranomalurus	M. Oligocene	E. Africa
7. Pedetidae	Megapedetes	L. Miocene	E. Africa
8. Cricetidae	Eucricetodon	U. Eocene	China
9. Spalacidae	Anomalomys	M. Miocene	Europe
10. Rhizomyidae	Kanisamys	M. Miocene	Pakistan
	Tachyoryctoides	M. Oligocene	Mongolia
11. Muridae	Antemus	M. Miocene	Pakistan
12. Zapodidae	Plesiosminthus	U. Eocene	China
13. Dipodidae	Protalactaga	M. Miocene	China
14. Platacanthomyidae			
15. Seleviniidae			
16. Gliridae	Eogliravus	M. Eocene	France
17. Ctenodactylidae	Saykanomys	L. Eocene	Kazhakstan
	Tamquammys	L. Eocene	Kazhakstan
18. Petromyidae			
19. Thryonomyidae	Paraphiomys	L. Miocene	E. Africa
(Phiomyidae)	Phiomys	L. Oligocene	Egypt
20. Bathyergidae	Bathyergoides	L. Miocene	E. Africa
21. Hystricidae	Sivacanthion	M. Miocene	Pakistan
22. Erethizontidae	Protosteiromys	L. Oligocene	Patagonia
23. Caviidae (Eocardiidae)	Chubutomys	L. Oligocene	Patagonia
24. Hydrochoeridae	Cardiatherium	Pliocene	Argentina
25. Dinomyidae	Branisamys	L. Oligocene	Bolivia
26. Octodontidae	Platypittamys	L. Oligocene	Patagonia
27. Echimyidae	Sallamys	L. Oligocene	Bolivia
28. Abrocomyidae	Protabrocoma	Pliocene	Argentina
29. Chinchillidae	Scotamys	L. Oligocene	Patagonia
30. Dasyproctidae	Cephalomys	L. Oligocene	Patagonia, Bolivia
31. Myocastoridae	Neoreomys	Miocene	Patagonia
32. Ctenomyidae	Actenomys	Pliocene	Argentina

During Mesozoic times, numerous nontherian multituberculates showed morphological convergence with rodents in possessing procumbent incisors and a diastema between the incisor and the cheek teeth. It has been suggested that, in a certain way, multituberculates occupied the niche of rodents in terrestrial communities (see Van Valen and Sloan, 1966).

Although Ameghino (1897) proposed possible phylogenetic relationships between the multituberculates and rodents, this hypothesis was rapidly abandoned. With these animals, we have a first look at a more general problem: does morphological convergence imply adaptive convergence? It has been emphasized by Bock (1977) that this problem is not easy, and a recent work by Krause (1982) has shown how we can possibly answer the question for multituberculates, with careful morphological and paleoecological analyses. In my opinion, one aspect of the question of the incredible success of the rodents during Cenozoic times can be answered by the study of Paleocene eutherian mammals. At that time, multituberculate diversity was decreasing, and numerous families of mammals had acquired large incisors as the main tool of their masticatory apparatus. These included: (1) Primates (Plesiadapidae, Carpolestidae, Paromomyidae); (2) Tillodontia (Esthonychidae); (3) "Insectivores" (Microsyopidae, Mixodectidae, Apatemyidae); (4) Mixodontia (Eurymyloidea); and (5) Rodentia.

It is also a striking fact that after the rise of rodents, enlarged incisors have developed only occasionally in nonrodent groups in "insular" conditions; for instance, *Vombatus* in the Australian marsupial community, the aye-aye *Daubentonia* in Madagascar, and *Myotragus*, an artiodactyl from the Baleares Islands . It appears as if rodents had "inhibited" the occurrence of this character in other orders of mammals during most Cenozoic periods. We can not give a clear and reasoned answer to this problem, and fortunately it is not the main question we are considering! However, this point allows me both to emphasize the importance of incisors in rodent evolutionary history, and, conversely, to indicate the limits of this important character; chisel-like incisors are not synonymous with Rodentia. The fossil record shows that this character has occurred in different groups at many times during mammalian history.

An important question needs to be cleared up: which upper and lower incisors have become enlarged and evergrowing in rodents (and lagomorphs)? When discussing this problem previously (Hartenberger, 1977), I noted that numerous authors were suggesting that for rodents they are I2/I1, and for lagomorphs I1-2/I1. Since that time, I have learned from Patrick Luckett that the question was resolved by Adloff (1898) a long time ago (see Luckett, this volume).

During the Symposium, the enamel structure of incisors was discussed at length (see the contributions of Koenigswald and Sahni, this volume), although there was disagreement as to which pattern represents the primitive rodent condition. There is only one minor point about the incisors that I will emphasize here - their size and extension in the jaw. Fossorial rodents have relatively larger incisors, probably because they use them as a pick for digging. In most rodents, the lower incisor extends distally beneath the last molar. However, in many advanced forms (murids mainly), there is a reduction of the incisors; their roots may extend only to M2 or M1. In this case, this character must be considered advanced. However, in the primitive genus Ailuravus (early Middle Eocene of Europe), Wood (1976) reported that the relatively short lower incisor only extends beneath M1. In the opinion of Wood, and I agree with him, this is a case of a primitive feature conserved by this lineage. For Lagomorpha, we note a regression of the lower incisor through time; it occurs beneath M2 in Eurymylus (Paleocene) and in the Oligocene lagomorph Palaeolagus (Wood, 1940) whereas it extends only beneath P3 in Recent leporids and P4 in ochotonids.

Determination of the polarity of this incisive character in an evolutionary analysis is not easy, if time and succession of forms are not taken into account. I have cited only this character, but examples of morphological modifications changing in both directions, with regard to time, are very frequent in rodents. This occurred particularly at the level of molar patterns. It must be added that in recent years cheek tooth studies have been the main material of paleontologists, and in fact the basis of most studies on rodent systematics. A direct consequence of these studies has been an excessive nomenclature for cusps and perhaps an overestimation of relationships at higher levels, because cusp homologies were not well established. As an example, the disagreement between Wood (1974) and Lavocat (1974) about the North American versus African origin of South American caviomorphs has as one of its bases the differing interpretations of the molar pattern of phiomyids and caviomorphs; is the middle ridge of the upper molar a mesoloph or a neoloph (also, see Butler, this volume)?

Molar nomenclature as used by many authors has only a practical purpose for descriptions and comparisons within given groups. Homologies or homoplasies are difficult to distinguish in cheek teeth of rodents. As a consequence, evolutionary analysis with only morphological considerations of cheek teeth data are of limited value (see also Table 8 and discussion concerning this problem). Molar occlusion in rodents has been rarely discussed. Butler (this volume) analyzes the evolutionary changes of molar cusp patterns, comparing early rodents to contemporaneous primitive groups.

THE GLIRES CONCEPT

In a broad sense, Glires designates small gnawing animals. It was proposed first by Linnaeus (1758) with clear taxonomic value, but gradually authors of the eighteenth century approached the problem in such a manner that serious doubt would exist about its taxonomic value. Indeed, the result of their studies was to divide Glires into two major groups: rodents and lagomorphs. Illiger (1811), Gervais (1848), and Brandt (1855) participated in the splitting of the Glires concept, but it was the monumental work of Tullberg (1899), followed by the very short but widely accepted note of Gidley (1912), that gave taxonomic expression to this idea. Table 5 lists the anatomical differences between rodents and lagomorphs taken into account by Gidley (1912), with an addendum by Wood (1957), which were the basis for the breakup of the Glires concept.

Why has the term Glires not completely disappeared from the taxonomic literature, as many others have? The roots of this conservatism have many different reasons. Ignorance of any paleontological ancestors for both groups was the reason invoked by Simpson (1945) for retaining Glires in his mammalian classification. From Mossman (1937) to Luckett (1977), embryologists have noted important similarities in the development of fetal membranes in both groups, and this was a strong argument for them to continue with the Glires concept. Functional anatomical studies like Turnbull's (1970) also emphasized similarities between the masticatory apparatus of rodents and lagomorphs. However, the main reason, in my opinion, for the persistence of Glires in the scientific literature is that Rodentology has been a protected field from the Hennigian revolution, and many investigators have uncritically followed Simpson's (1945) classification. Thus, specialists of rodents have not been engaged in an evolutionary analysis of the validity of the Glires concept. During the Symposium, in fact, it appeared that the problem was not whether to be or not to be a cladist, but rather the most important consideration in multidisciplinary discussions was how to communicate on matters of common interest. A great majority of workers in different fields of investigation, from anatomy to biochemistry, presented their results by referring to the primitiveness or derivedness of observed characters. This manner of presentation contributed greatly to mutual understanding of both methods and results. It should be added that the problem of homology is a general one, common to all disciplines, and it should not be considered as an obstacle, but rather as an opening to new and stimulating questions. By the end, it seemed to me that the Symposium provided new evidence for a revival of the Glires concept, but this is only a personal point of view (also, see Luckett, this volume; Novacek, this volume).

HYPOTHESES ABOUT THE ORIGIN OF RODENTS

As noted above, the gap between rodents and other mammals is so extensive that search for a hypothetical ancestor of rodents was never strongly attempted, and only imprecise hypotheses have been

Table 5. Main differences between Rodentia and Lagomorpha, as cited by Gidley (1912) and Wood (1957), to disprove the Glires concept.*

Lagomorpha		Rodentia
1. 4 upper incisors (+2 in young)	□ ■	1. 2 upper incisors
2. 3 functional upper premolars	□ ■	2. 1 functional upper premolar
3. 2 lower premolars	□ ■	3. 1 lower premolar
4. Broad palate (distance between upper tooth row > lower)	?	4. Narrow palate (distance between upper tooth row < lower)
5. Upper cheek teeth wider than lower	□ ■	5. Upper and lower cheek teeth equal in width
6. Glenoid fossa divided into 2 parts (lateral motion only for chewing)	□ ■	6. Glenoid fossa is a trough (lateral and anteroposterior motion for chewing)
7. Cheek tooth row in plane with ascending ramus of lower jaw	?	7. Cheek tooth row lying inside plane of ascending ramus of lower jaw
8. Caecum with spiral fold	?	8. Caecum without fold
9. Elbow joint modified, no rotary motion of the forearm	■ □	9. Elbow joint primitive, free rotary motion of the forearm
10. Fibula fused with tibia	■ ◩	10. Fibula fused or free
11. Fibula articulates with calcaneum	■ □	11. Fibula not articulated with calcaneum
12. Fenestration of lateral wall of snout	■ □	12. No fenestration of snout
13. Less complex jaw	□ ■	13. More advanced jaw
14. Fused symphysis	■ □	14. Unfused symphysis
15. Weak masseter and temporal muscles	□ ■	15. Strong masseter muscles
16. Incisive foramina large	?	16. Incisive foramina variable
17. Supraorbital processes developed	?	17. No supraorbital process
18. No os penis	■ □	18. Frequent os penis
19. Uncertain homologies of molar tooth pattern	■ □	19. Tribosphenic structure of upper molars
20. Short incisors	?	20. Long incisors
21. Enamel with one layer	□ ■	21. Enamel with 2 layers

* Symbols: Open square = more primitive condition; black square = more derived condition; half-blackened square = intermediately derived condition; ? = character state polarity uncertain.

proposed concerning the origins of rodents. In addition, two other reasons may explain the prudency and imprecision of some authors. Prior to the adoption of cladistic analysis by many workers as a research tool in the 1970's, a common method of research was to imagine what kind of strange animal looked like a protorodent. In addition, with regard to the known fossil record, as noted by Van Valen (1966), "the light from our narrow phyletic window does not yet illuminate much of the Cretaceous."

During the last 25 years, two not very different hypotheses have prevailed concerning rodent origins: (1) the primate hypothesis (McKenna, 1961; Wood, 1962, 1969; Van Valen, 1966, 1971), and (2) the palaeoryctid or eurymyloid hypothesis (Szalay and Decker, 1974; Szalay, 1977; Hartenberger, 1977, 1980). In fact, possible relationships between Rodentia and Lagomorpha were excluded in the first hypothesis, whereas they were possible in the second, with embryological arguments added by Luckett (1977).

Primate Hypothesis

At the end of his monumental review of Eocene paramyids from North America, Wood (1962) naturally speculated about the kind of mammal that could have been the ancestor of Rodentia. His approach was to imagine an animal possessing characters slightly more primitive than the known characters of Early Eocene paramyids. On this basis, Wood suggested that a protorodent would possess:

(1) Lower incisors relatively shorter (extending beneath P4);
(2) Enamel completely surrounding the incisors;
(3) Upper P3 conical, and transversely compressed;
(4) Upper P4 narrow, and protocone and metacone without crest; small anterior and posterior cingula;
(5) Upper M1 and M2 triangular, lacking a hypocone;
(6) Upper M3 less reduced than in paramyids;
(7) Lower P4 without protoconid; large metaconid; talonid small;
(8) Lower M1 and M2 comparable to those of paramyids; paraconid reduced or absent;
(9) Lower M3 with elongate talonid and strong hypoconulid.

In his conclusive reflections, Wood (1962) was influenced by the recent discovery of a good skull of the Paleocene primate _Plesiadapis_ by Russell (1959, 1964), and he concluded that his protorodent would not have been very different from plesiadapids, generally considered as primitive primates (although Wood excluded a direct phylogenetic relationship between Rodentia and _Plesiadapis_).

In my opinion, the main argumentation against this hypothesis was made by Gingerich (1976), who demonstrated that plesiadapids are not as primitive as previously considered. It appears that occurrence of common advanced characters in rodents and these primates is the consequence of heterochronic convergence. Finally, I must not forget to cite the "informal phylogeny" given by Van Valen (1966), in which by different and double pathways this author considered that his so-called group M (a wastebasket group including Pantolestidae, Mixodectidae, Leptictidae, Zalambdalestidae, Apatemyidae, Endotherium, and possibly Tupaiidae) was the ancestral group for Primates, Rodentia, and Lagomorpha, but without any documentation.

Palaeoryctoid Hypothesis

With this hypothesis, we arrive at the revival of the Glires concept, although the approaches of recent authors have been very different from earlier speculations, because they were the results of cladistic analyses. Authors did not imagine what kind of animal looked like the protorodent; instead, they analyzed the relationships of rodents with other mammals by evaluating the occurrence of advanced characters in fossil and living groups, following a more or less orthodox Hennigian approach. If I qualify these works as "more or less orthodox," I want to emphasize that all these authors have taken into account the geological time table and the occurrence of known fossil forms in this framework, their succession being considered as important data.

Most systematists are aware of the tentative classification of Mammalia by McKenna (1975), which was, if questionable, very stimulating. In his study, Rodentia were placed in the cohort Epitheria, but incertae sedis. McKenna did not consider them as possibly related to primates, and from a first draft of this paper he was inclined to put rodents in the Magnorder Ernotheria with Leptictidae, Anagalidae, Macroscelididae, and Lagomorpha. By 1977, independent and different data were used by Szalay, Luckett, and me to propose quite similar solutions to the problem of rodent origins. Szalay (1977), following previous ideas of Szalay and Decker (1974), took into account postcranial anatomy, dental morphology, and a reevaluation of the Paleocene fossil record from Asia and North America, and he proposed to unite in a cohort Glires the Leptictimorpha, Rodentia, Lagomorpha (= Duplicidentata, Macroscelidea, Anagalida, Eurymylidae), Zalambdalestidae, and Didymoconidae (Fig. 3). Just as the skull of Plesiadapis had been important in Wood's (1962) hypothesis, Asian Paleocene fossils recently studied by Szalay played a similar role in his conclusions.

Luckett (1977) evaluated the fetal membranes and placenta as a taxonomic tool, using common derived characters and reconstructing morphotypes of groups. In Table 6, I have summarized his and my own (Hartenberger, 1977) observations, although this is devoid of evaluation of additional skull characters. Figure 3 illustrates several of the phylogenetic trees proposed for relating rodents to other eutherians.

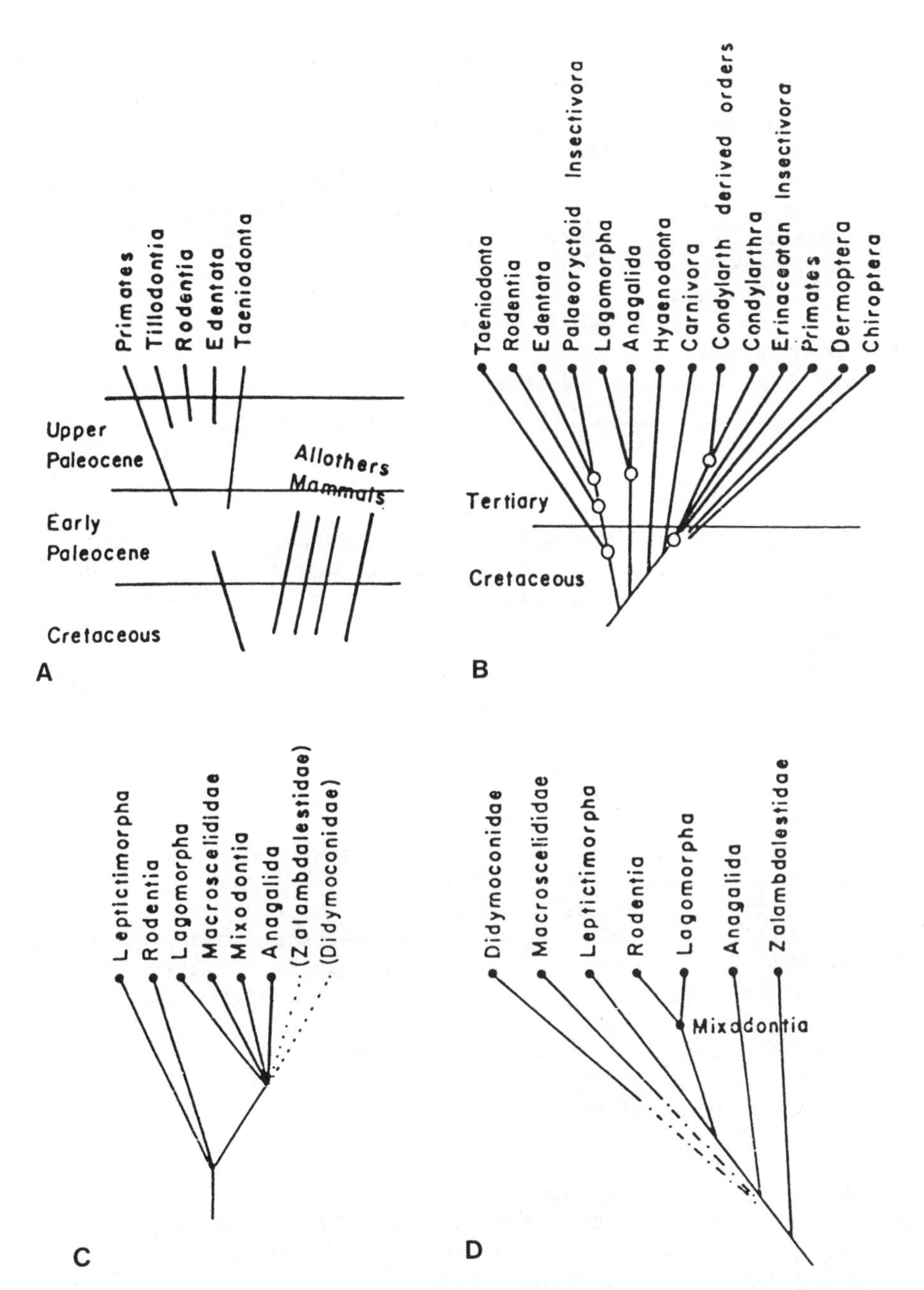

Figure 3. Proposed relationships of Rodentia with other mammals, as suggested by: (A) Matthew, in Gregory, 1910; (B) Szalay and Decker, 1974; (C) Szalay, 1977; and (D) Hartenberger, 1980.

Table 6. Synapomorphous characters noted by Luckett (1977) and Hartenberger (1977) to support the Glires concept.

1. Chorioallantoic placenta hemochorial
2. Bilaminar omphalopleure inverted or lost
3. Allantoic vesicle moderate sized or vestigial (moderate sized in Lagomorpha)
4. Loss of C/C, P1/1, P/2
5. Well developed premaxilla in contact with frontal
6. Long diastema
7. Well developed masseter muscles
8. Glenoid fossa elongated
9. Auditory bulla formed by ectotympanic

Eurymyloid Hypothesis

The discovery of *Eurymylus* and and its further study by Wood (1942) led this author to propose that it was a possible ancestor for Lagomorpha. He listed the following similarities between *Eurymylus* and Lagomorpha:

(1) Fenestration of maxilla, incipient as pitting in *Eurymylus*;
(2) Position of the anterior root of the zygoma;
(3) Position and size of the infraorbital foramen;
(4) Shape and position of the molars;
(5) Position of the anterior palatine fenestrae;
(6) Shape of the palate;
(7) Unilateral hypsodonty of the upper cheek teeth;
(8) Method of occlusion of the cheek teeth;
(9) Orientation of the lower cheek teeth;
(10) Large lingual basin in upper cheek teeth, and small buccal cusps;
(11) Possible postorbital process on the maxilla of *Eurymylus*;
(12) Pit on lateral face of maxilla above the masseteric fossa;
(13) Several mental foramina;
(14) Shape of masseteric fossa of mandible.

Some dissimilarities between the two taxa were also noted by Wood (1942):

(1) Absence of P2/ in *Eurymylus*;
(2) Greater length of lower incisor; beneath M/2 in *Eurymylus*;
(3) Anterior face of the zygoma slopes in *Eurymylus*, rather than being vertical as in lagomorphs;
(4) Pitting below infraorbital foramen in *Eurymylus*, rather than above.

One should add that at the time the number of upper incisors of *Eurymylus* was unknown, and it was only with the new material studied by Sych (1971) that it was possible to observe that *Eurymylus* has only one pair of upper incisors. Also, as noted above, Wood considered that Lagomorpha and Rodentia were two independent orders, following the conclusions of Gidley (1912).

In a later work, Wood (1957) suggested that periptychid condylarths were possible ancestors of eurymylids and lagomorphs. In the same paper, he reviewed the different hypotheses proposed for the origin of Lagomorpha. These can be summarized as follows:

(1) Marsupialia (Schlosser, 1884);
(2) Triconodonta (Gidley, 1906; Ehik, 1926);
(3) Ungulata (Gidley, 1912);
(4) Cainotheriid Artiodactyla (Hürzeler, 1936);
(5) Artiodactyla (Moody et al., 1949);
(6) Periptychid Condylarthra (Wood, 1957);
(7) Zalambdodont Insectivora (Russell, 1959);
(8) Hyopsodont Condylarthra (Van Valen, 1966).

Lopez Martinez (this volume) presents an extensive overview of the evolutionary history of the Lagomorpha, and I want only to recall here that new specimens of *Eurymylus* (Sych, 1971) and several other eurymyloids (Li, 1977; Li et al., 1979; Zhai, 1978) have been found since the time of Wood's (1942, 1957) analyses. New comments and hypotheses about these eurymyloids have been made, mainly by the authors of the first descriptions, but also by other investigators (Hartenberger, 1980; Patterson and Wood, 1982; Korth, 1984). Some of us had the opportunity to discuss these fossils at a meeting on rodent evolution in Pittsburgh, Pennsylvania during 1982*. Two years later at our Paris Symposium, many more fossil eurymyloids were available, and Li and Ting (this volume) have presented a survey of numerous cranial and dental features of these fossils and have discussed their possible affinities with other early Tertiary eutherians. In my opinion, one of the most striking facts about these abundant new finds is the indication that the Asiatic eurymyloids were very diversified at the generic and species levels during Paleocene and early Eocene times.

To facilitate discussions and comparisons, some of the primitive and derived characters that can be observed in several genera of eurymyloids and early rodents are listed in Table 7, and the distribution of these character states is illustrated in Figure 4. At first, it may appear from the figure that differences between *Heomys* and *Eurymylus* on the one hand, and *Paramys* and *Tamquammys* on the

---

* Sponsored by the Carnegie Museum of Natural History. Participants of this rodent symposium were: C. C. Black, M. R. Dawson, J.-L. Hartenberger, W. W. Korth, Li Chuan Kuei, A. Sahni, J. H. Wahlert, R. W. Wilson, A. E. Wood, and C. A. Woods.

Table 7. Character analysis of selected traits within primitive rodents and eurymyloids (see Fig. 4).

Primitive	Derived
1. Double mental foramina	1. Single mental foramen
2. Masseteric fossa beneath M/3	2. Masseteric fossa anteriorly shifted
3. Mandibular symphysis of insectivore type	3. Mandibular symphysis of rodent type
4. Coronoid apophysis well developed	4. Coronoid apophysis less developed
5. Angular process bent	5. Angular process parallel to tooth row
6. Lower diastema as long as upper diastema	6. Lower diastema shorter than upper diastema
7. Nasals short	7. Nasals extend posteriorly
8. Weak height of jaw beneath M/1 and M/2	8. High jaw beneath M/1 and M/2
9. Anterior root of zygoma even with M1/	9. Anterior root of zygoma more anteriorly situated
10. Broad palate	10. Narrow palate
11. Curved upper tooth row	11. Straight upper tooth row
12. Minute infraorbital foramen	12. Small infraorbital foramen
13. P2/ present	13. P2/ absent
14. P/3 present	14. P/3 absent
15. Bicuspidate trigone on P4/	15. Tricuspidate trigone on P4/
16. Hypocone present on P4/	16. Hypocone reduced on P4/
17. P/4 tuberculosectorial	17. P/4 not tuberculosectorial
18. P/4 short	18. P/4 as long as M/1
19. M1/ and M2/ transversely elongated	19. M1/ and M2/ not transversely elongated
20. Molar hypocone separated from protocone and as important as protocone	20. Molar hypocone reduced
21. Anterior cingulum does not reach molar protocone	21. Anterior cingulum reaches molar protocone
22. Narrow V-shaped molar trigone	22. Large molar trigone
23. M1/ larger than M2/	23. M1/ and M2/ equal sized
24. M3/ very reduced	24. M3/ less reduced
25. M/3 elongated	25. M/3 less elongated
26. Paraconid present on P/4	26. Paraconid absent on P/4
27. Molar hypoconulid important	27. Molar hypoconulid reduced

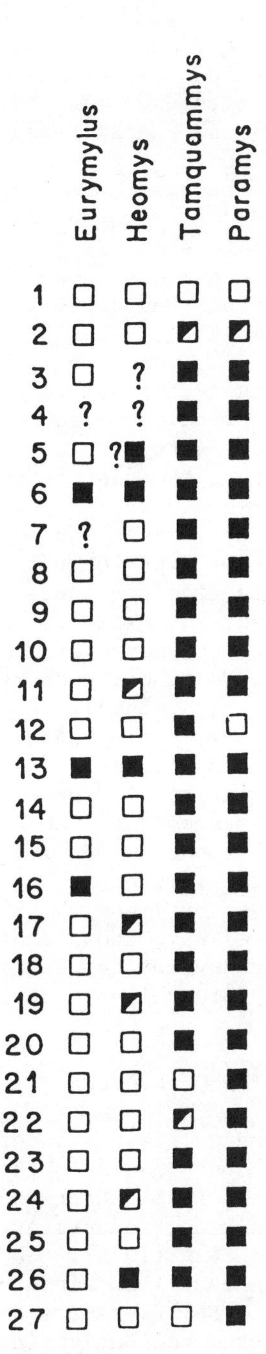

Fig. 4. Character analysis of some features observed within primitive rodents and eurymyloids (see also Table 7). Character symbols as in Table 5.

other, are extensive. However, this may be exaggerated; choice of characters has a direct influence in emphasizing differences, rather than pointing out similarities (see also Li and Ting, this volume). Nevertheless, it is clear that the differences between the two groups are important. Although it is useful to list the primitive and derived characters of the masticatory apparatus in lagomorphs and rodents, it is also important to understand how this apparatus functions in both groups. Thus, it is important to consider the question: is it possible to envision that the masticatory apparatus of both rodents and lagomorphs was derived from a common eurymyloid ancestor?

If we want to investigate the question of possible relationships between Rodentia and Lagomorpha, we need not only comparative anatomical analyses, but also points of view from other scientific fields. Thus, in introducing the discussion about the Glires concept, I have focused on paleontological and anatomical approaches. Nevertheless, many of our colleagues who participate in this volume have drawn attention to other types of biological data that can be useful for drawing phylogenetic inferences. Some of these data are biological in a broad sense; for example, many workers have long noted that caecotrophy was a common character of lagomorphs and herbivorous rodents. Other approaches, such as embryological methods and results (see Luckett, this volume), were ignored or overlooked by many paleontologists and neontologists. In recent years, observations at the cellular and molecular level have opened new fields of investigation for systematic relationships. In particular, biochemists have developed new tools and new methods for analyzing a variety of proteins. At the beginning, it was natural that Primates would be their main subject of interest. More recently, however, our colleagues Beintema, De Jong, Sarich, Shoshani and Goodman (see their papers in this volume) have shown that rodents and other mammals are also valuable for developing methodological approaches to biochemical analyses, and for evaluating evolutionary relationships.

RELATIONSHIPS AMONG FAMILIES OF RODENTS

As noted at the beginning of this paper (see Table 1), suprafamilial systematics of rodents has been the subject of numerous interpretations. One of the most recent and general reviews of evolutionary relationships among rodents at the suprafamilial level is the book by Chaline and Mein (1979). They emphasized "à juste titre" the most important results obtained by paleontologists in evaluating the phylogenetic relationships among rodents during the past 20 years. Following the foundation laid down by their great predecessors Drs. Lavocat and Wood, many workers have focused their efforts during this period on different groups of rodents (see the extensive bibliography in Chaline and Mein, 1979). A general remark can be made about these approaches: all are gradualistic in the sense that investigators have tried to establish ancestor-descendant lineages, establishing them step by step within a geological framework, and assessing phylogenetic and systematic relationships. These classical studies were very useful, followed a heuristic approach, and provided important biochronological results, as well as stimulating debates.

Many authors have ascertained from their studies that the evolution of rodents has shown a tremendous amount of convergence and parallelism (for discussion, see Wood, this volume). This reality has had poorly appreciated and even negative consequences for the authors themselves; because they believe homologous characters to be rare, homoplastic or analogous structures have been used extensively in the establishment of rodent systematics. An illustration of this perplexing problem is presented in Figure 5 and Table 8, which include some of the most commonly used characters by paleontologists and the distribution of these features in Recent families of rodents. With the exception of characters 7 and 21, and less probably 2 and 3, all of these occur as homoplastic characters in at least some instances; thus, they can not be used unequivocally to obtain systematic results, or at least to resolve some interesting questions about phylogenetic relationships among rodent families. Instead, we have only an indication of the level of complexity which characterizes these taxa.

Are homoplasies more frequent in rodents than in other mammals? I really do not know, but, if it is the case, Rodentology, more than any other taxonomic group, needs the collaboration of different research fields, from morphology to biochemistry, in order to determine evolutionary relationships. And, if I may summarize my point of view as a paleontologist, I would say that paleontology has many good questions to ask, but it is unable to provide answers by itself.

To conclude this introductory presentation, I wish to propose some hypotheses about rodent suprafamilial groupings as viewed by a paleontologist (Fig. 6 and Table 9). Some of us who participated in the Pittsburgh meeting will recognize some of the conclusions we reached during that meeting, although we did not share all of them! What I want to show in Figure 6 is mainly that chronology of occurrence of characters, as reflected by the fossil record, is important for paleontologists!

Group I (Ctenodactyloidea + Geomyoidea) is here considered as the earliest group of rodents to differentiate and probably also the most primitive. At first glance, this proposition would appear to be based mainly on the presence of several primitive characters in early ctenodactyloids, such as Tamquammys and Saykanomys. The Asian fossil record shows that the Heomys-Eurymylus group tooth pattern is not very different from that of Tamquammys-Saykanomys. This remark was made by some participants at the Pittsburgh meeting, and

Table 8. Occurrence of selected derived characters in Recent families of rodents (see Fig. 5).

1.	Short lower incisor	13.	Hypsodonty of molars
2.	Uniserial enamel	14.	Selenodonty
3.	Multiserial enamel	15.	Lophodonty
4.	Premolars lost	16.	Mesostyle
5.	P4/ reduced	17.	Mesoloph
6.	P4/ enlarged	18.	Neoloph
7.	Retention of dP4/	19.	High ectoloph (Langsträt)
8.	Enlarged dP4/	20.	Cement on molar crown
9.	M3/ reduced	21.	Hystricognathy
10.	M3/ enlarged	22.	Coronoid process low
11.	Homeodonty of molars	23.	Single masseteric crest
12.	High crowned molars	24.	Angular process developed

Korth (1984) has pointed out the following common characters:

(1) P4/ with single buccal cusp (paracone) and distinct hypocone;
(2) Upper molars with well developed, lingually situated hypocone; metaloph anterolingually situated;
(3) P/4 with small talonid;
(4) Lower molars with large hypoconulid.

I should add that I am impressed by the fact that the upper P4 of Cocomys lingchaensis is very similar to the upper P4 of Heomys (Dawson et al., 1984). It is possible that all these features of early ctenodactyloids are primitive, but this has not been demonstrated. At this moment, my point of view is that ctenodactyloids are the stem group of all other rodents.

The problem of the early differentiation of groups of rodents is difficult, because each shows only autapomorphic characters, and clear cladistic analysis is impossible! In addition, the number of characters available for analysis is low. In my opinion, an interesting way to analyze the evolutionary relationships of primitive rodents, where convergence and parallelism are more difficult to clarify, would be to perfect a sequence of successive cladistic analyses of contemporaneous taxa. In this way, we would obtain an early Eocene step (or cladogram), a middle Eocene step, etc. Up to now, the fossil record is too poor to attempt such an analysis. Usually, only teeth are available, and there is a scarcity of cranial and postcranial remains. Also, as noted above, synapomorphies are rare, and autapomorphies are more abundant, for the poor number of available characters. It is difficult to determine whether evolution occurred rapidly during the early radiation of rodents (illustrating a so-called punctuated model), or if this is only an artifact

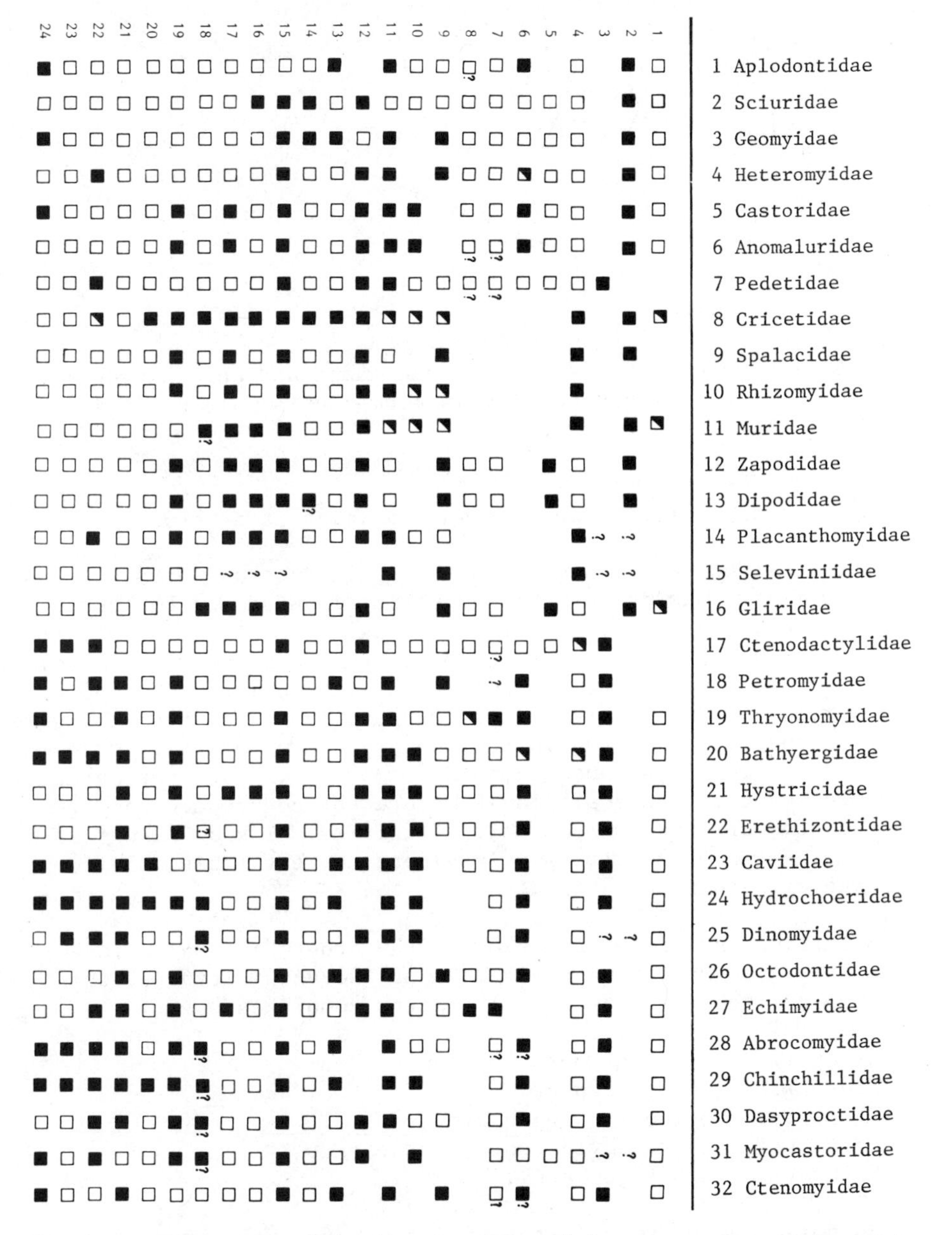

Fig. 5. Character analysis of selected features observed within families of rodents (see also Table 8).

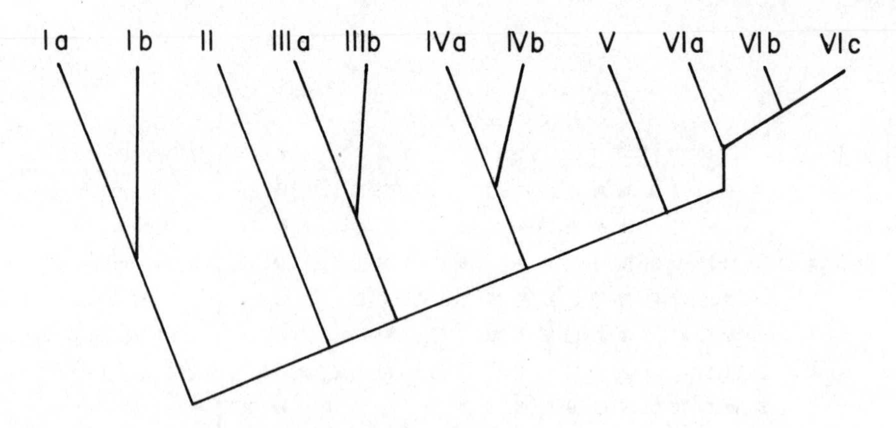

Figure 6. Proposed suprafamilial arrangement and chronological branching in Rodentia, as viewed by a paleontologist (see Table 9 for taxa).

of both the fossil record and the way in which it is studied.

The other comment I have about Group I concerns Geomyoidea. The evolutionary history of this group is discussed by Fahlbusch (this volume), and I only want to point out one aspect, namely its origin. Eocene sciuravids were probably the ancestors of geomyoids, and, in my opinion, it seems possible that the Sciuravidae and Ctenodactyloidea are related. This suggestion arises from the similarities that I have noted in the tooth patterns of early ctenodactyloids, sciuravids, and geomyoids. However, I can not determine at present whether these similarities are the consequence of primitiveness, convergence, or possible affinities. I know also that my colleagues Flynn et al. (this volume) and Luckett (this volume) do not agree with this suggestion!

Splitting and differentiation of the other five major rodent groups appear to have been the result of the acquisition of the following derived characters or character complexes:

Group II: gliroid tooth pattern;
Group III: sciuromorphy;
Group IV: cricetine tooth pattern;
Group V: theridomyid tooth pattern;
Group VI: hystricognathy.

As noted above, this is the viewpoint of a paleontologist, and the emergence of the five groups corresponds roughly to the chronological occurrence of the innovations represented by these five character complexes. I should also acknowledge that monophyly of Groups I, II, III, IV, and V, and their patterns as suggested in

Table 9. Relationships and suprafamilial systematics of Recent and fossil families of rodents depicted in Figure 6 (Arabic numerals refer to Recent families listed in Table 2).

Group	Taxa
Group Ia.	Ctenodactyloidea (17, Chapattimyidae, ?Cylindrodontidae)
Group Ib.	Geomyoidea (3, 4, Eomyidae, Sciuravidae)
Group II.	Gliroidea (16, ?15, Reithroparamyinae)
Group IIIa.	Sciuroidea (1, 2, Ischyromyidae, Mylagaulidae)
Group IIIb.	Castoroidea (5, Eutypomyidae, Rhizospalacidae)
Group IVa.	Dipodoidea (12,13)
Group IVb.	Muroidea (8, 9, 10, 11, 14)
Group V.	Anomaluroidea (6, ?Theridomyidae)
Group VI.	Hystricognathi
Group VIa.	Thryonomyoidea (19, 20, 21, ?18, Phiomyidae, Diamantomyidae)
Group VIb.	Erethizontoidea (22)
Group VIc.	Cavioidea (23, 24, Eocardiidae); Dinomyoidea (25); Octodontoidea (26, 27, 28, 32); Chinchilloidea (29); Dasyproctoidea (25, 30, 31, Heptaxodontidae)
Incertae sedis:	Pedetidae (Group I?)

Figure 6, are supported by results obtained from so-called gradualistic approaches, and I trust in this manner of practicing paleontology. For example, the evidence for separating the glirids from muroids was obtained mainly by the use of a historical approach (also see Vianey-Liaud, this volume).

Finally, I would like to make two remarks about the Sciurognathi and Hystricognathi. The first will be brief; it is clear now to virtually everyone that Sciurognathi is a wastebasket group, and there was no discussion of this "non-problem" during the symposium. Conversely, there were many discussions on the composition and monophyly of the suborder Hystricognathi. Wood (this volume) has presented his ideas on the subject and defended them with vigor, in particular, the inclusion of the North American Franimorpha within Hystricognathi. However, Lavocat and Parent (this volume) and others are not in complete agreement with him! Yet, it seems to me that this suprafamilial arrangement of African and South American hystricognathous rodents can now be considered as more than just a working hypothesis (see the contributions of Bugge, George, Luckett, and Woods, this volume). Search for the sister group of Hystricognathi seems to be the next logical direction in which research is heading.

Luckett (1980) has suggested that ctenodactylids and pedetids should be taken into account as potential candidates as sister groups of Hystricognathi. Although I have no specific comments on the affinities of pedetids (but I would be inclined to consider them as possibly related to ctenodactylids), Luckett's proposition about ctenodactylid relationships does not seem to me to be well founded. His argumentation was based primarily on the possession of four synapomorphies: (1) parietal endodermal layer of blastocyst fails to develop; (2) gestation period relatively long; (3) malleus and incus fused; and (4) sacculus urethralis present in penis. I have no comments to make about characters 1 and 4. As I have noted earlier, character 2 is shared also with castorids, and it is difficult to evaluate without precise information on litter weight in regard to maternal weight and other considerations. Character 3 is not uniformly present in hystricognaths, as noted by Wood (1974). In conclusion, because I consider ctenodactyloids as the stem group of all rodents, I can not imagine that Hystricognathi would be their only sister group (see Fig. 6).

REFERENCES

Adloff, P. 1898. Zur Entwickelungsgeschichte des Nagetiergebisses. Jena. Zeitsch. Naturw. 32: 347-411.

Ameghino, F. 1897. Mammifères crétacés de l'Argentine. Deuxième contribution à la connaissance de la faune mammalogique des couches à Pyrotherium. Bol. Inst. Geogr. Argent. 18: 406-429, 431-521.

Arata, A. A. 1975. The importance of small mammals in public health. In: Small Mammals: Their Productivity and Population Dynamics, F. B. Golley, K. Petrusewicz and L. Ryszkowski, eds., pp. 349-359, Cambridge University Press, Cambridge.

Asdell, S. A. 1964. Patterns of Mammalian Reproduction, 2nd Ed., Cornell Univ. Press, Ithaca, N. Y.

Blainville, H. M. D. de. 1816. Prodrome d'une nouvelle distribution systématique du règne animal. Bull. Sci. Soc. Philom. Paris, sér. 3, 3: 105-124.

Bock, W. J. 1977. Adaptation and the comparative method. In: Major Patterns in Vertebrate Evolution, M. K. Hecht, P. C. Goody and B. M. Hecht, eds., pp. 57-82, Plenum Press, New York.

Brandt, J. F. 1855. Beiträge zur nähern Kenntniss der Saügethiere Russlands. Mém. Acad. Imp. Sci. Pétersbourg, sér. 6-9: 1-375.

Chaline, J. and Mein, P. 1979. Les Rongeurs et l'Evolution. Doin, Paris.

Dawson, M. R., Li, C.-K., and Qi, T. 1984. Eocene ctenodactyloid rodents (Mammalia) of Eastern and Central Asia. Spec. Publ. Carneg. Mus. Nat. Hist. 9: 138-150.

Ehik, J. 1926. The right interpretation of the cheekteeth tubercles of Titanomys. Ann. Mus. Nat. Hungarici 23: 178-186.

Gervais, P. 1848. Zoologie et Paléontologie (Animaux vertébrés) ou nouvelles recherches sur les animaux vivants et fossiles de la France. Arthus Bertrand, Paris.

Gidley, J. W. 1906. Evidence bearing on tooth cusp development.

Proc. Wash. Acad. Sci. 8: 91-110.
Gidley, J. W. 1912. The lagomorphs an independent order. Science 36: 285-286.
Gingerich, P. D. 1976. Cranial anatomy and evolution of early Tertiary Plesiadapidae (Mammalia, Primates). Paper Paleont. Univ. Mich. 15: 1-141.
Gregory, W. K. 1910. The orders of mammals. Bull. Amer. Mus. Nat. Hist. 27: 1-524.
Hartenberger, J.-L. 1977. A propos de l'origine des Rongeurs. Géobios, Mém. Spéc. 1: 183-193.
Hartenberger, J.-L. 1980. Données et hypothèses sur la radiation initiale des Rongeurs. Palaeovertebrata, Mém. Jub. R. Lavocat: 285-301.
Hürzeler, J. 1936. Osteologie und Odontologie der Caenotheriden. Abh. Schweiz. Palaeont. Ges. 58: 1-89.
Illiger, C. 1811. Prodromus systematis mammalium et avium additis terminis zoographicis utriudque classis. C. Salfeld, Berlin.
Korth, W. W. 1984. Earliest Tertiary evolution and radiation of rodents in North America. Bull. Carneg. Mus. Nat. Hist. 24: 1-71.
Krause, D. W. 1982. Jaw movement, dental function, and diet in the Paleocene multituberculate Ptilodus. Paleobiology 8: 265-281.
Lavocat, R. 1974. What is an hystricomorph? Symp. Zool. Soc. Lond. 34: 7-20.
Li, C.-K. 1977. Paleocene eurymyloids (Anagalida, Mammalia) of Qianshan, Anhui. Vert. PalAsiat. 15: 103-118.
Li, C.-K., Chiu, C.-S., Yan, D.-F., and Hsieh, S.-H. 1979. Notes on some early Eocene mammalian fossils of Hengtung, Hunan. Vert. PalAsiat. 17: 71-82.
Linnaeus, C. 1758. Systema naturae per regna tria naturae, secundum classes, ordines, genera, species cum characteribus, differentiis, synonymis, locis, Vol. 1. Laurentii Salvii, Stockholm.
Luckett, W. P. 1977. Ontogeny of amniote fetal membranes and their application to phylogeny. In: Major Patterns in Vertebrate Evolution, M. K. Hecht, P. C. Goody, and B. M. Hecht, eds., pp. 439-516, Plenum Press, NY.
Luckett, W. P. 1980. Monophyletic or diphyletic origins of Anthropoidea and Hystricognathi: Evidence of the fetal membranes. In: Evolutionary Biology of the New World Monkeys and Continental Drift, R. L. Ciochon and A. B. Chiarelli, eds., pp. 347-368, Plenum Press, NY.
McKenna, M. C. 1961. A note on the origin of rodents. Amer. Mus. Novit. 2037: 1-5.
McKenna, M. C. 1975. Toward a phylogenetic classification of the Mammalia. In: Phylogeny of the Primates, W. P. Luckett and F. S. Szalay, eds., pp. 21-46, Plenum Press, NY.
Miller, G. S., Jr. and Gidley, J. W. 1918. Synopsis of the supergeneric groups of rodents. J. Wash. Acad. Sci. 8: 431-448.
Moody, P. A., Cochran, V. A., and Drugg, H. 1949. Serological evidence on lagomorph relationships. Evolution 3: 25-33.

Mossman, H. W. 1937. Comparative morphogenesis of the fetal membranes and accessory uterine structures. Contrib. Embryol. Carneg. Inst. 26: 129-246.

Nowak, R. M. and Paradiso, J. L. 1983. Walker's Mammals of the World, 4th ed. Johns Hopkins Univ. Press, Baltimore.

Patterson, B. and Wood, A. E. 1982. Rodents from the Deseadan Oligocene of Bolivia and the relationships of the Caviomorpha. Bull. Mus. Comp. Zool. 149: 371-543.

Prakash, L. 1975. The population ecology of the rodents of the Rajasthan desert, India. In: Rodents in Desert Environments, L. Prakash and P. K. Ghosh, eds., pp. 75-116, W. Junk, The Hague.

Romer, A. S. 1968. Notes and Comments on Vertebrate Paleontology. Univ. of Chicago Press, Chicago.

Russell, D. E. 1959. Le crâne de Plesiadapis. Bull. Soc. Géol. France (7th Ser.) 1: 312-314.

Russell, D. E. 1964. Les mammifères paléocènes d''Europe. Mém. Mus. Nat. Hist. Nat. Paris, sér. C 13: 1-324.

Schlosser, M. 1884. Die Nager des europäischen Tertiärs nebst Betrachtungen über die Organisation und die geschichtliche Entwicklung der Nager überhaupt. Palaeontographica 31: 1-143.

Simpson, G. G. 1945. The principles of classification and a classification of mammals. Bull. Amer. Mus. Nat. Hist. 85: 1-350.

Sych, L. 1971. Mixodontia, a new order of mammals from the Paleocene of Mongolia. Palaeont. Pol. 25: 147-158.

Szalay, F. S. 1977. Phylogenetic relationships and a classification of the eutherian Mammalia. In: Major Patterns in Vertebrate Evolution, M. K. Hecht, P. C. Goody, and B. M. Hecht, eds., pp. 315-374, Plenum Press, NY.

Szalay, F. S. and Decker, R. L. 1974. Origins, evolution and function of the pes in the Eocene Adapidae (Lemuriformes, Primates). In: Primate Locomotion, F. A. Jenkins, Jr., ed., pp. 239-259, Academic Press, NY.

Szalay, F. S. and Delson, E. 1979. Evolutionary History of the Primates. Academic Press, NY.

Thaler, L. 1966. Les rongeurs fossiles du Bas-Languedoc dans leurs rapports avec l'histoire des faunes et la stratigraphie du Tertiare d'Europe. Mém. Mus. nat. Hist. Nat. Paris, sér. C 17: 1-297.

Tullberg, T. 1899. Ueber das System der Nagetiere: eine phylogenetische Studie. Nova Acta Reg. Soc. Scient. Upsala 18: 1-514.

Turnbull, W. D. 1970. Mammalian masticatory apparatus. Fieldiana: Geology 18: 149-356.

Van Valen, L. 1966. Deltatheridia, a new order of mammals. Bull. Amer. Mus. Nat. Hist. 132: 1-126.

Van Valen, L. 1971. Adaptive zones and the orders of mammals. Evolution 25: 420-428.

Van Valen, L. and Sloan, R. E. 1966. The extinction of multituberculates. Syst. Zool. 15: 261-278.

Waterhouse, G. R. 1848. A Natural History of the Mammalia, Vol. 2, Rodentia. Hippolyte Baillière, London.
Winge, H. 1924. Pattedyr-Slaegter, Vol. 2, Rodentia, Carnivora, Primates. H. Hagerups Forlag, Copenhagen.
Wood, A. E. 1937. The mammalian fauna of the White River Oligocene. Part II. Rodentia. Trans. Amer. Phil. Soc. 28: 155-269.
Wood, A. E. 1940. The mammalian fauna of the White River Oligocene. Part III. Lagomorpha. Trans. Amer. Phil. Soc. 28: 271-362.
Wood, A. E. 1942. Notes on the Paleocene lagomorph, Eurymylus. Amer. Mus. Nov. 1162: 1-7.
Wood, A. E. 1957. What, if anything, is a rabbit? Evolution 11: 417-425.
Wood, A. E. 1962. The early Tertiary rodents of the family Paramyidae. Trans. Amer. Phil. Soc. 52: 1-261.
Wood, A. E. 1965. Grades and clades among rodents. Evolution 19: 115-130.
Wood, A. E. 1969. Rodents and lagomorphs from the "Chadronia Pocket," early Oligocene of Nebraska. Amer. Mus. Nov. 2366: 1-8.
Wood, A. E. 1974. The evolution of the Old World and New World hystricomorphs. Symp. Zool. Soc. Lond. 34: 21-60.
Wood, A. E. 1976. The paramyid rodent Ailuravus from the middle and late Eocene of Europe, and its relationships. Palaeovertebrata 7: 117-149.
Wood, B. J. 1971. Investigations of rats in rice fields demonstrating an effective control method giving substantial yield increase. Pest 17: 180-193.
Zhai, R.-J. 1978. More fossil evidences favoring an early Eocene connection between Asia and Nearctic. Mem. Inst. Vert. Pal. Paleoanthrop. 13: 107-115.

POSSIBLE PHYLOGENETIC RELATIONSHIP OF ASIATIC EURYMYLIDS AND RODENTS, WITH COMMENTS ON MIMOTONIDS

Li Chuan-kuei and Ting Su-yin

Institute of Vertebrate Paleontology and
Paleoanthropology, Academia Sinica
Beijing, People's Republic of China

RELATIONSHIP OF EURYMYLIDS AND RODENTS

Historical Review

The puzzling genus *Eurymylus*, from the latest Paleocene of the Gashato Formation, Mongolia, has been a subject of great scientific interest from the time of its original description (Matthew and Granger, 1925; Matthew et al., 1929). Despite the discovery of additional specimens (Sych, 1971), however, no consensus has been reached on the possible systematic affinities of eurymylids with rodents, lagomorphs, pseudictopids, zalambdalestids, anagalids, or leptictids (Wood, 1942; Van Valen, 1964; Szalay and McKenna, 1971; Sych, 1971; Hartenberger, 1977).

A more important advance came recently with the discovery of the eurymylid genus *Heomys* from the Middle and Late Paleocene of Qianshan, Anhui (Li, 1977), and the genus *Rhombomylus* from the early Middle Eocene of Turpan, Xinjiang (Zhai, 1978). According to Li, the eurymylid *Heomys* was thought to have affinities with rodents.

Several authors are now in agreement with Li (1977) concerning the relationship between *Heomys* and early rodents (Gingerich and Gunnell, 1979; Chaline and Mein, 1979; Hartenberger, 1980; Dawson et al., 1984; Li et al., in press). On the other hand, Patterson and Wood (1982) believed that eurymylids represented an independently evolving lineage which was close to the ancestry of lagomorphs, but lacked any special relationship with rodents.

Family Eurymylidae

Four genera, *Eurymylus*, *Heomys*, *Matutinia*, and *Rhombomylus*, are referred to the family Eurymylidae. Their history, horizons, associated mammals, and the materials are listed in Table 1.

General Characteristics of the Eurymylidae (Plates I and II)

(1) The dental formula is 1.0.2.3/1.0.2.3, similar to that of primitive rodents, except for the occurrence of one extra lower premolar (Figs. 1, 2).
(2) Gnawing-chewing are separated, with the incisors disengaging during cheek teeth occlusion, and *vice versa*. Correlated characters include: (a) upper diastema longer than lower; (b) glenoid fossa longitudinally elongate; and (c) when gnawing, predominantly anteroposterior (propalinal or palinal) movements of the mandible.
(3) Incisors are greatly enlarged and persistently growing (probably homologous to I2 or dI2; see Luckett, this volume); the roots of the upper incisors extend into the maxilla, and those of the lowers go as far back as M/2 or M/3.
(4) Enamel occurs only on the buccal surface of the incisors; two layers of incisor enamel are evident in the earlier forms (*Heomys*, *Eurymylus*), but only one layer is found in *Rhombomylus*.
(5) The upper cheek teeth are transversely elongated, slightly unilateral hypsodont, with obliteration of cusps relatively early in tooth wear (Figs. 1, 2).
(6) Upper P3 is large; upper premolars (P3/, P4/) possess only a single distinct cusp on the buccal side of the teeth, a morphology quite distinct from that of ischyromyid and paramyid rodents, but similar to that of the earliest ctenodactyloids, such as *Cocomys* (Dawson et al., 1984).
(7) Upper dP4 has a three-cusped trigon and a low, narrow talon; the talon in *Heomys* is restricted to the postero-lingual side, but is extended also to the buccal side in *Rhombomylus*. Upper dP3 is triangular and non-molariform. In *Heomys* sp. (V 4323), three upper deciduous teeth are present (dP2-dP4), but only two occur in later forms (dP3-dP4).
(8) The upper molars of *Heomys* have a hypocone, metaconule, and metaloph well developed, but no protoconule or protoloph. In *Matutinia* and *Rhombomylus*, a protoloph is distinct, the trigon is V-shaped, and the hypocone shelf is basin-shaped. In *Eurymylus*, the molars are compressed antero-posteriorly, and there is no hypocone shelf.
(9) The lower premolars exhibit little molarization, and are small and double-rooted (Fig. 2). Lower dP3 has a high pointed cusp in the trigonid; lower dP4 has a large paraconid and broad, low talonid.
(10) In the lower molars, the paraconid is weak or absent, and the trigonid is compressed antero-posteriorly. M/3 is the largest, with a distinct hypoconulid; there is an increase in height of the trigonid in later forms.
(11) A distinct shearing surface (surface 1 of Crompton, 1971) occurs on both upper and lower molars in *Matutinia* and *Rhombomylus*.
(12) The premaxillae are expanded backwards, contacting the frontals.

Table 1. Described specimens of the family Eurymylidae.

Genus	Materials	Age, Formation and Distribution	Associated Mammals
Eurymylus			
(Matthew and Granger, 1925)	1 maxilla and 1 mandible	Latest Paleocene, Gashato Formation, Gashato, Mongolia	Prionessus Phenacolophus Hyracolestes
(Matthew et al., 1929)	1 maxilla and 2 mandibles	Latest Paleocene, Gashato Formation, Gashato, Mongolia	Pseudictops Palaeostylops Prodinoceras
(Sych, 1971)	9 maxillae and mandibles	Earliest Eocene, Narun Buluk, Mongolia	Mongolotherium Hyracotherium
Heomys			
(Li, 1977)	1 skull and 1 mandible	Early Late Paleocene, Dou-mu Formation, Anhui, China	Hsiuannania Mimotona Sinostylops
(Li, 1977)	1 skull, fragmentary	Middle Paleocene, Wang-hu-dun Formation, Anhui, China	Huaiyangale Anictops Pappictidops
Matutinia			
(Li et al., 1979)	9 skulls and 30 jaws	Early Eocene, Lingcha Formation, Hunan, China	Cocomys Asiocoryphodon Propachynolophus
Rhombomylus			
(Zhai, 1978)	1 maxilla and mandible	Early Middle Eocene, Shi-san-jian-fang Formation, Xinjiang, China	Hyopsodus Coryphodon Heptodon
(Li, 1979)	19 skulls, 200 jaws and post-crania	Early Middle Eocene, Yu-huang-ding Formation, Hubei, China	Advenimus Asiocoryphodon

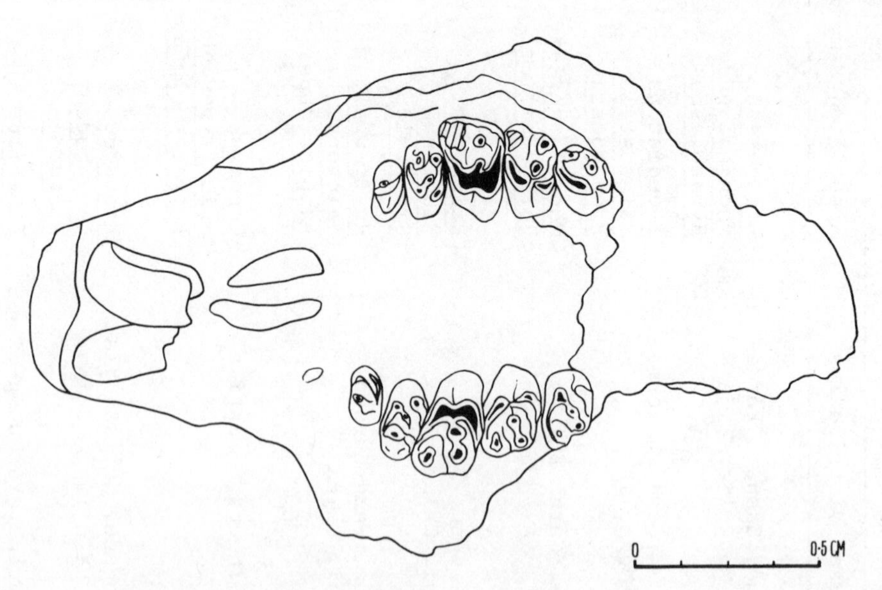

Fig. 1. Heomys orientalis Li, 1977, type (IVPP, V 4321), ventral view of anterior part of the skull, with upper P3-M3.

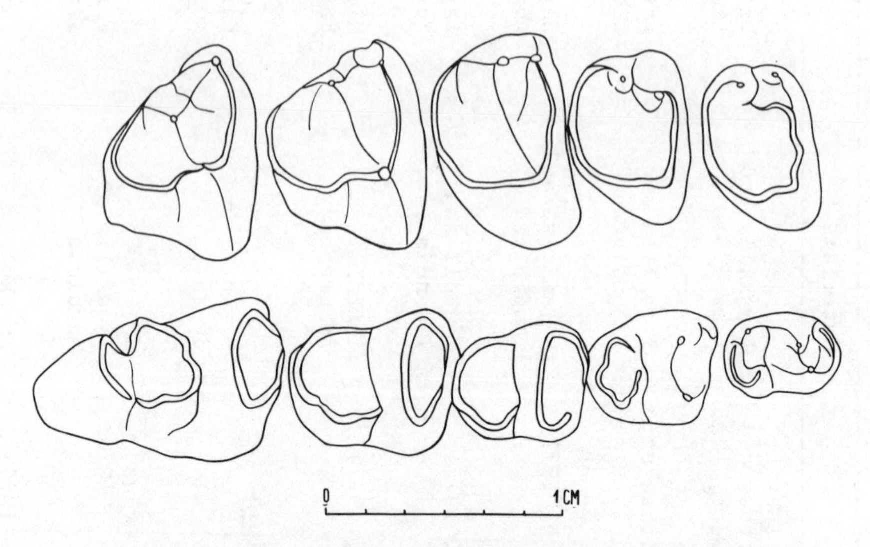

Fig. 2. Rhombomylus sp. (V 5293), showing a crown view of the upper cheek teeth (above), and lower cheek teeth (below). Anterior teeth (P3/3) are to the right.

(13) Incisive foramina show a decrease in size from moderate to small in the Heomys-Matutinia-Rhombomylus line (Fig. 3), but they are large in Eurymylus (Sych, 1971), extending posteriorly as far as P3/.
(14) The palatal bridge is wide, and the choanal openings are adjacent to M2/ (Rhombomylus) or M3/ (Heomys).
(15) The anterior root of the zygomatic arch lies adjacent to M1/ and M2/ in Heomys and Eurymylus, but in Rhombomylus it is adjacent to the posterior part of P4/. The masseter is restricted to the lateral surface of the zygoma; the infraorbital foramen is small and short (Figs. 3, 4).
(16) The orbital wing of the maxilla is extensively developed, and a lacrimal-palatine contact is lacking (Fig. 3).
(17) The optic foramina are confluent.
(18) The distance between the upper tooth rows is equal to that of the lowers; both sides of the tooth rows occlude simultaneously, as in rodents.
(19) In Rhombomylus, the inflated mastoid process extends not only on the occipital and ventral surfaces, but also on to the dorsal surface of the skull (Fig. 3); a cube-shaped inflation occurs on the cranial wall posterior to the tentorium of the cerebellum; many septae are present in the mastoid process.
(20) The auditory bulla is formed by the ectotympanic and mastoid (Ting and Li, 1984).
(21) The epitympanic recess is large and deep; the promontorium is well developed.
(22) There are no ossified canals present for the stapedial artery, promontory artery, or facial nerve.
(23) The coronoid process of the mandible is very large; both the condylar and coronoid processes are situated well above the level of the cheek teeth (Fig. 4); the angular process is inflected medially.
(24) The entepicondylar foramen of the humerus is large in Rhombomylus.
(25) In Rhombomylus, the third trochanter of the femur is situated about 1cm beneath the lesser trochanter; the patellar groove is narrow, not extending dorsally onto the anterior surface.
(26) There is no fusion of the tibia and fibula; the fibula is larger and more robust than in any rodents.
(27) A fibular facet is lacking on the calcaneum; the peroneal process is reduced.

Phylogenetic Significance of Eurymylid Characters

Items 1 and 2 above, related to the dentition and occlusal pattern, are the most important derived characters (synapomorphies) shared with Rodentia. Hartenberger (1977) and others have listed a number of fossil and extant mammalian forms, including multituberculates, plesiadapid primates, lagomorphs, and some insectivores, that possess characters similar to item 1. However, none of these groups share item 2 with Rodentia. Szalay (1971) made a reconstruction of the skull of Plesiadapis tricuspidens, in which the primary

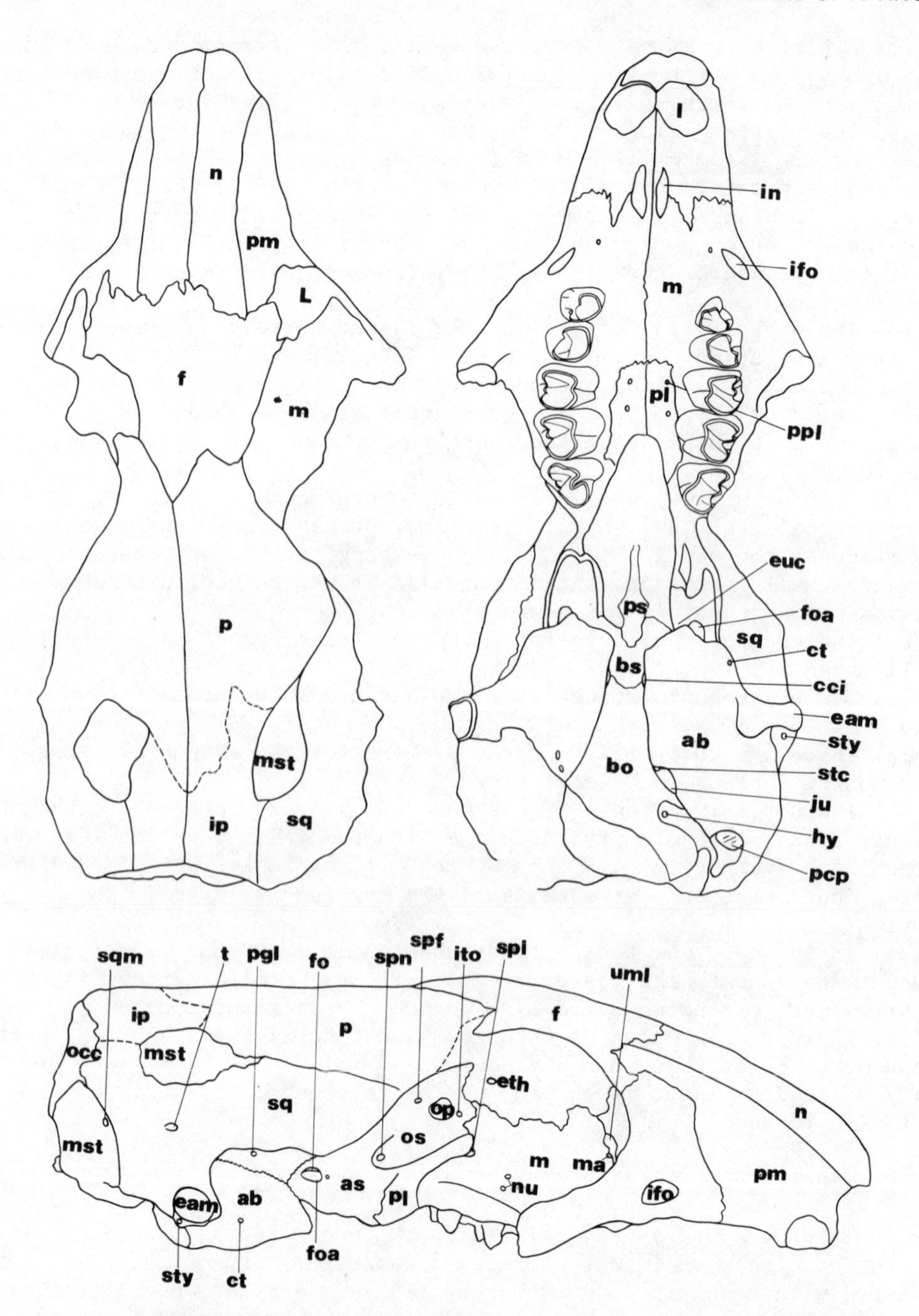

Fig. 3. Dorsal (upper left), ventral (upper right) and lateral (below) views of the skull of the middle Eocene eurymylid Rhombomylus (V 5289). See Table 2 for abbreviations.

Table 2. List of abbreviations used for skull structures and foramina illustrated in Figure 3 (adapted from Wahlert, 1974).

ab	auditory bulla
as	alisphenoid
bo	basioccipital
bs	basisphenoid
cci	internal carotid canal
ct	chorda tympani
eam	exterior acoustic meatus
eth	ethmoid foramen
euc	Eustachian canal
f	frontal
fo	foramen ovale
foa	foramen ovale accessorius
hy	hypoglossal foramen
i	incisor
ifo	infraorbital foramen
in	incisive foramen
ip	interparietal
ito	interior optic foramen
ju	jugular foramen
l	lachrymal
m	maxillary
ma	malar
mst	mastoid
n	nasal
nu	nutritive foramen
occ	occipital
op	optic foramen
os	orbitosphenoid
p	parietal
pcp	paracondyloid process
pgl	postglenoid foramen
pl	palatine
pm	premaxillary
ppl	posterior palatine foramen
ps	presphenoid
spf	sphenofrontal foramen
spl	sphenopalatine foramen
spn	sphenoidal fissure
sq	squamosal
sqm	squamoso-mastoid foramen
stc	stapedial artery canal
sty	stylomastoid foramen
t	temporal foramen
uml	unossified area in maxillary-lachrymal suture

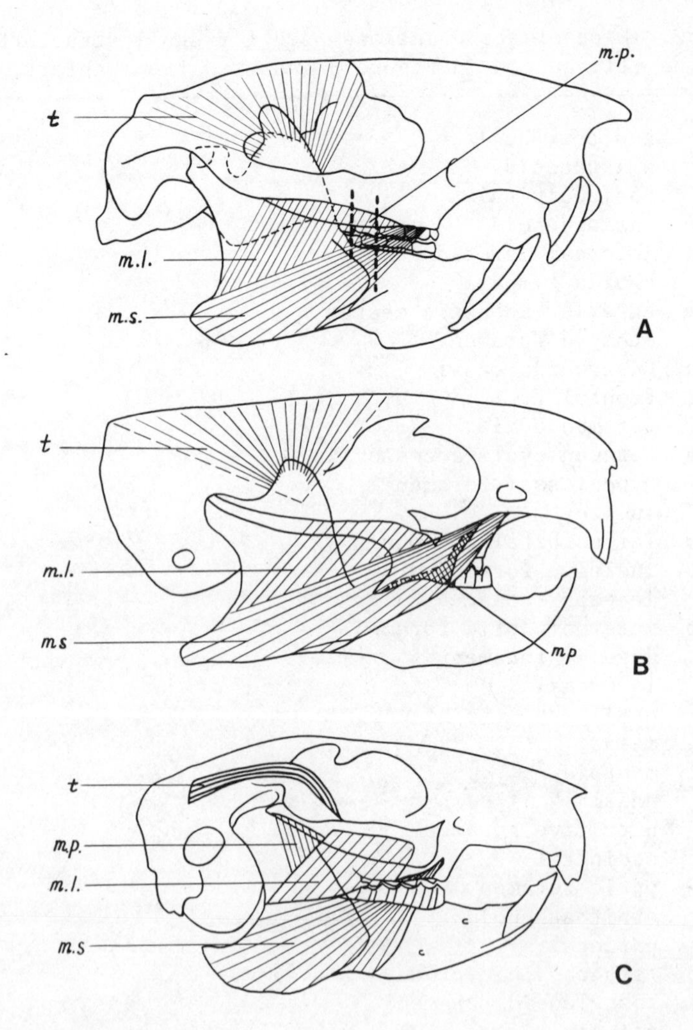

Fig. 4. The masticatory apparatus of: (A) an Eocene rodent, Ischyrotomus (after Wood, 1965); (B) an Eocene eurymylid, Rhombomylus; and (C) an Oligocene lagomorph, Palaeolagus (after Wood, 1940). Abbreviations of muscles: m.l., masseter lateralis; m.p., masseter profundus (within dashed lines); m.s., masseter superficialis; t, temporalis.

masticatory pattern shows a gnawing-chewing separation. However, Plesiadapis differs from rodents and eurymylids in the following: when chewing, the lower incisor remains in contact with the upper one; and the glenoid fossa is of a different shape. Krause (1982) has noted the differences between the masticatory system of rodents and the ptilodontid multituberculates. Both the jaw movement and cusp morphology of eurymylids differ obviously from those of the ptilodontids, but they are similar to those of rodents. Therefore, it should be emphasized that only the eurymylids share both items 1 and 2 above with rodents. Consequently, Li and Yan (1979) suggested a classification of rodents which includes eurymylids as follows:

Order Rodentia
  Suborder Mixodontia
    Family Eurymylidae
  Suborder Simplicidentata
    Infraorder Protrogomorpha, etc.

The following characters comprise the uniquely derived features (autapomorphies) of the family Eurymylidae: (1) a large molar hypocone shelf (or basin) situated lingually; and (2) compressed molar trigonids, with little transverse motion, such as is found in lagomorph mastication.

The following characters are plesiomorphous in comparison with those of rodents: (1) upper cheek teeth more transversely elongate, subhypsodont; (2) P/3 present; P3/ large; (3) coronoid process large; and (4) zygoma slightly expanded posteriorly (adjacent to M1/ or P4/).

The following characters are advanced or derived within eutherian mammals: (1) two layers of incisor enamel present (only one layer in *Rhombomylus*); (2) paraconid absent on lower molar trigonid; (3) premaxillae enlarged; and (4) No fibular facet on calcaneum.

The Family Eurymylidae and the Origin of Rodentia

Stehlin and Schaub (1951), McKenna (1961), and Wood (1962) proposed different hypotheses for the origin of rodents, and they suggested different reconstructions for the cheek tooth morphology of the ancestral rodents. Wood (1962) believed that the rodent ancestor may have been derived from some middle Paleocene *Plesiadapis*-like animal from North America. Based on calcaneal-astragalar morphology, Szalay (1977, this volume) concluded that the rodents originated from leptictids and, following Simpson (1945), referred both Rodentia and Lagomorpha to the Cohort Glires. Hartenberger (1980) arranged the Rodentia and Lagomorpha as descendants of the Mixodontia (= Eurymyloidea), and he suggested that Cretaceous leptictids (but not the Zalambdalestidae) gave rise to the Mixodontia. Li et al. (in press) have cited many characters aligning *Heomys* with rodents.

We would like to emphasize again the important value of the dentition and masticatory apparatus in assessing the phylogenetic relationships of the Eurymylidae. *Heomys* is, perhaps, very close to the ancestral stem of the Rodentia. We still believe that "there is no reason not to think that as yet unknown members of the Eurymylidae actually are ancestral" to rodents (Li et al., in press).

It remains a difficult task to determine the relationships between some Cretaceous Asiatic forms and the Eurymylidae. One lower jaw, described as Barunlestes butleri of the family Zalambdalestidae from the Barun Goyot Formation of Mongolia (Kielan-Jaworowska and Trofimov, 1980, Plate 7, Fig. 2, MgM-I/135), is very similar to the eurymylids in the following: (1) I/2 (?) extends posteriorly as far as M/2 (no other specimen of this genus or family possesses an incisor so enlarged); and (2) the trigonids are compressed anteroposteriorly. The main differences between this specimen and Paleocene eurymylids are that the former retain P/1 and P/2 and have a higher molar trigonid. We think that the jaw of MgM-I/135 should be referred to a new genus, until the upper dentition becomes known. Such an animal may be related to Paleocene eurymylids, and even might have given rise to the latter. Thus, we would like to suggest that some Asiatic Cretaceous proteutherian (similar to MgM-I/135, but not a true zalambdalestid) may be closely related to Asian eurymylids rather than to leptictids from other continents. If this hypothesis can be corroborated, the age of divergence of Eurymylidae from the Asiatic proteutherians would be earlier than ?middle Campanian, and the age of origin of rodents would perhaps also be much earlier (at least as early as the Cretaceous) than previously believed.

A character analysis of selected dental, cranial, and postcranial features of eurymylids, mimotonids, rodents, lagomorphs, and a variety of early Tertiary eutherians is presented in Figures 5 and 6, and in Tables 3 and 4. Data on taxa other than eurymyloids were obtained in part from Butler (1980), Evans (1942), Kielan-Jaworowska (1969, 1975, 1981), Kielan-Jaworowska et al. (1979), McKenna (1963, 1982), Novacek (1977, 1980), Simpson (1931), Sulimski (1969), Szalay (1977), Wahlert (1974), and Wood (1940, 1962).

COMMENTS ON THE POSSIBLE AFFINITIES OF THE MIMOTONIDS

Historical Review

In 1951, Bohlin described a rostrum and maxilla of the same individual from beds of probably Oligocene age from Gansu, Western China, as Mimolagus rodens, and he referred it to the Duplicidentata (= Lagomorpha). A hind limb collected from the same locality and certainly belonging to Mimolagus was referred to as a "rodent inc. sed." by Bohlin. Most later authors have considered Mimolagus as a eurymylid-like lagomorph (Wood, 1957; Van Valen, 1964; Dawson, 1967; Sych, 1971), because of a small, deep pit behind the large incisor, presumed to be the alveolus of a small posterior upper incisor.

The second genus, Mimotona, from the late Paleocene, Anhui, Eastern China, has been described as the type genus of the new family Mimotonidae, to which both Mimotona and Mimolagus belong (Li, 1977).

There are two other genera (Table 5) that can be referred to the family Mimotonidae. Gomphos (Shevyreva et al., 1975) is represented by mimotonid-like lower molars from the Gashato Formation of Mongolia and is possibly synonymous with Mimotona. Hypsimylus (Zhai, 1977) is represented by only two isolated teeth from the late Eocene of Beijing (Table 5).

Characteristics of the Family Mimotonidae (Plate I)

(1) The dental formula is 2(?).0.3.3/2-1.0.2-3.3 (Figs. 7, 8). I2/ is enlarged and extends slightly behind the premaxillary-maxillary suture; I3/ is small, with a single vertical root; I/2 extends far behind M/2; I/3 (when present) short and horizontally procumbent.
(2) Incisor enamel is distributed only on the anterior surface of the tooth; only one layer of enamel is present in Mimolagus and Mimotona.
(3) The upper premolars are not molariform (Fig. 7); there is a large cusp in the same position as the central cusp of lagomorph upper premolars (B-1 or metacone of Wood, 1942); protocone hollowed.
(4) The upper molars exhibit unilateral hypsodonty; the paracone and metacone are isolated and situated buccally; the stylar shelf is reduced; the protocone is V-shaped (Fig. 7).
(5) The anterior plane of the zygomatic arch is inclined anterodorsally, with the position of the anterior root of the arch slightly posterior to that of the Eurymylidae.
(6) The tibia and fibula are separate; there is no fibular facet on the calcaneum; the peroneal process is reduced in Mimolagus.

Mimotonidae and the Origin of the Lagomorpha

The origin of the order Lagomorpha has remained largely unknown, especially after Gidley (1912) concluded that lagomorphs share few similarities with rodents (also, see Hartenberger, this volume). Wood (1957) thought that lagomorphs were derived from periptychid condylarths; Russell (1959) suggested that the ancestor was a zalambdodont insectivoran; Van Valen (1964) stated that the Asian Paleocene genus Pseudictops was the closest to lagomorph ancestry; and McKenna (1975) believed that "Zalambdalestes is cladistically a Cretaceous lagomorph. A Zalambdalestes-like animal, possibly still with the tibia and fibula unfused, could have given rise to later lagomorphs ..." He united the Duplicidentata with the families Zalambdalestidae and Pseudictopidae within the order Lagomorpha, which was referred to the Grandorder Anagalida (also including Macroscelidea). However, Kielan-Jaworowska (1978) disagreed with McKenna, and she remarked that "in dentition, skull or postcranial skeleton ..., I see no reason to regard the Zalambdalestidae as a sister group of the Duplicidentata, except in the sense that both may have been derived from such generalized forms as the Early Cretaceous 'Prokennalestes.'"

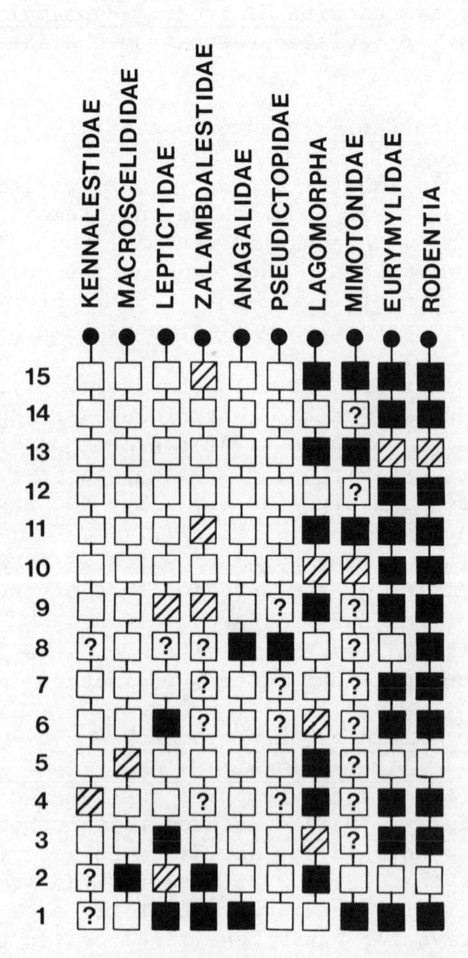

Fig. 5. Character state distribution of selected cranial and post-cranial features in eurymylids, mimotonids, rodents, and a variety of other extant and fossil eutherians. Derived character states are shown as black squares, open squares are primitive character states, and diagonally-lined squares are intermediately derived. Each taxon is represented by its reconstructed morphotypic condition. See Table 3 for the characters evaluated. Unknown character states are represented by (?).

Table 3. Character analysis of selected cranial and postcranial features in eurymyloids and other mammals.

Primitive	Derived
1. Calcaneal fibular facet present	1. Calcaneal fibular facet absent
2. Tibia and fibula unfused	2. Tibia and fibula fused
3. Infraorbital canal long	3. Infraorbital canal short
4. Ectotympanic not forming bulla	4. Bulla formed solely by ectotympanic
5. Temporal fossa and coronoid process well developed	5. Temporal fossa and coronoid process reduced
6. Alisphenoid orbital wing small	6. Alisphenoid orbital wing large, elongated
7. Accessory foramen ovale absent	7. Accessory foramen ovale present
8. Optic foramina confluent	8. Optic foramina not confluent
9. Maxillary orbital wing small or absent	9. Maxillary orbital wing well developed; lacrimal-palatine contact absent
10. Orbital fossae not notably shifted forward	10. Orbital fossae shifted forward relative to cheek teeth
11. Premaxilla not enlarged, no contact with frontal	11. Enlarged premaxilla contacts frontal
12. Glenoid fossa transverse	12. Glenoid fossa elongated longitudinally
13. Incisive foramina small	13. Incisive foramina large
14. Little or no dental diastema	14. Lower diastema shorter than upper diastema
15. Little or no diastema between incisors and premolars	15. Long diastema between incisors and premolars

Li (1977) cited several similarities between *Mimotona* and the Lagomorpha. These included: (1) dental formula; (2) I2 enlarged; (3) cheek teeth unilaterally hypsodont; (4) the pattern of the cheek teeth, especially the upper premolars; and (5) the presence of a single layer of enamel on the incisors (see Fig. 6). Consequently, he emphasized that the discovery of *Mimotona* is of the utmost importance to the study of the origin of the Lagomorpha. However, if the occurrence of I3/ in *Mimotona* is questionable (because the apparent "alveolus" of the tooth is located just at the suture between the premaxilla-maxilla; see Plate I, Fig. 6), then the possible relationship between *Mimotona* and the origin of the Lagomorpha would be more difficult to interpret. Nevertheless, the Oligocene genus *Mimolagus*, considered by us to be a mimotonid, did possess small alveoli in the correct position for I3/ (Bohlin, 1951, Fig. 3), and this suggests

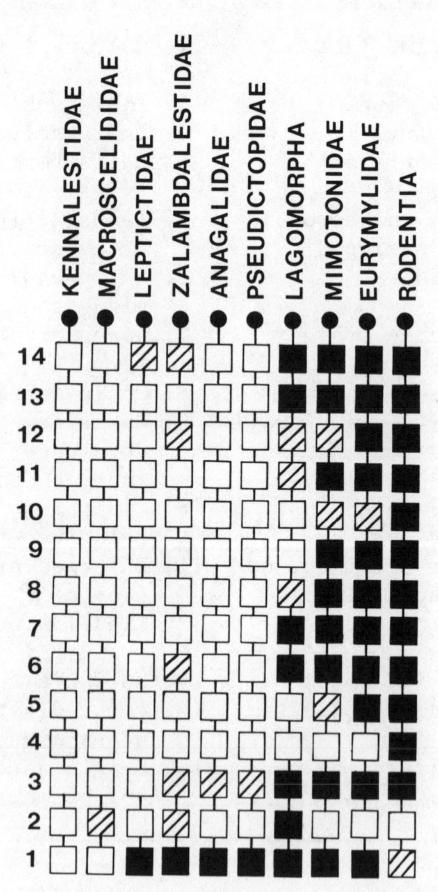

Fig. 6. Character state distribution of selected dental features in eurymylids, mimotonids, rodents, lagomorphs, and a variety of other extant and fossil eutherians. Derived character states are shown as black squares, open squares are primitive character states, and intermediately derived character states are shown as diagonally-lined squares. Each taxon is represented by its reconstructed morphotypic condition. See Table 4 for the characters evaluated.

Table 4. Character analysis of selected dental features in eurymyloids and other mammals.

Primitive	Derived
1. Molar trigonid not compressed	1. Molar trigonid compressed antero-posteriorly
2. M/3 not notably reduced	2. M/3 greatly reduced or absent
3. Molar paraconid well developed	3. Molar paraconid weak or absent
4. P/3 present	4. P/3 absent
5. P2/2 present	5. P2/2 absent
6. P1/1 present	6. P1/1 absent
7. Canines present	7. Canines absent
8. I/2 limited to anterior end of mandible	8. I/2 extends distally beneath M/2-M/3
9. I2/ restricted to premaxilla	9. I2/ extends distally into maxilla
10. Incisor enamel a single layer	10. Incisor enamel two layers
11. I2/2 surrounded by enamel	11. I2/2 enamel limited to buccal surface
12. I3/3 present	12. I3/3 absent
13. I2/2 not notably enlarged	13. I2/2 greatly enlarged
14. I1/1 present	14. I1/1 absent

Table 5. Described specimens of the family Mimotonidae.

Genus	Materials	Age, Formation and Distribution	Associated Mammals
Mimotona (Li, 1977)	5 jaws referred to 2 species	Early Late Paleocene, Dou-mu Formation, Anhui, China	_Heomys_ _Hyracolestes_ _Archaeolambda_
Mimolagus (Bohlin, 1951)	1 rostrum and 1 maxilla; 1 hindlimb	? Oligocene, Gansu, China	_Anagalopsis_
Gomphos (Shevyreva et al., 1975)	1 mandible with M1 and M2	Latest Paleocene, Gashato Formation, Gashato, Mongolia	_Eurymylus_ _Pseudictops_ _Khashanagale_
Hypsimylus (Zhai, 1977)	1 lower dP4 and M1	Late Eocene, Beijing, China	? _Eudinoceras_ _Imequincisoria_

PLATE I

Photographs of selected dental and cranial features from members of the Asian families Eurymylidae (Figs. 1-3, 7-8) and Mimotonidae (Figs. 4-6, 9). Anterior at top of all Figures.

Fig. 1. Heomys orientalis Li, 1977, type (IVPP Cat. No. V 4321). Ventral view of anterior part of the skull. x 4.

Fig. 2. Heomys orientalis Li, 1977 (V 4322). Crown view of right lower jaw. x 4.

Fig. 3. Heomys orientalis Li, 1977 (V 4331). Longitudinal section of upper incisor enamel, showing two distinct layers. x 150.

Fig. 4. Mimotona wana Li, 1977, type (V 4324). Crown view of left maxilla with P2/-M3/. x 5.

Fig. 5a. Mimotona wana Li, 1977 (V 4325.1). Crown view of left lower jaw with I/2-M/3. x 4.

Fig. 5b. Mimotona wana Li, 1977 (V 4325.2). Crown view of right lower jaw with I/2-M/2. x 4.

Fig. 6. Mimotona wana Li, 1977 (V 4326). Crown view of right premaxilla with I2/ and alveolus of I3/(?). x 3.

Fig. 7. Matutinia nitidulus Li et al., 1979 (V 5355). Crown view of right lower jaw with I/2-M/3. x 2.

Fig. 8. Matutinia nitidulus Li et al., 1979, type (V 5354). Ventral view of broken skull, with complete upper tooth row. x 2.

Fig. 9. Mimolagus rodens Bohlin, 1951, type (RV 51001.2). Crown view of right maxilla with P3/-M2/, and an alveolus for P2/. x 3.

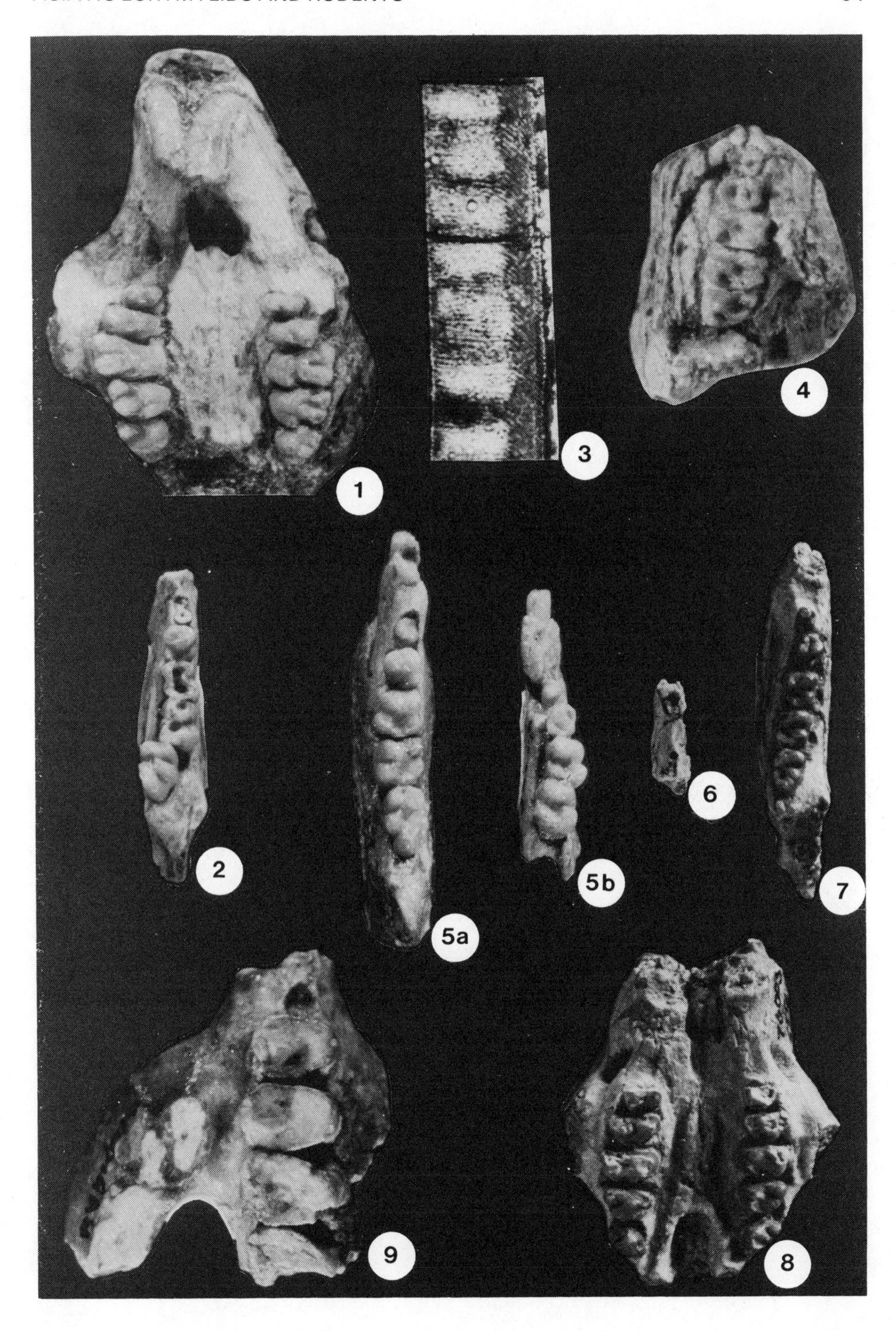

1
2
3
4
5a
5b
6
7
8
9

PLATE II

Photographs of various aspects of cranial morphology of the Eocene genus Rhombomylus (family Eurymylidae).

Fig. 1. Rhombomylus sp. (V 5289). The complete skull as seen in (a) dorsal, (b) lateral, and (c) ventral views. x 1.5. Compare with text Figure 3.

Fig. 2. Rhombomylus sp. (V 5293). Lateral view of the skull and lower jaw. x 1.

Fig. 3. Rhombomylus sp. (V 5263). Left middle ear cavity, showing cochlea. x 1.5.

Fig. 4. Rhombomylus sp. (V 5266-10b). Transverse section of the posterior part of the skull, showing the mastoid process, cochlea, malleus, and the ectotympanic bulla. x 2.

Fig. 5. Rhombomylus sp. (V 5262). Ventral view of posterior part of the skull, showing the septae of the mastoid process (right side) and the turns of the cochlea (left side). x 1.5

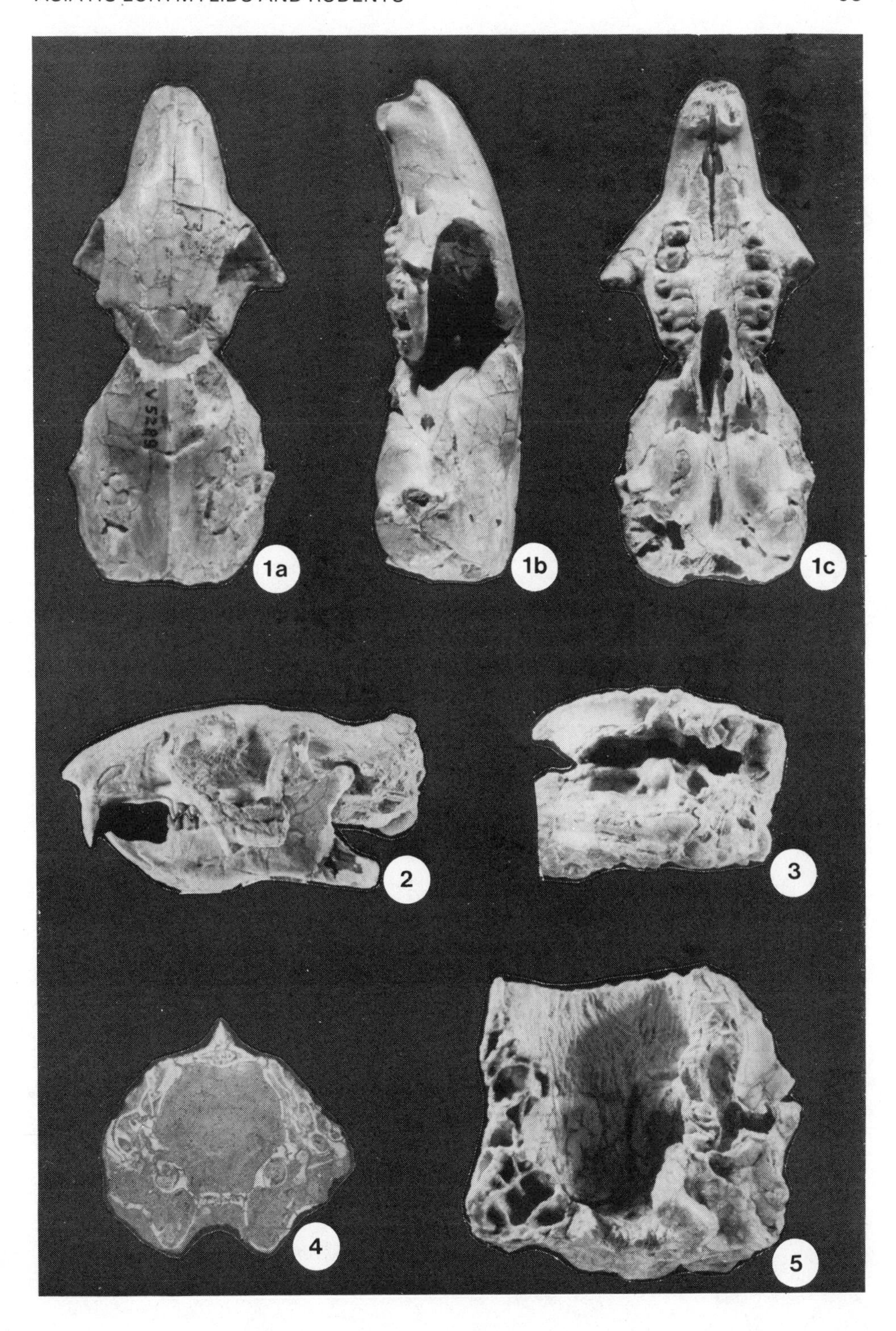
V 5289
1a
1b
1c
2
3
4
5

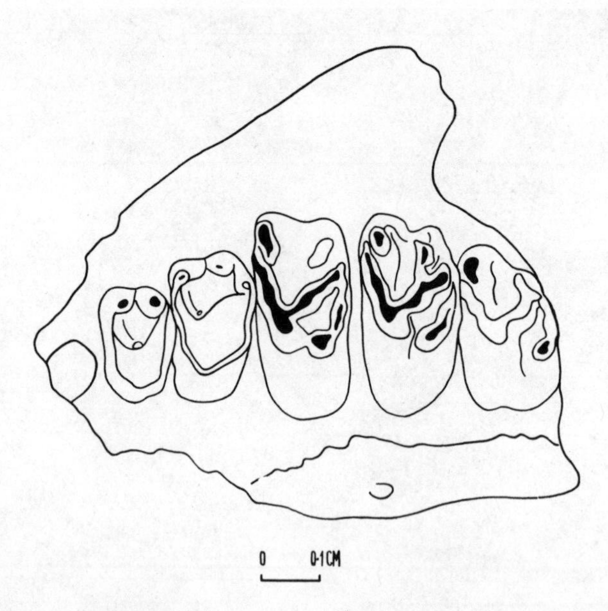

Fig. 7. Mimotona wana Li, 1977 (V 4324), showing crown view of left maxilla with upper P3-M3, and alveolus for upper P2. Anterior end is at left.

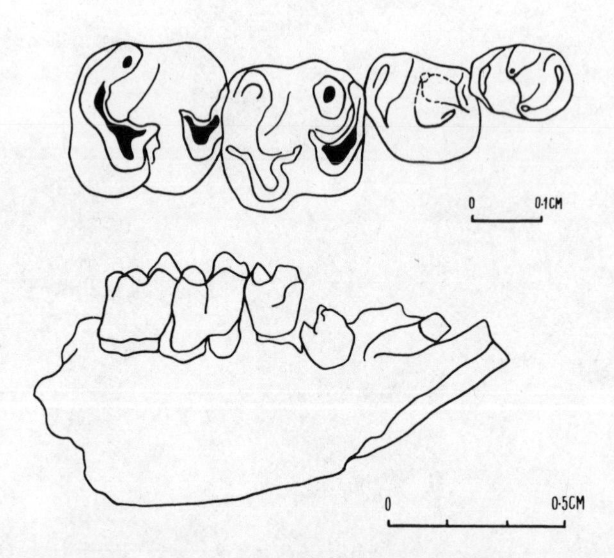

Fig. 8. Mimotona wana Li, 1977 (V 4324), showing crown view of right lower P3-M2 (above) and lateral view of right lower jaw with lower I2-M2 (below). Anterior end is at right.

that I3/ did occur in the morphotype of the mimotonids.

Based on the similarities of the pattern of the cheek teeth of Mimotonidae and Eurymylidae (Fig. 6, Table 4), we tentatively prefer to keep the two families within the superfamily Eurymyloidea. Our character analyses (Figs. 5, 6) show the close affinities among eurymyloids, rodents, and lagomorphs, but further data on the cranial and postcranial morphology of mimotonids are required for testing the earlier hypothesis (Li, 1977) that mimotonids may be closer to the lagomorphs, and that at least some eurymylids (*Heomys*) may be close to the ancestry of Rodentia.

ACKNOWLEDGMENTS

We are grateful to various colleagues who discussed the possible relationships of eurymyloids with the senior author at the NATO Rodent Phylogeny Symposium in Paris (July, 1984), notably Professors R. Lavocat and A. E. Wood, and Drs. J.-L. Hartenberger, P. Butler, J. H. Wahlert, N. Lopez Martinez, M. J. Novacek, F. S. Szalay, J.-P. Parent, and J. J. Jaeger. We would like especially to express our deep gratitude to Dr. W. P. Luckett for his helpful comments on the preliminary manuscript. We also thank Mr. Z. F. Wang for his photographs, and Dr. N. L. Martinez, Mrs. M. W. Yang, and Dr. Nancy Hong for their drawings.

REFERENCES

Bohlin, B. 1951. Some mammalian remains from Shih-ehr-ma-ch'eng, Hui-hui-p'u area, Western Kansu. Rep. Sino-Swedish Exped. N. W. Prov. China 35, Vert. Palaeont. 5: 1-47.

Butler, P. M. 1980. Functional aspects of the evolution of rodent molars. Palaeovertebrata, Mém. Jubil. R. Lavocat: 249-262.

Chaline, J. and Mein, P. 1979. Les Rongeurs et l'Evolution. Doin Editeurs, Paris.

Crompton, A. W. 1971. The origin of the tribosphenic molar. Zool. J. Linn. Soc. 50 (Suppl. 1): 65-87.

Dawson, M. R. 1967. Lagomorph history and the stratigraphic record. Univ. Kansas Dept. Geol. Spec. Publ. 2: 287-316.

Dawson, M. R., Li, C.-K. and Qi, T. 1984. Eocene ctenodactyloid rodents (Mammalia) of Eastern and Central Asia. Carneg. Mus. Nat. Hist. Spec. Publ. 9: 138-150.

Evans, F. G. 1942. The osteology and relationships of the elephant shrews (Macroscelididae). Bull. Amer. Mus. Nat. Hist. 80: 85-125.

Gidley, J. W. 1912. The lagomorphs an independent order. Science 36: 285-286.

Gingerich, P. D. and Gunnell, G. F. 1979. Systematics and evolution of the genus *Esthonyx* (Mammalia, Tillodontia) in the early

Eocene of North America. Contrib. Mus. Paleont., Univ. Mich. 25: 125-153.
Hartenberger, J.-L. 1977. A propos de l'origine des Rongeurs. Géobios, Mém. Spéc. 1: 183-193.
Hartenberger, J.-L. 1980. Données et hypothèses sur la radiation initiale des Rongeurs. Palaeovertebrata, Mém. Jubil. R. Lavocat: 285-301.
Kielan-Jaworowska, Z. 1969. Results of the Polish-Mongolian Palaeontological Expeditions. Part I. Preliminary data on the Upper Cretaceous eutherian mammals from Bayn Dzak, Gobi Desert. Palaeont. Pol. 19: 171-197.
Kielan-Jaworowska, Z. 1975. Preliminary description of two new eutherian genera from the Late Cretaceous of Mongolia. Palaeont. Pol. 33: 5-16.
Kielan-Jaworowska, Z. 1978. Evolution of the therian mammals in the Late Cretaceous of Asia. Part III. Postcranial skeleton in Zalambdalestidae. Palaeont. Pol. 38: 3-41.
Kielan-Jaworowska, Z. 1981. Evolution of the therian mammals in the Late Cretaceous of Asia. Part IV. Skull structure in *Kennalestes* and *Asioryctes*. Palaeont. Pol. 42: 25-78.
Kielan-Jaworowska, Z., Bown, T. M. and Lillegraven, J. A. 1979. Eutheria. In: Mesozoic Mammals, J. A. Lillegraven, Z. Kielan-Jaworowska and W. A. Clemens, eds., pp. 221-258, Univ. Calif. Press, Berkeley.
Kielan-Jaworowska, Z. and Trofimov, B. A. 1980. Cranial morphology of the Cretaceous eutherian mammal *Barunlestes*. Acta Palaeont. Pol. 25: 167-185.
Krause, D. W. 1982. Jaw movement, dental function, and diet in the Paleocene multituberculate *Ptilodus*. Paleobiology 8: 265-281.
Li, C.-K. 1977. Paleocene eurymyloids (Anagalida, Mammalia) of Qianshan, Anhui. Vert. PalAsiat. 15: 103-118.
Li, C.-K., Chiu, C.-S., Yan, D.-F. and Hsieh, S.-H. 1979. Notes on some Early Eocene mammalian fossils of Hengtung, Hunan. Vert. PalAsiat. 17: 71-80.
Li, C.-K., Wilson, R. W. and Dawson, M. R. In press. The origin of rodents and lagomorphs. J. Mammal.
Li, C.-K. and Yan, D.-F. 1979. Notes on the systematic position of eurymyloids and the origin of Rodentia. 12th Ann. Rep. Pal. Soc. China: 155-156. (In Chinese)
Matthew, W. D. and Granger, W. 1925. Fauna and correlation of the Gashato Formation of Mongolia. Amer. Mus. Novit. 189: 1-12.
Matthew, W. D., Granger, W. and Simpson, G. G. 1929. Additions to the fauna of the Gashato Formation of Mongolia. Amer. Mus. Novit. 376: 1-12.
McKenna, M. C. 1961. A note on the origin of rodents. Amer. Mus. Novit. 2037: 1-5.
McKenna, M. C. 1963. New evidence against tupaioid affinities of the mammaliam family Anagalidae. Amer. Mus. Novit. 2158: 1-16.
McKenna, M. C. 1975. Toward a phylogenetic classification of the

Mammalia. In: Phylogeny of Primates, W. P. Luckett and F. S. Szalay, eds., pp. 21-46, Plenum Press, New York.

McKenna, M. C. 1982. Lagomorph interrelationships. Géobios Mém. Spéc. 6: 213-223.

Novacek, M. J. 1977. A review of Paleocene and Eocene Leptictidae (Eutheria: Mammalia) from North America. Paleobios 24: 1-42.

Novacek, M. J. 1980. Cranioskeletal features in tupaiids and selected Eutheria as phylogenetic evidence. In: Comparative Biology and Evolutionary Relationships of Tree Shrews, W. P. Luckett, ed., pp. 35-93, Plenum Press, New York.

Patterson, B. and Wood, A. E. 1982. Rodents from the Deseadan Oligocene of Bolivia and the relationships of the Caviomorpha. Bull. Mus. Comp. Zool. 149: 371-543.

Russell, L. S. 1959. The dentition of rabbits and the origin of the Lagomorpha. Bull. Natl. Mus. Canada 166: 41-45.

Shevyreva, N. S. 1975. New data on the vertebrate fauna of Gashato Formation (Mongolia People's Republic). Bull. Acad. Sci. Georgian SSR 77: 225-228.

Simpson, G. G. 1931. A new insectivore from the Oligocene, Ulan Gochu Horizon, of Mongolia. Amer. Mus. Novit. 505: 1-22.

Simpson, G. G. 1945. The principles of classification and a classification of mammals. Bull. Amer. Mus. Nat. Hist. 85: 1-350.

Stehlin, H. G. and Schaub, S. 1951. Die Trigonodontie der simplicidentaten Nager. Schweiz. Pal. Abh. 67: 1-385.

Sulimski, A. 1969. Paleocene genus _Pseudictops_ Matthew, Granger and Simpson 1929 (Mammalia) and its revision. Palaeont. Pol. 19: 101-129.

Sych, L. 1971. Mixodontia, a new order of mammals from the Paleocene of Mongolia. Palaeont. Pol. 25: 147-158.

Szalay, F. S. 1971. Cranium of the Late Paleocene primate _Plesiadapis tricuspidens_. Nature 230: 324-325.

Szalay, F. S. 1977. Phylogenetic relationships and a classification of the eutherian Mammalia. In: Major Patterns in Vertebrate Evolution, M. K. Hecht, P. C. Goody and B. M. Hecht, eds., pp. 315-374, Plenum Press, New York.

Szalay, F. S. and McKenna, M. C. 1971. Beginning of the age of mammals in Asia: the Late Paleocene Gashato Fauna, Mongolia. Bull. Amer. Mus. Nat. Hist. 144: 269-318.

Ting, S.-Y. and Li, C.-K. 1984. The structure of the ear region of _Rhombomylus_ (Anagalida, Mammalia). Vert. PalAsiat. 22: 92-102.

Van Valen, L. 1964. A possible origin for rabbits. Evolution 18: 484-491.

Wahlert, J. H. 1974. The cranial foramina of protrogomorphous rodents: an anatomical and phylogenetic study. Bull. Mus. Comp. Zool. 146: 363-410.

Wood, A. E. 1940. The mammalian fauna of the White River Oligocene. Part III. Lagomorpha. Trans. Amer. Phil. Soc. 28: 271-362.

Wood, A. E. 1942. Notes on the Paleocene lagomorph, _Eurymylus_. Amer. Mus. Novit. 1162: 1-7.

Wood, A. E. 1957. What, if anything, is a rabbit? Evolution 11:

417-425.
Wood, A. E. 1962. The early Tertiary rodents of the family Paramyidae. Trans. Amer. Phil. Soc. 52: 1-261.
Wood, A. E. 1965. Grades and clades among rodents. Evolution 19: 115-130.
Zhai, R.-J. 1977. Supplementary remarks on the age of Changxindian Formation. Vert. PalAsiat. 15: 173-176.
Zhai, R.-J. 1978. More fossil evidences favouring an Early Eocene connection between Asia and Neoarctic. Mem. Inst. Vert. Pal. Paleoanthrop. 13: 107-115.

CRANIAL EVIDENCE FOR RODENT AFFINITIES

Michael J. Novacek

Department of Vertebrate Paleontology
American Museum of Natural History
New York, New York 10024 U.S.A.

INTRODUCTION

In his classic study of mammalian relationships, Gregory (1910) stated that a close grouping of rodents and lagomorphs seemed more compelling than most other proposals for higher eutherian categories. Nevertheless, the concept Glires, which unites these groups, soon drew influential criticism from Gidley (1912). Many workers acknowledged Gidley's arguments and looked elsewhere for rodent relationships. Some allied rodents with primates (e.g., McKenna, 1961; Wood, 1962). Lagomorphs, on the other hand, were related to primitive ungulates (Wood, 1957) or "zalambdodont insectivores" (Russell, 1959).

Despite these varied arguments, the case for rcdent-lagomorph relationships has not been abondoned altogether. Indeed, there has been, in recent years, a revival of support for Glires (Luckett, 1977; Szalay, 1977; Hartenberger, 1977; Novacek, 1982). Such support invokes features of the reproductive system, fetal membranes, and tarsal structure in addition to the more commonly cited evidence of the teeth.

Certain characters of the skull have, as well, been used in arguments for the rodent-lagomorph relationships. Notable among these are the presence of a well developed bulla consisting entirely of ectotympanic bone, a deep zygoma, modifications of the jaw articulations, and the relative positions of the orbital foramina. In this paper, I examine these and other features of the skull and jaws for their effectiveness as clues to the affinities of rodents. Comparisons of the osseous features cited here are based on first-hand examination of adult, neonatal and early postnatal skulls, relevant fossil material (where available), as well as the abundant literature on rodent and lagomorph phylogeny.

The methodology in this paper emphasizes the identity of homologous, shared specializations (synapomorphies) as evidence of close relationship (Hennig, 1966). The application of this approach to problems of higher-level relationships of mammals is discussed elsewhere (Novacek, 1980, 1982). I do wish, however, to raise one point here. There seems to be a growing concern that a simple "mapping exercise" that plots the distribution of special traits somehow diminishes our understanding of important morphological traits. Clearly, a firmer understanding of form, ontogeny, and function leads to a more reliable summary of character evidence and homology schemes. Nonetheless, a plot of character distributions does not conflict with the strategy of "in-depth" character analysis. The arrangement of attributes in a table does not preclude their further study. In this paper, certain characters mentioned are much better understood than others. Where there is a real disparity in relevant information, the problem is mentioned. Yet, I acknowledge that comparisons here are also guided by the need to consider a diversity of structures, and this requirement sometimes compromises the search for an equally profound understanding of every character.

RODENT AND LAGOMORPH MORPHOTYPES

Many rodent groups have highly specialized crania that reflect their diverse lifestyles. Reference to some of these groups as examples in a search of rodent affinities would be misleading. I follow Wood (1965), Wahlert (1974), and others in recognizing paramyids and sciuravids as having the most primitive skull morphology within rodents. Further, I agree with Lavocat (1967) and Parent (1980) that the Theridomorpha, the Sciuravidae, and the Paramyidae have the most primitive auditory construction for the group. I also endorse the suggestion of these authors that even Recent members of the Myomorpha are more conservative in auditory features than are most other rodents. Arguments that the most structurally primitive rodents may be early ctenodactyloids (e.g., Korth, 1984) pertain only to the dentition, jaw morphology, and one or two details of the cranial anatomy (e.g., the relative alignment of the nasals with the premaxilla and the relative anteroposterior position of the zygoma in relation to the cheek teeth - see Korth, 1984, p. 64). None of these comparisons seriously challenge the view that the paramyid skull provides a reasonable approximation of the primitive rodent morphotype. Published skulls of early ctenodactyloids are poorly known. For the present, the assumption concerning the "primitiveness" of certain paramyids seems valid.

The skull structure of forms like *Paramys* is, in fact, so generalized that it poses a problem for comparative studies. Early rodents are not distinguished from other eutherian groups by cranial features as readily as they are by dental features. By contrast, lagomorphs are markedly more derived in several cranial features than are morphotypical rodents. Thus, any theory invoking rodent-lagomorph relationships must rely on a small residuum of special features known in generally primitive skulls of rodents. It is noteworthy that Recent leporids and ochotonids do not differ significantly from early fossil members of the Lagomorpha (see Dawson, 1958, 1969; and Lopez-Martinez, 1978, for excellent syntheses on lagomorph structure and phylogeny). A better early fossil record of lagomorphs might reveal a more conservative cranial plan for this order. Unfortunately, various candidates for Late Cretaceous-Early Tertiary members or close relatives of lagomorphs (e.g., zalambdalestids, anagalids, mimotonids, pseudictopids, and others) are poorly represented by skull material. Many of the comparisons considered here are not possible for these early groups.

CHARACTER ANALYSIS

Table 1 describes cranial and jaw features that are compared in several eutherian groups listed in Table 2. The listing of characters is derived from an ongoing study of cranio-skeletal features in leptictids and other mammals (Novacek, in preparation). Only 36 features out of a larger sample of approximately 80 are considered here because these alone represent special traits in either or both lagomorphs and rodents. (One character, the suboptic foramen, may be represented by a primitive condition in lagomorphs and rodents, but polarity in this case is highly uncertain and the character is included in Table 1.) I acknowledge that this reference to numbers may seem artificial, that some features are perhaps highly correlated by functional or ontogenetic criteria, and that other authors would subdivide features of the cranium in different ways. The numbers do, however, provide a rough sense of the diversity of morphology sampled and the degree to which the character evidence supports one hypothesis over another.

Auditory features are not well represented in Table 1. This might seem unusual, as the auditory region is rich in characters and is accordingly given much attention in mammalian systematics. This system does not, however, offer much evidence bearing on rodent affinities. Aside from the development of the bulla and a few other features, the morphotypical condition of the rodent ear is, as noted above, remarkably impoverished in specializations. Lagomorphs are also unexpectedly conservative in many auditory features, although this group shows several autapomorphies of the tympanic roof and inner bulla not listed in Tables 1 and 2. These specializations help to define Lagomorpha but shed no light on their proposed affinities with rodents or other orders.

Table 1

Cranial Features of Selected Eutheria.
A, primitive eutherian condition; B, C, D, etc. derived condition.
A sequence A, B, C, does not necessarily imply a morphocline.

Feature	Primitive	Derived
1. Premaxilla	A) small, restricted to anterior portion of snout	B) large, nearly one half surface area of the snout or C) very large with postero-dorsal flange between nasal and frontal, D) very small
2. Infraorbital canal	A) long	B) short
3. Anterior nasal	A) recessed, truncated	B) slightly or markedly flared
4. Frontal-facial wing	A) weak or absent	B) projects as narrow process between premaxilla and nasal
5. Palatine size (on palate) relative to premaxilla	A) much larger	B) smaller
6. Incisive foramina	A) small, near anterior border of palate	B) larger, set more posteriorly on palate, or C) very large, elongate
7. Maxillary fenestrae or large foramen (on facial)	A) absent	B) present as a single or few large fenestrae or a complex lacework
8. Lacrimal	A) with facial process	B) confined to orbit

Feature	Primitive	Derived
9. Nasolacrimal foramen	A) small, opens posteriorly within orbit below antorbital rim or exits on pars facialis of lacrimal	B) larger, opens laterally at anterolateral edge of lacrimal, or C) opens laterally at lacrimal-frontal border, no antorbital rim of zygoma, or D) very large, otherwise as in B
10. Orbitosphenoid	A) small, low in orbit	B) large, high in orbit, or C) very large, contacts lacrimal
11. Maxilla	A) with small exposure in orbital wall	B) with large orbital exposure, restricts palatine, C) very large, excludes lacrimal-palatine contact
12. Palatine	A) with large orbital exposure, contacts lacrimal	B) narrowed at base of orbital wall between intrusive maxilla and alisphenoid, or C) very small, loses contact with lacrimal, confined to floor of orbit
13. Temporal fossa	A) distinct, extensive	B) obscured by lateral expansion of alisphenoid
14. Zygoma	A) moderate to slender in proportions	B) dorsoventrally deeper, more robust, or C) incomplete

(cont'd)

Table 1 (Continued)

Feature	Primitive	Derived
15. Optic foramen	A) moderate size, relatively low and posterior in orbital wall	B) large and C) high and anterior in orbit, and D) both right and left optic f. form continuous opening, or E) very small foramen
16. Sphenopalatine foramen	A) within palatine	B) at maxillary-palatine suture
17. Dorsal palatine foramen	A) distinct, behind sphenopalatine foramen	B) absent or confluent with sphenopalatine foramen
18. Suboptic foramen	A) opens in medial wall of sphenorbital	B) separate opening below optic foramen
19. Sphenorbital fissure	A) moderate size, crescentic	B) very large, dorsoventrally expanded, crescentic or slit-like
20. Alisphenoid canal	A) long, anterior opening confluent with sphenorbital fissure	B) shortened, anterior opening distinct, or C) very short
21. Sphenoidal foramina for masseteric and buccinator nerves	A) absent	B) present, diverge from common posterior opening
22. Accessory foramen ovale	A) absent	B) present
23. Pterygoid fossa	A) shallow, weak, or absent	B) deep, well-defined by strong pterygoid lamina
24. Middle lacerate foramen	A) small or absent	B) large, distinct

Feature	Primitive	Derived
25. Temporal (supra-meatal) foramen	A) present	B) absent
26. Auditory bulla	A) absent	B) present, comprising entotympanic C) entotympanic and ectotympanic, D) basisphenoid-petrosal, E) petrosal, F) ectotympanic, or G) composite of several elements
27. Ectotympanic	A) not expanded laterally	B) expanded laterally into meatal tube
28. Postglenoid foramen	A) present	B) vestigial or absent
29. Exit in tympanic wall for superior ramus of stapedial artery	A) present	B) absent
30. Roof of external auditory meatus	A) smooth	B) with distinct depression for epitympanic recess
31. Medial internal carotid artery	A) conveyed in a canal between medial petrosal and basisphenoid	B) conveyed in a tube that curves anterolaterally within bulla on tegmen tympani, or C) artery absent (or shifted to promontory position)
32. Coronoid process	A) well developed with semi-vertical anterior ridge	B) small with anterior margin inclined, or c) modified as mandibular condyle

(cont'd)

Table 1 (Continued)

Feature	Primitive	Derived
33. Jaw condyle	A) positioned relatively low	B) relatively high, or C) highest process on mandible
34. Postglenoid process	A) present, well developed	B) weak or absent, C) absent, with jaw articulation in glenoid region reoriented
35. Jaw adductors	A) temporalis muscle relatively more massive than masseter and pterygoid	B) masseter relatively more massive than temporalis, pterygoid relatively larger than in A), or C) pterygoid nearly half the mass of the large masseter
36. Posterior ventral zygomatic process	A) absent	B) present with strong posterior projection

By contrast, the nasal-facial region of the skull provides several specializations that support the Glires grouping. The enlarged premaxillary (Figs. 1, 3) is doubtless derived and may be correlated, as many workers have suggested, with the enlarged upper incisors (Fawcett, 1917). At early stages of ossification, the premaxilla in rodents and lagomorphs (Fig. 2A) has a frontal process that projects obliquely in a posterodorsal direction into the gap between the nasal and the maxilla. In lagomorphs the condition is further modified with the fully ossified premaxilla producing a distinct splint-like flange (Fig. 3B) that intrudes between the anteriorly projecting splint of the frontal and the nasal above. The overall configuration of the nasal-facial elements in adult lagomorphs is unique (see also Lopez-Martinez, 1978). Nonetheless, both rodents and lagomorphs depart from the typical mammalian condition, wherein the premaxilla arises as a small membrane outgrowth that lacks a strong posterodorsal process.

Lagomorphs also show marked enlargement of the incisive foramina, which function as openings for the duct of the Jacobson's organ. The condition contrasts in adult ochotonids and leporids. The latter show much longer, narrower incisive foramina. However, in juvenile skulls of both groups the condition of the palatal elements and foramina are strongly similar (Fig. 1). The palate is short and the posterior choanae are not anteriorly expanded. In the later ontogeny of the Ochotonidae, the incisive foramina increase in size, the rostrum and choanae less so. In the Leporidae, the much greater expansion of the choanae and incisive foramina is accompanied by the marked elongation of the rostrum. Still, the postnatal skull of either group shows a condition that contrasts with comparable stages in other mammals. It is noteworthy that in the Tertiary ochotonid Prolagus there are two pairs of incisive foramina, the posterior pair being the larger (Dawson, 1969, fig. 4). In modern ochotonids the two pairs appear to coalesce, although the anterior region of each foramen is bilaterally constricted and the opening partially filled with elements of the nasal septum.

Early stages of palatal development in rodents and lagomorphs are comparable (Voit, 1909; Fawcett, 1917; DeBeer, 1937). In the floor of the nasal capsule of the chondrocranium there appears a very elongate opening, the fenestra basalis, bordered by the lamina transversalis (anterior and posterior moieties), the paraseptal cartilage, and the lower margin of the pars nasi. Much of the fenestra basalis is blocked ventrally by the development of the palatal processes of the maxilla and palatine. The anterior end of the fenestra basalis remains open ventrally because the palatal wing of the premaxilla (incorrectly identified as the prevomer in early studies) develops with a broadly concave posterior border (Fig. 2B). A prevomerine process of the premaxilla projects posteriorly to cover ventrally the paraseptal cartilage (Figs. 1A, C; 2B). These outgrowths leave the incisive foramina as somewhat narrow or slit-like elongate openings continuous with the oral cavity. In other mammal groups the incisive foramina are usually very small openings resulting from a less extensive fenestra basilis and more marked expansion of the bony elements of the palate.

Thus, with respect to both ontogeny and adult configuration of palatal features, rodents and lagomorphs are notably similar. The morphotypical rodent condition includes enlarged, more posteriorly positioned incisive foramina (Wahlert, 1974) and greatly reduced palatine exposure on the palate. At the very least, the basic rodent condition is what might be expected in a lagomorph ancestor with a short snout, and clearly departs from the usual condition in mammals.

The orbital mosaic is most intriguing as evidence for higher-level mammalian relationships. Unfortunately, discrimination of specialized from primitive conditions in this region is exceedingly difficult. Needed are more detailed studies of ontogeny of the orbital elements in a diversity of taxa. Muller (1934) has contributed the most comparative information on the orbital-temporal

Table 2

Comparative Cranial Conditions in Selected Eutheria. Numbers refer to skull[1] features listed in Table 1. Letters denote traits also described in Table 1.

Features	RODENTIA	LAGOMORPHA	LEPTICTIDAE	MACROSCELIDEA	SCANDENTIA	SORICOMORPHA	TENRECOMORPHA	ERINACEOMORPHA	DERMOPTERA	"EUPRIMATES"	PLESIADAPIFORMES	CHIROPTERA	ANAGALIDAE
1. Premaxilla	B	C	A	A	A	A	A	A	A	A	B	D	A?
2. Infraorbital canal	B	B	A	A	A	B	B	A	A	A	A	B	A
3. Anterior nasals	B	B	A	A	A	A	A	A	A	A	A	A	-
4. Frontal-facial wing	A	B	A	A	A	A	A	A	A	A	A	A	A
5. Palatine-palatal exposure	B	B	A	A	A	A	A	A	B	A	B?	A	A
6. Incisive foramina	B	C	A	A	A	A	A	A	A	A	A	A	A
7. Maxillary fenestrae	A	B	A	A	A	A	A	A	A	A	A	A	A
8. Lacrimal	B	B	B	A	A	A	B	B	A	B?	A	A	-
9. Nasolacrimal foramen	B	D	A	A	A	B	C	A	A	A	A	A	A[2]
10. Orbitosphenoid	B	C	A	B	B	A	A	B	A	A	A?	A?	-
11. Maxilla in orbit	B	A	B	A	A	C	C	C	A	B?	B	A	-

Table 2 (continued)

Features	RODENTIA	LAGOMORPHA	LEPTICTIDAE	MACROSCELIDEA	SCANDENTIA	SORICOMORPHA	TENRECOMORPHA	ERINACEOMORPHA	DERMOPTERA	"EUPRIMATES"	PLESIADAPIFORMES	CHIROPTERA	ANAGALIDAE
12. Palatine in orbit	B	B	B	A	A	C	C	C	A	B?	B	A	-
13. Temporal fossa	B	B	A	A	A	A	A	A	B	B	B	B	B
14. Zygoma	B	B	A	A	A	C	C	A	B	B	B	A	A
15. Optic foramen	C	D	A	A	A	E	E	A	B	B	B	A	B
16. Sphenopalatine foramen	B	B	A	A	A	B	B	B	A	A	A	?	B?
17. Dorsal palatine foramen	A	B	A	A	A	A	B?	A	A	?	A	?	-
18. Suboptic foramen[3]	A	A	A	B	A	A	A	B	A	A	A	A	-
19. Sphenorbital fissure	B	B	A	A	B	B	B	B	B	B	A	A	B
20. Alisphenoid canal	B	C	B	A	B	A	A	B	A	A	A	A	B?
21. Sphenoidal foramina	B	B	A	B	A	A	A	A	A	?	?	A	A
22. Accessory foramen ovale	B	B	A	A	A	A	A	A	B	?	?	A	B?
23. Pterygoid fossa	A	B	B	A	A	A	A	B	A	A	A	A	A?

(continued)

Table 2 (continued)

Features	RODENTIA	LAGOMORPHA	LEPTICTIDAE	MACROSCELIDEA	SCANDENTIA	SORICOMORPHA	TENRECOMORPHA	ERINACEOMORPHA	DERMOPTERA	"EUPRIMATES"	PLESIADAPIFORMES	CHIROPTERA	ANAGALIDAE
24. Middle lacerate foramen	B	B	A	A	A	A	A	A	A	A	A	A	A
25. Temporal foramina	A	B	A	B	B	A	B	A	B	B	B	B	-
26. Auditory bulla	F	F	B	G	C	A	D	D	C	E	C?	C	F?
27. Ectotympanic	B	B	A	B	B	A	A	A	B	B	B	B	A
28. Postglenoid foramen	B	B	A	A	A	A	A	A	A[4]	A	A	A	A?
29. Superior ramus exit	B	B	A	A	A	A	A	A	B	A	A	A	?
30. Roof of auditory meatus	B	B	A	B	A	A	A	B	B	A?	A	A	?
31. Medial internal carotid a.	A	B	C	C	C	C	C	C	?	C	C	?	-
32. Coronoid process	B	C	A	B	A	A	A	A	B	A	A	A	A
33. Jaw condyle[6]	B	C	A	B	A	A	A	A	B	A	A	A	B[5]
34. Postglenoid process	C	C	A	B	B	B	B	B	A	A	A	B	B
35. Jaw adductors	B	C	A	?	A	A	A	A	?	A	A?	A	-

Table 2 (continued)

Features	RODENTIA	LAGOMORPHA	LEPTICTIDAE	MACROSCELIDEA	SCANDENTIA	SORICOMORPHA	TENRECOMORPHA	ERINACEOMORPHA	DERMOPTERA	"EUPRIMATES"	PLESIADAPIFORMES	CHIROPTERA	ANAGALIDAE
36. Posterior zygomatic process	A	B	A	A	A	A	A	A	A	A	A	A	A

[1] -, traits unknown; ?, trait uncertain, poorly preserved in fossil, not described, or morphotypical condition difficult to identify.

[2] In anagalids laterally open but below antorbital rim of zygoma.

[3] Polarity of presence or absence of suboptic foramen is highly tentative.

[4] Postglenoid foramen present but uniquely large, lenticular, cancellous, and located within anterolateral buttress of bulla.

[5] In _Anagale_, the jaw condyle is high, but not to the degree seen in rodents and lagomorphs.

[6] Features of the jaw joint, snout, and dentition in the Tertiary eurymyloids suggest relationships with rodents or lagomorphs. However, the Eurymyloidea is a problematic group showing a diversity of morphologies. Its affinities with the Glires clade remain uncertain. (See also Li and Ting, this volume)

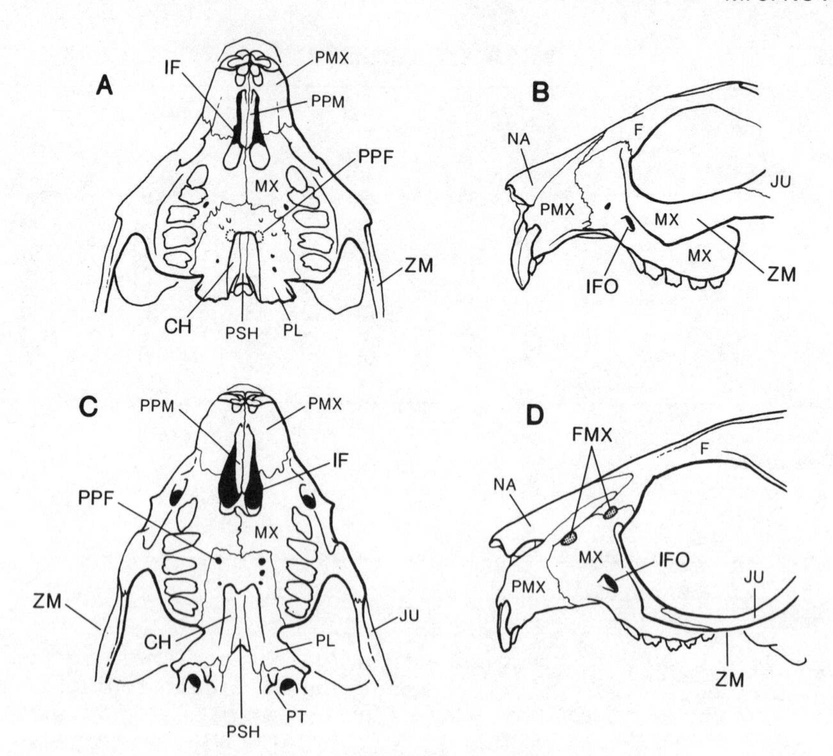

Fig. 1. Development of the anterior skull. (A) ventral (B) lateral views of AM 59737, *Ochotona* sp. (C) ventral (D) lateral views of AM 173732, *Lepus*, sp. For abbreviations, see text.

region. From Muller's important study, others (e.g., Butler, 1956; McDowell, 1958; Novacek, 1980) concluded that primitive eutherians had a large palatine in the orbital wall, whereas more specialized states showed the expanded maxilla and, sometimes, the alisphenoid "crowding out" the palatine in this region. Although this polarity scheme is supportable on several grounds (Novacek, 1980), it is by no means certain which condition is primitive for certain orders where both conditions exist (for example, primates, see discussion of Cartmill, 1975).

In rodents, there is a large orbital wing of the maxilla which, in some instances, forms a narrow lateral bridge (Fig. 3A) that contacts the forwardly expanded alisphenoid (Wahlert, 1974). The palatine is ventrally "pinched" by intrusion of its neighboring elements. The maxilla is a much less dominant element in the orbital wall of lagomorphs, a putatively primitive trait. Nevertheless, there is enough expansion of this element, and that of the alisphenoid, to effect a ventral constriction of the orbital process of the palatine (Fig. 3B). A possible correlative feature is the position of the sphenopalatine foramen at or near the maxillary-palatine suture (character 16B, Tables 1 and 2). The intrusion of the maxilla into the orbit is even more pronounced in various (lipotyphlan) insectivoran groups (Butler, 1956; McDowell, 1958).

In both rodents and lagomorphs there seems to be a similar modeling of the orbital wall. The temporal fossa is effectively lost or obscured by the lateral expansion of the alisphenoid that forms the back of the orbit. The orbitosphenoid is situated high in the orbital wall (Fig. 3). The ethmoidal, optic, and sphenopalatine foramina seem to be "shifted forward" relative to other reference points of the cranium (e.g., the cheek tooth row). Finally, in the orbital-temporal region there is a distinctive development of extra foramina and surface canals (Fig. 3) for passage of the mandibular branch of the trigeminal nerve (character 22B) and the masseteric and buccinator divisions of the maxillary nerve (character 21B). (See Wahlert, 1974, for excellent descriptions of these traits in rodents.) These attributes are conceivably correlated with refashioning of the jaw closing musculature and the cranio-mandibular joint, and they are not restricted to lagomorphs and rodents. Nonetheless, they are clearly derived conditions that corroborate other evidence for Glires.

The auditory bulla in rodents and lagomorphs has attracted special interest. Often emphasized is the fact that both rodents and lagomorphs have well developed, inflated bullae. This similarity hardly seems significant, as many mammals have inflated bullae. Possibly, in the most primitive rodents this structure was poorly developed. (The bulla is not preserved in many skulls of early rodents, but it is probable that these primitive forms had a bulla weakly attached to the basicranium; such a structure would be easily lost during fossilization - see Wood, 1962; Korth, 1984.)

What, then, is the significance of the bulla in lagomorphs and rodents? It is likely that the most primitive eutherians lacked a bulla and simply "covered" the auditory cavity with a tympanic membrane (Fig. 4, _Tachyglossus_) within a horizontally inclined, annular ectotympanic and with membranous or fibrous, dense connective tissue (Gregory, 1910; Archibald, 1977; Novacek, 1977; MacPhee, 1981). Subsequently, a well ossified bulla probably developed independently in a number of different lineages (Novacek, 1977). The important comparative distinctions involve the nature of the elements comprising the bulla rather than the degree of bullar inflation (Klaauw, 1931).

With respect to homology of bullar elements, lagomorphs and rodents clearly form a group. The bullae are made virtually entirely of an expanded ectotympanic (Fig. 4). This element contributes to the bullae of many other groups (Novacek, 1977; Table 2), but usually as a component with the entotympanic or some other bone. A com-

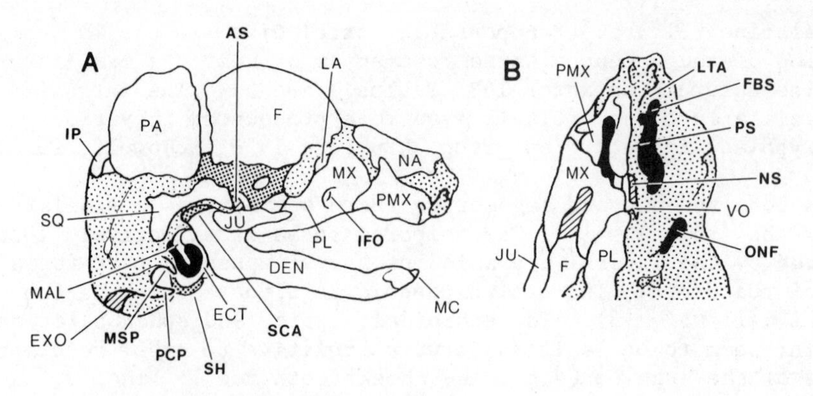

Fig. 2. Skull of Lepus sp., 45 mm. stage. (A) lateral view; (B) ventral view of nasal-palatal region (after Voit, 1909). For abbreviations, see text.

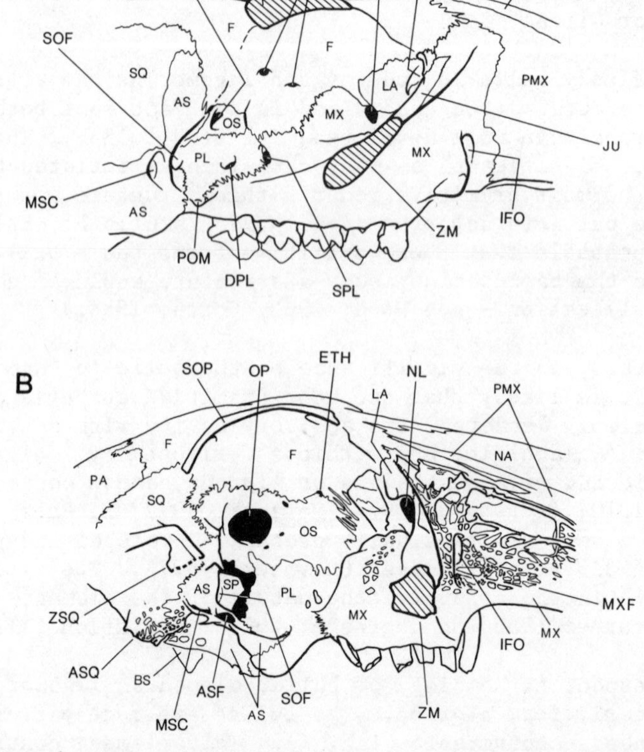

Fig. 3. Lateral views of orbital region. (A) C.A. 1004, Marmota sp.; (B) C.A. 1265, Lepus sp. For abbreviations, see text.

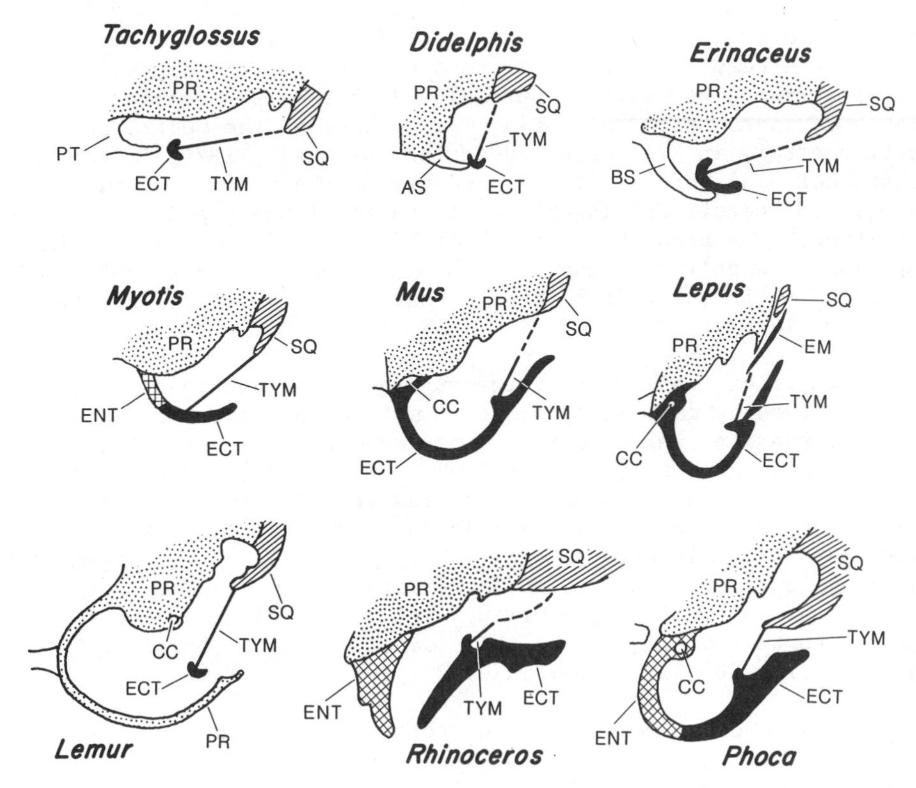

Fig. 4. Rostral views of transverse sections of the mid-tympanic region in several mammals (after Kampen, 1905). Dashed line represents pars flaccida. For abbreviations, see text.

pletely ectotympanic bulla is found only in rodents, lagomorphs, cetaceans, proboscideans, artiodactyls, and some perissodactyls (Klaauw, 1931). Association of rodents with any one of these non-lagomorph groups fails to offer an attractive alternative to Glires. It is by no means certain that ungulates primitively had a well developed ectotympanic bulla. The basic basicranial pattern for "ungulates" is, in fact, very conservative (MacIntyre, 1972), and rather like that of leptictids where the bulla is partial and formed from an entotympanic (for example, see mesonychid condylarths, Klaauw, 1931). The balance of cranial traits suggest that if any relationship of rodents with a certain ungulate group is likely, it entails the derivation of a monophyletic rodent-lagomorph lineage within the ungulate radiation. This alternative seems quite feasible but has not been explored sufficiently. For the present, the ectotympanic bulla must be recognized as a supportive, but not unique, character of Glires.

Aside from the condition of the bulla, the auditory region, as mentioned above, offers few clues to rodent affinities. The development of the ectotympanic meatal tube (Fig 4) and the depression for the epitympanic recess, are specialized, but rather common traits (Table 2, characters 27B, 30B). The pathway of the medial internal carotid artery in lagomorphs (described by Bugge, 1974) is anomalous (character 31B). However, the presence of the medial internal carotid in rodents and lagomorphs is quite likely a primitive condition. The promontory branch of the carotid clearly seen in tupaiids, insectivorans and other mammals possibly represents a secondary shift in the medial internal carotid to a lateral position (Presley, 1979).

Features of the jaw apparatus in lagomorphs and rodents have been reviewed by many workers. It is noteworthy that the jaw musculature orientation, relative mass, and function differ significantly between these groups (Lopez-Martinez, this volume). Yet, these traits form a constellation of features that distinguish lagomorphs and rodents from most other mammals whose affinities with the latter have been seriously entertained. The exceptions to this argument are the anagalids, eurymyloids, and a variety of other early Tertiary groups that have both a jaw arrangement and dentition somewhat similar to that of lagomorphs and rodents (McKenna, 1975; Li Chuan-kuei and Ting Su-yin, this volume).

These and other characters in Table 1 are considered in more detail in other publications (Wahlert, 1974; Novacek, 1980).

PHYLOGENETIC ALTERNATIVES

A simple tabulation of derived traits from the information provided in Table 2 clearly favors the lagomorph-rodent association. Consider the result:

Rodentia-Lagomorpha: 1B(C), 2B, 3B, 5B, 6B(C), 8B, 9B(D), 10B(C), 12B, 13B, 14B, 15C(D), 16B, 19B, 20B(C), 21B, 22B, 24B, 26F, 27B, 28B, 29B, 30B, 31B(C), 33B(C), 34C, 35B(C)

or 27 potential synapomorphies, where a more derived expression of a homologous trait is indicated by parentheses. Several of these traits are widely distributed among the groups listed in Table 2. Such homoplasious characters must be ruled out as evidence for the monophyly of Glires (Fig. 5). Nevertheless, the character data is less kind to alternatives. The next strongest association, that of Rodentia-Dermoptera (!), recruits only twelve shared-derived traits. Some groups, such as the leptictids, chiropterans, "euprimates," and tupaiids, fare poorly in these comparisons with less than six conditions shared with rodents in each case. Anagalids show only eight special similaries, but some of these features (auditory bulla) clearly imply a close relationship. The tally for anagalids is doubtless biased by incomplete knowledge of anagalid morphology, a

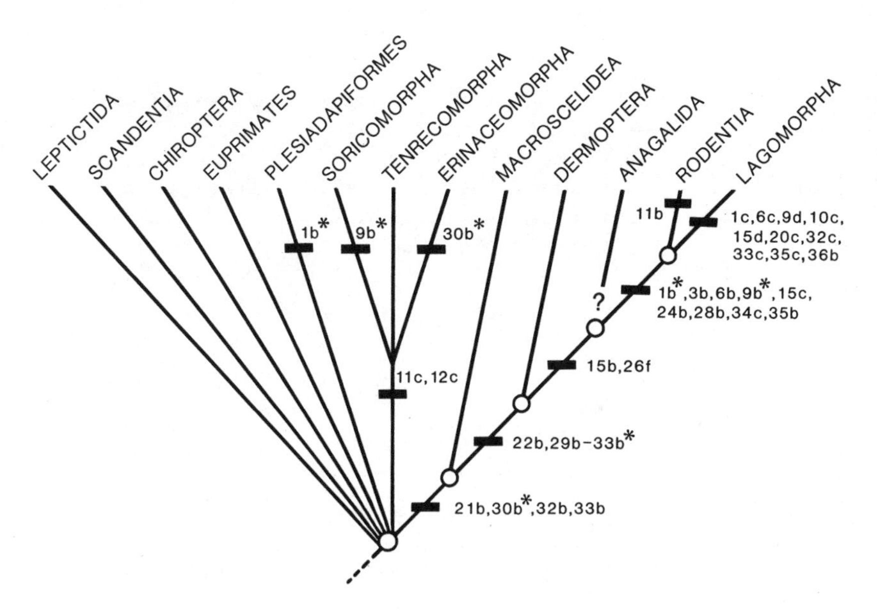

Fig. 5. A cladogram for cranial traits (Table 1) in selected Eutheria (Table 2). Asterisks indicate homoplasious (reversal or convergence) characters. Negative sign indicates reversal event. Some derived characters shared by lagomorphs and rodents show marked homoplasy and are omitted. Question mark indicates that many characters listed in Tables 1, 2 are not known for anagalids.

situation that will be improved upon (McKenna and Bleefeld, in preparation).

For those disturbed by this simple jostling of numbers, it is noted that at least six traits (anterior flaring of the nasals, incisive foramina, middle lacerate foramen, jaw articulation in glenoid region, and jaw closing musculature) are unique to lagomorphs and rodents, at least within the range of taxic diversity represented in Table 2. Obviously, the consideration of a broader sample would dilute this result (for example the middle lacerate foramen is present in some carnivorans). Thus, one is left with very little in the way of unique evidence for particular groupings of orders. A cladogram (Fig. 5) for the character matrix (Table 2) admits some homoplasious "noise." This pattern, however, clearly favors the monophyly of Glires. I conclude therefore that certain cranial features, when considered in combination with other biological traits of these taxa, allow one to muster a compelling case for Glires; at least, as Gregory (1910) noted, a case more convincing than most other associations of eutherian orders.

ACKNOWLEDGMENTS

I am grateful to W. P. Luckett, L. Flynn, N. Hong, M. C. McKenna, F. S. Szalay, A. Bleefeld, J. Wahlert, A. Wyss, N. Shubin and W. Maier for stimulating discussion before, during and after the NATO-CNRS symposium. To numerous other symposium participants I also extend thanks. L. Meeker, L. Lomauro, and C. Tarka prepared the figures. Partial support was provided by the Frick Laboratory Endowment Fund (American Museum of Natural History).

ABBREVIATIONS

AS alisphenoid
ASF anterior sphenoidal foramen
ASQ suture between alisphenoid and squamosal
BS basisphenoid
CC carotid canal
CH choanae
DEN dentary
DPL dorsal palatine foramen
ECT ectotympanic
EM external auditory meatus
ENT entotympanic
ETH ethmoidal foramen
EXO exoccipital
F frontal
FBS fenestra basilis
IF incisive foramen
IFO infraorbital foramen
IP interparietal
JU jugal
LA lacrimal
LTA lamina transversalis anterior
MAL malleus
MC Meckel's cartilage
MSC sphenoidal foramen (for masseteric nerve)
MSP mastoid process
MX maxilla
MXF maxillary fenestrae
NA nasal
NL nasolacrimal foramen
NS nasal septum
ONF orbitonasal fissure
OP optic foramen
OS orbitosphenoid
PA parietal
PCP paracondylar process
PL palatine
PMX premaxilla

POM posterior maxillary notch
PPF posterior palatine foramen
PPM prevomerine process of the premaxilla
PR petrosal
PS paraseptal cartilage
PSH presphenoid
PT pterygoid
SCA secondary cartilage
SH stylohyal cartilage
SOF sphenorbital fissure
SOP supraorbital process (of frontal)
SP sphenoid
SPL sphenopalatine foramen
SQ squamosal
TYM tympanic membrane
UML unossified area in maxillary-lacrimal suture
VO vomer
ZM zygomatic process of maxilla
ZSQ zygomatic process of squamosal

Institutions

AM American Museum of Natural History (Department of Mammalogy).
CA American Museum of Natural History (Department of Mammalogy, Comparative Anatomy Collections).

REFERENCES

Archibald, J. D. 1977. The ectotympanic bone and internal carotid circulation of eutherians in reference to anthropoid origins. Jour. Human Evol. 6:609-622.

Bugge, J. A. 1974. The cephalic arterial system in insectivores, primates, rodents, and lagomorphs, with special reference to systemic circulation. Acta Anat. 87 (suppl. 62):1-160.

Butler, P. M. 1956. The skull of *Ictops* and a classification of the Insectivora. Proc. Zool. Soc. London 125:453-481.

Cartmill, M. 1975. Strepsirhine basicranial structures and affinities of Cheirogaleidae, In: Phylogeny of the Primates, W. P. Luckett and F. S. Szalay, eds., pp. 313-354, Plenum Press, New York.

Dawson, M. R. 1958. Late Tertiary Leporidae of North America. Univ. Kansas Paleont. Contrib. 22:1-79.

Dawson, M. R. 1969. Osteology of *Prolagus sardus*, a Quaternary ochotonid (Mammalia:Lagomorpha). Palaeovertebrata 2:157-190.

DeBeer, G. R. 1937. The development of the vertebrate skull. University Press, Oxford.

Fawcett, E. 1917. The primordial cranium of *Microtus amphibus* (water rat), as determined by sections and a model of the 25 mm stage. With comparative remarks. J. Anat. 51:309-359.

Gidley, J. W. 1912. The lagomorphs, an independent order. Science N.S. 36(922):285-286.

Gregory, W. K. 1910. The orders of mammals. Bull. Amer. Mus. Nat. Hist. 27:1-524.

Hartenberger, J.-L. 1977. A propos de l'origine des Rongeurs. Geobios Mem. Spec. 1:183-193.

Hennig, W. 1966. Phylogenetic systematics. Univ. Illinois Press, Urbana.

Kampen, P. N. van 1905. Die Tympanalgegend des Säugetierschädels. Geg. Morph. Jb. 34:321-722.

Klaauw, C. J. van der 1931. The auditory bulla in fossil mammals, with a general introduction to this region of the skull. Bull. Amer. Mus. Nat. Hist. 62:1-352.

Korth, W. W. 1984. Earliest Tertiary evolution and radiation of rodents in North America. Bull. Carnegie Mus. Nat. Hist. 24:1-71.

Lavocat, R. 1967. Observations sur la region auditive des Rongeurs Theridomorphes. Colloq. Internatl. C.N.R.S. 163:491-501.

Lopez-Martinez, N. 1978. Cladistique et paleontologie. Application a la phylogenie des Ochotonides europeens (Lagomorpha, Mammalia). Bull. Soc. Geol. France 20:821-830.

Luckett, W. P. 1977. Ontogeny of amniote fetal membranes and their application to phylogeny. In: Major patterns in Vertebrate Phylogeny, M. K. Hecht, P. C. Goody and B. M. Hecht, eds., pp. 439-516, Plenum Press, New York.

MacIntyre, G. T. 1972. The trisulcate petrosal pattern in mammals. In: Evolutionary Biology, volume 6, M. K. Hecht, W. C. Steere and B. Wallace, eds., pp. 275-303, Plenum Press, New York.

MacPhee, R. D. E. 1981. Auditory regions of primates and eutherian insectivores. Morphology, ontogeny, and character analysis. Contrib. Primatol. 18:1-282.

McDowell, S. B. 1958. The Greater Antillean insectivores. Bull. Amer. Mus. Nat. Hist. 115:113-214.

McKenna, M. C. 1961. A note on the origin of rodents. Amer. Mus. Novit. 2037:1-5.

McKenna, M. C. 1975. Toward a phylogenetic classification of the Mammalia. In: Phylogeny of the Primates, W. P. Luckett and F. S. Szalay, eds., pp. 21-46, Plenum Press, New York.

Muller, J. 1934. The orbitotemporal region in the skull of the Mammalia. Archiv. Neerl. Zool. 1:118-259.

Novacek, M. J. 1977. Aspects of the problem of variation, origin and evolution of the eutherian auditory bulla. Mammal Rev. 7:131-149.

Novacek, M. J. 1980. Cranioskeletal features in tupaiids and selected Eutheria as phylogenetic evidence. In: Comparative Biology and Evolutionary Relationships of Tree Shrews, W. P. Luckett, ed., pp. 35-93, Plenum Press, New York.

Novacek, M. J. 1982. Information for molecular studies from anatomical and fossil evidence on higher eutherian phylogeny. In: Macromolecular Sequences in Systematic and Evolutionary Biology,

M. Goodman, ed., pp, 3-41, Plenum Press, New York.
Parent, J.-P. 1980. Recherches sur l'oreille moyenne des Rongeurs actuels et fossiles. Mem. Trav. EPHE, Inst. Montpellier, 11:1-286.
Presley, R. 1979. The primitive course of the internal carotid artery in mammals. Acta Anat. 103:238-244.
Russell, L. S. 1959. The dentition of rabbits and the origin of lagomorphs. Bull. Nat. Mus. Canada, 166:41-45.
Szalay, F. S. 1977. Phylogenetic relationships and a classification of the eutherian Mammalia. In: Major Patterns in Vertebrate Evolution, M. K. Hecht, P. C. Goody, and B. M. Hecht, eds., pp. 315-374, Plenum Press, New York.
Voit, M. 1909. Das Primordialcranium des Kaninchens unter Berücksichtigung der Deckknochen. Anat. Hefte 38:425-616.
Wahlert, J. H. 1974. The cranial foramina of protrogomorphous rodents; an anatomical and phylogenetic study. Bull. Mus. Comp. Zool. 146:363-410.
Wood, A. E. 1957. What, if anything, is a rabbit? Evolution 11:417-425.
Wood, A. E. 1962. The Early Tertiary rodents of the family Paramyidae. Trans. Am. Phil. Soc. N.S. 52(1):1-261.
Wood, A. E. 1965. Grades and clades among rodents. Evolution 19:115-130.

RODENT AND LAGOMORPH MORPHOTYPE ADAPTATIONS, ORIGINS, AND RELATIONSHIPS: SOME POSTCRANIAL ATTRIBUTES ANALYZED

Frederick S. Szalay

Department of Anthropology, Hunter College
City University of New York
New York, N. Y. USA

INTRODUCTION

The literature on postcranials, whether descriptive or analytical, paleontological or neontological, is amazingly scanty for the largest of the mammalian orders, the Rodentia. Although some excellent descriptions can be found on fossils (see especially Wood, 1937, 1962; Emry and Thorington, 1982), and equally important contributions have been made to comparative myology of the postcranium (e. g., Hildebrand, 1978; Woods, this volume, and references therein), this evidence has not been utilized convincingly to attempt an understanding of the origins and relationships, and the early postcranial adaptations of the order. My aim in this paper is to concentrate on these two closely interrelated goals of evolutionary analysis, primarily for the Rodentia, but with some definite observations on the Lagomorpha, *sensu lato*, as well.

Skeletal morphology of rodents is very unequally known. This is due both to the emphasis on descriptive or analytical studies of living species of *Rattus* and *Mus*, as well as to the discovery and study of different parts of fossil skeletons. It is without question that all aspects of the osteology and soft structures should be utilized to approximate the lofty goals of evolutionary morphology, as it was so clearly stated by Wood (1937). Yet my present short study is a mere expediency to maximize phylogenetic as well as adaptive input into the problem of rodent origins. I will attempt this by emphasizing a part of the skeletal anatomy which is complex, more widely known in the mammalian fossil record than other areas, and which is adaptively more clearly channelled for locomotor and substrate occupancy related mechanics than the rest of the skeleton.

I should also emphasize that the hand is an area of the skeleton which is comparable in its character complexity, and therefore in its potential evolutionary usefulness, to the foot. Knowledge of the carpus, however, either paleontological or comparative, lags far behind the modestly growing evolutionary accounts of the osseous foot. As I have discussed elsewhere (Szalay, 1984), the tarsus has adaptively less complex biological roles associated with it (inasmuch as it is rarely used for exploration and food manipulation) than the hand. Consequently, the significance of various tarsal character states in fossil taxa is probably more easily ascertained than that of the hand.

NOTES ON METHODS OF ANALYSIS

After the student identifies the broadly homologous features in the taxa studied, the morphocline, the analysis becomes imprecise and challenging. In which direction did evolutionary change occur, and is the most likely antecedent condition unrepresented in the morphocline?

The study of biological history still continues to be conducted by many students along the allegedly independent lines of either attempting to understand the genealogy of characters or taxa, or the nature of the mechanics or adaptations of specific entities throughout their entire ontogeny. What is remarkable about this dichotomous approach is that these events of the past, although defined as separate problems, are obviously inseparable in the actual evolutionary history of organisms (see especially Gutmann, 1977, for a thorough discussion of what phylogeny really is).

For the student of genealogies, the clues to the path of descent should be abstracted from all aspects of the characters studied (and from as great a variety of characters as it is possible to obtain), and, for the functionalist-adaptationist, the balanced analysis of the history of adaptive features is not likely without an attempt to see transformations in time. It is of some theoretical importance for this paper to consider, even briefly, the rationale for a combined approach to pursue these inseparable goals of evolutionary studies. In order to place this attempt in perspective, it is necessary to cast a brief critical glance at the reasons why cladistics, or neocladistics, or transformed cladistics, represent theoretical approaches, which, in spite of their glittering promise and appeal to the tidy in all of us, because of their biologically unrealistic, or wanting, assumptions, can not yield high probability approximations of actual biological history.

Cladists, starting from a sound principle defined more clearly than before by Willi Hennig (1966) (the "Hennig principle"), have developed a set of structuralist rules in order to generate genealogies. A genealogy is an abstract concept derived from the fact of evolutionary history, and genealogical knowledge must be based on the few features we have at our disposal from the diversity produced by evolution. This diversity of features, the morphoclines, must be interpreted to be consistent with an evolutionary hypothesis. Cladists, after their attempts to "eliminate" a number of axiomatic evolutionary assumptions from the study of history, pursue this activity by studying primarily, usually only, the taxic distribution of these characters, and consider this as the only and primarily acceptable activity for the understanding of genealogy. In order to construct a peculiar cladogram without any time value for taxa or characters (Stuffenreihe or Hennig combs, as called by some authors), the distribution of characters is analyzed by the so-called cladistic outgroup method (not to be confused by an uncircular and purely phenetic notion of the outgroup, meaning phena other than the one studied in detail), by the commonality principle, and when available, the sequence of appearances of characters in ontogeny.

The genealogy of sister-group relationships thus obtained, without weighting (or at least so claimed, and therefore without regard for the lesser or greater probability of parallelisms) is said to be the scientific basis for what has been called the scientifically less secure evolutionary hypothesis which may follow if one is inclined to consider an evolutionary scenario. Thus, through the writings and logic of a number of cladists this particular definition of systematics was slated to become the primary scientific base for all of the "less scientific" other activities of evolutionary science. The assumptions derived from evolutionary theory themselves have been considered by cladists as unacceptable in the practice of cladistic systematics, beyond the gentlemen's agreement that evolution really did take place. The weaknesses of this structuralist and essentially nonevolutionist view of biological systematics is considered elsewhere within the framework of a study of the adaptations, ecology, and character distribution of all fossil kangaroos (Szalay and Flannery, in preparation). The so-called primacy of a character distribution analysis in phylogenetics is true only inasmuch as it is necessary to take stock of the range of the available information one wants to analyze. The most important characters are often not even perceived until the investigation begins as a total transformation analysis considering, in addition to distribution, bioroles, form-function, ecology, temporal and spatial distribution, all as centrally relevant, rather than "more" and "less" scientific as various cladistic schemes would have it.

The recent ritualistic relegation of stratigraphic data, ancestor-descendant hypotheses, and functional-adaptive studies, all pejoratively labelled as "scenarios" to emphasize their distinction from "real science" practiced by cladists, and relegated to a second rate scientific status by post-Hennigean cladists (e. g., Eldredge and Cracraft, 1980), are a revealing symptom of the tenuous connection of cladistics to evolutionary theory. The ideas of the so-called transformed cladists, although vigorously under attack by the "real" cladists as having gone "too far" (to paraphrase the remarks of several colleagues of cladist persuasion), are merely the logical consequences of the very first attempts to decipher the genealogical meaning of biological features in a nonevolutionary, nonadaptive context. The various cladistic dictums appear to be manifestations of a curious "separatist" idealogy from evolutionary biology, and the now increasingly rule and procedure-bound field of cladistics, after having failed to come to grips with the probabilistic nature of evolutionary science, with its mixed use of induction, deduction, and prediction all as valid methods (see especially Van Valen, 1982), attempts to rid itself of any relationship to the study of adaptations.

Bock (1977, 1981), Gutmann (1977), Gingerich (1980), Godinot (1981), Szalay (1977a, 1981a, b), Mayr (1974, 1981), Peters (1972), Simpson (1975), Van Valen (1978, 1982), and many others have commented extensively on the theoretical unsoundness of the cladistic approach to evolutionary history, and subsequently to an aspect of this history, its genealogy.

In contrast to cladistic analysis, there is what may be referred to as evolutionary, or transformational analysis. Avenues of this approach have been outlined by Bock (1981), who has coined the term _functional-adaptive analysis_ for the study of character transformations, or, as it has been called, the analysis of polarity of morphoclines. Although I disagree with Bock on his rejection of the use of analogies in order to understand past adaptations (see Szalay, 1981a), I fully endorse his view that only a functional and adaptive appraisal of characters can yield bioscientifically (which should mean consistent with evolutionary theory) meaningful appraisal of morphoclines.

In the purposefully broadly defined sense, students who consider themselves phylogeneticists or evolutionists (their classificatory penchants notwithstanding) analyze the raw data derived from the distribution of characters and attempt, through transformational studies, to understand which are the most probable transformation sequences of the morphoclines. Because the form-functional and adaptive modifications of homologs are constrained, facilitated, and channelled by ancestor-descendant relationships (hence the necessity for vertical comparisons _sensu_ Bock, 1977; Szalay, 1977a), a fundamental assumption in all analyses involving evolution, this approach is consistent with evolutionary theory. If members of a morphocline can be shown to approximate a transformation series through the mechanics and associated bioroles, the latter often strengthened by the use of analogies from unrelated groups (Szalay, 1981a), and the reverse may be shown to be less probable, then a highly corroborated base for taxon phylogeny has been created. The fossil record, when it exists, can be enormously helpful in: (1) providing additional stages of morphology to facilitate both the erection and testing of transformation hypotheses; (2) suggesting a temporal, geographic, and sometimes ecological framework; and (3) increasing the probability of one hypothesis of transformation over others.

In contrast to cladistics, which is preoccupied with sister group hypothesizing, cladogram testing, and two, three (or more) taxon statements, in transformational analysis the effort is placed squarely on the understanding of character state transformations, both in terms of the constraints dictated by the developmental process (see Maier and Dierboch, this volume), as well as by the shaping forces of the adaptive process. Understanding of primitive character states is as important as the recognition of advanced ones, because it is absurd to contemplate one without the other within an evolutionary framework.

A character, or preferably a character complex, is advanced if it can be shown to have more probably evolved from its designated homolog than any other likely transformation sequence. It follows from this approach that proposals pertaining to the central concept of homology by cladists, which would restrict the concept to synapomorphies alone, lack evolutionary and therefore systematic significance. In such an artificially defined cladistic context, homologies could not potentially delimit the range of choices and potential points of origin for the variously derived characters.

In contrast to the cladist's most congruent synapomorphy scheme, based on the distribution of characters on a single time plane, irrespective of the actual temporal spread, which would represent the most corroborated Stuffenreihe, transformationists would construct a cladogram or tree based on the best understood, most complex characters (which are least likely to display parallel evolution), for which either sound stratigraphic, or functional and adaptive reasons, or both, exist to allow the differentiation of homologs into primitive and advanced states. Although selection may favor the same or mechanically equivalent form-function in a lineage as in a previous ancestral stage, the new solutions can be only those which will be available after the constraints of the antecedent stage. As Dollo (1893) recognized long ago, it is improbable that details of previous evolutionary stages ever reappear again. The appreciation of this is the cornerstone of transformational analysis and therefore phylogenetics, in contrast to the character distribution analysis of cladistics. Evolution of _Equus_ into _Hyracotherium_ becomes a rather remote possibility when the re-evolution of brachyodont cheek teeth bearing a limited number of cusps and ridges, along with the reacquisition of a pentadactyl foot while retaining the tarsal joint complex of the monodactyl "ancestor" must be contemplated.

I believe that no evidence of any sort exists which negates the general guiding principles (providing the analyzed evidence does not contradict it) that: (1) primitive characters are more likely to occur early in the history of a group; (2) parallelisms can be very common in simple and probably in more complex characters as well; and (3) evolutionary change is usually adaptive. It follows that to understand transformations one must consider stratigraphy (where judged adequate; see especially Simpson, 1975; Godinot, 1981), as well as the results of functional-adaptive analysis, in order to utilize high probability transformation information in accordance with Hennig's principle. Belief in the latter dictum alone is of questionable importance for any systematic undertaking without the bioscientifically (i. e., evolutionarily) oriented research program into the history of character complexes.

In characterizing aspects of the morphotypic conditions of the tarsus in the Rodentia I have used features gleaned from *Paramys*. This has nothing to do with notions of the cladistic outgroup method, but was based on the perception that all the known conditions of the skeleton among rodents (known superficially at best) can be derived from such a morphological complex more easily than from any other. Advanced characters, therefore, are established if they could be derived from the postulated primitive one more easily than the reverse. This transformation sequence is strongly corroborated, and consequently made even more probable by the stratigraphic position of paramyids. Increased knowledge of *Cocomys* and relatives (see Dawson et al., 1984) may undoubtedly alter present perceptions. It is particularly with a group like the Rodentia, whose ancestry or sister-group relationships are obscure, that the cladistic outgroup method would show itself to be a mere euphemism for "broad and meaningful comparisons." Although usually it is intended for more, this structuralist and circular dictum, when used as defined and prescribed by Eldredge and Cracraft (1980), can not produce biologically sound results without the study of characters. When this "method" is evoked alone, however, it invariably obfuscates the theoretically, conceptually, and practically important procedures necessary to arrive at the most probable transformation events of specific evolutionary histories.

SOME RELEVANT ASPECTS OF THE EUTHERIAN MORPHOTYPE

Before I begin to analyze the cruropedal attributes of protorodents, in order to shed some light on their origins and adaptive beginnings, I will touch on some pertinent aspects of the protoeutherian conditions.

A number of substantial disagreements can be found in the literature concerning the sequence of transformations of morphological characters of the pes in the Mammalia, and the nature and meaning of various mechanical solutions of specific form-function stages

(Matthew, 1904, 1909; Haines, 1958; Jenkins, 1974; Szalay and Decker, 1974; Szalay, 1977b, 1982; Lewis, 1980a, b, c). I have recently completed a study (Szalay, 1984), the results of which have close relevance, both to these differing views as well as the analysis attempted in this paper.

Lewis (1980a, b, c) has presented an elaborately documented thesis concerning the origins of placental arboreality. He stated that the arboreal marsupials of today represented both a form-functionally sound as well as behaviorally faithful antecedent condition for the first placentals, tupaiid-like in his estimation, and that the protoplacentals, as Matthew (1909) stated long ago, were arboreal as were their marsupial ancestors.

My reexamination of these ideas was presented in the framework of a number of distinct ancestral-descendant stages of the evolution of the therian tarsus. I attempted to show how the phylogenetic constraints on the articular morphology of the eutherian upper ankle joint (UAJ; for all abbreviations see Table 1, and the various figures) channelled the modifications dictated by the changing biological roles. I have emphasized that the evolution of the tenon-mortise UAJ in the last eutherian common ancestor (the protoplacental), because of the transverse stability and anteroposterior mobility that such a joint provides for a terrestrial animal, profoundly influenced all subsequent modifications of eutherian arborealists. Unlike the arboreal metatherians, which can both extensively plantarflex-dorsiflex as well as abduct-adduct the foot at this joint, those eutherians which have become arboreal accomplish the necessary inversion through the extensive helical movements in the lower ankle joint (LAJ). In these forms the LAJ axis becomes much more transversely aligned to the long axis of the calcaneus than in their terrestrial ancestors or their marsupial ecological vicars. Because of the extreme mobility of the UAJ in arboreal Metatheria, inversion is not associated with pronounced helical movements (i.e., translation in addition to rotation) in the LAJ.

Evidence from the fossil record unequivocally suggests that the primitive therian tricontact UAJ (in which the fibula has broad contact with the calcaneus) was independently transformed into the bicontact cruropedal joint of metatherians and eutherians.

It should be emphasized here, since I will discuss specific arboreal adaptations of rodents below, that the differences in those aspects of form-function that are intimately involved with arboreality were dissimilar in the morphotype of marsupials and the morphotypic condition of any group of eutherian arborealists. Rodents, primates, arboreal carnivorans, among others, develop their tarsal adaptations to their substrate occupancy, not only because of some unique demands of their arboreally related activities (as argued for the Rodentia below), but also because of the constraints of the protoeutherian adaptations which evolved in response to a fully terrestrial, probably at least semi-cursorial, commitment. That is the reason why no known eutherian, arboreal or not, resembles arboreal marsupials in the UAJ. In conclusion, there is no evidence or convincing argument that arboreality (and its associated morphological, functional, and behavioral adaptations) is anything but convergent in the Metatheria and Eutheria.

FORM AND FUNCTION OF THE PROTORODENT POSTCRANIUM

Before I comment on the possible biological roles associated with the form-function complex of the crus and pes, a brief digression about the rodent morphotype seems appropriate. Korth (1984), in his recent study of the early Paleogene rodents of North America, has stated his belief that the Asian ctenodactyloid evidence, dental in particular, represents a better approximation of rodent ancestry than the morphology embodied in the Paramyinae. Although he did not consider the Ischyromyidae (which stems, according to him, from his concept of the Reithroparamyinae) as representing the basal stock of all later rodents, he nevertheless entertained the possibility that sciurids, eutypomyids, castorids, and European glirids may have originated from them. The morphology of the Asiatic early Eocene *Cocomys* (Dawson et al., 1984), which I have examined, leaves little doubt that the earliest known Asiatic rodents were, at least dentally, very similar to *Paramys*, in spite of the slightly better developed hypocone in the former. Calling *Cocomys* a ctenodactyloid, which it may or may not have been, perhaps obscures the significant sharing of morphotypic similarities of these protrogomorphous forms. The postcranial features of early Paleogene Asiatic rodents, when found, will either contradict or strengthen the morphology proposed for the morphotype.

It may be noted here that most concepts of early Tertiary groups of rodents (and later ones as well) continue to be based on dental morphology as well as cranial criteria. This practice will undoubtedly continue as the prevalence of these fossils appears to dictate this approach. It seems appropriate to add, however, that the often abundant postcranial remains in wash collections, tarsals and other elements, can add a new dimension to the quests for the understanding of phylogeny and adaptations. This is unquestionably difficult because the extant forms have been barely studied from a postcranial perspective, and attempts to diagnose extant higher taxa so far completely omit the unquestionably unique postcranial features which families and higher categories possess. However, it is obvious, to me at least, that once the postcranium is more thoroughly considered, the increased diversity of the characters which will result will surely be followed by a deeper evolutionary appreciation of the order.

In a paper reviewing the evidence for the relationships of eutherians based on my understanding of the tarsal evidence at that time (Szalay, 1977b), I considered the basicranial as well as some of the postcranial evidence as indicative of leptictid origin of the rodents. In this paper I reaffirm the shared special similarities in the selected areas of the postcranium (for the cranial evidence and its interpretation see Novacek, this volume, and Parent and Lavocat, this volume), and propose some reasons for the form-function and adaptive aspects of paramyid skeletal remains, primarily the crus and pes. I do disagree with Parent and Lavocat (this volume), in that I consider the basicranial morphology of *Paramys*, particularly the fossa for the stapedial muscle, the most primitive among known rodents. *Cocomys* is as yet unstudied in this regard.

The similarity of the protorodent cruropedal complex (see below for an explanation for the bases of this concept) to Cretaceous and Paleocene leptictids can be best explained as the derivation of rodents from a leptictid, or from a form which homologously shared that morphology with leptictids. The reverse hypothesis is clearly an untenable one in light of the dental and cranial advances in the first rodents. Similarly, a sister-group relationship is merely a less precise version, in this instance, of the evolutionary conditions which may be hypothesized, based on known evidence. Either leptictids or rodents may be more recently related to other groups based on undiscovered special similarities, or else this latter fact may have become masked by the uniquely evolved features of these "crypto-sister groups". At the moment, it seems to me that the most precise statement of rodent origins calls for an ancestor-descendant hypothesis on the level of the Leptictidae and Paramyidae. To be sure, the derived fusion of the distal crus in a new late Paleocene leptictid under description by Novacek and Shubin would argue against this, if one did not stipulate the existence of more primitive Leptictidae without this fusion.

The earliest known rodents, from North America, are well described in Wood's (1962) monograph. More recent phylogenetically and adaptively analytical accounts, however, of the abundant rodent postcranial remains of the well washed quarries of the Eocene and the nearly complete skeletons of that epoch have not been attempted. This is no surprise, as the evolutionary skeletal morphology of the living rodents is known only in a most superficial way. Notable exceptions are such excellent contributions as those of Romankowowa (1960, 1963), Hildebrand (1978), Emry and Thorington (1982), among others.

Morphotypic features of the Rodentia, as often stressed by Wood (1962), are largely embodied by the composite one may gain of the genus *Paramys* and related, similar paramyids. These features, some specifically listed and analyzed, are considered ancestral rodent traits not because of any conceivably defined notion of outgroup comparison or principle of commonality. Concordant with the theory of evolution, the proposed transformations of hypothesized ancestral morphs, considering specific mechanisms, can be transformed into descendant morphs. These hypotheses of transformations can be corroborated by the ancestral constraints observed in the derived conditions. The stratigraphic position of the first known Paramys corroborates the primitiveness of that genus, with the general stipulation that a hypothesis of transformation based on stratigraphic information is always subject to stringent morphological testing. As suggested by Wilson (1949), Wood (1962), and others, there are probably no features in the cranial and dental anatomy of Paramys that would prohibit the derivation of craniodental attributes of other rodents from the former. For some possible exceptions, see Novacek (this volume) and Lavocat and Parent (this volume). The study of Cocomys will clearly add to this assessment.

There are significant gaps, not only in the fossil record, but also in the literature on the comparative, mechanical, and adaptational significance of mammalian skeletal morphology, and as noted above, particularly in the rodent literature. The most extensive effort of my present analysis, as stated in the introduction, is on the distal crus and the foot. Due largely to my own lack of familiarity with the evolutionary anatomy of the vertebral column, the finer details of the wrist, and due to lack of comparatively meaningful leptictid specimens, areas of paramyid postcranial anatomy that are not treated here should be gleaned from Wood's (1962) enduring informative accounts. The form-function of the selected features discussed below will be adaptively assessed in a later section.

Before I begin to concentrate on the lower hindleg, several important points have to be noted. In Paramys delicatus, AMNH 12506, as well as in other specimens and species of the genus, the proximal end of the ulna is bent forward, and the fovea capitis on the head of the femur for the ligamentum teres is very deep. The cranially bent olecranon and its flattened posterior edge for the transversely broad attachment of the triceps provides for optimal leverage for the triceps while the lower arm is flexed on the upper arm (Szalay et al., 1975). This was presumably the habitual condition. The hypertrophied ligamentum teres suggests resistance to unusual tensile loads.

The interpretation of the claw-bearing phalanges in paramyids is also significant. Like their homologs in phalangerid marsupials, tree squirrels, dermopterans, or in extinct Plesiadapis, these structures, transversely flattened, distally sharp and proximally deep, offer excellent resistance to tensile stresses that are generated when the falculae they bear are dug into a surface (Szalay et al., 1975).

The patellar groove of the paramyid femur, in spite of the generally greater robusticity of the entire skeleton than that of later rodents, particularly tree squirrels, is as elongated and deep as this structure is in the latter.

The crus and osseous foot of Paramys is well known (Figs. 1-12), and the distal portion of the tibia is clearly diagnostic of the order in the Eocene, as clearly as the rest of the lower leg or craniodental features. The robust fibula, a probably primitive eutherian attribute, is attached to the tibia by a syndesmosis (see in particular the contributions of Barnett and Napier, 1953a, b, on the form-function of the crus in therians). The distal end of the fibula is robust, with a well developed external malleolus, and the medially facing astragalofibular (AFi) facet is more vertical than this facet is in condylarths, fereans, or in archaic primates. This robust fibula is in striking contrast to the very reduced, splint-like, although heavily muscle-bearing bone of tree squirrels.

The paramyid tibia (Fig. 3) has a deep and robust medial malleolus and an equally deep, diagnostic, robust posterior process. On the distal crus of the late Paleocene leptictid under description by Novacek and Shubin, the presence of a modest posterior process is unmistakable. This is probably a leptictid specialization initially (although it is better developed in paramyids), as neither the process nor the corresponding articular surface on the astragalus can be seen in early condylarths, fereans, primates, or tupaiids. Although the noted Paleocene leptictid displays some degree of distal crural fusion, it is easy to envisage how an unfused leptictid crus represents a suitable ancestral stage for the Rodentia.

Reasons for the development of the tibial posterior process may lie in the mechanics of the articular contact with the astragalus, as well as in the advantage that the tendon of the flexor fibularis gains by its displacement posteriorly or by its direction or both (Fig. 4). It does appear from the manipulation of the UAJ in paramyids that the joint is exceptionally well braced medially during plantarflexion.

The striking shared and derived similarities between the astragali and calcanea of the late Cretaceous leptictids and Eocene rodents have been pointed out previously (Szalay, 1977b). Some of these details will now be examined. It is worthwhile to note that perhaps even more than the late Cretaceous specimens analyzed by Szalay and Decker (1974) and Szalay (1977b), some late middle Paleocene bones (see comparisons on Figs. 5 and 6) of leptictids that were recovered in 1969 by Hunter-CUNY parties are even more similar to paramyids than the bones of Procerberus and Cimolestes. There are no described or known rodents predating the Tiffanian Bear Creek Paramys atavus (see Wood, 1962; Korth, 1984). Nevertheless, the leptictid specimens from the Bison Basin bear perhaps a greater similarity to their homologs in paramyids than do the postcranial, dental, or cranial remains of any other nonrodent group.

Table 1. The following abbreviations, taken from Szalay (1984) with slight modifications, are used in the text and figures in connection with the lower leg and foot. Specific joints are abbreviated by the combination of the first letters in capitals of the names of those units which make up the joint, and the letter J for Joint. Abbreviations entirely in lower case designate bony processes, ligaments, tuberosities, grooves, tendons, etc., or anatomical directions.

ACJ	astragalocuboid joint
ACu	astragalocuboid
AFi	astragalofibular
ampt	astragalar medial plantar tuberosity
AmT	astragalomediotarsal
AN	astragalonavicular
ANJ	astragalonavicular joint
As	astragalus
at	anterior plantar tubercle
ATid	distal astragalotibial
ATil	lateral astragalotibial
ATim	medial astragalotibial
Ca	calcaneus
CaA	calcaneoastragalar
CaAd	distal calcaneoastragalar
CaAp	proximal calcaneoastragalar
CaCu	calcaneocuboid
CaFi	calcaneofibular
CAJ	cuboidoastragalar joint
CCJ	calcaneocuboid joint
CNJ	calcaneonavicular joint
Cu	cuboid
d	distal
dTs	distal tibial spine
Ec	ectocuneiform
EMIJ	entocuneiform-metatarsal I joint
EmT	entocuneiform-mediotarsal
En	entocuneiform
Fi	fibula
gtff	groove for the tendon of M. flexor fibularis
gtpb	groove for the tendon of M. peroneus brevis
gtpl	groove for the tendon of M. peroneus longus
l	lateral
LAJ	lower ankle joint
m	medial
Mc	mesocuneiform
mm	medial malleolus
mS	medial sesamoid
mT	medial tarsal (= mS1)
Mt	metatarsal
Na	navicular
NCJ	naviculocuboid joint
ne	neck of astragalus
NEn	naviculoentocuneiform
p	proximal
pp	peroneal process
sa	sulcus astragali
sc	sulcus calcanei
Su	sustentacular
Suh	sustentacular hinge
tca	tuber of calcaneus
tff	tendon of M. flexor fibularis
TFJ	tibiofibular joint
tft	tendon of M. flexor tibialis
tpp	tibial posterior process
UAJ	upper ankle joint

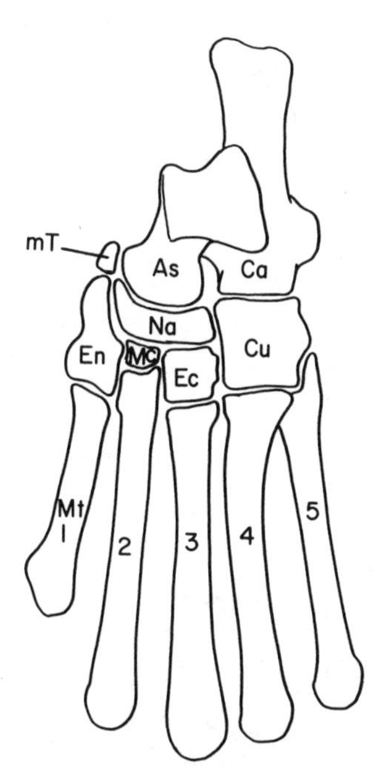

Figure 1. Schematic reconstruction of the general hypothesized relationships of the left protorodent foot skeleton in a dorsoventrally flattened, "exploded" view. Note in particular the large medial tarsal which braces the divergent head of the astragalus. There is medial contact of the entocuneiform with the navicular, but not with the astragalus. This arrangement may well be a uniquely derived feature of the protorodent. The entocuneiform was unlikely to have ever been part of the midtarsal joint. For abbreviations in this and later figures, see Table 1.

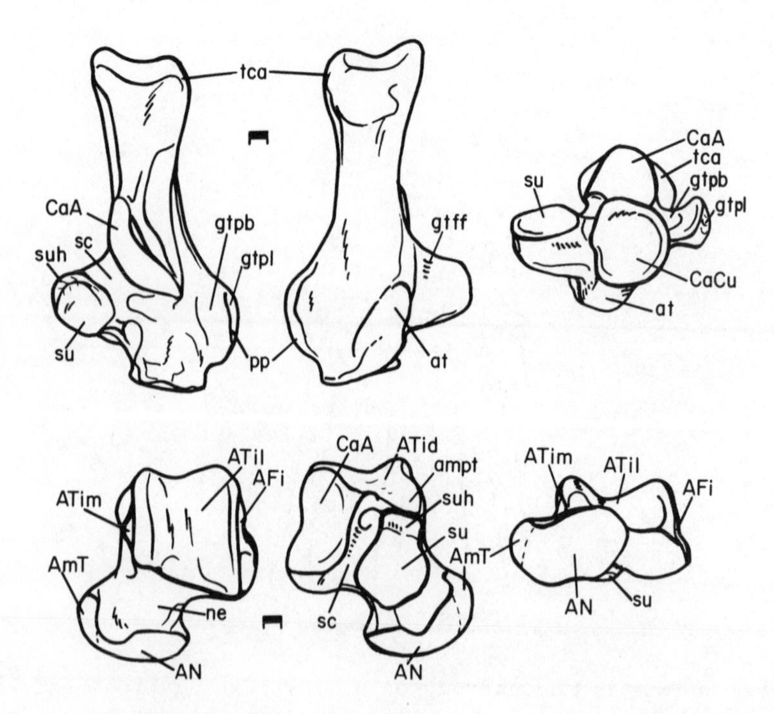

Figure 2. The anatomical terminology, as used in this paper, applied to the calcaneus (above) and astragalus (below), both from the left side, of the late Wasatchian Eocene Paramys copei, AMNH 2682. From left to right: dorsal, plantar, and distal views. For the morphology of the distal crus, see Figure 3. Scales represent one mm.

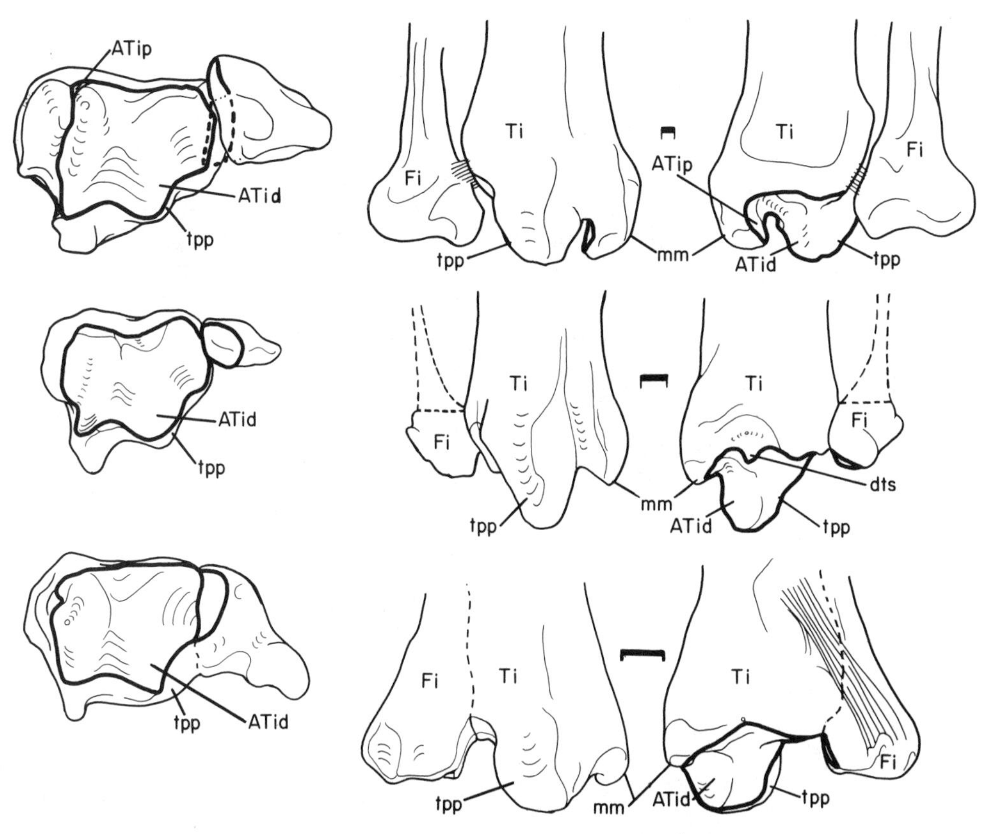

Figure 3. Comparisons of the left distal crus of the early Bridgerian Eocene Paramys delicatus, AMNH 12506 (above), Sciurus carolinensis (middle), and Rattus rattus (below). From left to right: distal, posterior, and anterior views, respectively. Note the robust fibula in the representative paramyid, the greatly reduced fibula and reduced tibial medial malleolus in the representative tree squirrel, and the reduced tibial medial malleolus and the fused and distally expanded fibula in a murid, a condition probably primitive for that group. Scales represent one mm.

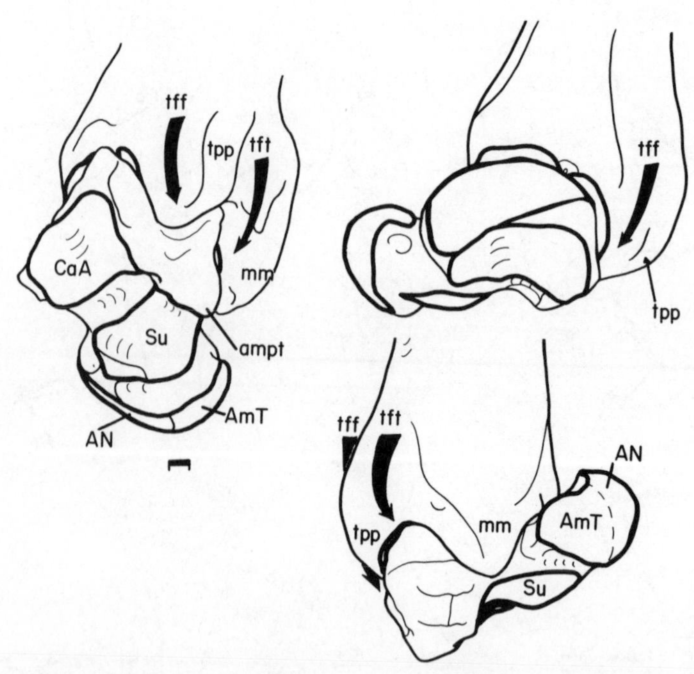

Figure 4. Distal (left), lateral (top right), and medial (lower right) views of the articulated left astragalus and tibia of the late Wasatchian Eocene Paramys copei, AMNH 2682, to illustrate the path of the tendon of M. flexor fibularis (=flexor digitorum fibularis; flexor digitorum longus) and the tendon of M. flexor tibialis (=flexor digitorum tibialis; flexor hallucis longus). Note that the large and diagnostic rodent tpp both guides the hypertrophied tff as well as braces the upper ankle joint, primarily against stresses directed toward the medial border of the foot (see text for details). Scale represents 1mm.

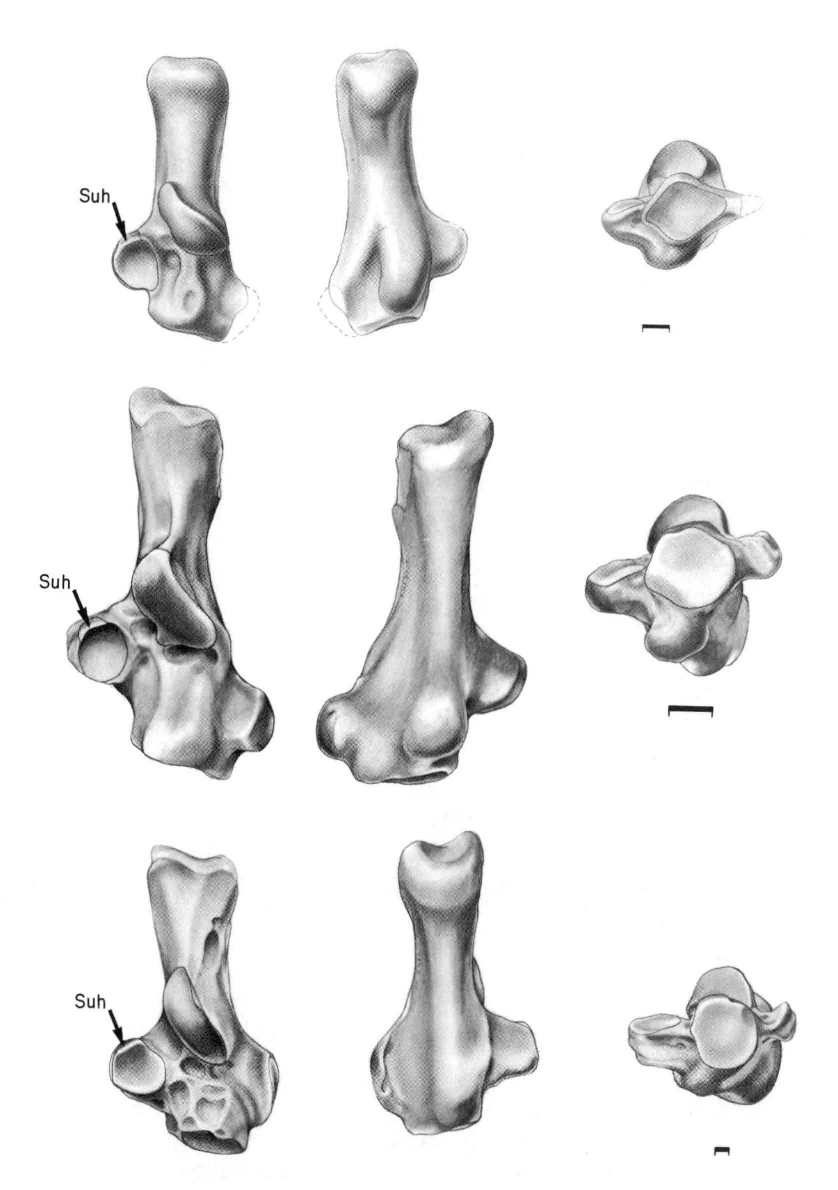

Figure 5. Comparisons of left calcanea of the late Cretaceous leptictid Procerberus formicarum, UMVP 1825 (above), a Paleocene (Bison Basin Saddle loc.) leptictid (middle), and a late Wasatchian Eocene Paramys copei, AMNH 2682 (below). From left to right: dorsal, plantar, and distal views, respectively. Scales represent one mm.

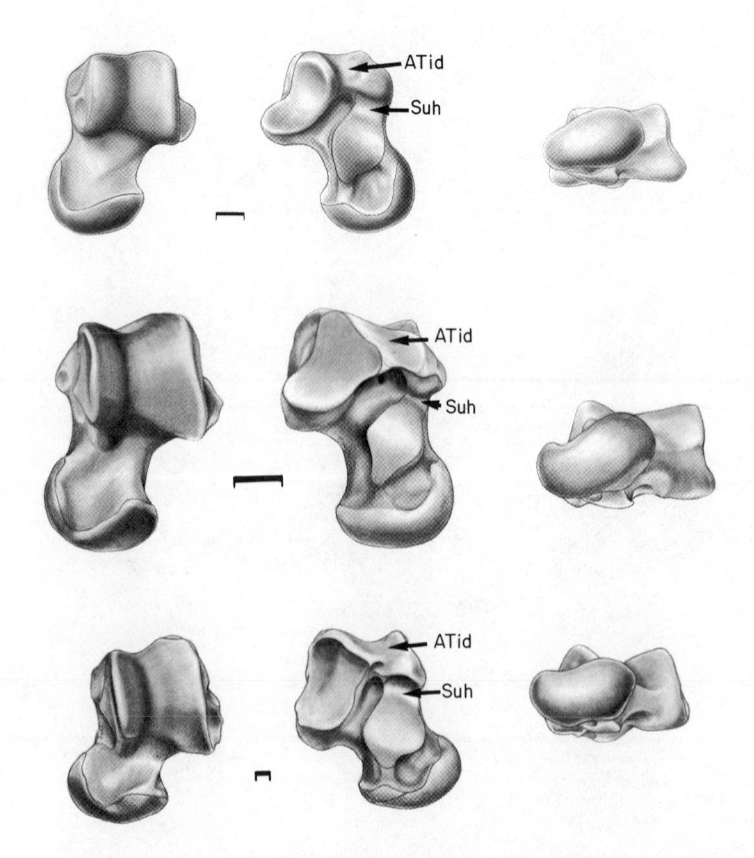

Figure 6. Comparisons of left astragali of the late Cretaceous leptictid Procerberus formicarum, UMVP 1806 (above), a Paleocene (Bison Basin Saddle loc.) leptictid (middle), and the late Wasatchian Eocene Paramys copei, AMNH 2682 (below). From left to right: dorsal, plantar, and distal views, respectively. Scales represent one mm.

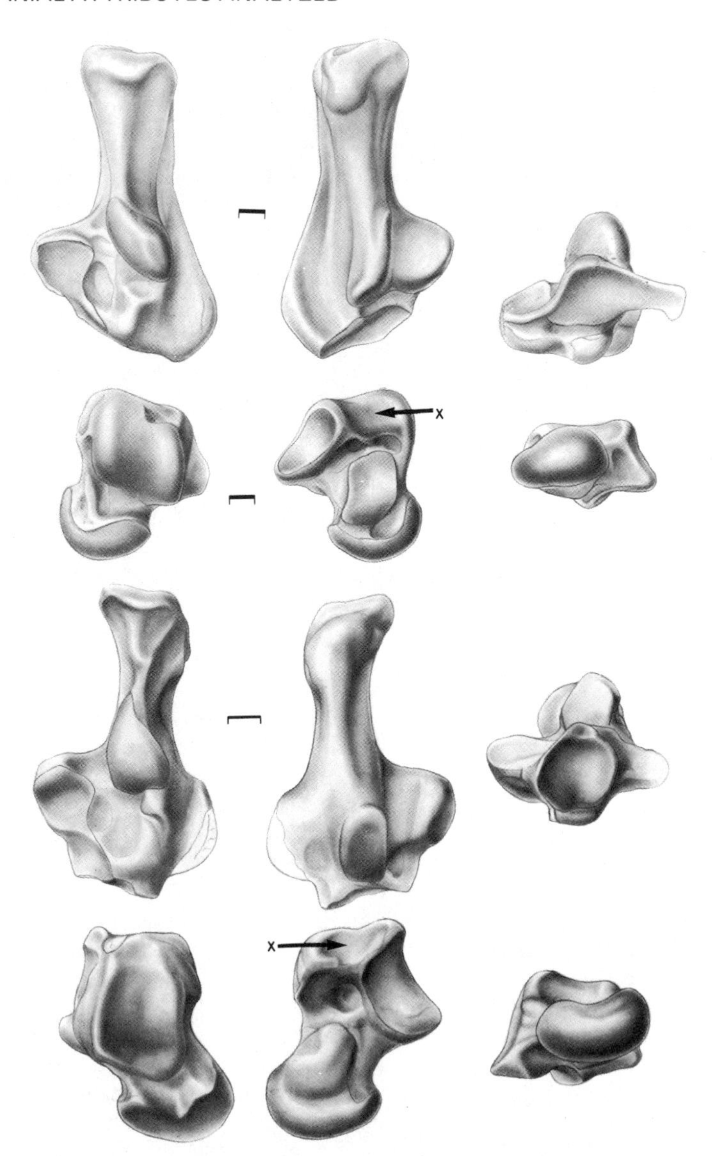

Figure 7. For comparison with leptictids and rodents, the calcanea and astragali of the late Cretaceous Protungulatum donnae, left side, based on UMVP 1914 and 1915 (above); and the Torrejonian Paleocene primate Plesiadapis gidleyi (right As), based on AMNH 17379 (below), are illustrated. From left to right: dorsal, plantar, and distal views, respectively. Note that the areas indicated with an X, which are grooves for the tendon of M. flexor fibularis, are transformed into the ATid facets of leptictids and paramyid rodents (see Figures 2-6 and text). Scales represent one mm.

The trochlear asymmetry of the presumably ancestral leptictid condition is diagnostically exaggerated by the first rodents. This diagnostic difference between the lateral and medial crests of the astragalar tibial trochlea (lapsus calami in Wood, 1962, p. 27) is even more exaggerated in the extant gliding sciurids. The high and sharp crests of the astragalar trochlea in the rodent morphotype, together with the corresponding crural articulations, sharply restrict all transverse movements in the UAJ, more so than the conditions displayed by the Cretaceous leptictids or by the earliest arctocyonids.

The head of the astragalus is an excellent predictor of the extent of habitual inversion, or its lack, in the positional repertoire of an eutherian species (see Szalay and Decker, 1974; Szalay, 1984). The medial enlargement of the head seen in Paramys, more so than in leptictids, just as the helical path of the calcaneal Su facet, strongly indicate all the mechanics associated with eutherian inversion. A quick visual comparison of the relative size of the head of the astragalus in arboreal and terrestrial forms is instructive (Fig. 9). The contrast between Rattus and Sciurus is striking.

The relatively greater size of the astragalar Su facet, compared to its calcaneal conarticular counterpart in Paramys, is suggestive of LAJ movements different than those in leptictids. Movement of the calcaneus, and therefore the foot, in a transverse direction is more facilitated in the protorodents than in the leptictids. These same functions are further exaggerated in the most arboreal of tree squirrels. Note the great transverse spread of the astragalar Su facet in Sciurus compared to the unaltered relative size of the calcaneal Su facet (Fig. 9). Significantly, in marmots there are osteological marks that depict a nonfunctional joint of the same relative form and proportions, but the actual contact area of the sustentaculum is clearly secondarily reduced (Figs. 8, 9).

Another significant qualitative difference between the morphology of primitive leptictids and paramyids lies in the greater mediolateral width of the calcaneal CaCu facet of leptictids than in Paramys, on which this facet is deeper than its width. This may be the result of the rotation of the calcaneus and cuboid laterally on their long axes. An extreme pedal position, hyperinversion in the posteriorly placed foot, which places great stress on the lateral border of the foot, is best resisted by the lateral rotation of the articular contact. The morphology of CaA facets of leptictids allowed less translation of the calcaneus on the astragalus than in Paramys; the axis of rotation in the former is much more closely aligned with the long axis of the calcaneus than in the latter. The peroneal process is larger in primitive leptictids than in paramyids.

The focal point in this discussion of the protorodent foot is the assessment of the origin and homology of the "split navicular" pervasive in the order. The mere presence of this structure does not appear to be correlated with a particular behavior. This bone has been called the "os sesamoideum tarsi proximale" by Romankowowa (1963), and the medial tarsal by Hildebrand (1978), whose designation I will follow (Fig. 1), with an elaboration on sesamoid bones appended below.

Nowhere is it more obvious than in attempting to understand the origins of the medial tarsal that any assessment of homology, mechanical function, and biological roles is intimately interrelated in a transformational analysis. Needless to say, widely based embryological studies on the pes of mammals would also be critically useful. Emry and Thorington (1982), in their careful and highly informative description of the early Oligocene Protosciurus skeleton, closely compared some aspects of that foot with those of Paramys spp. In so doing they have, inadvertently, prolonged a misunderstanding of a key area of the paramyid foot. Relying on Wood's (1962) landmark monograph, they stated (p. 31) that in Paramys "the contact of the astragalus and first cuneiform (= entocuneiform) is much more extensive than in any modern squirrel (many of which do not have an astragalar facet at all on the first cuneiform)." Unlike Wood's (1962, p. 27) statement that in Paramys the "... the entocuneiform ... articulates ... with the mesial side of the head of the astragalus," I believe that no known paramyid, or other rodent, or for that matter any other therian, has an entocuneiform-astragalus contact. This statement is clearly open to a simple falsification if specimens are known to show this to be otherwise.

I have examined and carefully fitted together the tarsus of Paramys delicatus (AMNH 12506) and Leptotomus parvus (AMNH 11591), paying particular attention to the critical EnNa contact, as that contact determines the extent to which the entocuneiform would extend proximally in the tarsus, and subsequently would or would not be in contact with the astragalus. The contact is illustrated in Figures 10-12. The small EmT facet on the proximal tip of the paramyid entocuneiforms could not be in contact with the astragalus, but, as its designation states, it articulated with a large medial tarsal.

When I examined specimens of Sciurus carolinensis with muscles and tendons in place, it became obvious that there are not two sesamoids, as Hildebrand (1978) reported, but that there are no less than four medially placed sesamoids embedded in both the spring ligament connecting the medial sides of the calcaneus and navicular and in the medial ligamentous binding of the entocuneiforms. The largest of these sesamoids, or S1 in the numbering used in this paper, is the medial tarsal (using Hildebrand's term), and it is in broad contact with the medial side of the astragalar head. This bone is very likely homologous in all major groups of rodents, although this problem has never been addressed (see Figs. 10-12). The large astragalar AmT facet becomes unmistakably recognizable once this relationship is understood. Although it is unlikely that medial tarsals of protorodents will be recovered, the morphology of the astragalar head leaves little doubt that this arrangement was in existence in the first paramyid. Whether or not this condition was present in more primitive eutherians predating the first rodent is perhaps impossible to assess at present.

The medial tarsal of rodents and the remaining sesamoids, numbered here as S2, S3, and S4, suggest the mechanically significant reasons for the fixation of these sesamoids in the first rodents, in light of the biological roles discussed below. Judged from the assumption widely regarded as plausible, namely that sesamoids, among other critical mechanical attributes, strengthen ligamentous and tendonous areas against damage and rupture, the medial sesamoid became hypertrophied to fulfill a mechanical role of resisting extreme stresses in the midtarsal region. Its hypertrophy and fixation were the result of selection emanating from a behavior which required an exceptionally strong medial binding of the astragalar head.

The ANJ and CCJ morphology of paramyids and squirrels is astonishingly similar (Fig. 11). The astragalar AN facet in *Paramys* and *Leptotomus* is almost uniformly deep and convex, from its lateral limits to the medially bordering AmT facet. This is reported to be similar or identical to the condition displayed by *Protosciurus* (Emry and Thorington, 1982), but unlike its homolog in sciurin tree squirrels, which have a highly derived sellar ANJ (see Fig. 9).

The hallucial mobility related morphology of the En-Mt1 joint of *Paramys* is discussed in detail in a paper in preparation on the evolution of this joint in therians. I will merely state here that in paramyids the En-Mt1 joint allowed a modest degree of hallucial abduction, nothing, however, comparable in degree to that seen in arboreal marsupials, tupaiids, or plesiadapiform or modern primates.

POSTURE AND LOCOMOTION OF THE PROTORODENT

The skeletal remains of Eocene paramyids represent relatively large species, such as *Paramys copei*, for example. As noted above, the earliest paramyid skeletons (and probably the teeth and skull as well, as emphasized by others) are a remarkably excellent structural as well as temporally justified actual ancestral stage for all later rodents. Judged from the impressive qualitative uniformity of the tarsus of both small and large representatives of such higher taxa as Perissodactyla, Artiodactyla, Borhyaenoidea, Phalangeroidea, etc., this relatively large size of the early skeletons is of little consequence for their postulated structural primitiveness.

Now I will attempt to explain some aspects of my understanding of the protorodent locomotor bioroles. As I have argued in detail

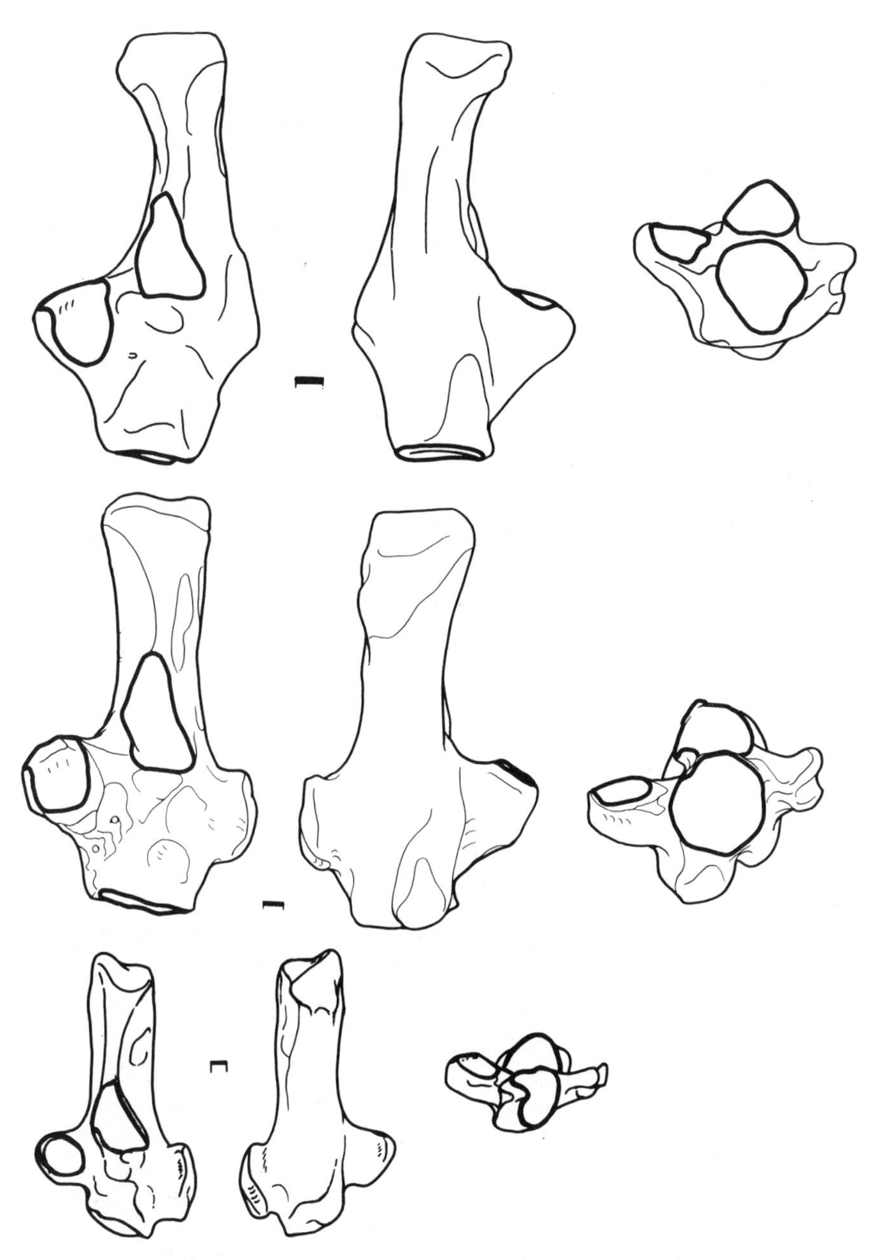

Fig. 8. Comparisons of the left calcanea of the representative tree squirrel, *Sciurus carolinensis* (above); a ground squirrel, *Marmota monax* (middle); and a representative of the possibly primitive murid morph, *Rattus rattus* (below). From left to right: dorsal, plantar, and distal views, respectively. Scales represent one mm.

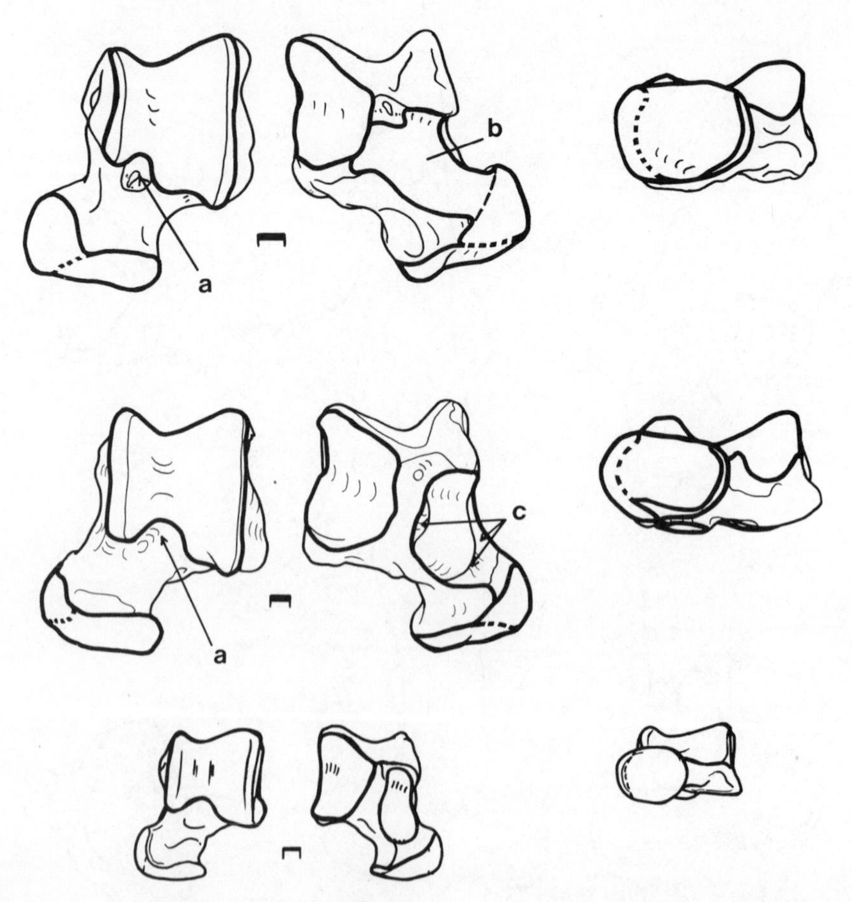

Fig. 9. Comparisons of the left astragali of the representative tree squirrel, Sciurus carolinensis (above); a ground squirrel, Marmota monax (middle); and a representative of the possibly primitive murid morph, Rattus rattus (below). From left to right: dorsal, plantar, and distal views, respectively. Note the relatively larger astragalar head and the medial tarsal articular facet in the tree squirrel compared to the two terrestrial forms. A pit (a) is for a spinous projection of the tibia (dTs, see Fig. 3) which probably stabilizes the sciurid foot in a dorsiflexed position. The depression persists in marmots. The astragalar Su facet (b), being much larger than its calcaneal conarticular counterpart, in Sciurus allows the extremes of hyperinversion or the transverse movements of the astragalus and crus, and therefore movement while the animal is hanging by hyperinverted feet. Some of this morphology (c) is retained in marmots, although this area can be distinguished from the articular part of the sustentaculum in these secondarily terrestrial forms.

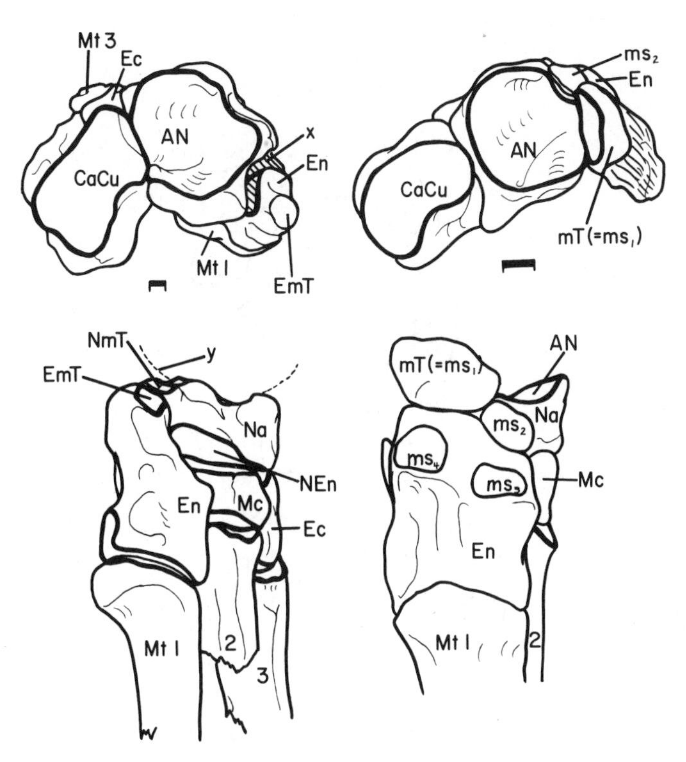

Fig. 10. Comparisons of the midtarsal joints in a proximal view (above) and the medial views of the tarsus (below) of the Eocene paramyid Leptotomus parvus, AMNH 11591 (left); and the sciurid Sciurus carolinensis (right). Note that in the fossil (left and above) the EmT facet on the entocuneiform was present, and as in squirrels it articulated with the medial tarsal as in the extant forms. If the tarsus of Leptotomus was correctly articulated in the lower left, which it is not, the NEn facet would be covered by the entocuneiform. The area designated with X (upper left) represents this artificial gap between the two bones. None of the four distinct sesamoids of the extant form is preserved in any fossils known to me. The curved broken line (Y) represents the approximate outline of the head of the astragalus, not in contact with the entocuneiform. The extreme proximal position of the latter, medial to the navicular, is probably diagnostic of the protorodent, compared to other therians. Scale represents one mm.

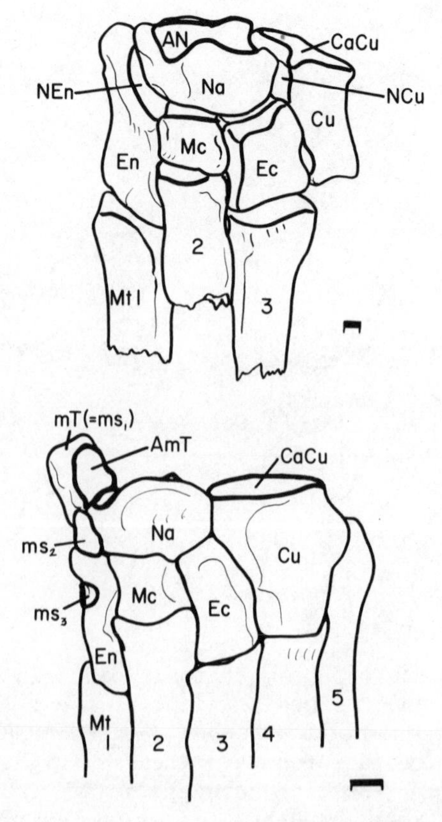

Fig. 11. Comparison of the tarsus in the paramyid Leptotomus parvus, AMNH 11591 (above), and Sciurus carolinensis (below), in dorsal view. The entocuneiform, prior to separation and preparation, is incorrectly articulated in Leptotomus, so the NEn facet is visible. In Sciurus, the medial tarsal is tightly articulated with the entocuneiform and medial to the navicular, forming a robust medial brace for the astragalar head. This was almost certainly the form-function pattern in paramyids also.

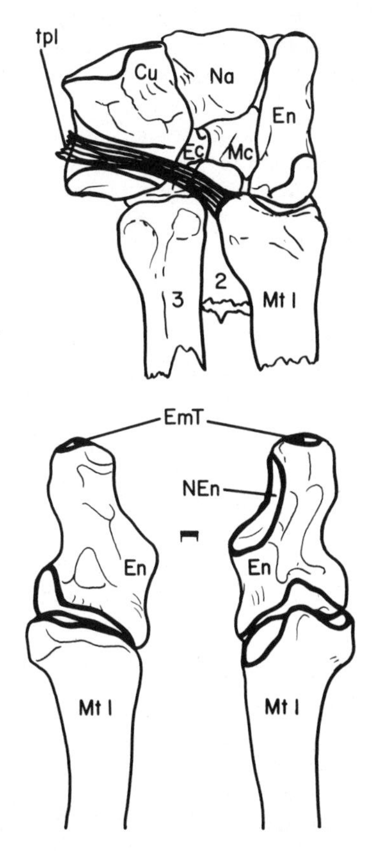

Fig. 12. Plantar view of the tarsus and the first three metatarsals (above); and lateral (below left) and the medial (below right) views of the articulated entocuneiform and first metatarsal of *Leptotomus parvus*, AMNH 11591. Note the large, reconstructed, tendon of *M. peroneus longus*. Scale represents one mm.

elsewhere (Szalay, 1981a), bone morphology tends to be viewed even by many functionally and adaptationally oriented morphologists within the conceptual confines of the "rubber hypothesis" of bone morphology The tacit assumption in most analyses of adaptations embodies the notion that bone morphology is largely the result of the forces generated during the ontogeny of mammals. The genetically rigorous control over bone morphology (initiated by the mechanical impulses from the developing musculature), down to the finest details of trabecular structure (Amtmann, 1979), the role of this morphology to generate and dissipate the forces which come to act on it, and the constraining influence of this inherited form-function in future evolution are largely ignored by many students of form, who consider phylogeny outside the confines of "science". This view is theoretically unsound, and consequently its practical manifestations often result in unsatisfactory attempts at solving evolutionary problems.

In interpreting the meaning of diagnostic morphotypic rodent characters, the evidence from paramyids, sciurids, murids, and representative members of caviomorphs and phiomyids figures largely. As a significant point in the discussion below, I want to note the following statement of the most recent and thorough students of sciurid osteology, Emry and Thorington (1982, p. 33): "Even the modern ground squirrels, which tend to have shorter and stouter limbs than arboreal squirrels, are more gracile than most of the paramyids. The differences that do exist between *Paramys* and *Protosciurus* are essentially the differences between *Paramys* and living squirrels and perhaps serve best to emphasize the extreme similarity between *Protosciurus* and living sciurines, particularly *S. niger*."

Protosciurus, in spite of its detailed similarity to arboreal squirrels, retained the primitively more robust build, although lightened, from a paramyid condition. Once the mechanics of a joint complex, such as the one we examined above, permits a particular locomotor and positional repertoire, and prohibits others, the robusticity of the bones becomes an adaptive factor which is honed not only by the consequences of the activities which the form-function allows, but also by the pressures exerted by the prey or predator, whichever the case may be. In the case of the earliest rodents, the speed of the predators or territorial competitors exerted selection on their opposites' skeletal biology. It is for these reasons that I believe that once the movements necessary for a specific locomotor or postural activity are facilitated in a joint complex, the relative robusticity of the bones themselves, out of their temporal context, adds little to the interpretive framework. The forces of selection are relative to the organisms and agents of selection, and they cannot be directly compared between ecologies separated by great gaps of time. As the ecosystem continued to evolve since the beginning of the Paleogene, the increase in the speed and agility of the mammalian hunters and the hunted (pursued by others besides mammals) significantly influenced the complex equation between body weight, skeletal robusticity, and of course the influence of the nervous system on the speed and efficiency of judgements of all movements.

Paleogene mammal bones, when compared to those of similar sized mammals of today, tend to be generally more robust, although this has not been adequately documented. Any study which would attempt to evaluate the skeletal biology of synchronous Paleogene mammals from the perspective of bone robusticity might result in important interpretive insights into species dynamics then and in later times. Another complicating factor of the differences noted by Emry and Thorington (1982) between the more slender bones of tree squirrels, compared to the terrestrial ones, involves bone dynamics as they relate to the stresses of compressive as opposed to tensile forces. The relatively greater magnitude of habitual tensile forces generated on the limbs of tree squirrels are best resisted by the tendons and ligaments of the skeletal system, and heavily built skeletons are unnecessary for such arborealists.

In spite of the general robusticity of known paramyid skeletons, the functionally critical attributes of the substrate reflecting distal phalanges (bearing similar shaped falculae), the ulna, the hip joint, the UAJ, LAJ, ANJ, and the medial tarsal should be appropriately weighted. Whether the first known paramyids were arboreal or not is not an issue here, as I believe they were capable arboreal forms, well beyond the primitive eutherian climbing abilities (Szalay, 1984). I will attempt to argue below that the derived morphotypic attributes of the rodents, well displayed in these paramyids, are the results of a special form of arboreal behavior practiced today by squirrels, _Potos flavus_ (which also has a prehensile tail), various marsupials, but particularly by secondarily arboreal dasyurids (which have reduced their pedal grasping abilities), such as _Phascogale tapoatafa_.

The following lines of evidence argue for a postural and locomotor repertoire in protorodents which was probably as fully committed to an arboreal substrate as arboreal squirrels or arboreal dasyurids are today. All of the characters or character modifications listed are advanced eutherian features, some attained in parallel with other arboreal eutherians.

The evidence cited from the morphology and mechanics of the proximal ulna, the relative size of the ligamentum teres, and the falcula bearing distal phalanges, by themselves as a functional complex, would strongly predict claw climbing arboreal activity. I will concentrate on the evidence of the cruropedal complex for a more precise hypothesis of posture and locomotion.

The character which invariably predicts the extensive ability of the foot to invert, and arboreality, is the long and helical CaA contact areas of the LAJ. As the foot inverts, given the rigidity of the tenon-mortise UAJ of eutherians, fully retained and further strengthened with the sharply keeled astragalar tibial trochlea in rodents, the foot is adjusted by the calcaneus moving forward, rollin outward and via the forces transmitted through the cuboid the foot inverts. The degree to which this modification is present in paramyids and squirrels, for example, is not any less than that seen in primates.

The advanced arboreal abilities of the protorodents are not questioned here. But what postural and locomotor repertoires should be predicted from the attributes reviewed and why? To any squirrel watcher, the agile claw climbing, the rapid spiral ascent and descent on a tree trunk, but particularly the head first descent and the ability to linger in that position and explore with the foreleg and hands while so suspended, are some of the most striking attributes of arboreal sciurid activities. In order to accomplish this, the animal uses its foot in a hyperinverted position (aided by the outward rotation of the hindleg). The forces on such a foot (and the ligamentum teres on which the animal is suspended) are relatively enormous at the UAJ and ANJ. This is the case because the extremely extended foot is also inverted in the LAJ and the anatomically medial border of the foot is now aligned lateral and behind (above) the hanging animal. The forces which must be resisted are directed anatomically medially, but in effect away from the animal's midsagittal plane. The hypertrophied sesamoid, the medial tarsal, the hypertrophied and medially located tibial posterior process articulating with the proximal surface of the astragalus, the exceptionally rigid UAJ articulation, and the relatively enormous tendon of the flexor fibularis all suggest that the arboreal descent by the hyperinverted clawed pes shaped the hindleg of the protorodent. For a discussion on hyperinversion in multituberculates, a group which solved the functional problem differently, due to its entirely different UAJ form and mechanics, see Jenkins and Krause (1983).

Thus, the evolution of the medial tarsal of rodents appears to be the result of selection to counteract the enormous stresses placed on the spring ligament, which ordinarily binds the medial head of the astragalus, during descent, mediated by the hyperinverted foot. Such a sesamoid might well have been the result of the selection generated from such forces as described above. Although this bone is present in all rodents, as far as known, the arboreal squirrels seem to have a relatively larger mT than various terrestrial murids. To expect identical modifications in all groups which have attained similar behaviors would be simplistic and phylogenetically naive. Eutherians metatherians, multituberculates and lizards which have come to deal with the problems of head first descent have different mechanisms, precisely because of the channelling influence of their mechanically distinctive heritage.

There are a number of differences between sciurin tree squirrels on the one hand and paramyids on the other, which would seem, if one pursued a strict total-analogy rooted methodology for biorole assessment, to prohibit the conclusion reached above that paramyids practiced the essentially squirrel-like arboreal activities. For example, why do modern squirrels have an exceptionally slender fibula in contrast to the robust paramyid one? It may be that the habitual stress distribution during hyperinversion, as discussed above, is all but reduced on the lateral side of the UAJ. This may well reflect the total arboreal commitment of the modern tree squirrels. This mode of life might not have been attained until after extensive radiation and competition of the Sciuridae for the arboreal, semi-arboreal, and fully terrestrial niches. The presence of a stout fibula, retained in paramyids from the non-rodent ancestry, does not invalidate any of the functional arguments advanced above.

Another possible objection to the biorole hypothesis for paramyids may center on the LAJ facets of sciurids (Fig. 9) being different from those of paramyids. In *Sciurus* the astragalar Su facet is transversely elongated, making contact laterally with the CaA facet and medially with the mT facet. The Suh facet persists, but the mediolateral movement of the foot in the LAJ (through the movement of the calcaneus) is clearly more extensive than in the more primitive condition. I interpret this sciurin character as the development of an additional safety factor for the integrity of the LAJ. While the animal is suspended by the claws of its hyperinverted foot, the forces resulting from its weight may be transversely adjusted in the LAJ. This arrangement allows a greater than before radius of mobility by the forequarters of the animal, increasing its reach while hanging. As Emry and Thorington (1982) discussed issues closely pertinent to this point, they noted that sciurin tree squirrels are highly derived in many ways, in contrast to other fully arboreal tree squirrels such as *Ratufa*.

On the anterolateral surface of the tibial medial malleolus, distinctly angled to the tibial ATi(m) facet, is an ATi(p) facet which made articular contact with the astragalus during the extreme of pedal dorsiflexion. In spite of the dorsiflexed accommodations of the paramyid cruropedal joint, the Eocene rodents lack an interesting cruropedal attribute of tree squirrels. This feature, the distal tibial spine (dTs, Fig. 3), possibly primitive for the Sciuridae (I have not examined *Protosciurus*, and this feature is neither mentioned nor figured by Emry or Thorington, 1982), consists of a spike-like process on the anterior portion of the medial ridge of the tibial ATi facet. This process fits into a correspondingly deep pit on the distal border of the astragalar tibial trochlea. The mechanical reason for this feature possibly lies in the stability it additionally confers to the UAJ during the extreme of pedal flexion. This specially realized stability may be needed when loading of this joint is rapidly increased while the foot is dorsiflexed. A tree squirrel, while keeping the dorsiflexed foot against the tree stable during ascent, increasingly and with varying speeds loads the UAJ by straightening the hindleg during vertical claw climbing. The absence of such a mechanism, if it really is that, does not prevent such an activity, it merely reduces the magnitude of such forces under which the dorsiflexed UAJ is likely to remain stable. The magnitude of these forces is directly correlated with the speed and the angular acceleration involved in the climb.

Wood (1962) gave an account of the caudals in Paramys as being indicative of a massive tail. Emry and Thorington (1982) supported Wood's idea that Paramys was not likely to have habitually performed the characteristic extreme dorsiflexion of the tail seen in sciurids. I am not aware of any study which links the sciurid scansorial locomotion with the probably partly social behavior related tail flexing exhibited by squirrels.

An additional final point relating to the scansorial and special claw climbing and hanging arboreal beginnings of the Rodentia concerns the well known dental morphology of paramyids. The cheek teeth of the earliest representatives are low crowned, crushing adapted like the cheek teeth of tree squirrels, and unlike the high crowned teeth of the terrestrial grazing squirrels, such as marmots (see especially Black, 1963).

FORM, FUNCTION, AND LOCOMOTION OF THE PROTOLAGOMORH

Another participant in this symposium, Lopez Martinez (this volume) has dealt extensively with the cranioskeletal morphology of the Lagomorpha. She contends that comparison of early representatives of orders is not reconcilable with the comparison of the "bauplan", or morphotype, of these orders. Concerning this last point, I would simply like to note that the use of ancient taxa as morphotypic representatives is only meaningful if these forms indeed embody the features which can either be recognized in younger taxa or if the younger character states can be shown to have probably transformed from the archaic conditions. Contrary to her assessment, no a priori use of fossils is intended if there is the least indication of unlikely transformation sequences. I do, therefore, sharply differ in my approach to the assessment of the most probable transformation sequence from the method espoused by Lopez Martinez, which, it appears to me, is a character distribution analysis.

Because of our different emphases and different approaches to the understanding of characters, some usefully contrasting interpretations of the same lagomorph features have resulted. Lopez Martinez maintains that the morphotypic cruropedal adaptations of the Lagomorpha have been ambulatory rather than cursorial as I have stated in Szalay (1977b).

The morphotypic lagomorph foot can be unequivocally interpreted as having evolved highly derived anteroposterior facilitation, severe mediolateral restriction of UAJ, LAJ, ANJ, and CCJ mobility, and therefore all the characters cited by Lopez Martinez as present in the "majority of the non-modified lagomorphs" point to the same morphotypic functional complex. In *Mimolagus* and in the other duplicidentate lagomorphs, the UAJ is deeply grooved into the astragalar body, consequently the medial and lateral crests are sharp and their outer walls steep, with no transverse movement and extreme close packing of this joint mediolaterally (see Figs. 15, 18). The CCJ is constructed in such a way that it is braced against transverse dislocation (the conarticular surfaces are sharply angled to the long axis of the calcaneus), yet its quasi-cylindrical surfaces allow a hinge-like movement. This mechanics was likely to have been adaptive in the necessity to adjust the digitigrade forefoot on the proximal tarsals through movements of the midtarsal joint, on an axis which was aligned anteroposteriorly, dorsoplantad, and slightly laterally. Incidentally, this quasi-cylindrical CCJ, as well as the deeply troughed UAJ, are characteristic of such undoubted cursors as perissodactyls and litopterns. There can be little doubt that the last common ancestors of *Mimolagus* and living lagomorphs possessed this suite of derived features. The strictly applied convergence method (level A correlations, see Szalay, 1981a) permits no other explanation than the establishment of advanced adaptations for running and leaping, an essentially cursorial mode of life, in that common ancestor. These characters are quintessentially associated with specialized cursors, not ambulatory forms, both among living metatherians and eutherians. A simple mechanical comparison (see Fig. 19) between *Mimolagus* and *Paleolagus* shows that an increase in the tarsal load arm in the latter over the former also suggests the adaptive trend for leaping, in addition to the cursoriality predicting aspects of the joints of the entire skeleton. There is little doubt that when these features are associated with non-cursorial species, they represent heritage characters of the cursorial ancestor.

The fact that *Paleolagus* (see Figs. 13-15) is early Oligocene in age detracts very little from its embodying the morphotypic characters of the cursorial protolagomorph. What compels my assessment of the lagomorph morphotype as a cursor is not that a fossil like *Paleolagus* displays cursorial mechanics (a form too young to be any kind of lagomorph, *sensu stricto*, ancestor), but the fact that the detailed mechanics of the foot and elbow, which unequivocally causally correlate with cursoriality and leaping, can be found in all modern lagomorphs, as either a habitus character or as a clear heritage of secondary noncursorial adaptations, as well as in the early Oligocene genus. If we knew of earlier, Eocene, taxa with these attributes identifying them as lagomorphs, then this assess-

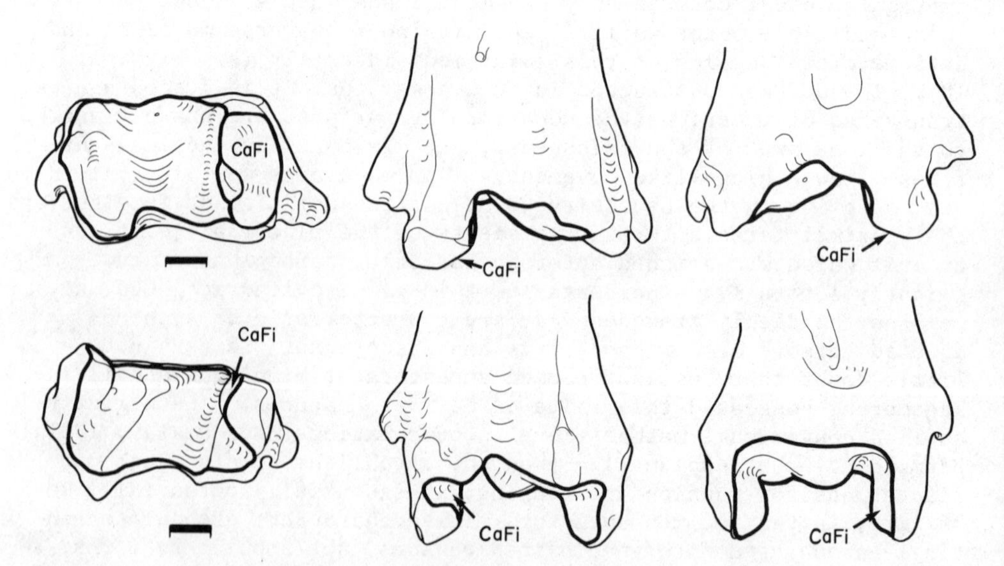

Fig. 13. Comparisons of the left distal crus of the early Oligocene lagomorph Paleolagus haydeni, AMNH 6275 (above), and Petrodromus tetradactylus, AMNH 203336 (below). From left to right: distal, posterior, and anterior views, respectively. Note the diagnostically contrasting manner in which the fibula contacts the calcaneus in the two groups. For comparisons of the proximal tarsals of these taxa, see Figs. 14-15. Scales represent one mm.

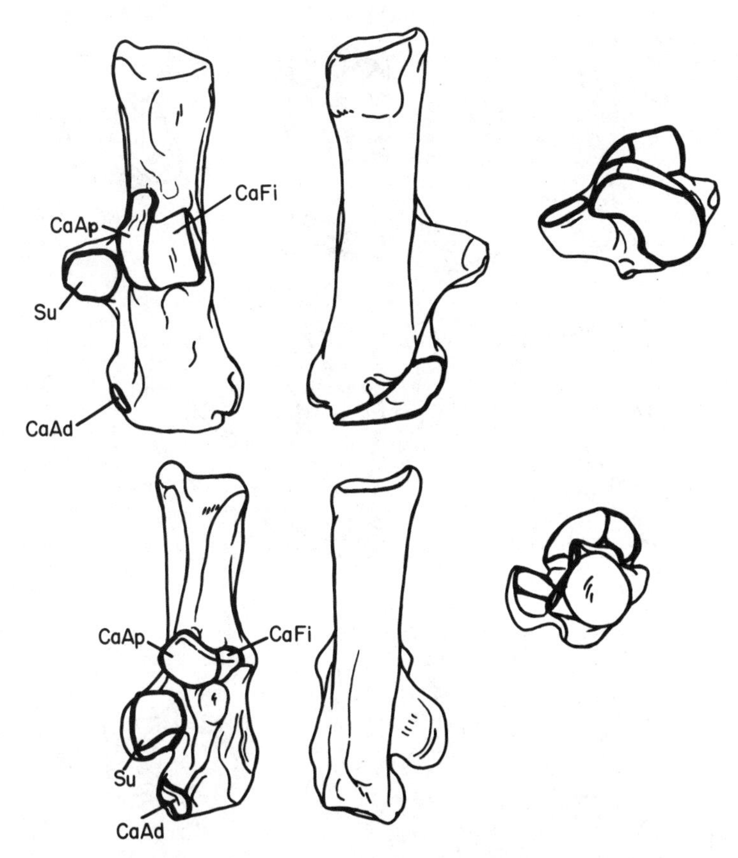

Fig. 14. Comparisons of the left calcanea of the Oligocene lagomorph Paleolagus haydeni, AMNH 6275 (above), and the macroscelidid Petrodromus tetradactylus, AMNH 203336 (below). Note the medially facing CaA facet and the broad, dorsally oriented CaFi facet in the lagomorph (and also in Pseudictops, as shown in Sulimski, 1968), in contrast to the dorsally oriented CaA facet and the small distally facing CaFi facet in macroscelidids. Unlike extant lagomorphs, Paleolagus lacks, primitively, a contact between the calcaneus and navicular.

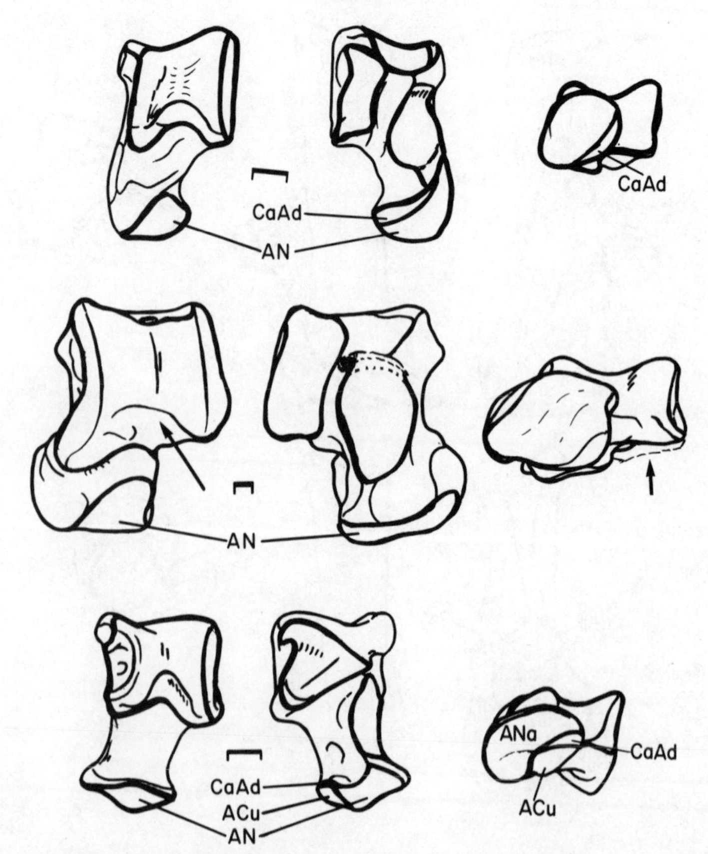

Fig. 15. Comparison of the left astragali of the Oligocene lagomorph Paleolagus haydeni, AMNH 6275 (above); the Paleocene pseudictopid Pseudictops lophiodon, AMNH 21755 (middle); and the macroscelidid Petrodromus tetradactylus, AMNH 203336 (below). Note the absence of a cuboid facet in the lagomorph, in contrast to its presence in the macroscelidid. Note the ovoid, ball and socket articulation of the tibial medial malleolus with the astragalus, as well as the pivot-like articulation of the ANJ, in macroscelidids. Scale represents one mm.

ment of the morphotype would be even more strongly corroborated.

ON THE POSSIBLE INTERRELATIONSHIPS OF THE GLIRES

The concept Glires has received vigorous support through Luckett's (this volume) important contributions on the development and suggested homologies of dI2 in both rodents and lagomorphs, and the developmental similarities of the fetal membranes in these two groups. There is clearly no way to evaluate the fetal membrane patterns in the various fossil groups which were undoubtedly somehow allied to the early rodents and lagomorphs. The Glires-type fetal membrane complex may or may not have been present in leptictids and Asiatic "anagalidans". This, then, leaves the problem of dI2 homologies open for future considerations. If it is found that enlargement of the anterior teeth in the Glires might have been constrained either phyletically (or ontogenetically) or adaptively, or both, and it is very important to consider that only four, but more realistically two, options are available to enlarge either of the first two incisors, then the cranial, dental, and postcranial morphology of fossils may have to be weighted differently. Assessing greater recency of relatedness between rodents and some "anagalidans" on the one hand, and lagomorphs and other "anagalidans" on the other, which may or may not have had enlarged incisors or non-dI2 enlargement, will become a matter of precarious analysis and judgement.

The vertical affinities of Heomys and Mimotona, considered to be members of the Eurymylidae and Mimotonidae, respectively, by Li and Ting (this volume), are problematic at best. Their great similarity to each other, however, as noted by Flynn (this volume and pers. com.), is an assessment I strongly concur with. So in spite of their different dental formulae, their cheek tooth similarities would suggest, at least at this time, inclusion in a single family. Following Butler's (this volume) assessment, I fail to detect any homologously shared derived occlusal similarities between the eurymylid Heomys on the one hand and Cocomys and paramyids on the other. Eurymylids do display a lagomorph-like, or a generally more primitive eutherian shear between the occluding molars than that envisaged in the protorodents. Unfortunately, the extensive postcranial material of the Eocene eurymylid Rhombomylus (Li and Ting, this volume) has not yet been described in detail.

Nothing in the evidence known so far compels one to conclude that the cruropedal complex originated synchronously with other character complexes at any of the nodes where the various groups split or originated. If some of the similarities of the paramyid cruropedal complex are homologously shared and derived with Cretaceous or Paleocene lepticid attributes, then this complex undoubtedly predates the protrogomorphous jaw and dI2 enlargement. It is clear that if a zalambdalestid and duplicidentate clade and a leptictid and rodent clade are two natural entities, then the Glires is an ancient group indeed which must be defined by non-incisor characters, in which incisors have come to be enlarged and modified independently. A functional-adaptive analysis of the feeding mechanisms will be clearly one of the acid tests of the likelihood of dI2 homology in the Glires.

It is important to note that Zalambdalestes is lagomorph-like, not only in some dental attributes, but also in aspects of its postcrania. The skeleton, which needs further preparation and study, was expertly described by Kielan-Jaworowska (1979), and one may point to some possible homologously shared and derived features with duplicidentates. The deep UAJ and the undeviated astragalar neck, along with the anteroposterior orientation of the ANJ, all on a highly digitigrade foot, point to advanced cursorial modifications. These attributes, as stated, are not necessarily homologous with those of Mimolagus and other duplicidentates. The latter and Zalambdalestes, and probably also Mimolagus, judged from the AN facet, share the presence of a posteroproximal extension of the navicular (tuber tibialis of Kielan-Jaworowska, 1979), which resists the great dorso-plantar bending movements at that joint. Macroscelidids have an entirely different solution for ANJ mobility and stability, in contrast to zalambdalestids and duplicidentates (see Fig. 15). It is interesting to contemplate whether this process may or may not be a transformed and ontogenetically incorporated medial tarsal into the navicular. To postulate and test such a homology between rodents and lagomorphs a great deal of embryological work must be done. The lack of a CaFi contact in Zalambdalestes, shared with Mimolagus, probably represents the primitive lagomorph condition (Table 2).

It is unknown whether the older alleged relatives of Mimolagus, such as Mimotona, Gomphos, and Hypsimylus (see Li and Ting, this volume), posessed a tarsal morphology similar to the younger Mimolagus for which an astragalus and calcaneus are known (Fig. 18). Although Paleolagus is primitive among the leporoid lagomorphs (including leporids and ochotonids, in juxtaposition to Mimolagus which should be placed in a family of its own, the Mimolagidae), in lacking a calcaneonavicular contact, it is considerably more advanced than Mimolagus in having generally more slender and less robust foot. Judged from the nature of the exceptionally broad, dorsally facing CaFi contact of the extant lagomorphs and their Oligocene family members, this condition is probably a secondary acquisition in the Lagomorpha. Incidentally, Pseudictops shares this feature with the Leporoidea, almost certainly convergently, or possibly parallel with them. Mimolagus, like leptictids, rodents, and Zalambdalestes completely lacks this contact (see also Li and Ting, this volume). The CaA facet is angled facing medially to brace the calcaneus and astragalus and not to allow the calcaneus to move on the ankle bone. This, of course, is the mechanical solution of the lagomorph LAJ, where the sustentaculum is immediately medial to this LAJ constraining

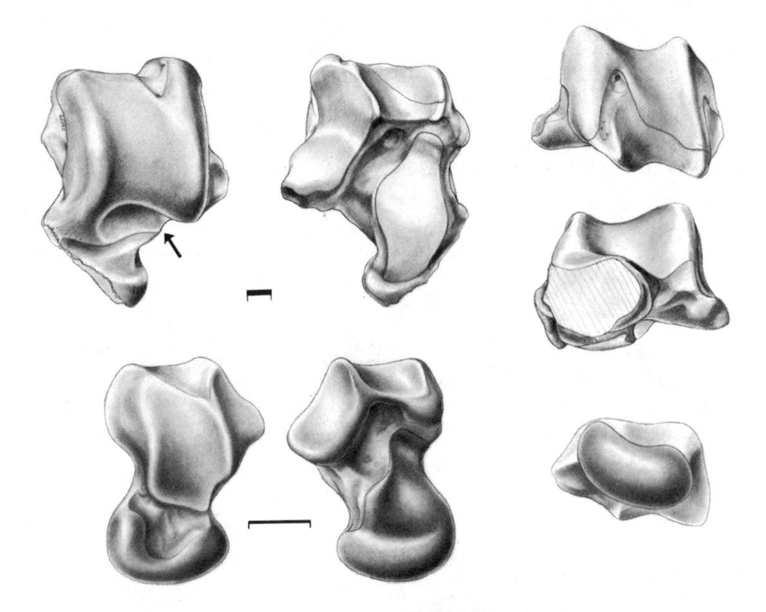

Fig. 16. Comparisons of the left astragali of the early Oligocene *Anagale gobiensis*, AMNH 26079, with its head broken off (above), and a Torrejonian Paleocene primate, AMNH 92015 (below). Dorsal, plantar, proximal, and distal views above. In the lower row the distal view is not shown. The arrow points to the area on the distal portion of the body in *Anagale*, indicating the strikingly different tibial contact compared to an arboreal primate. Note the similarity of *Anagale* to *Pseudictops* (Fig. 14) in this regard. Scales represents one mm.

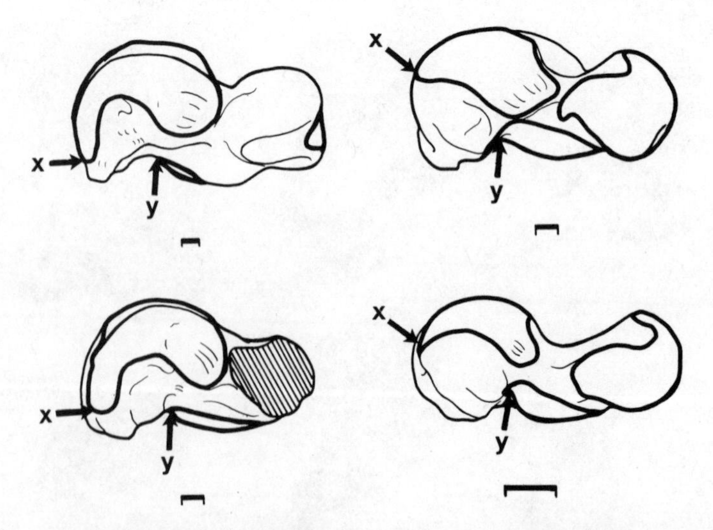

Fig. 17. Comparison of the medial views of the left astragali of *Pseudictops lophiodon*, AMNH 21755 (above left), *Anagale gobiensis*, AMNH 26079 (below left), *Paramys copei*, AMNH 2682 (above right), and a Bison Basin Paleocene lepticid (below right). Note the posterior and plantar extent of the AFi facet (X) of the first two taxa (on the left), characteristic of many terrestrial cursors (although it also occurs in arboreal leapers as well), and contrast this condition to those on the right. In the latter two, the sustentacular hinge facet, Suh (Y), is well pronounced, whereas in the Asiatic forms it is not. Scales represent one mm.

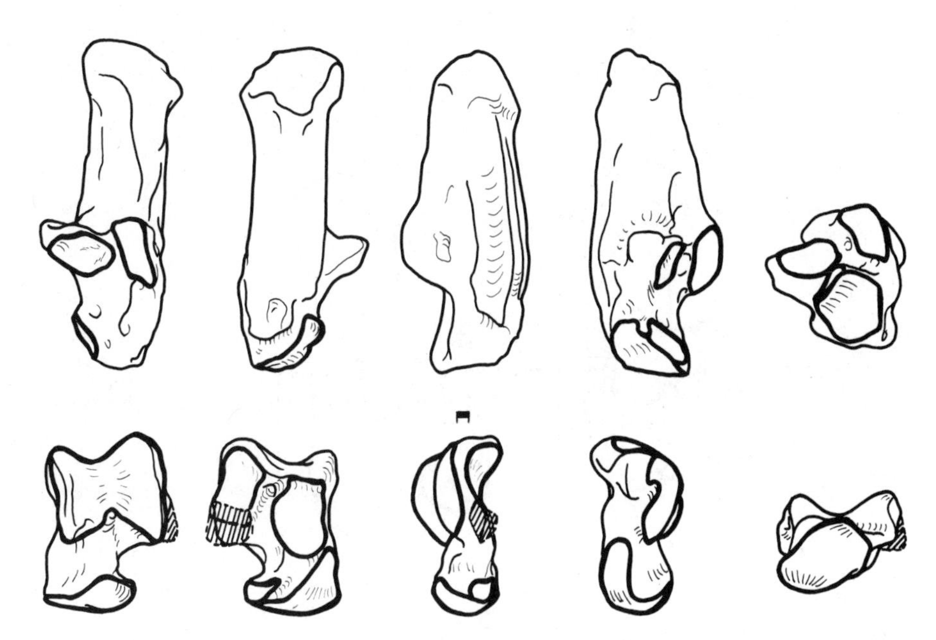

Fig. 18. Mimolagus rodens, calcaneus (above), and astragalus (below), based on casts in the AMNH, ?early Oligocene of Western China. From left to right: dorsal, plantar, lateral, medial, and distal views, respectively. Note the entirely duplicidentate appearance of the bones, but the lack of a CaFi facet. The immediate medial postion of the sustentaculum to the CaA surfaces, as well as the articulation of the latter, leaves no doubt that the LAJ was as immobile as it is in living leporid duplicidentates (I believe that the LAJ is secondarily modified in ochotonids). Note especially the proximoplantar extension of the ANJ articulation, probably for contact with the posteroproximal process of the navicular seen in lagomorphs. Hatched area is reconstructed. Scale represents one mm.

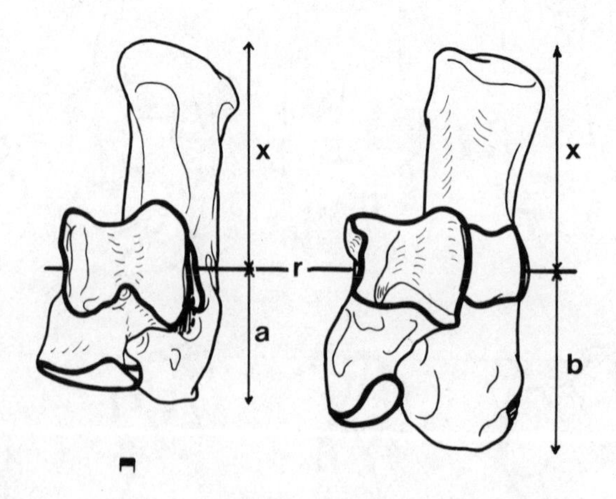

Fig. 19. Comparison of the articulated astragalus and calcaneus of Mimolagus (left) and Paleolagus (right). Note in particular the relatively broader astragalar tibial trochlea, the less anteroposteriorly oriented ANJ, and the lack of a calcaneal CaFi facet in the former in contrast to the latter. The lever arm (x) to load arm (a and b) ratios ($x/a > x/b$), shown in relation to a common axis of rotation for the UAJ (r), suggest a more highly advanced cursoriality in Paleolagus. Scales represent one mm.

CaA facet. In macroscelidids the calcaneal Su facet is proximal to the CaA facet. To conclude concerning Mimolagus, on the evidence which I examined, the casts of the proximal tarsals of this genus (see Fig. 18), I have no doubts that it is an archaic lagomorph, more recently related to the extant taxa than any other group which may be included in the Lagomorpha.

Table 2 gives a very conservative summary of a few characters which, in my view, lend themselves to this type of tabulation. No serious consideration, however, should be given to these isolated characters without consulting the text. The manner in which bone (and some other tissue) morphology is taxon specifically constructed around mechanically similar or identical joints is often the most significant expression of the underlying genetic code. Many subtle and important details do not lend themselves to the verbal simplification required for tabulation, yet these are some of the most important data bases on which systematic judgements should be based.

The living macroscelidids (see Evans, 1942, and Kielan-Jaworowska, 1979, for important postcranial details and discussions), like the morphotypic lagomorphs, are well adapted cruropedally to hop and scurry rapidly, and I have mistaken their adaptive similarity as indicative of a relatively recent common phylogeny (Szalay, 1977b). After a detailed examination of the cruropedal evidence, I believe that the similarities are only adaptive, yet the detailed solution for a hopping-cursorial hindleg are clearly independently and differently constructed. For example, the manner in which the tibia contacts the astragalus and the construction of the ANJ (see above) are highly distinctive between lagomorphs and mcroscelideans (Figs. 13-15). The astragalar ACu facet, well displayed in macroscelideans, is missing in lagomorphs, and the medially facing calcaneal CaA facet, as well as the broad, and dorsally oriented CaFi facet of lagomorphs, are in telling contrast to the dorsally oriented calcaneal CaA facet and the small distally facing CaFi facets of elephant shrews. These are convergent mechanical solutions to similar biorole demands.

I have compared a poorly preserved astragalus and a broken juvenile calcaneus (AMNH 21755, 21754) of Pseudictops lophiodon, consulting the excellent description and figures of the foot of this taxon by Sulimski (1968), with a partial astragalus of Anagale gobiensis (AMNH 26079), and compared these with all of the fossils and living taxa discussed in this paper. Neither of these Asiatic forms show any special similarity with the morphotypic conditions of rodents. There is no facet for a tibial posterior process, clearly because there either was not one or not one which homologously articulated with the astragalus. The diagnostic UAJ articulation of the astragalus in rodents does not resemble the well preserved bodies of the astragali of the Asiatic forms. I do detect, however, in spite of clearly diagnostic differences, whatever it is worth, significant similarities between Pseudictops and Anagale, more so

Table 2. A summary of some cruropedal characters in selected taxa.

Character	Protoeutherian	Protoleptictid	Protorodent	Protozalambdalestid	Protolagomorph	Protomacroscelidid
Lower TFJ	syndes-mosis	syndes-mosis	syndes-mosis	fused	fused	fused
tpp	absent	present	present	?	absent	present
CaFi facet	present	absent	absent	absent	absent	present
LAJ mobility	non-helical	slightly helical	helical	non-helical	non-helical	non-helical
CaCu facet	slightly ovoid, wide	slightly ovoid, wide	slightly ovoid, deep	cylin-drical, wide	cylin-drical, wide	ovoid, rounded
Posterior Na process	?	incip-ient	incip-ient	present	present	absent
Astragalar trochlea	(an important character complex, but as for many others, tabulation would merely dilute the meaning of the character states. These are compared and evaluated in the text.)					
Astragalar head	(the remarks under the astragalar trochlea apply here also.)					

than either one of these might resemble a primitive leptictid or a primitive rodent (Figs. 16-17). Pseudictops resembles the first lagomorphs in the construction of the LAJ, possibly a homologous feature, and in the conformation of the calcaneal CaFi facet, although probably convergently. Both Anagale and Pseudictops show restrictions and facilitations of the UAJ found in terrestrial cursors and leapers, well beyond the eutherian morphotypic adaptations. The digitigrady of the foot of Pseudictops described by Sulimski (1968) supports the assessment of the tarsals.

I see no pedal evidence which would bar the monophyly of the four groups, the Zalambdalestidae, Anagalidae, Pseudictopidae, and Lagomorpha, yet, contrary to my 1977 stand on classification, there seems to be no compelling evidence to unite these groups in a single order Lagomorpha.

SUMMARY AND CONCLUSIONS

The methodology for phylogenetic and adaptational reconstruction was a transformational analysis (i.e., a functional-adaptive analysis in a time context). The reason why cladistic analysis (i.e., character distribution analysis) and the cladistically based concept of a scenario are rejected is rooted in the unacceptable, biologically unrealistic assumptions on which cladistics as a structuralist school of systemetics is based, as well as the inability of cladistics to grapple with the significance of form-function, adaptation, time, the importance of their role in the assessment of character transformations, and therefore the descent of taxa. Assessment of the most probable morphocline polarities, i.e., transformation sequences, is dependent on the broadest understanding of the biology, mechanics, behavioral correlates, and ecological aspects of characters. Phylogenies which are not based on such a character analysis, but depend on character distribution analysis guided by biologically unfounded assumptions, are unlikely to reflect the most probable evolutionary history.

The first known postcrania of rodents display a morphology which is most similar in its derived attributes to the scantily known postcranials of the primitive members of the Leptictidae. These special similarities are largely restricted to the cruropedal remains until additional information about early leptictids becomes known.

The upper ankle joint in both groups is characterized by a distal tibial posterior process and a well developed medial malleolus which tightly restrict the sharp medial crest of the astragalar tibial trochlea. The more extensive but equally sharp lateral crest of the trochlea further ensures that this joint restricts all transverse mobility of the ankle. Unlike in other primitive eutherians, the articulation of the tibial posterior process is with the proximal border of the astragalus, on which all traces of the astragalar canal are obliterated. Although the lower ankle joint of the protorodents is distinctly more mobile than that of primitive leptictids, the conformation of the latter is an ideal structural antecedent for the former.

The skeletal specializations of the protorodents can be explained as solutions to claw climbing and head-first descent on tree trunks, as basically seen in the arboreal Sciuridae. This view is strongly corroborated by the mechanics of the mediolaterally compressed deep distal phalanges bearing sharp falculae, an "arboreal" ulna indicating a habitually flexed forearm, the tensile-load adapted hip joint, and the details of the cruropedal complex of the Eocene Paramyidae, in spite of the relatively greater robusticity of the bones of these compared to modern and fossil squirrels. The protorodent condition stands in sharp contrast with that of the protolagomorph form-function complex, which was modified from a hitherto unknown, possibly *Zalambdalestes*-like source, for a terrestrial cursorial and saltatorial mode of life.

There is no presently recognized postcranial evidence which would link the Rodentia with the Lagomorpha, or the former either with the Zalambdalestidae, Eurymylidae (including the Mimotonidae), Anagalidae, or Pseudictopidae. The latter families *may* have been part of a monophyletic group which was the source for the protolagomorph, *sensu stricto*, if the enlarged dI2 of the rodents and lagomorphs was not homologously acquired. The last common ancestor of *Mimolagus* and the living duplicidentates shared a complex of uniquely derived character combinations. The unequivocal mechanical interpretation of the relevant joints points to extreme facilitation of anteroposterior mobility and transverse restriction in the UAJ, coupled with great facility to plantarflex and dorsiflex the foot, not only at the UAJ, but also at the midtarsal articulation in the CCJ and ANJ. The morphology of the midtarsal joint, in addition to the advanced cursorial capabilities, suggests the ability to adjust the digitigrade foot to the unevenness of the substrate, or to favorably receive loads in a range of positions. This ancestral species, as stated before, was almost certainly a cursor-leaper, and not an ambulatory form. All relevant postcranial attributes of fossil and recent lagomorphs point to those types of restrictions and facilitations which widely occur in metatherian and eutherian terrestrial cursors-leapers.

The concept of Glires is discussed, and further work is suggested to understand the possible constraints on the limited options available in the development of enlarged incisors, either from ontogenetic or functional adaptive factors centered on the feeding mechanism, before the homology of the enlarged dI2 in the two orders is widely accepted.

My previous assessment of the cruropedal remains of macroscelidids and lagomorphs as indicative of monophyly were based on the mistaken judgement of similarities as homologies which, after a more detailed analysis, were reconsidered to be convergent features for cursorial adaptations.

In summary, with some noted departures from my previous (Szalay, 1977b) classification, based on the reanalyzed postcranial evidence, and on my appraisal of the important studies presented in this volume, the following evolutionary classification, if not a clearly understood phylogeny of the groups included, appears to be justified, as it creates no new categories, names no new taxa, yet allows the necessary means to discuss attributes and relationships of the known groups.

Cohort Glires Linnaeus, 1758
- Order Leptictimorpha Szalay, 1977
  - Family Leptictidae (including *Cimolestes* and relatives)
  - Family Pantolestidae
- Order Taeniodonta Cope, 1876
- Order Rodentia Bowdich, 1821
- Order Lagomorpha Brandt, 1855
- Order Anagalida Szalay and McKenna, 1971
  - Family Zalambdalestidae
  - Family Anagalidae
  - Family Pseudictopidae
  - Family Eurymylidae (including Mimotonidae)

ACKNOWLEDGEMENTS

I owe a debt of gratitude to Drs. J. L. Hartenberger and W. P. Luckett for providing the opportunity for so much learning by organizing the conference. I am also grateful to Dr. C.-K. Li and S.-Y. Ting for making so many of the important Asiatic Paleogene specimens relevant to the papers of many of us in Paris available for study. Their generosity has greatly added to the substance of this and other papers.

I thank Drs. M. C. McKenna, G. G. Musser, M. J. Novacek, and R. H. Tedford for permission to study specimens in their care. Support for the research presented in this paper was from PSC-CUNY grants nos. 6-62084 and 6-63094. The illustrations were skillfully prepared by Ms. Patricia Van Tassel.

Special thanks are given to Drs. Nancy Hong and Christiane Denys for their unfailing help before and during the conference.

REFERENCES

Amtmann, E. 1979. Biomechanical interpretations of form and structure of bones: role of genetics and function in growth and remodelling. In: Environment, Behavior, and Morphology, M. E. Morbeck, H. Preuschoft, and N. Gomberg, eds., pp. 347-366, G. Fischer, NY.

Barnett, C. H. and Napier, J. R. 1953a. The rotary mobility of the fibula in eutherian mammals. J. Anat. 87: 11-21.

Barnett, C. H. and Napier, J. R. 1953b. The form and mobility of the fibula in metatherian mammals. J. Anat. 87: 207-213.

Black, C. C. 1963. A review of the North American Tertiary Sciuridae. Bull. Mus. Comp. Zool. 130: 109-248.

Bock, W. J. 1977. Adaptation and the comparative method. In: Major Patterns in Vertebrate Evolution, M. K. Hecht, P. C. Goody, and B. M. Hecht, eds., pp. 57-82, Plenum Press, NY.

Bock, W. J. 1981. Functional-adaptive analysis in evolutionary classification. Amer. Zool. 21: 5-20.

Dawson, M. R., Li, C.-K., and Qi, T. 1984. Eocene ctenodactyloid rodents (Mammalia) of Eastern and Central Asia. Spec. Publ. Carneg. Mus. Nat. Hist. 9: 138-150.

Dollo, L. 1893. Les lois de l'evolution. Bull. Soc. Belg. Géol. Pal. Hydr. 7: 164-166.

Eldredge, N. and Cracraft, J. 1980. Phylogenetic Patterns and the Evolutionary Process. Columbia Univ. Press, NY.

Emry, R. J. and Thorington, R. 1982. Descriptive and comparative osteology of the oldest fossil squirrel, Protosciurus (Rodentia: Sciuridae). Smithsonian Contrib. Paleobiol. 47: 1-34.

Evans, F. G. 1942. The osteology and relationships of the elephant shrews (Macroscelididae). Bull. Amer. Mus. Nat. Hist. 80: 85-125.

Gingerich, P. D. 1980. Evolutionary patterns in early Cenozoic mammals. Ann. Rev. Earth Planet. Sci. 8: 407-424.

Godinot, M. 1981. Usefulness and meaning of the mammalian specific lineages which are used in European continental biostratigraphy. Int. Symp. Concpt. Meth. Paleont., Barcelona: 249-258.

Gutmann, W. F. 1977. Phylogenetic reconstruction: theory, methodology, and application to chordate evolution. In: Major Patterns in Vertebrate Evolution, M. K. Hecht, P. C. Goody, and B. M. Hecht, eds., pp. 57-82, Plenum Press, NY.

Haines, R. W. 1958. Arboreal or terrestrial ancestry of placental mammals. Quart. Rev. Biol. 33: 1-23.

Hennig, W. 1966. Phylogenetic Systematics. Univ. of Illinois Press, Urbana.

Hildebrand, M. 1978. Insertions and functions of certain flexor muscles in the hind leg of rodents. J. Morph. 155: 111-122.

Jenkins, F. A., Jr. 1974. Tree shrew locomotion and origins of primate arborealism. In: Primate Locomotion, F. A. Jenkins, Jr., ed., pp. 85-115, Academic Press, NY.

Jenkins, F. A., Jr. and Krause, D. W. 1983. Adaptations for climbing in North American multituberculates (Mammalia). Science 220: 712-715.

Kielan-Jaworowska, Z. 1979. Evolution of the therian mammals in the late Cretaceous of Asia. Part III. Postcranial skeleton in Zalambdalestidae. Palaeont. Pol. 38: 3-41.

Korth, W. W. 1984. Early Tertiary evolution and radiation of rodents in North America. Bull. Carneg. Mus. Nat. Hist. 24: 1-71.

Lewis, O. J. 1980a. The joints of the evolving foot. Part I. The ankle joint. J. Anat. 130: 527-543.

Lewis, O. J. 1980b. The joints of the evolving foot. Part II. The intrinsic joints. J. Anat. 130: 833-857.

Lewis, O. J. 1980c. The joints of the evolving foot. Part III. The fossil evidence. J. Anat. 131: 275-298.

Matthew, W. D. 1904. The arboreal ancestry of the Mammalia. Amer. Nat. 38: 811-818.

Matthew, W. D. 1909. The Carnivora and Insectivora of the Bridger Basin, Middle Eocene. Mem. Amer. Mus. Nat. Hist. 9: 289-567.

Mayr, E. 1974. Cladistic analysis or cladistic classification? Z. Zool. Syst. Evolut.-Forsch. 12: 94-128.

Mayr, E. 1981. Biological classification: toward a synthesis of opposing methodologies. Science 214: 510-516.

Peters, V. D. S. 1972. Das Problem konvergent entstandener Strukturen in der anagenetischen und geneologischen Systematik. Z. Zool. Syst. Evolut.-Forsch. 10: 161-173.

Romankowowa, A. 1960. The sesamoid bones of the autopodia of Insectivora and Rodentia. Zoologica Pol. 10: 225-256.

Romankowowa, A. 1963. Comparative study of the structure of the os calcaneum in insectivores and rodents. Acta Theriol. 7: 91-126.

Simpson, G. G. 1975. Recent advances in methods of phylogenetic inference. In: Phylogeny of the Primates, W. P. Luckett and F. S. Szalay, eds., pp. 3-19, Plenum Press, NY.

Sulimski, A. 1968. Paleocene genus _Pseudictops_ Matthew, Granger, and Simpson, 1929 (Mammalia) and its revision. Palaeont. Pol. 19: 101-133.

Szalay, F. S. 1977a. Ancestors, descendants, sistergroups, and the testing of phylogenetic hypotheses. Syst. Zool. 26: 12-18.

Szalay, F. S. 1977b. Phylogenetic relationships and a classification of the eutherian Mammalia. In: Major Patterns in Vertebrate Evolution, M. K. Hecht, P. C. Goody, and B. M. Hecht, eds., pp. 317-374, Plenum Press, NY.

Szalay, F. S. 1981a. Phylogeny and the problem of adaptive significance: the case of the earliest primates. Folia Primatol. 36: 157-182.

Szalay, F. S. 1981b. Functional analysis and the practice of the phylogenetic method as reflected by some mammalian studies. Amer. Zool. 21: 37-45.

Szalay, F. S. 1982. A new appraisal of marsupial phylogeny and classification. In: Carnivorous Marsupials, M. Archer, ed., pp. 621-640, Roy. Zool. Soc. NSW, Australia.

Szalay, F. S. 1984. Arboreality: is it homologous in metatherian and eutherian mammals? Evol. Biol. 18: 215-258.

Szalay, F. S. and Decker, R. L. 1974. Origins, evolution, and function of the tarsus in late Cretaceous eutherians and Paleocene primates. In: Primate Locomotion, F. A. Jenkins, Jr., ed., pp. 223-259, Academic Press, NY.

Szalay, F. S., Tattersall, I., and Decker, R. L. 1975. Phylogenetic relationships of Plesiadapis: postcranial evidence. Contrib. Primatol. 5: 136-166.

Van Valen, L. 1978. Why not to be a cladist. Evol. Theory 3: 285-299.

Van Valen, L. 1982. Why misunderstand the evolutionary half of biology? In: Conceptual Issues in Ecology, E. Saarinen, ed., pp. 323-343, D. Reidel Publ. Co., Dordrecht, Holland.

Wilson, R. W. 1949. Early Tertiary rodents of North America. Carneg. Inst. Wash. Publ. 584: 67-164.

Wood, A. E. 1937. The mammalian fauna of the White River Oligocene. Part II. Rodentia. Trans. Amer. Phil. Soc. 26: 157-279.

Wood, A. E. 1962. The early Tertiary rodents of the family Paramyidae. Trans. Amer. Phil. Soc. 52: 1-261.

ENAMEL STRUCTURE OF EARLY MAMMALS AND ITS ROLE IN EVALUATING RELATIONSHIPS AMONG RODENTS

Ashok Sahni

Centre of Advanced Study in Geology
Panjab University
Chandigarh-160014, India

INTRODUCTION

The evolution of mammalian enamels is a matter of multidisciplinary interest, and in recent years considerable attention has been paid to the subject (Grine, 1978; Fosse et al., 1973; Sahni, 1979). Work has concentrated on studying Recent reptilian and mammalian enamels (Cooper and Poole, 1973; Poole, 1957; Boyde, 1966), while other studies have centered on the micro- and ultrastructure of various "ancestral" groups such as the therapsid (mammal-like) reptiles in the hope of documenting evidence of the change from the thin, nonprismatic structure of reptilian enamels to the thicker, prismatic enamels of mammals (Poole, 1956; Moss, 1969; Osborn and Hillman, 1979). The enamels of various fossil and Recent rodents have invited special attention since the classic work of Tomes (1850). Rodent enamels are the most complex and highly organized within Mammalia. There are distinct subordinal variations among various groups; material from both fossil and Recent rodents is readily available and, furthermore, the process of amelogenesis (enamel secretion) can be observed throughout the life history of an individual because of the phenomenon of evergrowing incisors (Korvenkontio, 1934; Kiel, 1966).

However, in spite of the considerable progress made in the last two decades, there are several lacunae in our knowledge of the major steps leading to the evolution of mammalian enamels. These constraints include the adaptive responses in the arrangement and orientation of hydroxyapatite crystallites at the ultrastructural level, and the mechanism by which the process of amelogenesis proceeds in various taxa. Such inadequacies make the correlation of form and function difficult to establish. In addition, there is the omnipresent problem of recognizing parallel evolution of enamel elements, which tends to mask the evolutionary continuity of the enamel internal structure. Parallelism of features may arise more than once in the same group as well as in unrelated groups. Furthermore, some characters may be secondarily simplified. A good example of such degeneration is found in the odontocete whales which have typical Pattern One prisms without zone development, while their Eocene archaeocete ancestors have a similar pattern but with well developed zonation, a condition far removed from their supposed condylarthran or insectivore derivation. Another major hurdle in the application of enamel structure studies in taxonomy is the lack of data from well established evolutionary lineages of Mesozoic mammals, particularly those throughout the Jurassic and the early Cretaceous. The paucity of evidence tends to lay too much emphasis on the easily available and by now well studied upper Cretaceous mammals from Montana and ignores the large collections of Jurassic symmetrodonts and eupantotheres available for study. Lastly, the value of enamel structural elements in differentiating taxa and in determining phyletic lineages has yet to be firmly established and is still open to test.

The advent of scanning electron microscopy (SEM) has contributed greatly to the study of therapsid and primitive mammalian enamel ultrastructure, and it has revealed that the enamels of those early mammals and mammal-like reptiles, previously considered to be continuous and non-prismatic, are in fact prismatic (Moss, 1969; Moss and Kermack, 1967; Poole, 1967; Fosse et al., 1973; Grine, 1978; Sahni, 1979). Thus, while it was previously believed that prismatic enamel was a therian character, the earliest record of which was represented by early Cretaceous Pappotheriidae (Moss, 1969), recent work has shown that prismatic enamel is found in nontherian mammals, as well as in some therapsid reptiles (Grine et al., 1979; Poole, 1971). One extant reptile, the agamid lizard _Uromastyx_, also possesses prismatic enamel (Cooper and Poole, 1973). Conversely, Moss (1969) has suggested (but there is no SEM confirmation of this) that in one living mammal, the monotreme _Ornithorhynchus_, a reptilian type, non-prismatic enamel is present.

The present analysis is based on the author's published and unpublished work on traversodont, cynodont and dicynodont therapsids from Gondwana formations in India; Cretaceous sauropod and theropod teeth; Eocene crocodiles (including pristichampsines); upper Cretaceous allotherians, metatherians and eutherians from Montana, USA; Eocene archaeocetes, rhinocerotoids, tapiroids and brontotheres from Kashmir and Kutch; Indian Eocene ctenodactylid rodents, African phiomyids and several murids, rhizomyids and arvicolids from the Siwalik and Karewa Groups. The paper draws heavily on published accounts of therapsids, and of early mammal and rodent enamels cited herein. Taxa include cotylosaurs, _Dimetrodon_ (pelycosaur), and several therapsid genera (_Thrinaxodon_, _Probainognathus_, _Probelesodon_,

Diademodon, *Messetognathus*, *Pachygenelus*, and *Tritylodon*). Other taxa include *Oligokyphus* (ictidosaur), *Hargeria* (= *Hesperonis*) toothed birds; primitive mammals *Eozostrodon* (= *Morganucodon*), *Astroconodon*, dryolestids, plagiaulacid multituberculates, haramiyids, pappotheriids, and the monotreme *Ornithorhynchus*.

ENAMEL STRUCTURE OF REPTILES AND MAMMALS

Reptilian and mammalian enamels are of ectodermal origin and thus can be differentiated from the enameloid of fishes and amphibians, which is mesodermal in nature (Poole, 1971). Among the first to draw attention to the evolutionary changes from reptilian to mammalian enamels was the work of Poole (1956). He stressed the development of a pseudo-prismatic stage for therapsid reptiles, intermediate between the nonprismatic enamels of most reptiles and the prismatic stage of most mammals. Figure 1 illustrates the three broad evolutionary steps by which structureless, nonprismatic enamels changed to the prismatic condition via an intermediate pseudoprismatic (or preprismatic) stage. In most reptilian enamels, mainly as a result of the poor development of the Tomes process (Poole, 1957), the hydroxyapatite crystallites are oriented normal to the secretory plane of the ameloblasts, and the interprismatic phase is not developed. In most reptilian teeth, apart from the pronounced incremental lines, no other structure is discernible. In some reptiles, some irregular variation of the C axis of the apatite crystallites does occur, as for instance the "Säulengliederung" of crocodile enamel. In some therapsids and in some very primitive mammals, a more regular, repetitive change in the orientation of the crystallites has been noted (Fig. 1b). The condition is even present in the pelycosaur *Dimetrodon*. The optical effect of such repetitive, crystallite orientation changes in most therapsids (such as *Massetognathus*, *Diademodon*, and *Tritylodon*) is to give the appearance of a prismatic structure. However, the interprismatic phase (with differing crystallite orientation to that of the prisms) is not developed. This transitional stage, the pseudoprismatic, has been referred to as the "preprismatic" by Frank et al. (1984), based on their studies of Late

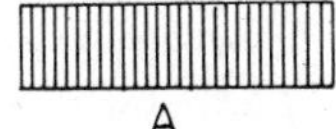

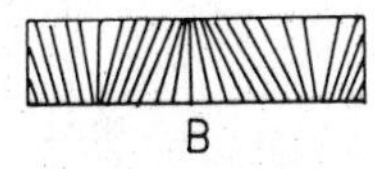

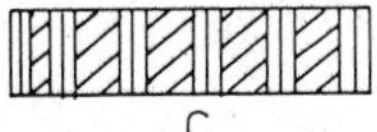

Fig. 1. Structureless and prismatic enamels.

triassic haramiyids. Prismatic enamel, which is currently documented in a few reptiles and in the majority of mammals, is marked by the development of an interprismatic phase with resultant sharp differences in crystallite orientation between the prism and interprism boundaries (Fig. 1c).

Most of the earlier work on the documentation of the reptilian-mammalian enamel transition was carried out by means of optical microscopy (Poole, 1956; Moss, 1969; Osborn and Hillman, 1979) and lacked the versatility and the resolution available in scanning electron microscopy. Consequently, recent ultrastructural work indicates that the clear-cut divisions visualized by optical microscopy do not exist. At least some reptiles do possess prismatic enamels: the agamid lizard _Uromastyx_ (Cooper and Poole, 1973); placodonts (Schmidt and Keil, 1971) and therapsids (Grine et al., 1979). These reptiles are known to have specialized, masticatory dentitions. This sample, albeit meager, lends credibility to Poole's (1956) contention that the main factor responsible for the development of prismatic enamels is the evolution of occluding dentitions with accompanying masticatory processes and the concomitant reduction in tooth replacement. However, exceptions to this rule exist. First, not all primitive mammals which are known to have occluding, non-multireplacement dentitions have prismatic enamels, for example, _Docodon_. This situation also applies to some therapsids (Grine, 1978) that are believed to possess occluding dentitions, namely _Tritylodon_ and _Diademodon_. On the other hand, some therapsids (the cynodont _Pachygenelus_) do appear to have prismatic enamel and are thought to have indulged in heavy mastication (Grine et al., 1979). Thus, at present, the status of enamel internal structure studies is hampered by poor documentation by electron microscopy. It is hoped that as more and more therapsid, non-therian and primitive therian mammalian enamels are studied, the evolutionary record of enamel development will improve.

STATUS OF ENAMEL ULTRASTRUCTURE STUDIES IN TAXONOMY

Recent work has stressed the potential of enamel ultrastructure studies in elucidating phylogenetic lineages and for differentiating forms at various taxonomic levels (Carlson and Krause, 1982; Fosse et al., 1978; Koenigswald, 1980). However, mainly because of inadequate sampling and poorly understood processes at the cellular level, the value of such studies has yet to gain universal acceptability for determining relationships in a consistent and systematic manner. At the ordinal level, unrelated groups possess similar enamel prism packing. Thus Pattern One prisms (Boyde, 1964) characterize the agamid lizard, _Uromastyx_, at least two genera of therapsids, _Eozostrodon_, ptilodontid multituberculates, most insectivores, some primitive primates, Chiroptera, Cetacea and Sirenia. The pattern is also found in the superficial sections of many other mammals. Prism Pattern Two is basically an ungulate pattern and is found in artiodactyls, perissodactyls, marsupials and, in a modified sense, in the rodents as well. Pattern Three prisms occur in carnivores, proboscideans and higher primates. In certain Cretaceous mammals such as *Stygimys* and *Protungulatum*, arcade-type prisms have been documented (Sahni, 1979). The prisms are completely surrounded by thick interprismatic regions and, in this sense, are not strictly comparable to typical Pattern Three prisms found in higher primates, where the interprismatic phase is greatly restricted and the prisms densely packed. At the subordinal level, enamel structural studies more closely support the results based on gross dental morphology and other anatomical features. This situation is broadly applicable in the multituberculates (Fosse et al., 1978; Sahni, 1979), rodents (Korvenkontio, 1934), Cetacea (Sahni, 1981; Ishiyama, 1984) and Primates (Boyde and Martin, 1984). The enamel structure of *Prozeuglodon* (Kozawa, 1984) and the Indian protocetids (Sahni, 1981) is similar to each other and well differentiated from that of various odontocete genera (Ishiyama, 1984). Similarly, it has been demonstrated that the strepsirhine primates possess a predominance of Pattern One enamel, while the haplorhine primates have generally a Pattern Three prism packing (Boyde and Martin, 1984). At the generic and specific levels, the utility of enamel structure studies is uncertain and inconsistent, mainly due to the lack of persistence of any one (or combination of) enamel characters at the familial level. It still has not been demonstrated why the genus *Vombatus* is the only exception to the marsupial condition by not having enamel tubules in its prismatic enamel; or why the Dall's porpoise (*Phocoenoides dalli*) has a nonprismatic, tubular, reptilian-type enamel (Ishiyama, 1984). Such exceptions, the reasons for which are not clearly understood today, make a comprehensive analysis of mammalian relationships a difficult task.

Enamel characters used for taxonomic and phylogenetic analyses include: the shape, size, pattern and density of transversely-sectioned prisms; layer differentiation; structure of the Hunter-Schreger Bands; the inclination, orientation, width, boundaries and development of prism zones; decussating and non-decussating prism patterns, and the distribution of enamel tubules. The distribution of characters in some representative reptilian and mammalian taxa is given in Table 1. The general conclusions from the table are: tubular, prismatic enamels are primitive in nature; Pattern One prisms are the least derived; some multituberculates have exceptionally large prisms; the acquisition of layered enamel is a specialized feature; and that rodent enamels constitute a distinctive group within the Mammalia.

Table 1. Distribution of enamel structural elements in various vertebrates. (P = Paleogene; R = Recent; T = Triassic; i = incisor; m = molar).

	R	T		Cretaceous					Tertiary						P	Neogene	
	Uromastyx	*Pachygenelus*	*Eozostrodon*	*Mesodma*	*Catopsalis*	*Protungulatum*	*Cimolestes*	*Gypsonictops*	ARTIODACTYLA	PERISSODACTYLA	MARSUPIALIA	CARNIVORA	PRIMATES	CETACEA	PRIMITIVE RODENTS	HYSTRICOGNATHI	MYOMORPHA AND SCIUROMORPHA
PRISM SHAPE (T. S.)																	
Pattern One	+	+	+	+	+	+	+	+	+	+	+	+	+	+	+	+	+
Pattern Two									+	+	+				+	+	+
Modified Pattern Three					+	+											
Pattern Three												+	+				
PRISM DIAMETER																	
4-5 micra	+	+	+	+i		+	+	+	+	+	+	+		+	+	+	+
8-10 micra				+m	+												
LAYERING				+i			+	+							+	+	+
ZONATION																	
Multiserial									+	+		+	+	+		+	
Other types															+		+
ENAMEL TUBULES	+	+	+	+	+						+					+	

PRISMATIC ENAMEL

Prism Shape

Figure 2 gives the temporal distribution of the various enamel patterns on the basis of transversely sectioned prism outlines. Circular prisms completely surrounded by interprismatic material (Pattern One) constitute the oldest type of prismatic enamel known and are present in some therapsids and the eotherian Eozostrodon since the late Triassic. In the earlier part of its record, this prism packing pattern is usually associated with numerous enamel tubules. In Tertiary and Recent insectivores, cetaceans, and sirenians (which are also characterized by Pattern One prisms), such

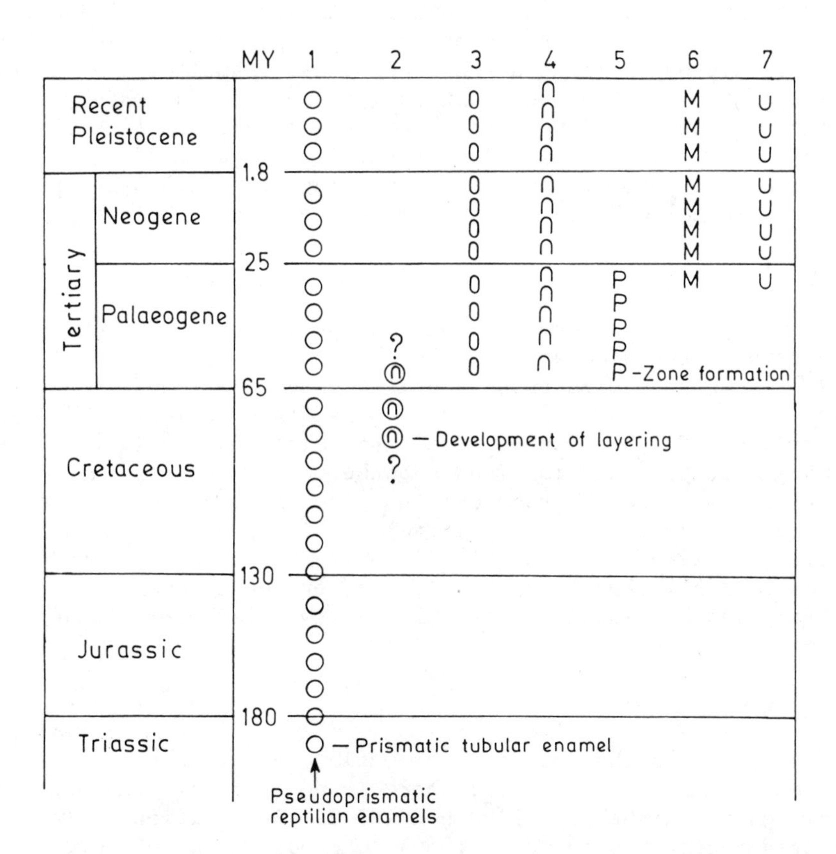

Fig. 2. The chronological distribution of transversely sectioned prisms. 1, Circular prisms completely surrounded by interprismatic material (Pattern One); 2, Arcade prisms surrounded by thick interprismatic material; 3, Oval prisms arranged in rows (Pattern Two); 4, Key-hole prisms with little interprismatic material (Pattern Three); 5, Pauciserial (P); 6, Multiserial (M); 7, Uniserial (U).

tubules are rare or absent in adult enamels. Some Cretaceous mammals, including the multituberculate Stygimys and the condylarthran Protungulatum, have arcade or horseshoe-shaped prisms that are surrounded by a thick interprismatic layer. This pattern does not appear to be strictly homologous to Pattern Three prisms as defined by Boyde (1964), as the latter have a poorly developed interprismatic phase. The former pattern is hence considered to be a modified version of Pattern Three. Pattern Two prisms of early Tertiary ungulates, with oval prisms arranged in rows, do not appear to be substantially different from those of their more modern descendants, and the same situation applies to Pattern Three prisms. Miocene to Recent canids, proboscideans and primates have basically the same enamel structure. The first record of enamel layering is in the Cretaceous, while that of zone development is in the early Paleogene.

Rodent enamels, with Pattern One and Pattern Two prisms, are clearly separated into three separate groups in chronological terms. The oldest group comprises the pauciserial enamel of diverse Paleogene rodents, and the other categories include two (mainly post-Oligocene) structural patterns, namely the multiserial and uniserial. During the Oligocene, a number of instances can be cited for the occurrence of forms transitional in structure between the pauciserial and multiserial or uniserial enamel patterns (see below).

Prism Size

In reptiles and mammals that do possess prismatic enamels, prism size remains the same and averages 4-5 micra. The size (= diameter) of ameloblasts of even those reptiles which are known to have structureless enamel averages the same as is found in metatherians and eutherians. Figure 3 shows models for amelogenesis in reptiles and mammals, with the special (and hypothetical) case for the multituberculates. In multituberculates with large prism diameters and circular prisms (more than 8 micra), it is conjectured that the ameloblasts were of comparable diameter, as had been proposed earlier by Fosse et al. (1973) on the basis of evidence demonstrated by Boyde (1969b), who has shown an ameloblast-prism size correspondence for Pattern One prisms.

Layering

In reptiles that do have prismatic enamel, as well as in the earliest mammal, Eozostrodon, there is no layering, and hence this is deemed to be the primitive condition. Acquisition of layering has been documented by Moss (1969), who has shown that Cretaceous multituberculates possess a 2-layered structure, whereas Jurassic plagiaulacids have an undifferentiated enamel. In the incisors of ptilodontid multituberculates a definite layered structure is caused by the bending over of the prisms in the central zone along the transverse plane. This results in the development of three layers:

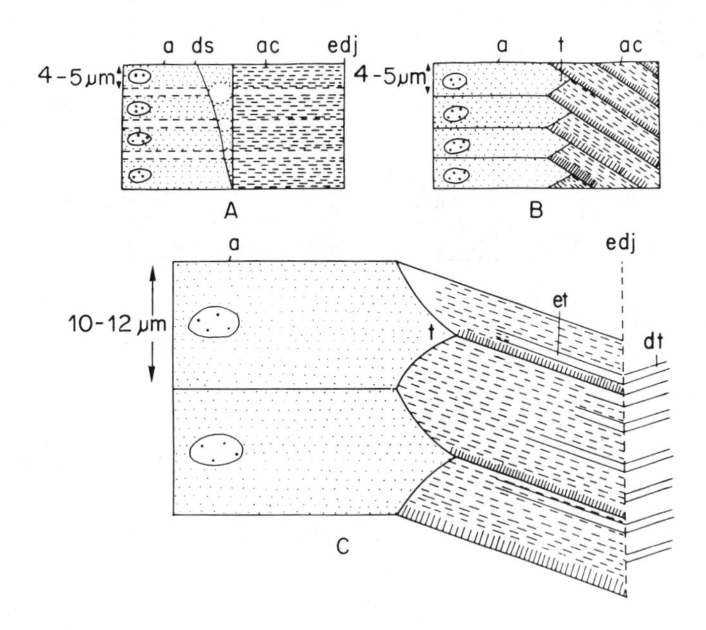

Fig. 3. Relationship of ameloblasts to developing enamel. (A) Reptilian condition (after Poole, 1957). (B) Eutherian mammal. (C) Catopsalis (taeniolabidid multituberculate). a, ameloblast; ac, apatite crystallites; dt, dentinal tubule; edj, enamel-dentin junction; et, enamel tubule; t, Tomes' process.

an innermost, consisting of transversely sectioned prisms, a central zone of apically directed prisms, and an outermost with structureless interprismatic enamel. This condition probably represents one of the first steps in increasing the biomechanical efficiency of the incisors. No pronounced layering could be detected in the few incisors studied of the taeniolabidid multiberculate Catopsalis. In horizontal sections of some multituberculate molars, a layered structure can sometimes be observed, consisting of an innermost prismatic layer and an outer structureless zone. A similar condition occurs in the early eutherian Cimolestes, but is not found in the condylarth Protungulatum. Layering (parallel to the enamel-dentin junction) is particularly well seen in rodents throughout their geological history (Fig. 4).

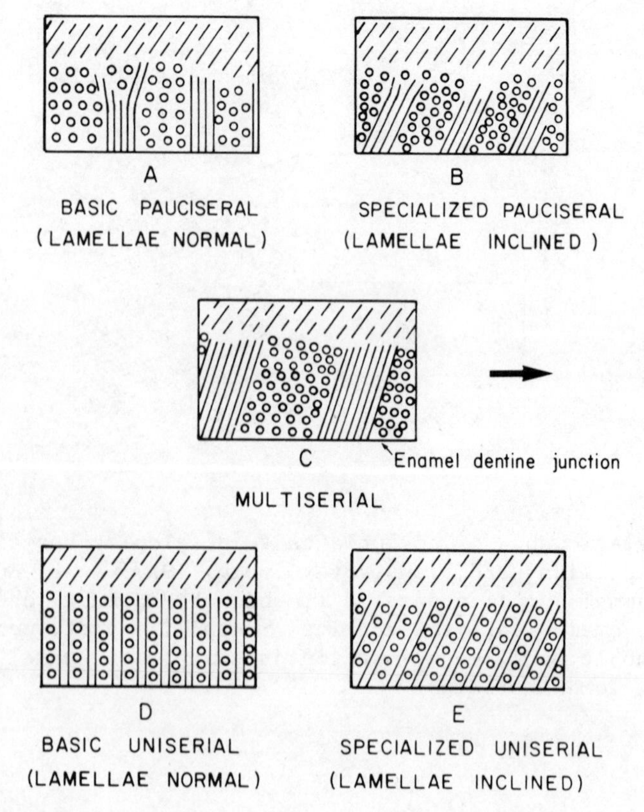

Fig. 4. Schematic diagrams of the variations within the three main patterns of rodent enamel.

Hunter-Schreger Bands

Hunter-Schreger bands are developed by bending or crossing-over of prisms, so that when enamel is sectioned, alternating zones are observed in which prisms lie parallel to the plane of sectioning, while others are cut obliquely or transversely. The structure of Hunter-Schreger bands is another important diagnostic character of mammalian enamels. In Pattern One type enamels, zones are not well developed, although they have been reported in one of the earliest mammals studied, Eozostrodon (Grine and Cruikshank, 1978). A slight apical bending of the prisms is observed in the lizard Uromastyx. The majority of mammals have Hunter-Schreger bands of the multiserial type (Fig. 4), in which alternating bands of longitudinally and transversely cut prisms (6 to 12 prisms wide) are seen to occur. Such zones are a common feature of most Tertiary and Recent mammals (Kumar, 1983). In rodents, the Hunter-Schreger bands assume a special significance (Koenigswald, 1980, this volume). The orientation, inclination, and uniformity of width of these bands (Fig. 4) help to group various rodent families at the subordinal level. When zones are formed by the relative sliding movement of the ameloblasts secreting the prisms, the prisms are said to decussate. This decussation is particularly well seen in rodents.

Enamel Tubules

An intriguing feature of reptilian and mammalian (mostly primitive) enamels is the presence of enamel tubules. This feature has been observed in the lizard Uromastyx, in most multituberculates (Fig. 3C), in all marsupials with the exception of Vombatus, and in some unspecified nontherians. In Recent eutherians, enamel tubules have been noted in insectivorans, bats, primates, developing pig enamel, and rodents (Boyde and Lester, 1967; Kozawa et al., 1981). Osborn and Hillman (1979) placed emphasis on this character and have tried to relate it to progressive evolutionary histories in three divergent lineages: (1) Dimetrodon (pelycosaur) to Messetognathus (traversodont); (2) a lineage to multituberculates; and (3) a lineage to therian mammals. These authors found a progressive increase in enamel tubules in each of the above lineages. In contrast, eutherian enamels from the Cretaceous to Recent do not show conspicuous enamel tubules (Sahni, 1979), and so the contentions of Osborn and Hillman (1979) have yet to be supported. Boyde and Lester (1967) have related the presence of enamel tubules in marsupials and rodents to a high rate of deposition (16 micra per day in rodents, as compared to 4 micra per day in man). It is possible that the same physiological controls applied for the densely tubular enamel of multituberculates.

RODENT ENAMEL CHARACTERISTICS

It is interesting to compare the enamel characteristics of multituberculates and rodents (Table 2). Evergrowing incisors are a characteristic feature of rodents, and this condition is thought to have occurred in taeniolabidid multituberculates as well (Kielan-Jaworowska, 1980), although probably not in ptilodontids (Van Valen and Sloan, 1966). The development of evergrowing incisors is func-

Table 2. Dental and enamel ultrastructural characters of rodent and ptilodontid and taeniolabidid multituberculate teeth.

CHARACTER	PTILODONTIDAE	TAENIOLABIDIDAE	RODENTIA
Evergrowing incisors	Absent	Present	Present
Self-sharpening edge	Absent	Present	Present
Enamel thickness	Even distribution	Restricted anterolabially	Restricted mainly anteriorly
Prism size	4-8 micra	8-10 micra	4 micra
Prism shape	Circular	Circular and arcade	Circular to oval
Layering	Poorly developed	Indistinct	Strongly developed
Zone formation	Absent	Absent	Strongly developed
Tubules	Common	Abundant	Uncommon
Function	Piercing and puncturing	Gnawing	Gnawing

tionally related and is found in some members of other mammalian orders (lagomorphs, primates, proboscideans, artiodactyls). Evergrowing incisors also developed in *Lystrosaurus*, a cynodont reptile (Poole, 1956). Taeniolabidid multituberculates and rodents possess self-sharpening edges which are maintained by restriction of incisor enamel only to the cutting edge; this results in differing physical properties of elasticity and rigidity in the inner and outer layers of the occluding incisors. In ptilodontids, enamel surrounds the incisor, although the thickness is greater on the buccal than on the lingual surface. Prism size also shows considerable variation withi[n] ptilodontids. Additional material has shown that prism diameters vary considerably in incisors, premolars (where they average 4 micra and in molars, where they are greater in diameter. In an earlier work (Sahni, 1979), uniformly large sized prisms were reported for different teeth of the same species. At the ultrastructural level, the enamels of rodent teeth are more function efficient than those of multituberculates. Rodent teeth (mostly incisors, but in some cases molars as well) have well differentiated layers (Fig. 4). The inner enamel lamellae have a meshwork of closely interdigitating prisms and interprisms in a three-dimensional framework, the Hunter-Schreger bands are function related and preferentially distributed within teeth, and the inner lamellae are inclined cuspally to prevent the extension of hairline cracks (Koenigswald, 1980, this volume). The outer layer of rodent enamel is arranged radially, regardless of the differences that occur in the inner layer.

Patterns of Hunter-Schreger Bands in Rodent Enamel

Rodent enamels have been classified as pauciseral, multiserial, and uniserial (Tomes, 1850; Korvenkontio, 1934). Pauciserial enamel was first recognized by Korvenkontio and was studied in a number of diverse families (Ischyromyidae, Paramyidae, Sciuravidae, Theridomyidae and Pseudosciuridae) of Eocene and Early Oligocene age (Wahlert, 1968). More recently, Sahni (1980) has shown that pauciserial enamel occurs as well in Eocene members of the Ctenodactylidae (*sensu* Hartenberger, 1982), which acquired a multiserial pattern by the Miocene. Similarly, Oligocene ischyromyids evolved a uniserial pattern from a pauciserial one in the Eocene. Pauciserial enamel represents a basic structural grade of enamel organization which cuts across taxonomic affinities and is irrespective of phyletic lineages established on megadental and skeletal features. Pauciserial enamel is most closely related to multiserial enamels of hystricognathous rodents.

Pauciserial enamel (Fig. 4) is characterized by the presence of Hunter-Schreger bands or zones oriented vertical to the enamel-dentin junction. The enamel is a 2-layered structure, consisting of an outer radial enamel with apically directed prisms, and an inner layer of irregularly decussating zones (Hunter-Schreger bands), 2-6 prisms in width. The zones are often irregularly branched, and the radial enamel is well developed. In the Indian Eocene ctenodactylids, molars also possess the pauciserial structure (Sahni, 1980). Multiserial enamels can be observed in a number of fossil and Recent rodent families, including all hystricognaths (Wahlert, 1968; Boyde, 1969b; Koenigswald, this volume). The enamel is two layered, with the Hunter-Schreger bands of 7-9 prisms in width (Fig. 4). The bands are inclined apically at a greater angle than those of pauciserial enamels. This is particularly well seen in incisor sagittal sections. Boyde (1969a) has pointed out that rodent enamels are quite similar to those of ungulates, except for the high degree of prism decussation in the enamels of the former group. Uniserial enamels are present in several extant rodent families including the murids, cricetids, sciurids and rhizomyids, and they represent the most specialized enamel structure known in mammals. Uniserial enamels are characterized by a two-layered structure, with the inner enamel represented by inclined zones of criss-crossing arrangement of Hunter-Schreger bands, one prism wide. The outer enamel layer consists of radial enamel as in the other rodent enamels.

There is as yet no satisfactory hypothesis to account for the different rodent enamel patterns or their evolutionary development.

One of the major controversies is the presence of pauciserial enamel in diverse Eocene rodents and their rather abrupt transition to a multiserial and uniserial structure. This phenomenon can be noted even at the infrafamilial level, as in the Ctenodactylidae and the Ischyromyidae by Oligo-Miocene times. Wahlert (1968) was one of the first to comment on the antiquity of pauciserial enamel and for its basis as a good structural predecessor for later multiserial and uniserial enamels. Sahni (1980) is in agreement with Wahlert (1968) but he considers that the multiserial enamels are less specialized than the uniserial ones. Multiserial enamels are present in most ungulates, both fossil and Recent, and in a variety of other therian (Koenigswald, this volume). Uniserial rodent enamels, on the other hand, have no parallel in any other mammal. It is possible to deriv both multiserial and uniserial enamels from pauciserial ones. A multiserial structure can be derived by the retention of zones in th upper prism width ranges found in pauciserial enamels, while uniserial enamels can be formed by keeping zones in the lower prism width range (one in this case). The relatively greater inclination of the inner lamellae in both the multiserial and uniserial enamels, in comparison to those of pauciserial ones, probably represents a response to greater mechanical efficiency and prevents the extension of hairlike fractures.

In contrast to the hypothesis which regards pauciserial enamel as a basic structural type, is the contention of Koenigswald (1980, this volume), who visualizes the multiserial arrangement as the most primitive, leading gradually to the pauciserial condition and then to the uniserial one. The main consideration for this hypothesis is that multiserial enamels, represented by zones up to 7-9 prisms in width, are the most common pattern found in other mammalian orders, particularly the ungulates. The smaller prism width band of pauciserial enamel is viewed as a transitional stage leading to the one prism wide zones of uniserial enamel. Although Koenigswald (1980) acknowledged the known geological antiquity of pauciserial enamels, he believed that multiserial enamels might be found in some Eocene rodents as well. He cited the work of Hussain et al. (1978) on the Eocene rodents of Pakistan, which were thought to have multiserial enamels. These rodents were assigned to the family Chapattimyidae by these authors, but they have been shown recently by Hartenberger (1982) to belong to the Ctenodactylidae. Sahni (1980) has shown that the Indo-Pakistan ctenodactylid rodents have, in fact, a pauciserial structure. Hence, as is so far known, rodent multiserial enamels do not predate the Oligocene (Wahlert, 1968).

The presence of pauciserial enamel in diverse, early rodents either reflects the rapid differentiation of rodents in the earliest Paleogene from a common ancestor possessing pauciserial enamel, so that this enamel condition was primitively retained, in comparison t other cranial and anatomical features, or else it implies that similar selective pressures were acting on various early rodents to produce the same enamel pattern. Although the three enamel patterns differ from each other in certain important aspects, they represent a unified category when compared to other mammalian enamels. The report by Li and Ting (this volume) of a two-layered condition in certain Paleocene eurymylids from China is interesting for identifying a possible precursor to the rodent enamel pattern. The acquisition of the rodent pauciserial condition has been attributed to various cranial and muscle tension modifications. Wood (this volume) states that the evolution of multiserial enamels (from the pauciserial stage) was not a modification that involved the lengthening of the anteroposterior tension of the muscles. On the other hand, Wood (1980) considered the evolution of uniserial enamel to be related to the change from a protrogomorphous to a sciuromorphous condition. In contrast, Wilson (1972) believed that the transition to uniserial enamels must have preceded the development of sciuromorphous musculature. Emry and Thorington (1982) supported Wilson's viewpoint, based on their study of the Oligocene genus *Protosciurus*, which is protrogomorphous and yet has uniserial enamel.

In conclusion, the present evidence suggests that the pauciserial condition is more primitive than the other rodent enamel patterns, contrary to the opinion of Koenigswald (this volume). The pauciserial enamels are chronologically the oldest known and are found exclusively in diverse early rodent families from different parts of the world. Furthermore, the transition from the pauciserial to the more advanced enamel conditions has been documented in several instances. Detailed study of the eurymylid enamel ultrastructure would throw additional light on possible ancestry of rodent enamel patterns. Future work should also concentrate on Late Eocene-Early Oligocene taxa to find out the steps leading to the evolution of multiserial and uniserial enamel. Korvenkontio's (1934) description of an Oligocene theridomyid, *Nesokerodon*, transitional between the pauciserial and uniserial patterns, deserves special attention. As yet, very little is understood about the biomechanical stresses which control the arrangement of prism pattern within a single tooth or between different teeth in the same dentition. The studies of Boyde (1969a) and Koenigswald (1982a, b) have shown the functional advantages of multiserial enamels and the possible genetic control for adaptive responses at the ultrastructural level. A better understanding of these processes is essential for comprehending the evolution of prism patterns in various rodents and other mammals.

REFERENCES

Boyde, A. 1964. The structure and development of mammalian enamel. Unpublished Ph. D. Thesis, Univ. of London, London.

Boyde, A. 1966. The development of enamel structure in mammals. In: Third European Symposium on Calcified Tissues, H. J. J. Blackwood and M. Owen, eds., pp. 276-280, Springer, Berlin.

Boyde, A. 1969a. Electron microscopic observations relating to the nature and development of prism decussation in mammalian dental enamel. Bull. Group Int. Rech. Sci. Stomat. 12: 151-207.
Boyde, A. 1969b. Correlation of ameloblast size with enamel prism pattern: use of scanning electron microscope to make surface area measurements. Z. Zellforsch. 93: 583-593.
Boyde, A. and Lester, K. S. 1967. The structure and development of marsupial enamel tubules. Z. Zellforsch. 82: 558-576.
Boyde, A. and Martin, L. 1984. A non-destructive survey of enamel prism packing pattern in primate enamel. In: Tooth Enamel Symposium IV, Odawara, Japan, R. W. Fearnhead, ed., preprint pp. 1-4.
Carlson, S. J. and Krause, D. W. 1982. Multituberculate phylogeny: Evidence from the tooth enamel ultrastructure. Abst. Geol. Soc. Amer. 2119: 460.
Cooper, J. S. and Poole, D. F. G. 1973. The dentition and dental tissues of the agamid lizard, Uromastyx. J. Zool. 169: 85-100.
Emry, R. J. and Thorington, R. W. 1982. Descriptive and comparative osteology of the oldest fossil squirrel, Protosciurus (Rodentia: Sciuridae). Smithsonian Contrib. Paleobiol. 47: 1-35.
Fosse, G., Eskidsen, O., Risnes, S. and Sloan, R. E. 1978. Prism size in tooth enamel of some Late Cretaceous mammals and its value in multituberculate taxonomy. Zool. Scripta 7: 57-61.
Fosse, G., Risnes, S. and Holmbakken, M. 1973. Prisms and tubules in multituberculate enamel. Cal. Tiss. Res. 11: 133-150.
Frank, R. M., Sigogneau-Russell, D. and Vogel, J. C. 1984. Tooth ultrastructure of Late Triassic Haramiyidae. J. Dent. Res. 63: 661-664.
Grine, F. E. 1978. Postcanine dental structure in mammal-like reptile Diademodon (Therapsida: Cynodonta). Proc. Electron Micr. Soc. S. Afr. 8: 123-124.
Grine, F. E. and Cruikshank, A. R. I. 1978. Scanning electron microscope analysis of postcanine tooth structure in the Late Triassic mammal Eozostrodon (Eotheria: Triconodonta). Proc. Electron Micr. Soc. S. Afr. 8: 121-122.
Grine, F. E., Gow, C. E. and Kitching, J. W. 1979. Enamel structure in the cynodonts. Proc. Electron Micr. Soc. S. Afr. 9: 99-100.
Hartenberger, J.-L. 1982. A review of the Eocene rodents of Pakistan. Contrib. Mus. Paleont. Univ. Mich. 26: 19-35.
Hussain, S. T., De Bruijn, H. and Leinders, J. M. 1978. Middle Eocene rodents from the Kala Chitta Range (Punjab, Pakistan). Proc. Kon. Ned. Akad. Wet., Amsterdam, ser. B, 81: 74-112.
Ishiyama, M. 1984. Comparative histology of tooth enamel in several toothed whales. In: Tooth Enamel Symposium IV, Odawara, Japan, R. W. Fearnhead, ed., preprint pp. 1-4.
Kiel, A. 1966. Grundzüge der Odontologie. Gebrüder Bornträger, Berlin.
Kielan-Jaworowska, Z. 1980. Absence of ptilodontoidean multituberculates from Asia and its palaeogeographic significance. Lethaia 13: 169-173.

Koenigswald, W. v. 1980. Schmelzstruktur und Morphologie in den Molaren der Arvicolidae (Rodentia). Abh. Senckenb. Naturforsch. Ges. 539: 1-129.

Koenigswald, W. v. 1982a. Zum Verständnis der Morphologie der Wühlmausmolaren (Arvicolidae, Rodentia, Mammalia). Z. Geol. Wiss. Berlin 10: 951-962.

Koenigswald, W. v. 1982b. Enamel structure in the molars of Arvicolidae (Rodentia, Mammalia), a key to functional morphology and phylogeny. In: Teeth: Form, Function, and Evolution, B. Kurten, ed., pp. 109-122, Columbia Univ. Press, New York.

Korvenkontio, V. A. 1934. Mikroskopische Untersuchungen an Nagerincisiven unter Hinweis auf die Schmelzstruktur der Backenzähne. Ann. Zool. Soc. Zool.-Bot. Fenn. Vanamo 2: 1-274.

Kozawa, Y. 1984. The development and evolution of mammalian enamel structure. In: Tooth Enamel Symposium IV, Odawara, Japan, R. W. Fearnhead, ed., preprint pp. 1-5.

Kozawa, Y., Tateishi, M., Akaishi, S. and Hirail, G. 1981. The fine projection of Tomes process in the pig ameloblast by electron microscopy. J. Oral Sci. Nihon Univ. 7: 223-228.

Kumar, K. 1983. Paleontological and paleohistological investigations of Subathu vertebrates from Jammu and Kashmir and Himachal Pradesh. Unpublished Ph. D. Thesis, Panjab Univ., Chandigarh.

Moss, M. L. 1969. Evolution of mammalian dental enamel. Amer. Mus. Novit. 2360: 1-39.

Moss, M. L. and Kermack, K. A. 1967. Enamel structure in two Upper Triassic mammals. J. Dent. Res. 46: 745-747.

Osborn, J. W. and Hillman, J. 1979. Enamel structure in some therapsids and Mesozoic mammals. Cal. Tiss. Int. 29: 47-61.

Poole, D. F. G. 1956. The structure of the teeth of some mammal-like reptiles. Quart. J. Micr. Sci. 97: 303-312.

Poole, D. F. G. 1957. The formation and properties of the organic matrix of reptilian tooth enamel. Quart. J. Micr. Sci. 98: 349-367.

Poole, D. F. G. 1967. Enamel structure in primitive mammals. J. Dent. Res. 46: 124.

Poole, D. F. G. 1971. An introduction into the phylogeny of calcified tissues. In: Dental Morphology and Evolution, A. Dahlberg, ed., Univ. of Chicago Press, Chicago.

Sahni, A. 1979. Enamel structure of certain North American Cretaceous mammals. Palaeontographica A, 166: 37-49.

Sahni, A. 1980. SEM studies of Indian Eocene and Siwalik rodent enamels. Geosci. J. 1: 21-30.

Sahni, A. 1981. Ultrastructure of fossil Mammalia: Eocene Archaeoceti from Kutch. J. Paleont. Soc. India 25: 33-37.

Schmidt, W. J. and Keil, A. 1971. Polarization Microscopy of Dental Tissue. Pergamon Press, Oxford.

Tomes, J. 1850. On the structure of the dental tissues of the order Rodentia. Phil. Trans. Roy. Soc. Lond. 1850: 529-567.

Van Valen, L. and Sloan, R. E. 1966. The extinction of the multituberculates. Syst. Zool. 15: 261-278.

Wahlert, J. H. 1968. Variability of rodent incisor enamel as viewed in thin section, and the microstructure of the enamel in fossil and recent rodent groups. Breviora, Mus. Comp. Zool. 309: 1-18.

Wilson, R. W. 1972. Evolution and extinction in Early Tertiary rodents. Proc. 24th Int. Geol. Cong. Montreal 7: 217-224.

Wood, A. E. 1980. The Oligocene rodents of North America. Trans. Amer. Philos. Soc. 70: 1-68.

RECONSTRUCTION OF ANCESTRAL CRANIOSKELETAL FEATURES IN THE ORDER LAGOMORPHA

Nieves Lopez Martinez

Departamento de Paleontologia, Facultad de Ciencias Geologicas, Universidad Complutense de Madrid
Madrid 3, Spain

INTRODUCTION

Living lagomorphs (hares, rabbits, and pikas) are not very different from their close fossil representatives of the Upper Eocene. Nevertheless, this conservative order of mammals has an astonishing number of modified features that can still be recognized after 45 million years.

Following Gidley (1912), the Lagomorpha were separated as an independent order from the Rodentia, which previously had been grouped with them in the cohort or order Glires. The similarities between the rodents and lagomorphs are considered generally to be the result of parallel evolution, and the sister group of lagomorphs has been sought among a variety of taxa, such as Mesozoic mammals (Ehik, 1926), zalambdodont insectivores (Russell, 1958), Condylarthra (Wood, 1940), Anagalida (Van Valen, 1964; Szalay and McKenna, 1971), and macroscelidids (Szalay, 1977).

Rodents and lagomorphs share an important number of features, and some authors (Luckett, 1977; Hartenberger, 1977) have supported the earlier idea of their close relationship (Simpson, 1945). It has never been analyzed in the class Mammalia as a whole whether these shared features are homologies or homoplasies. The difficulty of phylogenetic reconstruction in higher groups lies in the wide number of characters to be analyzed. Authors support their propositions with a limited number of characters, e.g., the molar crown pattern (Ehik, 1926; Wood, 1940; Russell, 1958; Van Valen, 1964; McKenna, 1975, 1982; each with a different interpretation); the morphology of the tarsus and orbital edges (Szalay, 1977); development of fetal membrane and placental morphology (Luckett, 1977); or the masticatory apparatus and auditory bulla (Hartenberger, 1977).

Reconstruction of the primitive conditions for each order of mammals may be an approach to simplify the problem. In that way, we may eliminate their variability, issued from successive apomorphies, and then we may compare each ancestral "Bauplan" or morphotype. This method is clearly different from that followed by Szalay (1977), who compared the older taxa of every order with each other. This implies that the older recorded taxon shows every plesiomorphic feature of its order, which may not be true.

I shall attempt here a character analysis of the Lagomorpha, in order to reconstruct their ancestral morphotype; this can then be compared with that of Rodentia and other mammals. This implies the identification of three different kinds of features: (1) Autapomorphic (uniquely derived) characters of primitive Lagomorpha in relation to other mammals; (2) plesiomorphic features of primitive Lagomorpha in relation to mammals and mammal-like reptiles; and (3) synapomorphic characters of primitive Lagomorpha, when compared to the primitive eutherian condition.

First, we will examine the adaptive strategies of living lagomorphs, in order to interpret the morphological cranioskeletal characters of both living and fossil lagomorphs.

METHOD OF CHARACTER ANALYSIS

Character analyses are the backbone of any evolutionary study. The morphological, ontogenetic, and functional analyses should provide a historical inference for the process of change, dividing the different character states into two groups; one of them will be primitive (plesiomorphic), the other or others will be derived (apomorphic). Cladistic systematics has developed some different methods of phylogenetic reconstruction, based on the distribution in taxa of both types of characters (see Hennig, 1966; Hecht and Edwards, 1977; Wiley, 1981).

The main difficulties of character analysis are of two kinds: (1) the selection of the characters; and (2) the inference of character state polarity. In the principles of cladistic systematics, selection of characters should be avoided. In almost every case, features should be included in the analysis without *a priori* differential weight. Nevertheless, in common practice, the selection of features introduces the first bias factor (Hecht and Edwards, 1977).

The historical interpretation of character states, and their attribution to a historical category (plesiomorphic/apomorphic) is mainly a matter of consensus founded on some more or less explicit criteria: (1) comparison with outgroups; (2) widespread distribution

(= commonality); (3) functional specialization; (4) ontogenetic precedence; and (5) geographical distribution. Another criterion, being one of the most criticized, is paleontological stratigraphic position. Each criterion by itself has some methodological difficulties. The agreement of two or more of them strengthens the historical inference (Hecht and Edwards, 1977; Sanchiz and G. -Valdecasas, 1980).

Within the framework of this study, the character distributions and their polarity were analyzed with the goal of recognizing the primitive features of the Lagomorpha. Therefore, an important criterion to be considered is commonality; the selected features are shared by almost every group of the order. We do not attempt an internal phylogenetic analysis of the lower taxa within Lagomorpha (see Lopez Martinez, 1978; McKenna, 1982).

The outgroup choice (the selection of the supposed ancestral or sister group of Lagomorpha) ia an apriorism that may be avoided only by choosing all Eutheria, Mammalia, or the mammal-like reptiles as the Lagomorpha outgroup. The choice by McKenna (1982) of the Eurymylidae and Anagalida as outgroups precludes their possible recognition _a posteriori_ as the ancestral groups for Lagomorpha, because this is circular reasoning.

Ontogenetic precedence is a rather ambiguous criterion, because either evolutionary recapitulation or reversed recapitulation are possible (see Alberch et al., 1979). As an example, we can consider the bony palate of lagomorphs (Major, 1899; Dice, 1933) to have two states: (1) bony palate mainly formed by the palatines (ancient leporids, ochotonids); or (2) bony palate formed mainly by the maxillary (Recent leporids). Both the paleontological and ontogenetic sequences support the polarity of 1 leading to 2 (Dice, 1933; Wood, 1940; Dawson, 1958), so that we can propose an evolutionary recapitulation. However, the outgroup Mammalia uniformly possesses character state 2, which implies a reverse ontogenetic recapitulation, and an evolutionary reversion for Recent leporids. When the criteria are not congruent, reversions and convergences with other taxa are not easy to detect.

The functional specialization criterion shows an advantage over the others. Functional analysis provides a meaning to the characters under study, which gives a historical interpretation following the Darwinian evolutionary theory (Dullemeijer and Barel, 1977). The meaning of the character is the most important question of this criterion, and it is not necessary to construct a complete engineering model to relate form and function (Gutmann, 1977).

The different meanings of a character state can be clustered in four groups: (1) adaptive characters related to a determined function; (2) secondary utilizable characters as a fortuitous effect of a feature (Vrba, 1983); (3) functionally neutral characters generated as a secondary effect of a developmental process (Hecht and Edwards, 1977; Gould and Vrba, 1982); and (4) residual characters of a degenerative structure. Only the second group does not imply a specialization and can be plesiomorphic. The same feature may have been transferred in the four categories along the evolutionary process (Williams, 1966).

As an example, the morphology of the bony palate mentioned above is the secondary effect of two other features often dissociated: the incisive foramina and the choanae which limit the bony palate anteriorly and posteriorly. The suture between maxillary and palatine bones is always placed in a stable position in front of P4/, and variation of the structure is not the suture displacement but the differential bony reduction: (1) anterior reduction of maxillary because of the development of the incisive foramina, related to Jacobson's organ (mainly derived in ochotonids); and (2) posterior reduction of the palatines because of the enlargement and advanced position of the choanae, related to respiratory improvement (mainly derived in Recent leporids). The primitive lagomorph condition (Major, 1899) is a longer palate with a forward position of the incisive foramina and a backward position of choanae, which have a lesser development as in <u>Megalagus</u>.

Functional evolutionary convergence, which can be argued against this criterion, is easily detected, because a functional adaptation has never been attained twice by the same steps or in the same way. It is only necessary to go a little forward in the character analysis to override the apparent similarity (see e. g., the masticatory apparatus below).

RECENT LAGOMORPHS AND THE HERBIVORE ADAPTATION

Dietary Adaptations

If we consider the Lagomorpha, both ochotonids and leporids, we can propose their common features of herbivorous diet and terrestrial mode of life as the primitive condition of this order. The herbivorous adaptation of lagomorphs is a very strict one. They are specialized for eating grass, leaves, juicy stalks, pulp from stems of tall grasses, tubers, and some barks of trees. Seeds are also eaten in a small proportion (Pielowski and Pucek, 1976; Myers and Bults, 1977; Soriguer, 1979; Ghose, 1981; Lopez and Cervantes, 1981; Smith, 1981a, b). The vegetable fibers in the lagomorph nutrition are an essential component, even in the artifically fed domestic rabbit. Nevertheless, efficiency in digesting cellulose is lower in the Lagomorpha than in the ungulates or herbivorous rodents, such as <u>Cavia</u> (Maynard and Loosli, 1969). Caecotrophy is the physiological behavioral mechanism that compensates for this.

The morphological adaptations of the Lagomorpha to grazing habits include: (1) the development of crests and high crowns on the cheek teeth; (2) the significant functional role of the premolars (lophodonty-hypsodonty-molarization); (3) the diastema between molariform teeth and the cutting incisors, which separates the two masticatory functions, biting and grinding; (4) the lateral (ecdental) main masticatory movements; (5) well developed pterygoid muscles; and (6) a very long intestinal caecum with symbiotic bacteria. The skeletal and dental features mentioned above are shared with all fossil lagomorphs since the Upper Eocene, and can be considered primitive for the order.

This set of features can also be found in most herbivorous extant mammals, such as macropodid marsupials, hyracoids, perissodactyls, and pecorans, and could be called a "grazing herbivorous adaptative set" (GHAS). This functional convergence, very useful for morphoecological purposes, disappears as we deepen the character analysis.

Other basically phytophagous mammals, such as Proboscidea, Sirenia, Dermoptera, and some Suina, Primates, and Rodentia, are not grazers. They feed mainly on the soft parts of vegetation, such as tree leaves, fruits, and seeds. We must therefore make a distinction between the browsing herbivores, frugivores, and granivores. The members of this latter category are mostly rodents (murids, hamsters, heteromyids, sciurids) but, in spite of its importance, the animal part of their nutritional habits has often been neglected (see Mares, 1980; Cheylan, 1982, for murids and glirids). These phytophagous and omnivorous mammals have some, but not all, of the GHAS features. They lack, for instance, the lophodont-hypsodont-molariform cheek teeth pattern, the well developed pterygoid muscles, and the very long caecum. Some rodent taxa, such as castorids, microtids, and Caviomorpha, have developed a secondary herbivorous-grazing adaptation, with a lophodont-hypsodont dental pattern and a long caecum. The first character was also present in the Oligocene theridomyids, but is absent in more primitive rodents and can be considered as a secondary adaptation, instead of a primary one as in Lagomorpha. Their similarity with lagomorphs is nevertheless of relevant interest.

Locomotion

The grazing herbivores are terrestrial and live in the open grassland, rarely in the forest. As size decreases, their mode of locomotion changes from cursorial to saltatorial, and finally fossorial. Szalay (1977) suggested, based on study of the tarsus, that cursorial locomotion was the primitive condition of the Lagomorpha. The larger lagomorphs, such as *Lepus*, have recently attained this veritable condition, but the remaining lagomorphs are rarely saltatorial or fossorial, and their ability for speed is too weak to escape from running predators. The movements of *Caprolagus*, for example, are so slow that they can be caught by hand (Ghose, 1981).

Lagomorphs always look for a shelter, and are rarely true burrowers. Their habitats are often areas of patchy bush shrubs and entangled high grass, or rocky-boulder areas, and their shelters are not very deep. Only Oryctolagus, Brachilagus, and perhaps Romerolagus are reported as habitual burrowers, as are the three steppe-dweller species of Ochotona. However, even Romerolagus seems to nestle above ground among rock shelters (Granados, 1981; Lopez and Cervantes, 1981; Cervantes, 1982). Lepus, Syvilagus, Caprolagus, Nesolagus, Pronolagus, and the majority of the Ochotona species are not diggers; instead, they hide in natural shelters (Nelson, 1909; Ghose, 1981; McNeely, 1981; Smith, 1981b). This mode of like can be found also in other small grazing herbivorous mammals, such as hyracoids and some caviomorphs (like Chinchilla, Lagidium, Dolichotis but the majority of rodents are true burrowers.

The mode of locomotion in lagomorphs is not a very specialized one; the locomotor skeleton of the majority of the non-modified lagomorphs seems very generalized, with small transformations: fissured ungual phalanges; terminal fusion of tibia and fibula; orthal mechanized facets of tarsus (fibula-calcaneum, tibia-astragalus, astragalus-navicular); reduced clavicle; very short or absent tail; small cervical region; hindlimb somewhat longer than forelimb, and both relatively short, especially the distal plantigrade segment. The head position is always higher than in rodents and primitive mammals, because of the more ventral orientation of the occipital foramen, and the basicranial-basifacial angle between presphenoids and basisphenoids.

There are two different modifications of the locomotion in some modern leporids. The mode of cursorial locomotion presents in Lepus an advanced phase towards an original kind of gait named leaping-gallop; the limbs become relatively longer, as well as the metapodial segment, whereas the elongated ulna becomes slender and is placed behind the radius. The saltatorial leporids, such as Oryctolagus and Sylvilagus, have a hindlimb clearly longer than the forelimb, and the lumbar region is relatively more developed.

Both specialized leporids have a closer basicranial-basifacial angle, a more ventrally oriented occipital foramen, a very enlarged nuchal region, the transverse and the hypapohyses developed on lumbar vertebra, a strong and oblique sacroiliac articulation, a more spherical astragalar trochlea, and a calcaneum-navicular joint. They often stand up on their hindlegs, and more by bounds and not by jerks Bipedal locomotion, found in some rodents, needs a strong tail, which has not been accomplished in the tailless lagomorphs.

The gnawing incisors of lagomorphs are not seizer-digger incisors (Hershkovitz, 1962), as in many burrowing rodents. The few burrowing lagomorphs only use the forefeet for digging; the larger flattened opisthodont incisors of lagomorphs are an apomorphic alimentary feature for cropping grass, such as horses do. The slender and more or less proodont rodent incisors have many different functions (Krumbach, 1904). They are used to open seeds, for cutting, grasping, biting, to dig a burrow, and to kill.

The primitive locomotory features of Lagomorpha can hardly be expressed as cursorial ones, and must be considered only as ambulatorial. Some similarities with the extant hyracoids can be found, but most of them may be plesiomorphies, considering the whole of Mammalia. The distal tibio-fibula fusion is shared with some rodents, macroscelidids, soricids, talpids, and solenodonts, whereas the fibulo-calcaneum joint is present also in Monotremata, Erinaceoidea, Artiodactyla, and Hyracoidea, though absent in the earlier named taxa, so that the order Lagomorpha has a unique combination. Lack or extreme reduction of tail is shared also with hyracoids and is a primitive feature of both orders, because the reduction of tail seems to be a secondary loss of some subterranean insectivores and rodents.

Size

The size of animals is an important feature and defines an ecological niche parameter. The lagomorphs are the larger micromammals of open-land communities in Holartic realms (Valverde, 1967; Mares, 1980). In the Mediterranean communities, they fill the gap in size distribution of secondary productors. *Oryctolagus* supports up to 40 predator species in this area (Soriguer and Roger, 1981; Delibes and Hiraldo, 1981). The lagomorph role is not so clear in the southern continents, where they are not an original component of the faunas. However, their widespread distribution after recent natural dispersal or artificial introduction in Australia, New Zealand, South America, and Africa allows one to predict the displacement by Lagomorpha of the autoctonous medium-sized terrestrial herbivore competitors, such as macropodid marsupials, hyracoids, or caviomorph rodents, if man lets nature take its course.

The primitive size of higher taxa is difficult to interpret, although increasing size is a general trend in communities (Bonner, 1965; Van Valen, 1973; Pianka, 1982; Margalef, 1982). The extant lagomorphs vary in size from 130-180g (200mm head-body length) to near 7kg (700mm length) with *Ochotona* in the lower range, rabbits in the middle range, and *Lepus* in the largest one. The mean of 22 extant species from which I could get reliable data is 2.014kg and 450mm head-body length. This is a rough index of the Recent lagomorph size; the primitive size is surely smaller, but to estimate it, fossil data must be introduced.

Fossil lagomorphs can provide only a size inference based mainly on dental and osteological measures. The size of the first lower molar has been used extensively for this purpose in other mammals (e.g., Gingerich, 1977; Michaux, 1983) in paleontological work. In lagomorphs, the teeth length and body size are not linearly correlated (Palacios, 1978), and the ancient taxa had relatively larger teeth. The ratio of lower M1/skull length varies from 3.1-4.4% in the extant species, and it reaches 5.1% in extinct Megalagus. The length of the skull is a better body size estimator, but, unfortunately, preserved skulls of fossil lagomorphs are scarce.

The well-known Oligocene lagomorph fauna of North America (Dice, 1933; Wood, 1940; Dawson, 1958; Gawne, 1978) has mainly three coexisting taxa. These primitive lagomorphs have a smaller size than extant leporids, and a rough estimation suggests that Megalagus, the largest one, with a skull measuring 67mm, was approximately the size of the recent Mediterranean rabbit (head-body length near 400mm), but with shorter limbs and larger teeth. Palaeolagus and Chadrolagus are smaller, about the size of Romerolagus (270mm head-body length), with a slight overlapping. This lagomorph size distribution is unknown in extant communities, where the body weight near 500g is rare (see Table 1 and 2). Before the Chadronian, the lagomorph genus Mytonolagus also had a size near Romerolagus (Dawson, 1970).

Following the Arikarean and the extinction of Megalagus in the Hemphillian, the whole lagomorph fauna became smaller, with the appearance of many small ochotonids (Oreolagus, Desmatolagus, Gripholagomys, Hesperolagomys) about the size of recent pikas. The leporids did not exceed the Oryctolagus size. In Hemphillian age, Hypolagus attained the Lepus size. We can summarize the main points:

(1) The primitive size of North American leporids was near Romerolagus.

(2) Taxa with morphological plesiomorphies were larger than the taxa morphologically advanced.

(3) The size of the ochotonids in North America was always smaller than that of contemporaneous leporid taxa. They occupied the smaller size class previously filled by small leporids. After their arrival, the leporids did not develop forms with small size.

The European lagomorph fossil record starts in the Oligocene with ochotonids, and the leporids appear in the Upper Miocene (Tobien, 1974a, 1975; Lopez Martinez, 1974, 1977; Lopez Martinez and Thaler, 1974, 1975; Heissig and Schmidt-Kittler, 1975, 1976; Sych, 1965). The first immigrant, "Shamolagus" franconicus is near Megalagus in size and grade. Later, from the end of the Oligocene to the Upper Miocene, many small ochotonids (Piezodus, Amphilagus, Titanomys, Prolagus, and Lagopsis), with a size similiar to that of Recent Ochotona, spread. There is overlapping in size, but competition is avoided by geographical displacement (Lopez Martinez, 1977). When they coexist, there are always slight size and grade differences as in North America (Tobien, 1974a), and as many as four related ochotonid taxa can coexist with very different abundances (see Lopez Martinez, 1977). Two larger ochotonids, "Amphilagus" ulmensis and

Table 1. The size classes and total number of taxa for Asian, European, and North American lagomorphs. See Table 2 for distribution of these classes.

NO.	SIZE CLASSES	NO.	SIZE CLASSES
2	Class I (e.g., Bohlinotona) length skull 25mm head-body length 100mm	37	Class IV (e.g., Oryctolagus) length skull 70mm head-body length 400mm
60	Class II (e.g., Ochotona) length skull 40mm head-body length 200mm	30	Class V (e.g., Lepus) length skull 90mm head-body length 500mm
26	Class III (e.g., Romerolagus) length skull 50mm head-body length 300mm		

Eurolagus have many plesiomorphic features, large size, and a short stratigraphic range.

The ochotonids in the European continent have never been larger than Oryctolagus, but in the Mediterranean insular areas they have developed all their size potential. Gymnesicolagus (Mein and Adrover, 1982) is the lagomorph with the largest teeth, and a rough size estimation suggests a skull length of 100-130mm, even larger than the largest arctic hare. Paludotona (Dawson, 1959) also has a large size, near that of the brown hare. After the Vallesian, the European lagomorph fauna consists of both leporids and ochotonids, combined in different abundances. At first, the ochotonids were more numerous; later, the leporids became most abundant. In Pliocene and Lower Pleistocene times, the leporids were dominant and presented the same size ratio as the Recent hare/rabbit one, but with a larger size. Thus, we can summarize:
(1) The primitive size of European ochotonids is near Ochotona.
(2) Also, as in North America, the relatively plesiomorphic taxa are larger than contemporary relatively apomorphic taxa.
(3) The leporids are larger than the contemporaneous ochotonids; after arrival of the former, the latter do not attain any further size increase.
(4) In the continental, as well as in insular, areas, lagomorphs are among the largest micromammals of the local fossil faunas.

Concerning this last point, it is interesting to note that insular Prolagus sardus lived in Quaternary Mediterranean islands, but it did not attain even half the size of ancient insular ochotonids.

The size of the Asiatic lagomorph fauna is clearly smaller than

Table 2. Distribution of Lagomorph taxa in size classes by chronological periods and continental regions. Class II is postulated as primitive for the order. A decreasing trend can be found in the Asiatic and North American Oligocene. At the bottom of the table, the eurymylids and anagalids (*) are tentatively classed by size for comparison. See Table 1 for the size of the classes.

	EUROPE					ASIA					NORTH AMERICA				
	I	II	III	IV	V	I	II	III	IV	V	I	II	III	IV	V
Recent				1	3		12		1	10		2	1	12	8
Pliocene/ Pleistocene		3		6	2		6	1	2	2		1	1	6	5
Miocene		12	3	2			6	1				4	7	4	
Oligocene		2			1	2	7	5				2	5	2	
Upper Eocene							3						2		
Paleocene							3*	1*	1*						

that in the other two continents (see Tables 1 and 2, and data from Matthew and Granger, 1923; Burke, 1936; Bohlin, 1942; Gureev, 1960, 1964; Dawson, 1961; Li, 1965; Erbajeva, 1970, 1981; Sych, 1975). Until the Villafranchian, we can not find a lagomorph surpassing the *Romerolagus* size, and the larger forms are very rare (*Gobiolagus major*, *Ordolagus*). The primitive fauna is near the *Ochotona* size (*Lushilagus*, *Shamolagus*, *Gobiolagus*, *Desmatolagus vetustus*). The first fauna of three coexisting lagomorphs (*Desmatolagus gobiensis*, *D. robustus*, and *Ordolagus*) shows a size distribution near the *Ochotona* (the former) and *Romerolagus* sizes, respectively. Later, we find even smaller forms like *Bohlinotona* and *Sinolagomys gracilis*, which are not represented in the Recent faunas. Nevertheless, most Asiatic lagomorphs are close to the *Ochotona* size throughout the Tertiary (*Sinolagomys*, *Desmatolagus radicidens*, *Agispelagus*, *Alloptox*, *Bellatona*), and they occurred mainly in monospecific associations. The modern fauna with *Ochotona* and Leporinae exhibited a larger size during the Villafranchian than it does in the Recent fauna. In summary:

(1) For Asiatic lagomorphs, the primitive size was near *Ochotona*.
(2) In contrast to the North American and European lagomorphs, the plesiomorphic taxa were smaller than apomorphic contemporaneous taxa, or nearly the same size.
(3) The size may be smaller than that of the smallest *Ochotona* species.

We can see in the Asiatic Paleocene record a group that may be related phylogenetically to Lagomorpha, the Eurymyloidea (Matthew et al., 1929; Wood, 1942; Van Valen, 1964; Sych, 1971; Xu, 1976; Li, 1977; Li et al., 1979; Li and Ting, 1982, this volume). In this group there are included some taxa (*Eurymylus*, *Heomys*, *Mimotona*) with sizes near to *Ochotona*, although larger than the oldest lagomorph *Lushilagus* (Li, 1965). Most of them are nevertheless rather large mammals, with a M/1 length exceeding four times that of primitive lagomorphs.

In Tables 1 and 2, we have summarized the number of lagomorph taxa filling each size-class in the three continents of the Holartic region (the insulary areas have not been taken into account). The diachronic and allopatric taxa are added, so the data are not representative of the local faunas. The classes I, IV, and V are clearly apomorphic, and most taxa are found in class II. The primitive size of Lagomorpha seems likely to be near that of *Ochotona*, with an increasing trend more frequent than a decreasing one. This fits with eurymyloid average taxa size, and is clearly larger than the primitive rodent size.

Life History Strategies

Extant Lagomorpha commonly live in colonies of many small family groups with a high degree of territoriality and social organization (Southern, 1940; Schneider, 1981; Daly, 1981; Smith, 1981a). As with many grazing herbivorous mammals, they do not have hibernation or estivation resting periods. Their common habitat is open land, both cold and warm deserts, steppes, and savannah. The populations of lagomorphs support a high rate of predation pressure and very important fluctuation cycles. They have an important ecological role as a basal link of secondary producers (Flinders and Hansen, 1973; Wagner, 1981; Keith, 1981).

The demographic turnover is high, with a size-related low longevity (Southern, 1979) and high reproduction-mortality rates. Unique physiological mechanisms like superfetation and induced estrous behavior, which can lead to continuing reproduction (Caillol and Martinet, 1981; Flux, 1981), favor the high reproduction rate. On the contrary, pseudopregnancy and embryo resorption are decreasing devices largely used in high density demographic conditions.

There are many reproductive differences in Recent lagomorphs. The gestation period is relatively short, and varies between 30 days (*Sylvilagus*, *Oryctolagus*, *Ochotona*) and 45 days (*Lepus*, *Romerolagus*). The first group has a high number of young per litter (5-7), and they are born blind and naked; the second group has a lower number of young per litter (2-3), and they are born well haired. The first strategy is postulated as plesiomorphic (Corbet, 1983; Cervantes, 1982).

Ecological competition is beginning to be understood and an important interspecific competition seems to be the rule among lagomorphs (Fraguglione, 1960; Lindlof, 1970; Flux, 1981; French and Heasley, 1981). The space distribution supports normally two parapatric Lepus species and a rabbit, or two rabbits and a hare. Thus, in extant lagomorph habitats, from one to three lagomorph species may coexist, but rarely in sympatry (never in synthopy). This distribution pattern can be found even in Oligocene fossil lagomorphs (see Gawne, 1978).

It is probable that social behavior, non-hibernation, and "r" strategy, with high demographic variations, are components of the original life strategy of the primitive lagomorphs. This can be associated with a relatively large number of individuals by fauna, communities with rather low species diversity, and small individual size. These features are shown by the majority of fossil lagomorph assemblages, and they may be retained as characteristics of the order. Rodents show demographic similarities, but they are mostly solitary, hibernating-estivating, smaller mammals with a different evolutionary strategy.

Several authors (Simpson, 1953; Eldredge, 1979; Vrba, 1983) have suggested that narrow environmental specialization (stenotopy) should be generally correlated with high speciation rates. This can not be applied to lagomorphs, which are stenotopic relative to rodents. Lagomorphs are more specialized in their habitat and diet, and are more evolved on the whole in social organization, physiological regulator mechanisms, and larger size, but can not override their interspecific competition.

MASTICATORY APPARATUS

This system is one of the most modified in Lagomorpha, and the one on which most attention has been focused. The dentition has been carefully analysed several times, because the dental pattern has been the starting point in studies of mammalian phylogeny. The dental pattern of lagomorphs has been the subject of many years of historical reconstruction; following Major's (1899) careful work, the homology of cusp pattern has been discussed by many authors, without agreement (Wood, 1957; Tobien, 1974b; McKenna, 1982). Therefore, we will use the neutral terminology of Lopez Martinez (1974). Before discussing the cusp homologies and masticatory muscles, we shall begin by surveying the occlusion pattern.

Dental Occlusion

The occlusion pattern of primitive mammals (Vandebroek, 1961; Crompton and Hiiemae, 1969; Crompton, 1971; Crompton and Jenkins, 1968; Bown and Kraus, 1979; Butler, 1980, this volume) provides a new line of evidence for assessing cusp homologies, as the wear facets of crowns on cheek teeth can also be easily homologized. This feature has never been studied in Lagomorpha, and we have found it a very interesting and fruitful area of study, as discussed below.

The occlusion of lagomorphs is modified in relation to common opposing occlusion, because of anisognathy of the skull and jaws. The lower tooth rows are nearer to each other than are the upper ones; thus, in the normal resting position, the lingual part of the upper crowns rests on the labial half of the lower crowns. The grinding movements are mainly transverse (ecdental), as in the majority of herbivores (GHAS feature), but lagomorphs differ from the others in their very strong anisognathy and the disjoint occlusion pattern.

In all mammals, occlusion implies the intercalation of the lower teeth with the upper ones. The trigonid has a major posterior wear facet against the paracone of the posterior upper tooth, and a wide anterior facet against the metacone of the anterior upper tooth. These have been named wear facets 1 and 2 after Crompton (1971). This author described and named 6 main wear facets in the primitive tribosphenic molars (Fig. 1).

Lagomorphs have an important wear facet in the same position as number 1. All the fossil and Recent taxa have a conspicuous thickening of the enamel band at the anterior edge of the upper molars and at the distal edge of the trigonids (Fig. 1). As hypsodonty increases, the sharp wear edge becomes more oblique, and a larger part of the vertical shaft of the crown is employed in the active grinding wear surface, which gets inclined backward in the upper molars and forward in the lower ones. The anterior part of talonid, which presented initially the wear facet 5 against the protocone, becomes incorporated into wear facet 1 in lagomorphs (Fig. 1).

The hypoconid is a thickened enamel cusp that runs over the middle part of the upper molars, and it produces a wear facet against the lagicone (see Fig. 4 for terminology) in primitive lagomorphs, or the enlarged hypoflexus (in Recent dental patterns). This is probably homologous to wear facet 4 of tribosphenic teeth. There are no wear facets on the talonids of most lagomorph teeth (Fig. 3). This occlusion model is shown by Paleogene fossil lagomorphs (*Megalagus*, *Palaeolagus*, *Shamolagus*) and many leporids (see Fig. 1), and it explains the peculiar lateral profiles of dental rows, resembling the steps of a staircase, which can be seen even in very ancient lagomorphs (Fig. 2).

The ochotonid and some leporid (*Romerolagus*) occlusion patterns, with new and important wear facets 4 and 7, are illustrated in Figures 1-3. All the distal parts of the talonid build a new sharp enamel edge and a posterior wear facet that occludes with the well

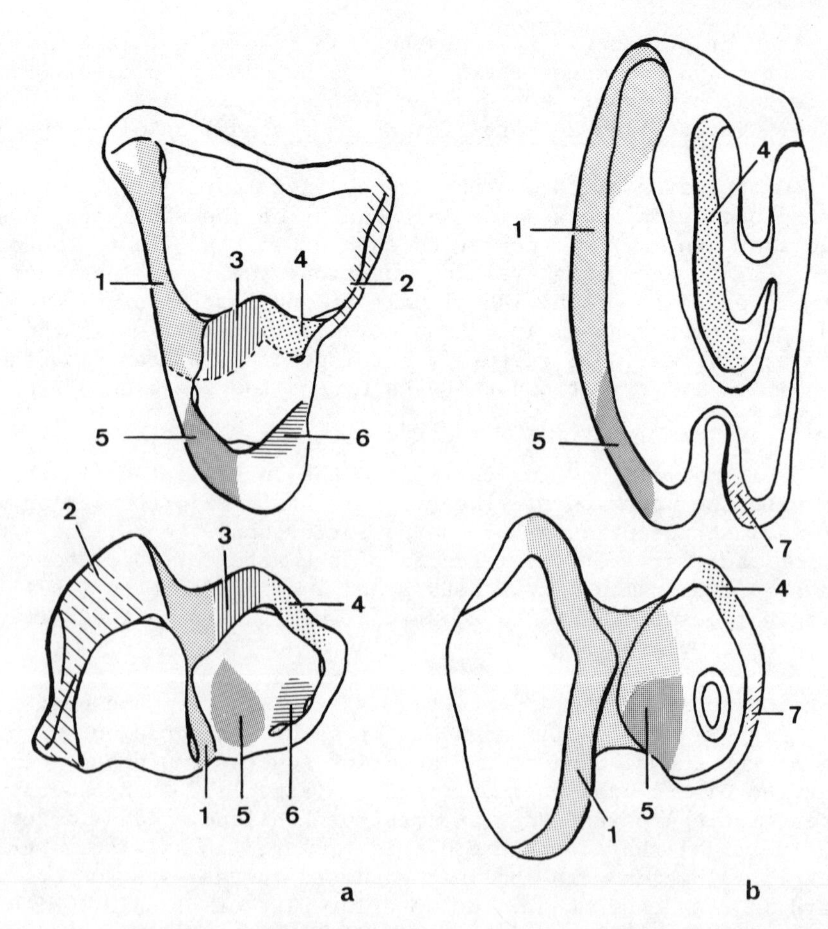

Fig. 1. Occlusal surface and wear facets (numbered) of upper (above) and lower (below) first molars in: (a) an early Tertiary eutherian with primitive tribosphenic molars (after Bown and Kraus, 1979); and (b) Lagomorpha. Facet 7 develops in ochotonids and a few leporids, such as Romerolagus, and it may become continuous with facet 4. The homology with tribosphenic wear facets is easily recognized, especially in the lower molars. Wear facet terminology after Crompton (1971). Mesial is to the left

developed anterior enamel crest of the lagicone (in primitive forms or the posterior part of the enlarged hypoflexus (in advanced ochotonids). This new facet adds an additional step to the stair-like profile, and it doubles the active grinding surface (Fig. 2). Consequently, the upper teeth are more separated from each other, and the talonids are as large as the trigonids (Fig. 3).

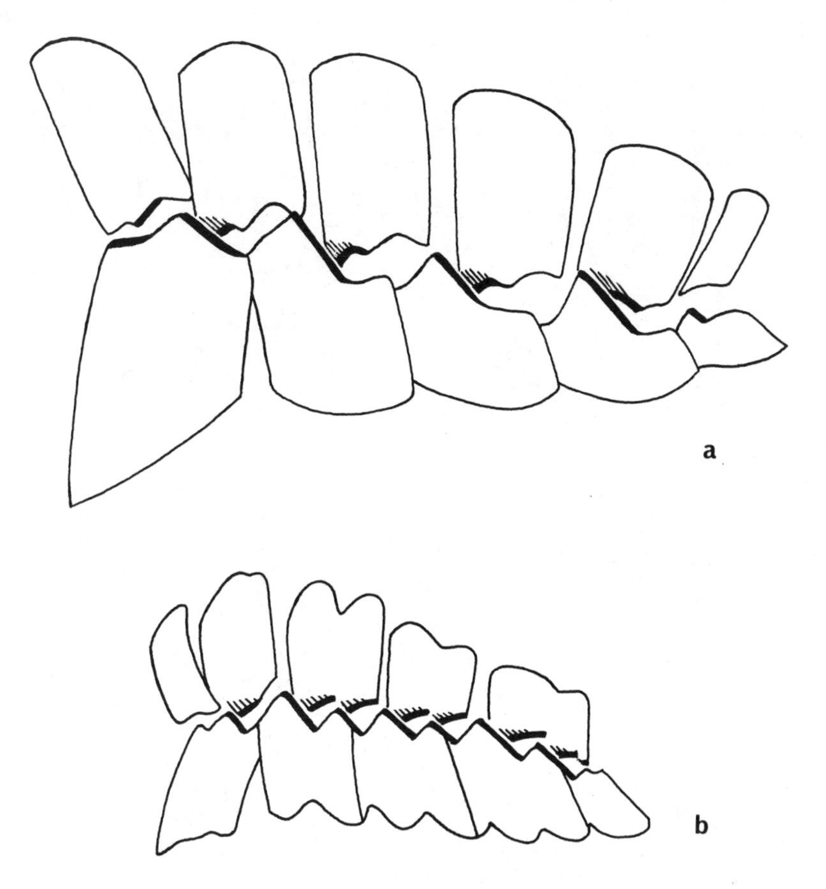

Fig. 2. Labial view of the wear facet orientation in lagomorph cheek teeth. (a) Leporids and primitive lagomorphs. (b) In some leporids and advanced ochotonids, the stair-like profile has added a new step series, with development of wear facets 4 and 7. Mesial is to the left.

These differences in the occlusal pattern can be observed from the Oligocene onwards in very primitive ochotonids such as *Piezodus*, *Amphilagus*, *Titanomys*, and "*Shamolagus*" *franconicus*, but, as far as can be inferred from the literature, they are not discernible in *Desmatolagus* or *Megalagus*. These latter genera usually do not show an anterior enamel thickening on the lagiloph-lagicone, such as occurs in the ochotonid occlusal pattern; instead, it is weak and disappears with wear (cf. Fig. 3a and 5c,d), as in the primitive leporid occlusion pattern. In advanced ochotonids, the enlarged posterior wall of the talonid occludes with the thickened posterior wall of the hypoflexus. On the contrary, the hypoconid of advanced leporids keeps wear facet 4 against the enlarged anterior wall of the hypoflexus (Fig. 3). This divergence of dental occlusion pat-

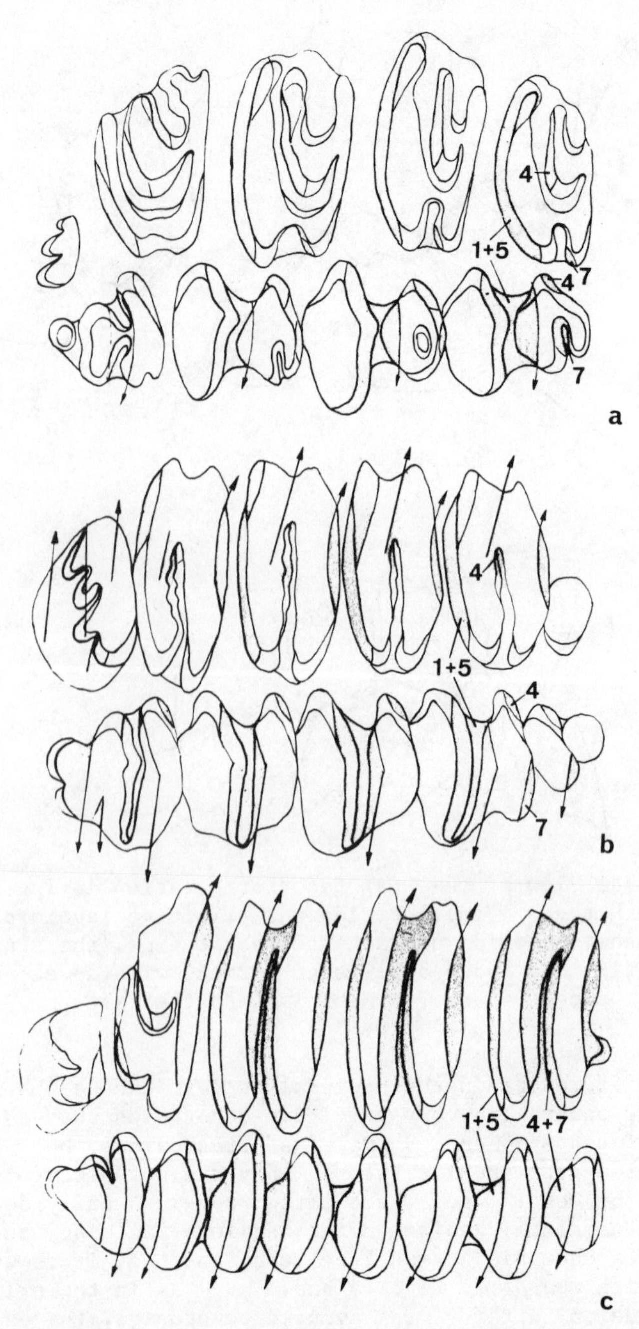

Fig. 3. Occlusal view of wear facets on lagomorph cheek teeth. (a) Primitive pattern (Piezodus). (b) Pattern in most leporids (Oryctolagus). (c) Pattern in advanced ochotonids (Ochotona) and Romerolagus. The rest position occurs when

terns in lagomorphs is also useful for internal phylogenetic and systematic studies of the order.

Molar Cusp Pattern

Analysis of the occlusion pattern can provide new evidence to aid the interpretation of cusp homologies. The relative position between the upper and lower cusps is very conservative in mammals, even though their initial position may have been changed (e.g., caenotheriids, Sudre, 1977).

The molar cusp pattern is eradicated in all lagomorphs because of the increased hypsodonty, and the upper molar pattern becomes unreliable for determining homologies with tribosphenic molar patterns. Nevertheless, the lower molars have retained the bilobed conservative pattern that is found in many other herbivorous mammals, and its homology with the tribosphenic lower molars is easily recognized. Thus, we should expect that the relative occlusion of the cusps has kept the same relationship.

As noted above, lagomorphs have a main wear facet between the metaconid and protoconid (combined facets 1 + 5) that corresponds with the proximal hyperloph wear facet on the upper molars (Fig. 3). The proximal hypercone lies on the talonid basin and takes part in this wear facet. Therefore, the proximal hypercone position and wearing pattern correspond to the protocone in the tribosphenic molar (Fig. 4a,b). The proximal hyperloph ends against the buccal precone, which is also involved in the wear facet 1 and may be related to the paracone (see Fig. 1).

The hypoconid lies under the proximal hypercone; it runs against the anterior part of the lagicone-lagiloph, to make a wear facet homologous with that of facet 4 of the tribosphenic molars at the anterior part of the metacone. The absolute position of the lagicone-lagiloph does not correspond with the posterobuccal metacone-premetacrista, but its relative position with the hypocone is the same. We can not interpret the nature of the postcone, because it has no wear facet. In advanced lagomorphs with a hypostria, we find the distal hypercone lying over the trigonid between the next two lower molars, and this occlusal position corresponds to the hypocone (Fig. 3).

This interpretation based on wear facets agrees with that of

---

the upper and lower wear facets touch each other. The wear facets are formed during the buccal phase of mastication, in the direction marked by the arrows. During the labial phase, no contact exists between upper and lower teeth. Not to scale.

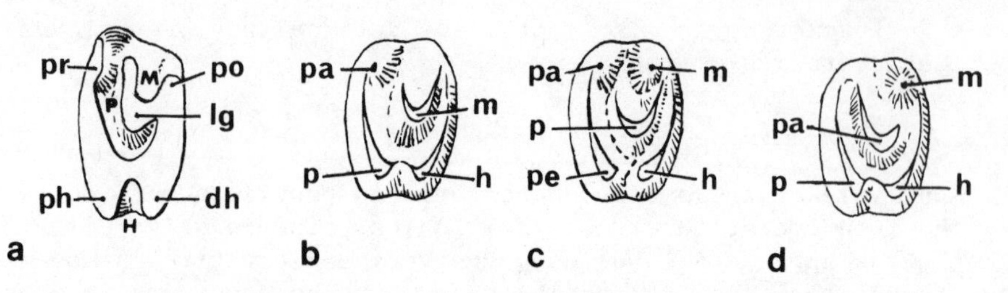

Fig. 4. Proposed homologies for upper molar cusp patterns in Lagomorpha. (a) Neutral descriptive terminology (after Lopez Martinez, 1974). dh, distal hypercone and hyperloph; h, hypoflexus and hypostriae; lg, lagicone and lagiloph; m, metaflexus; p, paraflexus; ph, proximal hypercone and hyperloph; po, postcone; pr, precone. (b) Cusp pattern according to our interpretation based on wear facet homologies, which coincides with Wood's (1940) and Bohlin's (1942) interpretations. h, hypocone; m, metacone; p, protocone; pa, paracone. (c) cusp pattern interpretation of McKenna (1982), based on comparison with anagalids. pe, pericone. (d) cusp homologies according to Burke (1934) and Tobien (1974b).

Wood (1940), based mainly on the "minimal modifications" and in comparisons with the condylarths (Wood, 1940, 1957), and that of Bohlin (1942).

Other interpretations were based on comparisons with different mammalian groups: primates (Major, 1899), triconodonts (Ehik, 1926), caenotherids (Hürzeler, 1936), Pseudictops, eurymylids, and anagalids (Van Valen, 1964; McKenna, 1982), and some others (Burke, 1936; Russell, 1958; Tobien, 1974b), all directed in a search for the ancestral group of the Lagomorpha. Nevertheless, all the interpretations can be consistent with several possible phylogenetic hypotheses, and they should be read only as polarity characters.

In Figures 4 and 5 we compare these propositions and suggested homologies, as well as some different lagomorph first upper molars. For comparison, molars of eurymyloids and anagalids are also shown. Our interpretation does not need a character reversal, as does that of McKenna (1982) (fusion, then subsequent splitting of the pericone-hypocone), and it is advantageous because it is based solely on the internal morphological evidence.

Hypsodonty and Molarization

Lagomorphs show the GHAS features in the dentition, which are present not only in the Recent members, but also in fossils since the Upper Eocene. The premolars become molariform in size and pat-

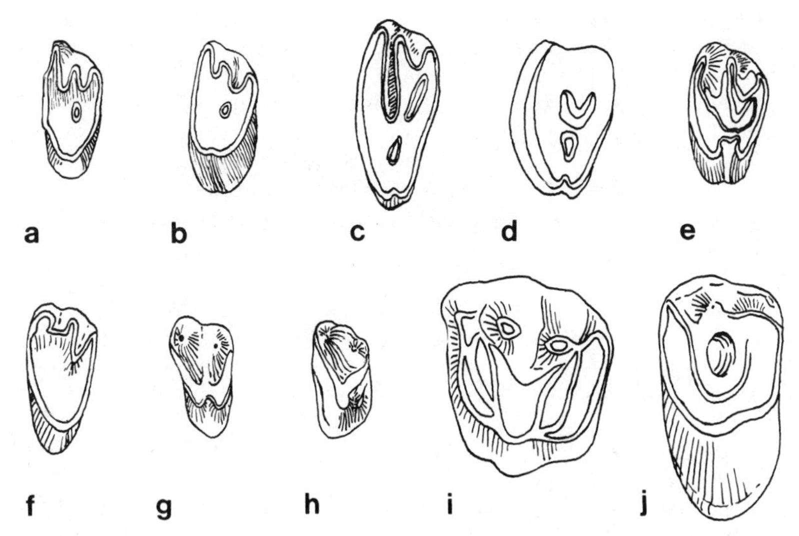

Fig. 5. Upper left first molars of: lagomorphs (a-e), eurymyloids (f-h), and anagalids (i-j). (a) Shamolagus, and (b) Gobiolagus (after Li, 1965). The fosset is interpreted by this author, correctly in my opinion, as a parafosset. (c) Megalagus brachiodon (= Desmatolagus dicei), and (d) M. turgidus (after Wood, 1940). The two fossets are the parafosset and the hypofosset (the labial one). (e) Eurolagus fontannesi (after Tobien, 1963). (f) Eurymylus (after Wood, 1942). (g) Heomys, and (h) Mimotona (after Li, 1977). (i) Hsiuannania, and (j) Huaiyangale (after Xu, 1976). x 5.

tern, and the main grinding activity in both jaws is modified and shifted from M1-M2 forward to P4-M1, while the M3 decrease and finally disappear in some taxa.

The premolars show a more molariform pattern as we advance distally in the jaws; thus, the anterior ones (P2/ and P/3) are never molarized. It is possible to find homologies with the molar cusp pattern, and earlier authors (Major, 1899; Burke, 1934; Wood, 1940; Bohlin, 1942) chose the premolar analogy. Molarization of premolars is a GHAS evolutionary trend in herbivores, and it must be considered for assessing the internal phylogenetic relationships of Lagomorpha. However, it apparently was not present in primitive lagomorphs (see Lushilagus, Li, 1965).

In contrast, hypsodonty is a primitive character for Lagomorpha, shared by all its members. Increasing hypsodonty is also a GHAS feature, and it is a major limiting factor in the longevity of individual herbivores. Primitive lagomorphs had rooted cheek teeth and a cusp-loph pattern that was worn very quickly. In every lineage there are different ways to attain rootless crowns (Bohlin, 1942; Tobien, 1974a). The upper cheek teeth have unilateral hypsodonty because of anisognathism, and the lingual root increases in relation to the labial ones, which are promptly eliminated (Tobien, 1974a, 1978). This hypsodonty pattern is similar to that of theridomyids (Vianey-Liaud, 1976), but in issiodoromyine rodents the buccal roots are usually conserved, probably because the masticatory movements are propalinal, more or less oblique relative to the enamel crests (Butler, 1980).

The degree of reduction of M3/3 is a variable feature in the order. Primitive lagomorphs had a three-lobed M/3. Leporids have a bilobed M/3, correlated with a weak M3/ for occlusal reasons (see Fig. 2). In ochotonids, M/3 is reduced to a single lobe, related to the lack of M3/, and they need a somewhat modified M2/ for occlusion (Oreolagus, Ochotona; Fig. 3c). In some genera lacking M3/3, such as Piezodus and Prolagus, M/2 develops a third lobe to occlude with the distal hyperloph of M2/ (Fig. 3a). A dental pattern with M3/ and a single-lobed M/3, as Muizon (1977) described for Bohlinotona pusilla, is not possible, and careful analysis shows it to be wrong (Teilhard de Chardin, 1926; personal observations).

The gnawing incisors of Lagomorpha are a primitive feature of the order. All known lagomorphs have a notched, cutting anterior upper incisor surrounded by enamel that, together with a small second upper incisor, both continuously growing, constitute unique features of the order (= Duplicidentata). The lower incisor seems to grow back initially beneath the cheek teeth in early forms (Wood, 1940; Li, 1965), but has become shifted progressively forward in more recent forms. The incisors have a transverse appearance, with a flat anterior face and a smaller antero-posterior dimension.

Many other mammalian groups develop enlarged, rootless incisors but only in Rodentia is this feature also the primitive state for the order. Multituberculates, megapodid and diprotodont marsupials, hyracoids, tillodonts, taeniodonts, eurymyloids, notungulates, plesiadapid and lemuriform primates, pyrotherians, proboscideans, and a few artiodactyls (Myotragus) have taxa with this character, which is not correlated necessarily with a herbivorous diet.

Enlarged incisors are useful for many tasks (see above), and we can find them in insectivores, frugivores, granivores, and omnivores, as well as in herbivores. Flattened incisors associated with a lophodont cheek tooth pattern are frequent in grazing herbivores, and we find them also in Lagomorpha and in some rodent groups (Castoridae, Caviomorpha, etc.). Rodent incisors can also be notched, but it is rare, and the enamel is always limited to the anterior face of the tooth. The lower incisors are always very long and chisel-like, while those of rabbits are short, because the lower incisors do not usually protrude in front of the upper ones, as they do in rodents (Ardran et al., 1958)

Masticatory Muscles

The masticatory apparatus has been studied in relation to its functions, and muscle development can be deduced by the osteological features found in fossils (Schumacher and Rehmer, 1959; Schumacher, 1961; Turnbull, 1970; Crompton and Parker, 1978). The lagomorphs, both fossil and Recent, shared a masticatory muscular apparatus (MMA) that can be assumed to be primitive for the order.

The jaws of lagomorphs have enlarged masseteric and pterygoid fossae, a high ramus and very reduced coronoid process, associated with the relative development of the three principal masticatory muscles. In modern lagomorphs, the relative mass of the masticatory muscles is: 12-15% temporalis, 56-63% masseter, and 25-29% pterygoids (Table 3). Mammalian MMAs can be clustered in three or four groups, based on the varying ratios of these muscles. These are: carnivores (temporalis dominant), ungulates (masseter-pterygoid dominant), rodents (masseter-temporalis dominant), and a fourth miscellaneous group (Turnbull, 1970). Turnbull's groupings have been used to support a synapomorphy of jaw musculature between Lagomorpha and Rodentia (Hartenberger, 1977).

On the contrary, with these same data we find a dissimilarity between rodents and lagomorphs, the latter being more similar to ungulates. In Table 3, the data of Turnbull's (1970) careful analysis clearly show the different pterygoid/temporalis ratios between lagomorphs and rodents, as well as a strong difference in the development of the internal pterygoid. In both features, lagomorphs are more similar to the ungulate herbivores of Turnbull's second group. Consequently, ungulates and lagomorphs can be grouped as herbivores characterized by a high pterygoid/temporalis ratio (> 1), and a greater development of the internal pterygoid in relation to the external, together comprising more than 25% of the total muscle mass. In contrast, rodents have a pterygoid/temporalis ratio near to or lower than 1, and the pterygoids, which are almost equally developed, are less than 20% of the total muscle mass.

The morphology of the MMA is also dissimilar between rodents and lagomorphs. According to the occlusal pattern described above, the main grinding movements of lagomorphs are transverse. However, the wear facets are forwardly inclined, so that the deep medial masseter is modified as a muscle posterior to the lateral masseter, for backward compression (see Fig. 6). It is attached to a long zygomatic projection that passes beyond the zygomatic arch behind the temporal process. This structure, misinterpreted by Wood (1940), was correctly identified by Bohlin (1942) and Schumacher and Rehmer (1959). The temporalis muscle is extremely reduced, and does not have the same function as in rodents, in which it is a powerful retractor. Rodents have developed very different propalinal movements, with an increasing importance of the forward projection of

Table 3. Percentage of relative muscular mass of masticatory apparatus. Data after Turnbull (1970).

Turnbull's groups	I carnivores			II ungulates					III rodents					
Proposed new groups	I carnivores			II herbivores							III omnivores			
	Bear	Cat	Nutria	Sheep	Bison	Roedeer	Pig	Horse	Rabbit	Hare	Sciurus	Rattus	Hystrix	Cavia
Muscles	Percentage of muscle mass													
Temporalis muscle	65	54	80	25	10	30	28	15	12	15	18	30	17	15
Masseter muscle	30	35	15	50	60	40	50	55	63	56	64	60	72	70
Pterygoid muscle	5	11	5	25	30	30	22	30	25	29	18	10	11	15
internal	4	10	4	22		26	17	25	20	23	10	7	6	8
external	1	1	1	3		4	5	5	5	6	8	3	5	7

the deep medial masseter, which becomes attached anterior to the lateral masseter (see Fig. 6).

The zygomatic arch is uniquely positioned in Lagomorpha, becaus the masseter is attached to its lateral face; it shows a large vertical plate grooved by a conspicuous fossa. It is placed more distally relative to rodents, and it bears a strong anterior tuberosity

CRANIAL FEATURES OF SENSORY ORGANS

Skull features have been described frequently in mammals for systematic and functional purposes, but there are many doubts about the polarity of the characters when comparing higher taxa (Roux, 1947; Starck, 1967; Kermack and Kielan-Jaworowska, 1971; Allin, 1975 Novacek, 1980). It is possible to summarize the major trends in lagomorphs, but fossil data are scarce, and analysis of the primitiv lagomorph features can only be speculative. Some characters are

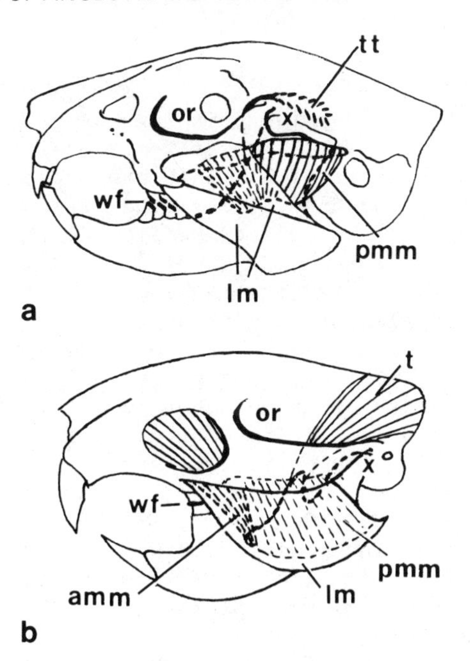

Fig. 6. Comparison of muscular masticatory apparatus in: (a) lagomorphs; and (b) rodents. Note for lagomorphs: the anterior and upward position of the condyle and the zygomatic process of the squamosal; reduction of the temporal fossa and coronoid process of the jaw; the peculiar lateral muscular insertion on the zygoma; and the posterior zygomatic process lateral to the mandibular ramus. amm, anterior medial masseter; lm, lateral masseter; or, orbit; pmm, posterior medial masseter; t, temporalis muscle; tt, tendinous temporalis; wf, wear facet orientation; x, condyle and glenoid fossa.

compared with those of other mammals, but there is not a "Bauplan" or morphotype of the mammalian skull for an outgroup comparison.

We have analyzed 68 craniomandibular characters used by various authors in lagomorph systematic studies (Major, 1899; Wood, 1940; Bohlin, 1942; Dawson, 1958, 1969; Lopez Martinez, 1974; Lopez Martinez and Thaler, 1975; Corbet, 1983), but the polarities are not clear for about 60% of these features. Many characters are correlated with one another, e.g., about 25 isolated characters belong to the masticatory apparatus, including the zygoma, dental tuberosity, jaws, etc. We have grouped these skull features by functional systems, retaining only those characters that are widely shared by most lagomorph taxa.

Olfactory System

The nasal region is very large in lagomorphs which are macrosmatic mammals. It is limited by very well developed nasal, premaxillary, and maxillary bones. The nasals are separated at the front from the premaxillary, and have a long area of insertion for the elevator nasii muscles that give great mobility to the rhinarium.

The nasals can taper forward (in leporids) or backward (in ochotonids), but they are always flanked laterally by two forward projections of the frontal bones, which seem to be an apomorphy of the order. There are also two very long posterior projections of the premaxilla, as in rodents and some other mammals. The nasal cavity is wide, with a large nasopalatine duct opening into the mouth by a great incisive foramen. The primitive feature of these openings seemed to be a state with four foramina, convergent into a large single opening (Major, 1899; Schreuder, 1936). There are no frontal or sphenoidal sinuses. The maxillary sinus is very small, and the maxillary nasal wall is lightened by a lacework of foramina that may converge onto a single large opening, increasing with ontogeny. The internal choanae have a forward position, anterior at least to M2/; this seems also to be an autapomorphy of the order Lagomorpha.

Optic System

This cranial region is also very large, with the orbits enlarged at the expense of the temporal fossa, even in the primitive taxa. The orbit is shifted forward and lies over the maxilla. The lacrimal bone is always very small, and does not leave the orbital cavity. The frontals are very narrow above the orbit and there is not always a "crestlike sculpturing of the orbital edges" (contra Szalay, 1977, p. 361); the supraorbital processes begin to appear in leporids, but ochotonids have a smooth orbital edge.

The internal orbital wall is mainly formed by the orbitosphenoid bones that converge in a unique axial wall provided with a single optic foramen connecting the orbits with each other. This is a character unique nowadays only in Lagomorpha; it could be a primitive feature, because of the similar disposition found in mammal-like reptiles and nontherian mammals, as well as in eurymylids (Li and Ting, this volume). The frontal enters dorsally to form a large part of the orbital wall, and below, the maxilla and mainly the palatines close the orbital vacuity. The basisphenoid does not appear in the orbit wall (Wood, 1940; Barone, 1966; Lopez Martinez, 1974), contrary to Dawson's (1969) interpretation. The most striking feature is the extreme reduction of the temporal fossae, which seem to be incorporated into the orbit; the posterior edge of the orbit is closed by the zygomatic process of the squamosal bone and the glenoid articulation of the jaw (Fig. 6). This forward position of the zygoma is typical autapomorphy of Lagomorpha and is correlated with a high mandibular ramus and a wide condyle, and with the posterior zygomatic process for the *M. masseter medialis*.

Auditory System

Lagomorphs have a well-developed ectotympanic bulla. In many fossil and recent genera (*Palaeolagus*, *Litolagus*, *Ochotona*), it attains the same relative size as in the very specialized desert rodents. The auditory bulla may be formed by the ectotympanic bone alone (leporids and some ochotonids), or also by the petrosal bone, as in *Ochotona*. In the latter, the cavity is filled by spongy bone; in some cases (e. g., *Prolagus*) an internal bony septum appears within the bulla, but normally the bulla is hollow. An entotympanic bone is absent, and many taxa show a more or less elongated meatal tube directed upward and backward.

The internal carotid arterial circulation has not been studied in fossil lagomorphs. Bugge (1974) compared this system in Recent Lagomorphs to the primitive stock of ungulates, near *Protungulatum*.

Cranial Foramina

The skull of lagomorphs shows many foramina, some with an unknown function (e. g., "pitted" bone in maxilla, frontals, alisphenoid, petrosal, occipital). Other foramina are present only in some taxa: retro-articular, ethmoidal, and premolar foramina. This last opening has been interpreted as an apomorphic feature of ochotonids (Bohlin, 1942; McKenna, 1982), where it is constant and sometimes very large (e. g., *Prolagus ibericus*). Nevertheless, this feature is also present in some recent leporids (*Poelagus*, some *Sylvilagus*) and may be a matter of intraspecific variation (e. g., *Nesolagus*) (see Corbet, 1983).

The most constantly present foramina in lagomorphs are common mammalian features: a single carotid-lacerum foramen, a sphenoidal fissure, and a round sphenoidal foramen. Others seem to be very primitive features, that do not exist in most mammals, such as the cranio-pharyngeal continuity, connecting the hypophysis and pharynx. In the jaws there are two primitive mental foramina and a large postero-internal canalis mandibularis, that opens dorsally in a foramen just behind the lower tooth-row; this seems to be an autapomorphic feature.

LOCOMOTION AND POSTCRANIAL SKELETON

As we have seen, the locomotory skeleton of lagomorphs does not show important modified features, except in some Recent leporids. Following a summary of postcranial characters in Recent lagomorphs, we shall study the primitive osteological features of fossil taxa.

Head Position

In this important character, some leporids are more derived than the ochotonids, but all lagomorphs show an apomorphic state of the head position in relation to rodents and primitive eutherians. The osteological features involved in this change are: basicranial-basifacial bending; ventral occipital foramen orientation and developed nuchal muscle attachments; and oblique orientation of the nucha plate. These characters show an increasing evolutionary trend in th fossil record (Dawson, 1958). This is apparently not related to the convex shape of the dorsal skull surface, which always shows a decreasing ontogenetic trend in mammals.

Postcranial Skeleton

The humerus and the principal bones of the tarsus are the only well-known postcranial elements of the main fossil lagomorphs, while the limb ratio, related to the segment lengths, can only be inferred in some well-preserved skeletons of a few fossil taxa. The main results of Major's (1899) valuable study are still accepted.

Numerous primitive eutherian characters, such as the carpal morphology, ulnar-radial mobility, presence of clavicle, and short and wide scapular neck, are retained in many lagomorphs.

In primitive lagomorphs, the humerus seems more rodent-like in some of its plesiomorphic features: the greater tuberosity is lower than the head and has an oblique orientation; the head is wide and evenly convex; the deltoid crest is more prominent and bent laterally; the distal epiphysis is larger, with entepicondylar and supratrochlear foramina; and the trochlea is asymmetric (Wood, 1940; Dawson, 1958; Li, 1965). All these features indicate a high degree of pronation-supination and lateral mobility of the forelimbs.

In the hindlimbs, the pelvic girdle and the femur exhibit the most important intraordinal differences. The pelvic girdle in primitive lagomorphs is narrow, almost parallel to the sacrum, and the ventral side of the iliac blade is weak, because of the ventral orientation of the acetabulum. There is always a strong ischial tuberosity.

The most common femoral pattern shows a very narrow neck, a dorsally expanded head, and a high third trochanter. These features may be primitive, and they are more frequent in terrestrial tetrapods than in scansorial-bipedal tetrapods. Rodents show the opposit characters, even in the terrestrial-fossorial taxa.

The tibio-fibula is fused distally. The tarsal structures show some interesting peculiarities. Included with several plesiomorphic eutherian features, such as the tibial bone (Major, 1899) and the fibular-calcaneal joint, there are some early specialized characters such as the longer calcaneum in relation to the astragalus, which provided even a calcaneum-navicular joint (the most frequent diarthrodial modification, derived from the serial arrangement found in primitive mammals, is an astragalus longer than the calcaneum). There is also a modified, L-shaped calcaneo-cuboid articulation, and two mechanized facets that Szalay (1977) compared with macroscelidids: the tibial-astragalar trochlea and the astragalo-navicular facet. The fibulo-calcaneum joint is also strongly mechanized; it attains a clear autapomorphic state. As noted above for the life habits of lagomorphs, there does not appear to be an efficient cursorial adaptation in primitive lagomorphs (for a different opinion, see Szalay, this volume).

DISCUSSION

We have listed in Tables 4-6 the postulated primitive (plesiomorphous) cranioskeletal features of the order Lagomorpha. These 56 characters show a wide distribution among the lower taxa of the order, and some of them are found in young ontogenetic stages of individuals.

These ancestral features of the order have been grouped into three categories: (1) plesiomorphic eutherian features (Table 4) that can be inferred after comparison with selected outgroups (other Mammalia and therapsid reptiles), and by ontogenetic precedence; (2) autapomorphic features of Lagomorpha (Table 5) that are rarely if ever present, as far as we know, in other mammalian orders; and (3) eutherian apomorphic features (Table 6), inferred by functional and outgroup analyses.

Plesiomorphic Characters

Most of the 16 plesiomorphic eutherian features of the reconstructed ancestral morphotype of Lagomorpha, as noted in Table 4, are mainly postcranial features, with only one dental and four cranial characters (numbers 1-5). In cladistic methodology, it is normally assumed that symplesiomorphies do not indicate phylogenetic relationships, because they can be retained from an earlier ancestor by any of the descendants. They are only indicative of the plesiomorphic grade of a taxon, and normally they are not given special weight in a cladistic phylogenetic analysis. Nevertheless, they could indicate reversal of evolution if the postulated plesiomorphic feature is not present in the entire outgroup.

As an example, characters 4 and 5 in Table 4 are shared with therapsid reptiles and with mammalian embryos, but these have been modified by other Mammalia. If we want to link lagomorphs with a mammalian group with two optic foramina and a closed basisphenoid, as is the rule for the Class, then lagomorphs must have had an

Table 4. Postulated primitive dental and cranioskeletal features of Lagomorpha. Part I. Plesiomorphic eutherian features retained in lagomorph morphotype.

1. Enamel surrounds the incisors	9. Humerus greater tuberosity lower than head
2. Four large incisive foramina	10. Entepicondylar and supra-trochlear foramina present
3. Two mental foramina	11. Weak femur neck
4. Optic foramina confluent between orbits	12. Dorsally expanded head of femur
5. Cranio-pharyngeal continuity	13. High third trochanter of femur
6. Short limbs related to body size	14. Central carpal bone free
7. Short metapodial segment of limbs	15. Tibial tarsal bone free
8. Parallel sacro-pelvic orientation	16. Fibula-calcaneum joint functional

evolutionary reversal in both characters. If we want to make this contingency disappear, it will be necessary to study in more detail the meaning of both features.

Plesiomorphic characters are important in the historical reconstruction of a taxon, because they are relative conditions, which do not always signify a general condition in the ancestral group or outgroup of the taxon.

Autapomorphic Characters

The important set of 17 autapomorphic characters of the ancestral condition of Lagomorpha (Table 5) attests to the high evolution ary grade of the order, but they are not used in the cladistic phylo genetic reconstruction. The high number of uniquely derived apomorphic characters supports the internal coherence of the group, and it justifies their systematic position as an independent order. Thus, the inclusion of eurymyloids and anagalids in the order Lagomorpha (Szalay, 1977; McKenna, 1982) would reduce their differential charac ters as an order to 0; they would have only a common apomorphic feature, orthal mechanized astragalar facets, shared also with macroscelidids. The dental pattern is not shared by this heterogeneous group (see Fig. 5).

Taxonomic groups are more coherent when more autapomorphies are present. Furthermore, autapomorphic features can also be useful in phylogenetic studies (Lopez Martinez, 1978). They can be compare with the internal variability of characters in a postulated sister group. These characters are not usually introduced in a phylogeneti analysis, but they have a high potential information content that ca

Table 5. Postulated primitive dental and cranioskeletal features of Lagomorpha. Part II. Autapomorphic features of the lagomorph morphotype.

1. Parallel and transverse dental wear facets	10. Lack of frontal and sphenoidal sinuses
2. Modified molar cusp pattern	11. Reduced coronoid process of jaw
3. Two pairs of upper incisors, one behind the other	12. High mandibular ramus capped by long condyle
4. Notched upper incisors	13. Dorsal opening of mandibular canal (behind tooth row)
5. Unilateral hypsodonty	
6. Lateral zygomatic fossa for outer muscle attachment	14. Frontals project forward at each side of nasals
7. Posterior zygomatic process for medial masseter	15. Reduced temporal fossa incorporated into the orbit
8. Choanae shifted forward in front of M2/	16. Orbitosphenoid builds a large part of the orbit
9. Maxilla lightened by lacework foramina	17. L-shaped facet of calcaneum-cuboid joint

serve as a test of a phylogenetic hypothesis. The sister group relationship is strengthened if the autapomorphies of one taxon are present in a variable state in the other taxon.

Synapomorphic Characters

We have listed in this category 23 derived characters that are shared by Lagomorpha and some other mammalian orders (Table 6). They are the most important data for phylogenetic reconstruction, but they must be compared with the ancestral features of the other mammalian orders, not only with some of their taxa, even assuming that they are plesiomorphic taxa. An isolated lower taxon rarely has the entire set of primitive characters of the order, and comparison with only a single taxon might lead us to overlook the homoplasies.

The ancestral morphotype of the order Rodentia has not been reconstructed completely, although it has been assumed that it is close to the Eocene family Paramyidae. Following Wood (1959), McKenna (1961), Hartenberger (1977), and Szalay (1977), the primitive rodents must have been very small scansorial mammals living in the forest and lacking a paraconid on the lower molars. They share with Lagomorpha the apomorphic features 3, 4, 7, 10, and ?13 in Table 6. Rodents seem to have derived character states instead of plesiomorphies for characters 1-6, 11-13, and 16 of lagomorphs listed in Table 4. On the contrary, rodents are plesiomorphic for characters 1, 5, 6, 8, 9, 11, and 14-23 of Table 6, which are apomorphic in Lagomorpha. A possible common ancestor of both orders would have to be plesiomorphous for all the characters in Tables 4 and 6, as well as

Table 6. Postulated primitive dental and cranioskeletal features of Lagomorpha. Part III. Apomorphic features of Lagomorpha shared with some other eutherian higher taxa.

1. Size near 200mm head-body length
2. Dental formula 2.0.3.3/1. 0.2.3.
3. Evergrowing incisors
4. Lower incisors grow beneath cheek teeth
5. Lophodont molar crown pattern
6. Molarized premolars
7. Dentary diastema long
8. High pterygoid/temporal ratio > 1
9. Inner pterygoid near 20% of total muscle mass
10. Orbit overlies tooth-row
11. Palatines included in the orbit wall
12. Very large auditory bullae
13. ?Bullae formed by ectotympanic bone
14. Ventral orientation of occipital foramen
15. Basicranial/basifacial bending
16. Oblique nuchal plate
17. Very short tail
18. Ventral orientation of acetabulum
19. Strong ischial tuberosity
20. Distal tibio-fibular fusion
21. Orthal mechanized astragular facets
22. Calcaneus longer than astragalus
23. Fissured ungual phalanges

for the high number of autapomorphic characters of rodents and lagomorphs.

Lagomorphs share some apomorphies with many herbivorous mammals (Characters 5-9, 13, 17-19 of Table 6 are shared with hyracoids, perissodactyls, and artiodactyls), but it is necessary to reconstruc the ancestral conditions of these orders for phylogenetic interpretation. Ungulate and condylarth relationships with lagomorphs have been suggested by Moody et al. (1949) on serological data, by Wood (1957) on dental evidence, and also by Bugge (1974) based on the similarity of the internal carotid arterial system.

Some insectivore groups also show synapomorphic features with the Lagomorpha; chrysochlorids share ?13; macroscelidids, soricids, talpids, and solenodontids have 20 and 23; and macroscelidids also have 21 of Table 6. Some other similarities are found among the plesiomorphic features.

Several eurymyloid genera, known principally by cranial masticatory characters, share with lagomorphs the the apomorphic features 1-7 of Table 6. Some genera have a dental pattern close to our interpretation of the lagomorph cusp pattern morphotype (see e.g., Mimotona, Li, 1977; also Fig. 5h). The dental formulae of this grou show a high variability (Li, 1977), and their systematic position is not clear. They have been grouped with the Anagalida (Li et al.,

1979), Lagomorpha (Wood, 1942; Szalay, 1977), or Rodentia (McKenna, 1982), or they have been considered as an independent order Mixodontia (Sych, 1971). Other characters of this interesting group, however, must be evaluated before reaching a conclusion about their relationship with Lagomorpha (also see Li and Ting, this volume).

CONCLUSION

Following an overview of numerous biological features of the order Lagomorpha, this mammalian group appears as a well characterized taxon with a unique set of morphological, physiological, ecological, and evolutionary strategies. The survey of cranioskeletal and dental features that can be postulated as ancestral characters of the order shows 56 traits that can be divided into three groups: 16 eutherian plesiomorphies; 17 lagomorph autapomorphies; and 23 apomorphies shared with some other mammalian groups.

Some plesiomorphic features of Lagomorpha have been modified in the whole of other Mammalia, and they must be studied carefully in order to analyze their evolutionary reversion. The third group of features, the only ones used in cladistic analysis, have been compared preliminarily with some supposed ancestral characters of the order Rodentia, and with some other mammalian taxa.

Rodents share four or five synapomorphies with lagomorphs, out of 23 ancestral lagomorph apomorphic features. The common ancestor of both groups would be plesiomorphic for all of these characters except the following: evergrowing incisors; lower incisors extending distally beneath cheek teeth; dentary diastema; orbits over the tooth row; and perhaps an ectotympanic bulla. Each of these features has appeared in many other mammals (multituberculates, hyracoids, macroscelidids, etc.), and they do not appear to be uniquely derived characters. Consequently, we can not corroborate the hypothesis of a sister group relationship between Rodentia and Lagomorpha (for a different viewpoint, see Luckett, this volume, and Novacek, this volume). Hyracoids, ungulates, and some insectivores share a few apomorphies with lagomorphs, but their ancestral character states are not yet known.

Eurymyloids share an important number of synapomorphies with lagomorphs, but they are still not a very well-characterized taxon. Some genera show as many as seven synapomorphous characters out of 15 available ones. They also share plesiomorphies 1 and 3 of Table 4, and some genera show a similarity in the autapomorphous molar characters 1, 2, and 5 of Table 5 (*Mimotona*, according to our suggested homology of molar cusp pattern). Eurymyloids may constitute a possible sister group of Lagomorpha, but their systematic position is not clear, however, and the higher taxa phylogeny of lagomorphs must be based on a more complete ensemble of characters (see Li and Ting, this volume).

ACKNOWLEDGMENTS

I am grateful to many people who have helped to prepare this work. P. Sevilla, F. Garcia Moreno, and W. P. Luckett corrected the preliminary manuscript. B. Sanchiz read the text and discussed the methods and principal results. M. Diaz Molina, M. A. Sacristan, and P. Sevilla carried on with the teaching and bureaucratic work at the University while I was occupied in preparing the manuscript. My husband and son provided support with their help at home. I want to thank the editors, Drs. Hartenberger and Luckett, for the invitation to attend the international meeting in Paris. Mrs. R. Ralomo, M. de Andres, and Dr. Nancy Hong typed the manuscript, and E. Marin took the photographs.

REFERENCES

Alberch, P., Gould, S. J., Oster, G. F. and Wake, D. B. 1979. Size and shape in ontogeny and phylogeny. Paleobiology 5: 296-317.

Allin, E. F. 1975. Evolution of the mammalian middle ear. J. Morph. 147: 403-438.

Ardran, G. M., Kemp, F. H. and Ride, W. D. L. 1958. A radiographic analysis of mastication and swallowing in the domestic rabbit Oryctolagus cuniculus (L.). Proc. Zool. Soc. Lond. 130: 257-274.

Barone, R. 1966. Anatomie comparée des Mammifères domestiques. Lab. Anat. E. Nat. Vet., Lyon.

Bohlin, B. 1942. The fossil mammals from the Tertiary deposit of Taben-Buluk, Western Kansu. Part I. Insectivora and Lagomorpha. Paleont. Sinica, N. S. C. 123: 1-113.

Bonner, J. T. 1965. Size and Cycle: An Essay on the Structure of Biology. Princeton Univ. Press, Princeton.

Bown, T. M. and Kraus, M. J. 1979. Origin of the tribosphenic molar and metatherian and eutherian dental formulae. In: Mesozoic Mammals, J. A. Lillegraven, Z. Kielan-Jaworowska, and W. A. Clemens, eds., pp. 172-181, Univ. of Calif. Press, Berkeley.

Bugge, J. 1974. The cephalic arterial system in insectivores, primates, rodents and lagomorphs, with special reference to the systematic classification. Acta Anat. 87 (Suppl. 62): 1-160.

Burke, J. J. 1934. Mytonolagus, a new leporine genus from the Uinta Eocene series in Utah. Ann. Carneg. Mus. 23: 399-420.

Burke, J. J. 1936. Ardynomys and Desmatolagus in the North American Oligocene. Ann. Carneg. Mus. 25: 135-154.

Butler, P. M. 1980. Functional aspects of the evolution of rodent molars. Palaeovertebrata Mem. Jubil. R. Lavocat: 249-262.

Caillol, M. and Martinet, L. 1981. Estrous behavior, follicular growth, and pattern of circulating sex steroids during pregnancy and pseudopregnancy in the captive brown hare. In: Proceedings World Lagomorph Conference, K. Myers and C. D. McInnes, eds., pp. 142-154, Univ. Guelph, Canada.
Cervantes Reza, F. A. 1982. Observaciones sobre la reproducción del zacatuche o teporingo Romerolagus diazi (Mammalia: Lagomorpha). Doñana, Acta Vert. 9: 416-420.
Cheylan, G. 1982. Les adaptations écologiques et morphologiques de Rattus rattus à divers environements insulaires méditerranées: étude d'un cas d'évolution rapide. D. E. A. U. S. T. L., Montpellier, 66 pág.
Corbet, G. B. 1983. A review of classification in the family Leporidae. Acta Zool. Fenn. 174: 11-16.
Crompton, A. W. 1971. The origin of the tribosphenic molar. Linn. Soc. Zool. J. 50: 65-87.
Crompton, A. W. and Hiiemae, K. M. 1969. Functional occlusion in tribosphenic molars. Nature 22: 678-679.
Crompton, A. W. and Jenkins, F. A. 1968. Molar occlusion in Late Triassic mammals. Biol. Rev. 43: 427-458.
Crompton, A. W. and Parker, P. 1978. Evolution of the mammalian masticatory apparatus. Amer. Sci. 66: 192-201.
Daly, J. C. 1981. Social organization and genetic structure in a rabbit population. In: Proceedings World Lagomorph Conference, K. Myers and C. D. McInnes, eds., pp. 90-97, Univ. Guelph, Canada.
Dawson, M. R. 1958. Late Tertiary Leporidae of North America. Univ. Kansas Paleont. Contrib. 22: 1-79.
Dawson, M. R. 1959. Paludotona etruria, a new ochotonid from the Pontian of Tuscany. Verh. Naturf. Ges. Basel 70: 157-166.
Dawson, M. R. 1961. On two ochotonids (Mammalia, Lagomorpha) from the later Tertiary of Inner Mongolia. Amer. Mus. Novit. 2061: 1-15.
Dawson, M. R. 1969. Osteology of Prolagus sardus, a Quaternary ochotonid (Mammalia, Lagomorpha). Palaeovertebrata 2: 157-190.
Dawson, M. R. 1970. Paleontology and geology of the Badwater Creek area, central Wyoming. 6: The leporid Mytonolagus (Mammalia, Lagomorpha). Ann. Carneg. Mus. 41: 215-230.
Delibes, M. and Hiraldo, F. 1981. The rabbit as prey in the Iberian Mediterranean ecosystem. In: Proceedings World Lagomorph Conference, K. Myers and C. D. McInnes, eds., pp. 614-622, Univ. Guelph, Canada.
Dice, L. R. 1933. Some characters of the skull and skeleton of the fossil hare, Palaeolagus haydeni. Michigan Acad. Sc. Art. Let. Pap. 18: 301-306.
Dullemeijer, P. and Barel, C. D. N. 1977. Functional morphology and evolution. In: Major Patterns in Vertebrate Evolution, M. K. Hecht, P. C. Goody, and B. M. Hecht, eds., pp. 83-118, Plenum Press, New York.

Ehik, J. 1926. The right interpretation of the cheek-teeth tubercles of Titanomys. Ann. Hist. Nat. Mus. Nat. Hung. 23: 178-186.
Eldredge, N. 1979. Alternative approaches to evolutionary theory. Bull. Carneg. Mus. Nat. Hist. 13: 7-19.
Erbajeva, M. E. 1970. Historiya Antropogenovoi fauni Zaüceobrazich (Lagomorpha) i grizunov (Rodentia) selenginskogo srednegoriya. Izdateistvo Nauka Moskva: 1-132.
Erbajeva, M. E. 1981. Miotsenovie pizhuji Mongolii. Iskopaiemie pozvonochnie Mongolii (Trudi, vip. 15). Moskva, Nauka: 86-95.
Flinders, J. T. and Hansen, R. M. 1973. Abundance and dispersion of leporids within a shortgrass ecosystem. J. Mammal. 54: 287-291.
Flux, J. E. C. 1981. Reproductive strategies in the genus Lepus. In: Proceedings World Lagomorph Conference, K. Myers and C. D. McInnes, eds., pp. 155-174, Univ. Guelph, Canada.
Fraguglione, D. 1960. Compétition interspécifique entre le lièvre commun (Lepus europaeus Pallas 1778) et le lapin de garenne (Oryctolagus cuniculus Linne 1758). Diana 10: 211-212.
French, N. R. and Heasley, J. E. 1981. Lagomorphs in the shortgras prairie. In: Proceedings World Lagomorph Conference, K. Myers and C. D. McInnes, eds., pp. 695-705, Univ. Guelph, Canada.
Gawne, C. E. 1978. Leporids (Lagomorpha, Mammalia) from the Chadronian (Oligocene) deposits of Flagstaff Rim, Wyoming. J. Paleont. 52: 1103-1118.
Ghose, R. K. 1981. On the ecology and status of the hispid hare. In: Proceedings World Lagomorph Conference, K. Myers and C. D. McInnes, eds., pp. 917-923, Univ. Guelph, Canada.
Gidley, J. W. 1912. The lagomorphs, an independent order. Science 36: 285-286.
Gingerich, P. D. 1977. New species of Eocene primates and the phylogeny of European Adapidae. Folia Primatol. 28: 60-80.
Gould, S. J. and Vrba, E. S. 1982. Exaptation - a missing term in the science of form. Paleobiology 3: 4-15.
Granados, H. 1981. Basic information on the volcano rabbit. In: Proceedings World Lagomorph Conference, K. Myers and C. D. McInnes, eds., pp. 935-942, Univ. Guelph, Canada.
Gureev, A. A. 1960. Zaitzeobrazyne (Lagomorpha) oligozena Mongolya y Kazakstana. Trylya Paleontology Instit. 77: 5-34.
Gureev, A. A. 1964. Zaitzeobrazyne. Die Hasenartigen (Lagomorpha) Fauna SSSR. Mammalia. Mlekop. III, 10 Moskow, Leningrad. N. S., 87: 1-276.
Gutmann, W. F. 1977. Phylogenetic reconstruction: theory, methodology and application to chordate evolution. In: Major Pattern in Vertebrate Evolution, M. K. Hecht, P. C. Goody, and B. M. Hecht, eds., pp. 645-670, Plenum Press, New York.
Hartenberger, J.-L. 1977. A propos de l'origine des Rongeurs. Geobios Mem. spéc. 1: 183-193.
Hecht, M. K. and Edwards, J. L. 1977. The methodology of phylogenetic inference above the species level. In: Major Patterns

in Vertebrate Evolution, M. K. Hecht, P. C. Goody, and B. M. Hecht, eds., pp. 3-52, Plenum Press, New York.
Heissig, K. and Schmidt-Kittler, N. 1975. Ein primitiver Lagomorphe aus dem Mitteloligozän Süddeutschlands. Mitt. Bayer Staatssamml. Paläont. hist. Geol. 15: 57-62.
Heissig, K. and Schmidt-Kittler, N. 1976. Neue Lagomorphen-Funde aus dem Mitteloligozän. Mitt. Bayer Staatssamml. Paläont. hist. Geol. 16: 83-93.
Hennig, W. 1966. Phylogenetic Systematics. Univ. Illinois Press, Urbana.
Hershkovitz, P. 1962. Evolution of neotropical cricetine rodents (Muridae) with special reference to the phyllotine group. Fieldiana Zool. 46: 1-524.
Hürzeler, J. 1936. Osteologie und Odontologie der Caenotheriden. Abh. Schweiz. Palaeont. Ges. 58/59: 1-111.
Keith, L. B. 1981. Population dynamics of hares. In: Proceedings World Lagomorph Conference, K. Myers and C. D. McInnes, eds., pp. 395-440, Univ. Guelph, Canada.
Kermack, K. A. and Kielan-Jaworowska, Z. 1971. Therian and non-therian mammals. Linn. Soc. Zool. J. 50 (Suppl. 1): 103-115.
Krumbach, T. 1904. Die unteren Schneidezähne der Nagetiere, nach Gestalt und Funktion betrachtet. Zool. Anz. 27: 273-290.
Li, C.-K. 1965. Eocene leporids of North China. Vert. PalAsiat. 9: 23-36.
Li, C.-K. 1977. Paleocene eurymyloids (Anagalida, Mammalia) of Qianshan, Anhui. Vert. PalAsiat. 15: 103-118.
Li, C.-K., Chiu, C.-S., Yan, D.-F., and Hsieh, S.-H. 1979. Notes on some early Eocene mammalian fossils of Hengtung, Hunan. Vert. PalAsiat. 17: 71-80.
Li, C.-K. and Ting, S.-Y. 1982. The Paleogene mammals of China. Bull. Carneg. Mus. Nat. Hist. 21: 1-96.
Lindlof, B. 1970. Ecological relations between mountain hare (*Lepus timidus*) and brown hare (*Lepus europaeus*). Zool. Revy. 32: 57-60.
Lopez-Forment, W. and Cervantes, F. 1981. Preliminary observations on the ecology of *Romerolagus diazi* in Mexico. In: Proceedings World Lagomorph Conference, K. Myers and C. D. McInnes, eds., pp. 943-949, Univ. Guelph, Canada.
Lopez Martinez, N. 1974. Evolution de la lignée *Piezodus-Prolagus* (Lagomorpha, Ochotonidae) dans le Cénozoique d'Europe sud-Occidentales. Thèse Univ. Sci. Tech. Languedoc Acad. Montpellier.
Lopez Martinez, N. 1977. Revisión sistemática y biostratigráfica de los Lagomorpha (Mammalia) del Terciario y Cuaternario inferior de España. Tesis Doctoral, Universidad Complutense de Madrid.
Lopez Martinez, N. 1978. Cladistique et paléontologie. Application à la phylogénie des Ochotonidés européens (Lagomorpha, Mammalia). Bull. Soc. géol. France 20: 821-830.
Lopez Martinez, N. and Thaler, L. 1974. Le plus ancien Lagomorphe d'Europe Occidentale. Réflexion sur la Grande Coupure.

Palaeovertebrata 6: 243-251.

Lopez Martinez, N. and Thaler, L. 1975. Biogéographie, évolution et compléments à la systématique du groupe d'Ochotonidés Piezodus-Prolagus (Mammalia, Lagomorpha). Bull. Soc. géol. France 17: 850-866.

Luckett, W. P. 1977. Ontogeny of amniote fetal membranes and their application to phylogeny. In: Major Patterns in Vertebrate Evolution, M. K. Hecht, P. C. Goody, and B. M. Hecht, eds., pp. 439-516, Plenum Press, New York.

Major, C. I. F. 1899. On fossil and recent Lagomorpha. Trans. Linn. Soc. Lond., Zool. 7: 433-520.

Mares, M. A. 1980. Convergent evolution among desert rodents: a global perspective. Bull. Carneg. Mus. Nat. Hist. 16: 1-51.

Margalef, R. 1982. La Biosfera, entre la Termodinámica y el Juego. Omega, Barcelona.

Matthew, W. D. and Granger, W. 1923. Nine new rodents from the Oligocene of Mongolia. Amer. Mus. Novit. 102: 1-10.

Matthew, W. D., Granger, W., and Simpson, G. G. 1929. Additions to the fauna of the Gashato Formation of Mongolia. Amer. Mus. Novit. 376: 1-12.

Maynard, L. A. and Loosli, J. K. 1969. Animal Nutrition. McGraw-Hill, New York.

McKenna, M. C. 1961. A note on the origin of rodents. Amer. Mus. Novit. 2037: 1-5.

McKenna, M. C. 1975. Toward a phylogenetic classification of the Mammalia. In: Phylogeny of the Primates, W. P. Luckett and F. S. Szalay, eds., pp. 21-46, Plenum Press, New York.

McKenna, M. C. 1982. Lagomorph interrelationships. Geobios Mém. Spec. 6: 213-223.

McNeely, J. A. 1981. Conservation needs of Nesolagus netscheri in Sumatra. In: Proceedings World Lagomorph Conference, K. Myers and C. D. McInnes, eds., pp. 926-929, Univ. Guelph, Canada.

Mein, P. and Adrover, R. 1982. Une faunule de mammifères insulaire dans le Miocène moyen de Majorque (Isles Balèares). Geobios Mém. Spec. 6: 451-463.

Michaux, J. 1983. Aspects de l'évolution des Muridés (Rodentia, Mammalia) en Europe sud-occidentale. In: Modalités, Rhytmes, Méchanismes de l'Evolution Biologique, J. Chaline, ed., pp. 195-200, Coll. Int. CNRS, Paris.

Moody, P. A., Cochran, V. A., and Drugg, H. 1949. Serological evidence on lagomorph relationships. Evolution 3: 25-33.

Muizon, C. de 1977. Révision des Lagomorphes des couches à Baluchitherium (Oligocène supérieur) de Sant-tao-lo (Ordos, Chine). Bull. Mus. Nat. Hist. Nat. 488: 265-292.

Myers, K. and Bults, H. G. 1977. A study of the biology of the wil rabbit in climatically different regions in eastern Australia. X. Measurement of changes in the quality of food. Aust. J. Ecol. 2: 215-230.

Nelson, E. W. 1909. The rabbits of North America. U. S. Dept. Agric. Bur. Biol. Surv. N. Amer. Fauna 29: 1-314.

Novacek, M. J. 1980. Cranioskeletal features in tupaiids and selected Eutheria as phylogenetic evidence. In: Comparative Biology and Evolutionary Relationships of Tree Shrews, W. P. Luckett, ed., pp. 35-94, Plenum Press, New York.

Palacios, F. 1978. Sistemática, distribución geográfica y ecológica de las liebres españolas. Situación actual de sus poblaciones. Tesis Doct. Univ. Politecn. (Ing. Montes), Madrid.

Pianka, E. 1982. Ecología Evolutiva (Translated by J. Ayala). Ed. Omega, Barcelona.

Pielowski, Z. and Pucek, Z. (eds.) 1976. Ecology and Management of European Hare Populations. State Publ. Agric. For., Warszawa.

Roux, G. 1947. The cranial development of certain Ethiopian "insectivores" and its bearing on the natural affinities of the group. Acta Zool. 28: 165-397.

Russell, L. S. 1958. The dentition of rabbits and the origin of the Lagomorpha. Bull. Nat. Mus. Canada 166: 41-45.

Sanchiz, F. B. de and G.-Valdecasas, A. 1980. Criterios metodológicos y glosario español de términos utilizados en sistemática cladística. Bol. R. Soc. Esp. Hist. Nat. (Biol.) 78: 223-244.

Schneider, E. 1981. Studies on the social behaviour of the brown hare. In: Proceedings World Lagomorph Conference, K. Myers and C. D. McInnes, eds., pp. 340-348, Univ. Guelph, Canada.

Schreuder, A. 1936. Hypolagus from the Tegelen Clay with a note on the recent Nesolagus. Arch. Néerl. Zool. II: 225-239.

Schumacher, G. H. 1961. Funktionelle Morphologie der Kaumuskulatur. Gustav Fischer Verlag, Jena.

Schumacher, G. H. and Rehmer, H. 1959. Morphologische und funktionelle Untersuchungen an der Kaumuskulatur von Oryctolagus und Lepus. Morph. Jahrb. 100: 678-705.

Simpson, G. G. 1945. The principles of classification and a classification of mammals. Bull. Amer. Mus. Nat. Hist. 85: 1-350.

Simpson, G. G. 1953. The Major Features of Evolution. Columbia Univ. Press, New York.

Smith, A. T. 1981a. Territoriality and social behavior of Ochotona princeps. In: Proceedings World Lagomorph Conference, K. Myers and C. D. McInnes, eds., pp. 310-323, Univ. Guelph, Canada.

Smith, A. T. 1981b. Population dynamics of pikas. In: Proceedings World Lagomorph Conference, K. Myers and C. D. McInnes, eds.. pp. 572-586, Univ. Guelph, Canada.

Soriguer, R. C. 1979. Biología y dinámica de una población de conejos, Oryctolagus cuniculus (L.) en Andalucía Occidental. Tesis Doctoral, Universidad de Sevilla.

Soriguer, R. C. and Rogers, P. M. 1981. The European wild rabbit in Mediterranean Spain. In: Proceedings World Lagomorph Conference, K. Myers and C. D. McInnes, eds., Univ. Guelph, Canada.

Southern, H. N. 1940. The ecology and population dynamics of the wild rabbit (Oryctolagus cuniculus). Ann. Appl. Biol. 27: 509-526.

Southern, H. N. 1979. Population processes in small mammals. In: Ecology of Small Mammals, D. M. Stoddart, ed., pp. 63-134, Chapman and Hall.

Starck, D. 1967. Le crâne des Mammifères. In: Traité de Zoologie, Vol. 16/1, P. P. Grassé, ed., pp. 404-549, Masson, Paris.

Sudre, J. 1977. L'évolution du genre Robiacina Sudre 1969, et l'origine des Caenotheriidae; implications systematiques. Geobios Mém. Spec. 1: 213-231.

Sych, L. 1965. Fossil Leporidae from the Pliocene and Pleistocene of Poland. Acta Zool. Cracov. 10: 2-62.

Sych, L. 1971. Mixodontia, a new order of mammals from the Paleocene of Mongolia. Paleont. Pol. 2: 147-158.

Sych, L. 1975. Lagomorpha from the Oligocene of Mongolia. Paleont. Pol. 33: 183-199.

Szalay, F. S. 1977. Phylogenetic relationships and a classification of the eutherian Mammalia. In: Major Patterns in Vertebrate Evolution, M. K. Hecht, P. C. Goody, and B. M. Hecht, eds., pp. 315-374, Plenum Press, New York.

Szalay, F. S. and McKenna, M. C. 1971. Beginning of the age of mammals in Asia: the late Paleocene Gashato fauna, Mongolia. Bull. Amer. Mus. Nat. Hist. 144: 269-318.

Teilhard de Chardin, P. 1926. Mammifères tertiaires de Chine et de Mongolie. Ann. Paléont. 15: 1-51.

Tobien, H. 1963. Zur Gebiss-Entwicklung Tertiärer Lagomorphen (Mamm.) Europas. Notizbl. Hess. L. -Amt. Bodenforsch. Wiesbaden 91: 15-35.

Tobien, H. 1974a. Zur Gebissstruktur, Systematik und Evolution der Genera Amphilagus und Titanomys (Lagomorpha, Mammalia) aus einigen Vorkommen in jüngeren Tertiär Mittel-und Westeuropas. Manizer geowiss. Mitt. 3: 95-214.

Tobien, H. 1974b. The structure of the lagomorphous molar and the origin of the Lagomorpha. Trans. First Int. Theriol. Congr. Moscow 2: 238.

Tobien, H. 1975. Zur Gebissstruktur, Systematik, und Evolution der genera Piezodus, Prolagus, und Ptychoprolagus (Lagomorpha, Mammalia) aus einigen Vorkommen im jüngeren Tertiär Mittel-und Westeuropas. Notizbl. Hess. L.-Amt. Bodenforsch. Wiesbaden 103: 103-186.

Tobien, H. 1978. Brachyodonty and hypsodonty in some Paleogene eurasian lagomorphs. Mainzer Geowiss. Mitt., Mainz 6: 161-175.

Turnbull, W. D. 1970. Mammalian masticatory apparatus. Fieldiana: Geology 18: 149-356.

Valverde, J. A. 1967. Estructura de una comunidad mediterránea de Vertebrados terrestres. Mon. C. Mod. C. S. I. C. 76. Est. Biol. Doñana 1: 1-218.

Vandebroek, G. 1961. The comparative anatomy of the teeth of lower and non specialized mammals. In: International Colloquium on the Evolution of Lower and Non Specialized Mammals, Part 1, G. Vandebroek, ed., pp. 215-320, Kon. Vlaamse Acad. Wetensch. Lett. Schone Kunst. Belgie, Brussels.

Van Valen, L. 1964. A possible origin for rabbits. Evolution 18: 484-491.

Van Valen, L. 1973. Body size and numbers of plants and animals. Evolution 27: 27-35.

Vianey-Liaud, M. 1976. Les Issiodoromyinae (Rodentia, Theridomyidae) de l'Eocene supérieur à l'Oligocene supérieur en Europe occidentale. Palaeovertebrata 7: 5-115.

Vrba, E. S. 1983. The evolution of trends. In: Modalitiés, Rythmes et Mechanismes de l'Evolution Biologique, J. Chaline, ed., pp. 239-246, Col. Int. CNRS, Paris.

Wagner, F. H. 1981. Role of lagomorphs in ecosystems. In: Proceedings World Lagomorph Conference, K. Myers and C. D. McInnes, eds., pp. 668-694, Univ. Guelph, Canada.

Wiley, E. O. 1981. Phylogenetics. The Theory and Practice of Phylogenetic Systematics. Wiley-Interscience, New York.

Williams, G. C. 1966. Adaptation and Natural Selection. Princeton Univ. Press, Princeton.

Wood, A. E. 1940. The mammalian fauna of the White River Oligocene. Part III. Lagomorpha. Trans. Amer. Phil. Soc. 28: 271-362.

Wood, A. E. 1942. Notes on the Paleocene lagomorph *Eurymylus*. Amer. Mus. Novit. 1162: 1-7.

Wood, A. E. 1957. What, if anything, is a rabbit? Evolution 11: 417-425.

Wood, A. E. 1959. Eocene radiation and phylogeny of the rodents. Evolution 13: 354-360.

Xu, Q. 1976. New materials of Anagalidae from the Paleocene of Anhui (A). Vert. PalAsiat. 14: 174-184.

A PHYLOGENY OF RODENTIA AND OTHER EUTHERIAN ORDERS: PARSIMONY ANALYSIS UTILIZING AMINO ACID SEQUENCES OF ALPHA AND BETA HEMOGLOBIN CHAINS

Jeheskel Shoshani[1], Morris Goodman[1,2], John Czelusniak[1,2], and Gerhard Braunitzer[3]

[1]Department of Biological Sciences, Wayne State University Detroit, MI, USA; [2]Department of Anatomy, School of Medicine, Wayne State University, Detroit, MI, USA; [3]Max-Planck-Institute für Biochemie, Abteilung Proteinchemie Martinsried bei München (Munich), West Germany

INTRODUCTION

Recent evidence of a mass extinction of eukaryotic life at the Cretaceous-Paleocene boundary, 65 million years ago, (e.g., Alvarez et al., 1984; Smit and Van der Kaars, 1984) strengthens the traditional view of a bush-like pattern of origin and radiation for the 18 or more orders in the infraclass Eutheria. Apparently several basal eutherian lineages survived the extinction and embarked 65 million years ago on a burst of cladogenetic change during which a series of dichotomous branchings occurred in rapid succession. Clearly it will not be easy to resolve such a bush-like pattern into a discriminating and accurate tree. Nevertheless, the molecular biological analysis of phylogeny promises to contribute significantly to this effort at resolving the branching pattern of this eutherian bush. Protein amino acid sequencing provides a major source of the molecular biological data. In this paper we focus on a genealogical reconstruction, carried out by the parsimony method, on alpha and beta hemoglobin sequences from 83 vertebrate species, 58 of which are placental mammals representing 13 of the 18 or so extant orders of Eutheria. Among the represented orders is Rodentia, the phylogeny of which is the topic of this symposium.

AMINO ACID SEQUENCES ANALYZED

The amino acid sequences analyzed in this study represent the complete alpha and beta hemoglobin chains of each of the 83 species listed in Table 1. The alpha and beta sequences of 55 of these species were employed in a previous study (Goodman et al., 1982). References to the sequences from the 28 new species are listed in Table 1. For purposes of constructing a species phylogeny from the 83 pairs of alpha and beta sequences, a tandemly combined alpha-beta alignment consisting of 289 amino acid or codon positions (143 for alpha sequences and 146 for beta) was employed. Although the typical amniote (mammal, bird, or reptile) alpha chains contain 141 positions, two single position gaps had to be inserted into the alpha portion of the alignment in order to maximize sequence matches between amniote and anamniote alpha chains. All 83 alpha sequences were provisionally assumed to be orthologues and, similarly, all 83 beta sequences. In contrast to paralogously related alpha and beta sequences which arose from a gene duplication that preceded the divergence of Elasmobranchii from the stem of Teleostei and Tetrapoda, the assumed orthologously related sequences - if true orthologues - would have diverged from one another simultaneously with the separations of their species-lineages.

Table 1. Vertebrate species list employed in this study.

Vertebrate species
ELASMOBRANCHII
Heterodontus portusjacksoni
TELEOSTEI
Cyprinus carpio
Carassius auratus
AMPHIBIA
Xenopus laevis[1]
REPTILIA
Alligator mississippiensis[2]
Crocodilus niloticus[2]
Caiman latirostris[2]
AVES
Struthio camelus
Rhea americana[3]
Aquila chrysaetos[4]
Phoenicopterus ruber ruber[5]
Anser anser
Anser indicus
Anseranas semipalmata[6]
AVES (cont'd)
Branta canadensis[7]
Cygnus olor[7]
Cairina moschata[8]
Anas p. platyrhynchos[9]
Phasianus c. colchicus[10]
Gallus gallus
Sturnus vulgaris[11]
MAMMALIA
MONOTREMATA
Tachyglossus aculeatus
Ornithorhynchus anatinus
MARSUPIALIA
Didelphis virginiana
Macropus cangoru
EDENTATA
Dasypus novemcinctus
PROBOSCIDEA
Elephas maximus[12]
HYRACOIDEA
Procavia capensis(=P. habessinica)[13]

Table 1 (continued)

ARTIODACTYLA
Sus scrofa

PERISSODACTYLA
Tapirus terrestris[14]
Ceratotherium simum[15]
Equus caballus
Equus asinus
Equus zebra[16]

ARTIODACTYLA
Hippopotamus amphibius[17]
Camelus dromedarius
Lama peruana

CETACEA
Tursiops truncatus[18]

ARTIODACTYLA
Bos taurus
Capra hircus
Ovis aries

RODENTIA
Cavia porcellus
Spalax ehrenbergi[19]
Ondatra zibethicus[20]
Mus musculus
Rattus rattus

INSECTIVORA
Talpa europaea[21]
Suncus murinus[22]
Erinaceus europaeus

CARNIVORA
Panthera leo
Felis catus
Procyon lotor
Nasua narica
Phoca vitulina[23]
Meles meles
Ursus (Thalarctos) maritimus
Urocyon cinereoargenteus

CARNIVORA (cont'd)
Canis latrans
Canis familiaris

SCANDENTIA
Tupaia glis

CHIROPTERA
Rousettus aegyptiacus[24]

LAGOMORPHA
Oryctolagus cuniculus

PRIMATES
Lemur fulvus
Nycticebus coucang
Loris tardigradus
Tarsius syrichta
Cebus apella
Ateles geoffroyi
Saguinus fuscicollis
Presbytis entellus
Colobus badius
Erythrocebus patas
Cercopithecus aethiops
Cercocebus torquatus
Theropithecus gelada
Papio cynocephalus
Macaca fuscata
Macaca mulatta
Pongo pygmaeus
Gorilla g. gorilla
Pan paniscus[25]
Pan troglodytes
Homo sapiens

Note: Species are listed in the order in which they appear on the complete phylogenetic tree (Figs. 1 and 2 combined), from the root of that tree (Elasmobranchii) to the most extreme major branch (Primates).

Species designated with superscript numbers are new; they are not included in Goodman et al., 1982. References for the hemoglobin sequences of these species include:

1. Kay et al., 1983; Patient et al., 1983.
2. Leclerq et al., 1981.
3. Oberthur et al., 1983b.
4. Oberthur et al., 1983c.
5. Godovac-Zimmermann and Braunitzer, 1984.
6. Oberthur et al., 1983a.
7. Oberthur et al., 1982.
8. Niessing, 1981; Erbil and Niessing, 1982.
9. Godovac-Zimmermann and Braunitzer, 1983.
10. Braunitzer and Godovac, 1982.
11. Oberthur and Braunitzer, 1984.
12. Braunitzer et al., 1982.
13. Kleinschmidt and Braunitzer, 1983a.
14. Mazur and Braunitzer, 1984.
15. Mazur et al., 1982.
16. Mazur and Braunitzer, 1982.
17. Braunitzer et al., 1983.
18. Kleinschmidt and Braunitzer, 1983b.
19. Kleinschmidt et al., 1984.
20. Bieber and Braunitzer, 1983.
21. Kleinschmidt et al., 1981.
22. Maita et al., 1981.
23. Matzuda, unpublished data.
24. Kleinschmidt and Braunitzer, 1982.
25. Goodman et al., 1983.

GENEALOGICAL TREE CONSTRUCTION BY THE PARSIMONY APPROACH

Our maximum parsimony method (Moore et al., 1973; Moore, 1976; Goodman et al., 1979; Czelusniak et al., 1982; Goodman et al., 1982) accounts for evolutionary descent of amino acid sequences by minimizing homoplasy, i.e., by seeking a genealogical arrangement which maximizes genetic likenesses associated with common ancestry, while minimizing incidences of parallel and back mutations. This is done by using the genetic code to represent amino acid sequences as mRNA sequences and by then seeking a tree with a minimum number of nucleotide replacements, or lowest NR length. Since common ancestry, as opposed to convergent evolution, is most likely responsible for extensive matches of nucleotide sequences between species, the tree reconstruction approach based on maximum parsimony is a sensible means of arriving at a preferred genealogical hypothesis. See also Beintema (this volume) and De Jong (this volume).

The strategy developed previously by Goodman et al. (1979) for finding the most parsimonious genealogic tree for a collection of related sequences was followed in the present study. In this strategy, as the first step a matrix of minimum mutation distances is calculated for the amino acid sequences by the method of Fitch and Margoliash (1967). Then an unweighted pair group tree is constructed from this matrix by the clustering algorithm of Sokal and Michener (1958). Another initial dendrogram, the distance Wagner tree, is constructed from the same matrix by the algorithm of Farris (1972). The unweighted pair group and distance Wagner trees provide a means to start the search for the parsimony genealogy without being biased by preconceived ideas on the phylogeny of the species from which the sequence data come. This search is conducted with a branch-swapping maximum parsimony algorithm (Goodman et al., 1979). Input data are the starting tree and the original file of amino acid sequences. The algorithm determines the Nucleotide Replacements(NR) length of the starting tree and of each alternative tree produced by the branch exchanges. The alternative having the lowest NR count is the start of the next round of exchanges. The rounds of exchanges continue until no new trees of lower NR count are found. Since the NR count reached may represent only a local minimum, trees which differ extensively from those previously examined and which test a wide range of phylogenetic possibilities, can now be submitted to the branch-swapping algorithm. This continuation of the heuristic search often reaches a deeper valley approaching the true minimum, in that further searching fails to reach a lower NR score. An important finding is that the lowest NR length trees contain fewer branching errors than the unweighted pair group and distance Wagner trees when judged in terms of accumulated knowledge on the phylogeny of the species represented by the sequences (Goodman et al., 1979).

In theory, if the statistical sample of aligned orthologous sequence positions representing the different species being compared is sufficiently large, the parsimony tree that minimizes only the NR count should yield the correct genealogic arrangement for these species. In the present study using a tandem alignment of just two polypeptide chains, the statistical sample was too meager to ensure reconstruction of the correct species phylogeny. Also, not all the alpha sequences were necessarily orthologues, nor all the beta. It is known that gene duplications occur among alpha genes and also among beta genes. Thus, minimizing NR length alone did not guarantee reconstruction of the correct species phylogeny. Indeed we noted, after finding the trees of lowest NR length, that occasional groupings of sequences violated some genealogical relationships among the species-lineages for which there was a strong *a priori* evidence. Such violations could be due to mistakes of the parsimony method whereby excesses of convergent amino acid residues cause those sequences containing them to be represented as showing a more recent common ancestor than they actually do. Alternatively, some of the presumed orthologues might have been, instead, paralogous descended from gene duplications which preceded the species-lineage splittings. To choose between these possibilities and construct a more accurate tree, we pursued the search for the most parsimonious genealogy as though we were constructing a gene phylogeny. To do so, the parsimony criterion was expanded to encompass not only NR's but also those categories of genic changes which can account for a different branching order of the lineages in a putative gene phylogeny (e.g., a tree of lowest NR length) from the branching order of the species-lineages. As discussed in detail elsewhere (Goodman et al., 1982), these categories are gene duplications (GDs) and deletions and regulatory mutations affecting gene expression, i.e., gene expression events (GEs). We then utilized this expanded parsimony criterion to find the tree or trees which minimized the sum of NRs+GDs+GEs. In practice for each hypothesized 1 GD needed to fit a putative gene phylogeny into the species phylogeny, an average of 3 GEs were also required. We emphasize here that our a priori assumed species phylogeny resembled, as far as orders of Eutheria go, a bush rather than tree, in that none of the eutherian orders were prespecified as being closer to one another than to other orders. Similarly, relatively few intra-ordinal relationships were considered sufficiently well established to be pre-specified. As will be apparent (see below, in connection with results presented in Tables 2 and 3), our "best" tree (the one of the lowest NR+GD+GE length) has an NR length (2298) only 4 NRs greater than the trees at lowest NR length (2294). Not only our "best" tree but also each 2294 or lowest NR length tree violated hardly any of the well established cladistic branching patterns of mammals and other vertebrates.

CLADISTIC RESULTS BASED ON ALPHA AND BETA HEMOGLOBINS

Major Vertebrate Branches and Ordinal Groupings within Eutheria

The 2294 NR tree as well as our "best" genealogical reconstruction (shown in Figs. 1 and 2) agree with the branching pattern of vertebrate phylogeny described by paleontologists (e.g., Romer, 1966 Colbert, 1982). In these trees, Aves and Crocodilia group with Mammalia as their sister class. This Amniota branch joins Amphibia; then the resultant Tetrapoda groups with Teleostei while Elasmobranchii serves as the outgroup of the tetrapods and teleosts. Within Mammalia, infraclass groupings are also in accordance with most workers (Gregory, 1910; Simpson, 1945; Romer, 1966; McKenna, 1975; Lillegraven et al., 1979): Euthera and Metatheria (Marsupialia) join together and the resultant Theria then groups with Prototheria (Monotremata). Phylogenetic relationships within the infraclass Eutheria are subject to controversies due to the "bush-like" splittings of most eutherian orders from one another at the end of the Cretaceous Period. Good reviews on the relationships among eutherian orders can be found in Gregory (1910), Simpson (1945), McKenna (1975), Szalay (1977), and Novacek (1982).

The 2294 NR tree and our "best" tree (the eutherian portion of which is shown in Fig. 2) depict six major branches within the infraclass Eutheria at increasing distances from the eutherian ancestral node: Edentata, condylarth derivatives, Rodentia, Carnivora-Insectivora, Lagomorpha-Chiroptera-Scandentia, and Primates. The condylarth derivatives represented here are Ungulata (Artiodactyla and Perissodactyla), Cetacea, and Paenungulata (Proboscidea and Hyracoidea). The position of the order Rodentia in Fig. 2 proved to

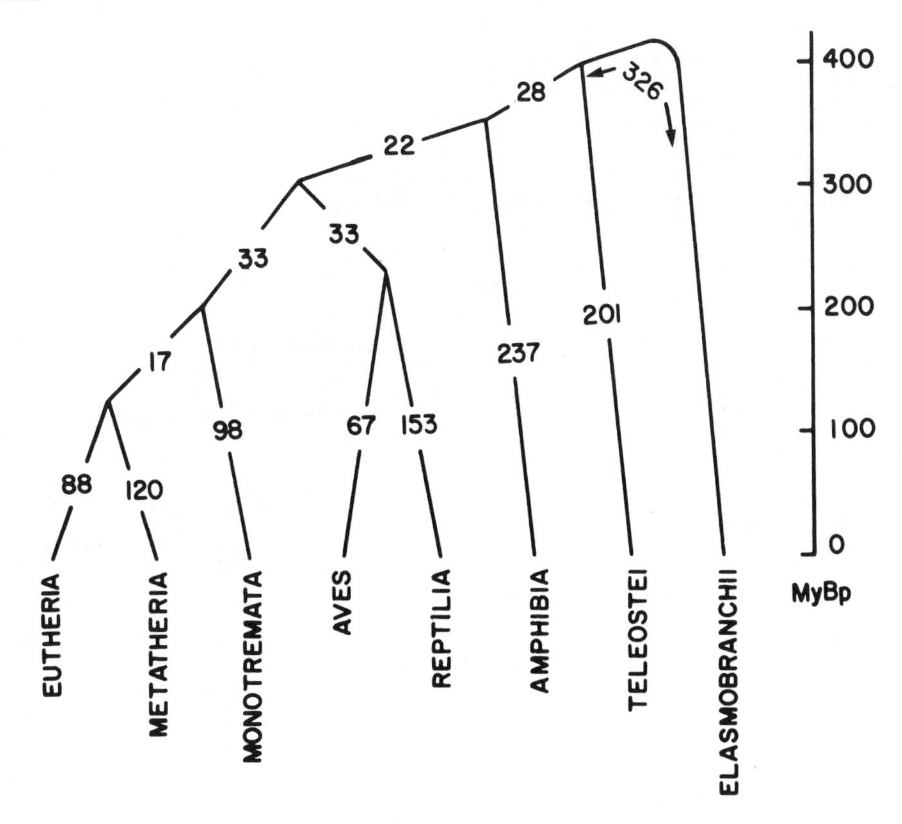

Fig. 1. Maximum parsimony tree based on amino acid sequences of alpha and beta hemoglobin chains of 83 vertebrate species. Augmented NR values are the numbers shown on the links. The mammalian infraclass Eutheria (58 species) is shown in detail in Fig. 2.

be weakly founded because the score of the tree was raised by only two or three NRs on testing other positions for Rodentia (see Table 2.A. and Fig. 3). (However, see Notes added in proof.)

The position of the order Edentata as the most ancient branch of Eutheria (Fig. 2) is a view agreed upon by recent workers (see McKenna, 1975, for cladistic analysis, and Honacki et al., 1982, for mammalian classification). The condylarth branch as presented in Fig. 2 is oversimplified; the complete tree depicts Cetacea (represented by Tursiops truncatus) closer to Bovidae than other artiodactylan species, and Sus scrofa being a sister-group to an Artiodactyla-Cetacea-Perissodactyla branch. Our results within the condylarth branch and those of the lowest NR length tree suggest that Cetacea is more closely related to Artiodactyla than to any other

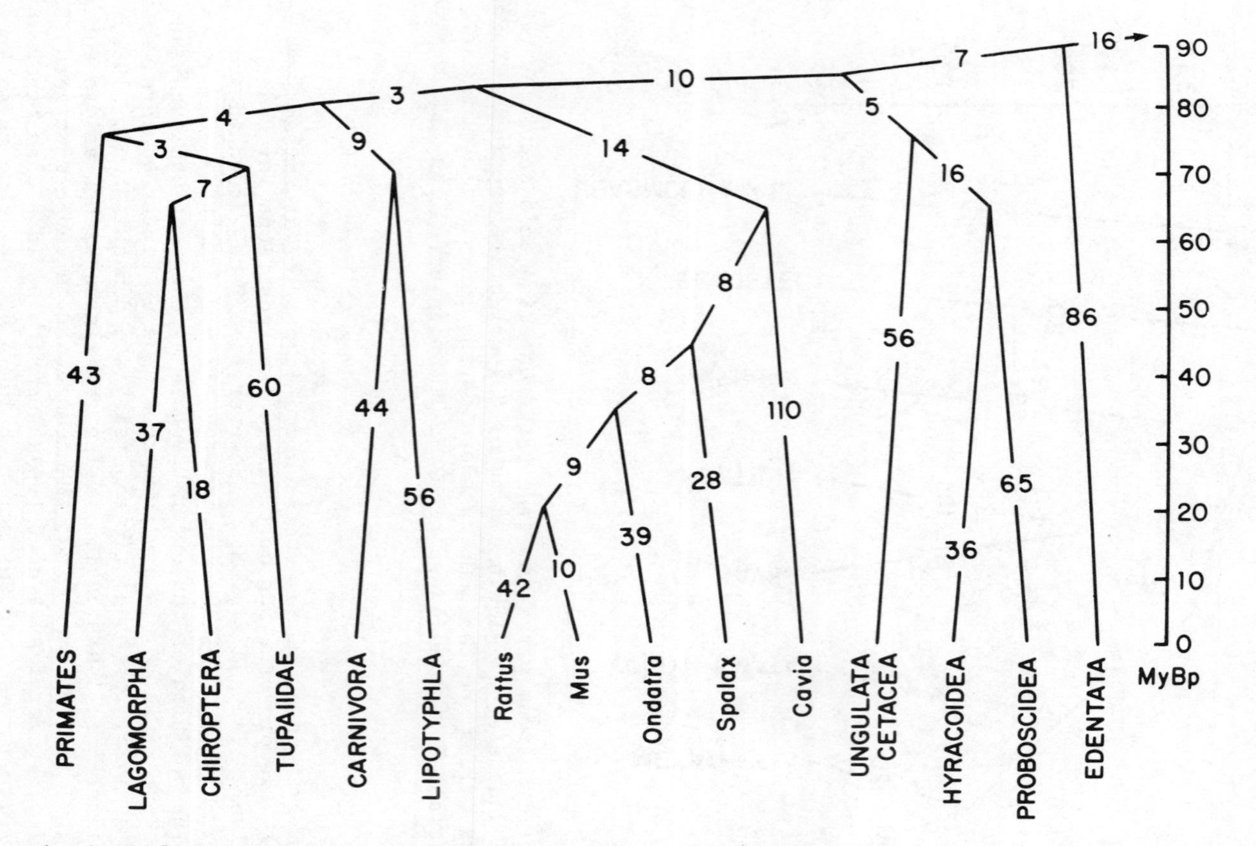

Fig. 2. A simplified eutherian portion of the tree shown in Fig. 1, numbers on links as in Fig. 1. Table 1 includes a complete species list. Ungulata = Artiodactyla and Perissodactyla.

order. Anatomical skeletal data further suggest that the order Cetacea has its origin within the mesonychid Condylarthra and is more closely related to Artiodactyla than to other mammalian taxa (see detailed discussion on the neontological and paleontological evidence in Barnes and Mitchell, 1978, and comments in Savage and Russell, 1983).

The NR scores of trees depicting alternative phylogenetic hypotheses to that of our "best" tree (Figs. 1 and 2) are listed in Tables 2 and 3. Hypotheses tested in Table 3 are based on a complete phylogenetic tree of the Eutheria (see Table 1 for list of species); an unabridged tree can be obtained upon request from the authors.

Table 2. NR lengths of trees representing hypotheses on the intra-ordinal and interordinal relationships of Rodentia.

RELATIONSHIPS	LENGTH
A. Among mammalian orders	
1. As represented in Fig. 2 and Fig. 3A	2298
2. As represented in Fig. 2 but, changes were made to join Rodentia to Edentata (Fig. 3B)	2300
3. As represented in Fig. 2 but, changes were made to join Rodentia to Lagomorpha (Fig. 3C)	2301
B. Within Rodentia	
1. As represented in Fig. 2 and Fig. 4A	2298
2. As represented in Fig. 2 but, with the following changes: <u>Rattus</u>-<u>Mus</u> branch joins <u>Ondatra</u>-<u>Spalax</u> branch then is joined by <u>Cavia</u> (Fig. 4B)	2305
3. As represented in Fig. 2 but, with the following changes: <u>Rattus</u>-<u>Spalax</u> branch joined successively by <u>Mus</u>, <u>Ondatra</u>, and <u>Cavia</u> (Fig. 4C)	2304
4. As represented in Fig. 2 but, with the following changes: <u>Rattus</u>-<u>Mus</u> branch joined <u>Spalax</u>, then is joined by <u>Cavia</u>-<u>Ondatra</u> branch (Fig. 4D)	2295

The paenungulate orders Proboscidea and Hyracoidea group together, forming a relationship supported by other workers (Le Gros Clark and Sonntag, 1926; Simpson, 1945; Romer, 1966; Thenius, 1969; Shoshani et al., 1981; De Jong et al., 1982; Novacek, 1982). It appears that this grouping of Proboscidea and Hyracoidea which is based on alpha and beta hemoglobin is firmly founded, for if we join Hyracoidea to Perissodactyla (a hypothesis proposed by, e.g., Whitworth, 1954 and McKenna, 1975), it costs 14 additional NRs. However, if we join the branch of Probos-idea and Hyracoidea, as a unit, to Perissodactyla, it costs 3 additional NRs (Hypotheses No.'s 3a and 3b in Table 3).

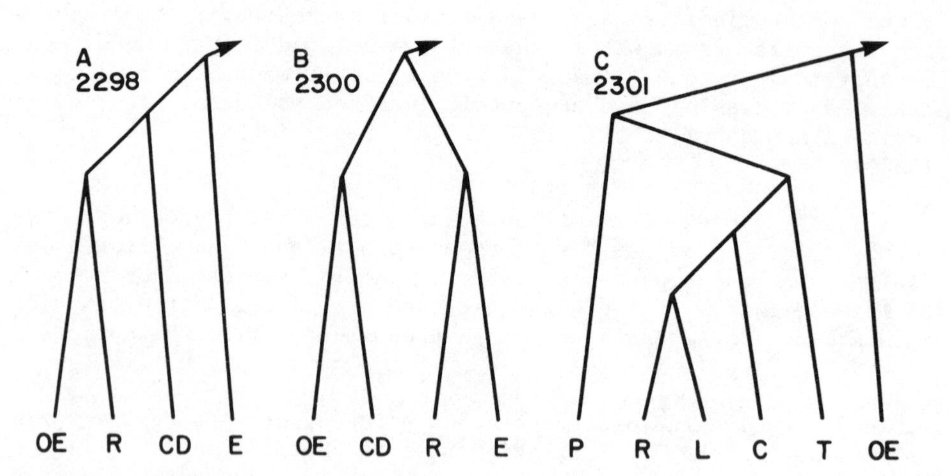

Fig. 3. Alternative phylogenetic position of Rodentia (R) within Eutheria. Fig. 3A is a condensed version of Fig. 2, and Figs. 3B and 3C are simplified relationships. OE = Other Eutheria, R = Rodentia, CD = Condylarth derivatives, E = Edentata, P = Primates, L = Lagomorpha, C = Chiroptera, T = Tupaiidae.

Relationships within the Carnivora were rearranged (in the complete tree) in a 2294 or lowest NR length tree (compare Hypothesis No. 2 to No. 1 in Table 3). The literature on relationships among carnivores (e.g., Leone and Wiens, 1956; Pauly and Wolfe, 1957; Romer, 1966; Sarich, 1975; Tedford, 1976) reveals, however, a wide range of possibilities for branch clustering within the Carnivora, provided members of Feloidea are not grouped with members of Canoidea.

Relationships within Rodentia

Among the five rodent species examined, the most parsimonious of phylogenetically acceptable trees (Fig. 2) depicts the *Rattus-Mus* branch as sharing a last common ancestor at increasingly older times with *Ondatra*, *Spalax* and *Cavia*, respectively. Testing other phylogenetic alternatives within the Rodentia shows that only one other alternative yields a tree of lower NR length (Fig. 4). In this alternative *Rattus-Mus* joins *Spalax*, and then this branch is joined by a *Cavia-Ondatra* branch (see Hypothesis B4 in Table 2 and Fig. 4D). This latter hypothesis, however, strongly violates generally accepted views on rodent phylogeny (Levine and Moody, 1939; Simpson, 1945; Romer, 1966; Carleton, 1984; Carleton and Musser, 1984) in which *Ondatra* (family Arvicolidae) is considered a myomorph, not a

Table 3. NR lengths of trees representing hypotheses on phylogenetic relationships among major mammalian groupings.

RELATIONSHIPS	LENGTH
1. As represented in Fig. 2	2298
2. As in Fig. 2 but, changes were made within Rodentia (see Table 2, B.4.), Carnivora and Aves	2294
3. As in Fig. 2 but with:	
a. Hyracoidea joined to Perissodactyla branch	2312
b. Paenungulata (Proboscidea and Hyracoidea) branch joined to the Perissodactyla branch	2301
4. As in Fig. 2 but with Lagomorpha (Oryctolagus) situated between Edentata and the branch of Ungulata-Cetacea-Paenungulata	2307

hystricognath rodent. Ondatra zibethicus (muskrat) which is traditionally placed in the family Cricetidae, subfamily Microtinae, tribe Microtini (Simpson, 1945), is now placed in a different family, Arvicolidae, along with other microtine rodents (Honacki et al., 1982). Our sample size among the Rodentia is small, and additional sequences from a denser range of rodent species are needed for more accurate parsimony reconstructions. However, based on the data available it appears that the divergence exhibited in the morphology of the microtine rodents may also be manifested in the amino acid sequences of alpha and beta hemoglobin. (See also Notes added in proof.)

COMPARISON TO OTHER MOLECULAR RESULTS

Data presented on amino acid sequences of ribonucleases and insulins (see Beintema, this volume), amino acid sequences of eye lens alpha crystallin A (see De Jong, this volume), and immunological results (see Sarich, this volume) provide no evidence for a Rodentia-Lagomorpha clade. Moreover, in our results with amino acid sequences of alpha and beta hemoglobins, the Lagomorpha joined Chiroptera and the combined branch is then joined by Scandentia (see Fig. 2). In the ribonuclease results of Beintema (this volume), the myomorph rodents separated from the hystricognaths. In the tree of De Jong (1982) the rodents, as a unit, share no synapomorphies. In contrast, our most parsimonious tree for alpha and beta hemoglobins grouped the rodents as a distinct monophyletic group, even though Ondatra joined Cavia (see Fig. 4 and Table 2). Generally speaking, in all four studies relationships within the Rodentia showed that murid (specially Rattus and Mus) and hystricognath rodents grouped in their respective clade. Of the four mentioned studies, only Sarich examined sciuromorph rodents.

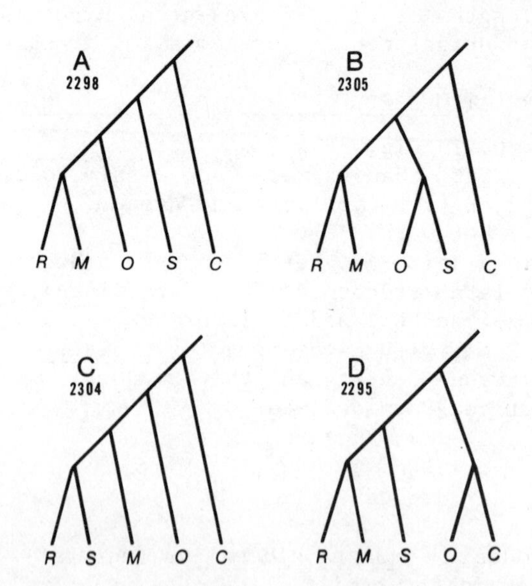

Fig. 4. Alternative relationships among five rodent species: R = Rattus rattus, M = Mus musculus, O = Ondatra zibethicus, S = Spalax ehrenbergi, C = Cavia porcellus. Numbers represent lengths of the total trees. Fig. 4A is the "best" phylogenetic branching arrangement, while Fig. 4D is the most parsimonious (cf. Table 2B).

Results form three out of the five proteins examined by De Jong (this volume), Sarich (this volume), and this paper demonstrated a monophyletic Paenungulata that groups Proboscidea, Hyracoidea, Sirenia, and Tubulidentata. Note, however, that results with alpha lens crystallin A (De Jong, this volume) and immunological results based on albumin (Shoshani et al., 1981), place Paenungulata at considerable distance from Ungulata (Artiodactyla and Perissodactyla), while our most parsimonious tree based on alpha and beta hemoglobins grouped Paenungulata close to Ungulata (Fig. 2). Immunological results (Sarich, this volume) do not provide information on the relationships between Paenungulata and Ungulata. Another relevant finding concerns the close relationship between Cetacea and Ungulata (De Jong, 1982, this volume; Sarich, this volume; Shoshani et al., 1981 and present data). Artiodactyla, Perissodactyla, Cetacea, and Paenungulata are all believed to evolve from the Condylarthra (Simpson, 1945; Romer, 1966; Barnes and Mitchell, 1978; Patterson, 1978).

We shall briefly mention additional results from maximum parsimony trees constructed for carbonic anhydrase sequences (Hewett-Emmett et al.,1984) and globin gene DNA nucleotide sequences (Czelusniak et al., 1982). These place ungulates (Bovidae, also for carbonic anhydrase II, Equus) rather than rodents (Mus) closer to Primates and Lagomorpha. Thus, the weak nature of the interordinal positioning of Rodentia within Eutheria by alpha and beta hemoglobin amino acid seqeunces (Table 2) is again apparent.

EVOLUTIONARY RATES

Evolutionary rates were calculated based on complete trees of Figs. 1 and 2; they are presented as NR%, i.e., NRs per 100 codons per 100 million years. The NR% values shown in Table 4 clearly demonstrate alternating fast and slow rates of hemoglobin evolution from the teleost-tetrapod ancestor along the line of descent to Homo sapiens. Table 5 also shows uneven rates of evolution for the Rodentia, with the line leading to Cavia porcellus evolving 2.3 times faster than the branch of muroid rodents. A larger contrast is observed when we compare evolutionary rates of the murids Rattus rattus and Mus musculus (73.17 vs 17.42, respectively); the former evolved 4.2 times faster than the latter (this faster rate resulted largely from a highly divergent Rattus alpha hemoglobin sequence). Our results and those of Beintema (this volume) also showed that evolutionary rates are faster in Muridae than in other mammals.

Table 4. Rates of evolution from teleost-tetrapod ancestor along the line of descent to Homo.

Evolutionary Period	Age my bp	$\alpha + \beta$ Hb NR%
Teleost-tetrapod ancestor to amniote ancestor	400-300	17
Amniote ancestor to Eutheria ancestor	300-90	11
Eutheria ancestor to Anthropoidea ancestor	90-40	37
Anthropoidea ancestor to Homo-Pan ancestor	40-5	8
Homo-Pan ancestor to Homo sapiens	5-0	0

Table 5. Rates of evolution from eutherian to rodent ancestor to Mus and Rattus.

Evolutionary Period	Age my bp	$\alpha + \beta$ Hb NR%
Eutheria ancestor to Rodentia ancestor	90-65	43
Rodentia along Cavia porcellus branch	65-0	59
Rodentia along Spalax ehrenbergi branch	65-0	19
Rodentia along Ondatra zibethicus branch	65-0	29
Rodentia along Myomorpha branch	65-0	26
Rodentia ancestor to Rattus-Mus ancestor	65-20	19
Rattus-Mus ancestor to Rattus rattus	20-0	73
Rattus-Mus ancestor to Mus musculus	20-0	17

Comparing the evolutionary rates among eutherian taxa as shown in Table 6, we find that the average NR% value for the time period from the ancestor of Eutheria to each ordinal ancestor is twice the average NR% value from each ordinal ancestor to its present day descendants (roughly 43 NR% compared to 23 NR%). Thus, a marked acceleration of rates in the early Eutheria (the preceding Amniota to Eutheria ancestor rate being only 11 NR%, see Table 4) is followed by sharply decelerated rates during the Tertiary. The highest NR% average rate for a eutherian order was that of Rodentia (32 NR%) and the lowest that of Perissodactyla (15 NR%) (Table 6). The steepest deceleration occurred within the Primates on the line to *Homo*; this rate fell during the last 40 million years from 8 NR% to 0 NR%.

Table 6. Rates of evolution from eutherian ancestor to ancestors of eutherian orders to the present.

Evolutionary Period	Age my bp	$\alpha + \beta$ Hb NR%
Eutheria ancestor to Perissodactyla ancestor	90-55	41
Perissodactyla ancestor to the present	55-0	15
Eutheria ancestor to Rodentia ancestor	90-65	43
Rodentia ancestor to the present	65-0	32
Eutheria ancestor to Insectivora ancestor	90-65	50
Insectivora ancestor to the present	65-0	26
Eutheria ancestor to Carnivora ancestor	90-65	59
Carnivora ancestor to the present	65-0	16
Eutheria ancestor to Primate ancestor	90-65	40
Primate ancestor to the present	65-0	20

CONCLUSION

The present body of hemoglobin amino acid sequence data is helping to elucidate the phylogeny of Primates and other eutherian orders but does not constitute a sufficient statistical sample of sequences and species to decisively place Rodentia within Eutheria. Nevertheless, the rate results demonstrate that accelerated evolution occurred during the initial burst of eutherian cladogenesis, and then uneven decelerated evolution occurred in most descending lineages with rodents tending to have faster rates than other mammals.

NOTES ADDED IN PROOF

Two additional alpha and beta hemoglobin sequences of rodents, *Mesocricetus auratus* and *Spermophilus mexicanus*, from one of us (G. Braunitzer) have now been included in the maximum parsimony analysis. The new results show that the tree of lowest NR length has the following branching arrangement: Mus-Rattus branch is joined by Mesocricetus-Ondatra (Cricetidae of Simpson, 1945) branch, and this resultant Muridae-Cricetidae branch is then joined successively by Spalax, Spermophilus, and Cavia. That is, Muroidea and Sciuridae grouped together, and this resultant branch is then joined with Caviidae. These findings are consistent with other molecular studies (Beintema, this volume; De Jong, this volume).

Results based on maximum parsimony analysis of 78 skull foramina characters of 137 mammalian and reptilian species (4 reptiles, 3 mammal-like reptiles, and 130 mammals) by one of us (J. Shoshani) support the Glires hypothesis, and, in contrast to the alpha and beta hemoglobin results presented above, show that Rodentia and Lagomorpha are the most closely related taxa among the 19 eutherian orders studied. Macroscelidea was found to be a sister-group to Glires. This finding (the support for the Glires hypothesis) is in agreement with Novacek's (this volume) analysis of cranial features and of Luckett's (this volume) findings based on dental and fetal membrane development. Evolutionary rates show that Rodentia display the fastest rate among the eutherian orders studied; similar results were obtained for alpha and beta hemoglobin sequences. Based on the above data, evolutionary relationships among the 44 rodent species examined corroborate certain traditional (Simpson, 1945) and most recent (Carleton, 1984) hypotheses on rodent phylogeny. Major branching arrangements within Rodentia show that, on the whole, all three major rodent groups of Simpson (1945) - Sciuromorpha, Myomorpha and Hystricomorpha -are well delineated (except for Pedetes). The hystricognath (N=21) and myomorph (N=13) branches join each other and the resultant branch is then joined by the sciuromorph rodents (N=9), with Pedetes being a sister-group to all other rodent species. Alternate arrangements among these major subgroupings are slightly less parsimonious. Of particular interest is the placement of the Geomyoidea (Geomyidae and Heteromyidae) as a sister-group to the branch of Sciuridae, Castoridae and Aplodontidae; this relationship is in agreement with traditional classifications (e.g., Simpson, 1945), and in disagreement with recent hypotheses, where Geomyoidea is closer to Dipodoidea and Muroidea (e.g., Carleton, 1984; Flynn et al., this volume; Luckett, this volume; Wahlert, this volume). Within the hystricognath rodents, the South American-West Indies complex appear to be more closely related to the African than to the North American genera. Limited time did not allow the senior author to present a full account here; it will be presented elsewhere.

ACKNOWLEDGMENTS

We thank Benjamin Koop and Sandra Lash for reading a draft of this paper and making constructive comments. This study was supported in part by National Science Foundation Grant No. NSF BRS 83-07336.

REFERENCES

Alvarez, W., Kauffman, E.G., Surlyk, F., Alvarez, L.W., Asaro, F., and Michel, H.V. 1984. Impact theory of mass extinctions and the invertebrate fossil record. Science 223:1135-1141.

Barnes, L.G., and Mitchell, E. 1978. Cetacea. In: The Evolution of African Mammals, V.J. Maglio, and H.B.S. Cooke, eds., pp. 582-602, Harvard Univ. Press, Cambridge, Massachusetts.

Bieber, F.A., and Braunitzer, G. 1983. Die Primaerstruktur des Haemoglobins von Bisam (Ondatra zibethica Rodentia). Hoppe-Seyler's Z. Physiol. Chem. 364:1527-1536.

Braunitzer, G., and Godovac, J. 1982. Hemoglobins, XLV: The amino acid sequence of pheasant (Phasianus colchicus colchicus) hemoglobins. Hoppe-Seyler's Z. Physiol. Chem. 363:229-238.

Braunitzer, G., Jelkmann, G.W., Stangl, A., Schrank, B. and Krombach C. 1982. Die primaere struktur des Haemoglobins des indischen Elephanten (Elephas maximus, Proboscidea): B2=Asn. Hoppe-Seyler's Z. Physiol. Chem. 363:683-691.

Braunitzer, G., Wright, P.G., Stangle, A., Schrank, B., and Krombach C. 1983. Amino acid sequence of haemoglobin hippopotamus (Hippopotamus amphibius, Artiodactyla). S. Af. J. Sci. 79: 411-412.

Carleton, M.D. 1984. Introduction to rodents. In: Orders and Families of Recent Mammals, S. Anderson and J.K. Jones, Jr., eds., pp. 255-265, John Wiley and Sons, New York.

Carleton, M.D., and Musser, G.G. 1984. Murid rodents. In: Orders and Families of Recent Mammals, S. Anderson and J.K. Jones, Jr., eds., pp. 290-379, John Wiley and Sons, New York.

Colbert, E.H. 1982. Evolution of the Vertebrates: a History of the Backboned Animals through Time (Third Edition). John Wiley and Sons, Inc. New York.

Czelusniak, J., Goodman, M., Hewett-Emmett, D., Weiss, M.L., Venta, P.J., and Tashian, R.E. 1982. Phylogenetic origins and adaptive evolution of avian and mammalian haemoglobin genes. Nature 298:297-300.

De Jong, W.W. 1982. Eye lens proteins and vertebrate phylogeny. In: Macromolecular Sequences in Systematic and Evolutionary Biology, M. Goodman, ed., pp. 75-114, Plenum Press, New York.

Ebril, G., and Niessing, J. 1982. The complete nucleotide sequence of the duck alpha A-globin gene. Gene 20:211-217.

Farris, J.S. 1972. Estimating phylogenetic trees from distance matrices. Am. Nat. 106:645-668.

Fitch, W.M., and Margoliash, E. 1967. Construction of phylogenetic trees. Science 155:279-284.

Godovac-Zimmermann, J., and Braunitzer, G. 1983. The amino acid sequence of Northern Mallard (Anas platyrhynchos platyrhynchos) hemoglobin. Hoppe-Seyler's Z. Physiol. Chem. 364:665-674.

Godovac-Zimmermann, J., and Braunitzer, G. 1984. The amino acid sequence of alpha A and beta chains from the major hemoglobin component of American Flamingo (Phonicopterus ruber ruber). Hoppe-Seyler's Physiol. Chemie. 365:437-443.

Goodman, M., Braunitzer, G., Stangl, A., and Schrank, R. 1983. Evidence on human origins from haemoglobins of African apes. Nature 303:546-548.

Goodman, M., Czelusniak, J., Moore, G.W., Romero-Herrera, A.E., and Matsuda, G. 1979. Fitting the gene lineage into its species lineage, a parsimony strategy illustrated by cladograms constructed from globin sequences. Syst. Zool. 28:132-163.

Goodman, M., Romero-Herrera, A.E., Dene, H., Czelusniak, J., and Tashian, R.E. 1982. Amino acid sequence evidence on the phylogeny of primates and other eutherians. In: Macromolecular Sequences in Systematic and Evolutionary Biology, M. Goodman, ed., pp. 115-191, Plenum Press, New York.

Gregory, W.K. 1910. The orders of mammals. Bull. Amer. Mus. Nat. Hist. 27:1-524.

Hewett-Emmett, D., Hopkins, P., Tashian, R.E., and Czelusniak, J. 1984. Origins and molecular evolution of the carbonic anhydrase isoenzymes. Ann. N.Y. Acad. Sci. 429:338-358.

Honacki, J.H., Kinman, K.E., and Koeppl, J.W. (eds.). 1982. Mammal species of the world: a taxonomic and geographic reference. Allen Press, Inc. and The Association of Systematic Collections, Lawrence, Kansas.

Kay, R.M., Harris, R., Patient, R.K., and Williams, J.W. 1983. Complete nucleotide sequence of a cloned cDNA derived from the major adult α-globin mRNA of *X. laevis*. Nucleic Acid Res. 11: 1537-1542.

Kleinschmidt, T., and Braunitzer, G. 1982. Die Primaerstruktur des Haemoglobins vom Aegyptischen Flughund (*Rousettus aegyptiacus*, Chiroptera). Hoppe-Seyler's Z. Physiol. Chem. 363:1209-1215.

Kleinschmidt, T., and Braunitzer, G. 1983a. Die Primaerstruktur des Haemoglobins vom Abessinischen Klippschlieferr (*Procavia habessinica*): Insertion von Glutamin in den a-ketten. Hoppe-Seyler's Z. Physiol. Chem. 364:1303-1313.

Kleinschmidt, T., and Braunitzer, G. 1983b. Die Primaerstruktur des Haemoglobins vom Grossen tuemmler (*Tursiops truncatus*, Cetacea). Biomed. Biochim. Acta 42:685-695.

Kleinschmidt, T., Jelkmann, W., and Braunitzer, G. 1981. Die Primaerstruktur des Haemoglobins des Maulwurfs (*Talpa europaea*). Hoppe-Seyler's Z. Physiol. Chem. 362:1263-1272.

Kleinschmidt, T., Nevo, E., and Braunitzer, G. 1984. The primary structure of the hemoglobin of the mole rat (*Spalax ehrenbergi*, Rodentia, Chromosome species 60). Hoppe-Seyler's Z. Physiol. Chem. 365:531-537.

Leclerq, F., Schnek, A., Braunitzer, G., Stangl, A., and Schrank, B. 1981. Direct reciprocal allosteric interaction of oxygen and hydrogen carbonate sequence of the hemoglobins of the caiman (*Caiman crocodylus*), the Nile crocodile (*Crocodylus niloticus*) and the Mississippi crocodile (*Alligator mississippiensis*). Hoppe-Seyler's Z. Physiol. Chem. 362:1151-1158.

Le Gros Clark, W.E., and Sonntag, C.F. 1926. A monograph of Orycteropus afer.--III. The Skull. The skeleton of the trunk and limbs. General summary. Proc. Zool. Soc. London 30:445-48

Leone, C.A., and Wiens, A.L. 1956. Comparative serology of carnivores. J. Mammal. 37:11-23.

Levine, H.P., and Moody, P.A. 1939. Serological investigation of rodent relationships. Physiol. Zool. 12:400-411.

Lillegraven, J.A., Kielan-Jaworowska, Z., and Clemens, W.A. (eds.). 1979. The First Two-thirds of Mammalian History: Mesozoic Mammals. University of California Press, Berkeley.

Maita, T., Matsuda, G., Takanaka, O., and Takahashi, K. 1981. The primary structure of adult hemoglobin of musk shrew (Suncus murinus). Hoppe-Seyler's Z. Physiol. Chem. 362:1465-1474.

Mazur, G., and Braunitzer, G. 1982. Haemoglobine, XLIV. Perissodactyla: Die sequenz der Haemoglobine vom Wildesel (Equus hemionus kulan) und zebra (Equus zebra). Hoppe-Seyler's Z. Physiol. Chem. 363:59-71.

Mazur,G., and Braunitzer, G. 1984. Perissodactyla: Die Primaerstruktur der Haemoglobine eines Flachlandtapirs (Tapirus terrestris), $Beta_2$=Glutaminsaeure. Hoppe-Seyler's Z. Physiol. Chem. 365:1077-1106.

Mazur, G., Braunitzer, G., and Wright, P.G. 1982. Die Primaerstruktur des Haemoglobins vom Breitmaulnashorn (Ceratotherium simum, Perissodactyla): $beta_2$ Glu. Hoppe-Seyler's Z. Physiol. Chem. 363:1077-1085.

McKenna, M.C. 1975. Toward a phylogenetic classification of the Mammalia. In: Phylogeny of the Primates: a Multidisciplinary Approach, W.P. Luckett, and F.S. Szalay, eds. pp. 21-46, Plenum Press, New York.

Moore, G.W. 1976. Proof for the maximum parsimony ("Red King") algorithm. In: Molecular Anthropology: Genes and Proteins in the Evolutionary Ascent of the Primates, M. Goodman and R.E. Tashian, eds., pp. 117-137, Plenum Press, New York.

Moore, G.W., Barnabas, J., and Goodman, M. 1973. A method for constructing maximum parsimony ancestral amino acid sequences on given network. J. Theor. Biol. 38:459-485.

Niessing, J. 1981. Molecular cloning and nucleotide sequence analysis of adult duck beta-globin cDNA. Biochem. Int. 2:113-120.

Novacek, M.J. 1982. Information for molecular studies from anatomical and fossil evidence on higher eutherian phylogeny. In Macromolecular Sequences in Systematic and Evolutionary Biolog M. Goodman, ed., pp. 3-41, Plenum Press, New York.

Oberthur, W., and Braunitzer, G. 1984. Haemoglobine vom Gemeinen star (Sturnus vulgaris, Passeriformes), die Primaerstruktur de A-, D-, and b-ketten. Hoppe-Seyler's Z. Physiol. Chem. 365: 159-173.

Oberthur, W., Braunitzer, G., Baumann, R., and Wright, P.G. 1983b. Die Primaerstruktur der A- and B- ketten der Hauptkomponeuten der Haemoglobine des Strausses (Struthio camelus) und des

Nandus (Rhea americana) (Struthioformes). Hoppe-Seyler's Z. Physiol. Chem. 364:119-134.

Oberthur, W., Braunitzer, G., Grimm, F., and Koesters, J. 1983c. Haemoglobine des steinadlers (Aquila chrysaetos, Accipitriformes): Die Aminosaeure-sequenz der A- and B-ketter der Hauptkomponente. Hoppe-Seyler's Z. Physiol. Chem. 364:851-858.

Oberthur, W., Godovac-Zimmermann, J., Braunitzer, G., and Wiesner, H. 1982. The amino acid sequence of Canada goose (Branta canadensis) and mute swan (Cygnus olor) hemoglobins: two different species with identical beta-chains. Hoppe-Seyler's Z. Physiol. Chem. 363:777-787.

Oberthur, W., Wiesner, H., Braunitzer, G. 1983a. Die Primaerstruktur der A- and B- ketten der Hauptkomponente der Haemoglobine der spaltfussgans (Anseranas semipalmata, Anatidae). Hoppe-Seyler's Z. Physiol. Chem. 364-51-59.

Patient, R.K., Harris, R., Walmsley, M.E., and Williams, J.G. 1983. The complete nucleotide sequence of the major adult beta globin gene of Xenopus laevis. J. Biological Chemistry 258: 8521-8523.

Patterson, B. 1978. Pholidota and Tubulidentata. In: Evolution of African Mammals, V.J. Maglio and H.B.S. Cooke, eds., pp. 268-278, Harvard University Press, Cambridge, Mass.

Pauly, L.K., and Wolfe, H.R. 1957. Serological relationships among members of the order Carnivora. Zoologica 42:159-166.

Romer, A.S. 1966. Vertebrate Paleontology (Third Ed.). University of Chicago Press, Chicago.

Sarich, V.M. 1975. Pinniped systematics: immunological comparisons of their albumin and transferrins. Amer. Zool. 15:826.

Savage, D.E., and Russell, D.E. 1983. Mammalian Paleofaunas of the World. Addison-Wesley Advanced Book Program, Reading, Mass.

Shoshani, J., Goodman, M., Barnhart, M.I., Prychodko, W., Vereshchagin, N.K., and Mikhelson, V.M. 1981. Blood cells and proteins in the Magadan mammoth calf: immunodiffusion comparisons of Mammuthus to extant paenungulates and tissue ultrastructure. In: Magadan Baby Mammoth, Mammuthus primigenius (Blumenbach), N.K. Vereshchagin, and V.M. Mikhelson, eds., pp. 191-220, "Nauka" Publishers, Leningrad.

Simpson, G.G. 1945. The principles of classification and a classification of mammals. Bull. Amer. Mus. Nat. Hist. 85:1-350.

Smit, J., and Van Der Kaars, S. 1984. Terminal Cretaceous extinctions in the Hell Creek area, Montana: compatible with catastrophic extinction. Science 223:1117-1179.

Sokal, R.R., and Michener, C.D. 1958. A statistical method for evaluating systematic relationships. Kansas Univ. Sci. Bull. 382:1409-1438.

Szalay, F.S. 1977. Phylogenetic relationships and a classification of the eutherian Mammalia. In: Major Patterns in Vertebrate Evolution, M.K. Hecht, P.C. Goody, and B.M. Hecht, eds., pp. 315-374, Plenum Publishing Co., New York.

Tedford, R.H. 1976. Relationship of pinnipeds to other carnivores (Mammalia). Syst. Zool. 25:363-374.

Thenius, E. 1969. Phylogenie der Mammalia: Stammesgeschichte der Saeugetiere (eischliesslich der Hominiden). Walter de Gruyter and Co., Berlin.

Whitworth, T. 1954. The Miocene hyracoids of East Africa with some observations on the order Hyracoidea. Fossil Mammals of Africa, British Museum (Nat. Hist.) 7:1-58.

SUPERORDINAL AFFINITIES OF RODENTIA STUDIED BY SEQUENCE ANALYSIS OF EYE LENS PROTEIN

Wilfried W. de Jong

Department of Biochemistry
University of Nijmegen
6525 EZ Nijmegen, The Netherlands

INTRODUCTION

The development of techniques to determine the amino acid sequences of proteins and the nucleotide sequences of genes has dramatically increased our insight into the structure and function of these macromolecules. These techniques also provide the ultimate tools to unravel evolutionary processes at the molecular level. Much is indeed known already about the evolutionary changes and mutational processes in proteins and their genes (Wilson et al., 1977; Sigman and Brazier, 1980; Dover and Flavell, 1982).

Because gene structure and regulation, and the function of proteins, are in fact so crucial in biological systems, the impression is sometimes given that the study of genes and proteins will eventually provide the complete understanding of life and evolution itself. Yet, within the context of this multidisciplinary volume, it is appropriate to recognize that virtually nothing is known about the causal relations between structure and function of genes and proteins on the one hand, and the anatomy, morphology, development, and behavior of organisms on the other. There still is a fundamental gap between the study of life and evolution at the molecular level and at the organismal level. Only minor advances are beginning to be made to bridge this gap (e.g., Slack, 1984). It is necessary for molecular biologists to realize that the selective forces in evolution act primarily at the organismal level.

Although our current knowledge of macromolecular sequences has indeed not yet contributed much to a better understanding of morphological evolution, this type of data is in fact ideally suited for for the analysis of phylogenetic relationships among organisms.

Given the fact that mutational events in the course of evolution have left their direct traces in the nucleotide sequences of the chromosomal DNA of present-day organisms, and hence in the amino acid sequences of their proteins, the comparison of homologous DNA and protein sequences in different species should, in principle, allow an accurate reconstruction of their phylogenetic relationships

In this contribution I will demonstrate the possibilities and limitations of protein sequence data in phylogenetic analysis, as exemplified by the comparative study of the eye lens protein α-crystallin in rodents and other mammals.

THE EYE LENS AND ITS PROTEINS

The vertebrate lens is an organ with many unique properties. It is fully transparent and devoid of blood vessels and innervation. It is enclosed in a thin and elastic collagenous capsule, and its plasticity allows the change of shape required for accommodation in many species. A single layer of epithelial cells covers the anterior side of the lens. At the equator of the lens these epithelial cells differentiate into greatly elongated fiber cells that form the body of the lens. In the process of fiber cell formation the nucleu and other organelles are lost. The soft cortex of the lens contains the most recently formed fiber cells, whereas the more solid nucleus of the lens is composed of the older fiber cells. Because no lens cells are ever shed or broken down during life, the embryonic lens cells are still present as the innermost part of the adult lens. All these properties make the eye lens an attractive object for studies in embryology, differentiation, ageing, protein structure, and gene expression (Bloemendal, 1981).

The eye lens is particularly rich in protein, of which the composition and properties have been studied in many vertebrate species (Clayton, 1974; De Jong, 1981). The water-soluble structural proteins of the eye lens are called the crystallins, of which three types can be distinguished in the mammalian lens, the α-, β- and γ-crystallins. Avian and reptilian lenses are characterized by the presence of the unrelated δ-crystallin, while they are lacking γ-crystallin (Piatigorsky, 1984). The different crystallins can be distinguished on the basis of molecular weight, charge, amino acid sequence, and subunit composition. Recently, the genes for α-, β-, γ-, and δ-crystallins have also been cloned and analyzed (reviewed by Piatigorsky, in press).

Of all lens proteins, α-crystallin is the best studied from a comparative and evolutionary point of view (De Jong, 1981). In most mammalian species it makes up approximately 30% of the total lens protein. It occurs as large aggregates, with an average molecular weight of 800,000. These aggregates are composed of two types of polypeptide chains: the acidic αA chain, and the more basic αB chain. The αA and αB chains have, in most vertebrates, a length of 173 and 175 amino acid residues, respectively. Their amino acid sequences show, in the ox, 57% homology, reflecting that the genes coding for αA and αB have originated early in vertebrate evolution by duplication of a common ancestral gene.

The evolutionary origin of the crystallin genes is, like that of the eye lens itself, shrouded in mystery. Only in the case of α-crystallin is there an intriguing sequence relationship with another group of proteins, the heat shock proteins (Ingolia and Craig, 1982). These ubiquitous proteins are induced in all eukaryotic and even prokaryotic cells under conditions of stress (like sudden elevation of temperature), and are thought to maintain cellular homeostasis to survive such conditions. The biological significance of this structural relationship between α-crystallin and the heat shock proteins is completely obscure. In fact, the function of α-crystallin itself in the eye lens is poorly understood, at best being described as contributing to the appropriate intracellular conditions, in order to guarantee transparency and proper diffraction of light by the lens fiber cells.

ANALYSIS OF α-CRYSTALLIN SEQUENCE DATA

A description of the biochemical techniques used to isolate α-crystallin and to determine the amino acid sequence of its composing αA and αB chains can be found in De Jong et al. (1984). The sequences of the αA chains of 43 mammalian species, 21 birds, two reptiles, two frogs, and a dogfish have been determined (De Jong, 1982; De Jong et al., 1984; Stapel et al., 1984).

These studies have shown that α-crystallin A evolves at an average rate of 3 amino acid replacements per 100 residues in 100 millon years. This is approximately as slow as the rate of evolution of cytochrome c, and considerably slower than the rate of change of hemoglobin, myoglobin, or pancreatic ribonuclease (Wilson et al., 1977). As in other proteins, the rate of replacements in α-crystallin evolution may be considerably accelerated or decelerated in individual lineages.

The relatively slow rate of evolution of α-crystallin makes it a suitable protein to study phylogenetic relationships at the level of higher taxonomic categories, like orders and subclasses. The results and significance of such phylogenetic analysis are obviously completely dependent on the number and types of mutational events which have occurred in the ancestral lines of the species under investigation. The interpretation of differences and similarities observed between the investigated amino acid sequences is seriously hampered by the frequent occurrence of parallel and back substitutions during protein evolution. Although several sophisticated computer programs using different algorithms, are available for the construction of phylogenetic trees from homologus sequence data (e.g. Fitch and Margoliash, 1967; Goodman et al., 1979; Hendy et al., 1978), these programs obviously cannot differentiate between genuine synapomorphous and parallel substitutions. Although indispensable in the handling of larger sets of sequence data, it is important to realize that such programs indiscriminatingly lump together all differences and similarities. This absolute objectivity is of course the strength, but also a weakness, of the method.

The additional information which is present in amino acid sequences, but not exploited by currently available computer programs, can be demonstrated by looking at some selected positions in mammalian α-crystallin A sequences. In Table 1 the species are arranged according to the generally accepted orders to which they belong. Full names of the investigated species and all sequence information (apart from Elephantulus) can be found in De Jong and Goodman (1982) In most cases the direction of a replacement can unambiguously be established. This can, for instance, clearly be seen in position 13 where Ala is primitive, present in almost all mammals, and Thr and Pro are derived.

It is also clear that not all replacements provide phylogenetic information of the same quality. The replacement 127 Ser → Thr seems to be a good character: it has been found exclusively in perissodactyls, and thus it is likely to be a real synapomorphous character for this order. Even more convincing are the replacements 51 Ser → Pro and 52 Leu → Val, which both are found only in the pinniped carnivores (seal and sea lion), lending strong support to their monophyly. Since deletions and insertions of amino acids are much rarer in evolution than amino acid replacements, the deletion at position 147 is a convincing synapomorphy of sloth and tamandua.

Poor indicators of phylogenetic relationships are the replacements 13 Ala → Thr and 61 Ile → Val, which both occur scattered among several orders, and thus by necessity must be due to independent, parallel mutational events. Similarly, at position 150 the residues Met, Leu and Val show a rather haphazardous distribution over the different orders.

Supraordinal relationships are more difficult to establish, both because the periods of common ancestry of mammalian orders during their evolutionary radiation must have been relatively short, and because such replacements which may have occurred in these periods of common ancestry could have been easily erased by subsequent superimposed replacements. Yet, the replacement 70 Lys → Gln is a good example of an apparent shared derived character which groups together the orders to which the paenungulates elephant, hyrax, manatee, and aardvark belong (De Jong and Goodman, 1982). Also, the replacements

3 Ile → Val and 4 Thr → Ala are valuable characters in the analysis of interordinal relationships.

In conclusion, the branch lengths (in numbers of unspecified nucleotide substitutions or amino acid replacements) usually provided in computer-constructed phylogenetic trees can become much more meaningful if one tries to analyze the actual nature of the mutational events on which the particular branching patterns are based. The more unique replacements or deletions that can be assigned to a particular branch in a phylogenetic tree, the stronger the probability that such a branch is an authentic reflection of common ancestry. Short branches based on ubiquitous replacements are of little value in phylogenetic analysis, and may just lead to false suggestions and increased confusion.

Obviously, and unfortunately, finding reliable indicators of phylogenetic relationships, i.e., real synapomorphies, becomes more and more unlikely if these relationships are the result of shorter periods of common ancestry in the distant evolutionary past. In those cases it will be a matter of good luck to pick up really convincing synapomorphies among the evolutionary noise caused by the mass of parallel and back replacements.

α-CRYSTALLIN SEQUENCES AND RODENT AFFINITIES

The total data sets of the sequence differences between the 68 investigated α-crystallin A chains can be found elsewhere (De Jong, 1982; De Jong et al., 1984; Stapel et al., 1984). Analysis of these data, using maximum parsimony procedures (Goodman et al., 1979) has provided valuable information about certain phylogenetic relationships, like those of the aardvark (De Jong et al., 1981) and the ratite birds (Stapel et al., 1984).

The αA chains of six rodent species are known (Table 1). Those of rat (*Rattus norvegicus*), gerbil (*Meriones unguiculatus*), hamster (*Mesocricetus auratus*), guinea pig (*Cavia porcellus*), and springhass (*Pedetes cafer*) have been determined by protein sequence analysis, while those of mouse (*Mus musculus*) and also the rat (*Rattus norvegicus*) have been deduced from the nucleotide sequences of the cloned copy-DNAs of their α-crystallin A mRNAs (Moormann et al., 1981; King and Piatigorsky, 1983). Actually, the αA chains of rat, mouse, gerbil, and hamster are all completely identical, and differ only at a single position from the αA chains of guinea pig and springhaas. The latter two have Gln at position 90, where the other rodents have Leu. This replacement 90 Gln → Leu is phylogenetically not very informative, since it occurs independently in several mammalian groups (Table 1).

In fact, the six investigated rodent αA sequences do not contain

Table 1

```
POSITION NR                          1111111111
IN ALPHA-A CHAIN                   15567990222444555
                                 3431210011237267038
                                    ***

MINKE WHALE                      IAASLIKQENSNSSVPMGA
PORPOISE                         |||||||||||||||T|||

HORSE                            ||||||||||||T|I||||
TAPIR                            ||T|||||||||T|||L||
RHINOCEROS                       ||T||V||||||T|I||||

PIG                              |||||V||||||||||V||
GIRAFFE, HIPPO                   ||||||||||||||I|V||
OX                               ||T|||||||||||I|V||
CAMEL                            |||||||L||||||I|V||

DOG, CAT                         |||||||L||||||||V||
BEAR                             |||||||L|||||||||||
MINK                             |||||V|LQ|||||||V||
SEAL, SEA LION                   |||PV||L||||||||V||

PIKA                             VT|||||||S|||||QL||
RABBIT                           VTT||||||||||||QL||

TUPAIA                           VT|||||L|||||||QL||

ELEPHANT SHREW                   ..||||.LD||||.|Q|S|

RAT, HAMSTER, GERBIL             VT|||||L|||||||QL||
GUINEA PIG, SPRINGHAAS           VT|||||||||||||QL||

HEDGEHOG                         VT|||||L|SPSA||AL||

LEMUR                            VTP||V|||||||||QL||
GALAGO, POTTO                    VTP||V|||||||||QL||
RHESUS MONKEY                    VTT|||||D|||||IQL-|
HUMAN                            VTT|||||D||||CIQL-|

ELEPHANT                         VT||||Q|D||||CIQ|S|
HYRAX                            VT||||QLD||||C|Q|S|
MANATEE                          VT||||QLD||||C|Q|S|
AARDVARK                         VT||||QLD||||C|Q|||

2-TOED SLOTH                     VT|||||LD|TA||I-VST
TAMANDUA                         VT||||RLD|TA||L-VST

KANGAROO                         |T|||||LDS||||IH|SS
OPOSSUM                          |T||||RLDS||||IH|SS
```

a single synapomorphous replacement which would group them together as rodents. As a consequence, the most parsimonious cladograms that can be constructed on the basis of the mammalian αA sequences fail to resolve the order Rodentia from the other mammalian orders. These cladograms group the rodents in a cluster together with the investigated representatives of the primates, lagomorphs, insectivores, and tree shrews (Fig. 1). Although the topologies of these cladograms differ considerably, they have as a common feature that the cluster of these orders is set apart from the ungulates, whales, carnivores, pangolins, and bats on one hand, and the edentates, paenungulates, and marsupials on the other. The α-crystallin sequences thus support the view that the closest relatives of the rodents are to be found among the primates, lagomorphs, and insectivores, either as a clade, or as a grade in the lineage to carnivores and ungulates.

The most parsimonious trees shown in Figure 1 obviously violate several well-established opinions about mammalian phylogeny. Most disturbing indeed is the mixing up of species belonging to different orders, causing for instance the separation of rabbit from pika, and rat from guinea pig.

This mainly is a reflection of the fact that the αA sequences just have not accepted enough evolutionary changes in these lineages to be informative. The few replacements that have occurred are moreover of the recurrent type, like 13 Ala ↔ Thr and 90 Leu ↔ Gln. Thus, α-crystallin has not the resolving power to unravel the relationships between the rodents and their relatives.

Goodman et al. (1979) have developed a computer program which combines the protein sequence data with other biological evidence about phylogenetic relationships. Because there is overwhelming evidence that the order Rodentia is a natural, monophyletic group, and so are the orders Lagomorpha and Primates, one can limit the freedom

---

Table 1. Replacements in mammalian αA sequences which are informative for the study of interordinal relationships. Only those positions in the αA chains are included (apart from 51, 52, and 61) at which different residues occur in at least three different orders. Species with identical αA sequences are shown on the same line. The vertical lines indicate where residues in the sequences of the mammalian αA chains are identical to those in the topmost sequence. The one-letter notation for amino acids has been used: A, Ala; C, Cys; D, Asp; E, Glu; G, Gly; H, His; I, Ile; K, Lys; L, Leu; M, Met; N, Asn; P, Pro; Q, Gln; S, Ser; T, Thr; V, Val; (•) not determined; (-) a deletion.

* Position 51, 52, and 61 are included, not for their interordinal informational content, but to illustrate some aspects of protein evolution as discussed in the text.

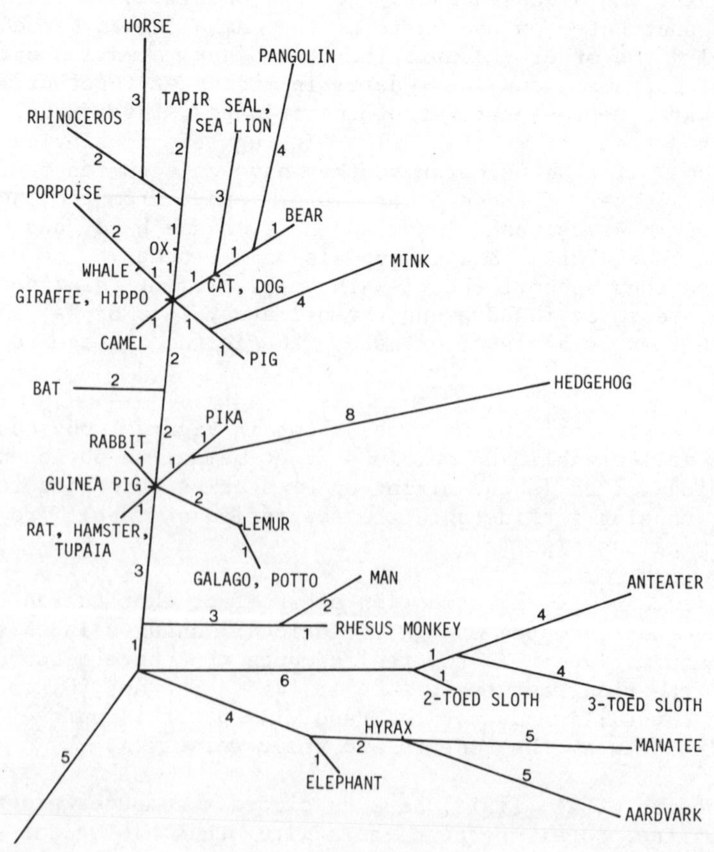

Fig. 1. Three most parsimonious cladograms (a, b, c) constructed on the basis of the known eutherian α-crystallin A sequences. These cladograms each require 152 nucleotide substitutions, including the outgroup sequences of the αA chains of kangaroo, opossum, chicken, and frog (not shown in these trees). The computer methods used to construct these trees have been described in Goodman et al. (1979) and De Jong and Goodman (1982). Springhaas and gerbil are not shown in this figure, but are identical in position to guinea pig and rat, respectively. Branch lengths are given in numbers of inferred nucleotide substitutions. These three cladograms are representative examples of the major equally parsimonious branching arrangements that can be obtained from the data set of α-crystallin A sequences. They emphasize that such most parsimonious topologies can show considerable differences, but demonstrate also the consistent grouping of certain species.

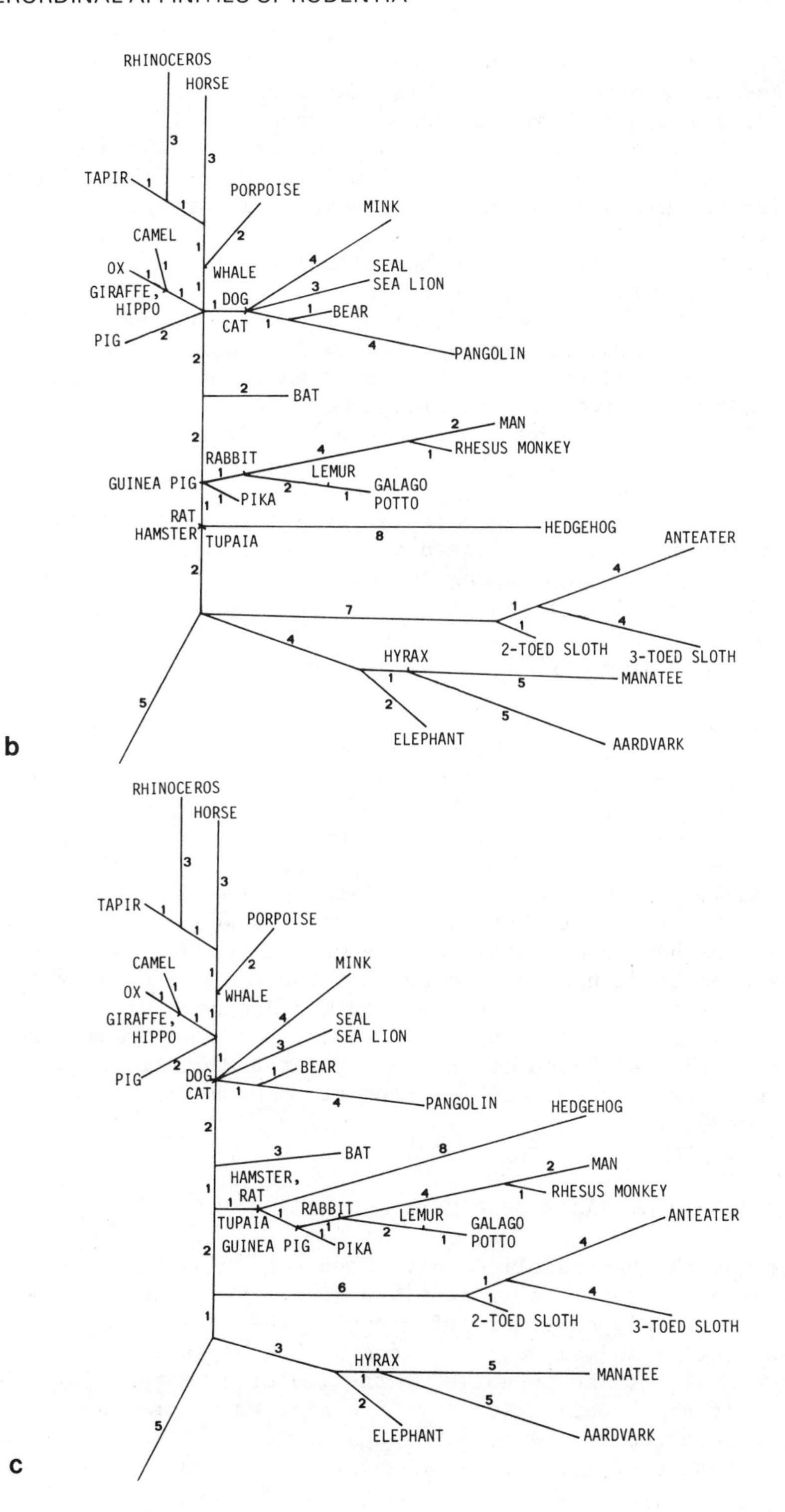
RHINOCEROS
HORSE
TAPIR
PORPOISE
MINK
CAMEL
OX
WHALE
SEAL
SEA LION
GIRAFFE, HIPPO
DOG
CAT
BEAR
PIG
PANGOLIN
BAT
MAN
RHESUS MONKEY
RABBIT
LEMUR
GUINEA PIG
PIKA
GALAGO
POTTO
RAT
HAMSTER
TUPAIA
HEDGEHOG
ANTEATER
2-TOED SLOTH
3-TOED SLOTH
HYRAX
MANATEE
ELEPHANT
AARDVARK
b
RHINOCEROS
HORSE
TAPIR
PORPOISE
CAMEL
MINK
OX
WHALE
GIRAFFE, HIPPO
SEAL
SEA LION
BEAR
PIG
DOG
CAT
PANGOLIN
HEDGEHOG
BAT
MAN
HAMSTER, RAT
RHESUS MONKEY
RABBIT
TUPAIA
LEMUR
GALAGO
POTTO
GUINEA PIG
PIKA
ANTEATER
2-TOED SLOTH
3-TOED SLOTH
HYRAX
MANATEE
ELEPHANT
AARDVARK
c

of the tree-construction program by forcing the species together in their respective orders. A most parsimonious tree constructed under this restriction, which is biologically more acceptable, is shown in Figure 2. Because of the forceful separation of branches, which are joined in Figure 1, this tree requires five additional replacements to realize the grouping of the mammals in their orders.

While Figure 1 illustrates the actual phylogenetic information that can be deduced from the mammalian αA sequences, Figure 2 provides a more realistic representation of the evolutionary history of this protein. It emphasizes that the evolutionary change in rodent α-crystallin has indeed been much less than in other taxa, such as primates and insectivores. It also demonstrates that the αA sequences nicely resolve the divergence of the Strepsirhini and Anthropoidea within the primates (although they fail again to group them in a monophyletic order). The single replacement 150 Val → Leu which in this tree holds together the rodents, primates, lagomorphs, insectivores, and tree shrews, is indeed a rather unique shared derived character for this group (among the other vertebrates 150 Leu has been found only in tapir and frog), but position 150 is a variable one (Table 1), which makes it a rather weak synapomorphy. As stated already, also residue 90 Leu or Gln is not a very reliable character.

Obviously, only a few representatives of the approximately 32 families of Recent rodents have been included in our study. The mouse, rat, hamster and gerbil moreover belong to the related families Muridae and Cricetidae. It thus may be of interest to analyze the αA sequences of some species of other major rodent groups, like Gliroidea, Sciuroidea, Castoroidea, and Dipodoidea (Hartenberger, this volume), to confirm whether the lack of replacements in αA is indeed a general rodent character, or to find whether other rodent groups are perhaps characterized by specific replacements. Also among carnivores it has been found that the αA chains of the representatives of the families Felidae and Canidae have not accepted a single replacement since their divergence from the common carnivore stem, while the αA chains of the investigated Mustelidae, Ursidae, and the pinnipeds do contain a number of informative substitutions (De Jong, 1982).

NO RELATIONSHIP ON BASIS OF RODENT AND MACROSCELID αA SEQUENCES

Because the Macroscelidea are often suggested to be related to lagomorphs and rodents, we recently have analyzed the αA chain of an elephant shrew (_Elephantulus rufescens_). The lenses of a single specimen of this animal were provided by Dr. Michael J. Novacek. Electrophoresis of the subunits of the isolated α-crystallin showed a remarkably high mobility of the αA chain, which previously had only been observed for αA of hyrax and African elephant. Unfortunately the amount of isolated _Elephantulus_ αA chain was not suffi-

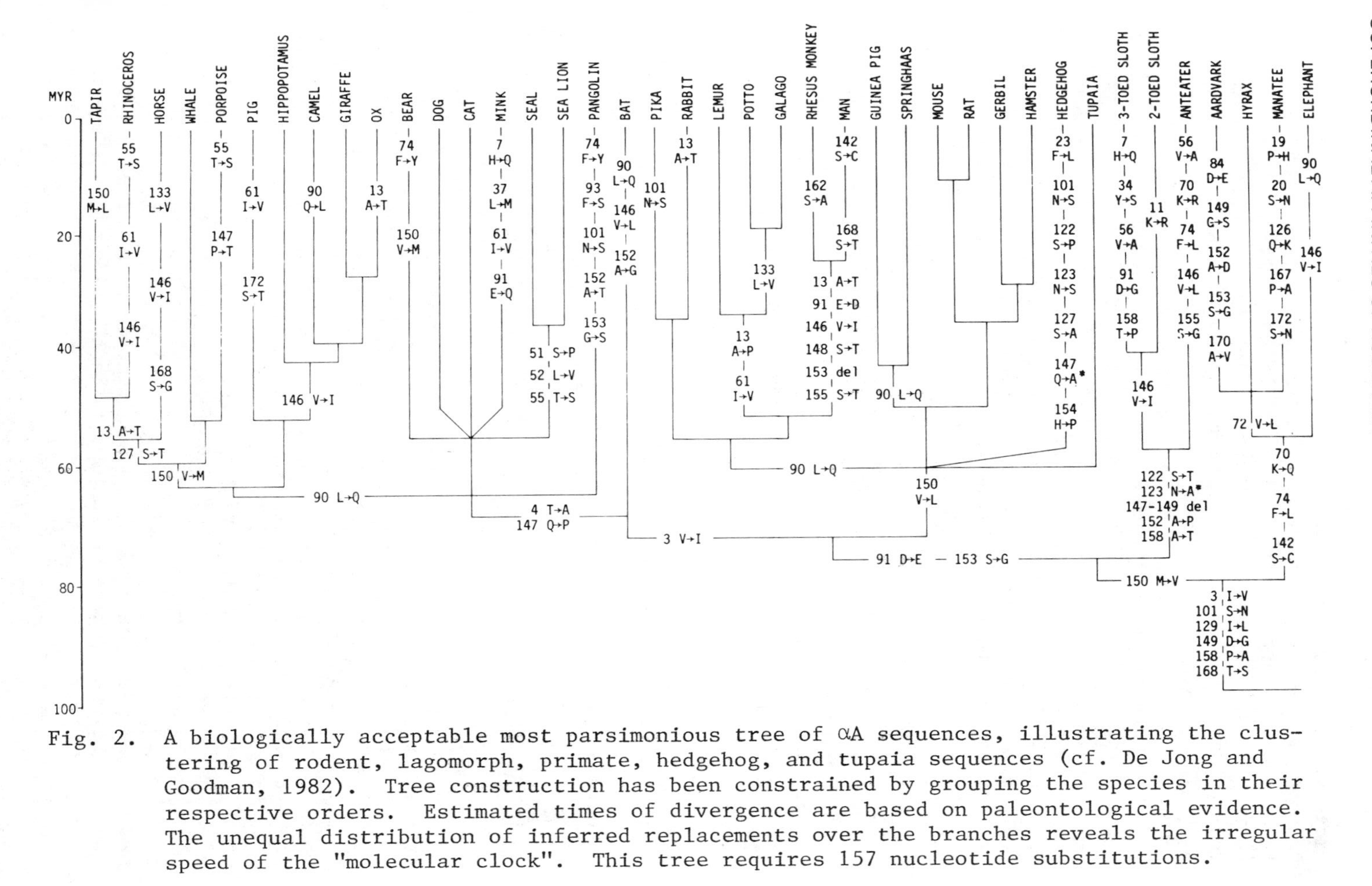

Fig. 2. A biologically acceptable most parsimonious tree of αA sequences, illustrating the clustering of rodent, lagomorph, primate, hedgehog, and tupaia sequences (cf. De Jong and Goodman, 1982). Tree construction has been constrained by grouping the species in their respective orders. Estimated times of divergence are based on paleontological evidence. The unequal distribution of inferred replacements over the branches reveals the irregular speed of the "molecular clock". This tree requires 157 nucleotide substitutions.

cient to determine the complete sequence. Although fragmentary protein sequences are not much appreciated in the biochemical literature, they are not out of place in a volume dealing largely with bone and skull fragments. Moreover, like paleontological fragments, this partial *Elephantulus* αA sequence does provide some information about its possible relationships.

The results of this incomplete sequence determination are incorporated into Table 1, which presents those sequence positions in the mammalian αA chains which have proved to be informative for the resolution of the relationships between the orders (De Jong and Goodman, 1982). It turns out that the *Elephantulus* αA chain lacks the derived replacements 91 Glu, 150 Leu, and 153 Gly, which are present in the cluster of lagomorphs, rodents, primates, hedgehog, and tupaia. Instead, the *Elephantulus* αA chain contains the supposedly primitive conditions 91 Asp, 150 Met, and 153 Ser; these are also present in marsupials, edentates (not 150 Met), and subungulates. It can, however, be seen that back-mutations to 91 Asp and 150 Met also occurred, in several species, which weakens their diagnostic value. The elephant shrew lacks in its αA sequence any synapomorphous replacement with other specific mammalian orders, nor does this chain have any autapomorphies (as far as determined). Also, a more thorough comparison with the total data set of vertebrate αA sequences leads us to conclude that on this basis no specific relationship between Macroscelidea and rodents and/or lagomorphs can be inferred. Our limited sequence data suggest instead that the Macroscelidea represent one of the earliest offshoots of the main eutherian stem, before the radiation of most other orders (including rodents and lagomorphs) took place.

AN ELONGATED α-CRYSTALLIN A CHAIN IN MUROID RODENTS

Whereas the molecular evolution of the αA chain itself in rodents has been quite uneventful, a very startling observation of a related chain has been made in a number of muroid rodents. Initially in rat, and later on in mouse, golden hamster, Chinese hamster, and Mongolian gerbil, it was noticed that they possessed, apart from the normal α-crystallin A and B chains, an additional minor α-crystallin chain. This minor component has been purified and characterized from rat and golden hamster lenses (Cohen et al., 1978). It turned out that this chain was identical in sequence to the normal muroid αA chain, apart from the fact that it contained an additional sequence of 23 amino acids inserted between residues 63 and 64. King and Piatigorsky (1983) have shown that this elongated chain, now called αA^{Ins}, is the result of a unique aberration in the processing of the messenger-RNA transcribed from the single αA gene. Probably due to one or a few nucleotide substitutions in the αA gene of these rodents, the primary mRNA is processed in such a way that in approximately 10% of the final mRNA an additional internal sequence of 69 nucleotides is present. In the remaining 90% of the mRNA molecules this sequence, which is part of a so-called noncoding intervening sequence (or intron), is removed in the normal way. Translation of this elongated mRNA then results in the synthesis of αA^{Ins} chain, which thus occurs in a 1:10 ratio as compared to normal αA.

This phenomenon of differential processing of mRNA, resulting in the production of two different polypeptide chains from the same gene, is unique in molecular biology. In cladistic terms it constitutes an extremely rare and unlikely synapomorphous character, which sets apart the muroid rodents among all vertebrates. Since we have studied only very few of the rodent families, it may be that we will find the αA^{Ins} chain also outside the Muridae and Cricetidae. In that case it would provide firm evidence for a monophyletic relationship with Muridae and Cricetidae, and thus might contribute to the resolution of rodent intraordinal relationships.

CONCLUDING REMARKS ON THE APPLICATION OF SEQUENCE DATA TO PHYLOGENY

As compared to the other biochemical methods used in phylogenetic studies, such as electrophoresis, immunology, and DNA-DNA hybridization, the most detailed information about relationships can, in principle, be obtained from protein and DNA sequence determinations. The sequence data, moreover, are presented in a discrete number of distinct character states (20 in the case of amino acids, and 4 for nucleotides), and this "digital" information is ideally suited for computer handling. Also, the direction of changes in sequences can often reliably be inferred, which is an obvious advantage in cladistic analyses, because it enables the distinction between primitive and derived character states.

Nevertheless, the results from comparative sequence analysis in studies over the past years have not yet been very numerous and impressive; only few problems have conclusively been solved by this approach. The reasons for this are obvious:

(1) Relative to the enormous number of established protein and DNA sequences, only a few comparative sequence studies have been aimed directly at solving specific phylogenetic problems. Almost all available sequences have been determined in the context of other molecular biological studies. Moreover, the techniques involved are expensive and laborious, and they require specialized equipment and expertise, which are not easily available at a taxonomic institute. From a phylogenetic point of view, the choice of species in sequence studies has generally been clearly inappropriate.

(2) Because different proteins have different rates of evolutionary change, it is often found that protein sequences show either too few or too many differences among the investigated species to be

informative. More closely related species should be studied, using rapidly evolving proteins (such as fibrinopeptides or pancreatic ribonuclease), whereas the more slowly evolving proteins (such as cytochrome c or α-crystallin) are more likely to give good results in higher taxonomic categories. Yet, even then the success will be unpredictable, because of the great variations in evolutionary rates in different lineages of the same protein.

(3) Some intrinsic properties of the evolutionary process itself also present fundamental obstacles in phylogenetic studies based on sequence comparisons. A major problem is the frequent occurrence of parallel and back replacements in protein evolution, which often make it impossible to decide whether a given similarity reflects common ancestry or is due just to parallel mutations. Also, the fact that many proteins are encoded by multiple copies of a gene can give rise to problems. There is the danger of comparing paralogous* instead of orthologous gene products in different species, in which case the inferred cladogram will not reflect the species phylogeny. Moreover, in the case of closely linked duplicated genes, the process of gene conversion (the exchange of DNA sequences between such genes) may seriously hamper the correct interpretation of comparative sequence data.

(4) In many cases, the periods of common ancestry that must be traced for the proper reconstruction of phylogenetic relationships will have been too short to have allowed the fixation of sufficient and characteristic replacements. The more distant these periods of common ancestry have been in the geological past, the more difficult it will be to pinpoint the synapomorphous replacements among the preponderance of subsequent and possibly superimposed replacements.

(5) The presently available computer programs for the construction of phylogenetic trees from sequence data still fail to extract the maximum possible information from these data. They indiscriminately lump together all sequence similarities, and they are unable to distinguish the more informative replacements from the evolutionary noise of back-and-forth replacements. The length, in terms of nucleotide substitutions, of the connecting branches in the phylogenetic trees that are based on such computer programs, are no reliable measure of the relative strength of such branches. To that end, it is necessary to estimate the informational content of the mutational events that are inferred to be responsible for a particular connection. The strength and quality of an inferred relationship should depend on the number and uniqueness of the specific replacements (and insertions or deletions) combined in the line of common ancestry. The need for such an extension of the computer-derived

* Paralogy refers to sequence homology originating from gene duplication; orthology indicates homology arising via speciation.

information, which has also been recognized by others (Joysey, 1981) will certainly lead to future refinements in the analysis of comparative sequence data.

SUMMARY

In this paper an attempt has been made to demonstrate the possibilities and limitations of the application of comparative protein sequence analysis to phylogenetic studies. This has been illustrated with examples from the eye lens protein α-crystallin A in mammals and, more specifically, rodents. Because of its slow rate of evolution, the αA chain is unable to clearly resolve the relationships among most mammalian orders. The αA sequences do, however, confirm that the closest relatives of the rodents have to be found among the lagomorphs, primates, insectivores, and tree shrews. The concept Glires is thus not contradicted by the αA seqence data. These data also tentatively indicate that the Macroscelidea do not belong to the just mentioned cluster of mammalian orders. In the rodent families Muridae and Cricetidae an elongated αA chain is present in small amounts, along with the normal αA chain, due to a unique mutational event.

ACKNOWLEDGEMENTS

I am grateful to Dr. Michael J. Novacek for providing Elephantulus lenses, to Anneke Zweers and Marlies Versteeg for technical assistance, and to Els van Genne for secretarial help. This work was supported in part by the Netherlands Foundation for Chemical Research (SON), with financial aid from the Netherlands Organization for Pure Research (ZWO).

REFERENCES

Bloemendal, H. (ed.) 1981. Molecular and Cellular Biology of the Eye Lens, Wiley, New York.

Clayton, R. M. 1974. Comparative aspects of lens proteins. In: The Eye, Volume 5, H. Davson and L. T. Graham, eds., pp. 399-494, Academic Press, New York.

Cohen, L. H., Westerhuis, L. W., De Jong, W. W., and Bloemendal, H. 1978. Rat α-crystallin A-chain with an insertion of twenty-two residues, Eur. J. Biochem. 89: 259-266.

De Jong, W. W. 1981. Evolution of lens and crystallins. In: Molecular and Cellular Biology of the Eye Lens, H. Bloemendal, ed., pp. 221-278, Wiley, New York.

De Jong, W. W. 1982. Eye lens proteins and vertebrate phylogeny. In: Macromolecular Sequences in Systematic and Evolutionary Biology, M. Goodman, ed., pp. 75-114, Plenum Press, New York.

De Jong, W. W. and Goodman, M. 1982. Mammalian phylogeny studied by sequence analysis of the eye lens protein α-crystallin. Z. Säugetierkunde 47: 257-276.

De Jong, W. W., Zweers, A., and Goodman, M. 1981. Relationship of aardvark to elephants, hyraxes and sea cows from α-crystallin sequences. Nature 292: 538-540.

De Jong, W. W., Zweers, A.. Versteeg, M., and Nuy-Terwindt, E. C. 1984. Primary structures of the α-crystallin A chains of twenty-eight mammalian species , chicken, and frog. Eur. J. Biochem. 141: 131-140.

Dover, G. A. and Flavell, R. B. (eds.) 1982. Genome Evolution. Academic Press, New York.

Fitch, W. M. and Margoliash, E. 1967. Construction of phylogenetic trees. Science 155: 279-284.

Goodman, M., Czelusniak, J., Moore, G. W., Romero-Herrera, A. E., and Matsuda, G. 1979. Fitting the gene lineage into its species lineage, a parsimony strategy illustrated by cladograms constructed from globin sequences. Syst. Zool. 28: 132-163.

Hendy, M. D., Penny, D., and Foulds, L. R. 1978. Identification of phylogenetic trees of minimal length. J. Theor. Biol. 71: 441-452.

Ingolia, T. D. and Craig, E. A. 1982. Four small Drosophila heat shock proteins are related to each other and to mammalian α-crystallin. Proc. Natl. Acad. Sci. USA 79: 2360-2364.

Joysey, K. A. 1981. Molecular evolution and vertebrate phylogeny in perspective. Symp. Zool. Soc. Lond. 46: 189-218.

King, C. R. and Piatigorsky, J. 1983. Alternative RNA splicing of the murine αA-crystallin gene. Cell 32: 707-712.

Moormann, R.J. M., Van der Velden, H. M. W., Dodemont, H. J., Andreoli, P. M., Bloemendal, H., and Schoenmakers, J. G. G. 1981. An unusually long non-coding region in rat lens α-crystallin messenger RNA. Nucl. Acids Res. 9: 4813-4822.

Piatigorsky, J. 1984. δ-Crystallins and their nucleic acids. Molec. Cell. Biochem. 59: 33-56.

Piatigorsky, J. in press. Lens crystallins and their gene families. Cell.

Sigman, D. S. and Brazier, M. A. B. (eds.) 1980. The Evolution of Protein Structure and Function. Academic Press, New York.

Slack, J. 1984. A Rosetta stone for pattern formation in animals? Nature 310: 364-365.

Stapel, S. O., Leunissen, J. A. M., Versteeg, M., Wattel, J. and De Jong, W. W. 1984. Ratites as oldest offshoot of avian stem: evidence from α-crystallin A sequences. Nature 311: 257-259.

Wilson, A. C., Carlson, S. S., and White, T. J. 1977. Biochemical evolution. Ann. Rev. Biochem. 46: 573-639.

SUPERORDINAL AND INTRAORDINAL AFFINITIES OF RODENTS: DEVELOPMENTAL EVIDENCE FROM THE DENTITION AND PLACENTATION

W. Patrick Luckett

Department of Anatomy
University of Puerto Rico
San Juan, Puerto Rico 00936, USA

INTRODUCTION

The origin and subsequent evolutionary radiation of the order Rodentia were based primarily on the development of a pair of enlarged, evergrowing incisors in both jaws, and concomitant specializations of the cheek teeth and associated musculoskeletal features of the jaws for gnawing and chewing. The earliest known fossil rodents from the Late Paleocene of North America (Wood, 1962), the Early Eocene of Europe (Michaux, 1968), and the Early Eocene of Asia (Shevyreva, 1976; Hartenberger, 1980) already exhibit elongated incisors, and the search for possible ancestors or sister-groups of rodents has focused on other eutherian groups that show some degree of incisor enlargement. These taxa have included the Lagomorpha and a number of early Tertiary groups: plesiadapiform Primates, Tillodontia, Mixodectidae, Microsyopidae, Apatemyidae, and, most recently, the Paleocene eurymyloids of Asia (see Li, 1977; Hartenberger, 1977, 1980; Li and Ting, this volume).

Uncertainties regarding the homologies of the large, evergrowing incisors of rodents create a major difficulty in assessing the possible ancestral or sister-group relationships of rodents with other eutherians. As an example, several authors (Wood, 1962, 1977a; McKenna, 1969; Van Valen, 1971) have suggested a possible origin of rodents from Paleocene plesiadapiform primates, based primarily on the presence of a pair of enlarged, "rodent-like" incisors in these early primates. None of these investigators, however, discussed the evidence for homology of enlarged incisors in plesiadapids and rodents. Early embryological studies by Adloff (1898) suggested that the gliriform incisors of rodents and lagomorphs are I2, based in part on the development of small, abnormal dI1 in the prenatal jaws of both groups. This important work was cited by Tullberg (1899) in his analysis of rodent-lagomorph similarities, but it has been overlooked by most subsequent investigators. On the other hand, it is generally believed that the enlarged incisors of plesiadapiform primates are I1 (Simons, 1972; Gingerich, 1976; Szalay and Delson, 1979; Butler, 1980).

If incisor enlargement occurred at different tooth loci in rodents and plesiadapids, then large incisors in the two taxa would be an example of convergent evolution, and incisor morphology would provide no evidence to corroborate a hypothesis of close relationship between rodents and primates. Progress in evaluating the origin and the interrelationships of rodents has been retarded by the widespread failure to assess the evolutionary basis of shared similarities among biological traits, such as enlarged incisors. Thus, there has been little discussion of the criteria used to distinguish among shared primitive retentions, shared derived traits, and shared similarities that may be due to convergence or parallelism. Such character analyses are particularly important in assessments of rodent phylogeny, because it is commonly believed that many adaptive changes in the teeth and skull for increased masticatory efficiency have evolved in parallel several times within unrelated higher taxa of rodents (Simpson, 1945, 1959; Lavocat, 1956; Wood, 1965, 1980).

Most assessments of rodent evolution have been restricted to studies of cheek tooth patterns, infraorbital-zygomasseteric relationships, and mandibular structure; these are all components of a single, functionally interrelated complex - the masticatory system, as emphasized by Wahlert (1974). The apparent occurrence of widespread parallelism among these features during rodent evolution underscores the necessity for phylogenetic analyses of other organ systems and functional complexes, in order to test conflicting hypotheses of rodent phylogeny. It should be evident that phylogenetic hypotheses that can be corroborated by a wide range of biological evidence, obtained from several unrelated organ systems, are more likely to approximate or reflect the actual phylogeny of a group, than are those based solely on analysis of a single organ system.

THE VALUE OF ONTOGENETIC DATA IN PHYLOGENETIC RECONSTRUCTION

Character analysis is essential for evaluation of both ancestral-descendant and cladistic relationships during phylogeny. Fundamental tasks for the phylogenetic analysis of characters are: (1) determination of the homoplastic (parallelism, convergence) or homologous nature of shared biological similarities among organisms; (2) identification of all possible character states of homologous traits, and their arrangement in a transformation series or morphocline; and (3) assessment of the relatively primitive or derived nature of each character state in a hypothesis concerning the pattern and direction of evolutionary change of characters (determination of morphocline polarity). It is safe to conclude that most disagreements concerning rodent and mammalian phylogeny are related directly to differing interpretations of character homology and morphocline polarity.

Despite the fact that character analysis is an essential prerequisite for phylogeny reconstruction, the methodology of character analysis is the least discussed aspect of systematic inquiry (Bock, 1977). Homologous character states are defined by their inheritance from a common ancestor, but the recognition of homologous features is based on detailed similarities in structure and ontogeny (Simpson, 1961; Bock, 1977; Gaffney, 1979). The most common procedure for character analysis is to compare "presumably homologous" traits and to determine their relatively primitive and derived character states, usually by a consideration of commonality, ontogeny, and outgroup comparisons (Hennig, 1966; Gaffney, 1979; Eldredge, 1979; Cracraft, 1981). Considerable disagreement exists with regard to whether the evaluation of character homologies should be the first or last step in phylogenetic analyses (see contrasting views of Bock, 1977, and Cracraft, 1981). Many systematists rely on the parsimony principle for distinguishing between homologous and homoplastic traits, when there are incongruences or contradictions in the distribution of synapomorphies. This method is highly suspect when there are an insufficient number of taxa and characters incorporated into a phylogenetic hypothesis. Others have stressed the central importance of homology in character analysis, and have urged a more biological approach to the assessment of homology (Bock, 1977; Szalay, 1981). This entails a detailed study of ontogeny and form-function interrelationships, with an emphasis on character complexes and organ systems, rather than relatively simple "counts" of characters.

Given the low level of resolution for methods of testing hypotheses of homology, it seems naive to ignore any type of data that might bear on phylogenetic analysis. Ontogenetic studies provide valuable evidence for the assessment of homology and character state polarity, especially when considered from a broad comparative viewpoint (Simpson, 1961; Hennig, 1966; Luckett, 1975, 1982; Martin, 1975; Nelson, 1978; MacPhee, 1981). As emphasized by MacPhee (1981), Hennig's criterion of "ontogenetic character precedence" should not be used in a strict recapitulation sense, but rather in conjunction with other aspects of homology assessment, including commonality, outgroup comparisons, and form-function considerations.

USE OF FETAL MEMBRANE DATA FOR ASSESSING MAMMALIAN PHYLOGENY

Developmental features of the fetal membranes and placenta comprise a functionally interrelated organ system that offers a number of advantages for assessment of mammalian phylogeny (Mossman,

1937, 1953, 1967; Luckett, 1975, 1977). The mammalian fetal membranes (amnion, chorion, yolk sac, and allantois) develop from all three embryonic germ layers and exhibit a fundamental homology in their ontogenetic relationships among all amniotes. Data on fetal membrane morphogenesis are available for all orders and suborders of eutherian mammals, as well as for many families, and these developmental features can be compared with those of marsupials and monotremes for character analyses and phylogenetic reconstruction.

Compared with dental and cranial traits, fetal membranes have remained relatively conservative in their patterns of evolutionary change during mammalian phylogeny; this may be related in part to their relative isolation from selective effects of the external environment. Individual fetal membrane features, as well as their entire developmental pattern, vary rarely at the generic and family levels. Evolutionary changes are more evident above the family level, and the frequent occurrence of intermediately derived character states facilitates the analysis of cladistic relationships among mammalian higher taxa. Mossman (1967) emphasized that the entire developmental pattern of the fetal membranes and placenta is available for assessing mammalian phylogeny, rather than only the mature or definitive condition of these traits. Such ontogenetic analyses are particularly useful for recognition of homologous, parallel, or convergent evolution among shared and derived similarities. As an example, it is highly improbable that the shared similarity of a hemochorial placenta in rodents and haplorhine primates is homologous, because these structures develop as the result of different ontogenetic pathways in the two taxa. In summary, the genetic complexity and relative conservatism of fetal membrane traits, as well as the availability of their entire developmental pattern for study, provide a valuable organ system for the assessment of cladistic relationships among rodents and other mammals.

USE OF DENTAL ONTOGENY FOR ASSESSING MAMMALIAN PHYLOGENY

Although the dentition is the character complex most often used for evaluating evolutionary relationships among fossil and extant mammals, its developmental features are infrequently incorporated into such analyses. Histological studies of prenatal dental development are rarely complete for any species of rodent or other mammals. Commonly, only a few developmental "stages" are included in such reports (Freund, 1892; Adloff, 1898), or else they focus upon only selected aspects of molar (Santoné, 1935; Gaunt, 1955, 1961; Butler, 1956) or incisor development (Fitzgerald, 1973; Moss-Salentijn, 1978; Ooë, 1980). Rodents (especially the mouse) are commonly studied as experimental models for early aspects of dental development, especially for investigations of pattern formation and epithelial-mesenchymal interactions during early morphogenesis (Kollar, 1972; Ruch et al., 1973; Lumsden, 1979; Thesleff and Hurmerinta, 1981). These studies have provided valuable insight into tissue interactions during early phases of dental development, but they have not contributed as yet to our understanding of rodent evolution.

Developmental studies have proved valuable in clarifying the deciduous or successional nature of gliriform incisors in lagomorphs (Moss-Salentijn, 1978; Ooë, 1980), as well as for identifying the tooth loci occupied by normal and vestigial incisors. These homologies are less clear for rodent incisors, as noted by Moss-Salentijn (1978). Ontogenetic analyses have also helped to clarify homologies of cusps and crests in cricetid and murid molars (Gaunt, 1955, 1961), but such studies are lacking for most rodent families.

EARLY DENTAL DEVELOPMENT IN RODENTS

Cuspidate, low-crowned molars and a primitive rodent dental formula of $\frac{1\text{-}0\text{-}2\text{-}3}{1\text{-}0\text{-}1\text{-}3}$ are found in most Eocene rodents and are retained in the family Sciuridae among extant taxa. On the other hand, premolars are reduced in number or lost in most other Oligocene-Recent families. Previous embryological studies (Freund, 1892; Adloff, 1898; Fitzgerald, 1973; Moss-Salentijn, 1978) suggest that the occurrence of one or more "vestigial" incisors in each jaw quadrant is a common feature in prenatal sciurids, whereas their presence is more variable in the investigated cricetids, murids, and caviomorphs. Because of their retention of several relatively primitive (plesiomorphous) dental features, the sciurids were selected for more careful analysis of early phases of tooth development in the present study. These developmental features will be compared with those of lagomorphs, primates, tupaiids, macroscelidids, and lipotyphlous insectivores, with emphasis placed on determining the homologies of the normal and vestigial incisors of rodents and lagomorphs.

The sciurids studied were of unknown gestational age, although in most cases either the general length (GL) or head length (HL) of the fetus was recorded. Careful examination of the developmental features of each tooth facilitates arrangement of the fetuses into an ontogenetic sequence or series, and this allows comparisons to be made with developmental stages from other mammals. It should be evident that developmental "stages" are erected for the convenience of the embryologist; nevertheless, they aid us to recognize major patterns of developmental change for any organ system. Early phases of eutherian dental morphogenesis are fundamentally similar, regardless of the individual tooth locus or species examined (Fig. 1). These morphologically recognized phases include: (1) differentiation of a dental lamina from the oral epithelium; (2) proliferation and invagination of a tooth bud from the dental lamina; (3) indentation of the inferior surface of the epithelial bud by a proliferation and aggregation of mesenchymal cells to form a cap stage; (4)

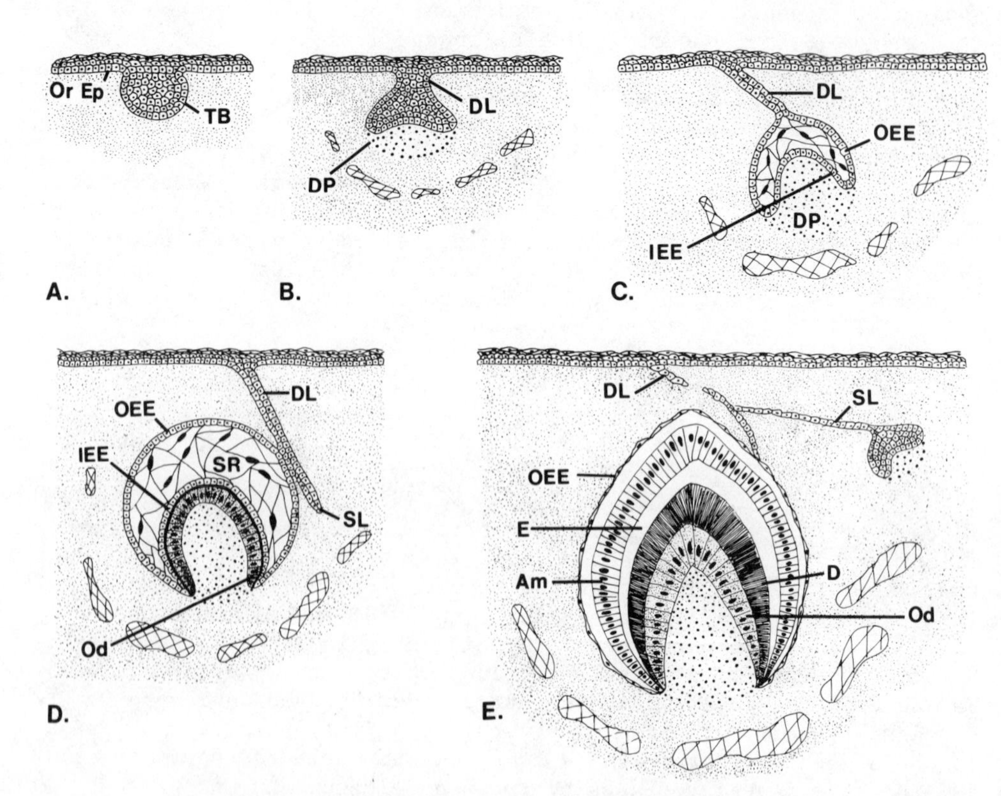

Fig. 1. Overview of the major features of early dental development in eutherians. (A) Tooth bud is proliferated from the oral epithelium. (B) Mesenchyme of dental papilla indents tooth germ to form early cap stage. (C) The enamel organ envelops the dental papilla to form a bell stage. (D) Odontoblasts differentiate in the late bell stage, and stellate reticulum separates inner and outer enamel epithelium. A successional lamina becomes evident on the lingual surface of the enamel organ. (E) Dentin is secreted by odontoblasts, and this induces the inner enamel epithelial cells to differentiate into ameloblasts and secrete enamel. The successional lamina differentiates into a successional tooth germ, which undergoes the same developmental stages as its deciduous predecessor. Abbreviations: Am, ameloblasts; D, dentin; DL, dental lamina; DP, dental papilla; E, enamel; IEE, inner enamel epithelium; Od, odontoblasts; OEE, outer enamel epithelium; Or Ep, oral epithelium; SL, successional lamina; SR, stellate reticulum; TB, tooth bud.

continued increase in envelopment of the mesenchymal aggregation (dental papilla) by the growth of the oral epithelium (enamel organ) to form a bell stage; (5) differentiation of odontoblasts from the mesenchymal cells of the dental papilla, adjacent to the enamel organ; (6) deposition of predentin matrix by odontoblasts; (7) differentiation of the inner enamel epithelial cells of the enamel organ into ameloblasts; and (8) deposition of enamel matrix by ameloblasts, at the dentin-enamel junction. This developmental sequence characterizes the "normal" deciduous or primary dentition of all eutherians examined.

Another constant developmental feature is the differentiation of a successional lamina in continuity with the outer enamel epithelium of the deciduous enamel organ, at its lingual margin. Subsequently, the successional lamina differentiates into a successional or secondary tooth bud at each antemolar locus (Fig. 1d, e), and this bud then undergoes the same sequence of development (bud, cap, bell, formation of dentin and enamel) as does its deciduous predecessor. Molar teeth develop from a secondary, distally directed wave of differentiation of the primary dental lamina, but they do not develop distinct secondary tooth germs from their undifferentiated, lingual "successional" lamina. Thus, developmental evidence corroborates the hypothesis that molars are unreplaced members of the primary dentition (Leche, 1895; Ziegler, 1971; Ooë, 1979).

A summary of the major features of sciurid dental development found during the present study is presented in Tables 1 and 2. This includes several additional specimens described and illustrated by Adloff (1898). The most significant features of prenatal dental development in sciurids are: (1) the constant occurrence of three deciduous incisors in the upper jaw, although two of these are abnormal and do not erupt; the constant occurrence of two deciduous incisors in the lower jaw (one normal, one abnormal), and the frequent development of a third, minute, deciduous incisor germ distal to these; (3) the lack of development of secondary or successional incisors beyond the bud-early cap stage; and (4) the common development of an abnormal upper dP2 and lower dP3, even though these teeth and their successors are unknown in postnatal extant and fossil rodents.

In the earliest specimen examined, a 23mm GL embryo of _Sciurus vulgaris_, the three deciduous incisors of the upper jaw arise by separate connections from the primary dental lamina, and all are associated with the developing premaxillary ossification. The largest of these tooth germs is the second (dI2), and it is flanked mesially and distally by a smaller dI1 and dI3 (Fig. 2). Despite its large size, dI2 is the least differentiated of the premaxillary incisors; it has attained only the early cap stage. The most differentiated tooth of the upper jaw is the small dI1 (Table 1; Fig. 2). This tooth bears a thin layer of dentin (or predentin) over its developing cusp, although odontoblasts are not very distinct. The small dI1 differs from "normal" tooth germs in the virtual absence of stellate reticulum in its enamel organ, as well as by the absence of differentiating ameloblasts. These developmental abnormalities are even more evident in older fetuses, in which the dentin forms a spherical clump or knot, and both enamel and ameloblasts are lacking (Fig. 3). A small dI3 occurs near the distal end of the premaxillary ossification, and it is somewhat more advanced developmentally than dI2 in the 23mm embryo, being in an early-middle bell stage. Subsequent development of the abnormal dI3 is similar to that for dI1, although odontoblasts are somewhat better developed in a 33mm fetus with early dentin formation (Fig. 3).

The large upper dI2 differs from the small abnormal incisors in its normal development of stellate reticulum within the enamel organ, and in the normal appearance of its odontoblasts during dentinogenesis (Fig. 3). The large dI2 develops in association with a deep bony alveolus of the premaxilla in the 33mm fetus, but later stages were not available to determine when the tooth penetrates into the maxilla. Adloff (1898) reported that dI2 is erupted in a 3-4 week juvenile _Sciurus_, and that the base of the incisor extended distally into the premolar region of the maxilla.

Dental development in the lower jaw of sciurids is slightly in advance of the upper jaw, similar to the condition in most eutherians studied. A small, abnormal deciduous incisor, with a dentinal knot, develops at the cranial end of the dental lamina in the 23mm embryo, and this is followed distally by a larger, less differentiated, but normally developing, deciduous incisor in the early bell stage (Table 2). This developmental pattern parallels that of the upper jaw, and it is almost certain that the tooth germs in the lower jaw are dI1 and dI2. A minute, abnormal tooth that contains dentin or dentin-like material was also identified distal to dI2 in many, but not all, the fetal mandibles examined by Adloff (1898) and during the present study. The proximity of this tooth to dI2 suggests that it is an abnormal dI3, rather than an abnormal deciduous canine.

A small, bud-like thickening on the lingual successional lamina of the large lower dI2 was evident in a 33mm GL fetus of _Sciurus_, and a slightly advanced early cap stage for I2 was described and illustrated by Adloff (1898) in a 21mm HL fetus of _Spermophilus_ (Table 2). There is no evidence for the further differentiation of a successional tooth germ for I2 in either jaw of sciurids (or in any other rodent). This abortive development of a successional tooth corroborates the hypothesis that the gliriform incisors of fetal and postnatal rodents are highly specialized and unreplaced deciduous teeth.

Early developmental stages of dP3 and dP4 are present in both jaws of a 23mm GL _Sciurus_ embryo, and their further differentiation

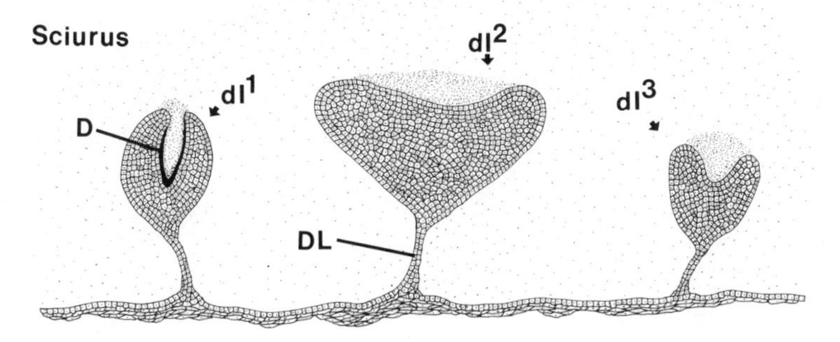

Fig. 2. Diagram of tooth germs in the premaxillary region and their relationship to the oral epithelium of the upper jaw in a 23mm GL Sciurus embryo. Note that dI1/ is the most advanced tooth developmentally; it contains abnormal dentin (D), but lacks stellate reticulum. The dental laminae (DL) of dI1/ and dI2/ are closely associated in histological sections, and they are separated here only for clarity.

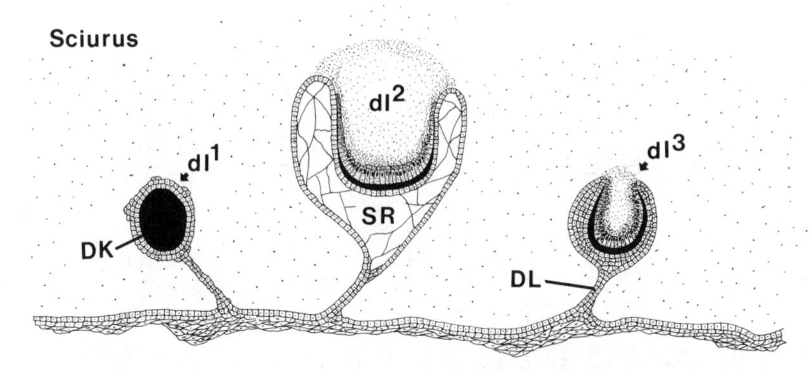

Fig. 3. Diagram of tooth germs in the premaxillary region of the upper jaw in a 33mm GL Sciurus fetus. An abnormal dentinal knot occurs in dI1/, but there is no evidence of enamel. The small dI3/ is developing abnormal dentin. Note the normal appearance of stellate reticulum (SR) in dI2/; this is the largest but also the least differentiated incisor at this stage.

Table 1. Dental developmental sequence in upper jaw of sciurids (* = data from Adloff, 1898).

Genus	dI^1	dI^2	dI^3	dP^2	dP^3	dP^4	M^1
Sciurus 23mm GL	Tiny, late bell; early dentin	Large, early cap	Small, early bell	Tiny bud	Late bud-early cap	Late cap-early bell	Possible early bud
Callosciurus 10mm HL *	Late bell; dentin	Early cap	Early bell	Early cap	Early cap	Early bell	Distinct bud
Spermophilus 15mm HL *	Abnormal; dentin knot	Middle bell	Early bell	Early cap	Middle bell	Late bell	Early bell
Sciurus ±30mm GL	Abnormal; dentin knot	Late bell; early dentin	Late bell	Absent	Early bell	Large, late bell	Middle bell
Sundasciurus 15mm HL *	Abnormal; dentin knot	Late bell; early dentin	Dentin ring	Early cap	Middle bell	Middle-late bell	Early bell
Sciurus 33mm GL	Abnormal; dentin knot	Late bell; early dentin	Early dentin	Tiny bud	Middle bell	Late bell	Middle bell
Spermophilus 21mm HL *	Abnormal; dentin knot	Large bell; dentin and ameloblasts	Late cap	Small cap	Middle-late bell	Late bell	Middle bell
Sciurus 44mm GL	Tiny dentin rudiment	Large; dentin, thin enamel	Small, no dentin	Absent	Middle bell	Late bell, odonto-blasts	Late bell
Sciurus 58mm GL * (3-4 wks)	Absent	Erupted; extends to premolars	Absent	Absent	Calcified; unerupted; bud for P3	Calcified; unerupted; cap for P4	Calcified; unerupted

Table 2. Dental developmental sequence in lower jaw of sciurids (* = data from Adloff, 1898).

Genus	dI_1	dI_2	dI_3	dP_3	dP_4	M_1
Sciurus 23mm GL	Small bell; dentin knot	Early-middle bell	?Small bell, early dentin	Late bud-early cap	Early bell	Early bud?
Callosciurus 10mm HL *	Small bell; dentin	Late cap-early bell	Small bud?	Late bud-early cap	Early mid-bell	Distinct bud
Spermophilus 15mm HL *	Abnormal; dentin ring	Late bell; dentin	Not detected	Not detected	Middle-late bell	Middle-late bell
Sciurus ±30mm GL	Abnormal dentin	Large, late bell; thin dentin	Not detected	Abnormal; small bell; early dentin	Large, late bell	Middle bell
Sundasciurus 15mm HL *	Abnormal; dentin	Late bell; dentin	Rudimentary; dentin	Small bud	Middle-late bell	Middle bell
Sciurus 33mm GL	Small, abnormal dentin	Late bell; dentin; small bud for I2	Small; dentinal knot	Small, late bell; dentin	Large, late bell	Middle-late bell
Spermophilus 21mm HL *	Small; dentinal knot	Late bell, dentin; I_2 bud ?	Small; dentinal knot	Bell-like epithelial knot	Late bell	Middle-late bell
Sciurus 44mm GL	Tiny; dentinal knot	Dentin; buccal enamel	-	Small, late bell; dentin cap	Late bell, dentin on protoconid	Large, late bell
Sciurus 58mm GL * (3-4 wks)	Absent	Erupted, extends to M3	Absent	Absent	Erupting; late cap P4	Calcified; unerupted

is described in Tables 1 and 2. The fate of the small lower dP3 is similar to that of the abnormal deciduous incisors. It exhibits little or no stellate reticulum during the bell stage, lacks ameloblasts, and forms only a small amount of abnormal dentin.

The homologies of a small, vestigial tooth germ that develops sometimes between dP3 and dI3 in the upper jaw are more difficult to determine. The dental lamina is greatly reduced at the cranial end of the maxillary ossification in a 23mm embryo, and there is no tooth germ in this normal locus of a deciduous canine. There is a small bud from the dental lamina somewhat distal to this region, and a similar bud or early cap stage has been identified in most later fetuses (Table 1). The location of this tooth germ in relation to the maxilla suggests that it is a rudimentary dP2, rather than a deciduous canine as suggested by Adloff (1898). This vestigial tooth apparently does not develop beyond the early cap stage, and there is no evidence that it forms dentin. In the lower jaw, the dental lamina is absent in the diastema between the region of dI3 and dP3 in all stages examined, and there is no indication of a rudimentary bud for dP2.

Similar to the condition in other eutherians, the primary dental lamina of sciurids grows distally from the site of dP4 to give rise to the bud for M1, and in subsequent stages, the process is repeated to form M2 and M3. Distinct buds for M1 are not initiated until dP4 have attained the early bell stage (Tables 1 and 2). This pattern has also been observed in a wide variety of other eutherians, including primates, tupaiids, lagomorphs, and insectivorans (Luckett, unpublished), as well as in the caviomorph *Cavia* (Berkovitz, 1972).

The fate of the small, abnormal incisors and premolars of postnatal sciurids is unclear, although they are probably either resorbed or shed into the oral cavity. Adloff (1898) reported that there was no trace of these abnormal rudiments in a juvenile *Sciurus* of 3-4 weeks age (see Tables 1 and 2). Vestigial dI1 are also found in castorids (Heinick, 1908) and in many murid fetuses (Fitzgerald, 1973; Moss-Salentijn, 1978), but there is apparently no report of dI3 or vestigial premolars in other rodents examined. Vestigial incisors have not been identified yet in *Cavia* (Freund, 1892; Adloff, 1898; Tims, 1901; Berkovitz, 1972), but other hystricognathous rodents remain to be examined. The small dI1 of *Mus* are still detected during the early postnatal period (up to 15 days), whereas those of *Rattus* are not found postnatally (Fitzgerald, 1973; Moss-Salentijn, 1978).

Corroboration for the identification of three deciduous incisor loci in the upper jaw of fetal sciurids is provided by studies on the initial phases of dental lamina differentiation in the mouse. Strassburg et al. (1971) identified three distinct swellings of the dental lamina at the anterior end of each upper jaw quadrant in

11.5 day mouse embryos. During days 12-13, the first and third dental buds regressed, whereas the second anlage (=dI2) continued to enlarge and differentiated into a bell stage enamel organ by the 14th day. A similar developmental pattern occurred in the lower jaw, where the second bud at the anterior end of the dental lamina also differentiated into a normal bell stage. Further development of the normal dI2 and vestigial dI1 in older mouse fetuses was described by Hay (1961) and Fitzgerald (1973), respectively.

DENTAL HOMOLOGIES AS CLUES TO THE AFFINITIES OF RODENTS

Only a single enlarged incisor is known in each jaw quadrant of early Eocene-Recent rodents, although ontogenetic evidence suggests that the last common ancestor of sciurids, castorids, and murids possessed three incisor primordia in the premaxillary segment, and probably three in the lower jaw as well. Thus, it seems likely that small, abnormal or "vestigial" deciduous incisors have been developing in fetal rodents for the last 50-55 million years, despite the fact that they do not erupt or become functional. The same may be true for the abnormal lower dP3 and upper dP2 found in sciurids.

It seems likely that prenatal development of abnormal deciduous teeth in rodents and other mammals represents an intermediate stage in the evolutionary loss of individual tooth loci (Moss-Salentijn, 1978; Luckett and Maier, 1982). The reasons for retention of these dental rudiments remain obscure; nevertheless, their occurrence aids the reconstruction of the dental pattern in the ancestors of rodents. Thus, developmental evidence suggests that the ancestral stock of rodents had already lost the deciduous and permanent canines, dP1, and probably also lower P2 and dP2. On the other hand, enlargement of I2 and/or dI2 probably occurred in jaws that still retained reduced I1 and I3. Such a pattern of incisor relationships is evident in the lower jaw of early Eocene tillodonts (Cope, 1888; Gazin, 1953; Gingerich and Gunnell, 1979), although they still retained moderately developed canines and lower P2 (see Table 3 and Fig. 4).

A survey of numerous features of the dentition in rodents and other eutherians, with emphasis on the presence and size of different tooth loci, serves to narrow the search for the possible sister group or ancestral taxon of rodents. Considering only these traits, the greatest number of dental similarities to rodents is detected in lagomorphs (Table 3 and Fig. 4). Character analysis, based principally on commonality and outgroup comparisons of fossil and extant mammals, indicates that most of these shared similarities are derived or apomorphic features. In contrast, several other groups that have been suggested as close relatives of rodents, such as plesiadapiform primates and mixodectids, share only a few derived similarities with rodents. Moreover, I1/1, rather than I2/2, are the most enlarged incisors in plesiadapids and mixodectids.

Table 3. Character analysis of selected eutherian dental traits.

Primitive	Derived
1. P4 premolariform	1. P4 molariform
2. Molar paraconid present	2. Molar paraconid absent
3. Lower P3 present	3. Lower P3 absent
4. Lower P2 present	4. Lower P2 reduced or absent
5. Upper P2 present	5. Upper P2 reduced or absent
6. P1 present	6. P1 absent
7. Lower canine present	7. Lower canine reduced or absent
8. Upper canine present	8. Upper canine reduced or absent
9. Lower I3 present	9. Lower I3 reduced or absent
10. Upper I3 present	10. Upper I3 reduced or absent
11. (d)I2 surrounded by enamel	11. (d)I2 enamel restricted buccally
12. Lower (d)I2 not notably enlarged	12. Lower (d)I2 greatly enlarged, evergrowing
13. Upper (d)I2 not notably enlarged	13. Upper (d)I2 greatly enlarged, evergrowing
14. Lower I1 not greatly enlarged	14. Lower I1 greatly enlarged
15. Lower I1 present	15. Lower I1 reduced or absent
16. Upper I1 present	16. Upper I1 reduced or absent

Recently, several authors (Szalay and McKenna, 1971; McKenna, 1975; Szalay, 1977) have suggested that lagomorphs are more closely related cladistically to anagalids, macroscelidids, leptictids, pseudictopids, and zalambdalestids than they are to rodents. This hypothesis was based in part on shared similarities of the postcranial skeleton; it is not supported, however, by analysis of most dental characters (Fig. 4). In lagomorphs, as in rodents, the large evergrowing incisors are dI2/2 (Moss-Salentijn, 1978; Ooë, 1980; personal observations). In the upper jaws of Asian Cretaceous zalambdalestids, I1 and I3 are small, whereas I2 is moderately enlarged (Kielan-Jaworowska, 1975). On the other hand, I/1 is greatly enlarged and extends distally beneath P/3 in the lower jaw of zalambdalestids, while I/2 and I/3 are small or reduced. I/1 is slightly enlarged in anagalids (Simpson, 1931), whereas neither I/1 nor I/2 are notably enlarged in macroscelidids (Patterson, 1965), pseudictopids (Sulimski, 1968), or leptictids (Novacek, 1977). Given the dental homologies determined by ontogenetic studies, it seems improbable that the ancestral stock of rodents or lagomorphs would have exhibited noticeable enlargement of I1/1 or reduction of I2/2.

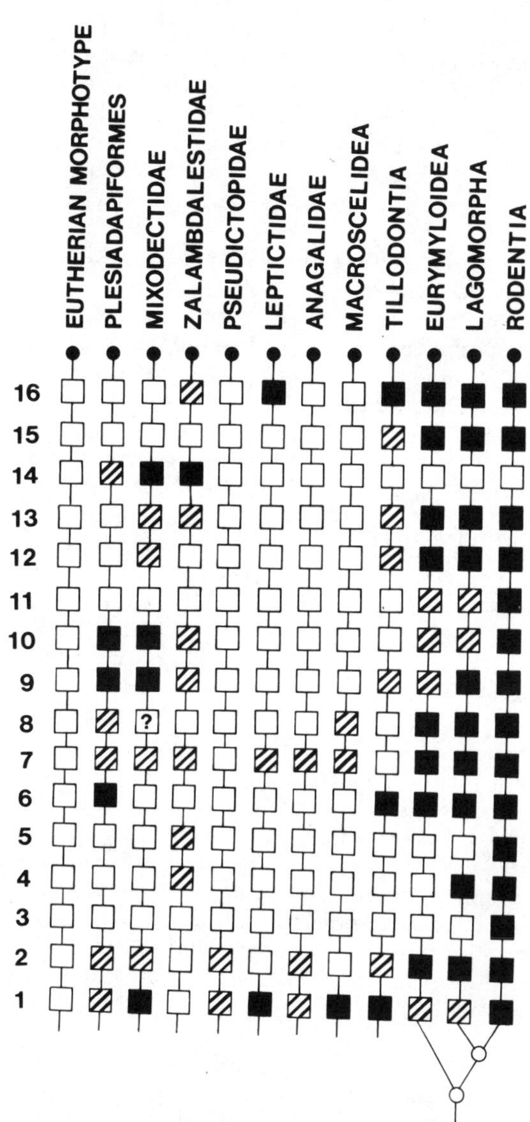

Fig. 4. Character phylogeny of selected dental traits in rodents and other eutherians. Derived character states are shown as black squares, open squares are primitive character states, and diagonally-lined squares are intermediately derived, in this and other figures. Each taxon is represented by its reconstructed morphotypic condition. See Table 3 for characters evaluated. Note: for character 1, recent evidence suggests that the rodent morphotype would have included a non-molariform P4/4 (Dawson et al., 1984).

There is increasing fossil evidence, based on analysis of both dental and cranial features, that the Asiatic Paleocene eurymyloids lie near the ancestry of both rodents and lagomorphs (see Li and Ting, this volume). For this reason, I have interpreted the enlarged incisors in the upper and lower jaws of eurymyloids as I2/2 (or possibly dI2/2), rather than I1/1, as interpreted previously by Li (1977). Li and Ting (this volume) have also accepted this suggested homology for the enlarged incisors of eurymyloids.

Examination of early dental development in lagomorphs reveals a striking similarity to the pattern in sciurid rodents (Table 4). There are three deciduous incisor loci in the upper jaw, and two in the lower jaw, corroborating previous reports (Moss-Salentijn, 1978; Ooë, 1980). The developmental pattern and relationships of dI1 and dI2 in both jaws are fundamentally identical in the rabbit and sciurids. The small dI1/1 develop rapidly and become abnormal, forming a dentinal knot after the bell stage is attained (Figs. 5, 6). The large dI2 develop normally, with stellate reticulum differentiating during the bell stage. There is no replacement of the evergrowing dI2/2, although the successional lamina may differentiate into a small bud or early cap stage (Moss-Salentijn, 1978; personal observations), similar to the condition in sciurids.

A notable difference in the premaxillary region is the normal development of a small dI3/ in lagomorphs, in contrast with its abnormal appearance in sciurids (cf. Fig. 6 and Fig. 3). This distinction becomes evident during the bell stage with the appearance of stellate reticulum in the normal dI3/ of lagomorphs, whereas this tissue is absent in the abnormal tooth of sciurids. As noted elsewhere (Moss-Salentijn, 1978; Luckett and Maier, 1982), the absence of stellate reticulum during the bell stage of differentiation is correlated with the subsequent abnormal or "vestigial" appearance of deciduous teeth in therian mammals, and this pattern was evident in all abnormal teeth of rodents and lagomorphs examined during the present study. An abnormal dI/3 has not been detected in lagomorphs (ochotonids are unstudied).

Premolar development follows a normal pattern in lagomorphs (Table 4) for both deciduous and successional teeth, and there is no trace of a rudimentary dP/2. The chronology of normal tooth replacement for the deciduous premolars and dI3/ in the rabbit has been summarized previously (Hirschfeld et al., 1973; Michaeli et al., 1980).

Character analysis of dental developmental features provides further corroboration for the hypothesis of a cladistic relationship between rodents and lagomorphs, when compared to other extant eutherians, including macroscelidids and primates (Fig. 7, Table 5). These

Table 4. Dental developmental sequence in upper jaw of leporid lagomorphs.

Genus	dI^1	dI^2	dI^3	dP^2	dP^3	dP^4	M^1
<u>Oryctolagus</u> 14mm GL	Short, late bud-early cap	Large, elongate bud	Small bud	Indistinct	Late bud-early cap	-	-
<u>Oryctolagus</u> 17.5mm GL	Small, middle-late bell; no odontoblasts	Large, elongate bud	Flattened bud	Small, early bud	Late bud	Large, early bud	-
<u>Oryctolagus</u> 20.5mm GL	Small, late bell; odontoblasts; thin dentin	Large, elongate bud	Early cap	Small, middle bud	Middle cap	Late bud	-
<u>Oryctolagus</u> 28mm GL	Small, abnormal dentin	Large, middle bell	Middle cap	Middle cap	Middle-late bell	Middle bell	Early bud
<u>Oryctolagus</u> ±35mm GL	Small, abnormal; thick dentin	Large, late bell; early cap for I2	Small, middle-late bell	Early-middle bell	Middle-late bell	(Not seen)	(Not seen)

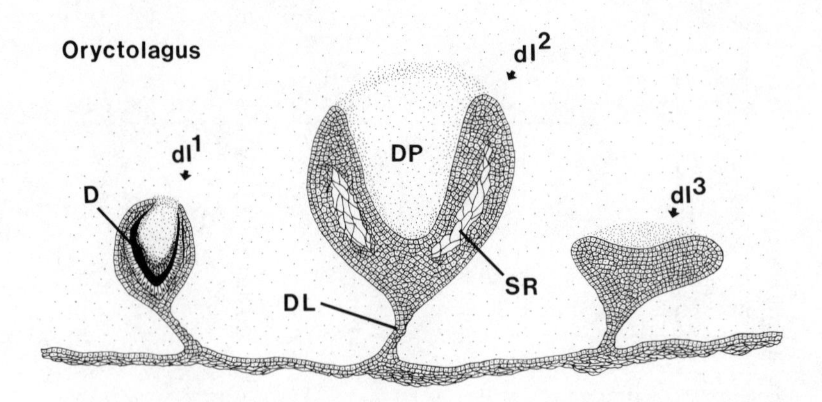

Fig. 5. Diagram of deciduous incisors in the premaxillary region of the upper jaw in a 28mm GL fetus of the lagomorph Oryctolagus. The small dI1/ exhibits a cap of abnormal dentin (D) and lacks stellate reticulum. The larger, but less differentiated dI2/ is a normal tooth in the middle bell stage, and it contains early stellate reticulum (SR). The small dI3/ is the least differentiated incisor; it has only attained the cap stage.

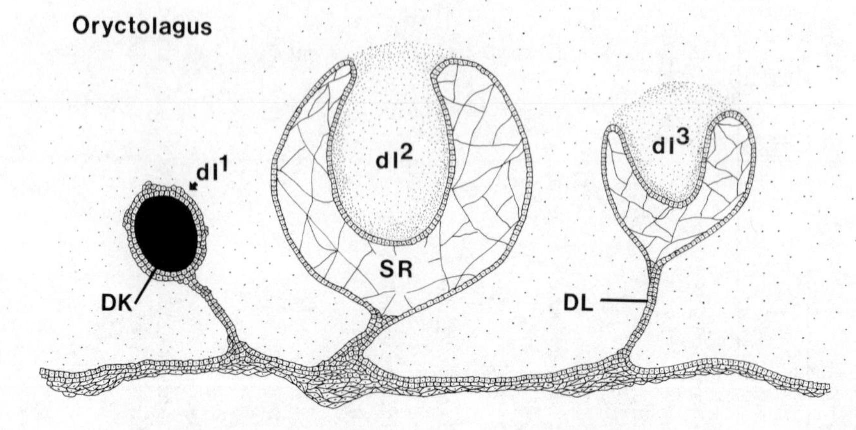

Fig. 6. Diagram of deciduous incisors in the premaxillary region of the upper jaw in a 35mm GL fetus of the lagomorph Oryctolagus. The small dI1/ consists of a dentinal knot (DK). The large dI2/ is in the late bell stage and has well developed stellate reticulum (SR). The smaller, middle bell dI3/, like dI2/, has a normal appearance and contains stellate reticulum.

developmental similarities are even more striking when the abnormal incisors and premolars are included in such a comparison. Thus, numerical differences that occur in the adult dentitions between rodents and lagomorphs are less notable when their deciduous teeth are compared (Fig. 8). If the homologies suggested for the small, abnormal dP2/ and dP/3 of sciurids are correct, then the developmental pattern of the incisors and premolars is virtually identical during early ontogenetic phases of the two taxa. Sciurids (and probably the rodent morphotype) are more derived by virtue of the abnormal development and postnatal disappearance of dI3/, dP2/, and dP/3, in contrast to normal development and subsequent replacement of these teeth in lagomorphs. Considered as individual characters, as well as an ontogenetic complex, the dentition provides a uniquely derived character complex that corroborates the sister-group relationship of rodents and lagomorphs (Fig. 7) among extant eutherians. In contrast, early dental development in macroscelidids presents no unusual features (Kindahl, 1957), and all deciduous antemolars except dP1 are replaced by successional teeth (Fig. 8).

Homology or Homoplasy of Gliriform Incisors

Incisor enlargement in mammals can occur at either the first or second tooth position in the jaws, and, in some cases, it is associated primitively with retention of the canines and anterior premolars (tillodonts, soricids, zalambdalestids, multituberculates, mixodectids, and plesiadapid primates). Thus, there is no reason to suppose, *a priori*, that the entire developmental suite of characters of the rodent and lagomorph dentitions might have evolved convergently, as a concomitant of differentiation of enlarged, gliriform incisors.

Of course, the hypothesis of homology or homoplasy of shared, derived dental features in rodents and lagomorphs can (and should) be tested by evaluating other organ systems, such as the fetal membranes (see below), or cranial morphology (see Novacek, this volume). Character analysis of these systems provides further support for the hypothesis of a monophyletic sister-group relationship for rodents and lagomorphs, and, in doing so, corroborates the suggested homologies of shared dental traits. In addition, assessment of dental and cranial features also corroborates the hypothesis of a monophyletic relationship among rodents, lagomorphs, and Paleocene-Eocene eurymyloids (Fig. 4; Li and Ting, this volume).

Use of Dental Developmental Data for Assessing Intraordinal Relationships of Rodents

Because of the limited available data on early dental development for most rodent families, these features are of little use at present for evaluation of relationships within the order. However,

Table 5. Character analysis of developmental features of the dentition and fetal membranes in rodents and other eutherians.

Primitive	Derived
1. Blastocyst attachment non-invasive	1. Blastocyst attachment invasive
2. C-A placenta epitheliochorial	2. C-A placenta hemochorial
3. Embryonic disc antimesometrial	3. Embryonic disc mesometrial
4. Blastocyst attachment at paraembryonic pole	4. Blastocyst attachment at abembryonic pole
5. C-A placenta at site of initial attachment	5. C-A placenta at opposite pole from attachment
6. Allantoic diverticulum large	6. Allantoic diverticulum small
7. Yolk sac splanchnopleure not inverted	7. Yolk sac splanchnopleure completely inverted
8. Amniogenesis by folding	8. Amniogenesis by cavitation
9. C-V placenta present	9. C-V placenta absent
10. I1 present	10. I1 absent; abnormal dI1
11. I2 present	11. I2 absent; dI2 large
12. (d)I2 smaller than (d)I1	12. (d)I1 smaller than (d)I2
13. Upper I3 present	13. Upper dI3 and I3 absent
14. Lower I3 present	14. Lower I3 absent
15. dI2 surrounded by enamel	15. Enamel restricted to labial surface of dI2
16. Canines present	16. Canines absent; large diastema
17. dP1 present	17. dP1 absent
18. Upper P2 present	18. Upper P2 absent; abnormal upper dP2 in fetus
19. Lower P2 present	19. Lower P2 absent
20. Lower P3 present	20. Lower P3 absent; abnormal dP3 in fetus

the pattern and sequence of molar cusp differentiation provide valuable insight into the probable homologies of molar cusps in cricetids and murids. Embryological studies on the mouse and hamster (Gaunt, 1955, 1961) corroborate the hypothesis, based on the paleontological record (see Jacobs, 1977 for discussion), that the lingual, longitudinal row of cusps is the neomorphic feature of murid upper molars.

Studies of dental ontogeny can also be invaluable for resolving disagreements concerning the deciduous or successional nature of premolars in the adult dentition. Wood (1974) has argued that an

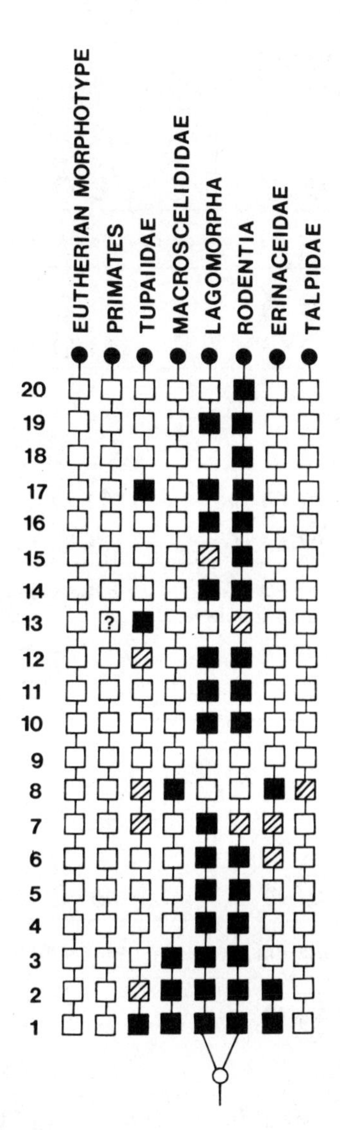

Fig. 7. Character phylogeny of developmental features of the dentition and fetal membranes in rodents, lagomorphs, and selected eutherians. Symbols as in Figure 4. See Table 5 for the characters evaluated.

important distinction between thryonomyoids and hystricids among the African hystricognaths is the retention of dP4/4 in adults of the former taxon, in contrast to normal replacement of these teeth by P4/4 in the latter. This is based in part on the apparent absence of postnatal replacement of these teeth in the extant genera

MACROSCELIDIDAE

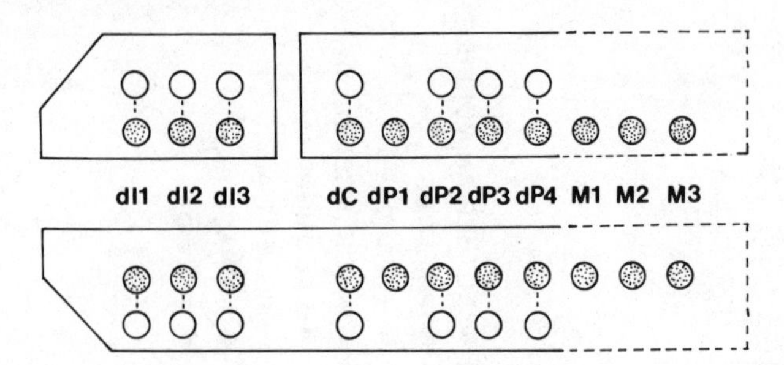

LAGOMORPHA

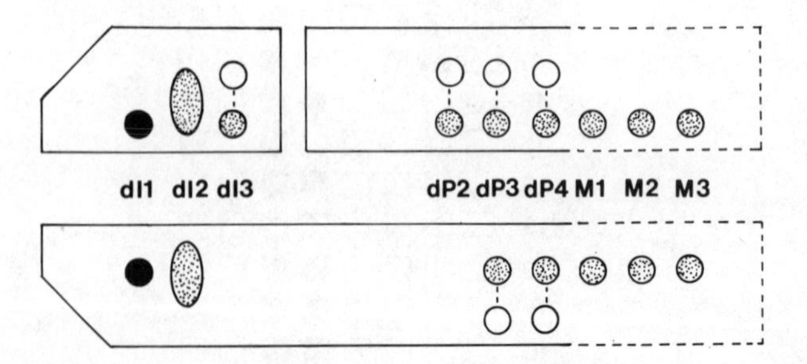

RODENTIA

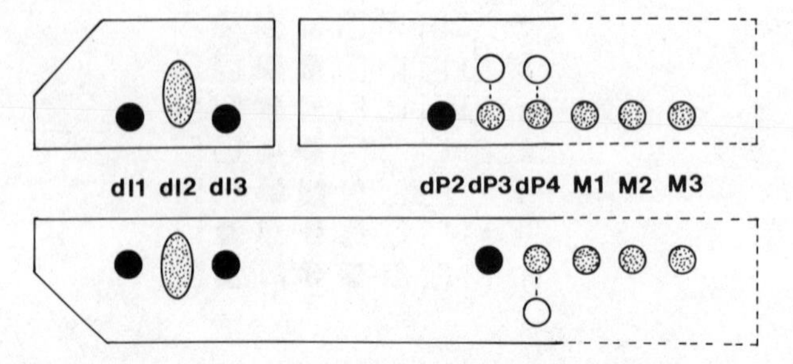

Fig. 8. Diagram of the deciduous (stippled) and successional (clear circle) tooth loci that develop in both jaws of prenatal and postnatal sciurid rodents, lagomorphs, and macroscelidids. Molars are considered as unreplaced members of the primary dentition. Abnormal deciduous teeth that do not erupt are shown in black. The large, evergrowing incisors of rodents and lagomorphs are dI2 in both jaws.

Thryonomys and *Petromus* (Friant, 1945; Wood, 1962b), and in most Oligocene fossil thryonomyoids (Wood, 1968). However, one should be cautious in interpreting such "observations" as evidence for retained deciduous premolars. It is well established that in the guinea pig (Cavia), dP4/4 develop, "erupt," are partially resorbed, and shed into the oral cavity during prenatal life (Adloff, 1904; Harman and Smith, 1936; Berkovitz, 1972; personal observations), and that the successor P4/4 also develop and erupt before birth. Therefore, the premolars are not "replaced" postnatally, although these teeth in the neonate, juvenile, and adult animals are homologous with P4/4, rather than dP4/4 of other eutherians.

Thomas (1894) reported (but did not illustrate) the occurrence of small, rudimentary dP4/4 overlying the P4/4 in a late fetus of Thryonomys. On the other hand, Woods (1976) could find no evidence of such vestigial dP4 in fetal Thryonomys jaws that he x-rayed and dissected grossly. Again, caution should be exercised in considering this as corroboration for the suppression of successional premolars and the retention of dP4 in postnatal animals. X-ray analysis was not successful for identifying the tiny (0.09-0.18mm length) abnormal dI1/1 of rabbits (Horowitz et al., 1973), even though these dental rudiments are consistently identified in histological sections. The question of the deciduous or successional nature of premolar loci in postnatal extant hystricognaths, as well as in Pedetes (see Wood, 1965b) can be resolved by careful histological analysis of fetuses. A similar controversy over premolar homologies in the prosimian primate Tarsius was settled by such a study (Luckett and Maier, 1982).

USE OF FETAL MEMBRANE DATA FOR ASSESSMENT OF RODENT RELATIONSHIPS

Mossman's (1937) monumental study on comparative mammalian placentation marked a turning point in the evolutionary analysis of placental features, because he emphasized the central importance of evaluating the entire developmental pattern of the fetal membranes and placenta, rather than only selected individual traits. Comparative aspects of fetal membrane development in lagomorphs and the major subordinal groups of rodents played a central role in Mossman's analysis of homologies and patterns of evolutionary change in placental traits among eutherians.

Morphogenetic relationships of the fetal membranes and placenta are quite varied among rodents, although four general developmental patterns can be recognized: (1) a sciurid-aplodontid-castorid pattern; (2) a geomyoid-dipodoid pattern; (3) a muroid pattern; and (4) an hystricognath pattern. Some variation occurs within each of these groups, and other taxa (such as the families Ctenodactylidae, Pedetidae, and Anomaluridae) do not associate readily with any of these patterns.

No attempt will be made here to review the extensive literature on fetal membrane development in rodents; this can be found in several syntheses of comparative rodent placentation (Mossman, 1937;

Fischer and Mossman, 1969; Luckett, 1971, 1980). Instead, a brief summary will be presented for the major developmental features that are most useful for assessing phylogenetic relationships within the order. In addition, reconstruction of the morphotypic conditions of the order will facilitate analysis of the superordinal affinities of rodents. For convenience of description, this ontogenetic pattern can be subdivided into five major phenomena: (1) implantation; (2) amniogenesis; (3) yolk sac relationships; (4) allantoic vesicle and body stalk; and (5) chorioallantoic placentation. The relatively primitive or derived nature of character states for these developmental traits is assessed on the basis of ontogeny, commonality, outgroup comparisons, and form-function analysis of individual characters and of the entire ontogenetic complex (see Luckett, 1977, 1980 for further discussion).

Implantation

The processes of fertilization, cleavage, and early blastocyst formation are fundamentally similar among eutherians, and significant differences become evident only as the blastocyst becomes prepared for attachment (= implantation) to the uterine wall. The embryonic knot or disc is oriented toward the mesometrial pole of the uterus at the time of initial implantation in all rodents (Fig. 9). The initial attachment of the blastocyst occurs by local invasive activity of its abembryonic trophoblast into the antimesometrial wall of the uterine endometrium. Following this common pattern of initial attachment in rodents, the later phases of implantation may take one of three pathways. The first, superficial implantation, is characteristic of sciurids and aplodontids. Abembryonic invasion remains limited, and the blastocyst expands to occupy a moderately large, eccentric implantation chamber of the uterine lumen (cf. Fig. 9A and 10A). This general arrangement is widespread in eutherians with noninvasive and moderately invasive implantation, and it doubtlessly represents the primitive rodent condition. A similar process of implantation is found in castorids, geomyoids, dipodoids, pedetids, and ctenodactylids. This pattern probably occurs also in anomalurids, although the critical early stages have not yet been examined.

The initial attachment phase is identical in muroids, but this is soon followed by secondary closure and fusion of the uterine epithelium, so that the antimesometrial implantation chamber becomes separated secondarily from the main portion of the uterine lumen (Fig. 9C). This pattern of secondary interstitial implantation occurs in cricetids (including gerbillines and arvicolines) and murids, whereas spalacids, rhizomyids, and glirids have not yet been investigated.

The most derived pattern of rodent implantation occurs in Old and New World hystricognaths (Fig. 9D). Following initial invasive attachment of the abembryonic trophoblast, the small blastocyst

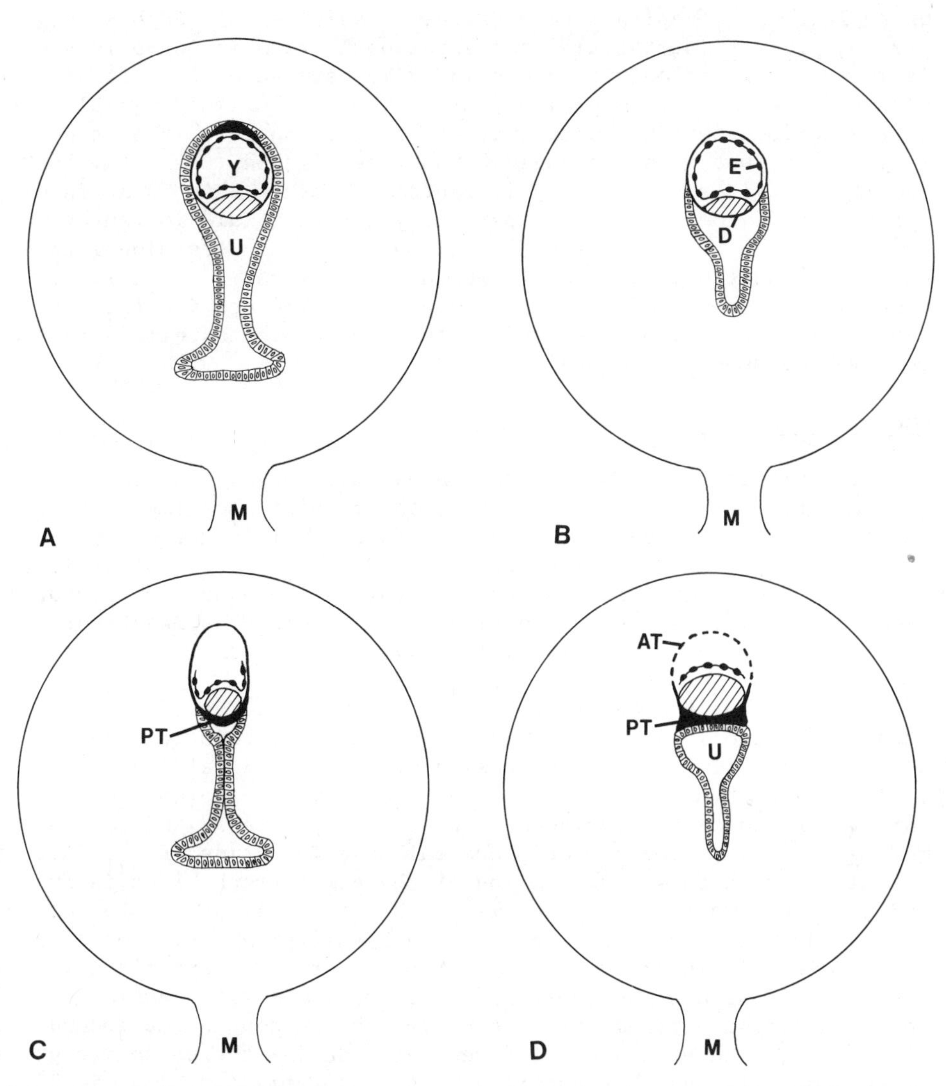

Fig. 9. Schematic diagram of pregnant uteri in transverse section, showing the four major patterns of blastocyst implantation in rodent higher taxa. (A) Sciurid-aplodontid pattern, approximating the rodent morphotype. (B) Geomyoid-dipodoid pattern. (C) Muroid pattern. (D) Hystricognath pattern. See text for description. Abbreviations: AT, Abembryonic trophoblast; D, Embryonic disc; E, Yolk sac endoderm; M, Mesometrium of uterus; PT, Polar trophoblast; U, Uterine lumen; Y, Yolk sac cavity.

penetrates beneath the uterine epithelium, with subsequent "healing" or replacement of epithelium over the invasion site. Such a mechanism of primary interstitial implantation is rare in eutherians; it is known elsewhere only in hominoid primates, where it is attained by a different developmental pattern. The single implanted blastocyst described for the North American caviomorph Erethizon is notabl larger than that of other caviomorphs investigated (Perrotta, 1959), and it is still superficially implanted. Whether this is an intermediate condition leading to "partially interstitial" implantation, as suggested by Perrotta (1959), is impossible to determine without access to additional and later implantation stages. Virtually all other known aspects of erethizontid placentation are identical to those of other caviomorphs, and it seems probable that implantation follows the same fundamental process.

Amniogenesis

The process of amnion formation exhibits several different patterns in rodents, related in part to the mechanism of implantation and the fate of the polar trophoblast that overlies the embryonic knot during the early implantation period. Polar trophoblast is los during early implantation phases in sciurids, aplodontids, castorids geomyids, and dipodoids, and amniogenesis occurs by somatopleuric folding in all these taxa (Fig. 10A, B).

In muroids, the polar trophoblast is thin at the time of initia attachment, and it is soon brought into intimate contact with the uterine epithelium, following fusion of the epithelial lips of the implantation chamber (Fig. 9C). Consequently, the polar trophoblast thickens to form an ectoplacental plate in cricetids, and an even more elongate, thickened ectoplacental cone in murids (Fig. 10C). Associated with this modification of the polar trophoblast in muroids, the embryonic cell mass elongates and invaginates into the developing yolk sac. This is followed by differentiation of a separate cavity within the ectoplacental plate or cone (epamniotic cavity), and within the embryonic epiblast (primordial amniotic cavity). These extraembryonic cavities become continuous secondaril (Fig. 10C). The exocoelom differentiates during the presomite phase of development, and this contributes to the formation of amniotic folds that fuse to separate the definitive amniotic and epamniotic cavities.

Although definitive amniogenesis occurs by folding in sciurids and muroids, there are notable differences in the developmental pathways leading to this condition. To a considerable degree, the intermediate stages between these differing patterns of folding are found in the amniogenesis process of geomyids and dipodoids. Polar trophoblast disappears in early implantation stages of these taxa (Fig. 9B); nevertheless, they exhibit invagination of the embryonic mass into the yolk sac (Fig. 10B). Subsequently, somatopleuric amni

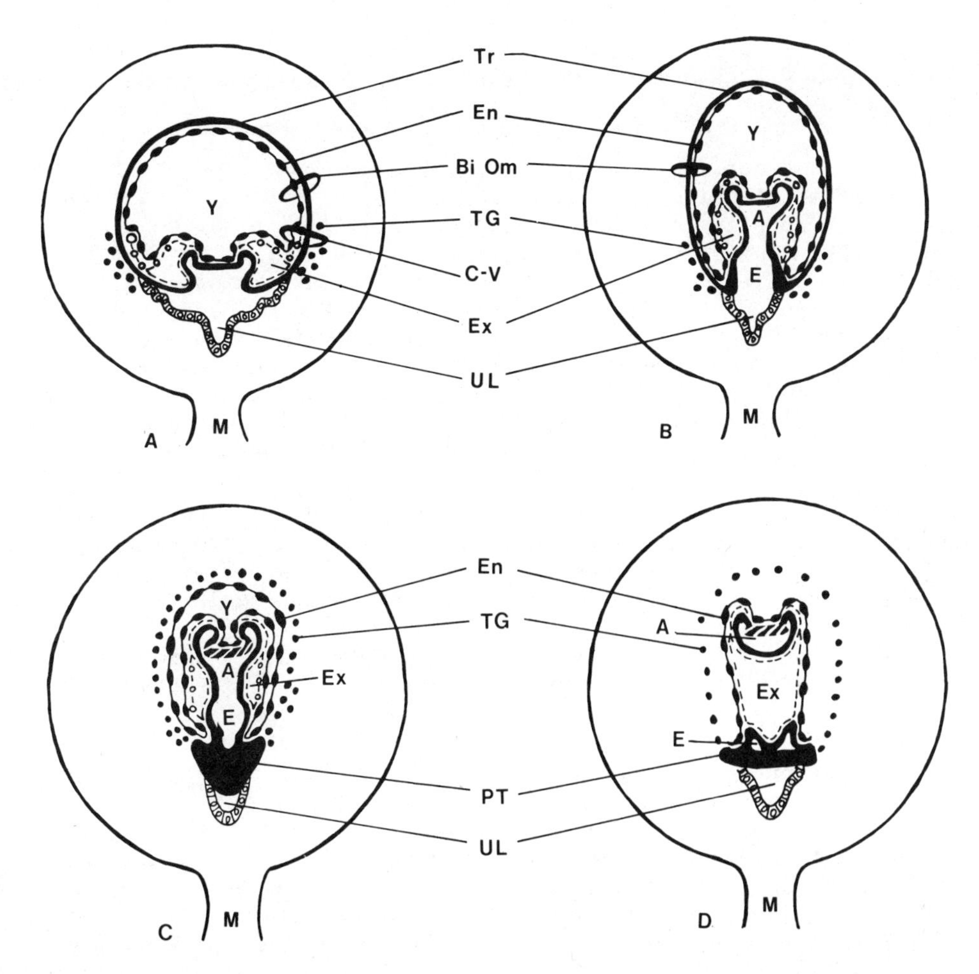

Fig. 10. Schematic diagram of four major patterns of postimplantation development in rodent higher taxa. (A) Sciurid-aplodontid pattern. (B) Geomyoid-dipodoid pattern. (C) Muroid pattern. (D) Hystricognath pattern. Note the appearance of the yolk sac, amnion, and primitive streak. See text for description. Abbreviations: A, Amniotic cavity; Bi Om, Bilaminar omphalopleure; C-V, Chorio-vitelline placenta; E, Epamniotic cavity; En, Endoderm; Ex, Exocoelom; PT, Preplacental trophoblast; TG, Trophoblastic giant cells; Tr, Trophoblast; UL, Uterine lumen; Y, Yolk sac cavity.

otic folds separate the amniotic cavity from an incomplete epamniotic cavity. The roof of the latter cavity remains incomplete in geomyids and dipodoids, due to the absence of polar trophoblast. The intermediate nature of amniogenesis and other aspects of fetal membrane morphogenesis in geomyids and dipodoids have been discussed ably by Mossman and his associates (Mossman, 1937; Mossman and Strauss, 1963; King and Mossman, 1974).

The most derived pattern of rodent amniogenesis occurs in Hystricognathi. A primordial amniotic cavity develops by cavitation within the epiblast of the early interstitially implanted blastocyst (Fig. 10D). This cavity persists to form the definitive amniotic cavity, in contrast to its transitory nature in muroids. An epamniotic cavity originates also by cavitation within the moderately developed ectoplacental plate, and it remains separated from the primordial amniotic cavity by an elongate, temporary proexocoelomic cavity. Following differentiation of the primitive streak, the proexocoelom is converted into an exocoelom (Fig. 10D) by the spread of extraembryonic mesoderm. The initial phases of amniogenesis have not been described in the African hystricognathous families Hystricidae or Bathyergidae. Later phases of implantation and the nature of the amnion and epamnion in these families are virtually identical, however, to those of caviomorphs (Luckett and Mossman, 1981), and it is almost certain that the pattern of initial amniogenesis is the same.

Pedetids and ctenodactylids develop a transitory primordial amniotic cavity, but the polar trophoblast is soon disrupted, exposing the embryonic epiblast to a temporary trophoepiblastic cavity whose roof is formed by the differentiating ectoplacental or preplacental trophoblast. Definitive amniogenesis occurs by somatopleuric folding which is initiated in late presomite stages, in conjunction with exocoelom differentiation. In contrast to the condition in geomyoids, dipodoids, and muroids, the embryonic disc does not invaginate noticeably into the yolk sac during the process of amnion formation; consequently, there is no development of an epamniotic cavity in conjunction with the definitive amniotic cavity. This is clearly a primitive retention, but it occurs in association with derived development of a preplacental proliferation of trophoblast at the future mesometrial placental site.

Yolk Sac and Choriovitelline Placentation

During later implantation phases, extraembryonic endoderm spreads peripherally along the inner surface of the trophoblast to form the endodermal lining of the yolk sac in most rodent families (Fig. 10A, B, C). A notable and rare exception to this primitive mammalian condition is the failure of peripheral endodermal spread in early implantation phases of all Old and New World hystricognaths (Fig. 10D); this feature is known elsewhere in rodents only in the family Ctenodactylidae (Luckett, 1980). Despite this difference, all rodents are characterized by failure of extraembryonic mesoderm to spread from the primitive streak to the peripheral half of the yolk sac wall. Consequently, only the embryonic half of the yolk sac can become vascularized during pregnancy (Fig. 11).

The distal or abembryonic wall of the modified yolk sac (= bilaminar omphalopleure) persists throughout pregnancy in sciurids, aplodontids, pedetids, and anomalurids (Fig. 11A); this apparently represents the primitive rodent condition. Loss of both the trophoblastic and endodermal layers of the bilaminar omphalopleure takes place in most other rodent families as gestation progresses (Fig. 11B, C, D). This leads to the inversion of vascular splanchnopleure of the persisting proximal portion of the yolk sac against the uterine endometrium. This inversion of the yolk sac occurs relatively late during gestation in castorids, geomyoids, and dipodoids, whereas inversion takes place earlier in muroids.

In hystricognaths, the abembryonic trophoblast degenerates soon after interstitial implantation is completed, and, coupled with absence of peripheral yolk sac endoderm, this leads to very early and complete inversion of the yolk sac during pre-primitive streak stages (Fig. 9D, 10D). This highly derived pattern of early yolk sac development is unique to hystricognaths. The abembryonic trophoblast persists throughout gestation in ctenodactylids, but the parietal or peripheral layer of yolk sac endoderm fails to develop, as noted above. Both the bilaminar omphalopleure and the inverted splanchnopleuric yolk sac of rodents have been shown to play an important functional role in the uptake of macromolecular substances, including immunoglobulins and other proteins, from maternal tissues (King, 1977; Carpenter, 1980).

Peripheral spread of extraembryonic mesoderm into the paraembryonic region of expanded, postimplantation blastocysts, and its subsequent vascularization, result in differentiation of a choriovitelline placenta in sciurids, aplodontids, castorids, anomalurids, pedetids, and ctenodactylids (Fig. 10A). This is clearly a primitive retention in rodents, as well as in therian mammals in general. In other rodents (geomyoids, dipodoids, muroids, and hystricognaths), in which the embryonic disc invaginates into the roof of the yolk sac during presomite stages, the vascular mesoderm of the yolk sac splanchnopleure is prevented from contacting the chorion (Fig. 10B, C, D). Therefore, these taxa can not develop a choriovitelline placenta during ontogeny.

Allantoic Vesicle

An endodermal allantoic vesicle differentiates in continuity with the hindgut during early somite phases of development in sciurids, aplodontids, castorids, anomalurids, pedetids, and ctenodactylids. The vascularized allantoic vesicle fuses with the mesometrial chorionic or preplacental trophoblast during early limb bud stages in these taxa; this initiates the development of the chorioallantoic placenta. The small allantoic vesicle persists at the fetal surface of the placental disc throughout gestation in these families (Fig. 11A); this is the primitive condition for the order.

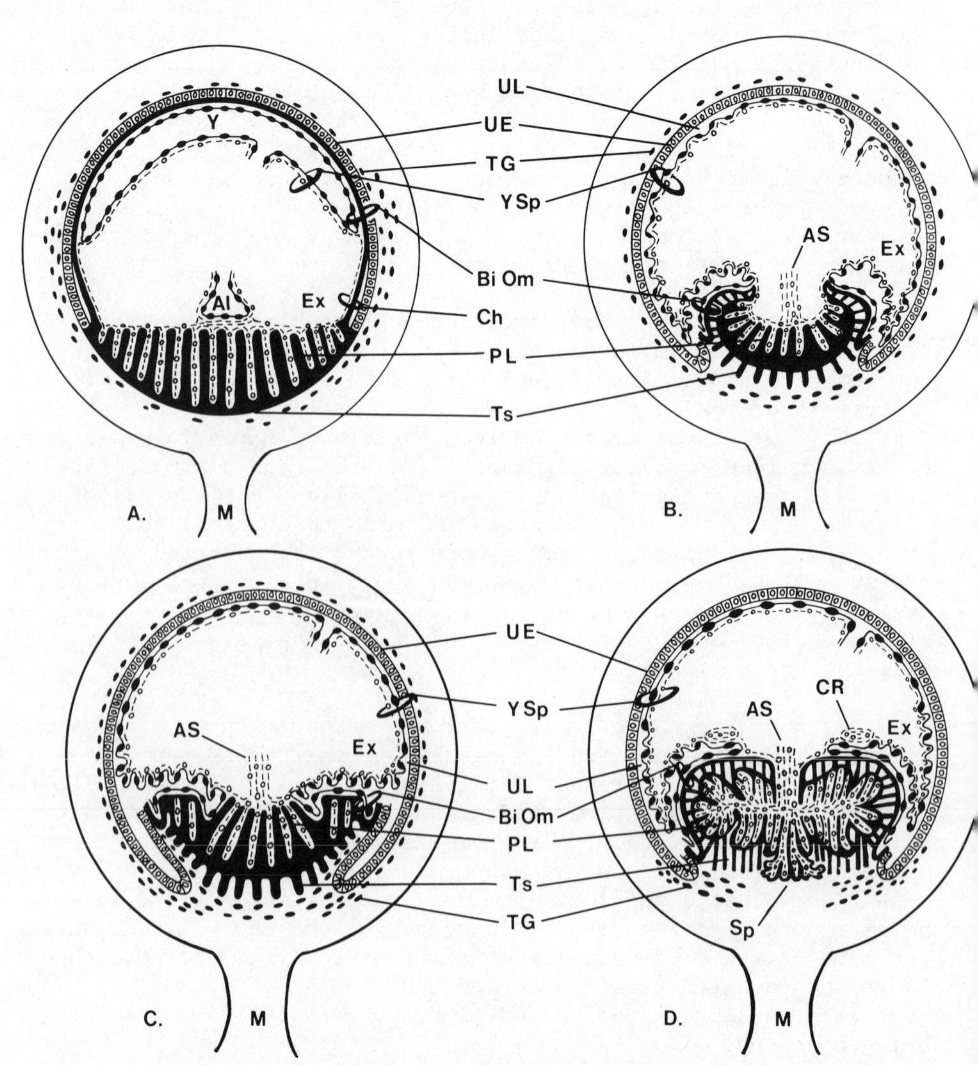

Fig. 11. Schematic diagram of four major patterns of the definitive fetal membranes and placenta in rodent higher taxa. (A) Sciurid-aplodontid pattern. (B) Geomyoid-dipodoid pattern. (C) Muroid pattern. (D) Hystricognath pattern. See text for description. Abbreviations: Al, Allantoic vesicle; AS, Allantoic stalk; Ch, Chorion; CR, Capillary ring; PL, Placental labyrinth; Sp, Subplacenta; Ts, Trophospongium; UE, Uterine epithelium; Y Sp, Yolk sac splanchnopleure. Other abbreviations as in Figure 10.

A bud of allantoic mesoderm differentiates from the caudal end of the primitive streak in late presomite-early somite embryos of geomyoids, dipodoids, muroids, and hystricognaths. With further growth, this mesodermal bud expands into the exocoelom, but it lacks an endodermal allantoic diverticulum. Continued growth of the vascularized mesodermal bud or body stalk brings it into contact and subsequent fusion with the chorionic trophoblast to initiate chorioallantoic placentation (Fig. 11B, C, D). The body stalk, devoid of an endodermal allantoic diverticulum, contacts the chorion during early limb bud stages in geomyoids and hystricognaths (Nielson, 1940; Mossman and Strauss, 1963; Luckett and Mossman, 1981), at a comparable developmental period to that of sciurids and aplodontids. In contrast, chorioallantoic fusion is initiated at an earlier developmental period in dipodoids and muroids (7-8 somite stage); this precocious vascularization represents a derived eutherian condition.

Chorioallantoic Placenta

A discoidal, hemochorial placenta differentiates at the mesometrial pole of the uterus in all rodents (Fig. 11). Despite this basic similarity in structural plan, differences are evident in the histological organization of the placental disc in some rodent families. A characteristic feature of geomyids and heteromyids is the development of a cup-shaped placental hillock in early stages, and the subsequent differentiation of a pileate or inflected placental disc (Mossman and Strauss, 1963). A somewhat similar inflected disc is evident in late stages of ctenodactylids (Luckett, unpublished), but the developmental pattern leading to this condition is very different from that in geomyoids, suggesting that this is an example of convergent evolution.

In all hystricognath rodents, the definitive placental disc is highly lobulated (Perrotta, 1959; Roberts and Perry, 1974; Luckett and Mossman, 1981). It is likely that this lobulation develops in part as the result of proliferative activity of another unique hystricognath feature - the subplacenta. The hystricognath subplacenta is a specialized growth zone of cytotrophoblast (Fig. 11D), and it may also function in placental endocrine secretion (Davies et al., 1961), or in secretion of progesterone-binding globulin (Heap and Illingworth, 1974).

Placental lobulation also occurs in pedetids (Fischer and Mossman, 1969) and castorids (Fischer, 1971), but it is not associated with a cytotrophoblastic subplacenta. Lobulation of the placental disc in these families is probably associated with large body size, and it provides a mechanism for increasing the area of efficient interchange between fetal and maternal blood. Unfortunately, the term "subplacenta" has been used in castorids to describe a specialized region of syncytiotrophoblast found in the junctional zone between placental disc labyrinth and the degenerating uterine epithelium. Fischer (1971) acknowledged that this "subplacenta" is unlike that of hystricognaths, both histologically and developmentally. Therefore, continued use of this term for the castorid placenta can only lead to confusion and imprecision for functional and evolutionary studies.

In addition to the subplacenta, another uniquely derived feature of Old and New World hystricognaths is the occurrence of a capillary ring or fibrovascular ring that surrounds the site of attachment of the yolk sac splanchnopleure to the fetal surface of the placental disc (Fig. 11D). The functional significance of this unique structure is unknown, and no precursor or homologous feature has been found in other rodent taxa.

At the ultrastructural level, there are differences in the number of trophoblastic layers that participate in the interhemal membrane or zona intima (= placental barrier), which separates fetal and maternal blood in the placental disc. Mossman (1937) originally believed that the definitive interhemal membrane of geomyoids, muroids, and caviomorphs lacked an intact trophoblastic layer, and he classified this as a "hemoendothelial" type of placenta. However, electron microscopic studies have revealed that not only is the trophoblast layer intact, but that it may also consist of two or three layers of trophoblast in some rodent families (Enders, 1965). Three trophoblastic layers (a hemotrichorial condition) occur in all murids and cricetids (including cricetines, arvicolines, and sigmodontines) examined (Enders,1965; King and Hastings, 1977); this derived condition has not been detected in any other eutherians. It seems clear that these three layers are derived from the chorionic cytotrophoblast, the roof of the epamnion (= ectoplacental trophoblast or trager), and the trophoblastic giant cells that arise from the bilaminar omphalopleure (Fig. 11C, D), in close association with the ectoplacental trophoblast (Enders, 1965; Mossman and Fischer, 1969; Carpenter, 1972).

In dipodoids, a highly derived pattern occurs, in which the chorionic cytotrophoblast degenerates during early chorioallantoic differentiation, resulting in the development of a hemomonochorial interhemal membrane, formed only by trophoblastic giant cells of bilaminar omphalopleure origin (King and Mossman, 1974). The trophoblastic roof of the epamniotic cavity is not complete in this superfamily; this accounts for the absence of the "middle" layer of trophoblast that occurs in muroids. Relationships in geomyoids are more difficult to evaluate, because of the scarcity of ultrastructural studies. In the heteromyid Dipodomys, two layers separate the fetal endothelium from the maternal blood, but it is unclear whether this is an endotheliochorial condition (unknown elsewhere in rodents), or a hemodichorial relationship (King and Tibbitts, 1969). A consideration of the possible homologies of the trophoblastic layers in the hemodichorial placentae of pedetids and castorids, and the single layer in the hemomonochorial placentae of sciurids and hystricognaths, is beyond the scope of the present report.

Character Analysis of Fetal Membrane Data for Assessment of Intraordinal Relationships among Rodents

No assumptions were made at the onset of this study concerning the cladistic or sister-group relationship of Rodentia to other eutherian mammals. Therefore, the "outgroup" used for assessing the relatively primitive and derived character states of fetal membrane development in rodents was all other Eutheria. Such a broad-based character analysis facilitates morphotype reconstruction of fetal membrane and placental features in Rodentia and other mammalian orders. Following morphotype reconstruction for all eutherian orders (see Luckett, 1977), a search for the sister group of rodents can be undertaken, based on the identification of shared, derived similarities in individual traits, as well as in the entire developmental complex of the fetal membranes and placenta (see below).

There is general agreement (Mossman, 1937; Fischer and Mossman, 1969; Luckett, 1971; Luckett and Mossman, 1981) that the fetal membrane complex of sciurids and aplodontids most closely resembles the primitive rodent pattern. This developmental complex includes: superficial, eccentric implantation of the blastocyst by its abembryonic pole; loss of polar trophoblast during the peri-implantation period; absence of a primordial amniotic cavity; amniogenesis by folding; absence of an epamniotic cavity; occurrence of a complete endodermal lining of the early yolk sac; presence of a transitory choriovitelline placenta; development of a small, permanent allantoic vesicle; and persistence of the abembryonic trophoblast throughout gestation (see Table 6). Some of these primitive rodent features, especially those related to amniogenesis and the yolk sac, are doubtlessly retentions of the primitive eutherian morphotype (see Table 5; also, Luckett, 1977).

Other developmental traits that occur in all rodent taxa studied can also be attributed to their last common ancestor. These include: invasive blastocyst attachment by its abembryonic pole; mesometrial orientation of the embryonic disc at the time of implantation; establishment of the chorioallantoic placenta at the opposite (mesometrial pole) region of the uterus from the site of initial attachment; and development of a hemochorial placental disc. These morphotypic features are not useful for assessing the intraordinal relationships among rodents, but they provide valuable evidence for evaluating the eutherian affinities of the order (see Table 5, Fig. 7, and discussion below).

The most derived pattern of fetal membrane development among

Table 6. Character analysis of selected rodent traits.

Primitive	Derived
1. Mandible sciurognathous	1. Mandible hystricognathous
2. Subplacenta absent	2. Subplacenta present
3. Pars reflexa of superficial masseter muscle small	3. Pars reflexa of superficial masseter well developed
4. Pars posterior, deep division of masseter lateralis profundus indistinct	4. Pars posterior, deep division of masseter lateralis profundus well developed
5. Fibrovascular ring absent	5. Fibrovascular ring present
6. Implantation superficial	6. Implantation primary interstitial
7. Transitory proexocoelomic cavity absent	7. Transitory proexocoelomic cavity present
8. Pterygoid fossa does not open anteriorly	8. Pterygoid fossa opens anteriorly into orbit
9. Complete endodermal lining develops in early blastocyst	9. Parietal endodermal layer fails to develop
10. Gestation period short	10. Gestation period long
11. Malleus and incus unfused	11. Malleus and incus fused
12. Sacculus urethralis absent	12. Sacculus urethralis present
13. Abembryonic trophoblast persists	13. Abembryonic trophoblast disappears precociously
14. No primordial amniotic cavity; amniogenesis by folding	14. Primordial amniotic cavity develops by cavitation; persists to form definitive amniotic cavity
15. True chorion permanent and paraplacental	15. True chorion temporary and placental
16. Incisor enamel multiserial	16. Incisor enamel uniserial
17. Masseter muscle complex protrogomorphous	17. Masseter muscle complex myomorphous or hystricomorphous
18. Allantoic vesicle small, permanent	18. Allantoic vesicle absent
19. Choriovitelline placenta present	19. Choriovitelline placenta absent
20. Epamniotic cavity absent	20. Epamniotic cavity develops by cavitation

rodents (indeed, within Eutheria) is found in the South American and African hystricognaths (Fig. 12), as reported previously (Luckett, 1980). This is true not only for individual traits, but also for the entire developmental complex. This developmental pattern includes a number of uniquely derived features, such as a subplacenta,

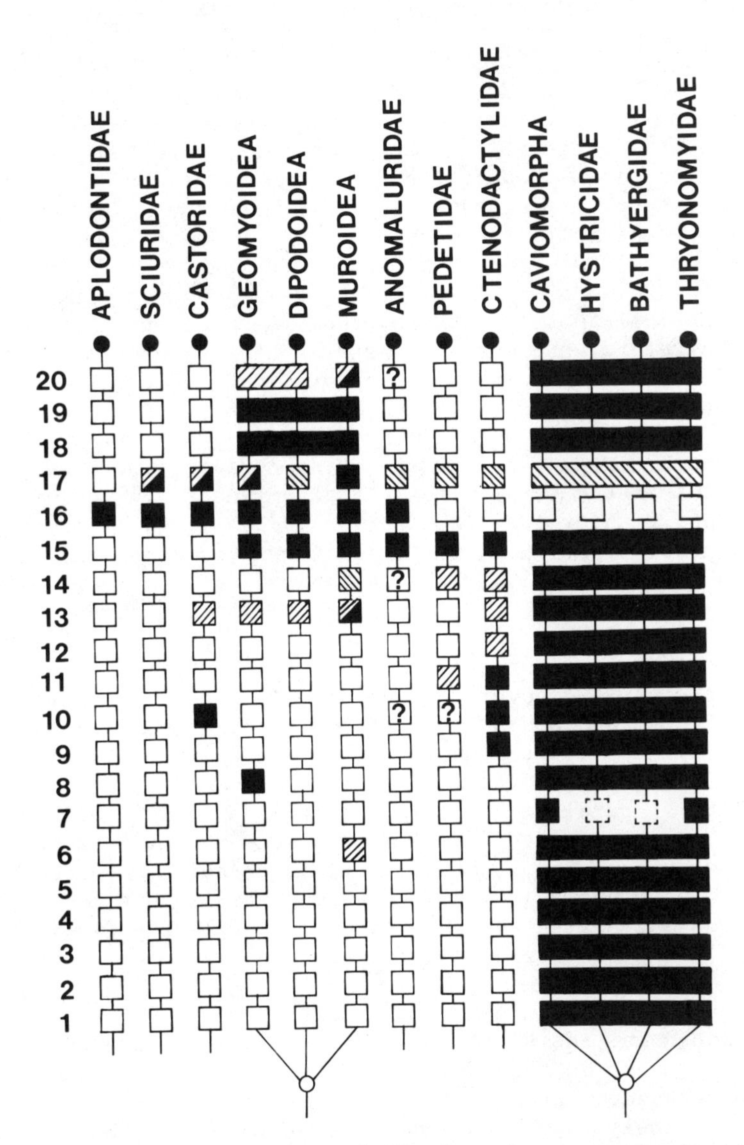

Fig. 12. Character phylogeny of fetal membrane, reproductive, and other anatomical features of rodents. See Table 6 for characters evaluated. Monophyly of the suborder Hystricognathi is strongly corroborated, and that of Myomorpha is more weakly supported. Broken-lined squares represent unknown character states.

primary interstitial implantation, and occurrence of a fibrovascular ring on the inverted yolk sac (Table 6 and Fig. 12).

Recently, placentation in the African hystricognathous family Thryonomyidae has been described by Oduor-Okelo and Gombe (1982) as differing in several important aspects from the developmental pattern in other hystricognaths, despite the presence of a subplacenta and fibrovascular ring. However, careful examination of their illustrations suggests that they have misinterpreted these features in Thryonomys, especially the mechanism of implantation, the nature of the inverted yolk sac, and the fate of the parietal trophoblast. Instead, the developmental pattern in this genus appears to be virtually identical to that of Bathyergus (see Luckett and Mossman, 1981). Therefore, the character states of Thryonomyidae included in Figure 12 are based on my re-interpretation of the data and illustrations provided by Oduor-Okelo and Gombe (1982).

The occurrence of numerous shared, derived, and homologous character states of the fetal membranes and placenta in Caviomorpha, Hystricidae, Bathyergidae, and Thryonomyidae (Fig. 12) corroborates the hypothesis of a monophyletic suborder Hystricognathi, in agreement with Lavocat (1973, 1974). Because of the virtually identical ontogenetic pattern of the highly derived fetal membrane complex in extant hystricognaths, it is reasonable (and parsimonious) to conclude that this developmental complex was also present in their last common ancestor. Moreover, it is impossible to distinguish clearly between African and South American hystricognaths, or among hystricids, bathyergids, and thryonomyids, on the basis of fetal membrane features. As emphasized elsewhere (Luckett and Mossman, 1981), the possibility of parallel evolution for this uniquely derived and functionally interrelated complex of developmental features within Hystricognathi appears remote. It is interesting to note that neither albumin immunological data (Sarich, this volume) nor myological data (Woods and Hermanson, this volume) can clearly distinguish between Old and New World hystricognaths, in agreement with the fetal membrane observations.

Although muroids share a number of derived features with hystricognaths, such as absence of an allantoic vesicle and choriovitelline placenta, and development of an epamniotic cavity (Fig. 12), it is likely that these shared similarities have evolved convergently in the two taxa. This is based on a consideration of the entire developmental complex of fetal membrane features in muroids, as well as on evidence for the origin of the muroid ontogenetic pattern from an intermediate condition similar to that which occurs in geomyoids and dipodoids (see Mossman, 1937; Nielson, 1940; Mossman and Strauss, 1963; King and Mossman, 1974). Indeed, fetal membrane data provide corroboration for the hypothesis (Wilson, 1949; Wood, 1959; Wahlert, 1978) of a sister-group relationship among Geomyoidea, Dipodoidea, and Muroidea, as members of a monophyletic taxon Myomorpha (Fig. 12). Unfortunately, data on fetal membrane development are not available for the myomorphous families Rhizomyidae, Spalacidae, and Gliridae. This would be particularly valuable for the latter family, because of the evidence which suggests that myomorphy was acquired convergently in glirids and muroids (Vianey-Liaud, this volume). A more detailed analysis of fetal membrane and reproductive features among myomorphs does not provide a clear picture of the branching sequence among muroids, geomyoids, and dipodoids (Fig. 13, Table Table 7).

In contrast to the ontogenetic evidence for intermediate pathways of evolution for the myomorph fetal membrane patterns, there are few developmental clues available for the evolutionary origin of the highly derived hystricognath pattern. Thus, the fetal membrane data provide little indication for the sister-group relationship of Hystricognathi to other rodents (Fig. 12). The family Ctenodactylidae exhibits a lack of development of the distal (abembryonic) portion of the yolk sac endoderm; this unusual condition is known elsewhere in rodents only in hystricognaths. Other ctenodactylid features, such as the development of a primordial amniotic cavity and differentiation of preplacental trophoblast, would also be expected to occur in the sister group of Hystricognathi. However, most of the developmental features of the ctenodactylid fetal membranes are retentions of the reconstructed rodent morphotype (Fig. 12); as such, they provide no evidence for assessing the intraordinal affinities of the family.

When fetal membrane, reproductive, and musculoskeletal features of rodents are evaluated together (Luckett, 1980), a larger number of shared, derived traits (Fig. 12, Table 6) can be identified for ctenodactylids and hystricognaths (also, see George, this volume). While none of these features alone provides convincing evidence for affinities between the two taxa, the combination of derived traits is consistent with a working hypothesis of a sister-group relationship between Ctenodactyloidea and Hystricognathi (for an opposing view, see Wood, this volume). Preliminary analysis of dental features in the newly discovered late Eocene phiomyids from North Africa (Jaeger et al., this volume) provides additional support for a close ctenodactyloid-hystricognath relationship, as did the earlier study of middle Eocene ctenodactyloids from Pakistan (Hussain et al., 1978). It is unfortunate that molecular data are unavailable as yet to test this hypothesis.

Affinities of pedetids and anomalurids, based on fetal membrane data, are even less clear than those of ctenodactylids. Fischer and Mossman (1969) argued for a close relationship between pedetids and ctenodactylids, and for their higher taxa affinities with "Sciuromorpha." However, this was based primarily on the retention of numerous shared primitive features of the fetal membranes and placenta (see Fig. 12), and this is also true for their similarities shared with anomalurids (Luckett, 1971).

Table 7. Character analysis of selected fetal membrane and reproductive features of myomorph rodents.

Primitive	Derived
1. True chorion permanent, paraplacental	1. True chorion temporary, placental
2. Choriovitelline placenta present	2. Choriovitelline placenta absent
3. Small allantoic vesicle present	3. Allantoic vesicle absent
4. Amniotic folds fuse during early somite stage	4. Amniotic folds fuse during presomite stage
5. Bilaminar omphalopleure persists	5. Bilaminar omphalopleure disappears early
6. Epamniotic cavity absent	6. Epamniotic cavity roofed by trophoblast
7. Initial chorioallantoic fusion during early limb bud stage	7. Initial chorioallantoic fusion during 7-10 somite stage
8. Implantation superficial, eccentric	8. Implantation secondarily interstitial
9. Polar trophoblast transitory	9. Polar trophoblast persists
10. Primordial amniotic cavity indistinct or absent	10. Primordial amniotic cavity secondarily continuous with epamniotic cavity
11. Preplacental trophoblast absent	11. Preplacental trophoblast well developed
12. Hemochorial placenta monochorial or dichorial	12. Hemochorial placenta trichorial
13. Ectoplacental cone absent	13. Ectoplacental cone present
14. Placental disc lacks inflected rim	14. Placental disc bears inflected rim
15. Ovarian thecal gland moderately developed	15. Ovarian thecal gland extensively developed
16. Ovarian bursa present	16. Ovarian bursa absent
17. Placental labyrinth formed in part by chorionic trophoblast	17. Placental labyrinth trophoblast formed solely by giant cells

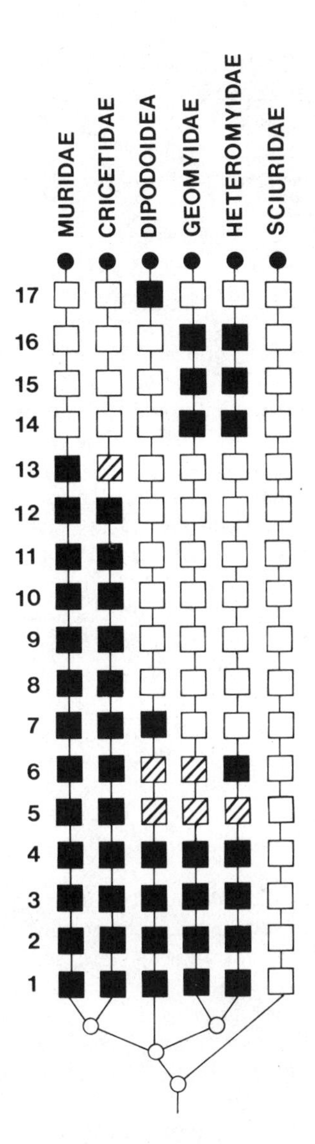

Fig. 13. Character phylogeny of selected fetal membrane and reproductive features in muroid, dipodoid, and geomyoid rodents. The family Sciuridae is included in the analysis as an outgroup for comparison. Monophyly of a suborder Myomorpha that includes Dipodoidea and Geomyoidea is corroborated by this analysis, but relationships between the three myomorph superfamilies are less clear. See Table 7 for the characters evaluated.

The hypothesis of a sister-group relationship between Sciuridae and Aplodontidae, supported by analysis of cranial and basicranial features in fossil and extant rodents (Vianey-Liaud, this volume; Lavocat and Parent, this volume), is neither corroborated nor refute by their sharing of a common suite of fetal membranes characters. This is due to the fact that both families retain a developmental pattern that approximates the rodent morphotype (Fig. 12).

Character Analysis of Fetal Membrane Data for Assessment of Superordinal Relationships of Rodents

Following the proposed separation of rodents and lagomorphs int distinct orders (Gidley, 1912), most paleontologists and comparative anatomists chose to emphasize the differences rather than the similarities between the two taxa. Thus, Hartman (1925) cited distinct differences between the rabbit and common laboratory rodents (mouse, rat, guinea pig) in the pattern of implantation and subsequent fetal membrane differentiation. At that time, however, there was little information available on placental development in sciurids, and subsequent investigations (Mossman, 1937) of this family revealed strik ing similarities between sciurids and lagomorphs in the pattern of fetal membrane morphogenesis. Indeed, Mossman (1937) emphasized tha the placental characters of lagomorphs differ in only minor details from those of relatively primitive rodents such as sciurids, and he retained lagomorphs (= Duplicidentata) as a suborder within the orde Rodentia. Subsequent studies on fetal membrane development in ochotonid lagomorphs and aplodontid rodents (Harvey, 1959a, b) provided further evidence for detailed similarities of these features in lago morphs, sciurids, and aplodontids.

As with other cases of shared similarities, these developmental resemblances might be due to symplesiomorphy, synapomorphy, or homoplasy. A prerequisite for comparing the patterns of fetal membrane development between rodents and lagomorphs is reconstruction of the ancestral (= morphotype) condition for each order. The developmenta patterns of leporids and ochotonids are fundamentally identical, and this character complex can be taken to represent the ancestral condition in the last common ancestor of extant lagomorphs. As noted above, the morphotypic condition of the major developmental features of the fetal membranes in rodents is represented by the aplodontid-sciurid pattern.

When the morphotype of the rodent fetal membrane pattern is compared to that of all other eutherian orders (and to the reconstructed eutherian morphotype), there is a striking degree of near identity with the pattern of Lagomorpha (Fig. 7, Table 5). This is true for individual features, as well as for the entire ontogenetic complex. The rodent-lagomorph pattern of invasive attachment at the abembryonic pole of the blastocyst, associated with orientation of the embryonic disc mesometrially, and the differentiation of the placental disc mesometrially at the pole opposite the initial attachment site, is a uniquely derived developmental complex that occurs in no other eutherian taxon. This developmental complex of the fetal membranes and placenta provides strong corroboration for a sister-group relationship of Rodentia and Lagomorpha as a superorder (or cohort) Glires, as suggested previously (Luckett, 1977).

Examination of early implantation stages in lagomorphs provides valuable clues to the evolutionary origin of the highly derived and unique pattern of implantation in rodents. The preimplantation blastocyst of lagomorphs expands considerably within the uterus, in contrast to rodents; this is a primitive eutherian condition shared with most taxa that subsequently develop an epitheliochorial or endotheliochorial placenta. By the time of initial attachment, the lagomorph blastocyst is considerably larger than that of rodents (Fig. 14). The trophoblastic cones or knobs of the rabbit blastocyst are distributed over the abembryonic and paraembryonic surface, instead

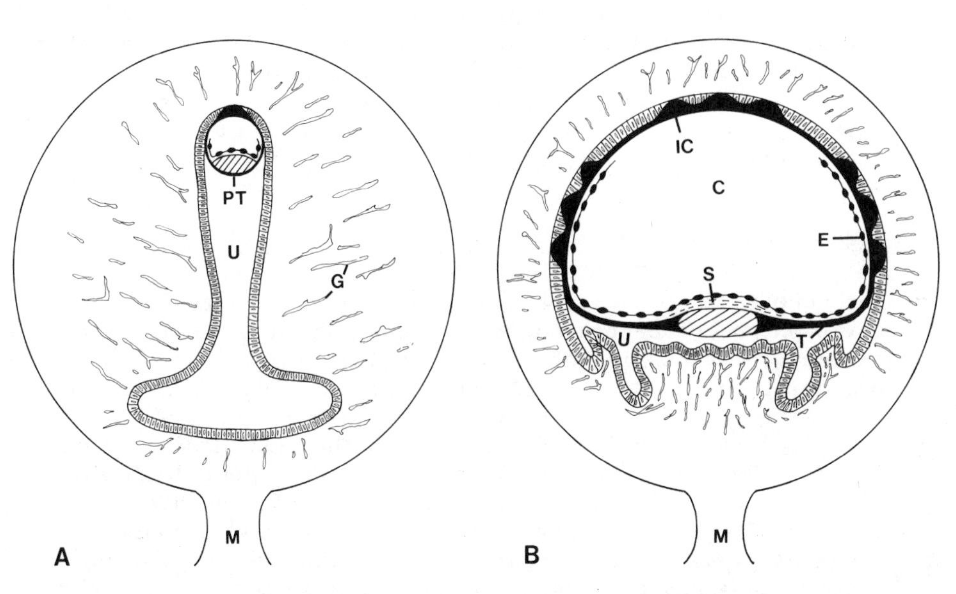

Fig. 14. Comparison of early implanted blastocysts in sciurid rodents and lagomorphs. (A) In sciurids, the small blastocyst is attached at its abembryonic pole by one or more implantation cones. (B) In lagomorphs, the more expanded blastocyst exhibits multiple implantation cones (IC) over its abembryonic and paraembryonic poles. The blastocyst is also more differentiated and possesses an early primitive streak (S). C, Blastocyst cavity; G, Uterine glands; T, Trophoblast. Other abbreviations as in Figure 9.

of being restricted to the abembryonic pole, as in rodents. Several hundred of these implantation knobs may develop over the blastocyst surface during the first 12 hours after initial trophoblastic invasion in the rabbit (Boving, 1962). This pattern of multiple attachment sites may be an intermediate condition in the development of the rodent abembryonic attachment cone from a more primitive eutherian condition of diffuse, paraembryonic attachment of the implanting blastocyst.

In contrast to the fetal membrane evidence for a rodent and lagomorph sister-group relationship, there are no shared, derived similarities between rodents and primates. In addition, fetal membrane data provide no corroboration for a hypothesis (Szalay, 1977) of special lagomorph-macroscelidid affinities. Those derived features shared by lagomorphs and macroscelidids are also shared with rodents (mesometrial orientation of the disc, invasive attachment, hemochorial placenta). As indicated above, consideration of the _entire_ developmental pattern suggests that these shared features in rodents and lagomorphs are homologous, whereas they are clearly convergent in macroscelidids. For instance, blastocyst attachment occurs at the same pole as the placental disc in macroscelidids (Horst, 1950), leading to a completely different interrelationship (Fig. 7). This emphasizes the value of considering the functional and ontogenetic interrelationships of characters that comprise an organ system, when assessing the probable homology or homoplasy of shared similarities in a phylogenetic analysis.

PHYLOGENETIC CONCLUSIONS

Ontogenetic data from two completely unrelated organ systems, the dentition and placentation, corroborate the hypothesis of a sister-group relationship between Rodentia and Lagomorpha, as a superordinal taxon Glires. This hypothesis is also corroborated by analysis of cranial features (Novacek, this volume). This concept of Glires is virtually identical with that espoused previously by Tullberg (1899) and Gregory (1910), based on a compilation of shared dental, cranial, neural, reproductive, and placental traits. The findings of the present study provide additional support for the paleontological evidence that Asiatic Paleocene eurymyloids may be near the ancestry of both rodents and lagomorphs (Li, 1977; Hartenberger, 1977, 1980; Li and Ting, this volume). In contrast, these data provide no support for the hypothesis of a close relationship between rodents and primates, _contra_ Wood (1962a, 1977) and McKenna (1969). Nor is there any support for the hypothesis that lagomorphs and macroscelidids shared a more recent common ancestry with each other than either group did with rodents, _contra_ McKenna (1975) and Szalay (1977). This is in agreement with the recent assessment of eutherian phylogeny by Novacek (1982), based on cranioskeletal, dental, and soft anatomical data. The Macroscelidea are considered as the sister group of Glires (among extant mammals) by Novacek (1982), but some of the characters used to support this hypothesis, such as the nature of the reduced canines and the composition of the auditory bulla, are of questionable homology and/or significance.

Consideration of numerous dental (Fig. 4) and cranial features (Li and Ting, this volume; Novacek, this volume) in leptictids, rodents, lagomorphs, and eurymyloids provides little support for Szalay's (this volume) hypothesis that the ancestry of the order Rodentia can be found in the Cretaceous-Paleocene leptictids.

Developmental evidence indicates that the large, gliriform incisors of both rodents and lagomorphs are retained, evergrowing dI2. Therefore, the ancestral stock or sister group of these taxa can be expected to show enlargement of I2 and concomitant reduction of I1 and I3, probably associated with loss of the canines and P1. In contrast, taxa that enlarged I1 and reduced I2 are unlikely to be close to the ancestry of Glires.

Within the order Rodentia, there is strong developmental support for the monophyletic clades Hystricognathi and Myomorpha, whereas other rodent groups retain numerous primitive fetal membrane traits. This makes it difficult to corroborate any hypotheses for the affinities of Sciuridae, Aplodontidae, Castoridae, Anomaluridae, Pedetidae, and Ctenodactylidae with either of the two large monophyletic clades, based on the available developmental data.

A final comment is warranted concerning the cladistic methodology used in this paper. The "character phylogenies" or "character state distributions" of developmental features presented in this report are simplified methods for summarizing and communicating large amounts of biological data to a diverse scientific community that shares a common interest in evolutionary biology, regardless of their primary anatomical, molecular, or paleontological interests. These character phylogenies are not a substitution for careful study of the developmental, functional, or "biological" meanings of characters that comprise an organ system. However, presentation of character state distributions in this manner encourages the investigator to evaluate the same suite of features in all taxa under consideration, and to make decisions concerning the morphotypic condition for each taxon. Mistakes may be made in such character analyses, but this method allows other biologists to clearly understand the data on which our phylogenetic hypotheses are based, and it provides an alternative to the "weight of authority" pronouncements, which are rarely accompanied by a careful elucidation of the relevant supporting data. Character analyses serve as tests for phylogenetic hypotheses, and the present analysis of developmental data is offered with this goal.

ACKNOWLEDGMENTS

The author is grateful to Dr. H. W. Mossman, Professor Emeritu Department of Anatomy, University of Wisconsin, Madison, for making his extensive collection of rodent embryological material available for study, as well as for his extensive discussions of rodent and mammalian placentation with me over the years. Most of the studies of dental development in rodents and lagomorphs were carried out at the Hubrecht Laboratory, Utrecht, the Netherlands, as were the exam nations of fetal membrane development in African rodents. Special thanks are due to Dr. Elze Boterenbrood of that Laboratory for her continued support and encouragement. Several colleagues in West Germany also made specimens available for study: Drs. W. Maier, R. Schneider, and A. Dierbach (Frankfurt); and Prof. H.-J. Kuhn (Göttingen). Finally, a special debt of gratitude is owed to Dr. Nancy Hong, not only for the illustrations for this paper, but also for her editorial assistance, advice, and support during all phases of this study. Financial support for research in the Netherlands and West Germany was provided in part by the University of Puerto Rico School of Medicine, through the active support of Dean Pedro Santiago Borrero, M. D.

REFERENCES

Adloff, P. 1898. Zur Entwickelungsgeschichte des Nagetiergebisses Jena. Zeitsch. Naturwiss. 32: 347-411.

Adloff, P. 1904. Ueber den Zahnwechsel von Cavia cobaya. Anat. Anz. 25: 141-147.

Berkovitz, B. K. B. 1972. Ontogeny of tooth replacement in the guinea pig (Cavia cobaya). Archs. Oral Biol. 17: 711-718.

Bock, W. J. 1977. Foundations and methods of evolutionary classification. In: Major Patterns in Vertebrate Evolution, M. K. Hecht, P. C. Goody, and B. M. Hecht, eds., pp. 851-895, Plenum Press, New York.

Böving, B. G. 1962. Anatomical analysis of rabbit trophoblast invasion. Carneg. Inst. Wash., Contrib. Embryol. 37: 33-55.

Butler, P. M. 1956. The ontogeny of molar pattern. Biol. Rev. 31: 30-70.

Butler, P. M. 1980. The tupaiid dentition. In: Comparative Biolog and Evolutionary Relationships of Tree Shrews, W. P. Luckett, ed., pp. 171-204, Plenum Press, New York.

Carpenter, S. J. 1972. Light and electron microscopic observations on the chorioallantoic placenta of the golden hamster (Cricetus auratus). Days seven through nine of gestation. Amer. J. Anat 135: 445-476.

Carpenter, S. J. 1980. Placental permeability during early gestation in the hamster. Anat. Rec. 197: 221-238.

Cope, E. D. 1888. The mechanical causes of the origin of the dentition of the Rodentia. Amer. Nat. 22: 3-13.

Cracraft, J. 1981. The use of functional and adaptive criteria in phylogenetic systematics. Amer. Zool. 21: 21-36.

Davies, J., Dempsey, E. W. and Amoroso, E. C. 1961. The subplacenta of the guinea pig: Development, histology and histochemistry. J. Anat. 95: 457-472.

Dawson, M. R., Li, C.-K. and Qi, T. 1984. Eocene ctenodactyloid rodents (Mammalia) of Eastern and Central Asia. Carneg. Mus. Nat. Hist. Spec. Publ. 9: 138-150.

Eldredge, N. 1979. Cladism and common sense. In: Phylogenetic Analysis and Paleontology, J. Cracraft and N. Eldredge, eds., pp. 165-198, Columbia Univ. Press, New York.

Enders, A. C. 1965. A comparative study of the fine structure of the trophoblast in several hemochorial placentas. Amer. J. Anat. 116: 29-68.

Fischer, T. V. 1971. Placentation in the American beaver (Castor canadensis). Amer. J. Anat. 131: 159-184.

Fischer, T. V. and Mossman, H. W. 1969. The fetal membranes of Pedetes capensis, and their taxonomic significance. Amer. J. Anat. 124: 89-116.

Fitzgerald, L. R. 1973. Deciduous incisor teeth of the mouse (Mus musculus). Archs. Oral Biol. 18: 381-389.

Freund, P. 1892. Beiträge zur Entwicklungsgeschichte der Zahnanlagen bei Nagethieren. Arch. Mikr. Anat. 39: 525-555.

Friant, M. 1945. La formule dentaire des Rongeurs de la famille des Thryonomyidae. Rev. Zool. Botan. Afr. 38: 200-205.

Gaffney, E. S. 1979. An introduction to the logic of phylogeny reconstruction. In: Phylogenetic Analysis and Paleontology, J. Cracraft and N. Eldredge, eds., pp. 79-111, Columbia Univ. Press, New York.

Gaunt, W. A. 1955. The development of the molar pattern of the mouse (Mus musculus). Acta Anat. 24: 249-268.

Gaunt, W. A. 1961. The development of the molar pattern of the golden hamster (Mesocricetus auratus W.), together with a re-assessment of the molar pattern of the mouse (Mus musculus). Acta Anat. 45: 219-251.

Gazin, C. L. 1953. The Tillodontia: An early Tertiary order of mammals. Smith. Misc. Coll. 12: 1-110.

Gidley, J. W. 1912. The lagomorphs an independent order. Science 36: 285-286.

Gingerich, P. D. 1976. Cranial anatomy and evolution of early Tertiary Plesiadapidae (Mammalia, Primates). Univ. Mich. Pap. Paleont. 15: 1-141.

Gingerich, P. D. and Gunnell, G. F. 1979. Systematics and evolution of the genus Esthonyx (Mammalia, Tillodontia) in the Early Eocene of North America. Univ. Mich. Contrib. Mus. Paleont. 25: 125-153.

Gregory, W. K. 1910. The orders of mammals. Bull. Amer. Mus. Nat. Hist. 27: 1-524.

Harman, M. T. and Smith, A. 1936. Some observations on the development of the teeth of Cavia cobaya. Anat. Rec. 66: 97-111.

Hartenberger, J.-L. 1977. A propos de l'origine des Rongeurs. Géobios, Mém. Spéc. 1: 183-193.

Hartenberger, J.-L. 1980. Donnees et hypotheses sur la radiation initiale des rongeurs. Palaeovertebrata Mém. Jubil. R. Lavocat: 285-301.

Hartman, C. G. 1925. On some characters of taxonomic value appertaining to the egg and the ovary of rabbits. J. Mammal. 6: 114-121.

Harvey, E. B. 1959a. Placentation in Ochotonidae. Amer. J. Anat. 104: 61-85.

Harvey, E. B. 1959b. Placentation in Aplodontidae. Amer. J. Anat. 105: 63-90.

Hay, M. F. 1961. The development *in vivo* and *in vitro* of the lower incisor and molars of the mouse. Archs. Oral Biol. 3: 86-109.

Heap, R. B. and Illingworth, D. V. 1974. The maintenance of gestation in the guinea-pig and other hystricomorph rodents: Changes in the dynamics of progesterone metabolism and the occurrence of progesterone-binding globulin (PBG). Symp. Zool. Soc. Lond. 34: 385-415.

Heinick, P. 1908. Uber die Entwicklung des Zahnsystems von *Castor fiber* L. Zool. Jahrb. Abt. Anat. 26: 355-402.

Hennig, W. 1966. Phylogenetic Systematics. Univ. Ill. Press, Urbana.

Hirschfeld, Z., Weinreb, M. M. and Michaeli, Y. 1973. Incisors of the rabbit: Morphology, histology, and development. J. Dent. Res. 52: 377-384.

Horowitz, S. L., Weisbroth, S. H. and Scher, S. 1973. Deciduous dentition in the rabbit (*Oryctolagus cuniculus*). A roentgenographic study. Archs. Oral Biol. 18: 517-523.

Horst, C. J. van der 1950. The placentation of *Elephantulus*. Trans. Roy. Soc. S. Afr. 32: 435-629.

Hussain, S. T., de Bruijn, H. and Leinders, J. M. 1978. Middle Eocene rodents from the Kala Chitta Range (Punjab, Pakistan). Proc. Kon. Ned. Akad. Wetensch., Amsterdam, Ser. B. 81: 74-112.

Jacobs, L. L. 1977. A new genus of murid rodent from the Miocene of Pakistan and comments on the origin of the Muridae. Paleobios 25: 1-11.

Kielan-Jaworowska, Z. 1975. Results of the Polish-Mongolian palaeontological expeditions. Part VI. Preliminary description of two new eutherian genera from the Late Cretaceous of Mongolia. Palaeont. Pol. 33: 5-16.

Kindahl, M. 1957. Some observations on the development of the tooth in *Elephantulus myurus jamesoni*. Ark. Zool. 11: 21-29.

King, B. F. 1977. An electron microscopic study of absorption of peroxidase-conjugated immunoglobulin G by guinea pig visceral yolk sac *in vitro*. Amer. J. Anat. 148: 447-456.

King, B. F. and Hastings, R. A. II 1977. The comparative fine structure of the interhemal membrane of chorioallantoic placentas from six genera of myomorph rodents. Amer. J. Anat. 149: 165-180.

King, B. F. and Mossman, H. W. 1974. The fetal membranes and unusual giant cell placenta of the jerboa (Jaculus) and jumping mouse (Zapus). Amer. J. Anat. 140: 405-432.
King, B. F. and Tibbitts, F. D. 1969. The ultrastructure of the placental labyrinth in the kangaroo rat, Dipodomys. Anat. Rec. 163: 543-554.
Kollar, E. J. 1972. Histogenetic aspects of dermal-epidermal interactions. In: Developmental Aspects of Oral Biology, H. C. Slavkin and L. A. Bavetta, eds., pp. 125-149, Academic Press, New York.
Lavocat, R. 1956. Réflexions sur la classification des Rongeurs. Mammalia 20: 49-56.
Lavocat, R. 1973. Les Rongeurs du Miocène d'Afrique orientale. I. Miocène inférieur. Mém. Trav. E.P.H.E., Inst. Montpellier 1: 1-284.
Lavocat, R. 1974. What is an hystricomorph? Symp. Zool. Soc. Lond. 34: 7-20.
Leche, W. 1895. Zür Entwicklungsgeschichte des Zahnsystems der Säugethiere. I. Ontogenie. Bibl. Zoologica 17: 1-160.
Li, C.-K. 1977. Paleocene eurymyloids (Anagalida, Mammalia) of Qianshan, Anhui. Vert. PalAsiat. 15: 103-118.
Luckett, W. P. 1971. The development of the chorio-allantoic placenta of the African scaly-tailed squirrels (Family Anomaluridae). Amer. J. Anat. 130: 159-178.
Luckett, W. P. 1975. Ontogeny of the fetal membranes and placenta: Their bearing on primate phylogeny. In: Phylogeny of the Primates, W. P. Luckett and F. S. Szalay, eds., pp. 157-182, Plenum Press, New York.
Luckett, W. P. 1977. Ontogeny of amniote fetal membranes and their application to phylogeny. In: Major Patterns in Vertebrate Evolution, M. K. Hecht, P. C. Goody and B. M. Hecht, eds., pp. 439-516, Plenum Press, New York.
Luckett, W. P. 1980. Monophyletic or diphyletic origin of Anthropoidea and Hystricognathi: Evidence of the fetal membranes. In: Evolutionary Biology of the New World Monkeys and Continential Drift, R. L. Ciochon and B. Chiarelli, eds., pp. 347-368, Plenum Press, New York.
Luckett, W. P. 1982. The uses and limitations of embryological data in assessing the phylogenetic relationships of Tarsius (Primates, Haplorhini). Géobios, Mém. Spéc. 6: 289-304.
Luckett, W. P. and Maier, W. 1982. Development of deciduous and permanent dentition in Tarsius and its phylogenetic significance. Folia Primatol. 37: 1-36.
Luckett, W. P. and Mossman, H. W. 1981. Development and phylogenetic significance of the fetal membranes and placenta of the African hystricognathous rodents Bathyergus and Hystrix. Amer. J. Anat. 162: 265-285.
Lumsden, A. G. S. 1979. Pattern formation in the molar dentition of the mouse. J. Biol. Buccale 7: 77-103.
MacPhee, R. D. E. 1981. Auditory regions of primates and eutherian

insectivores. Contrib. Primat. 18: 1-282.

Martin, R. D. 1975. The bearing of reproductive behavior and ontogeny on strepsirhine phylogeny. In: Phylogeny of the Primates, W. P. Luckett and F. S. Szalay, eds., pp. 265-297, Plenum Press, New York.

McKenna, M. C. 1969. The origin and early differentiation of therian mammals. Ann. N. Y. Acad. Sci. 167: 217-240.

McKenna, M. C. 1975. Toward a phylogenetic classification of the Mammalia. In: Phylogeny of the Primates, W. P. Luckett and F. S. Szalay, eds., pp. 21-46, Plenum Press, New York.

Michaeli, Y., Hirschfeld, Z. and Weinreb, M. M. 1980. The cheek teeth of the rabbit: morphology, histology and development. Acta Anat. 106: 223-239.

Michaux, J. 1968. Les Paramyidae (Rodentia) de l'Eocène inférieur du Bassin de Paris. Palaeovertebrata 1: 135-194.

Moss-Salentijn, L. 1978. Vestigial teeth in the rabbit, rat and mouse; their relationship to the problem of lacteal dentitions. In: Development, Function and Evolution of Teeth, P. M. Butler and K. A. Joysey, eds., pp. 13-29, Academic Press, London.

Mossman, H. W. 1937. Comparative morphogenesis of the fetal membranes and accessory uterine structures. Contrib. Embryol. Carneg. Inst. 26: 129-246.

Mossman, H. W. 1953. The genital system and the fetal membranes as criteria for mammalian phylogeny and taxonomy. J. Mammal. 34: 289-298.

Mossman, H. W. 1967. Comparative biology of the placenta and fetal membranes. In: Fetal Homeostasis, Vol. 2, R. M. Wynn, ed., pp. 13-97, N. Y. Acad. Sci., New York.

Mossman, H. W. and Fischer, T. V. 1969. The preplacenta of _Pedetes_ the Träger, and the maternal circulatory pattern in rodent placentae. J. Reprod. Fert., Suppl. 6: 175-184.

Mossman, H. W. and Strauss, F. 1963. The fetal membranes of the pocket gopher illustrating an intermediate type of rodent membrane formation. II. From the beginning of the allantois to term. Amer. J. Anat. 113: 447-478.

Nelson, G. 1978. Ontogeny, phylogeny, paleontology, and the biogenetic law. Syst. Zool. 27: 324-345.

Nielson, P. E. 1940. The fetal membranes of the kangaroo rat, _Dipodomys_, with a consideration of the phylogeny of the Geomyoidea. Anat. Rec. 77: 103-127.

Novacek, M. J. 1977. Evolution and Relationships of the Leptictida (Eutheria: Mammalia). Unpublished Ph. D. Thesis, Univ. Calif., Berkeley.

Novacek, M. J. 1982. Information for molecular studies from anatomical and fossil evidence on higher eutherian phylogeny. In: Macromolecular Sequences in Systematic and Evolutionary Biology M. Goodman, ed., pp. 3-41, Plenum Press, New York.

Oduor-Okelo, D. and Gombe, S. 1982. Placentation in the cane rat (_Thryonomys swinderianus_). Afr. J. Ecol. 20: 49-66.

Ooë, T. 1979. Development of human first and second permanent molar, with special reference to the distal portion of the dental lamina. Anat. Embryol. 155: 221-240.

Ooë, T. 1980. Développement embryonnaire des incisives chez le lapin (Oryctolagus cuniculus L.). Interprétation de la formule dentaire. Mammalia 44: 259-269.

Patterson, B. 1965. The fossil elephant shrews (family Macroscelididae). Bull. Mus. Comp. Zool. Harv. 133: 295-335.

Perrotta, C. A. 1959. Fetal membranes of the Canadian porcupine, Erethizon dorsatum. Amer. J. Anat. 104: 35-59.

Roberts, C. M. and Perry, J. S. 1974. Hystricomorph embryology. Symp. Zool. Soc. Lond. 34: 333-360.

Ruch, J. V., Karcher-Djuricic, V. and Gerber, R. 1973. Les déterminismes de la morphogenèse et des cytodifférenciations des ébauches dentaires de souris. J. Biol. Buccale 1: 45-56.

Santoné, P. 1935. Studien über den Aufbau, die Struktur und die Histiogenese der Molaren der Säugetiere. I. Molaren von Cavia cobaya. Z. Mikr. Anat. Forsch. 37: 49-100.

Shevyreva, N. S. 1976. Paleogene rodents of Asia. Acad. Sci. U. S. S. R., Trans. Palaeont. Inst. 158: 1-115. (In Russian)

Simons, E. L. 1972. Primate Evolution. An Introduction to Man's Place in Nature. Macmillan, New York.

Simpson, G. G. 1931. A new insectivore from the Oligocene, Ulan Gochu horizon, of Mongolia. Amer. Mus. Novit. 505: 1-22.

Simpson, G. G. 1945. The principles of classification and a classification of mammals. Bull. Amer. Mus. Nat. Hist. 85: 1-350.

Simpson, G. G. 1959. The nature and origin of supraspecific taxa. Cold Spring Harbor Symp. Quant. Biol. 24: 255-271.

Simpson, G. G. 1961. Principles of Animal Taxonomy. Columbia Univ. Press, New York.

Strassburg, M., Peters, S. and Eitel, H. 1971. Zur Morphogenese der Zahnleiste. 2. Histologische Untersuchungen über die frühesten Differenzierungsphasen der Zahnleiste bei der Maus. Dtsch. Zahnärztl. Z. 26: 52-57.

Sulimski, A. 1968. Paleocene genus Pseudictops Matthew, Granger and Simpson 1929 (Mammalia) and its revision. Palaeont. Pol. 19: 101-129.

Szalay, F. S. 1977. Phylogenetic relationships and a classification of the eutherian Mammalia. In: Major Patterns in Vertebrate Evolution, M. K. Hecht, P. C. Goody and B. M. Hecht, eds., pp. 315-374, Plenum Press, New York.

Szalay, F. S. 1981. Functional analysis and the practice of the phylogenetic method as reflected by some mammalian studies. Amer. Zool. 21: 37-45.

Szalay, F. S. and Delson, E. 1979. Evolutionary History of the Primates. Academic Press, New York.

Szalay, F. S. and McKenna, M. C. 1971. Beginning of the age of mammals in Asia: The Late Paleocene Gashato Fauna, Mongolia. Bull. Amer. Mus. Nat. Hist. 144: 269-318.

Thesleff, I. and Hurmerinta. 1981. Tissue interactions in tooth development. Differentiation 18: 75-88.

Thomas, O. 1894. Description of a new species of reed-rat (Aulacodus) from East Africa, with remarks on the milk-dentition of the genus. Ann. Mag. Nat. Hist. 13: 202-204,

Tims, H. W. M. 1901. Tooth-genesis in the Caviidae. J. Linn. Soc. (Zool.) 28: 261-290.

Tullberg, T. 1899. Ueber das System der Nagetiere: eine phylogenetische Studie. Nova Acta Reg. Soc. Scient. Upsala, Ser. 3, 18: 1-514.

Van Valen, L. 1971. Adaptive zones and the orders of mammals. Evolution 25: 420-428.

Wahlert, J. H. 1974. The cranial foramina of protrogomorphous rodents; an anatomical and phylogenetic study. Bull. Mus. Comp. Zool. 146: 363-410.

Wahlert, J. H. 1978. Cranial foramina and relationships of the Eomyoidea (Rodentia, Geomorpha). Skull and upper teeth of Kansasimys. Amer. Mus. Novit. 2645: 1-16.

Wilson, R. W. 1949. Early Tertiary rodents of North America. Carneg. Inst. Wash. Publ. 584: 67-164.

Wood, A. E. 1959. Eocene radiation and phylogeny of the rodents. Evolution 13: 354-361.

Wood, A. E. 1962a. The early Tertiary rodents of the Family Paramyidae. Trans. Amer. Phil. Soc. 52: 1-261.

Wood, A. E. 1962b. The juvenile tooth patterns of certain African rodents. J. Mammal. 43: 310-322.

Wood, A. E. 1965a. Grades and clades among rodents. Evolution 19: 115-130.

Wood, A. E. 1965b. Unworn teeth and relationships of the African rodent, Pedetes. J. Mammal. 46: 419-423.

Wood, A. E. 1968. Early Cenozoic mammalian faunas, Fayum Province, Egypt. II. The African Oligocene Rodentia. Bull. Peabody Mus. Nat. Hist. 28: 23-105.

Wood, A. E. 1974. The evolution of the Old World and New World hystricomorphs. Symp. Zool. Soc. Lond. 34: 21-60.

Wood, A. E. 1977. The Rodentia as clues to Cenozoic migration between the Americas and Europe and Africa. Milwaukee Public Mus., Spec. Publ. Biol. and Geol. 2: 95-109.

Wood, A. E. 1980. The origin of the caviomorph rodents from a source in Middle America: A clue to the area of origin of the platyrrhine primates. In: Evolutionary Biology of the New World Monkeys and Continental Drift, R. L. Ciochon and A. B. Chiarelli, eds., pp. 79-91, Plenum Press, New York.

Woods, C. A. 1976. Deciduous premolars in Thryonomys. J. Mammal. 57: 370-371.

Ziegler, A. C. 1971. A theory of the evolution of therian dental formulas and replacement patterns. Quart. Rev. Biol. 46: 226-249.

POSSIBLE EVOLUTIONARY RELATIONSHIPS AMONG EOCENE AND LOWER OLIGOCENE RODENTS OF ASIA, EUROPE AND NORTH AMERICA

Monique Vianey-Liaud

Institut des Sciences de l'Evolution
U. S. T. L., Place E. Bataillon
34060 Montpellier, France

INTRODUCTION

A previous review by Dawson (1977) summarized the late Eocene rodent faunas of Asia, Europe and North America. Numerous new data provide information about the origin of rodents, and their evolution and radiation within the different continents. Many of these new findings will be discussed elsewhere in this volume.

First, I want to emphasize some methodologic aspects that explain a part of the difficulties for establishing phylogenetic relationships among Paleogene rodents from such diverse regions as Asia, Europe, and North America. The fossil record of Paleogene rodents in the three continents has been analyzed by very different paleontological traditions. Until recently, these could be characterized as follows:

(1) North America was rich in "beautiful" fossils, discovered in numerous localities from large sedimentary basins; however, in many of the studies there was little analysis of stratigraphically delineated large populations. Evolutionary lineages among these rodents have rarely been described by paleontologists, paleoanatomists, or taxonomists. Thus, there was no precise biostratigraphy based on evolutionary stages of rodents, and no clear analyses of possible faunal events.

(2) In western Europe, systematic research has identified numerous fossiliferous localities in small sedimentary basins, and in numerous, rich fissure fillings. Many studies have focused on population studies and on the establishment of evolutionary lineages. As a consequence, a precise biostratigraphy (mainly a scale of standard levels) has led to clear analyses of faunal variation.

(3) In Central Europe and Asia, however, the autochtonous paleontologists were not numerous, and therefore the fossiliferous local-

ities discovered have been relatively few. Thus, paleontological finds have been very scattered.

At present, knowledge is changing in varying degrees in the three continents. For Asia, mainly in China, the number of fossils and localities is increasing greatly. This is leading to considerable progress in knowledge of the origin of numerous mammalian taxa, particularly for Rodentia and Lagomorpha. In Europe, the improvement in methods leading to the refining of evolutionary lineages allows a better understanding of their evolutionary modes. In North America on the contrary, the methods used for tracing evolutionary lineages for Paleocene-Eocene Primates (Gingerich, 1976, 1977) have not been applied as yet to Eocene-Oligocene rodents, perhaps because of the abundance of "interesting" rodent specimens. However, "the time is not far off when detailed correlations of the continental Tertiary will be based primarily on the time ranges of rodent genera, species, or even of stages within species," as postulated by Wood (1980b) in his work on North American Oligocene rodents.

Table 1 shows the present state of lower Paleogene rodent biostratigraphy in the three continents. Taking into account the differences in progress of studies on Paleogene rodents, precise correlation are now nearly impossible, and any discussion must consider this limitation. Discussions about the evolutionary origin of rodent families must also consider data obtained from each of these three continental regions. To establish his argumentation on the origin of rodent families, the paleontologist generally searches for the earliest species of the family studied, in order to determine the place where the family may have arisen (and, if possible, it is better for him to find it in his own native land!). Of course, this localization is important; however, we must be very careful with the stratigraphic correlations between different continental regions, if we wish the notion of "earliest species" to remain meaningful. In this regard, Cavelier and Pomerol (1983) have recently dated the European Eocene-Oligocene boundary between 37 and 40.5 million years ago (MYA), and this appears more congruent with the scale used for North America than the classical one (International Stratigraphical Lex.) that I have used previously (for example, Vianey-Liaud, 1979). Therefore, for this discussion on the possible relationships among Eocene and lower Oligocene rodents from the different continents, I will follow the time scale of Cavelier and Pomerol (see Table 1).

Heeding these methodological cautions, we can begin to discuss the possible relationships among early Paleogene rodents. Emphasis will be placed on those families that I have studied directly; for the others, data from the literature will be analyzed.

ORIGIN OF THE SCIURIDAE: RELATIONSHIPS WITH THE APLODONTIDAE

I have previously described (Vianey-Liaud, 1974a, b, 1975) the earliest European sciurid (Palaeosciurus goti) from Mas de Got, Quercy. This species also occurs in the lower Oligocene locality of Aubrelong 1 (Quercy). Thus, sciurids appeared in Western Europe immediately after the "Grande Coupure," with the immigrant's wave (Lopez and Thaler, 1974; Vianey-Liaud, 1979; Hartenberger, 1983). Wood (1980b) denied the existence of any true Oligocene sciurids (sciuromorphous rodents with uniserial enamel), either in North America or in Europe. Later, Emry and Thorington (1982) described "the oldest fossil squirrel," Protosciurus jeffersoni, from the lower Oligocene of the United States. The earlier reports on fossil sciurids by Vianey-Liaud seem to have been unknown to the above three authors. Thus, up to now, two publications have described the "oldest" sciurid from the lower Oligocene of Europe and North America, but a third author (Wood) did not agree. To understand these divergent opinions, it seems necessary to provide first a definition of this family, as used now. I will discuss here only some dental, cranial, and skeletal features. The data are taken from various authors (Bryant, 1945; Grassé, 1955; Black, 1963; Vianey-Liaud, 1974a, b, 1975; Parent, 1980; Emry and Thorington, 1982).

In sciurids, the axial skeleton is quite conservative. The number of vertebrae is: 7 cervical, 12 thoracic, 7 lumbar, 3-4 sacral, and 15-26 caudal. The number of sternebrae varies from 4 to 5, and the number of ribs varies from 8 true ribs and 4 false ribs in tree squirrels, to 7 true ribs and 5 false ribs in the other sciurids. There are different ratios between the bones of the limbs and of the digits in different groups of sciurids (arboreal and ground squirrels). Tibia and fibula are incompletely fused. In the carpus, the scaphoid and lunate are fused. In the pelvic girdle, the iliac wings are generally strongly curved outward, and the tuber ischii are well marked. A subscapular spine is present on the scapula.

The skull is sciuromorphous, with a reduced infraorbital foramen, and the insertion of the M. masseter lateralis anticus, lateral to this foramen, is clearly localized on the masseteric tubercle. The rostrum is generally short. There are postorbital processes preceded by small notches. The auditory region of sciurids is relatively specialized (Parent, 1980), althought it does retain some primitive characters, such as the tensor tympani muscle, which is not covered. The presence of a distinct meato-cochlear bridge (a derived feature), and of a functional stapedial artery, are characteristic of sciurids. The mandible is sciurognathous, with the anterior end of the masseteric fossa reaching the level of P/4-M/1 junction. The angle of the mandible is generally curved inward.

The dental formula is primitively 1-0-2-3/1-0-1-3. The incisors are generally compressed, moreso in tree squirrels than in other sciurids, and possess uniserial enamel. The dental pattern is tritubercular, with transverse lophs linking the main cusps. The teeth

Table 1. Biostratigraphic distribution of early Paleogene rodents in North America, Europe, and Asia (after Wilson, 1980; Rose, 1981; Russell et al., 1982; Cavelier and Pomerol, 1983). Correlations are difficult between the continents, and radiometric dates for the beginning of the different stages are variable.

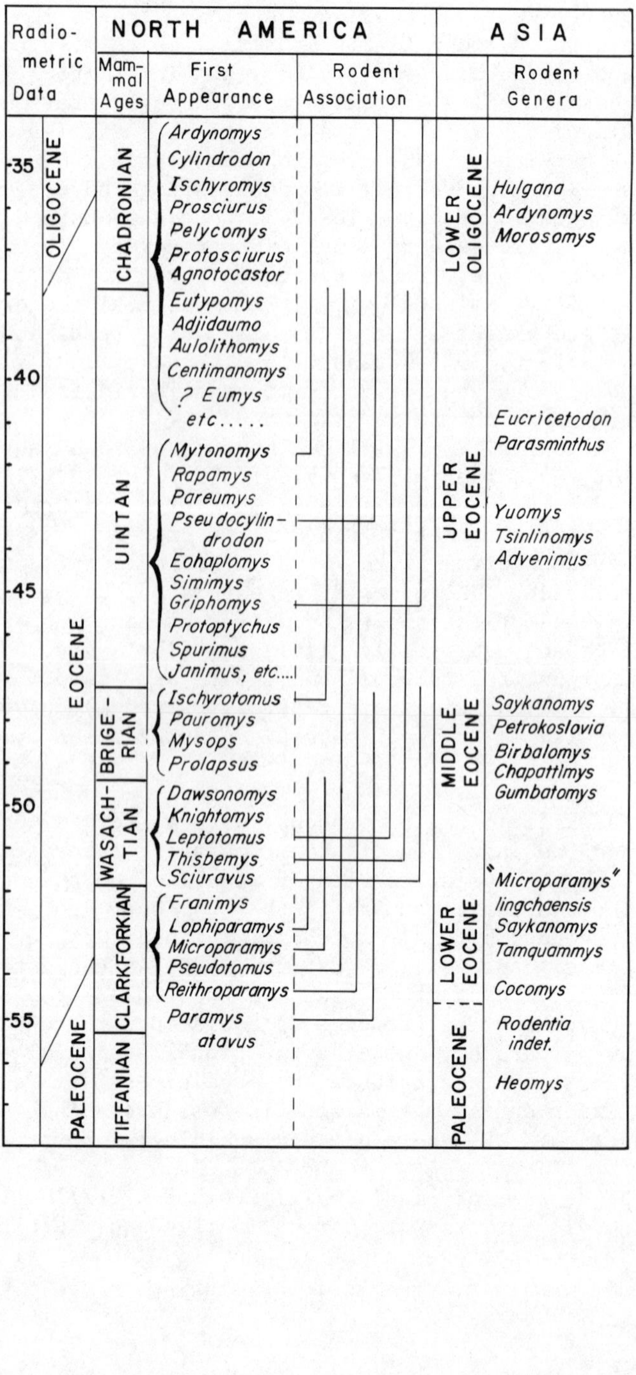

Radiometric Data	Epoch	NORTH AMERICA: Mammal Ages	NORTH AMERICA: First Appearance	NORTH AMERICA: Rodent Association	ASIA	ASIA: Rodent Genera
35	OLIGOCENE	CHADRONIAN	*Ardynomys*, *Cylindrodon*, *Ischyromys*, *Prosciurus*, *Pelycomys*, *Protosciurus*, *Agnotocastor*		LOWER OLIGOCENE	*Hulgana*, *Ardynomys*, *Morosomys*
40	EOCENE	UINTAN	*Eutypomys*, *Adjidaumo*, *Aulolithomys*, *Centimanomys*, *? Eumys* etc.....		UPPER EOCENE	*Eucricetodon*, *Parasminthus*
45		UINTAN	*Mytonomys*, *Rapamys*, *Pareumys*, *Pseudocylindrodon*, *Eohaplomys*, *Simimys*, *Griphomys*, *Protoptychus*, *Spurimus*, *Janimus*, etc....		UPPER EOCENE	*Yuomys*, *Tsinlinomys*, *Advenimus*
		BRIGERIAN	*Ischyrotomus*, *Pauromys*, *Mysops*, *Prolapsus*		MIDDLE EOCENE	*Saykanomys*, *Petrokoslovia*, *Birbalomys*, *Chapattimys*, *Gumbatomys*
50		WASACHTIAN	*Dawsonomys*, *Knightomys*, *Leptotomus*, *Thisbemys*, *Sciuravus*		LOWER EOCENE	"*Microparamys*" *lingchaensis*, *Saykanomys*, *Tamquammys*, *Cocomys*
		CLARKFORKIAN	*Franimys*, *Lophiparamys*, *Microparamys*, *Pseudotomus*, *Reithroparamys*		LOWER EOCENE	
55	PALEOCENE	CLARKFORKIAN	*Paramys atavus*		PALEOCENE	*Rodentia indet.*, *Heomys*
	PALEOCENE	TIFFANIAN			PALEOCENE	

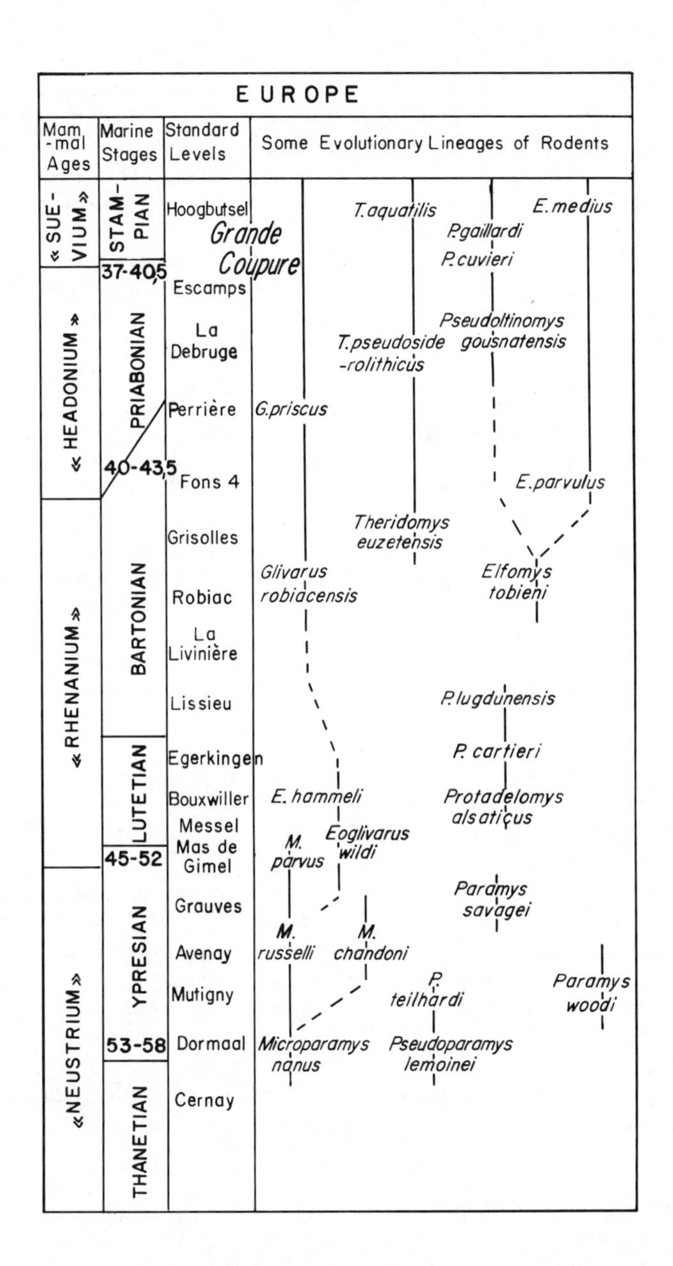
EUROPE
Mam-mal Ages
Marine Stages
Standard Levels
Some Evolutionary Lineages of Rodents
« SUEVIUM »
« HEADONIUM »
« RHENANIUM »
« NEUSTRIUM »
STAMPIAN
PRIABONIAN
BARTONIAN
LUTETIAN
YPRESIAN
THANETIAN
37-40,5
40-43,5
45-52
53-58
Hoogbutsel
Escamps
La Debruge
Perrière
Fons 4
Grisolles
Robiac
La Livinière
Lissieu
Egerkingen
Bouxwiller
Messel
Mas de Gimel
Grauves
Avenay
Mutigny
Dormaal
Cernay
Grande Coupure
T. aquatilis
P. gaillardi
P. cuvieri
E. medius
Pseudoltinomys gousnatensis
T. pseudosiderolithicus
G. priscus
E. parvulus
Theridomys euzetensis
Elfomys tobieni
Glivarus robiacensis
P. lugdunensis
P. cartieri
E. hammeli
Protadelomys alsaticus
Eoglivarus wildi
M. parvus
Paramys savagei
M. russelli
M. chandoni
P. teilhardi
Paramys woodi
Microparamys nanus
Pseudoparamys lemoinei

are generally low-crowned, except in the graminivorous genera (Marmota, Citellus, Cynomys). These genera show a triangular shape of the upper cheek teeth, with a narrow protocone. More generally, the upper cheek teeth show a protocone more or less elongated, a protocone and a metaloph more (high-crowned genera) or less high (the other genera), and a parastyle and mesostyle generally are present. The teeth tend to be selenodont in flying squirrels. When present, P3/ is either a conical and functional cutting tooth, or a peg-like one, as in chipmunks and tree squirrels. The lower cheek teeth show a reduced trigonid, except in the high-crowned genera, where it is enlarged (as long as the talonid). P/4 is molariform in these genera; in others, the protoconid and paraconid are close together. In flying squirrels, the floor of the trigonid basin is rough.

Actually, there is a very specialized family, the Aplodontidae, that shows some affinities with the Sciuridae. The genus Aplodontia is specialized for a fossorial habit, with fossorial adaptations of the skeleton, such as the curious apophyses of the mandible, or of the skull. The teeth are hypsodont and selenodont. However, in spite of its peculiar adaptations, the living member of the Aplodontidae shows similarities with the Sciuridae. Although the stapedial, promontory, and internal carotid arteries have been lost in Aplodontia, it still shares with sciurids many specialized characters of the auditory region. The petrosal and ectotympanic are fused, the facial canal is ossified, the stapedius muscle is reduced, and the bent cochlea is quite swollen. Both families possess a meatocochlear bridge, the high fenestra rotunda twisted, and the progressively enlarged ostium tubae continued in the fossa of the tensor tympani muscle (Parent, 1980; Lavocat and Parent, this volume). Also, the bullae are not very partitioned.

If the family Aplodontidae is at present restricted to a single genus, this has not always been the case. They were a diversified group, known since the end of Eocene times in North America. The axial and appendicular skeletons of the early genera are not well known. If we consider their variety of tooth patterns, they were probably adapted to different habits, perhaps with arboreal and/or terrestrial adaptations. Thus, the cranial and dental characteristics are better known than the postcranial ones. These fossil aplodontids were protrogomorphous and sciurognathous. The Eocene and Oligocene species had brachydont and bunolophodont to bunoselenodont teeth (such as Eohaplomys, Spurimus, Plesispermophilus, and Pelycomys). Some of these genera gave rise to brachydont and selenodont to hypsodont and selenodont forms (such as Plesispermophilus ernii, Allomys, Entoptychus, Sciurodon, Promylagaulus, and Meniscomys).

Thus, we have at present two families, the Sciuridae and Aplodontidae, easily separated by their skeletal adaptations, their dental specializations, and their infraorbital regions. However, if we also include the Tertiary fossil aplodontids, the separation is not so easy. The dental formulae of the two families are the same; fossil aplodontids are brachydont and buno-selenodont, as are some sciurids, and flying squirrels are selenodont. We do not know the real diversity of skeletal adaptations in aplodontids; thus, it remains that the most striking difference between the two families is that of the infraorbital regions. The sciuromorphous sciurids and the evolved protrogomorphous aplodontids probably originated from primitive protrogomorphous rodents. It would be logical to find a protrogomorphous "squirrel," but: how would we know if it was a squirrel, if, at that moment, it did not have the characteristics of the family? Because of the similarities between Sciuridae and Aplodontidae, it is no longer possible to distinguish an archaic sciurid from an aplodontid, if they really do belong to the same ancestral stock. Because of the nature of the evolutionary process, there are limits to the systematics of taxa higher than the species! However, as we need to use a family taxon, the family Aplodontidae, including protrogomorphous rodents, seems to be the most appropriate for including a protrogomorphous "pre-sciurid," or "para-sciurid," or "para-aplodontid."

Now, we return to the beginning of this section. Considering the available data, and adding some others about *Palaeosciurus goti*, it is possible to provide some answers to three questions:
(1) Is *Palaeosciurus goti* a true sciurid?
(2) Is *Protosciurus jeffersoni* a true sciurid or an aplodontid?
(3) What is the respective antiquity of these two species, and is there a direct phyletic relationship between them?

It is clear that *Palaeosciurus goti* belongs to the family Sciuridae. Its infraorbital region is undoubtedly sciuromorphous (Vianey-Liaud, 1975). The insertion of *M. masseter lateralis anticus*, lateral to the infraorbital foramen, is clearly localized on the masseteric tubercle. On the mandible, the anterior end of the masseteric fossa reaches the level of P/4-M/1 junction, as in modern sciurids. All other known cranial features are similar to those of Sciuridae, such as the postorbital processes preceded by a small notch (Vianey-Liaud, 1975), the very compressed incisors (Fig. 1b), and the appendicular bones. A sagittal section of incisor enamel, observed by scanning electron microscopy, shows a fairly uniserial pattern (Fig. 1a). The auditory region of *Palaeosciurus goti* is not known; however, *P. feignouxi* from the lower Miocene shows the characteristic sciurid specializations (Parent, 1980).

Thus, the earliest Oligocene European "squirrel" is clearly a sciurid; all the characters enumerated above, except the enamel structure, were described and published 10 years ago. Some characteristics of the zygomatic plate and of the limb bones of *Palaeo-*

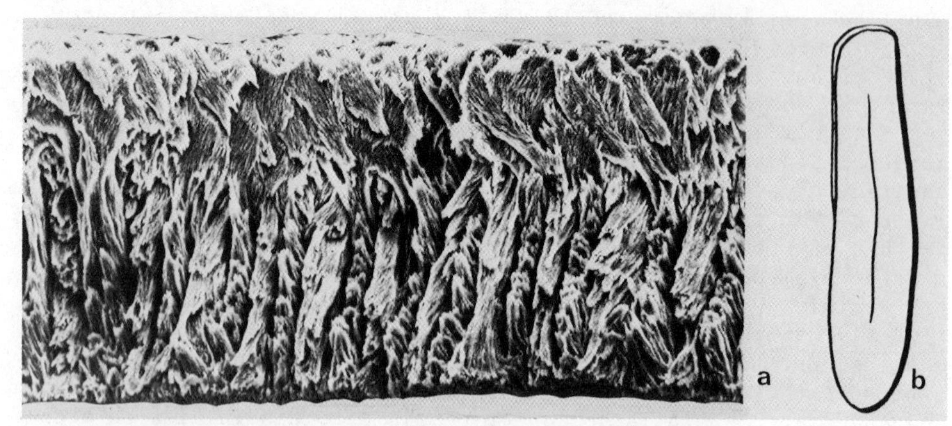

Fig. 1. Incisor morphology of the Oligocene sciurid *Palaeosciurus goti*. (a) Scanning electron micrograph of a sagittal section through the lower incisor, showing the uniserial enamel pattern. x 1000. (b) Occlusal view of lower incisor. x 25.

sciurus goti are probably associated with a terrestrial habit (Vianey-Liaud, 1975). If *Palaeosciurus goti* is the true ancestor of *P. feignouxi*, this evolutionary lineage showed the change from a terrestrial life-habit to a more specialized arboreal life-habit. Indeed, known skeletal features of *P. feignouxi* show that this species is a tree squirrel, like *Sciurus vulgaris*. This hypothesis has to be corroborated following a current study on evolutionary lineages of European Oligocene sciurids, and a more detailed anatomical study.

Protosciurus jeffersoni is not a member of the family Sciuridae, according to Wood (1980b). This species is protrogomorphous, with no evidence that the *M. masseter lateralis anticus* had spread along the side of the snout, lateral to the infraorbital foramen. Nevertheless, Emry and Thorington (1982) remarked that the anterior end of the masseteric fossa of the mandible is more anterior than in protrogomorphs such as paramyids, although it is less so than in the modern sciurids. However, if we compare *Protosciurus* with contemporaneous "advanced" protrogomorphous rodents, such as the Oligocene aplodontids *Plesispermophilus* or *Sciurodon* (Schmidt-Kittler and Vianey-Liaud, 1979), the position of the anterior end of the masseteric fossa is the same in both groups. And, at least for *Plesispermophilus*, the infraorbital foramen is fairly protrogomorphous. Thus, this mandibular character is not useful to argue a real step toward sciuromorphy for *Protosciurus*. It is probably a derived trait, attained in parallel by these lineages, like those of the auditory region, or the compressed incisors, or the fourth toe longer than the third, all mentioned by Emry and Thorington

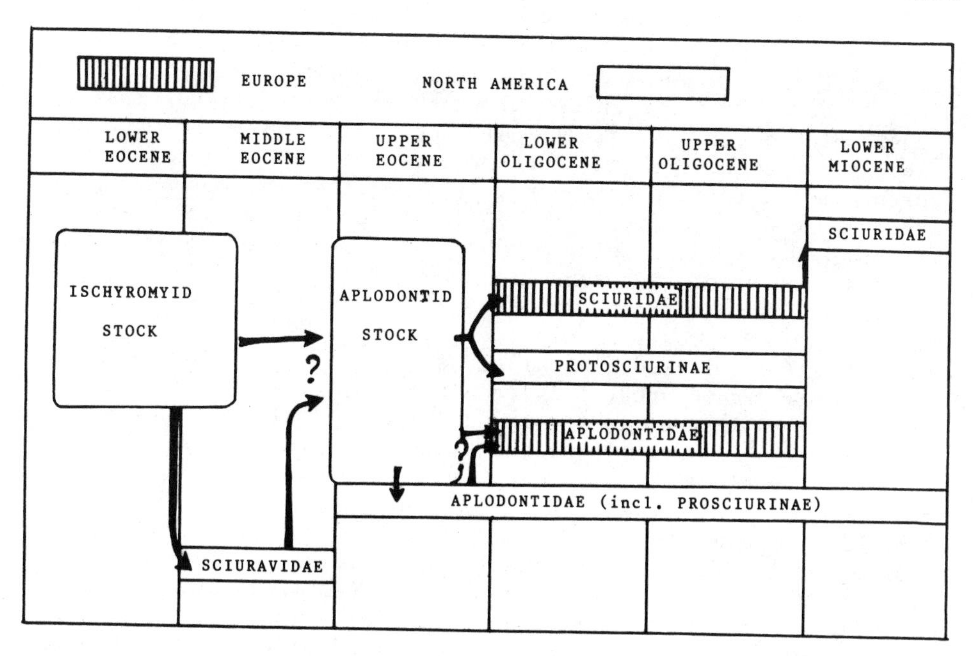

Fig. 2. Suggested relationships between early Aplodontidae and Sciuridae.

(1982). Derived characters of the auditory region, shared by *Protosciurus* and the Sciuridae, are also shared by the Aplodontidae and Sciuridae (Parent, 1980). The occurrence of a stapedial artery in both *Protosciurus* and the Sciuridae is a primitive character; they are more conservative for this than is the modern *Aplodontia*. The only special character shared by *Protosciurus* and the Sciuridae, remaining of those mentioned by Emry and Thorington (1982), is the development of the subscapular spine. However, the authors noted that this character also occurs in *Anomalurus* (and perhaps in other anomalurids that they have not yet studied). It may perhaps be another derived character, eventually linked to arboreal adaptations.

Emry and Thorington (1982) did not compare the dental pattern of *Protosciurus* with that of primitive aplodontids, but with flying squirrels and some paramyids. Distinct conules on upper cheek teeth are also aplodontid characters, and a hypolophid is present in primitive aplodontids. These authors also considered that the postorbital processes, observed in *Protosciurus* and Sciuridae, represent a derived character. But this character exists in all Lagomorpha, in some ischyromyids, and in *Cocomys*. I suspect that it is a primitive character, retained in modern Lagomorpha and Sciuridae.

The sharing of derived characters between *Protosciurus* and the

Sciuridae, on one hand, and between Sciuridae and Aplodontidae on the other hand, seems to show parallelism in the evolution of these three groups, from one ancestral stock, instead of true close parental relationships (Fig. 2). Although considering the cautions given above about the precise correlation among Europe, Asia, and North America for the Paleogene, it seems that Palaeosciurus goti and Protosciurus jeffersoni were nearly contemporaneous. The former was already a typical sciuromorphous sciurid, with a possible terrestrial skeleton. The contemporaneous Protosciurus jeffersoni, protrogomorphous with an arboreally-specialized skeleton, can not be a direct ancestor for Palaeosciurus goti. The genus Protosciurus seems to have followed a parallel way of specialization with the Aplodontidae, throughout Oligocene times in North America (if time "zero" was an instant in the lower Oligocene, the contemporaneous zoologists would have classified Protosciurus close to the prosciurine aplodontids).

Protosciurus became extinct in the lower Miocene. It was replaced in fossiliferous localities by undeniable sciurids. This arrival was accompanied by that of other families, probably Eurasiatic immigrants (Fahlbusch, 1973; Engesser, 1979). From a systematic point of view, this genus could form a subfamily Protosciurinae. Because it shares with the aplodontids numerous derived cranial and mandibular characters, as well as the primitive protrogomorphous infraorbital region, and evolved in the same area from common ancestors (to occupy various ecological niches), it seems logical to include the subfamily Protosciurinae within the family Aplodontidae.

The Aplodontidae originated from early Prosciurinae, although their precise origin is unclear. In North America, typical aplodontids first appear in the upper Oligocene, with Haplomys (Rensberger, 1975). During Oligocene times, only primitive genera are known (Prosciurus, Pelycomys, Cedromus). Morphologically, one of these genera could be ancestral to Haplomys (Rensberger, 1975), although the fossil record is lacking between these two steps. However, after studies on the European Oligocene aplodontids (Schmidt-Kittler and Vianey-Liaud, 1979), and especially by the demonstration of an evolutionary lineage with intermediate stages between the lower Oligocene primitive species (Plesispermophilus angustidens) and the upper Oligocene advanced species (P. ernii), we can suppose a similar evolutionary process in North America during the Oligocene.

In North American upper Eocene localities, two specialized genera are considered as a Prosciurinae for one (Spurimus), and as an Aplodontidae for the other (Eohaplomys). Both show too many derived characters to be the ancestors for Oligocene aplodontids. They are probably the remnants of the "aplodontid stock" that spread in several lineages. Typical primitive aplodontids, already diversified, appeared in Europe by the time of the "Grande Coupure"

(*Plesispermophilus atavus*, *P. angustidens*, *Sciurodon cadurcensis*). At the same time, three genera of prosciurines occurred in North America. The family is recorded in Asia by the middle Oligocene *Prosciurus lohiculus*.

Because of the arrival of immigrants in Europe and Asia toward the Eocene-Oligocene boundary, of the occurrence of the diversified Eocene "protrogomorphous stock" in North America, and of the lack of any Eocene protrogomorphous rodents in Asia, the origin of the Aplodontidae is probably North American, associated with that of the Sciuridae. The ancestral group may be found either among the middle Eocene sciuravids, or among the lower Eocene ischyromyids. Since upper Eocene, after the great faunal replacement, the Sciuridae appears as the sister group of the Protosciurinae. The latter group remained protrogomorphous, whereas the former evolved sciuromorphy. Because there is a typical sciurid in the European lower Oligocene, we must recognize an earlier development of sciuromorphy during the Eocene. Thus, *Protosciurus* does not illustrate a stage in the change between protrogomorphy and sciuromorphy, and *Palaeosciurus* was already a Sciuridae.

ORIGIN OF THE CASTORIMORPHA

Hugueney (1975) has shown that the Castoridae appeared at the same time (lower Oligocene) on the three northern continents: *Steneofiber butselensis* in Europe, *Propalaeocastor kazachstanensis* and *P. habilis* in Asia, and *Agnotocastor galushai* in North America. These apparently contemporaneous species showed similar evolutionary states, and similar dental patterns, although some dental and mandibular details indicate that they belong to different lineages.

In North America, the Eutypomyidae evolved in parallel with the Castoridae (Emry, 1972; Wahlert, 1977; Wood, 1980b). As with the case of the Sciuridae, it is only after the "Grande Coupure" that the Castoridae occurred in Europe, and at nearly the same time elsewhere. The origin of these sciuromorphous rodents is not well recorded. One North American upper Eocene genus (*Janimus*) shows some dental similarities with the eutypomyids. Yet, Wood (1974) emphasized the stratigraphic proximity between *Janimus* and *Eutypomys*, and he suggested that, if there is a possible relationship between them, it is not a direct ancestral one.

Thus, at present, we do not have many indications about the pre-Oligocene history of Castorimorpha. The occurrence of *Janimus* and that of two lineages (Eutypomyidae, Castoridae) are signs for locating their place of origin in North America. Their further dispersal may have followed the same routes as for the other migrants of the "Grande Coupure."

ORIGIN OF THE GEOMYOIDEA

The sciuromorphous Geomyoidea are a specific North American group. The oldest family, the Eomyidae, is well diversified as earl as the lower Oligocene (four or five genera), with typical pentaloph odont eomyids (lacking P3/), or with trends toward bilophodonty (Meliakrouniomys), as well as yoderimyines (which retained P3/). Th fossil record clearly shows phyletic relationships between the eomyi and geomyoids (see Fahlbusch, this volume). This group seems to hav had a history completely independent from that of the muroids.

During the middle Eocene, at least two protrogomorphous species of sciuravids show evolutionary trends indicating geomyoid character Pauromys has peculiar molars tending to bilophodonty. A "sciuravid sp." from the same locality shows some similarities with Pauromys, and its morphology could be compatible with that of Simimys (Dawson, 1968). In fact, it is no more contradictory with that of the Eomyidae; two teeth show a trend toward pentalophodonty.

An important diversification occurred between middle and upper Eocene times. Along with unquestioned eomyids (Protadjidaumo, Namatomys), there are a number of evolving geomyoid species (Floresomys) and bilophodonty appears several times (Griphomys, Presbymys). Thus according to many authors (Wood, 1974, 1980b; Dawson, 1977), the Eomyidae appear to originate from the Sciuravidae. Their appearance in European localities occurred at the time of the "Grande Coupure." Until now, the first migration, presumably from North America, has not left any remains in the Oligocene fossil record of Asia.

ORIGIN OF THE HYSTRICOGNATHI: IS THIS GROUP MONOPHYLETIC?

I can not avoid writing some words about the suborder Hystricognathi. This suborder, characterized by the hystricognathous condition of the mandible, includes five infraorders recognized by Patterson and Wood (1982): Franimorpha, Phiomorpha, Bathyergomorpha, Hystricomorpha, and Caviomorpha. The discussion between Wood (1975, 1977, 1980b; also Patterson and Wood, 1982) and Lavocat (1973, 1974a b) concerning the origin of the Caviomorpha, and their relationships with African Phiomorpha, which has continued since the 1960's, is far from over (Fig. 3). I do not have any new solid arguments, base on fossils. Nevertheless, in light of the new Asian rodent fossil record, and of the re-evaluation of the meaning of parallelism in regards to hystricomorphy and hystricognathy, I wish to bring my small stone to this monument. Like my famous predecessors, I continue to mix phyletic relationships based on morphology and paleogeography in the discussion.

Because of the evidence of "incipiently" or fully hystricognath ous rodents in Central (North) America, and because of numerous ana-

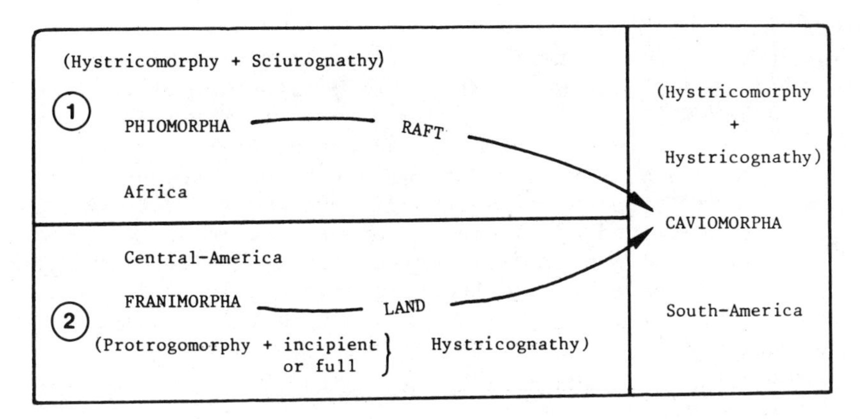

Fig. 3. Summary of the opinions of Lavocat (1) and Wood (2) on the origin of Caviomorpha.

tomical and dental characters, Wood (1975) and Patterson and Wood (1982) placed the North American franimorphs at the origin of the Caviomorpha (Fig. 3). The South American caviomorphs, clearly hystricomorphous since the lower Oligocene, would have quickly attained their infraorbital pattern from the still protrogomorphous Franimorpha of the upper Eocene. Hystricomorphy exists in numerous families that, according to Wood (1980a), have no close phyletic relationships. He mentioned the Theridomyidae, Anomaluridae, Ctenodactylidae, and Pedetidae. According to the fossil record, we can also add the Cricetidae, Dipodidae, Protoptychidae, and Simimyidae. However, it is not so clear that all or some of these families have no close phyletic relationship (see below). Wood (1980a) wrote that "parallelism is rampant.... in the development of hystricomorphy." But, he did not mention that hystricognathy may also have been attained independently in several families of rodents. Actually, hystricognathy appears fully developed only five times: during Eocene times in _Prolapsus_; during Oligocene times in the Caviomorpha and Phiomorpha; and during Miocene times in the Hystricidae and Bathyergomorpha. This specialization did not happen frequently; but, must we think that it happened only once? Should we exclude (_a priori_) any possibility of parallelism?

Hartenberger (1980) restricted the number of early North American hystricognathous rodents to the unique genus _Prolapsus_. Korth (1984) has observed that most of the "incipiently" hystricognathous franimorphs are badly crushed and distorted specimens, including the mandible of _Prolapsus_. According to Wood (1977), this genus, although it appears clearly protrogomorphous, may show a slightly enlarged infraorbital foramen. "Dans l'état actuel, les seuls territoires hors l'Amérique du Sud ou l'on connaissait des Hystricognathes indubitables restent l'Afrique et l'Asie (Hartenberger, 1982). This opinion is not contradictory with that of Wood (1980a), who agreed with Dawson (1977) when she wrote that the incipient hystricognathy on the sciurognathous mandible of Reithroparamys is recognized only a posteriori: "If full hystricognathy were not known, the slight hystricognathy of Reithroparamys would not be worthy of note." Wood suggested that this incipient hystricognathy characterized all North American franimorphs and the cylindrodontids, and he proposed (Wood, 1980b) that cylindrodontids should be included in the Franimorpha. This hystricognathy is at first incipient; "it is inconceivable that hystricognathy appeared instantaneously" (Patterson and Wood, 1982). The primitive cylindrodontids were believed to have given rise to the more evolved Asiatic Tsaganomyinae, now considered by Wood to be in the Bathyergomorpha, and to the bathyergoids; all of these forms are fully hystricognathous (Wood, 1980b; Patterson and Wood, 1982). Among them, no sign of true hystricomorphy appears. As yet, there has been no illustration of the slightly enlarged infraorbital foramen of Prolapsus, and we are not certain that the only true North American hystricomorphous rodent (Protoptychus) is really a franimorph. Dawson (1977) mentioned the possibility of an Asiatic (ctenodactylid) origin for this genus. This hypothesis should be considered more carefully; it is perhaps just as likely to originate Protoptychus from the Asiatic hystricomorphous forms, as to include it in the franimorphs. North America was not completely isolated from the other areas of rodent differentiation (Asia and Europe). Thus, migrations did occur from North America to Eurasia (Sciuridae, Aplodontidae, Cylindrodontidae, Castoridae, Eomyidae) during the upper Eocene; exchanges in the opposite direction were also possible (Cricetidae, Dipodidae, Simimyidae, Protoptychidae?). All these possible Asiatic migrants are hystricomorphous and sciurognathous.

Returning to the suggestion of Wood (1980a) that there is "no reason why hystricomorphy may not have arisen independently several times in the hystricognaths as well as in the sciurognaths," it seems to me that we can first reverse this proposal, and second make a remark about the significance of sciurognathy and hystricognathy. Thus, there is no reason why hystricognathy could not have arisen independently several times among rodents showing various infraorbital structures. If we examine the general rodent fossil record, it shows that hystricomorphy appears at about the same time as hystricognathy (middle Eocene for Protadelomys, and, doubtfully, Prolapsus), and probably earlier (if the Asian cocomyids are more than incipiently hystricomorphous). Also, we can not consider hystricognathy and sciurognathy at the same level. At our present state of knowledge, sciurognathy is primitive among rodents. From this condition, several degrees of sciurognathy, more or less specialized, evolved, as well as the hystricognathous (incipient to complete) conditions. Therefore, hystricognathy had less chance than sciurognathy to be associated with various infraorbital patterns. Indeed, the fossil record shows that hystricognathy is only associated with protrogomorphy and hystricomorphy.

Thus, generally speaking, sciurognathy is primitive and hystricognathy is one among several specialized conditions; hystricomorphy seems to appear before hystricognathy; protrogomorphy and hystricomorphy have evolved in parallel. For Caviomorpha, Wood (1980a) supposed hystricomorphy to originate later than hystricognathy. However, there is no undoubted proof that incipient or fully hystricognathous and protrogomorphous rodents evolved toward an enlargement of the infraorbital foramen (cylindrodontids and tsaganomyines remained protrogomorphous).

Consequently, and to summarize, an alternative hypothesis would be that three groups have independently attained the hystricognathous condition. The first group (protrogomorphous, and incipient to fully hystricognathous) would be the franimorphs and their relatives, the cylindrodontids (and perhaps the bathyergoids). The second and third groups (hystricomorphous and hystricognathous) include the caviomorphs and the hystricids. The caviomorphs, as strongly supported by Wood (1974, 1977, 1980a, b; also Patterson and Wood, 1982), may have originated from some middle or upper Eocene franimorph. However, there are one or two problems for this suggested relationship: the possible ctenodactyloid origin for the protoptychids, and the sudden and general rise to full hystricomorphy in caviomorphs. It may also be possible for them to originate from hystricomorphous ctenodactyloids, because there were already groups in the upper Eocene of North America that are probably of Asian origin (cricetids, dipodids, and perhaps protoptychids). Furthermore, it is not necessary to postulate a "franimorph population" (Patterson and Wood, 1982) in Asia for the origin of the third group (family Hystricidae) and also for the hystricognathous Phiomorpha. The early Asian ctenodactyloids may have provided ancestral populations for these taxa. However, the problem of the origin of the various African rodents needs new data (fossils) to be resolved, and this symposium has provided important new information in that regard (see Jaeger et al., this volume).

Therefore, I suspect that the taxon "Hystricognathi" is probably not monophyletic, especially if the Protoptychidae can be excluded from the Franimorpha, and if the ctenodactyloids appear as likely candidates for the origin of some infraorders or families of the Hystricognathi.

ORIGIN OF THE GLIRIDAE

Hartenberger (1971) has shown that the glirids may have originated from the middle Eocene European *Microparamys*. His arguments were based on dental morphology. The ancestral genus of the Eocene glirids, *Eogliravus*, showed one species (*E. hammeli*) with a paramyid-

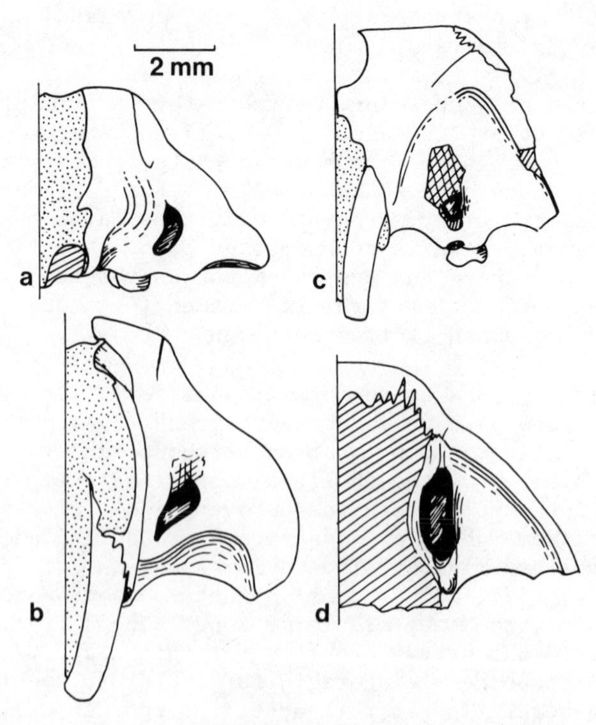

Fig. 4. Frontal view of the rostrum in four species of Gliridae, showing differences in the infraorbital foramen and zygomatic arch. (a) Gliravus majori, QP 625, Type, Anc. Coll., Oligocene from Quercy. (b) G. species indet., ITD, middle Oligocene from Itardies, Quercy. (c) "Gliravus" priscus, MGB 99, middle Oligocene from Mas de Got, Quercy. (d) Vasseuromys sp., Anc. Coll., Miocene from St.-Gérand-le-Puy. Note the "pseudomyomorphous" condition in this genus.

like tooth pattern (cusps higher than those of later glirids, low and interrupted crests, and a well marked hypocone), and another species (E. wildi) that was more glirid-like (lower main cusps, and regular transverse lophs). In addition, Hartenberger noted that Gliravus majori, the only species for which the skull was known, had a protrogomorphous infraorbital foramen. This species is recorded in lower and middle Oligocene localities from Quercy. Another primitive species, now under study, from the middle Oligocene of Itardies (Quercy) is also protrogomorphous. However, the infraorbital foramen seems to be more specialized, slightly contracted at its lower part, and perhaps a little larger than that of Gliravus majori. The skull fragment of Gliravus priscus (Vianey-Liaud, 1974b) from the upper Eocene and lower Oligocene shows a rather different infraorbital and masseteric region (Fig. 4). The infraorbital foramen tends to be myomorphous. The masseteric edge is restricted on the lower margin of the anterior part of the zygomatic arch in G. majori, slightly curved upward in the species from Itardies, but lies completely above the infraorbital foramen in G. priscus and in Vasseuromys from the Miocene of St.-Gérand-le-Puy (Fig. 4). In these two species, the infraorbital foramen is laterally compressed, and a low masseteric tubercle is well developed, as in myomorphous rodents. Thus, at least since the end of the Eocene and the beginning of the Oligocene, the glirids were already diversified, both in their dental morphology and in their infraorbital structure. The similarity of tooth patterns in G. priscus and G. majori has allowed us to put them in the same genus Gliravus. However, it seems clear now that their infraorbital regions are quite different. Therefore, it will be necessary to re-evaluate the taxonomy of this group during an extensive study of Paleogene Gliridae.

Thus, it seems clear that the "myomorphy" of the Recent Gliridae had a different origin from the myomorphy of the Cricetidae and other muroids, because it did not exist in at least two protrogomorphous lower Oligocene glirid lineages. According to Wood (1980b), the similarity between gliroids and muroids, emphasized by Wahlert (1978) in his cladogram is certainly the result of parallelism. I agree with this opinion, and consider that the "myomorphy" of glirids is only a condition of "pseudo-myomorphy," attained convergently with the pattern in muroids.

ORIGIN OF THE THERIDOMYIDAE

Hartenberger (1980) has discussed the ideas on the origin of this group, which appeared in Europe during middle Eocene times (level of Bouxwiller). He considered two possibilities: either an origin from a European ischyromyid stock, or from an Asiatic immigration. Hartenberger believed that there were no data at present to support either hypothesis. Nevertheless, the occurrence of primitive ctenodactyloids in the Asiatic lower and middle Eocene seems to support the second hypothesis. The tooth pattern of the early theridomyid Protadelomys cartieri is certainly different from that of the ctenodactyloids Birbalomys, Chapattimys, and Tamquammys. For example, a number of characters, such as the lack of an anteroconid, the break of the longitudinal crest, the swell of the hypoconulid, and the lingual hypocone, are found together only in ctenodactyloids. The importance of these differences has to be evaluated by detailed dental studies on the earliest ctenodactyloids. Yet, we know that the earliest theridomyids are clearly hystricomorphous and sciurognathous, like ctenodactylids. We also know that cricetids, because of their primitive hystricomorphy and their antiquity on the Asian continent, could also have originated from primitive ctenodactyloids, despite the originality of their dental pattern.

Protadelomys is clearly hystricomorphous since the middle Eocene. It occurred in Europe at the time of a great faunal replacement, due partly to an autochtonous evolution, and partly to an allochtonous invasion (Sige, 1976). After this initial migration (if this hypothesis is true), the Theridomyidae spread in Europe, where they became the characteristic Paleogene rodents (Thaler, 1966; Schmidt-Kittler, 1971; Hartenberger, 1973; Vianey-Liaud, 1979; Vianey-Liaud and Hartenberger, 1983).

One of the theridomyid subfamilies, the Columbomyinae, occurred only at the time of the "Grande Coupure." Sciuromys, its most primitive genus, is clearly a true theridomyid, based on cranial and mandibular morphology and on its dental formula. However, it also shows some peculiar characters, such as the lack of a mesolophid and the small deciduous teeth. The Columbomyinae seem to have been absent from western Europe until the beginning of the Oligocene. In this case, they must have originated outside of this region, separate from other European theridomyids since the middle Eocene. A sister group of the other theridomyids, they would have joined them by the time of the closing of the Turgai Straits. Or, they could have evolved somewhere in northern or eastern Europe, where they have not yet been found. Their dispersal could have followed the late Eocene climatic change, and the faunal environmental change of the "Grande Coupure." At present, we can not choose between these two hypotheses.

ORIGIN OF THE CTENODACTYLOIDEA, AND THE FIRST RADIATION OF RODENTIA

Recent analyses of new findings of early Paleogene Asiatic rodents (Shevyreva, 1972; Dawson, 1977; Sahni and Srivastava, 1977; Hartenberger, 1977, 1980, 1982; Hussain et al., 1978; Li et al., 1979; Dasheveg, 1982; De Bruijn et al., 1982; Dawson et al., 1984) show that all described species belong to the Ctenodactyloidea. This conclusion (Hartenberger, 1982) is supported mainly by dental arguments (small premolars, hypocone originating from the lingual end of the posterior cingulum, and heavy hypoconulid), and also by the shape of the infraorbital foramen, when available. In the upper Eocene genus Yuomys (Li, 1975), this foramen is clearly hystricomorphous. Petrokoslovia, from the middle Eocene, is related to Yuomys because its dental morphology. Terrarboreus (Shevyreva, 1972), an hystricomorphous middle Oligocene rodent, is related to the Tamquammys-Advenimus-Saykanomys group, from the lower and middle Eocene. The infraorbital morphology of these genera is not yet known, except for the recent report that Tamquammys is hystricomorphous (Dawson et al., 1984). Nevertheless, the hypothesis of an hystricomorphous foramen, or at least a pre-hystricomorphous condition, with an enlarged foramen, is more likely than that of a protrogomorphous structure.

In Asiatic upper Eocene localities, the first cricetids and

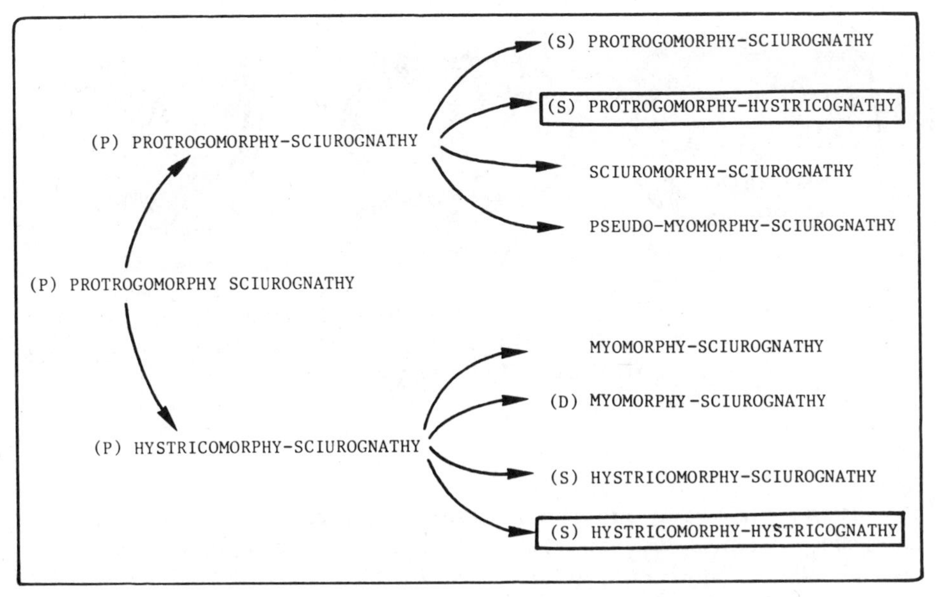

Fig. 5. Possible relationships between infraorbital and mandibular morphology among Paleogene rodents. P, Primitive; S, Specialized; D, Dipodid type. See text for details.

dipodids are known only by teeth (see below). At least for cricetids, their immediate relative at the beginning of the Oligocene (Eucricetodon) showed a large infraorbital foramen (Vianey-Liaud, 1974c).

Cocomys, the oldest known ctenodactyloid from the lower Eocene of China, has a protrogomorphous foramen, but one that is fairly larger than that of contemporaneous North American protrogomorphous ischyromyids (Dawson et al., 1984; Li, pers. comm.). The similarities between Cocomys and the Paleocene genus Heomys have been emphasized by Hartenberger (1982), and they are discussed in more detail by Dawson et al. (1984). Heomys had a rounded, protrogomorphous infraorbital foramen. The transition from Heomys to the cocomyids, if true, involved a slight enlargement of the infraorbital foramen. The dichotomy in the evolutionary history of the Rodentia, suggested by Hartenberger (1980), could have occurred at that time (Fig. 5) by: (1) retaining the relatively small infraorbital foramen in the Ischyromyidae of North Atlantic continents, with subsequent specializations, such as "evolved" protrogomorphy, sciuromorphy (Geomyoidea, Sciuridae, or Castoridae), or "pseudo-myomorphy" (Gliridae); or (2) enlargement of the infraorbital foramen in the Ctenodactyloidea in Asia, with further specializations, such as "pre-hystricomorphy" in the cocomyids, hystricomorphy (Ctenodactylidae or Dipodidae), or myomorphy (advanced Cricetidae).

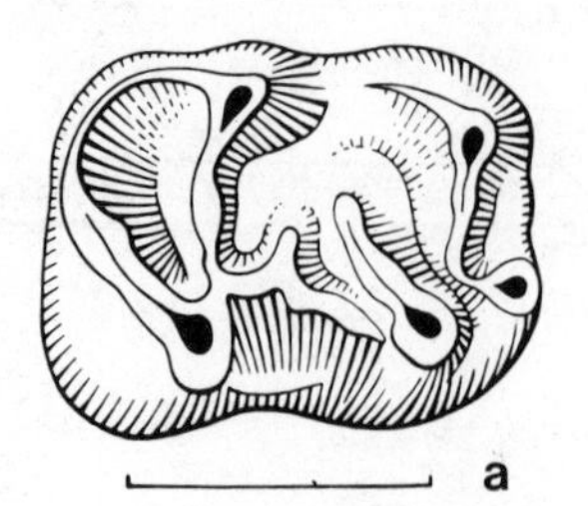

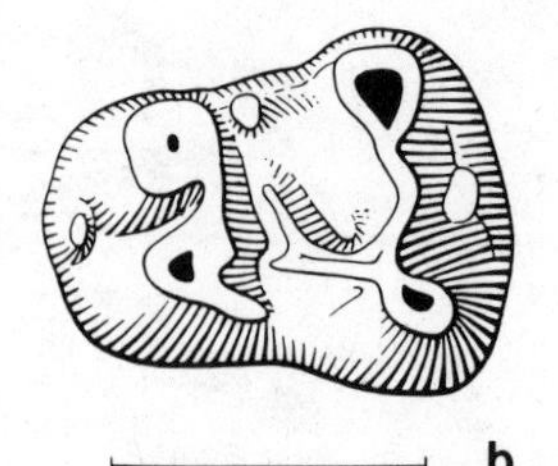

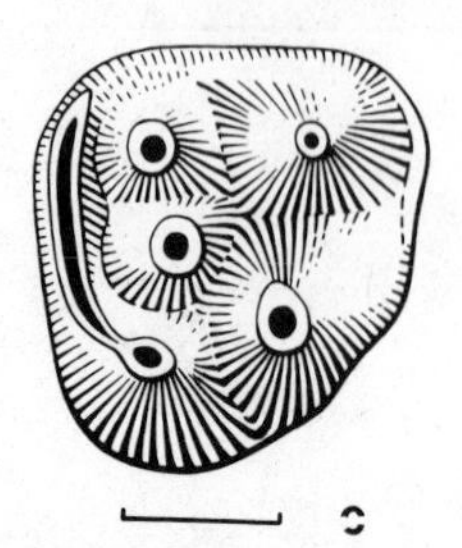

Fig. 6. Rodent cheek teeth from the River Section locality, upper Eocene, China (Uppsala Coll.). (a) *Eucricetodon schaubi*. (b) ? *Parasminthus*. (c) Ctenodactylidae indet.

These cocomyids and their diversified relatives may have been near the origin of several Asiatic families (Ctenodactylidae, Cricetidae, Dipodidae) and Euramerican families (Protoptychidae, Simimyidae, Theridomyidae ?). The first "evolved" protrogomorphous rodents occurred in Asiatic faunas during the Oligocene, with the Tsaganomyinae, *Hulgana* (Ischyromyidae indet.), and the Aplodontidae (*Prosciurus lohiculus*). This probable migration from North America is the Asiatic witness of the "Grande Coupure."

ORIGIN OF THE CRICETIDAE AND DIPODIDAE

The first typical cricetid (*Eucricetodon schaubi*) occurred in China during upper Eocene (Zdansky, 1930), and in Europe and North America during the lower Oligocene (*Eucricetodon atavus* and *Eumys pristinus*, respectively). The occurrence of *E. atavus* seems to be earliest in the Suevian (Northeast of Rhenanian "trench," latest Eocene) (Schmidt-Kittler and Vianey-Liaud, 1975). In the locality of River Section, China, *Eucricetodon* is associated with a true dipodid (? *Parasminthus*; Hartenberger et al., 1975; Hugueney and Vianey-Liaud, 1980), and a ctenodactylid (Fig. 6). Based on tooth morphology, the separation of the Cricetidae and Dipodidae took place at least in the upper Eocene. Is this also true for the infraorbital region (hystricomorphous foramen, with independent lower neurovascular groove in dipodids; myomorphous foramen in cricetids)? The infraorbital region of European and North American cricetids is known; *Eucricetodon atavus* and *Pseudocricetodon montalbanensis* (from the middle Oligocene) have a large infraorbital foramen of the hystricomorphous type, whereas *Eumys* is myomorphous. The study of one *Eucricetodon* lineage (*E. atavus* to *E. collatus*) shows the change from the hystricomorphous pattern to the typical myomorphous condition (Vianey-Liaud, 1974c, 1979). The myomorphy of cricetids has been attained at least twice in parallel, and earlier in *Eumys*. *Cricetops* from the middle Oligocene of Mongolia still retained a large hystricomorphous infraorbital foramen.

The dipodoid infraorbital pattern appeared in *Simimys* in North America during the upper Eocene. Because of its dental formula (1.0.0.3), its molar pattern, and the erroneous interpretation of its orbital region (Lindsay, 1977), *Simimys* was long considered as a possible ancestor for the Cricetidae. Emry (1981) has shown that this genus had a typical dipodid infraorbital region. Wood (1980b) finally decided that he "should cut the Gordian knot, and formalize a conclusion reached by Wilson thirty years ago: *Simimys* can perhaps be viewed as a more or less primitive survivor into the Late Eocene of a stalk which was ancestral to both cricetids and the Dipodoidea." Thus, he created the family Simimyidae. However, *Simimys* is not the only specialized genus with a dipodid infraorbital structure. This condition also occurs in the lower Oligocene *Nonomys* (Emry, 1981). These two genera also have very peculiar dental specializations. The last premolars (P4/4) have disappeared in the two forms; the molars show well developed cingula (lower labial and upper lingual) in *Nonomys*, whereas those of *Simimys* are more generally of dipodid-cricetid type. Because of their specializations, these two lineages can not be ancestral to the classical Dipodidae. As suggested by Wood (1980b) and others for *Simimys*, these two genera could be witnesses of the ancestral radiation of the Dipodidae (for a different interpretation, see Flynn et al., this volume). They can be placed together in the family Simimyidae. Separation of the Simimyidae from the Cricetidae probably happened earlier, because the two families were distinct since the upper Eocene (Fig. 7). This separation could have originated from hystricomorphous forms. At present, the earliest known hystricomorphous rodents are the Asiatic ctenodactyloids. Even though the earliest Asiatic cricetid showed the typical cricetid tooth pattern, the earliest dipodid (? *Parasminthus*) showed a well developed molar hypoconulid. This character agrees with a ctenodactyloid origin for dipodids.

The earliest American dipodoids (Simimyidae) disappeared at the beginning of the Oligocene. There are no further American representatives of this group before the Miocene (*Schaubeumys*). This genus is a typical dipodid, included in the European upper Oligocene genus *Plesiosminthus* by Green (1977). The origin of this genus seems to be the Asiatic *Parasminthus*.

For the family Cricetidae, the oldest known species is Asiatic; the oldest primitive infraorbital structures known for the family are European and Asiatic. We can suppose a dispersal from an Asiatic center during the upper Eocene (Fig. 7). A migration could have taken place toward Europe during the close of the Turgai Straits, and another one toward North America. The Cricetidae and Simimyidae seem to be the only certain immigrants into North America during the late Eocene and the beginning of the Oligocene. The arrival of *Simimys* was earlier than that of *Eumys* and *Nonomys*. It was contem-

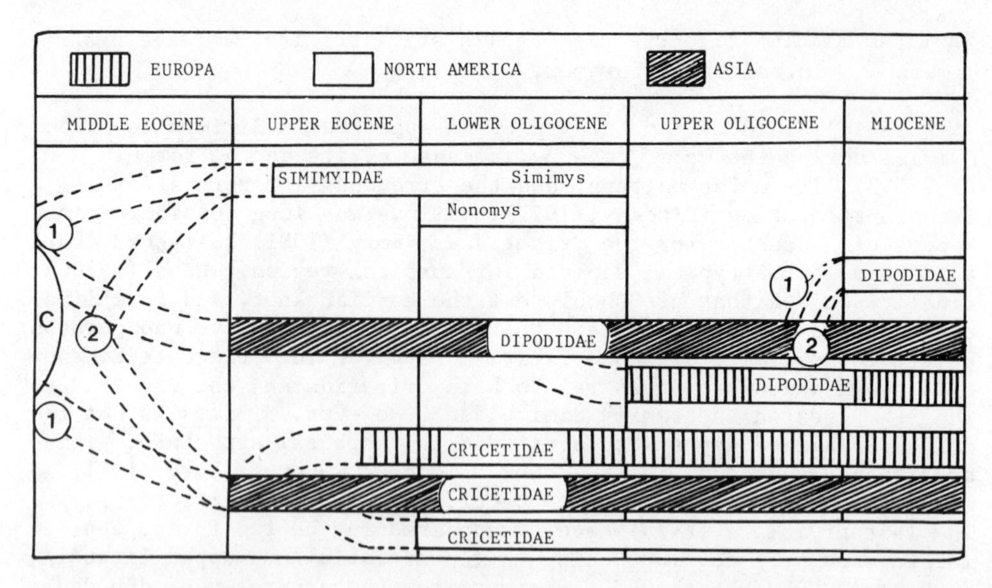

Fig. 7. Suggested relationships between early Cricetidae and Dipodidae, and their possible ancestry from a ctenodactyloid stack (C). Different hypotheses of relationships are indicated at (1) and (2). See text for details.

porary with the appearance of the only true early North American hystricomorphous family, the Protoptychidae. A more detailed analysis of the entire North American fauna from this time is necessary in order to clarify this faunal change, which is reminiscent of the European "Grande Coupure."

TENTATIVE PALEOGEOGRAPHIC AND PHYLOGENETIC RELATIONSHIPS SUGGESTED FOR EARLY PALEOGENE RODENTS

It remains difficult to correlate the Asian, European, and North American faunas, until the detailed local biostratigraphical analyses are completed and calibrated with global scales. Keeping this in mind, discussions about some major problems can be quickly ended. The first, and not the least, is that of dating the oldest rodents in North America and Asia. Although Paramys atavus is well dated as later Paleocene (early Clarkforkian), and correlated with European biostratigraphy (Rose, 1981), the relative age of the first Asiatic rodents (Rodentia indet., from Naran-Bulak; Dasheveg, 1982) is actually more difficult to determine. These Asian rodents apparently came from upper Paleocene deposits, but no other details are known. In addition, Heomys orientalis, found in Paleocene deposits earlier than these, is considered here to be a primitive rodent; it is associated with eurymylids (also see Li and Ting, this volume).

Thus, in spite of imprecise correlations, the dispersal area of these rodents (Hartenberger, 1980) can be localized without too much doubt in Asia during the Paleocene (Fig. 8). From there, a first radiation during the late Paleocene could have given rise to the Ischyromyidae (= Paramyidae) in the North Atlantic continents, while the Asiatic rodents proceeded with their evolution in Asia. This episode in rodent evolution is still not well recorded (Paramys atavus on the one hand, and Rodentia indet. on the other). The data are more abundant for the lower Eocene. The Ischyromyidae spread and evolved in North America and reached Europe. At the same time, the cocomyids began their radiation in Asia (Fig. 8).

During middle Eocene, faunal exchanges between these three Northern continental regions are recorded, mainly be the appearance of the Theridomyidae in Europe, with the beginning already of the specializations that are going to characterize their endemic evolution during the upper Eocene. In North America, the radiation of the protrogomorphous Ischyromyidae took place (Fig. 9), giving rise to several families (Aplodontidae, ? Sciuridae, Eomyidae, Cylindrodontidae, Castoridae). The fossil record is fairly good, but few evolutionary lineages have been described up to now. In Europe, the hystricomorphous Theridomyidae are the dominant rodents; they soon replace most of the ischyromyids, except for the Gliridae, which originated from Microparamys, and the large Plesiarctomys. The fossil record is still insufficient, but it permits us to describe several evolutionary lineages in Europe. Asiatic rodents proceeded with their diversification; during upper Eocene, at least three new genera of ctenodactyloids appeared, as well as one cricetid and one dipodid. Clearly, the Eocene fossil record is still poor, but the findings are increasing!

Until now, the beginning of the rodent story in South America and Africa is known only in the lower Oligocene (Fig. 8). Because of the development of hystricomorphous ctenodactylids in Asia, I will add just a small contribution to the two strong and divergent hypotheses concerning the relationships among the Caviomorpha, Phiomorpha, and Franimorpha, and their occurrence in North America, Africa, and perhaps Asia. Why should we assume, according to Wood (1980b) that only hystricomorphy must have arisen several times independently within Rodentia? Why cannot hystricognathy have arisen twice (or more) among the originally hystricomorphous (ctenodactyloid) rodents, or among the protrogomorphous (franimorph) rodents? We know an earlier stock of hystricomorphous Asiatic rodents, but we have neither the transition between a hystricomorphous-sciurognathous form and a hystricomorphous-hystricognathous one, such as the caviomorphs, nor the place where this evolution would have taken place. However, up to now there is no more evidence that incipient or fully hystricognathous, protrogomorphous rodents evolved toward a hystricomorphous condition.

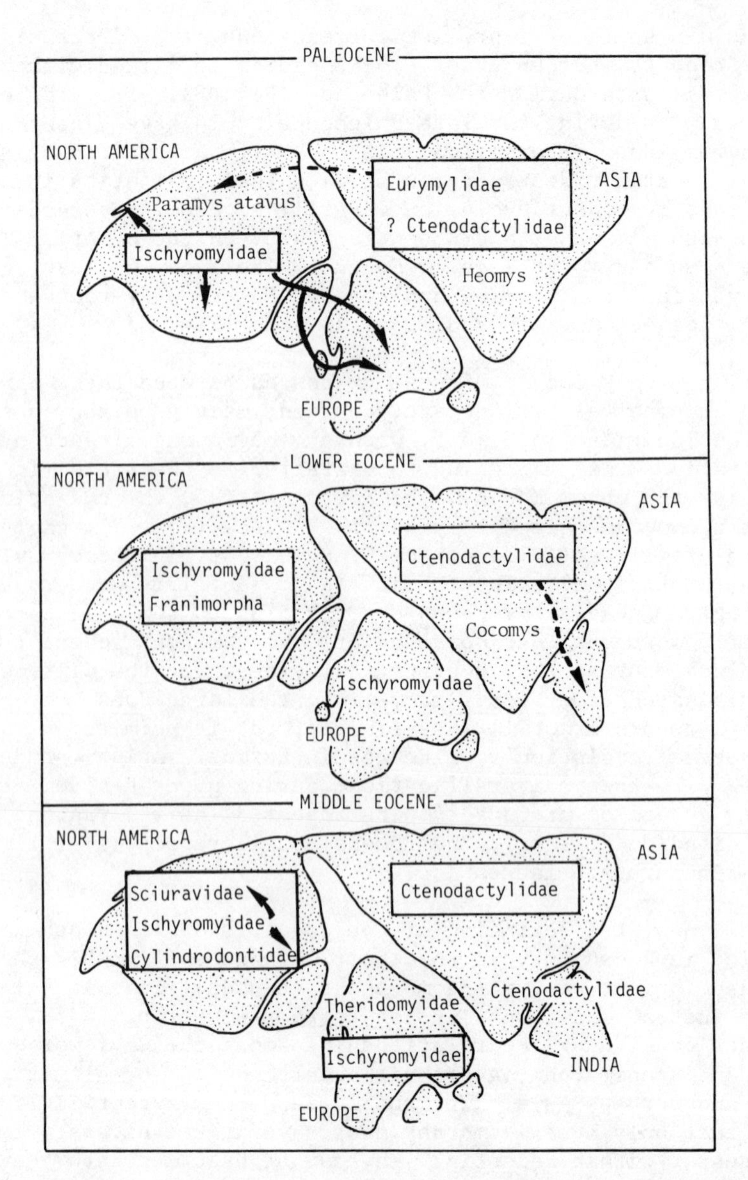

Fig. 8. Diagrammatic representation of paleogeographic relationships among early Paleogene rodents of North America, Europe, and Asia (Africa and South America included for the lower Oligocene). North polar projections of the continents are simplified. Adapted from Smith and Briden (1977).

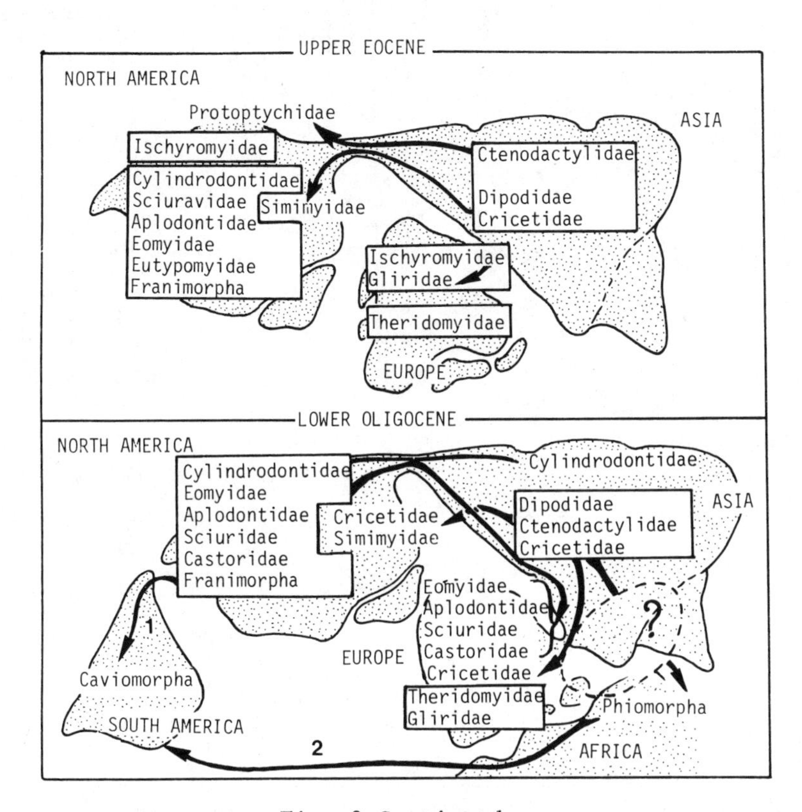

Fig. 8 Continued

There is a distinct change between the upper Eocene and lower Oligocene rodent faunas from the Northern continents. This is the "Grande Coupure," recognized in Europe by Stehlin (1909). Since Stehlin's time, this important event has received strong support. The method of evolutionary lineages and the determination of standard levels have allowed us to define the very details of this "revolution." Numerous investigators, working on large mammals (Stehlin, 1909; Brunet, 1977) or on rodents and other micromammals (Lopez and Thaler, 1974; Schmidt-Kittler and Vianey-Liaud, 1975; Hartenberger, 1973, 1977, 1983; Schmidt-Kittler, 1977; Vianey-Liaud, 1979; Sigé

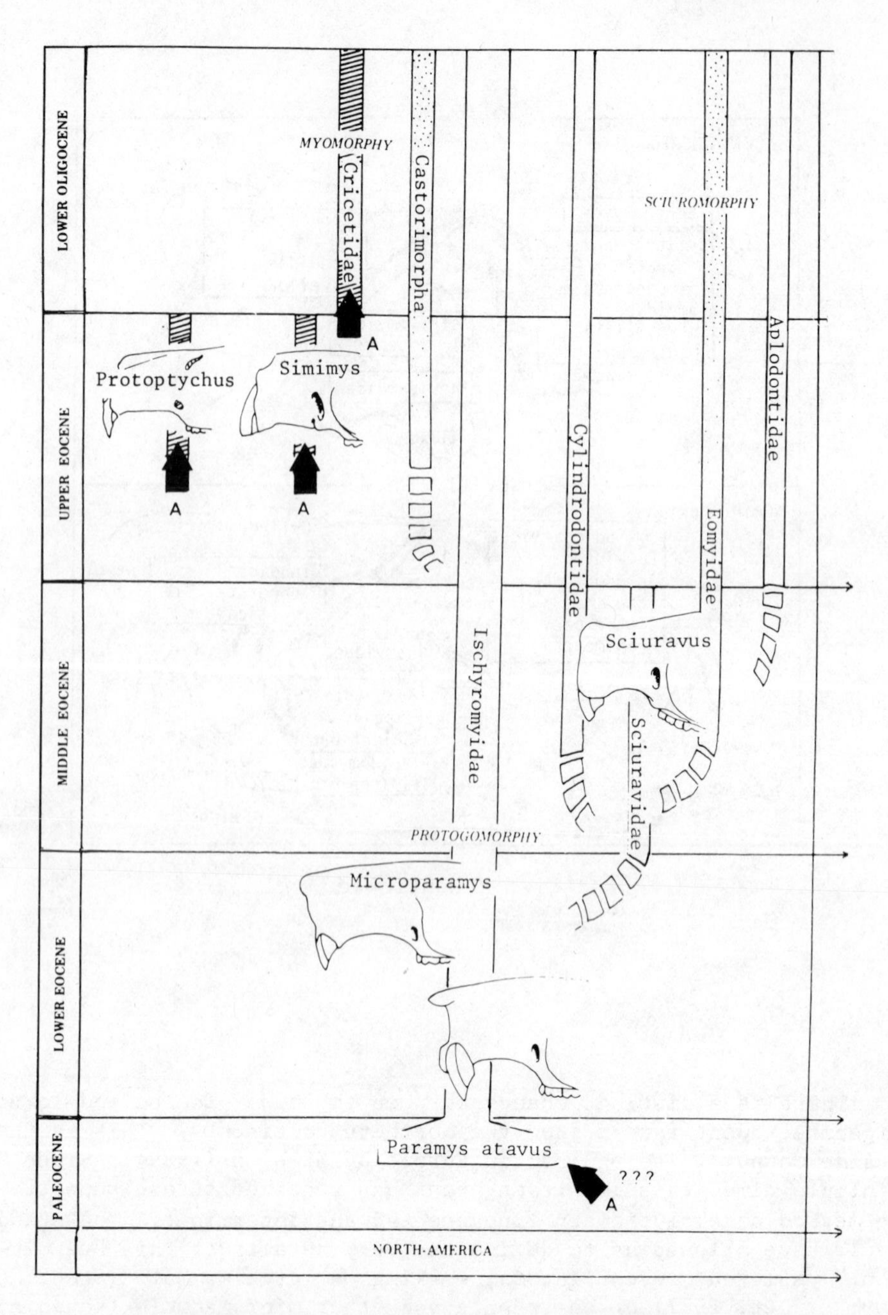

Fig. 9. A diagrammatic representation of suggested phyletic relationships among Asia, European, and North American early Paleogene rodent families. The snout and infraorbital region are illustrated for some families. Hystricomorphy is indicated by lined columns, protrogomorphy by white

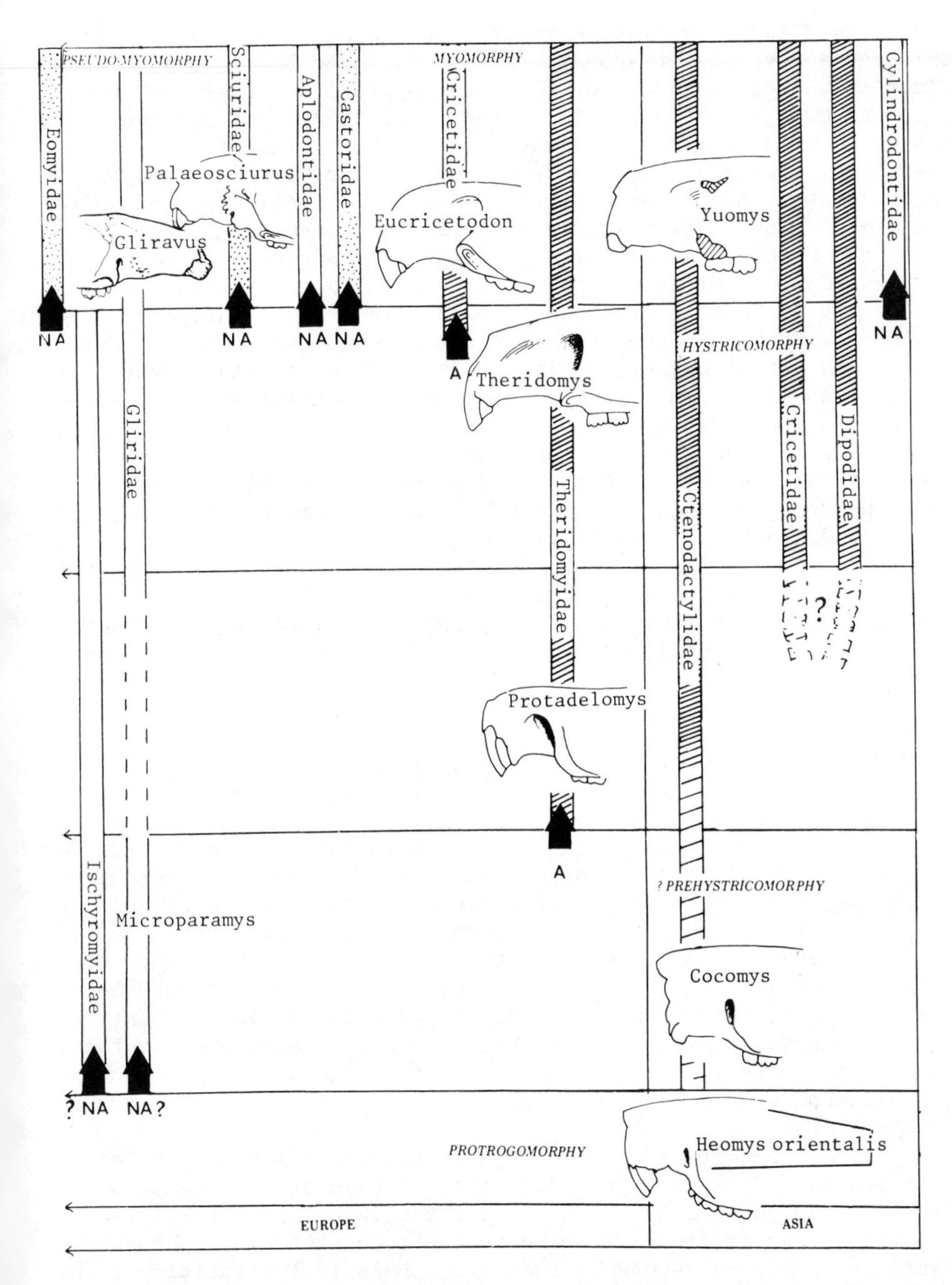

columns, and sciuromorphy by stippled columns. Myomorphy and pseudomyomorphy are noted when they appear on the appropriate columns. A, Asia; NA, North America; black arrows show suggested immigrations.

and Vianey-Liaud, 1979) have contributed to our understanding of this faunal event. It appears that this faunal break is very distinct in western Europe, but, like any migration it seems to have occurred over a period of time. The first immigrants reached southern Germany earlier (late Eocene - it took some time to cross the Rhenanian "trench") than western Europe (lower Oligocene). The origin of this migration was related to general climatic and paleogeographic changes, leading to the closing of the Turgai Straits (Southern Oural). A considerable break is observed between upper Eocene and lower Oligocene rodent faunas in North America (Wilson, 1972). To explain this change, Wilson used evolutionary and climatic arguments. He considered the possibility of migrations from Eurasia, but he did not strongly support this. For him, the influence of climatic change did not seem to have been important in Europe, whereas it was more important in Asia. It now appears, after a bibliographical analysis of North American faunas, that Asiatic immigrations have really occurred (Cricetidae and Dipodidae), and that these migrations lasted from the late Eocene (Simimyidae, and perhaps, Protoptychidae) to early Oligocene (Cricetidae). The basic argumentation for these hypotheses must be accomplished by a careful analysis of the various evolutionary lineages. Several immigrant families in Europe (Aplodontidae, Sciuridae, Eomyidae, Castoridae) and in Asia (Aplodontidae, Castoridae, Cylindrodontidae), on the other hand, apparently came from North America during the lower Oligocene.

Thus, the time from upper Eocene to lower Oligocene showed a great global paleogeographic change (Fig. 9), related to strong climatic changes. The "Grande Coupure" is no longer a limited local event, but rather is worldwide. We have to analyze the records for this change on all the Northern continents, with as much detail as in Europe, and moreso if possible, in order to estimate its various steps in time and space, and the possible causes for these changes. Following such an analysis, the phyletic relationships among rodents of North America, Asia, Europe, and perhaps Africa and South America should be more clearly understood. For earlier phyletic relationships, as noted above, the analysis is not so clear, and the fossil record is not so good. We need to improve this situation, especially the biochronology and correlations.

Another important conclusion concerns the origin of the major infraorbital patterns of the Rodentia. I have original data mainly for sciuromorphy and myomorphy. If the sciuromorphy of sciurids apparently did originate from North American protrogomorphous rodents, there is no record of such a morphological transition. The oldest true Sciuridae, which is clearly sciuromorphous) is an Oligocene immigrant in Europe, _Palaeosciurus_ _goti_. The possible North American evolutionary lineage that gave rise to this sciurid remains to be described.

Myomorphy is shared by gliroids and muroids. Based on the upper Eocene and Oligocene fossil record of the Gliridae and Cricetidae, however, this condition is not homologous in the two groups. For the cricetids, the transition from a large hystricomorphous infraorbital foramen (Eucricetodon atavus) to a compressed myomorphous infraorbital foramen (E. collatus) is observed in a single evolutionary lineage. The attainment of myomorphy is heterochronous in various cricetid lineages (Eumys, Eucricetodon, Cricetops).

An evolutionary transition also occurred in the Gliridae. In the upper and lower Oligocene, two different forms are recorded. One is clearly protrogomorphous (Gliravus majori), whereas the second ("Gliravus" priscus) is already myomorphous. However, this "myomorphy" is associated with a relatively small, compressed infraorbital foramen. A specialized genus from the Miocene (Vasseuromys) showed a more compressed foramen, and this is even moreso in Recent glirids.

Thus, the myomorphy of the Cricetidae took origin from a large (hystricomorphous) infraorbital foramen, whereas that of the Gliridae originated from a small (protrogomorphous) foramen. These two derived conditions were attained convergently by the two families, from different ancestral conditions. If we can speak of myomorphy for the cricetids and their muroid relatives, we can speak of "pseudo-myomorphy" for the glirids. The functional aspects of this morphological convergence remain to be studied.

I shall add just a brief comment on the possible relationships between Dipodoidea and Muroidea. The meager fossil record suggests an Asiatic origin from a primitive hystricomorphous group. The first dipodid skull is recorded in North America (Simimys), whereas the earliest cricetid skulls have been found in Eurasia. It would be reasonable to consider these families as sister groups, which have originated from a common Asiatic ancestor. In upper Eocene Asiatic localities there are isolated teeth that have been attributed to both families (Eucricetodon schaubi to Cricetidae; Parasminthus sp. to Dipodidae). Did they belong to the same ancestral family, or have they been separated for a long time? At present, we can not answer this question. Other evidence from skulls, dentition, and fetal membranes (Flynn et al., this volume; Luckett, this volume) supports the concept of a sister-group relationship between Dipodoidea and Muroidea. This hypothesis is not inconsistent with the fossil record.

Finally, I wish to emphasize that, if this review presents some concrete data, and supports some hypotheses about evolutionary relationships among early Paleogene rodents, it also asks more questions than it provides answers!

REFERENCES

Black, C. C. 1963. A review of the North American Tertiary Sciuridae. Bull. Mus. Comp. Zool. 130: 109-248.

Brunet, M. 1977. Les Mammifères et le problème de la limite Eocène Oligocène en Europe. Geobios, Mem. spéc. 1: 11-27.

Bruijn, H. de, Hussain, S. T., and Leinders, J. J. M. 1982. On som Early Eocene rodent remains from Barbara Banda, Kohat, Pakistan and the early history of the order Rodentia. Proc. Kon. Ned. Akad. Wetensch., B, 85: 249-258.

Bryant, M. D. 1945. Phylogeny of Neartic Sciuridae. Amer. Midl. Nat. 33: 257-390.

Cavelier, C. and Pomerol, C. 1983. Echelle de corrélation stratigraphique du Paléogène. Stratotypes, étages standards, biozones, chimiozones et anomalies magnétiques. Géologie de la France, BRGM, Orléans 3: 261-262.

Dasheveg, D. 1982. La faune de Mammifères du Paléogène inférieur de Naran-Bulak (Asie Centrale) et ses corrélations avec l'Europ et l'Amérique du Nord. Bull. Soc. Géol. France, (7), XXIV, 2: 275-281.

Dawson, M. 1968. Middle Eocene rodents (Mammalia) from Northeastern Utah. Ann. Carnegie Mus. 20: 327-370.

Dawson, M. 1977. Late Eocene rodent radiations: North America, Europe and Asia. Geobios, Mem. spéc. 1: 195-209.

Emry, R. 1972. A new species of Agnotocastor (Rodentia, Castoridae from the Early Oligocene of Wyoming. Amer. Mus. Novit. 2485: 1-7.

Emry, R. 1981. New material of the Oligocene muroid rodent Nonomys and its bearing on muroid origins. Amer. Mus. Novit. 2712: 1-14.

Emry, R. and Thorington, R. W., Jr. 1982. Descriptive and comparative osteology of the oldest fossil squirrel, Protosciurus (Rodentia, Sciuridae). Smithsonian Contrib. Paleobiol. 7: 1-35

Engesser, B. 1979. Relationships of some insectivores and rodents from the Miocene of North America and Europe. Bull. Carnegie Mus. Nat. Hist. 14: 1-68.

Fahlbusch, V. 1973. Die stammesgeschichtlichen Beziehungen zwische den Eomyiden (Mammalia, Rodentia) Nordamerikas und Europas. Mitt. Bayer. Staatssamml. Paläont. Hist. Geol. 11: 141-175.

Gingerich, P. D. 1976. Cranial anatomy and evolution of early Tertiary Plesiadapidae (Mammalia, Primates). Univ. Mich. Mus. Paleont. Papers Paleont. 15: 1-140.

Gingerich, P. D. 1977. Systematics, phylogeny, and evolution of Early Eocene Adapidae (Mammalia, Primates) in North America. Contr. Mus. Pal., Univ. Mich. 24: 245-279.

Grassé, P. P., (ed.) 1955. Traité de Zoologie, Vol. 17 (2). Masso et Cie, Paris.

Green, M. 1977. Neogene Zapodidae (Mammalia, Rodentia) from South Dakota. J. Paleont. 51: 996-1015.

Hartenberger, J.-L. 1971. Contribution à l'étude des genres Gliravus et Microparamys (Rodentia) de l'Eocène d'Europe. Palaeovertebrata 4: 97-135.
Hartenberger, J.-L. 1973. Etude systématique des Théridomyoidés (Rodentia) de l'Eocène supérieur. Mém. Soc. Géol. Fr. (N.S.) 52 (1-5), 117: 1-76.
Hartenberger, J.-L. 1977. A propos de l'origine des rongeurs. Geobios, Mem. spéc. 1: 183-193.
Hartenberger, J.-L. 1980. Données et hypothèses sur la radiation initiale des rongeurs. Palaeovertebrata, Mém. Jubil. R. Lavocat: 285-301.
Hartenberger, J.-L. 1982. Exemples de données géophysiques et paléontologiques antinomiques dans le Tertiaire ancien. Bull. Soc. Géol. Fr., 7e ser., XXIV (5-6): 927-934.
Hartenberger, J.-L. 1983. La Grande Coupure. Pour La Science 67: 26-38.
Hartenberger, J.-L., Sudre, J. and Vianey-Liaud, M. 1975. Les mammifères de l'Eocene supérieur de Chine (gisement de River Section); leur place dans l'histoire des faunes eurasiatiques. 3ème R.A.S.T., Montpellier: 186.
Hugueney, M. 1975. Les Castoridae (Mammalia, Rodentia) dans l'Oligocène d'Europe. Coll. Intern. CNRS 218: 791-806.
Hugueney, M. and Vianey-Liaud, M. 1980. Origine et évolution des Dipodidae (Rodentia) paléogènes d'Europe Occidentale. Palaeovertebrata, Mém. Jubil. R. Lavocat: 303-342.
Hussain, S. T., Bruijn, H. de and Leinders, J. M. 1978. Middle Eocene rodents from the Kala Chitta Range (Punjab, Pakistan). Proc. Kon. Ned. Akad. Wetensch., Amsterdam, B, 81: 74-112.
Korth, W. W. 1984. Earliest Tertiary evolution and radiation of rodents in North America. Bull. Carnegie Mus. Nat. Hist. 24: 1-71.
Lavocat, R. 1973. Les rongeurs du Miocène d'Afrique Orientale. I. Miocène inférieur. Mém. Trav. E.P.H.E., Montpellier 1: 1-284.
Lavocat, R. 1974a. The interrelationships between the African and South American rodents and their bearing on the problem of the origin of South American monkeys. J. Human Evol. 3: 323-326.
Lavocat, R. 1974b. What is an hystricognath? Symp. Zool. Soc. Lond. 34: 7-19.
Li, C.-K. 1975. Yuomys, a new ischyromyid rodent genus from the Upper Eocene of North China. Vert. PalAsiat. 13: 58-70.
Li, C.-K., Chiu, C.-S., Yan, D.-F. and Hsieh, S.-M. 1979. Notes on some Early Eocene mammalian fossils of Hengtung, Hunan. Vert. PalAsiat. 17: 71-82.
Lindsay, E. H. 1977. Simimys and origin of the Cricetidae (Rodentia: Muroidea). Geobios 10: 597-623.
Lopez, N. and Thaler, L. 1974. Sur le plus ancien Lagomorphe européen et la "Grande Coupure" Oligocène de Stehlin. Palaeovertebrata 6: 243-251.

Parent, J. P. 1980. Recherches sur l'oreille moyenne des Rongeurs actuels et fossiles. Anatomie. Valeur systématique. Mem. Trav. E.P.H.E., Montpellier, 11: 1-286.
Patterson, B. and Wood, A. E. 1982. Rodents from the Deseadan Oligocene of Bolivia and the relationships of the Caviomorpha. Bull. Mus. Comp. Zool., Harvard, 149: 371-543.
Rensberger, J. 1975. Haplomys and its bearing on the origin of the aplodontoid rodents. J. Mammal. 56: 1-14.
Rose, K. 1981. The Clarkforkian Land-mammal age and mammalian faunal composition across the Paleocene-Eocene boundary. Univ. Mich. Pap. Paleont. 26: 1-189.
Russell, D. E., Hartenberger, J.-L., Pomerol, C., Sen, S., Schmidt-Kittler, N. and Vianey-Liaud, M. 1982. Mammals and stratigraphy: The Paleogene of Europe. Palaeovertebrata, Mem. extr.: 1-77.
Sahni, A. and Srivastava, M. C. 1977. Eocene rodents of India: their palaeobiogeographic significance. Geobios, Mem. spéc. 1: 87-95.
Schmidt-Kittler, N. 1971. Odontologische Untersuchungen an Pseudosciuriden (Rodentia, Mammalia) des Altertiärs. Abh. Bayer. Akad. Wiss., München, math.-naturw. Kl., N.F., 150: 1-133.
Schmidt-Kittler, N. 1977. Some aspects of evolution and provincialism of rodent faunas in the European Paleogene. Geobios, Mem. spéc. 1: 97-106.
Schmidt-Kittler, N. and Vianey-Liaud, M. 1975. Les relations entre les faunes de rongeurs d'Allemagne du Sud et de France pendant l'Oligocène. C. R. Acad. Sci. Paris, D, 281: 511-514.
Schmidt-Kittler, N. and Vianey-Liaud, M. 1979. Evolution des Aplodontidae oligocène européens. Palaeovertebrata 9: 33-82.
Shevyreva, N. 1972. New rodents in the Paleogene of Mongolia and Kazakhstan. Paleont. J., Moscow, 3: 134-145.
Sigé, B. 1976. Les insectivores et chiroptères du Paléogène moyen d'Europe dans l'histoire des faunes de mammifères de ce continent. Paleobiologie continentale, Montpellier 7: 1-25.
Sigé, B. and Vianey-Liaud, M. 1979. Impropriété de la Grand Coupure de Stehlin comme support d'une limite Eocène-Oligocène. Newsl. stratigr. 8: 79-82.
Smith, A. G. and Briden, J. C. 1977. Mesozoic and Cenozoic Paleocontinental Maps. Cambridge Univ. Press, Cambridge.
Stehlin, H. G. 1909. Remarques sur les faunules de Mammifères des couches éocènes et oligocènes du Bassin de Paris. Bull. Soc. Géol. Fr., Paris, 9: 488-520.
Thaler, L. 1966. Les Rongeurs fossiles du Bas-Languedoc dans leurs rapports avec l'histoire des faunes et la stratigraphie du Tertiaire d'Europe. Mém. Mus. Nat. Hist. Nat., Paris, C, (Sci. Terre) 17: 1-295.
Vianey-Liaud, M. 1974a. Palaeosciurus goti nov. sp., écureuil terrestre de l'Oligocène moyen du Quercy. Données nouvelles sur l'apparition des Sciuridés en Europe. Ann. Pal. (Vert.), Paris 60: 103-122.

Vianey-Liaud, M. 1974b. Les Rongeurs de l'Oligocène inférieur d'Escamps. Palaeovertebrata 6: 197-241.

Vianey-Liaud, M. 1974c. L'anatomie crânienne des genres Eucricetodon et Pseudocricetodon (Cricetidae, Rodentia, Mammalia); essai de systématique des Cricétidés oligocènes d'Europe Occidentale. Géologie Méditerranéenne, Marseille 1: 111-132.

Vianey-Liaud, M. 1975. Caractéristiques évolutives des Rongeurs de l'Oligocène d'Europe Occidentale. Coll. Internat. C.N.R.S., Paris 218: 765-776.

Vianey-Liaud, M. 1979. L'évolution des Rongeurs à l'Oligocène en Europe occidentale. Paleontographica 166: 136-236.

Vianey-Liaud, M. and Hartenberger, J.-L. 1983. Modalités évolutives des rongeurs paléogènes européens. Coll. Internat. C. N.R.S. 330: 225-237.

Wahlert, J. H. 1977. Cranial foramina and relationships of Eutypomys (Rodentia, Eutypomyidae). Amer. Mus. Novit. 2626: 1-8.

Wahlert, J. H. 1978. Cranial foramina and relationships of the Eomyidae (Rodentia, Geomorpha). Skull and upper teeth of Kansasimys. Amer. Mus. Novit. 2645: 1-16.

Wilson, R. W. 1972. Evolution and extinction in Early Tertiary rodents. 24th I.G.C., Sect. 7: 217-224.

Wilson, R. W. 1980. The stratigraphic sequence of North American rodent faunas. Palaeovertebrata, Mem. Jubil. R. Lavocat: 273-284.

Wood, A. E. 1974. Early Tertiary vertebrate faunas, Vieja Group, Trans-Pecos Texas: Rodentia. Texas Mem. Mus. Bull. 21: 1-112.

Wood, A. E. 1975. The problem of the hystricognathous rodents. Univ. Mich. Papers Paleont. 12: 75-80.

Wood, A. E. 1977. The Rodentia as clues to Cenozoic migrations between the America and Europe and Africa. Milwaukee Pub. Mus. Spec. Publ. Biol. Geol. 2: 95-109.

Wood, A. E. 1980a. Problems of classification as applied to the Rodentia. Palaeovertebrata, Mem. Jubil. R. Lavocat: 263-272.

Wood, A. E, 1980b. The Oligocene rodents of North America. Trans. Amer. Phil. Soc. 70: 1-68.

Zdansky, O. 1930. Die altertiären Säugetiere Chinas stratigraphischen Bemerkungen. Paleontologia Sinica, C, VI (2): 1-87.

CRANIAL FORAMINA OF RODENTS

John H. Wahlert

Department of Vertebrate Paleontology
American Museum of Natural History
New York, New York, USA

INTRODUCTION

A paleontologist does not have the luxury of designing experiments to test hypotheses; the experiments, termed evolution, have already occurred. It is the job of the paleontologist to use the sparse evidence of fossil remains, along with the surviving results of evolution, in order to discover the experiments. Any aspect of morphology, from molecules to entire animals, can yield valuable evidence.

Dental and cranial evidence have been used extensively in evaluating proposed relationships among rodents, and of rodents to other mammals. The teeth and jaw musculature, both parts of the masticatory system, have not provided sufficient evidence to support a single, best hypothesis. The cranial foramina provide an additional source of comparative information. They are clear evidence for the presence or absence of certain blood vessels and of the course of vessels and nerves. The positions of arteries and nerves offer reference points for assessing the relative attitudes of bones. This important aspect of the foramina has not been considered in studies of rodent crania.

Merriam (1895) described and figured crania of living geomyid rodents in excellent detail. Howell (1932) and Hill (1935, 1937) described the cranial foramina in rodents and found differences among taxa that suggested the utility of these apertures as characters in systematic analysis. Their lead was not quickly followed; Landry (1957, p. 8) was "... convinced that the cranial foramina of rodents are not very reliable characters", chiefly because homologies were uncertain. Since then, Wood (1962) has described paramyid skulls and those of Oligocene rodents (1974, 1976) in detail. Vianey-Liaud (1976) has carefully presented theridomyid cranial remains. Patterson and Wood (1982) described skulls of early Caviomorpha. Musser (1981a, b, 1982a, b) and Musser and Newcomb (1983) have recorded and figured a wealth of information on cranial foramina in murids; their analyses of primitive and derived character states are especially helpful. My own research has covered the protrogomorphous rodents (Wahlert, 1974); subsequent papers dealt with *Eutypomys* and the early Geomorpha (Wahlert, 1977, 1978, 1983). Description of the Geomyoidea is in press, but that of other sciuromorphous rodents is available only in my Ph.D. dissertation (Wahlert, 1972). I have begun a study of the Myoxoidea and have completed a preliminary survey of the Myomorpha. In this paper I draw on the observations made on all of these groups.

Certain works have been extremely helpful in the study of cranial foramina. Greene's Anatomy of the Rat (1935) is a thorough treatment on which to base dissection; initially the illustrations are confusing, because the system that is the subject of a figure is often disproportionately enlarged. Hyman (1942) presented excellent verbal directions for dissection; these are especially helpful when the specimen is very different from the rat. A third source, valuable for its completeness in anatomical description and terminology, is Miller's Anatomy of the Dog (1964). McDowell's (1958) glossary of foramina, although it is for the Lipotyphla, is the best set of definitions available.

The carotid circulation in rodents has been studied in detail. Tandler (1899) examined a variety of rodents. Guthrie (1963, 1969) and Guthrie and De Long (1977) extended the work; Bugge (1970, 1971a, b, c, 1974) used acrylic-resin injection followed by corrosion to produce a remarkable set of photographs and comparative diagrams. The cranial nerves present few problems in dissection, because homologies are easy to determine, and the pattern of their distribution is consistent. Veins are the least studied part of the head and appear to have great variation in some regions.

The most accessible investigation of cranial architecture is by Buckland-Wright (1978) on skulls of *Felis catus*; direct measurement of strain and cracking patterns of resin coating in experimental simulation of a functioning masticatory system and microradiography were used to determine the patterns of force distribution in bones. The morphological contrast between cat and rodent skulls provokes thought. Bateman (1954) presented valuable information on the shaping of rodent skulls during development. A bone by bone analysis of remodeling in the mouse skull illustrates possible alternative scenarios for development of different cranial characteristics in rodents.

Little information about cranial foramina can be gleaned from the literature other than from the works cited. The exact positions of foramina and sutures are unclear in most published illustrations of skulls; the orbital morphology is usually hidden by the zygomatic arch; and often features described in the text are not tied in with the figures by labels. Fossil skulls present special problems, because they require extremely careful preparation and thoughtful mental restoration. Illustrations of skulls in their damaged state are misleading unless the true edges of bones are distinguished from broken ones. Bilateral symmetry, direction of apertures, and molding of edges are important clues to identification of foramina.

The problem of defining the primitive condition of any character in rodents is difficult, since the earliest representatives are clearly distinct from other mammals. There is often no sure basis for comparison. I accept Wood's view (1965) that the extinct paramyids are the most primitive specimens available, and I add the sciuravids to this category. For the purpose of my analysis, it is not necessary that these taxa be ancestors of the other rodents studied.

The chief criterion for a taxon to be used in the study of foramina is that well preserved, complete skulls be available. *Paramys copei*, *Paramys delicatus*, and *Sciuravus nitidus* fulfill this requirement. All of these taxa are protrogomorphous--the origins of the lateral and medial masseter are restricted to the zygomatic arch. Of all the arrangements of the masseter that occur in rodents, the protrogomorphous condition is the most like that in other mammals. The anterior extension of lateral and medial masseter in other rodent groups is accepted as modifications that improve the effectiveness of the chisel-like incisors, the basic characteristic of rodents. Any modification in the masticatory musculature rearranges the pattern of chewing stresses in the skull; the accompanying changes in cranial architecture alter positions and shapes of the foramina, which mark the passage of nerves and vessels. It is not surprising, therefore, that the foramina in protrogomorphous rodents, which have a primitive masticatory system, can be compared with those of other mammals more easily than can the foramina of other rodents. *Paramys* and *Sciuravus* are the standards of primitiveness used in the following descriptions.

This paper is no more than a progress report, since the number of taxa examined is still far short of a thorough survey of the Rodentia. My chief purpose is to tell other investigators what features can be seen in a skull, so that descriptions of cranial remains can be presented amply and consistently. I have reexamined data in an attempt to sort derived and primitive character states, because this is a first step in discovering both the origin of rodents and the pattern of their diversification.

I have not begun to investigate the cranial foramina of the

Hystricognathi (Hystricomorpha of many authors), and I do not pretend familiarity with them. The few occasions on which I have attempted to interpret hystricognath morphology convinced me that dissections will be a necessary beginning. For these reasons the Hystricognathi are largely omitted from this progress report.

CRANIAL FORAMINA

The foramina are grouped below for convenience in viewing, for their structural relationships, and by common breakage points in fossils. The palate is thick and commonly preserved, often with the rostrum attached. The orbit is less frequently seen and is usually broken through the optic foramen and anterior part of the pterygoid region. In some few specimens the zone of breakage is at the front of the auditory region, and the pterygoid fossae are preserved. Complete skulls are the rarest remains. The palate is of primary importance, because it usually retains enough teeth to identify the taxon.

Figures 1 through 4 illustrate crania of a paramyid, sciurid, heteromyid, and murid rodent, and these show most of the foramina described. The comments that accompany the figures permit comparison of important features without reference to the text. The incisive foramina in *Paramys* and *Sciuravus* are moderately large and occupy from 41 to 47 percent of the diastemal length, the distance from the back of the incisor alveolus to the root of the first cheek tooth on the same side. Within ischyromyoids some taxa, including *Ischyrotomus*, *Ischyromys*, and *Manitsha*, have much lower ratios; these are considered to be specializations as are other features in these taxa. Low ratios occur in many living rodents, most notably in geomyoids (Fig. 3), where the extremely small size of the foramina is certainly derived. In all of these taxa, however, the premaxillary-maxillary suture intersects the lateral margins of the foramina near the back, a common mammalian condition. Long incisive foramina that occupy much of the diastema occur in muroids, but a great range of ratios is encompassed by the group. The suture intersects the elongated foramina near the middle, and this is a derived condition. The aperture remains in the primitive position; in a few recent specimens, in which cleaning stopped short of completion, the posterior extension of the foramina is covered by membrane (Fig. 4).

The major pair of posterior palatine foramina, which transmit the descending palatine artery and a vein, is within the palatine bones in *Paramys* and *Sciuravus*, eomyids, and some taxa of sciurids and muroids. The foramina are in the maxillary-palatine suture in most rodents. The common position of this pair of foramina is medial to a zone that extends from the middle of the first molar to the middle of the second molar. The foramina are farther posterior in prosciurids, aplodontids, some sciurids, and some heteromyids, and they are farther anterior in *Ischyrotomus*, *Ischyromys*, and cylindrodontids. The primitive position relative to the suture and teeth is not established.

A notch behind the posterior point of the maxilla marks the passage of the descending palatine vein in Paramys and Sciuravus (Fig. 1). Broad union of the maxilla and alisphenoid, a derived condition in many rodent groups, encloses the passage as a posterior maxillary foramen (Fig. 2). In many muroids the foramen is lacking (Fig. 4); Greene (1935) showed that in the rat, which lacks the foramen, the vein accompanies the descending palatine artery.

The infraorbital canal is very short in Paramys and Sciuravus; this is a special derived condition of the Rodentia that is associated with the steep angle at which the maxillary root of the zygoma stands relative to the axis of the skull. The canal transmits the infraorbital nerve, artery, and vein. Recently Eastman (1982; see also Coues, 1877, p. 549) described a slip of the medial masseter that passes through the infraorbital canal of Aplodontia, a living protrogomorphous rodent; he proposed that this condition, hystricomorphy, is primitive in rodents and probably occurred in the extinct protrogomorphs such as Paramys. The character occurs in the Hystricognathi, the Myomorpha, and a few other groups. If hystricomorphy is a primitive and apparently useful adaptation, as its extreme development in many rodents testifies, then true sciuromorphy would not have been likely to evolve. Sciuromorphy has arisen independently at least three or four times: in the sciurids, castorids, ischyromyids, and the Geomorpha. In these the origin of the anterior part of the lateral masseter extends onto the rostrum, and the medial division does not penetrate the infraorbital canal. In the Muroidea the origin of the lateral masseter extends forward on a part of the of the maxillary root, termed the zygomatic plate (Fig. 4); origin of the medial masseter on the rostrum apparently prevented attachment of the lateral division there. Klingener (1964) proposed that the myomorphous masseter was derived from the hystricomorphous type, and there is some fossil evidence to support this hypothesis (see Vianey-Liaud, this volume).

The long infraorbital canal in many sciuromorphous rodents (Fig. 2) looks similar to the common mammalian condition. Here, however, it is a derived condition associated with the increased anterior origin of the lateral masseter. The canal remains short in the sciurid genera Protoxerus, Eutamias, and Tamias, although they are fully sciuromorphous. The canal stands lateral to the rostral wall in the Castoridae and Sciuridae. The fact that its anterior end is depressed into the rostrum in the Geomorpha (Fig. 3) is evidence of independent derivation. The infraorbital foramen is the anterior aperture of the infraorbital canal. Its shape is strongly influenced by impinging musculature. In some instances the lacrimal canal causes a bulge in its medial wall. Heteromyids have a unique rostral perforation in the maxilla medial and anterior to the

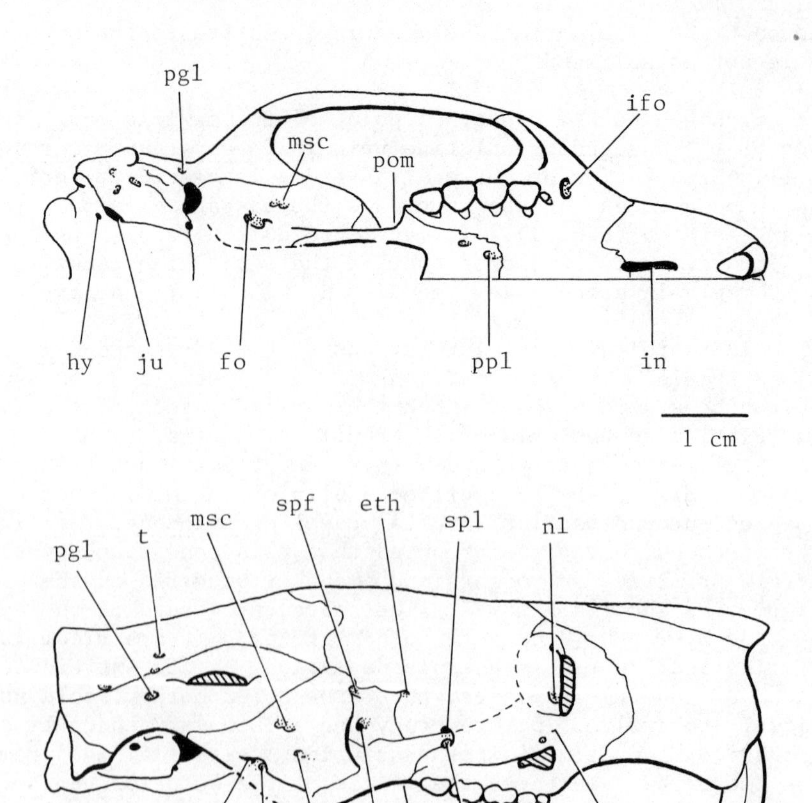

Fig. 1. Paramys copei (Paramyidae) (after Wahlert, 1974). Abbreviations for foramina and other apertures as noted: ac, alisphenoid canal; bu, buccinator; cc, carotid canal; cHu, canal of Huguier; dpl, dorsal palatine; eth, ethmoid; fo, foramen ovale; foa, accessory foramen ovale; hy, hypoglossal; ifo, infraorbital; in, incisive; ito, interorbital; ju, jugular; mlf, middle lacerate foramen; msc, masticatory; nl, nasolacrimal; op, optic; paf, post-alar fissure; pgl, postglenoid; pom, posterior maxillary notch or foramen; ppl, posterior palatine; rp, rostral perforation; spf, sphenofrontal; spl, sphenopalatine; spt, sphenopterygoid canal; spv, sphenopalatine vacuity; sqm, squamosomastoid; st, stapedial; sty, stylomastoid; t, temporal; uml, unossified area between maxilla and lacrimal; vf, venous foramen. Cross hatching indicates imagined cut through bone.

Comment: Paramys copei is taken as the standard of primitiveness among rodents.

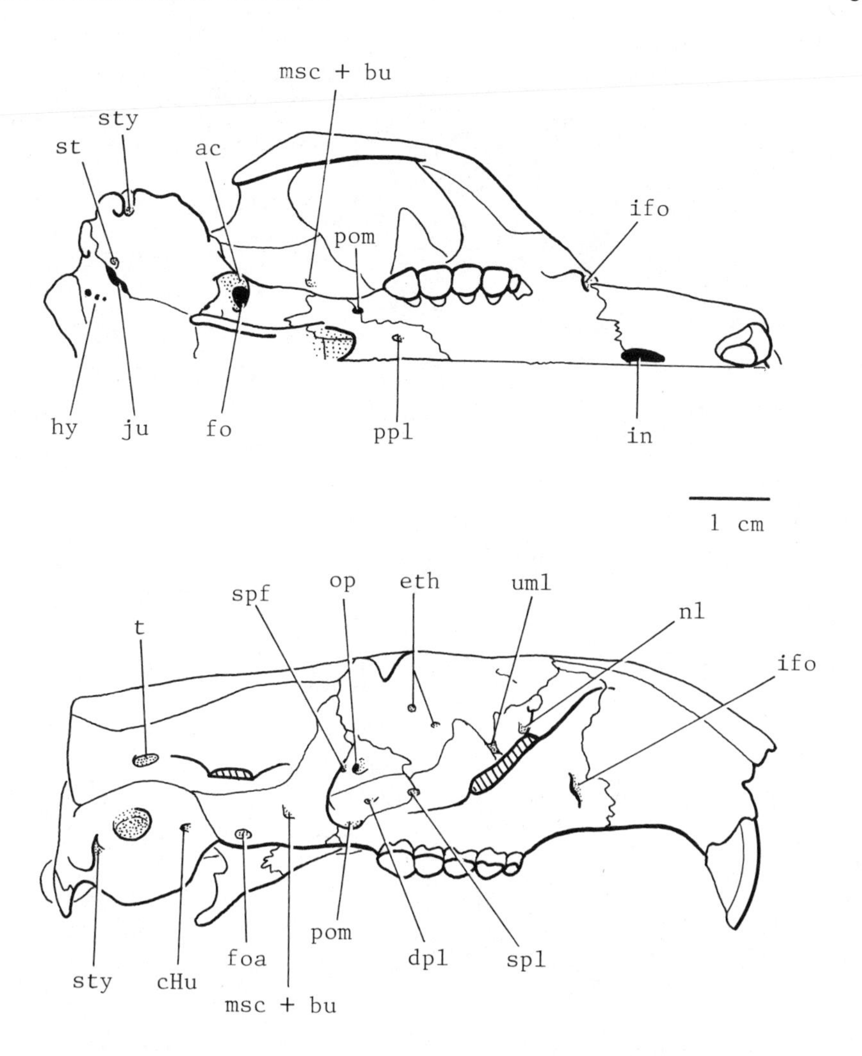

Fig. 2. Marmota monax (Sciuridae) (after Wahlert, 1974). See fig. 1 for abbreviations.

Comment: Anterior root of zygomatic arch inclined; infraorbital canal long and opening lateral to rostrum; sphenopalatine foramen anterior to dorsal palatine; double ethmoid foramen dorsal to cheek teeth; sphenofrontal and stapedial foramina present; maxilla and alisphenoid just touching; posterior maxillary foramen enclosed; distance from third molar to auditory region shorter than in Paramys; pterygoid fossa long and completely roofed; lateral pterygoid flange strong and enclosing accessory foramen ovale; auditory bulla in contact on all sides with skull; cranium deep.

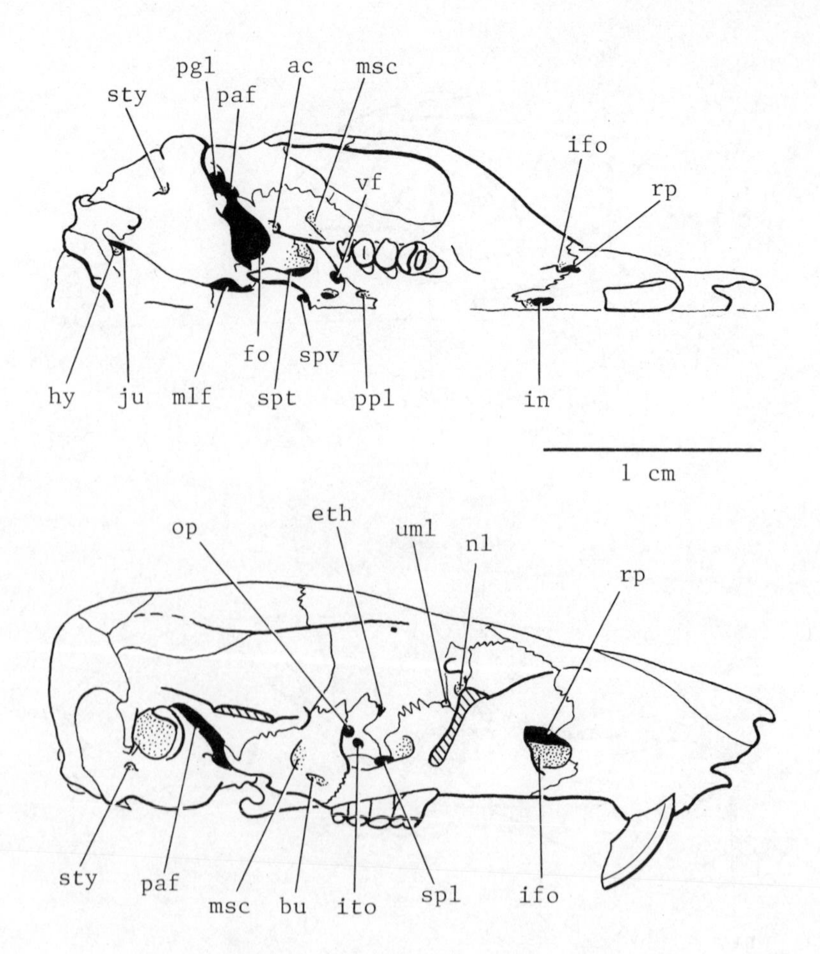

Fig. 3. Liomys pictus (Heteromyidae), AMNH (American Museum of Natural History) 190271. See fig. 1 for abbreviations.

Comment: Anterior root of zygomatic arch inclined; infraorbital canal long and opening depressed into rostrum; rostrum perforated; sphenopalatine foramen anterior to dorsal palatine foramen; all orbital foramina dorsal to cheek teeth; sphenofrontal and stapedial foramina absent; maxilla and alisphenoid joined in broad suture; posterior maxillary foramen lacking; distance from third molar to auditory region short; pterygoid fossa broad; sphenopterygoid canal forming passage from pterygoid fossa to orbit; lateral pterygoid flange strong but not enclosing accessory foramen ovale; broad gap anterior and anteromedial to bulla; cranium deep.

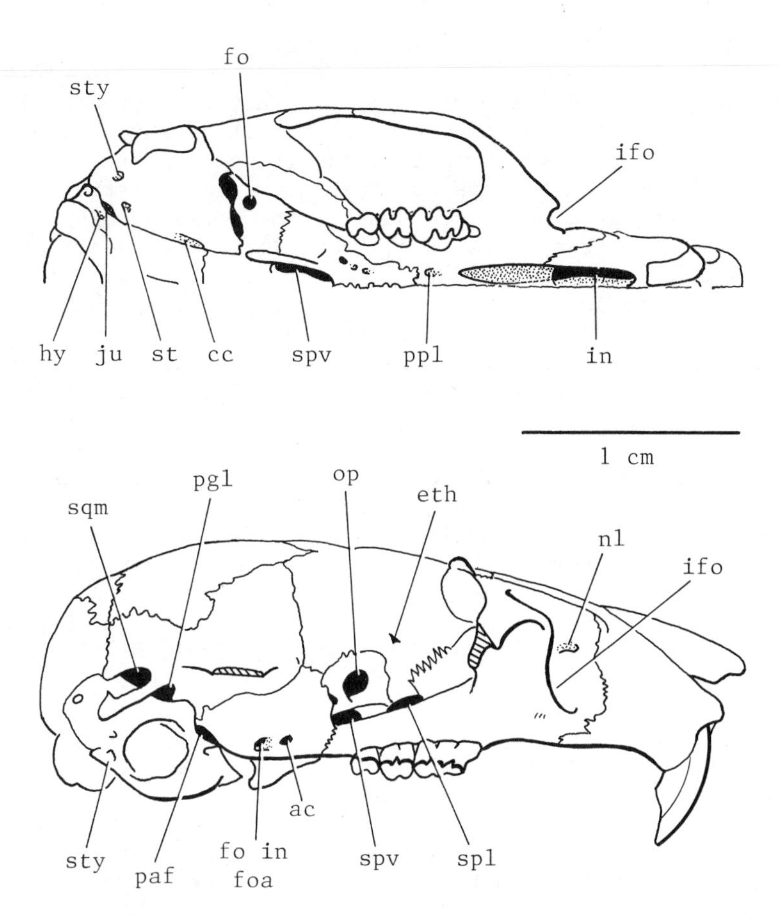

Fig. 4. Phyllotis magister (Muridae), AMNH 74152. See fig. 1 for abbreviations.

Comment: Anterior root of zygomatic arch inclined as plate-like structure; infraorbital canal short with dorsal part penetrated by medial masseter muscle; sphenopalatine foramen large and including dorsal palatine foramen; all orbital foramina above cheek teeth; sphenofrontal foramen absent, but furrow on inner surface of alisphenoid indicating that superior ramus present and emerging at dorsal end of anterior-alar fissure; stapedial foramen present; maxilla and alisphenoid broadly joined; posterior maxillary foramen lacking; distance from third molar to auditory region shorter than in Paramys; pterygoid fossa broad and completely roofed; lateral pterygoid flange enclosing accessory foramen ovale; entrance to alisphenoid canal anterior; series of gaps anterior to bulla; squamosal embayed posteriorly; cranium deep.

foramen (Fig. 3); its function is unknown.

The nasolacrimal foramen is usually in the anterodorsal part of the orbital wall; the lacrimal bone bounds it on three sides and the maxilla anteriorly in the primitive condition. The region is incomplete in specimens of *Paramys* and *Sciuravus*, but existing remains suggest that the design was the same in them. In muroids, myoxoids, and dipodoids the foramen is in the maxillary bone anteromedial to the infraorbital foramen, a derived condition (Fig. 4). The lacrimal canal actually begins in the usual orbital position, but the first part of its course is covered only by connective tissue. A gap in the orbital wall occurs at the junction of the lacrimal, frontal, and maxillary bones in rodents of many groups (Figs. 2, 3); the lack of ossification is a derived character.

The three chief foramina of the orbit, the optic, ethmoid, and sphenopalatine, form a triangular constellation that is easy to recognize. The most interesting feature of these foramina is that they are farther anterior relative to the cheek teeth than in other mammals. Apparently the basic rodent design not only emphasizes anterior extension of the masseter muscle but also a more posterior placement of the dentition under the orbit. The orbital foramina are farthest posterior in the early protrogomorph *Paramys* (Fig. 1); the optic foramen is posterior to the third molar; the ethmoid, posterodorsal or dorsal to that tooth; the sphenopalatine dorsal to the junction of the second and third molars or to the second molar. The foramina in *Sciuravus* are somewhat farther anterior and similar to the positions seen in a few groups of living rodents. In early rodents the optic and sphenopalatine foramina are far apart and the ethmoid foramen is low, so that the triangle they form has a broad base and little height.

The optic foramen is surrounded by the orbitosphenoid bone with few exceptions. In extremely rare instances its ventral margin appears to be defined by the orbital process of the palatine. In some rodents the foramen is united with an unossified area dorsal and posterior to the orbitosphenoid. The position of the foramen, however, can be found in most cases by locating the saddle on the medial part of the orbitosphenoid where right and left optic nerves diverged. The optic foramen is posterior to the third molar in *Paramys*, dorsal to the back of the third molar in *Sciuravus* and farther anterior in many rodents. The foramen is usually oval, and its most common lesser dimension is 1.0 mm; this occurs in *Paramys*. The large size of the foramen in prosciurids, sciurids, florentiamyids, and many myomorphs is derived as is its small size in geomyids and some other rodents. The optic foramen has often been confused with the interorbital foramen which is anterior, anteroventral, or ventral to it in the orbitosphenoid bone. Hill (1935) reported that this foramen transmits a sinusoid vein between the orbits. The key to sure identification of the interorbital foramen is thorough cleaning of fossil specimens; it never leads into the braincase as the optic foramen must.

The sphenopalatine foramen transmits the sphenopalatine nerve, artery, and vein. The maxilla always participates in its margin. The palatine and frontal bones commonly bound it posteriorly and dorsally. In a few rodents the orbitosphenoid reaches the foramen posteriorly; in others it is entirely surrounded by the maxilla; both of these conditions are derived. Occasionally the ethmoturbinal is exposed on the dorsal side of the aperture. The sphenopalatine foramen is primitively dorsal to the junction of the second and third molars as in Paramys. It is farther anterior in Sciuravus and is as far forward as the junction of the fourth premolar and the first molar in some rodents such as eomyids.

The ethmoid foramen is at or close to the suture between the anterodorsal edge of the orbitosphenoid and the frontal bones. In sciurids (Fig. 2) it has two divisions, and both are within the frontal. It can be found in fossils with careful scrutiny, because a small groove often leads to it across the orbitosphenoid. Since it faces ventrally, a tiny bit of matrix usually hides it in fossils. In some rodents the ethmoid foramen is within an unossified area that separates the orbitosphenoid and frontal bones.

The sphenofrontal foramen is situated in the alisphenoid-orbitosphenoid suture near its intersection with the frontal bone. It transmits the superior ramus of the stapedial artery. Bugge (1974) proposed that this condition is the primitive pattern, and he demonstrated that derived conditions which entail the loss of the foramen are brought about by anastomoses between the orbital portion of the superior ramus and other arteries of the head. He pointed out that the term ophthalmic artery is incorrect for this vessel. There is strong evidence that Paramys and Sciuravus had large stapedial arteries, and each preserves a sizable sphenofrontal foramen. This foramen is easily missed in fossils, because its margin is frequently damaged. It can also be combined with the anterior lacerate foramen at the dorsal end of the anterior-alar fissure. The sphenofrontal foramen is absent in aplodontids, castorids, geomyines, some myoxoids, and some muroids.

The dorsal palatine foramen in the floor of the orbit is the aperture through which the descending palatine artery enters a canal that leads to the posterior palatine foramen on the palate. The position of the dorsal palatine foramen is more closely tied to the dentition than to the upper part of the cranium. It is always in or next to the suture between the maxilla and the orbital lamina of the palatine and is dorsal to the junction of the second and third molars. The foramen is immediately ventral to the sphenopalatine foramen in Paramys (Fig. 1) and Sciuravus; this association is also seen in cylindrodontids and Ischyromys in which the posterior palatine foramina are anteriorly situated. The same association can be seen in some insectivores (McDowell, 1958). The dorsal palatine foramen usually retains its primitive position relative to the cheek teeth, whereas the sphenopalatine is more anterior. The situation in cylindrodontids and Ischyromys is unusual.

The alisphenoid bone is extended anteriorly and hides the anterior lacerate foramen (equals sphenorbital or sphenoidal fissure) exactly as in insectivores (McDowell, 1958). I have termed the leading edge of the alisphenoid in rodents the anterior-alar fissure (Wahlert, 1983), because it is the anterior end of the alisphenoid canal. This canal transmits the internal maxillary artery. The alisphenoid canal in rodents may be open in part or entirely on its medial side into the cranial cavity. Many mammals lack an alisphenoid canal, and the maxillary artery runs lateral to the bone; this condition appears to be primitive, and alisphenoid canals in different mammalian orders are not homologous.

In addition to a long alisphenoid canal, rodents, but not insectivores, have canals in the alisphenoid bone for the masseteric and buccinator divisions of the maxillary nerve. The canals have a common posterior aperture and diverge to separate anterior openings. The masseteric nerve emerges from the masticatory foramen and ascends in a broad groove in the alisphenoid bone before turning laterally to pass between the coronoid and condyloid processes of the jaw. The buccinator nerve is smaller and runs anterolaterally. When the two foramina are combined (Fig. 2) the edge of the opening is L-shaped and faces anteriorly, since the nerves are at right angles to each other. The fact that the nerves are enclosed in bone suggests that the alisphenoid bone is situated farther laterally in rodents than in other mammals. Separation or union of the masticatory and buccinator foramina may reflect the different positions of the bone relative to a constant disposition of the diverging nerve pathways. In many murids and other rodents the masticatory and buccinator foramina are united with the accessory foramen ovale. The entrance to the alisphenoid canal is at the anterior end of this opening in many murids (Fig. 4) (Musser and Newcomb, 1983). It is possible that the arteries and nerves follow their usual courses, but the bone is situated more medially than in Paramys and Sciuravus.

The unusual extent of the alisphenoid in rodents is illustrated also by the presence of the accessory foramen ovale which perforates the alisphenoid bone lateral to the foramen ovale. The mandibular branch of the trigeminal nerve passes through a foramen only because bone stands in its course. The foramen has been found in specimens of Paramys in addition to those described by Wahlert (1974). I believe the accessory foramen ovale is a derived character shared by all rodents. It is lost again in groups such as the geomyoids (Fig. 3) in which the posteroventral part of the alisphenoid apparently lacks a structural role and does not develop.

The pterygoid fossa is always bounded medially by a prominent plate of bone that is anteroposteriorly oriented alongside the nasal passage. In *Paramys* and *Sciuravus* the lateral boundary is defined by a low ridge that anteriorly approaches the medial plate; the fossa is very shallow. A prominent lateral ridge and broad fossa are characteristic of many later rodents. The fossa in *Paramys* is an acute V-shape; in many later rodents the angle of the V is less acute. The posterior end of the fossa is bounded by the auditory bulla in most living rodents; together with the medial and lateral flanges it frames a triangle.

The foramen ovale is lateral to the posterior part of the base of the hamular process; the usual position of the foramen in mammals is posterolateral to the process. The foramen ovale and the base of the hamular process are a substantial distance anterior to the auditory region in *Paramys*, slightly closer in *Sciuravus*, and adjacent to the bulla in many recent rodents. The entire pterygoid region appears proportionally shorter in most rodents than in *Paramys*, but no comparison of dimensions has been made. The foramen ovale may be surrounded by bone, bounded posteriorly by the bulla, or open posteriorly into a gap.

In *Paramys copei* the foramen ovale and the alisphenoid, masseteric, buccinator, and transverse canals open onto a common depression. The depression is somewhat more elongated in *Sciuravus*. Close association of the foramen ovale and alisphenoid canal is common when the latter structure is present as it is in insectivores and *Canis*. In many taxa of rodents the alisphenoid canal is best seen in lateral view; it often begins at the anterior end of the accessory foramen ovale. Musser and Newcomb (1983) described this condition and the relationship of the morphology of the pterygoid region to the cranial arteries in muroids. McDowell (1958) suggested that the presence of the transverse canal, which traverses the basisphenoid, is primitive in insectivores; in rodents it permits connection of the two internal maxillary veins. Primitively it opens onto the posteromedial wall of the alisphenoid canal; it is more medially situated, near the base of the hamular process in all myomorph rodents including the Geomorpha and the Myoxoidea.

A complete roof of the pterygoid fossa is primitive. In many myomorphs there is a gap, because the posterior edge of the palatine fails to overlap the anterior part of the pterygoid. This gap may be the first stage in evolution of a sphenopterygoid canal (Wahlert, 1978), which allows anterior extension of the internal pterygoid muscle. In living geomyoids (Fig. 3) the muscle arises in part from the walls of the canal and extends into the orbit ventral to the optic foramen. An open fossa involves failure of ossification and is associated frequently with unossified areas along the dorsal border of the orbitosphenoid in the wall of the orbit.

Restriction of the origin of the internal pterygoid muscle to a fossa posterior to the cheek teeth is a characteristic of rodents that is unusual among mammals. A similar condition may be seen in Turnbull's (1970) illustration of Echinosorex. In Felis, Odocoileus, and Ovis (ibid.) the origin of the internal pterygoid extends beside the alisphenoid bone into the orbit, where it is attached chiefly to the broad flange of the palatine. In both rodents and insectivores the orbital exposure of the palatine is small; Novacek (1980) interpreted this condition in insectivores as derived. A broad palatine wing in the orbit is probably associated with orbital origin of the internal pterygoid muscle. Extension of the muscle medial to the alisphenoid wall in geomyoids, Spalax, and many hystricognaths (Landry, 1957) is a reinvention of the orbital origin; it is associated with propalinal chewing.

The middle lacerate foramen for passage of the internal carotid artery is at the posteromedial apex of the triangular pterygoid fossa. It is simply a notch in the pterygoid bone that is bounded posteriorly by the auditory bulla. I have previously called this the carotid foramen. Bugge (1971a) showed that in sciurids the internal carotid artery is obliterated distal to the departure of the stapedial artery. He (1974) proposed that Paramys had both stapedial and internal carotid branches. The petrobasilar canal in sciurids resembles the region termed carotid canal in Paramys, so the presence or absence of the anterior continuation of the internal carotid cannot be determined in the fossil. The promontorium of Sciuravus bears clear grooves that mark divergence of stapedial and promontorial arteries within the middle ear. Presley (1979) presented embryologic evidence that indicates the internal carotid continuation and promontory artery are the same vessel; he considered the promontorial position of the artery in the middle ear of the adult to be a derived condition.

There is a gap, which I have previously termed the middle lacerate foramen, between the bulla and the back of the pterygoid fossa in many rodents; the middle lacerate foramen for the carotid artery is continuous with it. This gap is variously developed in different groups. Another gap occurs in many rodents between the alisphenoid and squamosal and the bulla; I have called it the post-alar fissure (Figs. 3, 4). Primitively the postglenoid foramen is in the squamosal bone posterior to the glenoid fossa (Fig. 1). It is between the squamosal and the auditory bulla in aplodontoids, geomyoids, myoxoids, some dipodoids, and some muroids (Figs. 3, 4). The middle lacerate foramen, gap anterior to the bulla, foramen ovale, post-alar fissure, and postglenoid foramen can form a continuous fissure, or can be united in a variety of combinations that are often variable within genera. The gaps around the auditory region may reduce sounds of chewing that reach the ears by conduction through bone. In Dipodomys there is even a gap between the petrosal and basioccipital.

Temporal foramina in or near the squamosal-parietal suture are common in rodents and occur in Paramys and Sciuravus. They are connected via the venous system to the postglenoid foramen. The muroid rodents usually lack these apertures. In muroids the posterior part of the squamosal contains an embayment in the edge adjacent to the mastoid and occipital bones (Fig. 4). This embayment may be analogous to the squamosomastoid foramen that is occasionally present in a small variety of rodent taxa (Musser, 1982a). The mastoid foramen between the dorsal part of the mastoid bone and the occipital is usually a long slit, but in some taxa it is very small or absent.

The sphenopalatine vacuities in the roof of the nasal passage are between the orbitosphenoid bone and the palatine flanges that flank it. They appear to be lacking in Paramys, Sciuravus and other early North American taxa. The vacuities are present in most of the Myomorpha (Figs. 3, 4). In some rodents they extend so far anteriorly that they can be seen in lateral view of the orbit.

The stapedial foramen, which is usually between the tympanic and petrosal bones at the posteromedial curve of the bulla, transmits the artery of the same name (Figs. 2, 4). Its size gives a good measure of the size of the artery. The foramen is easily overlooked in rodents in which it is deep within the jugular foramen. Presence of the artery is primitive, and it occurs in both Paramys and Sciuravus. It is lost independently in a variety of rodent groups, including ischyromyids, cylindrodontids, mylagaulids, some aplodontids, some heteromyids, geomyines, some myoxoids, and some muroids.

The jugular foramen transmits the jugular vein and allows passage of nerves and other vessels. Its shape is lenticular, and it is connected anteriorly with the petrobasilar canal. The hypoglossal foramen is posteromedial to the jugular foramen; it often has more than one opening, and these are commonly included in a single depression. The stylomastoid foramen is posteroventral to the bony external auditory meatus and allows the facial nerve to pass between this tube and the mastoid process. I have found no features of these foramina that are of taxonomic interest in rodents.

DISCUSSION

Cranial foramina can be used as evidence in support of certain phylogenetic hypotheses and as indicators of changes in skull design in the course of evolution. The data that are being accumulated have been useful in limited phylogenetic problems but are not yet sufficient for a larger view. Tabulation of data for major groups revealed that some members of most groups, but not all members, possess any particular derived character. As Wood (1935) demonstrated, parallelism is common in rodents. Foramina have not been used to describe the supporting architecture of rodent skulls.

Two basic problems exist in rodent phylogeny. The first is the relationship of rodents to other mammals, and the second concerns initial branching within the Rodentia. Any close relative of the Rodentia should possess the basic features of rodent design: ever-growing, chisel-like incisors with enamel concentrated on the anterior surface, and a single pair of incisors dominant in the upper and lower dentitions; long glenoid fossa sloping anteroventrally and shorter diastema in the lower dentition than in the upper which, together, permit occlusion of incisors while cheek teeth are disengaged. In addition to these traditional characters the foramina indicate another important set: very short infraorbital canal with maxillary root of zygoma at nearly a right angle to the wall of the rostrum; cheek teeth posterior relative to orbital foramina; alisphenoid bone anteroposteriorly wide and situated far enough laterally to enclose passages of internal maxillary artery and masseteric and buccinator nerves; origin of internal pterygoid muscle restricted to fossa posterior to cheek teeth; palatine bone with limited orbital exposure. In side view a rodent skull is box-shaped rather than wedge-chaped like the skulls of most mammals; this difference probably reflects specializations of the masticatory system.

The similarity of rodents and insectivores may reflect common origin or merely the fact that comparative information about mammal skulls is rare. Recent discovery of eurymyloid skulls in China (Li, 1977; Ting and Li, 1984) gives credence to the theory that rodents and lagomorphs share common ancestry. Li (this volume) describes several rodent-like characters of eurymyloids; perhaps the most significant among them is the physical separation of gnawing and chewing. Debate continues about the earliest diversification of rodents (Patterson and Wood, 1982; Lavocat, 1974; Hartenberger, 1980; Hussain et al., 1978). I have not yet had the opportunity to dissect hystricognath rodents and find myself unable to interpret many of their cranial foramina. This difficulty, itself, may point to early separation of the group from the well known North American protrogomorphs and the sciruomorphs and myomorphs. The accumulated data on cranial foramina are still inadequate to lend support to any particular hypothesis. The problem is compounded by the fact that very few rodents are represented by adequate cranial remains.

I have used characteristics of the cranial foramina to support certain hypotheses of relationship among protrogomorphous and sciuromorphous rodents. The Ischyromyidae and Cylindrodontidae retain primitive association of palatine and sphenopalatine foramina in the orbit, although the orbital foramina are farther anterior relative to the cheek teeth than in other protrogomorphous rodents. In both families the orbitosphenoid bone reaches the sphenopalatine foramen, the sphenofrontal foramen is lost, and the arrangement of foramina in the pterygoid fossa appears to be unique (Wahlert, 1974).

The prosciurids, aplodontids, and sciurids form a group with regard to their foramina (Wahlert, 1972). For the most part the number and positions of the foramina are primitive. The ethmoid foramen, however, is in the frontal bone; it is double in many living sciurids. Members of the three families also have a postorbital process; it is lacking in Aplodontia, itself. This process of the frontal bone projects over the orbit, and it may well be a derived feature. The postorbital process is seen only in Franimys (Wood, 1962) among early rodents.

I have placed the Muroidea, Dipodoidea, Myoxoidea, and Geomorpha in the suborder Myomorpha, because they share certain derived features (Wahlert, 1978). The carotid canal is short, and the entrance to the transverse canal is medial and separated from the alisphenoid canal (Wahlert, 1983). The Muroidea, Dipododoidea, and Myoxoidea share enlargement of the infraorbital foramen and have the aperture of the nasolacrimal canal medial or anteromedial to this foramen. The variety of shapes of the infraorbital foramen and morphologies of the zygomatic plate indicate that the history of this region is complex. Vianey-Liaud (This volume) suggests on the basis of fossils that dipodoids and myoxoids achieved myomorphy independently.

Cranial foramina suggest possible structural roles of the elements of rodent skulls. The most frequently described cranial character is the size of the infraorbital foramen. A large foramen is usually accompanied by clear evidence that anterior fibers of the medial masseter extended through the foramen and attached to the side of the rostrum--hystricomorphy. In other rodents the origin of the lateral masseter extends onto the rostrum alongside the maxillary root of the zygoma. In this condition, termed sciuromorphy, the infraorbital foramen is usually constricted, and a long canal permits covered passage of the artery, vein, and nerve. In myomorphous rodents origins of both the medial and lateral masseter have extended anteriorly. Anterior extension of the masseter's insertion on the mandible is part of the same transformation. The results of these modifications are two. First, the cheek teeth are within the muscular sling, and the incisors are close to it; the force available in both chewing and gnawing is thereby increased. Second, more muscle is oriented in a direction similar to that of the superficial masseter , and the anterior component of chewing force is increased.

Other foramina indicate further complexity of this transformation. In most rodents the chief orbital foramina, the sphenopalatine, ethmoid, and optic, are farther anterior relative to the cheek teeth than in Paramys, and the distance between the back of the third molar and the foramen ovale appears proportionally less. The teeth themselves appear to have been shifted posteriorly in the muscular sling. The position of the dorsal palatine foramen gives support to this view. The foramen usually retains its position relative to the cheek teeth and is posterior to the sphenopalatine foramen, whereas the two apertures are closely associated and the sphenopalatine is farther posterior in *Paramys*. The nature of the passage for the descending palatine vein presents further evidence. In *Paramys* the vessel appears to have run posterior to the point of the maxilla and alongside the stout extension of the palatine from the palate to the pterygoid region. In most rodents a foramen is enclosed in this position by a connection of the maxillary point and the anteroventral part of the alisphenoid. The palatine extension is covered by this bridge, and it is also very short. Again, the tooth-bearing part of the maxilla appears to be farther posterior.

In any view the skulls of rodents differ from those of other mammals. In ventral view the attitude of the maxillary root of the zygoma is striking. It stands out from the skull so that the zygomatic arch is nearly parallel to the anteroposterior axis. In side view the rostrum is deep to accommodate the incisor alveoli, and the skull is box-shaped. The primitive form of the face in mammals is a wedge.

Buckland-Wright (1978) has shown the chief structural elements that resist the stress of biting and chewing in *Felis catus*, which has a wedge-shaped skull. The dentition is situated under the face, and the chief structural girders of the skull join the teeth directly to the anterior zygomatic root and through the face to the temporal region. The strength of the maxillary root of the zygoma is also apparent in rodents; it is frequently preserved in fossils and is a very tough part of decalcified skulls. I believe that the facial brace between the dentition and the temporal region is unimportant in rodents compared with other mammals. The box-shaped skull would make it a circuitous route. Instead, the posterior part of the dentition is close to the palatine. The posterior extension of the palatine to the pterygoid region is stout; it appears to be the chief structural connection between the cheek teeth and the temporal region. This view is suported by the fact that in many rodent lineages the maxilla is joined in a broad suture directly to the anterior edge of the alisphenoid. The posterior brace is thereby strengthened.

The pterygoid region and alisphenoid bone that forms its lateral side are particularly interesting. Insectivoran breadth of the alisphenoid would be a possible precursor for the condition in rodents. Anteroposterior elongation of the glenoid fossa may have necessitated widening in the adjacent ventral element, the alisphenoid bone. Lateral position of the alisphenoid wall, indicated by the presence of alisphenoid, masseteric, and buccinator canals, may have been a precondition for subsequent union of the maxilla and alisphenoid. The importance of such a union is demonstrated in geomyoid rodents. In them the origin of the internal pterygoid muscle is extended anteriorly. Instead of passing lateral to the alisphenoid bone, as in other mammals, the internal pterygoid muscle extends through a canal, and the alisphenoid-maxillary union is not disrupted. The sphenopterygoid canal is created by incomplete ossification of the roof of the pterygoid fossa. Unossified areas adjacent to the orbitosphenoid bone in these rodents may be a consequence of this lack of ossification. Gaps in the medial wall of the orbit indicate that the wall does not resist compressive stress or torsion.

CONCLUSION

The special nature of rodent skulls must be understood in order to determine which other mammalian groups may be closely related to rodents. Cranial foramina provide evidence in addition to the host of special characters associated with gliriform incisors and gnawing. The story of rodent diversification appears to be one that encompasses a limited variety of improvements on the basic design. These improvements are influenced by special adaptation to specific environments, and parallelism is common. The cranial foramina, which serve as landmarks of nerves and vessels, yield information useful in determining the relationships of taxa and in understanding the function of cranial elements. I have found the evidence of cranial foramina to be the most helpful at the familial level in the groups that I have studied. The foramina are not the final arbiter of phylogenetic questions, but they are equal in value to the evidence of dental morphology.

ACKNOWLEDGMENTS

My participation in this conference was supported by the Early Tertiary Mammal Fund of the Department of Vertebrate Paleontology, American Museum of Natural History. I thank both the departments of Vertebrate Paleontology and Mammalogy for making research facilities and collections available to me.

REFERENCES

Bateman, N. 1954. Bone growth: A study of the grey-lethal and microphthalmic mutants of the mouse. J. Anat. 88: 212-262.

Buckland-Wright, J. C. 1978. Bone structure and the patterns of force transmission in the cat skull (*Felis catus*). J. Morphol. 155: 35-62.

Bugge, J. 1970. The contribution of the stapedial artery to the cephalic arterial supply in muroid rodents. Acta Anat. 76: 313-336.

Bugge, J. 1971a. The cephalic arterial system in mole-rats (Spalacidae) bamboo rats (Rhizomyidae), jumping mice and jerboas (Dipodoidea) and dormice (Gliroidea) with special reference to the systematic classification of rodents. Acta Anat. 79: 165-180.

Bugge, J. 1971b. The cephalic arterial system in sciuromorphs with special reference to the systematic classification of rodents. Acta Anat. 80: 336-361.
Bugge, J. 1971c. The cephalic arterial system in New and Old World hystricomorphs, and in bathyergoids, with special reference to the systematic classification of rodents. Acta Anat. 80: 516-536.
Bugge, J. 1974. The cephalic arterial system in insectivores, primates, rodents and lagomorphs, with special reference to the systematic classification. Acta Anat. 87(suppl. 62): 1-160.
Coues, E. 1877. Family Haplodontidae. In: Monographs of North American Rodentia, E. Coues and J. A. Allen, pp. 543-600, Rep. U. S. Geol. Surv. Terr. 11.
Eastman, C. B. 1982. Hystricomorphy as the primitive condition of the rodent masticatory apparatus. Evol. Theory 6: 163-165.
Greene, E. C. 1935. Anatomy of the rat. Trans. Amer. Philos. Soc., n.s., 27: 1-370.
Guthrie, D. A. 1963. The carotid circulation in the Rodentia. Bull. Mus. Compar. Zool. 128: 455-481.
Guthrie, D. A. 1969. The carotid circulation in Aplodontia. J. Mammal. 50: 1-7.
Guthrie, D. A., and N. De Long. 1977. Carotid arteries in the rodent genera Pappogeomys, Geomys and Thomomys (family Geomyidae). Bull. S. Calif. Acad. Sci. 76: 63-66.
Hartenberger, J.-L. 1980. Données et hypothèses sur la radiation initiale des Rongeurs. Palaeovertebrata, Mém. Jubil. R. Lavocat: 285-301.
Hill, J. E. 1935. The cranial foramina in rodents. J. Mammal. 16: 121-129.
Hill, J. E. 1937. Morphology of the pocket gopher mammalian genus Thomomys. Univ. Calif. Publ. Zool. 42: 81-172.
Howell, A. B. 1932. The saltatorial rodent Dipodomys: The functional and comparative anatomy of its muscular and osseous systems. Proc. Amer. Acad. Arts Sci. 67: 375-536.
Hussain, S. T., H. de Bruijn, and J. M. Leinders. 1978. Middle Eocene rodents from the Kala Chitta Range (Punjab, Pakistan) (1). Proc. Koninkl. Nederland. Akad. Wetensch., Amsterdam, ser. B, 81: 74-112.
Hyman, L. H. 1942. Comparative Vertebrate Anatomy. Univ. Chicago Pr., Chicago.
Klingener, D. 1964. The comparative myology of four dipodoid rodents (Genera Zapus, Napaeozapus, Sicista, and Jaculus). Misc. Publ. Mus. Zool. Univ. Michigan, no. 124: 1-100.
Landry, S. O., Jr. 1957. The interrelationships of the New and Old World hystricomorph rodents. Univ. Calif. Publ. Zool. 56: 1-118.
Lavocat, R. 1974. What is an hystricomorph? Symp. Zool. Soc. Lond. 34: 7-20.
Li, C. K. 1977. Paleocene eurymyloids (Anagalida, Mammalia) of Qianshan, Anhui. Vert. PalAsiat. 15: 103-118.

McDowell, S. B., Jr. 1958. The Greater Antillean insectivores. Bull. Amer. Mus. Natur. Hist. 115: 113-214.

Merriam, C. H. 1895. Monographic revision of the pocket gophers family Geomyidae (exclusive of the species of Thomomys). North Amer. Fauna 8: 1-258.

Miller, M. E. 1964. Anatomy of the Dog. W. B. Saunders Co., Phila.

Musser, G. G. 1981a. Results of the Archbold Expeditions. No. 105. Notes on the systematics of Indo-Malayan murid rodents, and descriptions of new genera and species from Ceylon, Sulawesi, and the Philippines. Bull. Amer. Mus. Natur. Hist. 168: 225-334.

Musser, G. G. 1981b. The giant rat of Flores and its relatives east of Borneo and Bali. Bull. Amer. Mus. Natur. Hist. 169: 67-176.

Musser, G. G. 1982a. Results of the Archbold Expediations. No. 108. The definition of Apomys, a native rat of the Philippine Islands. Amer. Mus. Novit. 2746: 1-43.

Musser, G. G. 1982b. Results of the Archbold Expeditions. No. 110. Crunomys and the small-bodied shrew rats native to the Philippine Islands and Sulawesi (Celebes). Bull. Amer. Mus. Natur. Hist. 174: 1-95.

Musser, G. G., and C. Newcomb. 1983. Malaysian murids and the giant rat of Sumatra. Bull. Amer. Mus. Natur. Hist. 174: 327-598.

Novacek, M. J. 1980. Cranioskeletal features in tupaiids and selected Eutheria as phylogenetic evidence. In: Comparative Biology and Evolutionary Relationships of Tree Shrews, W. P. Luckett, ed., pp. 35-93, Plenum Publ. Corp., New York.

Patterson, B., and A. E. Wood. 1982. Rodents from the Deseadan Oligocene of Bolivia and the relationships of the Caviomorpha. Bull. Mus. Compar. Zool. 149: 371-543.

Presley, R. 1979. The primitive course of the internal carotid artery in mammals. Acta Anat. 103: 238-244.

Tandler, J. 1899. Zur vergleichenden Anatomie der Kopfarterien bei den Mammalia. Denkschr. kaiserlich. Akad. Wiss., Math.-Natur. Cl. 67: 677-784.

Ting, S., and C. K. Li. 1984. The structure of the ear region of Rhombomylus (Anagalida, Mammalia). Vert. PalAsiat. 22: 92-102 (in Chinese, English summary).

Turnbull, W. D. 1970. Mammalian masticatory apparatus. Fieldiana: Geol. 18: 147-356.

Vianey-Liaud, M. 1976. Les Issiodoromyinae (Rodentia, Theridomyidae) de l'Eocène supérieur à l'Oligocène supérieur en Europe occidentale. Palaeovertebrata 7: 1-115.

Wahlert, J. H. 1972. The cranial foramina of protrogomorphous and sciuromorphous rodents; an anatomical and phylogenetic study. Unpubl. Ph.D. thesis, Harvard Univ., Cambridge.

Wahlert, J. H. 1974. The cranial foramina of protrogomorphous rodents; an anatomical and phylogenetic study. Bull. Mus. Compar. Zool. 146: 363-410.

Wahlert, J. H. 1977. Cranial foramina and relationships of

Eutypomys (Rodentia, Eutypomyidae). Amer. Mus. Novit. 2626: 1-8.

Wahlert, J. H. 1978. Cranial foramina and relationships of the Eomyoidea (Rodentia, Geomorpha). Skull and upper teeth of Kansasimys. Amer. Mus. Novit. 2645: 1-16.

Wahlert, J. H. 1983. Relationships of the Florentiamyidae (Rodentia, Geomyoidea) based on cranial and dental morphology. Amer. Mus. Novit. 2769: 1-23.

Wahlert, J. H. in press. Skull morphology and relationships of geomyoid rodents. Amer. Mus. Novit.

Wood, A. E. 1935. Evolution and relationships of the heteromyid rodents. Ann. Carnegie Mus. 24: 73-262.

Wood, A. E. 1962. The early Tertiary rodents of the family Paramyidae. Trans. Amer. Philos. Soc., n.s., 52(pt. 1): 1-261.

Wood, A. E. 1965. Grades and clades among rodents. Evolution 19: 115-130.

Wood, A. E. 1974. Early Tertiary vertebrate faunas, Vieja Group, Trans-Pecos Texas: Rodentia. Bull. Texas Memorial Mus. 21: 1-112.

Wood, A. E. 1976. The Oligocene rodents Ischyromys and Titanotheriomys and the content of the family Ischyromyidae. In: Athlon: Essays on Paleontology in Honour of Loris Shano Russell, C. S. Churcher, ed., pp. 244-277, Royal Ontario Mus., Life Sci., Misc. Publ.

PHYLOGENETIC ANALYSIS OF MIDDLE EAR FEATURES

IN FOSSIL AND LIVING RODENTS

René Lavocat[1] and Jean-Pierre Parent[2]

[1]Ecole Pratique des Hautes Etudes, USTL,
34060 Montpellier, France, and [2]Institut Supérieur
d'Agriculture, 59046 Lille, France

INTRODUCTION

The anatomy of the middle ear of mammals is certainly well known and has recently been redescribed by Parent (1980). Nevertheless, it seems advisable to recall the main features of this anatomy, which can be used to evaluate the probable affinities of the various groups of rodents.

ANATOMY OF THE MIDDLE EAR

The promontorium forms the external bony wall of the cochlea, producing a more or less prominent swelling on the medioventral part of the petrosal. The variations of shape and inflation of the promontorium are very significant. It is well known that two openings can be seen on or near the promontorium: the fenestra ovalis, latero-externally, closed by a membrane to which is applied the stapes; and posteriorly, the fenestra rotunda, also closed by a membrane, the elastic power of which has the effect to limit the pressure of the liquid inside the cochlea. The shape, orientation, relative positions, and dimensions of each fenestra are highly significant.

In latero-external position to this promontorium, mainly dorsally, except for its posterior part, is found the facial canal (= Aquaeductus Fallopii) for the facial nerve. This aqueduct can be an open sulcus or groove (primitive character), or it can also be a true tube, ossified partially or throughout its length (advanced character). Anteriorly, in the region where it joins the primitive alisphenoid foramen crossing through the petrosal, it can give way to the stapedial artery, when this is present, after its crossing of the fenestra ovalis. Posteriorly, the facial canal usually bends ventrally to reach the secondary alisphenoid canal, the so-called stylomastoid foramen, in the posterior corner of the skull.

Anteriorly, on the roof of the tympanum (or tympanic cavity), between the facial canal and the promontorium's swelling, is the place of insertion for the tensor tympani muscle. This insertion may be entirely exposed (primitive character), or inside a fissure of the bone, or even in a tubular structure (derived character).

Posteriorly, medial to the facial canal, near to the fenestra rotunda, and posterior to it, is the pit of insertion for the stapedius muscle. This pit can be such that the muscle extends out toward the posteroventral region of the skull (primitive character), eventually by the benefit of an enlargement of the facial canal, giving way both to the nerve and the muscle. This surface can also be a depression, strictly preventing the muscle from passing outside (advanced character).

Laterally, the epitympanic recess, dorsally excavated and enlarged to provide room for the malleus and allowing it to hinge, is sometimes overlying the external auditory meatus (primitive character), or sometimes entirely distinct from it (advanced character). The external auditory meatus itself can be a simple opening, or it can become a more or less developed osseous tube. The ear drum or tympanic membrane is inserted on the tympanic ring. This ring may be nearly horizontal (primitive character), or in any position up to nearly vertical (advanced character). The tympanic membrane closes the external auditory meatus at the entrance of the tympanic cavity, and it transmits the auditory vibrations to the manubrium of the malleus.

According to the early classic interpretation (Matthew, 1909), not supported by Presley (1979), three arterial vessels could be found primitively in the ear region, or immediately near it: (1) the medial internal carotid artery; (2) the promontory artery; and (3) the stapedial artery (any disappearance being an advanced character).

The medial internal carotid circulates generally through a cana situated more or less between the basioccipital and the medial borde of the petrosal and the auditory bulla. The stapedial artery separates from the internal carotid to enter into the tympanic cavity, runs close to the fenestra rotunda, crosses the fenestra ovalis (frequently over an osseous bridge) between the two crura of the stapes, then enters the facial canal and exits the tympanic cavity by several possible routes. A promontory artery, which imprints a groove on the ventral surface of the promontorium, separates from the stapedial artery (in some very rare, ancient, and primitive rodents) and enters into the skull by an anterior opening, situated in line with the axis of the promontorium.

According to Presley (1979), embryological studies demonstrate that the medial internal carotid artery and promontory artery are actually the "same" artery, which may follow one or the other course. Thus, both could not exist simultaneously in the same animal. The existence of three arteries in the prosimian primate *Cheirogaleus* and other genera described by Saban (1963) is not in opposition to the findings of Presley, because it has also been demonstrated that the so-called "anterior carotid" artery of Saban (1963) is not a true carotid artery, but instead a highly modified ascending pharyngeal artery (see MacPhee, 1981). Such a homology had been suggested as possible by Saban himself.

In any case, it is important to recognize that three arteries are simultaneously present around the auditory region of some living mammals. Two of these are clearly the promontory and stapedial arteries, and the third, the ascending pharyngeal, seems to maintain relationships with the auditory and basicranial regions similar to those of a true internal carotid artery, according to the descriptions of Saban (1963). Consequently, if anatomical data favor it (a new examination of these data will be necessary), we have no positive reason to reject the possibility of finding three arteries in the auditory region, at least in Creodonta and in some ancient rodents, although probably not with all the homologies proposed by Matthew (1909). Thus, it remains a problem; its solution can not be that suggested by Matthew, and another solution is needed.

The mammalian tympanic cavity does not remain "open" ventrally, although its covering may be only membranous or cartilaginous. However, among rodents, usually if not always, this ventral covering is a bony structure, generally rather inflated and shaped like a ball - the ectotympanic auditory bulla. This bulla may be attached loosely only (primitive character) to the petrosal, mainly upon a medial and moderately developed expansion of the petrosal. The bulla can also be very strongly and firmly fused with the petrosal (advanced character).

These are the main auditory structures of evolutionary interest. Of course, many shared details in common give a family likeness to ear regions, which, for other biological reasons, we consider to belong to related species, genera, and families. We can not describe these here in detail; these similarities have to be studied in themselves, and they are easier to recognize than to describe. Their perception results from the integration by the eye and brain of an enormous number of details, practically impossible to quantify, whose meaning is perceived only as the result of a great number of repeated observations, as is also true for skulls, teeth, or postcranial features (also see Szalay, this volume).

PRIMITIVE AUDITORY STRUCTURES OF EARLY RODENTS

Keeping in mind all of these preliminary considerations, it is possible to distinguish, among rodents, auditory regions whose structure can be called primitive, either because they are similar to the structures of other ancient mammals, most of them Cretaceous or Paleocene, or to the structures of the insectivorans *Echinosorex* and *Erinaceus*. These latter genera are Recent indeed, but generally they are assumed to be very primitive, in part, because it is clear from comparative anatomical studies that they have retained many archaic traits during their evolution.

The fossil rodents with the most primitive auditory regions are: the Theridomorpha, from the Eocene and Oligocene of Europe; the Sciuravidae, known until now only from North America; and the Paramyidae, found in both Europe and North America, although really well known only in the latter continent.

All these forms have in common a promontorium not notably swollen; a fenestra rotunda opening more toward the back of the skull than latero-externally; a facial canal or groove not ossified; a vertical fenestra ovalis; an open area for the insertion of the tensor tympani muscle; a stapedial artery; a groove for the promontory artery (not seen in *Paramys*, but clearly seen in some other Paramyidae) (Fig. 1). The Paramyidae and Theridomorpha, but not *Sciuravus*, show a canal normally characteristic of the medial internal carotid artery. This is not absolute proof for the presence of the artery, because other observations have shown convincingly that identical osseous structures may sometimes receive the carotid, sometimes not. Nevertheless, it seems probable to us that, if these osseous structures are existing, it is at least because the internal carotid artery had been coursing this way in some ancestor.

However, an important character definitely separates the American Paramyinae of Lower Eocene, for which we know the auditory region, from the Sciuravidae and Theridomorpha, so clearly that it is impossible to accept the known Paramyinae as ancestors to these latter groups. In *Sciuravus* and Theridomorpha, the auditory region is located at the very posterior part of the skull, and the stapedius muscle, according to the shape of its insertion, was extending out of the tympanic cavity onto the surface of the posterior region of the skull, in the same manner as in many Recent Myomorpha, some archaic Insectivora (*Echinosorex*, *Erinaceus*, *Tenrec*), or *Tupaia*.

In contrast, Paramyinae and *Franimys* show a very long horizontal mastoid exposed on the ventral part of the skull. This mastoid separates the tympanic cavity from the posterior part of the skull, while the pit of insertion for the stapedius muscle, being more vertical, results in the fact that this muscle was certainly restricted to the inside of the bulla. This double advanced character of the

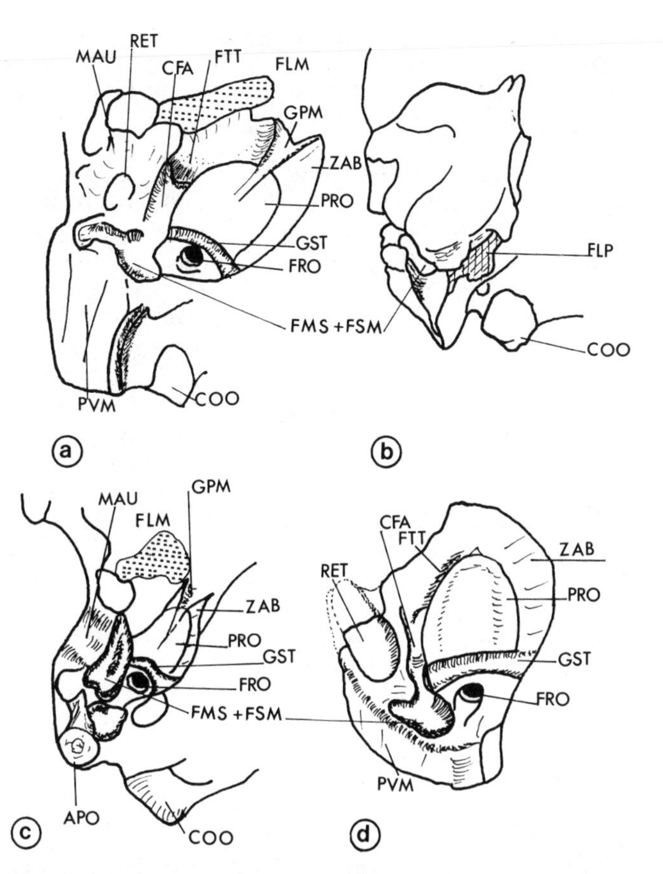

Fig. 1. Right auditory regions of Paleogene rodents. (a) *Adelomys* (Theridomyidae) with bulla removed. (b) *Adelomys* with bulla intact. (c) *Paramys* (Paramyidae). (d) *Pseudocricetodon* (Cricetodontidae). Abbreviations for Figures 1-3: AMA, Mastoid process; APO, Paraoccipital process; CCH, Semicircular canal; CFA, Canal for facial nerve; COO, Occipital condyle; EOS, Ear ossicles; FAN, Facial nerve; FLM, Foramen lacerum medium; FLP, Foramen lacerum posterior; FMS, Fossa for stapedius muscle; FOV, Fenestra ovalis; FRO, Fenestra rotunda; FSM, Stylomastoid foramen; FTT, Fossa for tensor tympani muscle; GPM, Groove for promontory artery; GST, Groove for stapedial artery; MAE/MAU, External auditory meatus; OSB, Osseous meato-cochlear bridge; PBA, Part of auditory bulla; PRO, Promontorium; PVM, Vertical part of mastoid; RET, Epitympanic recess; RHT, Hypotympanic recess; STA, Stapes; STC, Osseous canal for small stapedial artery; TEU, Eustachian tube; TGL, Glenoid foramen; TR, Osseous canal for large stapedial artery; TRA, System of trabeculation; ZAB, Contact between petrosal and bulla.

Paramyinae is enough to separate them from a group which retains more primitive structures. This group includes not only the ancient Sciuravidae and Theridomorpha, but also the Myomorpha (Fig. 2), with most of their Recent genera, as well as the Sciuroidea, Gliroidea (Fig. 3), and the genus *Eutypomys*.

More studies will be necessary to determine whether the structure of the mastoid region is or is not in favor of eventual relationships of the Paramyinae with the Caviomorpha, or their ancestors. All in all, it seems that the American Paramyinae are already very diversified, and too specialized on several points to be the ancestors of all the families previously suggested to have taken origin from them.

The most primitive rodent auditory region presently known by us is that of the Theridomorpha. They are the only group to have preserved the previously supposed "three" branches of the primitive internal carotid artery. They show a clear groove for the promontory artery (Fig. 1), as well as a primitive stapedius muscle. Although known from Europe only, they are not, even there, the most ancient rodents known.

Should we suppose the Microparamyinae, more anciently recorded in Europe, to be the ancestors of theridomorphs? It is not impossible; we do not know the auditory region of any microparamyine, thus there is no proof that it was identical to that of the American paramyines. Moreover, we know that Hartenberger (1971) believed from a study of the teeth that the Gliroidea, whose ancestors are known up to now only in Europe, are closely related to the Microparamyinae. Parent (1980) has also demonstrated that the Gliroidea show rather specialized and peculiar characters of the stapedial artery. This crosses the fenestra rotunda as in Sciuroidea, whereas it runs around the inferior rim of the same fenestra in the Muroidea. The cochlea is also bent in Gliroidea and Sciuroidea, but not in Muroidea. These observations strongly support the conclusions of Vianey-Liaud (this volume) that glirids are not closely related to the muroids. In addition, the Gliroidea have a peculiar specialization of a fundamentally primitive character; their stapedius muscle passes out of the bulla to be inserted on the back of the skull, but through an ossified tubular canal (Fig. 3). Clearly, its ancestor had this external insertion of the muscle, and if it was really a microparamyine, there is nothing to prevent this group from having given rise to the Theridomorpha.

FROM THERIDOMORPHA TO CAVIOMORPHA?

Are the Gliroidea and Theridomorpha the only relatives of the ancient European rodents? This is not certain at all. Earlier, one of us (Lavocat, 1967) had placed emphasis, too much indeed in our present opinion, on the very important differences in structure of the middle ear between the Theridomorpha and the African Phiomorpha. These differences are real, and they certainly forbid us from putting these two groups into a single systematic unit. But we must consider that the existing differences are essentially those separating an advanced group (Phiomorpha) from a more primitive one. In the opinion of Lavocat, there are good reasons to think that it is from a general skull structure similar to that of the Theridomorpha, in fact from a more or less primitive group related to them and being ancestral to them, that originated the Phiomorpha. At present, the auditory region of the Phiomorpha, not changed since the Miocene (Parent, 1980), shows the following characters: (1) strongly pronounced swelling of the promontorium; (2) cochlea more whorled, bent, with the posterior whorl strongly developed and in such a way that the fenestra rotunda is then facing laterally and not backward; (3) ossified facial canal; (4) ossified covering for the insertion of the tensor tympani muscle; (5) stapedius muscle clearly restricted to the inside of the bulla; (6) vestigial remains (osseous bridge crossing the fenestra ovalis) of the stapedial artery, itself lost in the adult; (7) osseous canal for the internal carotid artery, perhaps functional in fossils with some evidence in Recent forms of the ancestral condition.

Another important difference between Phiomorpha and Theridomorpha is the hystricognathy of the Phiomorpha, while the Theridomorpha remained sciurognathous. However, hystricognathy is certainly a character derived from sciurognathy.

The findings of the Algerian team with Jaeger and the Montpellier team in the Eocene of Brezina (Mahboubi et al., 1984), show that the problem of the faunal relationships and supposed isolation of Africa in Early Tertiary is far from being clear.

Following his study of the Auvergne Theridomorpha, Lavocat (1951) noted many structural affinities with *Anomalurus*, and he thought it reasonable to provisionally maintain *Anomalurus* rather near to the Theridomyoidea. It is well known that up to now the anomalurids are strictly African endemics. Thus, it is highly interesting that Parent (1980) reported that the auditory region of Anomaluridae is strikingly intermediate between the theridomorph structure and the phiomorph-caviomorph condition: the cochlea, the exit of the Eustachian tube, and the fossa for the tensor tympani muscle are identical to those of the genus *Echinoprocta*, a South American porcupine, which shows many primitive caviomorph structures.

During this Symposium, Jaeger et al. (see this volume) described one tooth of an anomaluroid-like rodent, from the Eocene of the Nementcha mountains, Algeria. After an examination of this specimen subsequent to the Symposium, Lavocat believes the presence of anomalurids possible at that time in Africa. Now new findings should elucidate the anatomical patterns of the immigrants and their relationships with the pretheridomorphs.

PHIOMORPHA AND CAVIOMORPHA

Whatever the given interpretation may be, there is no doubt that the structures of the middle ear of Phiomorpha and Caviomorpha are strikingly similar (Lavocat, 1973, 1976; Parent, 1980, 1983). Indeed, Lavocat has been arguing for many years in favor of the African origin of the Caviomorpha. A North American origin of Caviomorpha would not seem to fit very well with the mastoid structures found in Paramyinae, but we agree completely that a more detailed study of this particular point is required.

We will not review here all the arguments that have been proposed in favor of the African origin of caviomorph ancestors (Lavocat, 1973, 1976). We will discuss only one point which was perhaps not completely clear previously. It has been objected (Wood, 1974) that the Phiomorpha could not have given rise to the Caviomorpha, in part because the Erethizontidae retain an internal carotid artery, a feature that is absent in Recent African phiomorphs (i.e., *Thryonomys*, *Petromus*). It can be answered that, even if there is no possibility of knowing whether the osseous structures that are found in Miocene Phiomorpha were functional or not, these structures did occur, and they certainly mean something. One of us (RL) recently re-examined the beautiful stereophotos of the skull of the Miocene phiomorph *Diamantomys*, published previously (Lavocat, 1973, plate 4), and also a cast of this skull. These observations reaffirm the striking similarities of this skull, noted previously, to that of the Recent South American porcupine, *Coendu insidiosus*. The triangular fossa observed between the posteromedial region of the bulla, the occipital condyle, and the basioccipital, is nearly identical in shape and proportions in the two genera, one Miocene and one Recent. The positions of the condylar foramen, the foramen lacerum posterius, and the carotid foramen are also the same in both. There is, however, some sort of intermediate space at the mouth of the foramen lacerum posterius in *Coendu*, which is not present in *Diamantomys*. Thus, the triangular fossa is deeper in the former genus. The carotid foramen, proportionally less wide in *Diamantomys*, is in exactly the same relative position to the bulla and the basioccipital as it is in *Coendu*. There is no doubt that this foramen is strictly homologous to the functional carotid foramen of *Coendu*. The discovery of an Oligocene phiomorph skull would not provide more certainty about the ancestral anatomy of this feature in Phiomorpha. It is not possible to prove that the carotid foramen of *Diamantomys* was still functional, but its similarity to the foramen in *Coendu* proves at least that there were remnants of a functional structure in the Miocene African phiomorphs, a structure retained at present only in the family Erethizontidae among the Hystricognathi.

To end this point, we may also note that the carotid canal is also absent in the Sciuridae and Aplodontidae, which lack an internal catotid artery (also see, Bugge, this volume).

The Hystricidae, which appear first in the Miocene of Southern Asia, without known ancestors, have a middle ear morphology strikingly similar to that of Erethizontidae, and both retain some archaic features, such as absence of lengthening of the cochlea. Moreover, hystricids retain the normal replacement of dP4/4 by permanent premolars. This takes us back earlier than the Fayum Oligocene of Africa, where this replacement pattern was in the process of disappearing in phiomorphs (Wood, 1968), or to another, unknown part of the Oligocene population, less advanced in this feature. Remarkably, the shared primitive traits of Hystricidae and Erethizontidae would agree with the idea of an African center of hystricognath dispersal, and possibly with the idea of an initial migration of archaic rodents, followed a little later in South America by a second migration. Perhaps these features are even more consistent with the view of an ancient African population that was more heterogeneous (while nevertheless of a single origin) than was supposed previously.

Numerous primitive and derived auditory features shared by the African Phiomorpha and South American Caviomorpha are summarized in Figure 4 and Table 1.

PHIOMORPHA AND CTENODACTYLOIDEA

Our previous discussion considered the possible relationships among Anomaluroidea, Theridomorpha, Protheridomorpha, and Phiomorpha, but there are also important similarities between the ear region of the Ctenodactyloidea and that of the Phiomorpha. These include: the pattern of the cochlea; ossification of the facial canal; ossification and trabeculation of the bulla; and the reduced carotid circulatory system (Fig. 4, Table 1). However, it is true that a more complete and detailed analysis also reveals important differences, mostly regarding the muscular attachments. The insertion of the tensor tympani is not covered in ctenodactylids, and there is no pit of insertion for the stapedius muscle. These differences are not absolutely conclusive; however, study of the dentition also argues against a very close relationship. No other fossil ctenodactylid ear region has been available to us except *Sayimys* from Taben-Buluk, China, previously described by Bohlin (1946). This ear region is not very well preserved; what can be seen suggests strongly that the structures known in the Recent genera were already present, with the exception of the mastoid sinus.

Discussing the origin of Phiomorpha, Hussain et al. (1978) suggested that it may be found, not in European pretheridomorphs, but instead in the Eocene Chapattimyidae of the Indian subcontinent.

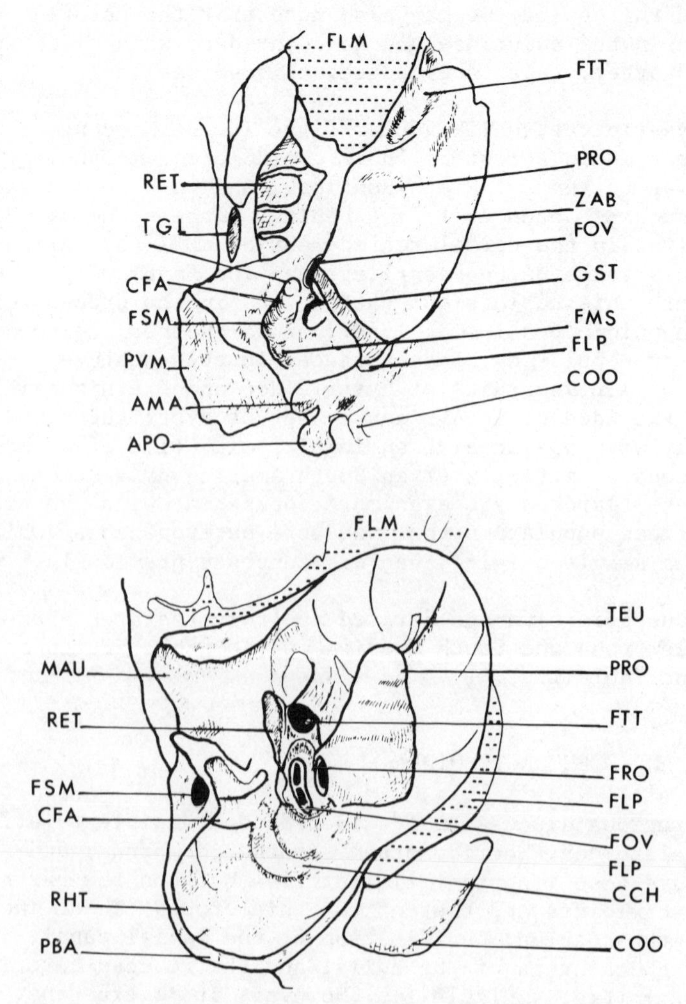

Fig. 2. Primitive and advanced features of the middle ear in the arvicolid Ondatra (Muroidea) (above), and the caviid Cavia (Caviomorpha) (below). See Fig. 1 for abbreviations.

It is clear that this family is closely related to or included in the Ctenodactyloidea (Hartenberger, 1982). Following their recent findings of late Eocene phiomyids in North Africa, Jaeger et al. (this volume) suggest the probable derivation of Phiomorpha from the Chapattimyidae. After an examination of these primitive phiomyid teeth, Lavocat now agrees with such a possible early relationship. Since at least late Eocene times, Ctenodactyloidea and Phiomorpha have had distinct and divergent evolutionary paths.

In summary, it seems that there are numerous good reasons to consider the hypothesis that a primitive rodent stock migrated into Africa and gave rise to the Anomaluridae on the one hand, and the Phiomorpha on the other. Looking at the ear region only (Fig. 4), one could easily place Pedetes near to this ancestral group also, if the dental structures (Lavocat and Michaux, 1966) were not such a real problem.

MUROIDEA

One of the most striking paradoxes in the history of the Rodentia is the fact that, while for several suborders or families we can see the results of an intense evolution of middle ear structures, the superfamily Muroidea has retained, up to the present, a very primitive pattern, to be compared most closely with that of the Theridomorpha (Fig. 4, Table 1). This is true, not only for some isolated groups, but for most of its representatives, including the most abundant of them, the Recent Muridae and Cricetidae. The fossorial forms are a notable but not unique exception to this rule. Except for them, the muroid cochlea is generally rather low, not very prominent ventrally, and not uncoiled; the fenestra rotunda has its opening facing backward; the stapedius muscle exits the tympanic cavity posteriorly, to insert partly in an external fissure on the base of the back of the skull; the facial canal is not completely ossified; and the insertion of the tensor tympani muscle is not completely covered by bone. There is no promontory artery, but the stapedial and medial internal carotid arteries are both present and functional, although the stapedial artery is not always well developed (also, see Bugge, 1974, this volume). The bulla is only weakly attached to the petrosal and can be isolated without real damage.

Thus, study of the ear region of fossil muroids completely supports the phylogenetic conclusions reached previously by specialists of the dentition. The initial, complex evolution of the Muroidea took place mainly in Asia, and from there began several distinct waves of migration. The initial radiation was toward Europe and North America, as early as the Oligocene, then later, in the Miocene, toward Africa.

Eurasiatic Muroidea

Two muroid genera were already distinct in Europe during the Oligocene. The first, Eucricetodon, is notably more advanced in the lower Oligocene than are most of the Recent Muroidea. Its hemicylindrical promontorium is notably swollen; the fenestra rotunda is high and narrow; the facial canal is completely ossified; and the stapedius muscle is restricted inside the bulla. In contrast, the Oligocene genus Pseudocricetodon (Fig. 1) retains many primitive characters of the group, and it also shows many similarities with the Asiatic genus Cricetops, a form which exhibits somewhat peculiar cranial structures.

This simultaneous presence in the European Oligocene of two quite distinct groups of cricetodontids can be found also in the Miocene. The famous site of La Grive Saint Alban has yielded two distinct types of petrosals. One (type B) is quite similar to the Oligocene genus Eucricetodon, but differs from it, not only by the derived loss of the stapedial artery, but also by the completely normal shape of the fenestra rotunda. This last character probably prohibits the closely related Eucricetodon from being the direct ancestor of "type B", because the fenestra rotunda was already more specialized in the Oligocene genus.

The other petrosal specimen at La Grive (type A) shows many similarities to the Oligocene genus Pseudocricetodon, although several minor differences occur. Moreover, a major difference in the course of the stapedial artery refutes the idea of a direct relationship. The Oligocene genus, with an anteriorly ossified canal for the artery, is clearly more derived in this feature than the Miocene form, in which the artery crosses the nerve and immediately leaves the tympanic cavity.

These observations are in complete agreement with the conclusions of other investigators. They have proposed that the European populations of Oligocene and Miocene muroids are the result of two successive migrations, each originating from Asiatic centers, and not the result of local evolution. On the contrary, it appears that the muroids present in Europe during the Miocene were capable of being ancestral to the Recent Cricetinae and Arvicolidae, showing only slight modifications. The North American Hesperomyinae, while originating probably from North American Eumyinae, also appear very similar in their ear region to the European Miocene forms, although with a tendency to lose the stapedial artery, absent in at least four genera. The Hydromyinae of Australasia, with a very primitive ear region, seem to be a group of relatively more ancient origin than supposed by Ellerman (1940).

African Muroidea

The Miocene Cricetodontidae of Africa, particularly Afrocricetodon, although certainly taking their origin from Miocene Asiatic forms, are especially interesting because their cranial structures are still morphologically at the level of Oligocene European cricetodontids. The ancestral stock probably originated in Southern Asia in close proximity to Africa. One of us (Lavocat, 1973) proposed the hypothesis that the Miocene population of European cricetodontids, much more advanced, resulted from a separate migration from northern Asia, where prevailing environmental conditions would have resulted in severe selection, leading to a more rapid adaptive evolution than in Southern Asia. In addition, Parent (1980) has shown that, with a few more derived features (bulla strongly fused to petrosal), the middle ear of the Afrocricetodontinae is basically at the level of the most primitive European Oligocene cricetodontids. The facial canal is not ossified, and an osseous rim can be seen behind the fenestra rotunda; this latter is reminiscent of the osseous rim present in primitive forms of certain other groups, such as Sciuravidae and Theridomorpha.

Considered as a whole, the Oligocene Cricetodontidae of Eurasia and the African Miocene Afrocricetodontinae are related to an ancestral Asiatic group, itself more primitive, originating probably near the base of the ancestral rodent stock. Some of the African muroids, as well as the Nesomyidae of Madagascar, are certainly derived from the afrocricetodontines. The Recent _Cricetomys_ has an ear region nearly identical to that of _Afrocricetodon_, and the Malagasy nesomyids are very near to _Protarsomys_ of the East African Miocene. The ear region of _Otomys_, rather conservative, also suggests cricetodontid affinities, notwithstanding the possibly erroneous indications suggested by the dentition of _Euryotomys_ (Pocock, 1976). _Mystromys_, certainly a cricetodontid, also needs further investigation.

Myocricetodontidae and Gerbillidae

The history of the Muroidea is quite complex; in addition to the families that show a primitive ear region, the group also includes the Gerbillidae, in which nearly all the characters are advanced or derived. The gerbillid bulla is greatly enlarged and strongly fused to the petrosal; the promontorium is swollen, with the apex of the cochlea well exposed; the tensor tympani muscle is inserted in a covered fossa; the facial canal is ossified; and the stapedius muscle remains inside the bulla.

According to Jaeger (1977), the Gerbillidae should be descendants of the Myocricetodontidae, well known from the Miocene of Africa, and whose ancestors were certainly living in Asia. Associated with the characters noted above, which can be interpreted as mostly advanced or adaptive, others are uniquely found among the Rodentia, for example, the complete dissociation of the stapedius muscle from the facial nerve. This suggests that the Myocricetodontidae must have had, at least in the Miocene, an advanced ear region, and that they were an original group, separated probably for a very long time from other muroids. Unfortunately, the ear region of myocricetodontids is still unknown.

Large Fossorial Muroidea

Four families or subfamilies of mono- or paucispecific fossorial rodents, of large to medium size, are muroids: The Rhizomyidae, Tachyoryctinae, Spalacidae, and Myospalacinae. All have a very advanced ear region: bulla strongly fused to the petrosal; cochlea very high and lightly bent; fenestrae advanced; tensor tympani muscle reduced or absent, inserted in a covered fossa; stapedius muscle greatly reduced, remaining inside the bulla; facial canal ossified; and loss of the stapedial artery. These characters, with several other advanced features, found also in the gerbillid Tatera, raise the question of whether they provide some indication that these four fossorial taxa should perhaps be allied with the Myocricetodontidae. This hypothesis requires careful testing, especially to determine whether it is compatible with evidence derived from the dentition. The discovery of an ear region from Myocricetodon would greatly aid the analysis of this hypothesis.

Because of the numerous advanced or derived characters, these fossorial muroids show an ear region rather similar to that of Phiomorpha, Ctenodactyloidea, and Caviomorpha. However, the internal carotid artery is retained in the fossorial muroids, whereas it is lost in most members of the above taxa (also see Bugge, this volume).

In summary, the history of the Muroidea seems to confirm the great stability of the structures of the middle ear, as soon as these features are established. Thus, the family Muridae, although now in full expansion, is not fundamentally different in the ear region from some Oligocene Cricetodontidae.

SCIUROMORPHA

While the Hystricognathi (sensu Lavocat, 1973) and the Muroidea appear to be natural or monophyletic groups, the Sciuromorpha, in contrast, seems to be a composite or paraphyletic group. In this section we will discuss those rodent taxa with a sciuromorphous jaw, as well as the protrogomorphous Aplodontidae, a group with possible affinities to one sciuromorphous family.

Sciuroidea

The Sciuroidea (family Sciuridae) are certainly a homogeneous and natural group, easily recognized by two rare auditory characters that are always associated: (1) the unusual position of the functional stapedial artery, crossing the middle part of the fenestra rotunda within a bony canal, as in the Gliridae, instead of following the inferior rim of this fenestra; and (2) the existence of a strong osseous bridge connecting the promontorium to the auditory meatus and hiding the ossicles in ventral view (Fig. 3). The cylindrical and globular bulla is strongly fused to the petrosal, but the suture remains distinct. Other bullar characters show various stages of evolution. The cochlea is prominent, but with a twisted fenestra rotunda (advanced character); the facial canal is completely ossified (advanced character); the tensor tympani muscle is inserted openly (primitive character). The stapedius muscle is reduced in Recent forms.

As early as the Oligocene, Palaeosciurus feignouxi possessed the essential ear region features of Recent sciurids. On the contrary, some forms from the Miocene of La Grive show a posterior opening, which suggests that the stapedius muscle in these species was passing out of the bulla to insert on the posterior part of the skull. Thus, it is not possible to derive Sciuroidea from North American Paramyinae, in which this muscle was already restricted to the internal part of the bulla.

The Sciuroidea appear simultaneously in the Oligocene of North America, Europe, and Asia, suggesting a previous Asiatic history. However, we have no information to connect them to earlier forms, except the primitive character of the insertion of the tensor tympani muscle (retained in Recent taxa), and the apparently primitive pattern of insertion of the stapedius muscle (preserved in some Miocene forms). It is evident that the sciuroid auditory region is quite original, and far from primitive in many characters. This is in agreement with their advanced sciuromorphous infraorbital foramen, and with the derived features of their internal carotid circulation. These findings are contrary to the opinion that sciuroids are relatively archaic forms, based on their cuspidate dentition. For further discussion of sciurid features, see Vianey-Liaud (this volume).

Aplodontidae

It seems likely that data on the auditory region can provide some insight into the systematic affinities of Aplodontia. Curiously enough, this group exhibits closest affinities with Sciurus and other sciurids in the ear region (Fig. 4, Table 1). These shared similarities include: (1) cochlea bent at nearly a right angle; the peculiar and unique feature of an osseous meato-cochlear bridge, not known elsewhere in Rodentia; (3) a high fenestra rotunda twisted upon itself, with its inferior half covered by a thin osseous plate; (4) a large uncovered insertion for the tensor tympani muscle; (5) a short stapedius muscle, completely separated from the posterior part of the skull; and (6) absence of the internal carotid artery (Fig. 3). The middle ear of Aplodontia is more advanced than that of sciurids in two features: the loss of the stapedial artery and loss of the stapedial bony tube crossing the fenestra rotunda. In one character, the subhorizontal position of the tympanic ring, Aplodontia remains more primitive. Based on these shared similarities of the auditory region, a close ancestral relationship between Sciuroidea and Aplodontidae seems likely (also see Vianey-Liaud, this volume). On the other hand, a possible relationship between Sciuroidea and Gliroidea, on the basis of the similar position of their stapedial arteries, is more dubious.

Castoroidea

These rodents, at least the Recent ones, seem to be one of the

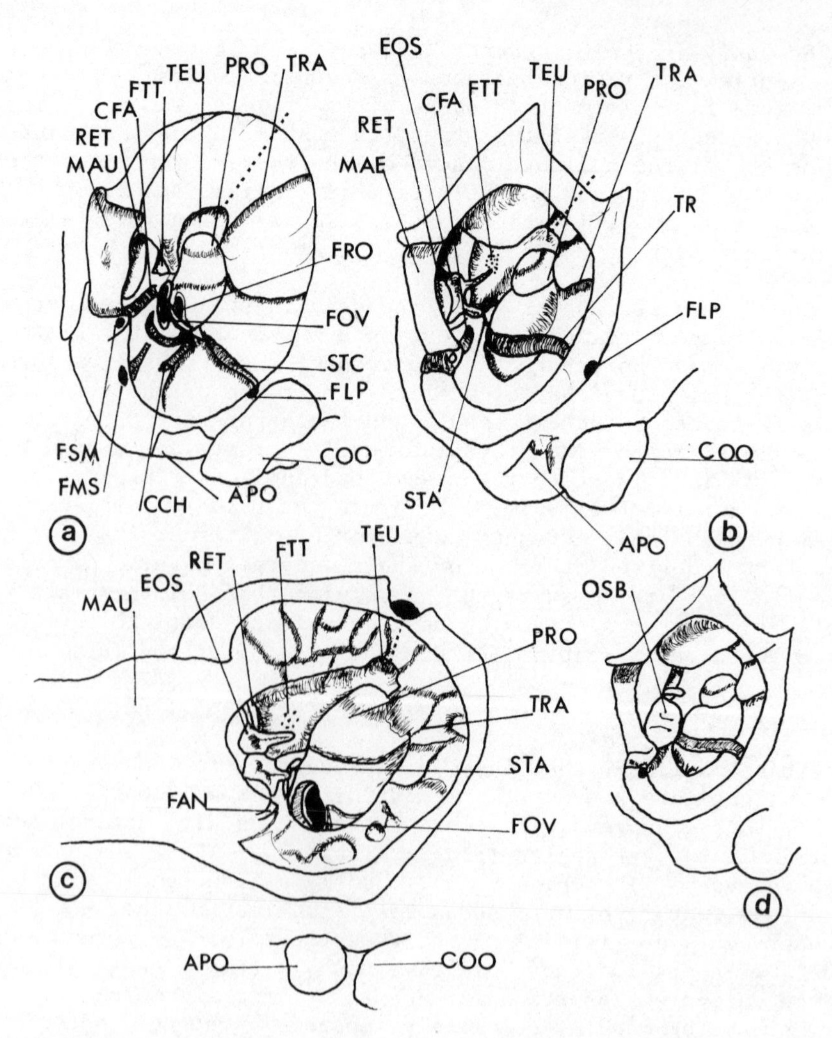

Fig. 3. Middle ear features in (a) the glirid Eliomys, (b) the sciurid Sciurus, with the osseous bridge removed, (c) the aplodontid Aplodontia, and (d) the sciurid Sciurus with the osseous bridge (OSB). See Fig. 1 for abbreviations.

most isolated groups, near to no other Recent family. Among rodents, Castor is the only genus that shows a globular cochlea without an anterior ridge, associated with a continuous petrosal roof where the soft tissues remain unseen. There is no stapedial artery, but the internal carotid artery persists. Comparisons of the auditory regions of Castor and the Oligocene genus Eutypomys do not provide much support for the opinion (Wahlert, 1977; Wood, 1980) that Eutypomys should be part of the ancestral group of the Castoridae. However, our data do not falsify this hypothesis, because Eutypomys is very near the primitive pattern found in Sciuravidae, Theridomorpha, and primitive Muroidea. Evidently, there must be some way in which Castor is connected with this primitive pattern, and this may well be through Eutypomys, because there is other evidence for such a relationship, as noted by Wahlert (1977) and Wood (1980). However, a relationship of castorids to Paramyinae would then have to be rejected, because the auditory region of Eutypomys (except for loss of the promontory artery) is quite similar to that of Sciuravus, and not close to that of Paramys.

Geomyoidea

The Geomyoidea are rodents of rather uncertain affinities. Their Recent representatives are highly specialized, in a way different from that seen in other groups. Notably, the promontorium remains primitively low, in association with derived features such as ossification of the facial canal and loss of the stapedius muscle. The fossa for the tensor tympani muscle lies beneath the facial canal.

Three fossil geomyoid ear regions are known for Sanctimus, Kansasimys, and Paradjidaumo from the middle Oligocene (Wahlert, 1978, 1983). They retain numerous primitive features, such as a non-uncoiled cochlea, an unossified facial canal, presence of the stapedial artery, and a pit for the stapedius muscle at the rear of the skull, but they also show the beginning of specialization. Another auditory region is known, that of the eomyid Viejadjidaumo (Wahlert, 1978). It may have lost the stapedial artery, a feature retained in Recent Geomyidae and Heteromyidae, as well as in other eomyids, on the promontorium surface. If true, this could be an indication that the separation of the Eomyidae was already begun before the Oligocene.

The fossil Eomyidae could be related to the group Sciuravus - Adelomys - ancient Muroidea. The ear region does not provide evidence of a special relationship for Eomyidae with the Geomyidae and Heteromyidae. (See Fahlbusch, this volume, for a discussion of the dental evidence for an eomyid-geomyoid relationship,)

CONCLUSIONS

A character analysis of many of the auditory features discussed in this report is presented in Figure 4 and Table 1. This general account shows us that data from the auditory region are in good agreement with most phylogenetic conclusions drawn from the dentition. From all these data, we can draw the following conclusions.

As early as the Eocene, even though known rodents remained near to the probable primitive eutherian pattern of ear morphology, found

Table 1. Primitive and derived characters of the auditory region of rodents.

Primitive	Derived
1. Promontorium low	1. Promontorium swollen
2. Cochlea uniformly coiled	2. Cochlea with its last part uncoiled
3. Cochlea uniformly coiled	3. Cochlea with its last part more important
4. Auditory bulla weakly connected to petrosal	4. Auditory bulla fused to petrosal
5. Auditory bulla without septae	5. Auditory bulla with septae
6. Promontory artery present	6. Promontory artery absent
7. Stapedial artery present	7. Stapedial artery reduced or absent
8. Internal carotid artery present	8. Internal carotid artery reduced or absent
9. Fenestra ovale vertical	9. Fenestra ovale horizontal
10. Fenestra ovale small	10. Fenestra ovale large
11. Fenestra rotunda small	11. Fenestra rotunda large
12. Fenestra rotunda regular	12. Fenestra rotunda twisted
13. Fenestra rotunda above the stapedial artery	13. Fenestra rotunda crossed by the stapedial artery
14. Posterior part of the stapedius muscle outside the bulla	14. All parts of the stapedius muscle inside the bulla
15. Stapedius muscle large	15. Stapedius muscle reduced or absent
16. Stapedius muscle uncovered	16. Stapedius muscle in a closed fossa
17. Tensor tympani muscle uncovered	17. Tensor tympani muscle in a closed fossa
18. Facial canal not ossified	18. Facial canal ossified
19. Epitympanic recess overlies the roof of the auditory meatus	19. Epitympanic recess distinct from the auditory meatus
20. No hypotympanic recess	20. Hypotympanic recess present
21. Petrosal horizontal	21. Petrosal tilted more vertical
22. Auditory region at the posterior part of the skull	22. Auditory region not at the posterior part of the skull
23. No osseous meato-cochlear bridge	23. Osseous meato-cochlear bridge present

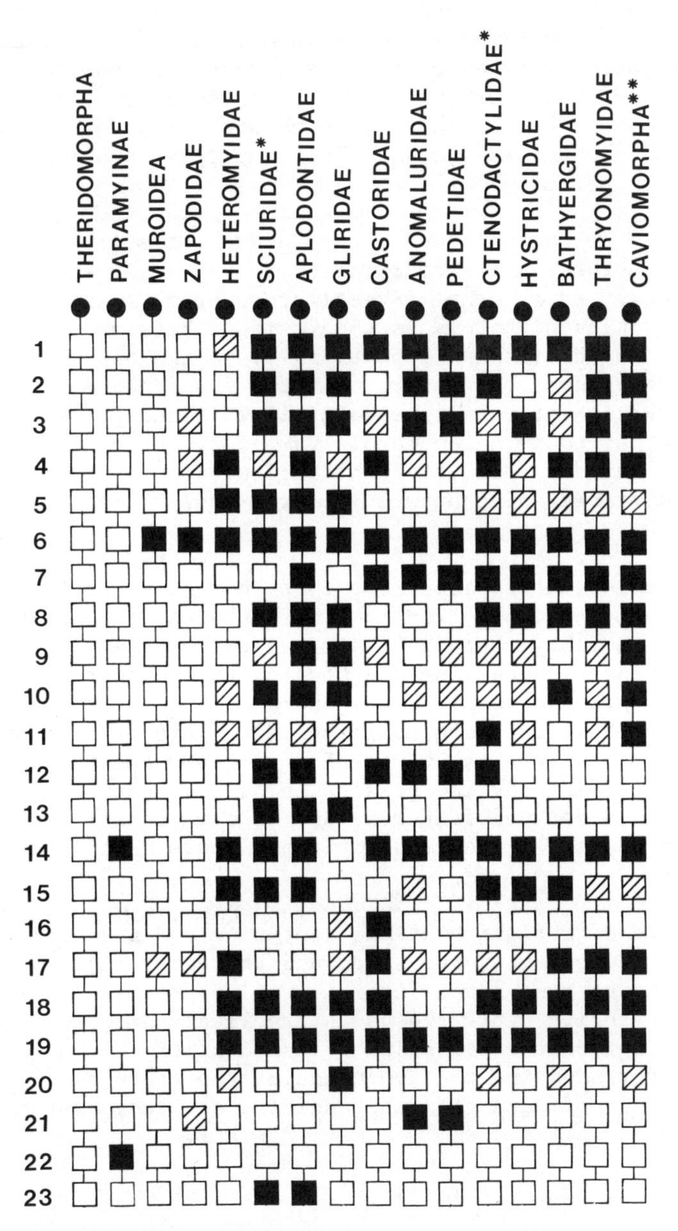

Fig. 4. Character distribution of selected middle ear features in rodents. Open squares = primitive character states; black squares = derived character states; diagonally-lined squares = intermediately derived characters. See Table 1 for characters evaluated. Note: * = only Recent members of family studied; ** = analysis exclusive of family Erethizontidae.

also in Eocene Carnivora and Insectivora, a clear dichotomy in auditory structures was already established in Rodentia. The important and varied North American Paramyinae already possessed some derived or specialized characters, which exclude the possibility of considering this group as ancestors of most modern groups of rodents. In contrast, Eocene Sciuravidae and Theridomorpha retained a more primitive auditory pattern.

As early as the Oligocene, very characteristic and diagnostic auditory structures are found; these are generally specific to taxa already recognized by other investigators from dental and cranial data. A striking dichotomy is then evident between Ctenodactyloidea and Hystricognathi (sensu Lavocat, 1973) on the one hand, possessing already a very advanced pattern, and the Muroidea, which retain a very conservative or primitive condition. Although derived features have been attained in a few muroid families, these families nevertheless retain numerous primitive characters. Thus, there is no confusion in distinguishing between a large-bodied fossorial muroid and a caviomorph.

It is important to emphasize that, not only the fundamental advanced characters, but also the fine details, are strikingly similar in the auditory regions of African Phiomorpha and South American Caviomorpha. Indeed, some genera of these geographically separated groups are more similar to genera on the other continent than to genera in their own geographic region. This is an important factor in favor of the existence of a single ancestral hystricognath group, characterized by the possession of numerous well developed and established derived features, and from which all genera showing this assemblage of derived auditory characters must be the direct descendants.

The rate of evolution of the dentition and the auditory region has not been necessarily the same, and the ear region has not been subjected to the same adaptive or selective pressures. Therefore, the auditory region can serve as an excellent guide for following the meanderings of the evolution of the dentition. The continued importance for studying the evolution of the teeth is not diminished by the addition of data from the auditory or other systems.

REFERENCES

Bohlin, B. 1946. The fossil mammals from the Tertiary deposits of Taben-Buluk, Western Kansu. Part II. Paleont. Sinica, New Series C (8b): 1-259.

Bugge, J. 1974. The cephalic arterial system in insectivores, primates, rodents and lagomorphs, with special reference to the systematic classification. Acta Anat. 87 (Suppl. 62): 1-160.

Ellerman, J. R. 1940. The Families and Genera of Living Rodents. Vol. I. Rodents Other than Muridae. British Museum (Nat. Hist.), London.

Hartenberger, J.-L. 1971. Contribution à l'étude des genres Gliravus et Microparamys (Rodentia) de l'Eocène d' Europe. Palaeovertebrata 4: 97-135.

Hartenberger, J.-L. 1982. A review of the Eocene rodents of Pakistan. Contrib. Mus. Paleont. U. Mich. 26: 19-35.

Hussain, S. T., de Bruijn, H. and Leinders, J. M. 1978. Middle Eocene rodents from the Kala Chitta Range (Punjab, Pakistan). Proc. Kon. Ned. Akad. Wetensch., Ser. B, 81: 74-112.

Jaeger, J.-J. 1977. Les rongeurs du Miocène moyen du Maghreb. Palaeovertebrata 8: 1-144.

Lavocat, R. 1951. Révision de la faune des mammifères Oligocènes d'Auvergne et du Velay. Editions Sciences et Avenir, Paris.

Lavocat, R. 1967. Observations sur la région auditive des Rongeurs Théridomorphes. Col. Inter. C. N. R. S., Problèmes actuels de Paléontologie 163: 491-501.

Lavocat, R. 1973. Les Rongeurs du Miocène d'Afrique Orientale. I. Miocène inférieur. Mém. Trav. E. P. H. E. (Montpellier) 1: 1-284.

Lavocat, R. 1976. Rongeurs caviomorphes de l'Oligocène de Bolivie. II. Rongeurs du bassin de Salla-Luribay. Palaeovertebrata 8: 15-90.

Lavocat, R. and Michaux, J. 1966. Interprétation de la structure dentaire des rongeurs africains de la famille des Pedetidae. C. R. Acad. Sci., Paris 262D: 1677-1679.

MacPhee, R. D. E. 1981. Auditory regions of primates and eutherian insectivores. Contrib. Primatol. 18: 1-284.

Mahboubi, M., Ameur, R., Crochet, J.-Y. and Jaeger, J.-J. 1984. Earliest known proboscidean from early Eocene of North-west Africa. Nature 308: 543-544.

Matthew, W. D. 1909. The Carnivora and Insectivora of the Bridger Basin, middle Eocene. Mem. Amer. Mus. Nat. Hist. 9: 289-567.

Parent, J.-P. 1980. Recherches sur l'oreille moyenne des Rongeurs actuels et fossiles. Anatomie. Valeur systématique. Mém. Trav. E. P. H. E. (Montpellier) 11: 1-286.

Parent, J.-P. 1983. Anatomie et valeur systématique de l'oreille moyenne des rongeurs actuels et fossiles. Mammalia 47: 93-122.

Patterson, B. and Wood, A. E. 1982. Rodents from the Deseadan Oligocene of Bolivia and the relationships of the Caviomorpha. Bull. Mus. Comp. Zool. Harvard 149: 371-543.

Pocock, T. N. 1976. Pliocene mammalian fauna from Langebaanweg. A new fossil genus linking the Otomyinae with the Muridae. S. Afr. J. Sci. 72: 58-60.

Presley, R. 1979. The primitive course of the internal carotid artery in mammals. Acta Anat. 103: 238-244.

Saban, R. 1963. Contribution à l'étude de l'os temporal des Primates. Mém. Mus. Nat. Hist. Nat., Paris, sér. A, 29: 1-377.

Wahlert, J. H. 1977. Cranial foramina and relationships of Eutypomys (Rodentia, Eutypomyidae). Amer. Mus. Novit. 2626: 1-8.

Wahlert, J. H. 1978. Cranial foramina and relationships of the Eomyoidea (Rodentia, Geomorpha). Skull and upper teeth of Kansasimys. Amer. Mus. Novit. 2645: 1-16.

Wahlert, J. H. 1983. Relationships of the Florentiamyidae (Rodentia, Geomyoidea) based on cranial and dental morphology. Amer. Mus. Novit. 2769: 1-23.

Wood, A. E. 1962. The early Tertiary rodents of the family Paramyidae. Trans. Amer. Philos. Soc. 52: 1-261.

Wood, A. E. 1968. The African Oligocene Rodentia. Bull. Peabody Mus. Nat. Hist. 28: 23-105.

Wood, A. E. 1974. The evolution of the Old World and New World hystricomorphs. Symp. Zool. Soc. Lond. 34: 21-60.

Wood, A. E. 1980. The Oligocene rodents of North America. Trans. Amer. Philos. Soc. 70: 1-68.

SYSTEMATIC VALUE OF THE CAROTID ARTERIAL PATTERN IN RODENTS

Jørgen Bugge

Department of Anatomy
Royal Dental College
DK-8000 Aarhus C, Denmark

INTRODUCTION

The only important early description of the cephalic arterial system in mammals is that of Tandler (1899), which has been referred to in most of the subsequent literature. Tandler found no specific connection between the cephalic arterial pattern and the systematic position of the species he investigated. Recently, however, in connection with a comprehensive comparative anatomical investigation of the carotid arterial pattern in many mammalian orders, it has been found that the pattern of carotid circulation may serve as a character complex of taxonomic importance. Some studies have involved insectivores and primates (Bugge, 1972, 1974b), carnivores (Bugge, 1978), and edentates and pangolins (Bugge, 1979), but most have concerned the order Rodentia (Guthrie, 1963; Bugge, 1970, 1971a, b, c, 1974a, b).

It is not sufficient, however, merely to observe homologies, because not all similarities have the same systematic value. Therefore, that classification which accords with the largest number of organ systems must be given priority. In this respect, it is important to note that "characters of common inheritance" (Le Gros Clark, 1959) or symplesiomorphic characters (Hennig, 1950, 1966) cannot be accorded special systematic value, whereas "characters of independent acquisition" (Le Gros Clark, 1959) or synapomorphic characters (Hennig, 1950, 1966) are significant, i. e., only synapomorphic homology (Martin, 1968) has taxonomic value in evolutionary analyses. The same synapomorphic characters can, however, be developed in different lines independently by convergence, just as common synapomorphic characters may be acquired separately by related groups through parallelism. Therefore, osteological and dental features which can be traced in fossil material will be of special taxonomic importance, reflecting the probable evolutionary history and relationships of mammals, together with those soft organ systems, for instance the cephalic arterial system, which can be wholly or partly traced back, owing to their leaving traces in hard tissues.

My original research (Bugge, 1974b) focused especially on a description of the cephalic arterial pattern in a large and comprehensive section of the order Rodentia and on a discussion of the results in an ontogenetic and phylogenetic perspective, with special reference to taxonomy, including a comparison of the classification thus obtained with former groupings. It is my intention here to discuss the systematic value of the carotid arterial pattern compared with other organ systems used as a basis for systematic classification, emphasizing in particular the taxonomically ambiguous groups, where earlier investigations of different organ systems have resulted in a varying systematic allocation.

MATERIALS AND METHODS

The investigation is primarily based on the examination of corrosion casts of the cephalic arterial system, prepared by means of a special plastic injection and corrosion technique, developed by the author (Bugge, 1963, 1974b, 1975), and comprising 78 species and 478 specimens (detailed in Bugge, 1974b). Some of the species are represented in the material by a rather large number of specimens, with a view to testing the variation of the pattern of cephalic arterial system within the individual species. Besides the corrosion casts, the investigation comprised a few dissected animals.

BASIC PATTERN OF CEPHALIC ARTERIES

The cephalic arterial supply pattern in Recent rodents (and all other mammals) is assumed to have been derived from a primary basic pattern (Fig. 1A), comprising the internal carotid artery connected with the vertebral-basilar artery system by means of the posterior communicating artery, the external carotid artery, and the stapedial artery with its supraorbital, infraorbital, and mandibular branches.

There is evidence from ontogenetic studies that the primary basi pattern employed (Fig. 1A) as a basis for the following description of the cephalic arterial system in rodents represents the most primitive cephalic arterial supply pattern in all mammals (see Grosse 1901; Tandler, 1902; Hofmann, 1914; Struthers, 1930; Padget, 1948). Also paleontological studies indicate that the basic pattern employed (Fig. 1A) corresponds fairly closely to the cephalic arterial pattern in the early Tertiary mammals (Matthew, 1909). McKenna (1966) and Van Valen (1966) accepted Matthew's point of view that the internal carotid artery was possibly originally divided into a lateral and a medial stem, whereas Presley (1979) has recently emphasized that there is no evidence for separate vessels in any mammalian embryo studied that could give rise to non-homologous internal carotid arteries in the two positions. Instead, the single embryonic internal carotid artery assumes a more lateral or more medial position during early ontogeny. The lateral course has apparently been retained in certain mammals, such as insectivores and primates, and the medial in others, like lagomorphs and those rodents which have not entirely lost an internal carotid artery. The primary basic pattern is therefore considered to be the original cephalic arterial pattern in the following, and the patterns in Recent mammals are interpreted as modifications of this.

The primary area supplied by the internal carotid artery is the brain (with the assistance of the vertebral-basilar artery system) and the eyeball, while the external carotid artery system, especially supplies the tongue and the face. The rest of the head (i.e., the upper and lower jaws, the dura, and the extrabulbar part of the orbit), is supplied from the stapedial artery and its three branches (ramus supraorbitalis, ramus infraorbitalis, and ramus mandibularis).

Alterations to this basic pattern (Fig. 1A) are presumed to have occurred primarily by obliteration of the internal carotid artery and/or certain parts of the stapedial system (st, rs, ri, rm), in connection with the presence of a varying number and combination of six anastomoses, designated a 1 - a 6 (Fig. 1B), all of which occur as permanent or temporary anastomoses during human embryogenesis (see Padget, 1948; De la Torre and Netsky, 1960). Apart from the possible existence of anastomosis a 1 (Fig. 1B), hardly any of these anastomoses could have been developed in the primitive mammals from the early Tertiary. The anastomosis a 1, however, is present in all insectivores, primates, rodents (Bugge, 1974b), and carnivores (Bugge, 1978) investigated and is only lacking in edentates (Bugge, 1979). (For further details concerning the system employed, see Bugge, 1974b.)

Possible Anastomoses

Of the six anastomoses shown in Fig. 1B, the most important in rodents are anastomosis a 3, connecting the distal part of the external carotid artery with the mandibular branch (rm) and forming the first part of the maxillary artery, and anastomosis a 2 between the infraorbital branch (ri) and the orbital part of the supraorbital branch (rs) forming the external ophthalmic artery. By means of these anastomoses the external carotid artery system annexes the whole stapedial area of supply, that is, the upper and lower jaws (ri and rm), the dura (mm) and the extrabulbar part of the orbit (1, f, e); the stapedial artery stem and the central part of the supraorbital branch (rs) becoming obliterated. Anastomosis a 1,

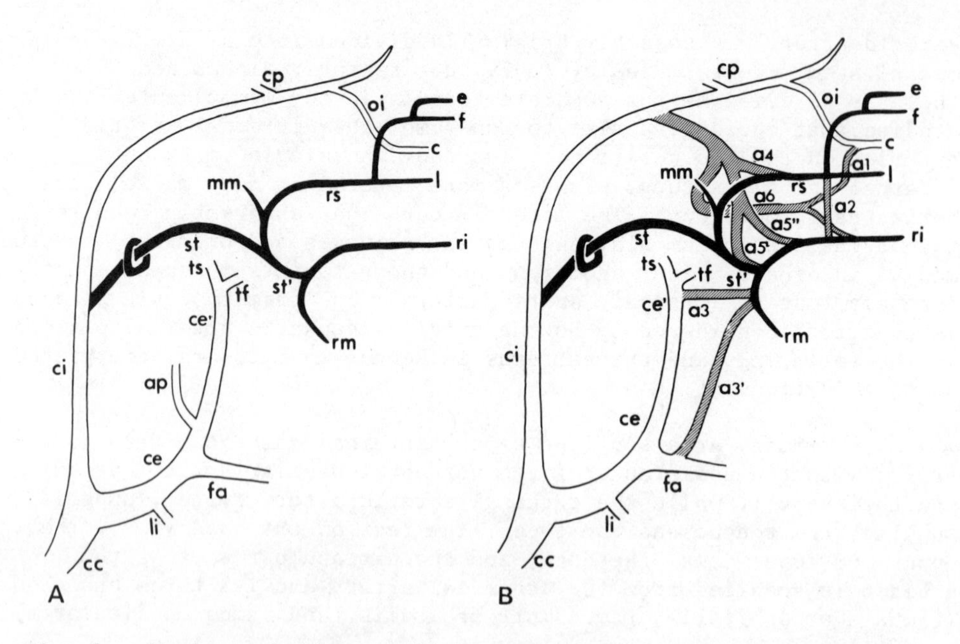

Fig. 1. Basic cephalic arterial pattern and possible anastomoses. (A) Basic pattern; (B) basic pattern and six possible anastomoses (a 1 - a 6). White, internal-external carotid artery system; black, stapedial artery system; hatched, anastomoses; dashed, only temporarily developed. Abbreviations (Figs. 1-7): ap = a. auricularis posterior; c = a. ciliaris; cc = a. carotis communis; ce = a. carotis externa (proximal part); ce' = a. carotis externa (distal part); ci = a. carotis interna; cp = a. communicans posterior; e = a. ethmoidalis; f = a. frontalis; fa = a. facialis; l = a. lacrimalis; li = a. lingualis; mm = a. meningea media; oi = a. ophthalmica interna; ri = r. infraorbitalis; rm = r. mandibularis; rs = r. supraorbitalis; st = a. stapedia; st' = distal part of the stapedial artery stem; tf = a. transversa faciei; ts = a. temporalis superficialis. (Reproduced from Bugge, 1974b.)

connecting the extrabulbar arteries (Fig. 1B, l, f, e) with the bulba arteries (c) of the orbit, is also important. By means of this anastomosis the external carotid artery also annexes the supply of the eyeball (c), and the internal ophthalmic artery (oi) from the circulus arteriosus in most cases disappears. An anastomosis a 3' sometimes develops between the proximal part of the external carotid artery and the mandibular branch (rm). Anastomosis a 4, or anastomotic artery, and anastomosis a 5, or anastomotic branch, which connect the distal end of the internal carotid artery with the supraorbital (rs) and infraorbital (ri) branches, respectively, occur irregularly among rodents and are sometimes very important for systematic classification. Anastomosis a 6, the distal end of which forms the Vidian artery, connects the distal end of the internal carotid with the ciliary artery (c) or the infraorbital branch (ri), or both. It occurs in a few hystricognaths but contributes only slightly to the cephalic arterial supply. The common trunk for anastomoses a 4, a 5, and a 6 is thought to represent the vestige of a "persistent primitive maxillary artery" (De la Torre and Netsky, 1960).

RODENT ARTERIAL PATTERNS

Muroidea: Families Cricetidae and Muridae

In the rodents examined, the most primitive pattern of cephalic arterial supply is as a rule found in the muroids, where the internal carotid and stapedial arteries are well developed. The most primitive supply pattern is found in hamsters (*Cricetini*) and certain New World mice (*Hesperomyini*), namely grasshopper mouse (*Onychomys torridus*) and deer mouse (*Peromyscus boylei*), where the complete stapedial artery system (Figs. 2A, st, rs, ri, rm) is intact, and only the most common anastomosis (a 1) is developed. The cephalic arterial supply pattern differs from the primitive basic pattern only with regard to the supply of the eyeball (c), which is annexed by the stapedial artery system by means of anastomosis a 1, the internal ophthalmic artery (oi) obliterating.

In certain New World cricetids (*Hesperomyini*), namely rice rat and cotton rat (*Oryzomys palustris* and *Sigmodon hispidus*, Fig. 2B) and wood rat (*Neotoma albigula*, Fig. 2C), the general cephalic arterial pattern is still relatively primitive, with most of the original system intact, whereas the *supply* pattern is entirely altered by the additional development of three of the following four anastomoses (Fig. 1B): a 2, a 3, a 4 and a 5''. The external carotid artery (Fig. 2B-C) supplies the upper jaw (rm) via anastomosis a 3, while the internal carotid artery assumes the supply of the entire orbit (1, f, e) and upper jaw (ri) by means of a different combination of anastomoses a 5'', a 2, a 1, and a 4.

In voles and lemmings (Microtinae), the most primitive pattern of cephalic arterial supply is found in the former group (Microtini, Fig. 2D), where the stapedial artery system has assumed the supply of the entire orbit (1, f, e, and c) by means of anastomoses a 2 and a 1. The cephalic supply pattern in lemmings (Lemmini), most gerbils (Gerbillinae), and most Murinae differs from this rather primitive pattern only in the presence of anastomosis a 3 (Fig. 3A), via which the external carotid artery system assumes the supply of the lower jaw (rm).

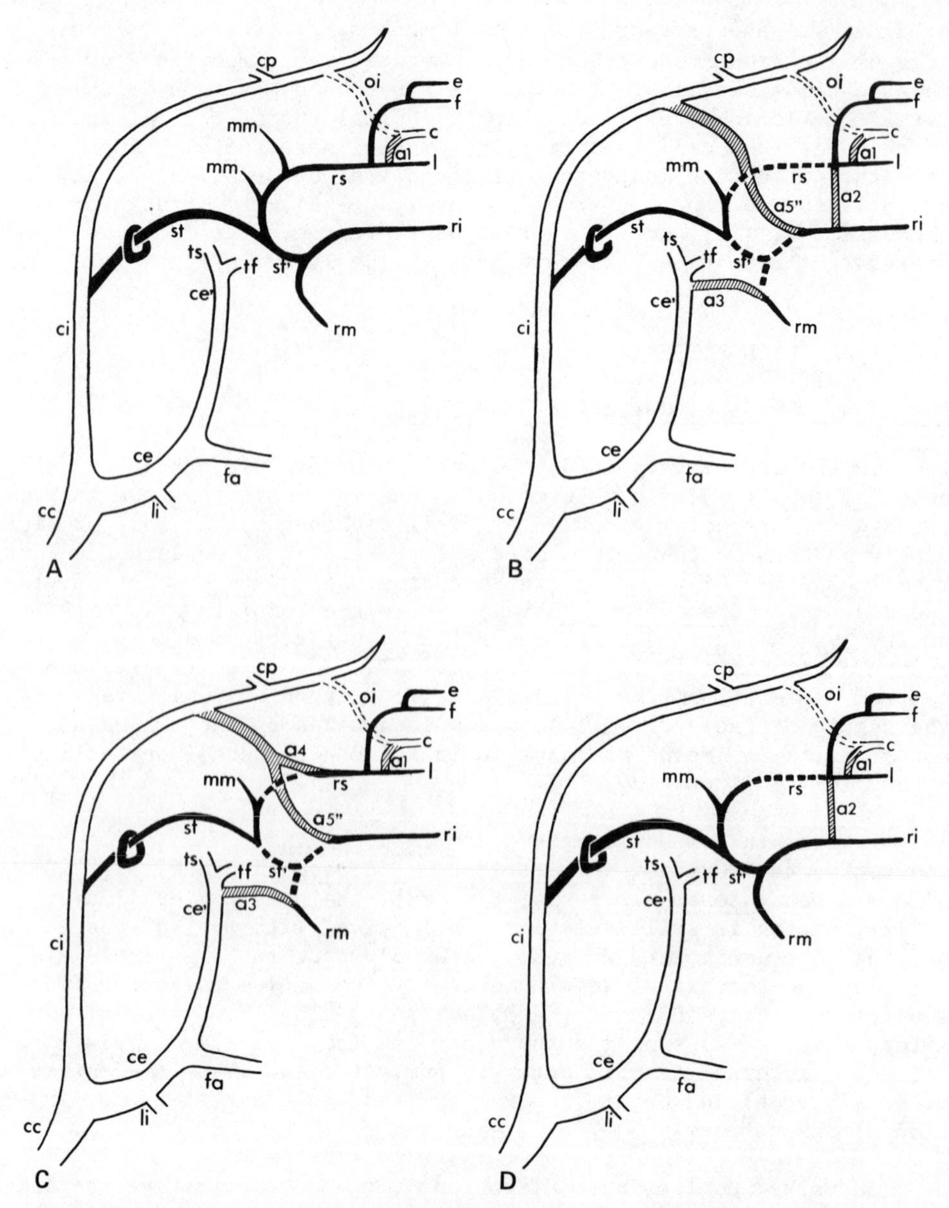

Fig. 2. The cephalic arterial pattern in certain muroids (A-D). (A) Cricetini, e.g. *Cricetus cricetus*; and certain Hesperomyini, viz. *Onychomys t. torridus* and *Peromyscus b. rowleyi*. (B) Certain Hesperomyini, viz. *Oryzomys palustris* and *Sigmodon hispidus*. (C) certain Hesperomyini, viz. *Neotoma a. albigula*. (D) Microtini, e.g. *Microtus agrestis*. (Reproduced from Bugge, 1974b.)

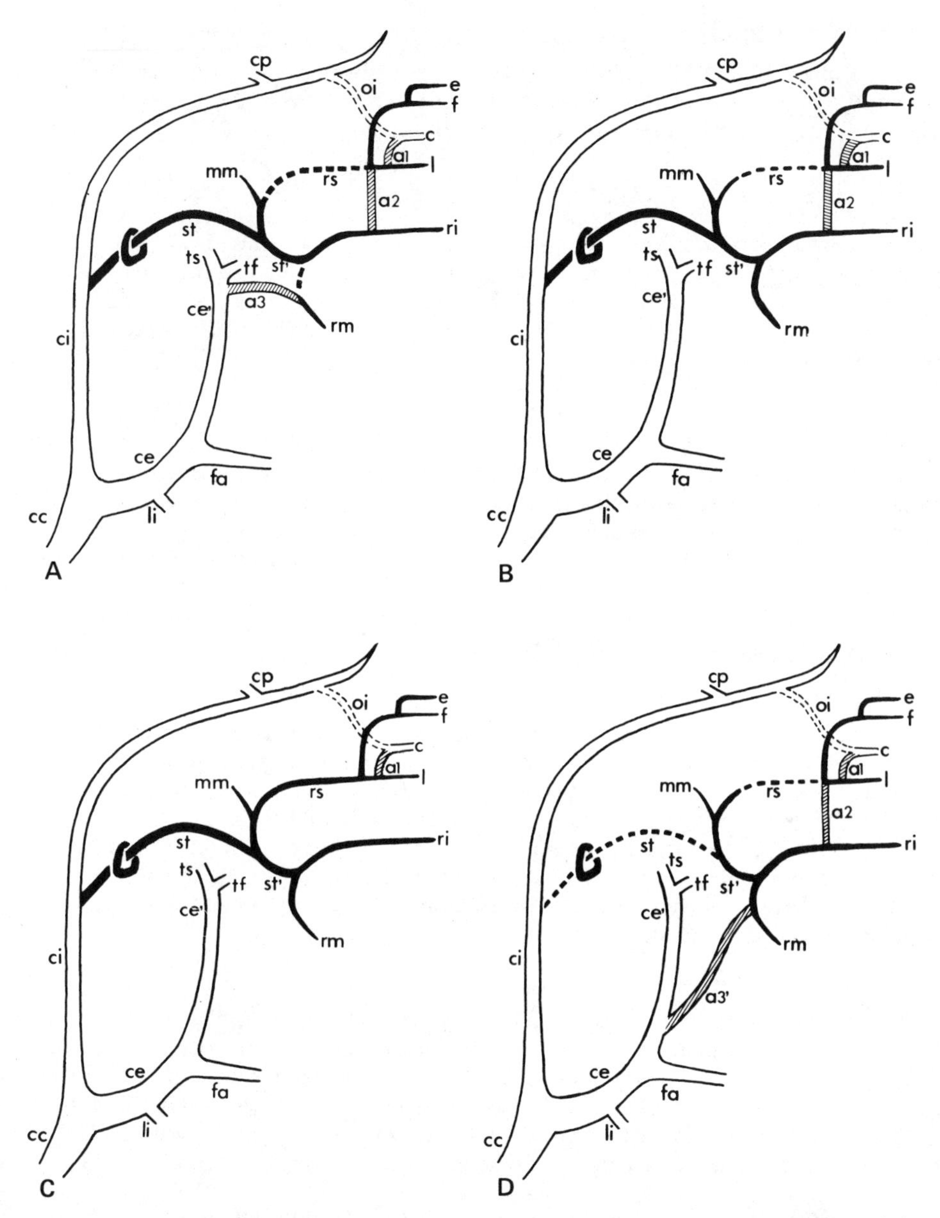

Fig. 3. The cephalic arterial pattern in certain muroids (A); dipodoids (B); and geomyoids (C, D). (A) Lemmini, viz. Lemmus lemmus; most Gerbillinae, e.g. Gerbillus pyramidum; and most Murinae. e.g. Rattus norvegicus. (B) Zapodidae, viz. Sicista betulina; and Dipodidae, e.g. Jaculus blanfordi. (C) Heteromyidae, e.g. Perognathus b. baileyi. (D) Geomyidae, viz. Geomys bursarius. (Reproduced from Bugge, 1974b.)

Dipodoidea: Families Zapodidae and Dipodidae

In the dipodoids investigated (birch mice; Sicistinae) and jerboas; Dipodinae, the cephalic arterial pattern (Fig. 3B) is likewise primitive and similar to that of voles (Fig. 2D).

Geomyoidea: Families Heteromyidae and Geomyidae

In pocket mice and kangaroo rats (Heteromyidae), the cephalic arterial pattern is extremely primitive (Fig. 3C), with only anastomosis a 1 developed, via which the stapedial artery system assumes the supply of the eyeball (c), with the internal ophthalmic artery (oi) obliterating.

In pocket gophers (Geomyidae), the cephalic arterial pattern is more advanced (Fig. 3D). The internal carotid artery is still well developed and supplies the brain, together with the vertebral-basilar artery system, while the entire stapedial area of supply is taken over by the external carotid artery system, by means of anastomoses a 3', a 2, and a 1, with the stapedial artery stem obliterating.

Families Spalacidae and Rhizomyidae

In the spalacids and rhizomyids investigated (Fig. 4A, B), the cephalic arterial pattern differs markedly from the pattern found in muroids and dipodoids. The internal carotid artery (Fig. 4A-B) is well developed and supplies the brain, assisted by the vertebral-basilar artery system, while the entire stapedial area of supply is assumed by the internal-external carotid artery system, by means of four anastomoses out of a total of six anastomoses in various combinations (see below), with the stapedial artery stem and its supraorbital (rs) and mandibular (rm) branches being partly obliterated.

In spalacids (lesser mole-rats (Spalax leucodon)), the external carotid artery system (Fig. 4A), supplies the mandible (rm) via anastomosis a 3, while the internal carotid artery has taken over the supply of the dura (mm), the upper jaw (ri), and the entire orbit (1, f, e, and c) by means of anastomoses a 5' between the distal ends of the internal carotid artery and the stapedial artery stem, a 2, and a 1, with the internal ophthalmic artery (oi) becoming obliterat

In the rhizomyids investigated (e.g. lesser bamboo rat (Cannomy badius minor)), the external carotid (Fig. 4B, ce) has annexed the supply of the lower jaw via anastomosis a 3' from the proximal part the facial artery (fa), while the remainder of the stapedial area of supply, i.e. the dura (mm), the upper jaw (ri) and the entire orbit (1, f, e, and c), is taken over by the internal carotid artery (ci) by means of anastomoses a 4 between the distal end of the internal carotid artery (ci) and the orbital part of the supraorbital branch (rs), a 2 and a 1.

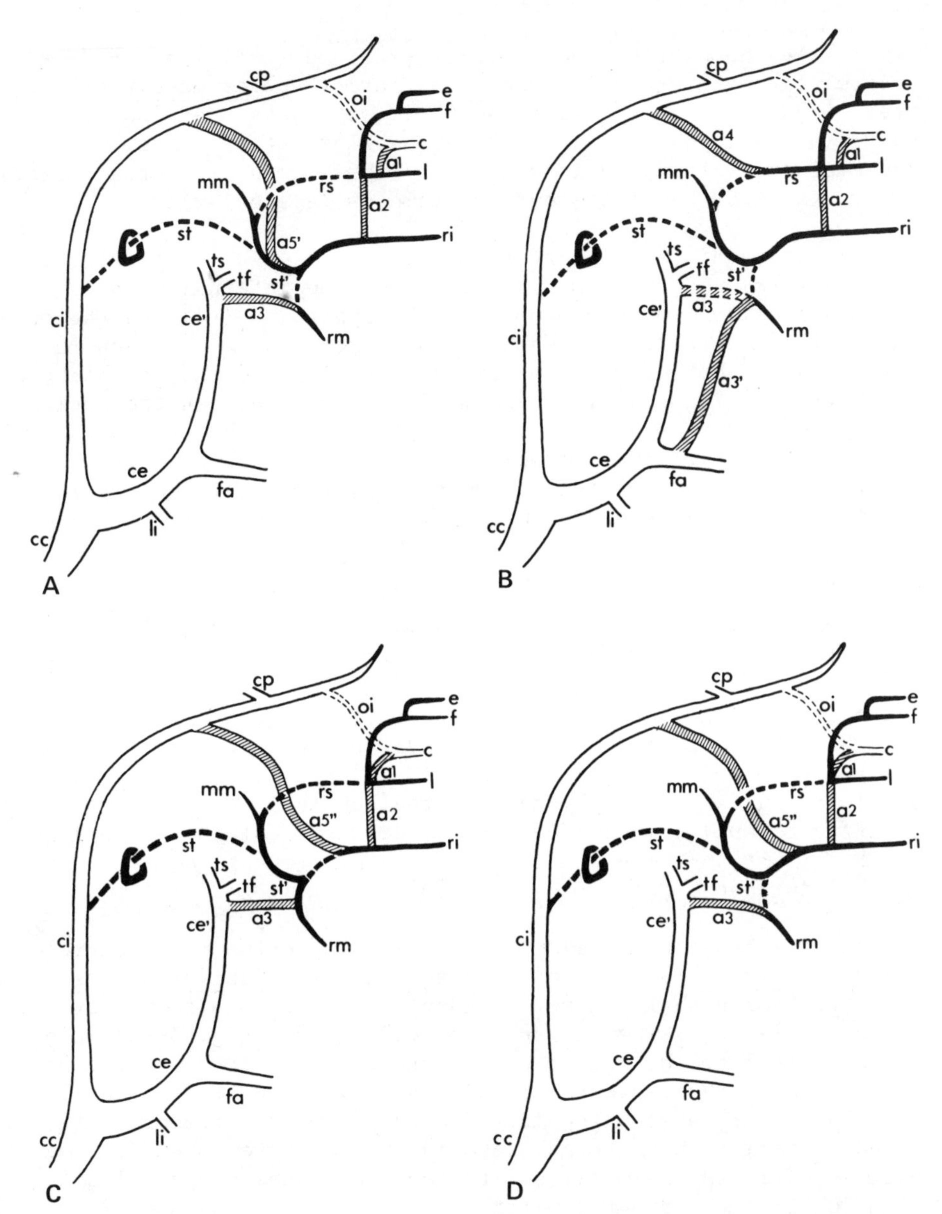

Fig. 4. The cephalic arterial pattern in spalacids (A); rhizomyids (B); anomalurids (C); and pedetids (D). (A) Spalacidae, viz. *Spalax leucodon*. (B) Rhizomyidae, e.g. *Cannomys badius minor*. (C) Anomaluridae, viz. *Anomalurus peli*. (D) Pedetidae, viz. *Pedetes capensis*. (Reproduced from Bugge, 1974b.)

Families Anomaluridae and Pedetidae

In the anomalurids and pedetids examined (scaly-tailed "flying" squirrel (*Anomalurus peli*) and spring hare (*Pedetes capensis*)), the internal carotid artery (Figs. 4C-D) is well developed and supplies the brain, assisted by the vertebral-basilar artery system, while the stapedial area of supply is annexed by the internal-external carotid artery system by means of four anastomoses (see below).

The external carotid artery system (Fig. 4C-D) supplies the lower jaw (rm) via anastomosis a 3, while the internal carotid artery system has taken over the supply of the upper jaw (ri) and the entire orbit (l, f, e, and c) by means of anastomoses a 5'' between the distal end of the internal carotid artery (ci) and the infraorbital branch (ri), a 2 and a 1. The only difference between the cephalic arterial patterns of anomalurids and pedetids is that the dura (mm) in scaly-tailed "flying" squirrels (Fig. 4C) is supplied from the external carotid artery but in the spring hares (Fig. 4D) from the internal carotid artery system.

Family Castoridae

In castorids (*Castor fiber*), a well-developed internal carotid artery (Fig. 5A), with the vertebral-basilar artery system, supplies the brain. On the other hand, the entire stapedial area of supply, i.e. the upper and lower jaws (ri and rm), the dura (mm) and most of the orbit (l, f, e, and c) is taken over by the external carotid artery system by means of anastomoses a 3' from the base of the facial artery, a 2 and a 1, the internal ophthalmic artery (oi) contributing to only a modest degree to the supply of the eyeball (c).

Family Sciuridae

In all the sciurids examined, for instance the red squirrel (*Sciurus vulgaris*) and eastern red squirrel (*Tamiasciurus hudsonicus*) the pattern of cephalic arterial supply is fairly uniform (for detail see Bugge, 1974b). The internal carotid artery (Fig. 5B-C) is obliterated distal to the origin of the stapedial artery, and the brain is supplied by the vertebral-basilar artery system (cp). The stapedial artery system supplies the dura (mm) and the orbit (l, f, e, and c), while the external carotid artery system supplies the lower jaw (rm) and, sometimes with the aid of the stapedial artery (cf. Fig. 5B), the upper jaw (ri).

Family Aplodontidae

In the mountain beaver (*Aplodontia rufa*), both the internal carotid artery (Fig. 5D) and the stapedial artery system are lacking

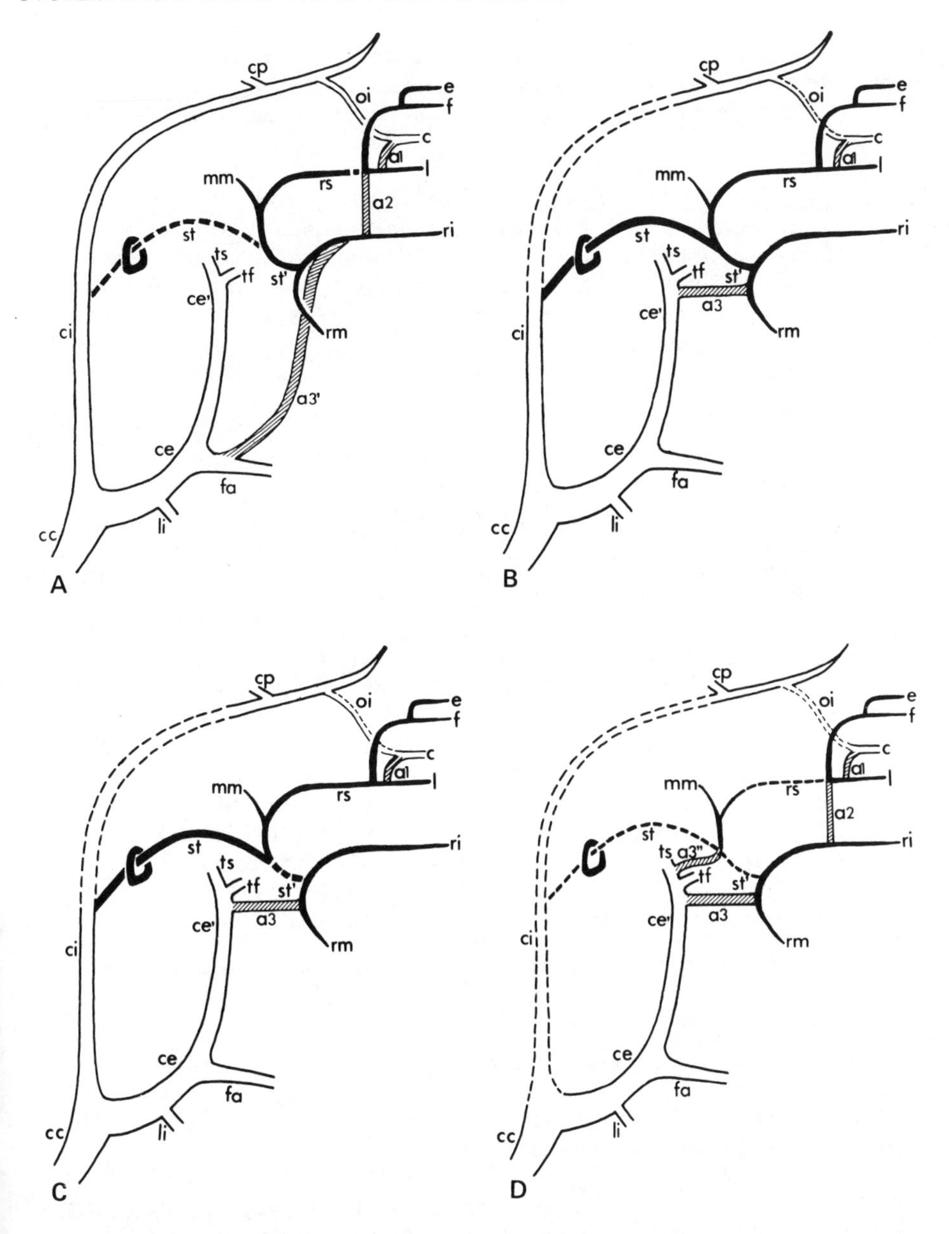

Fig. 5. The cephalic arterial pattern in castorids (A); in typical sciurids (B, C); and in aplodontids (D). (A) Castoridae, viz. Castor fiber. (B) Sciurini, viz. Sciurus vulgaris. (C) Tamiasciurini, viz. Tamiasciurus hudsonicus; and certain Funambulini, viz. Funambulus pennanti. (D) Aplondotidae, viz. Aplodontia rufa. (Reproduced from Bugge, 1974b.)

The brain is supplied by the vertebral-basilar artery system (cp), while the entire stapedial area of supply is taken over by the external carotid artery system by means of four anastomoses (a 1, a 2, a 3, a 3'').

The external carotid artery (Fig. 5D) annexes the supply of the upper and lower jaws (ri and rm) via anastomosis a 3, and also that of the entire orbit (l, f, e, and c) by means of anastomoses a 2 and a 1, the internal ophthalmic artery (oi) aborting. For the supply of the dura (mm) there is developed a very specific anastomosis (a 3'') from the superficial temporal artery (ts).

Family Gliridae

In the glirids examined, the most primitive pattern of cephalic arterial supply is found in the garden dormouse (*Eliomys quercinus*) and the African dormouse (*Graphiurus murinus*), representing the subfamilies Glirinae and Graphiurinae, respectively. The stapedial artery system (Fig. 6A) is well developed and supplies the dura (mm) and the orbit (l, f, e, and c) and contributes to the supply of the upper jaw (ri), while the external carotid artery system has assumed the supply of the lower jaw (rm) and, together with the stapedial artery, the upper jaw (ri), only two anastomoses (a 1, and a 3) being developed. In other glirids (common dormouse (*Glis glis*) and hazel mouse (*Muscardinus avellanarius*)), there is an increasing curtailment of the stapedial artery stem (Fig. 6B-C) in favor of the external carotid artery system, and three anastomoses (a 1, a 2, and a 3) are developed. The internal carotid artery is lacking in all glirids, aborting after the departure of the stapedial artery (Fig. 6A-B) or completely (Fig. 6C), and the brain is supplied by the vertebral-basilar artery system (cp) alone.

Caviomorpha (New World hystricognaths)

Superfamily Cavioidea; Families Caviidae and Dasyproctidae. In all cavioids examined, for instance guinea pigs (*Cavia porcellus*), the internal carotid artery (Fig. 6D) is lacking, and the brain is supplied by the vertebral-basilar artery system, assisted by the external carotid system (see below).

The external carotid artery assumes the supply of the upper and lower jaws (ri and rm) and the dura (mm) via anastomosis a 3' from the facial artery (fa), most of the stapedial artery stem obliterating. Also the entire orbit, including the extrabulbar part (l, f, e originally supplied by the stapedial artery system, and the bulbar part (c), originally supplied by the internal carotid artery, is annexed by the external carotid artery by means of anastomoses a 2 and a 1, respectively. The external carotid artery system has not only annexed the entire stapedial area of supply, but contributes to a considerable degree to the supply of the anterior part of the

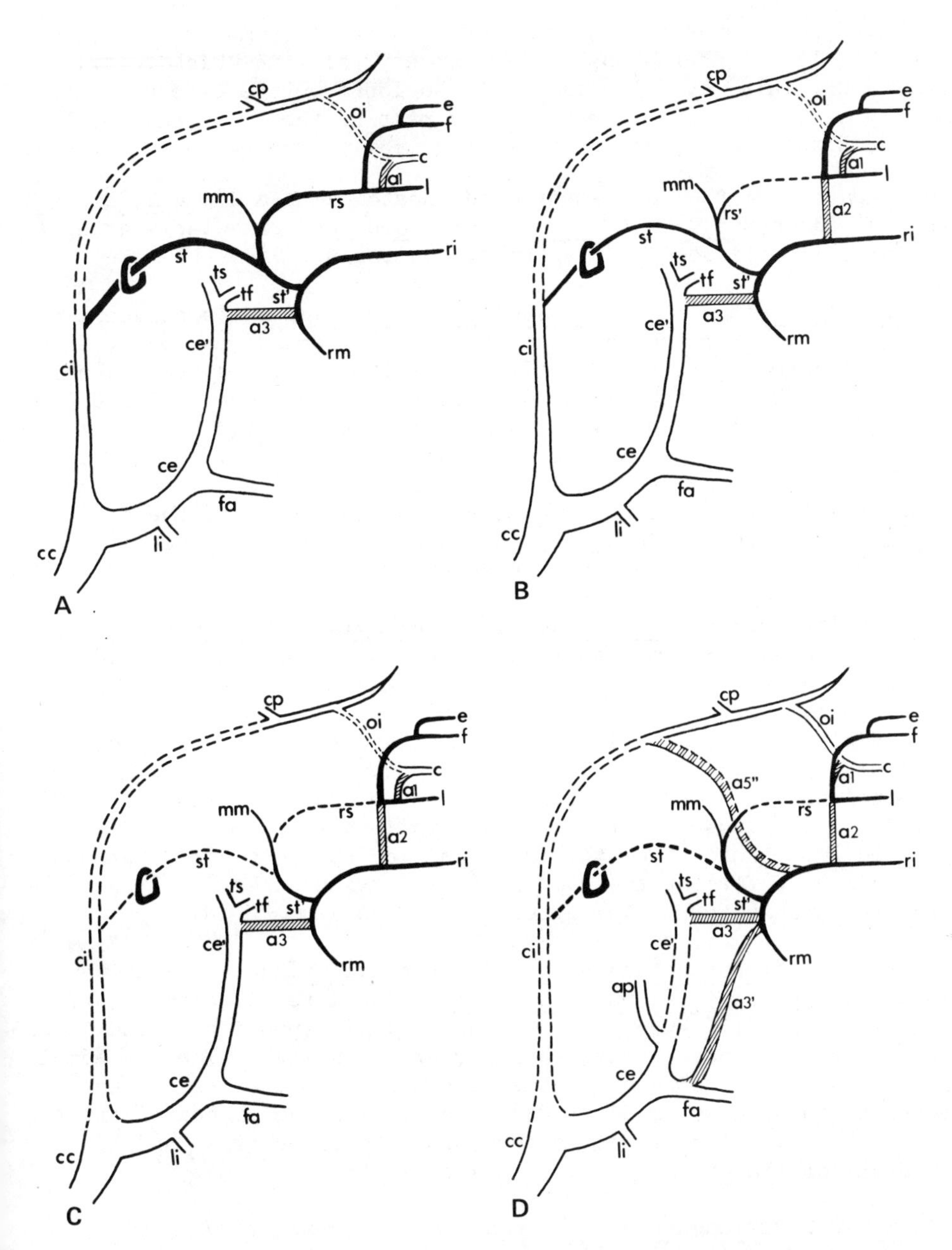

Fig. 6. The cephalic arterial pattern in glirids (A–C); and in certain caviomorphs, viz. cavioids (D). (A) certain Gliridae, viz. Graphiurus murinus and Eliomys quercinus. (B) certain Gliridae, viz. Glis glis. (C) certain Gliridae, viz. Muscardinus avellanarius. (D) Caviidae, viz. Cavia porcellus. (Reproduced from Bugge, 1974b.)

circulus anteriosus via the extremely short anastomosis *a 1* and the well developed internal ophthalmic artery (oi), in which the flow is reversed. This very specific contribution from the external carotid artery to the supply of the anterior part of the brain has not been found in any other mammals and is peculiar to the cavioids.

Apart from the above-mentioned anastomoses (*a 3'*, *a 2*, and *a 1*), the anastomoses *a 3*, *a 4*, *a 5''*, and *a 6* occur irregularly among the cavioids investigated (for details, see Bugge, 1974a, b).

Superfamily Chinchilloidea: Family Chinchillidae. In chinchilla (*Chincilla laniger*) and plains viscacha (*Lagostomus maximus*), the internal carotid artery (Fig. 7A) is obliterated, and the brain is supplied by the vertebral-basilar artery system alone, while the external carotid artery system has assumed the entire stapedial area of supply by means of anastomoses *a 3'*, *a 2*, and *a 1*. The most marked difference from the pattern in cavioids is that there is no contribution from the external carotid artery to the supply of the brain, the internal ophthalmic artery (oi) aborting in the chinchilla and being attenuated in the plains viscacha.

Superfamily Octodontidea: Families Capromyidae, Octodontidae, Ctenomyidae, and Echimyidae. In all the octodontoids examined, the internal carotid artery (Fig. 7B) is lacking, and the brain is supplied by the vertebral-basilar artery system (cp) alone. The stapedial artery system is obliterated, and the entire stapedial area of supply is annexed as in cavioids and chinchilloids by the external carotid artery system, by means of anastomoses *a 3'*, *a 2*, and *a 1*. The cephalic arterial pattern in octodontoids is distinguished from that of chinchilloids only by the development of anastomosis *a 3*, forming the common trunk for the superficial temporal (ts) and transverse facial artery (tf).

Superfamily Erethizontoidea: Family Erethizontidae. In the New World porcupines (*Coendou prehensilis*), the internal carotid artery (Fig. 7C) is well developed, and the brain is supplied by the internal carotid artery and vertebral-basilar artery system. The stapedial area of supply is assumed by the external carotid artery system by means of anastomoses *a 3'*, *a 2*, and *a 1*, with the stapedial artery stem, the central part of the supraorbital branch (rs), and the internal ophthalmic artery (oi) all obliterating.

Old World Hystricognaths

Superfamily Hystricoidea: Family Hystricidae. In the Old World porcupines investigated (*Hystrix leucura* and *Hystrix cristata*), the internal carotid artery (Fig. 7D) is obliterated, and the brain is supplied by the vertebral-basilar artery system (cp) alone. The stapedial artery stem and the supraorbital branch (rs) are also obliterated to the same extent as in the caviomorphs, and the entire

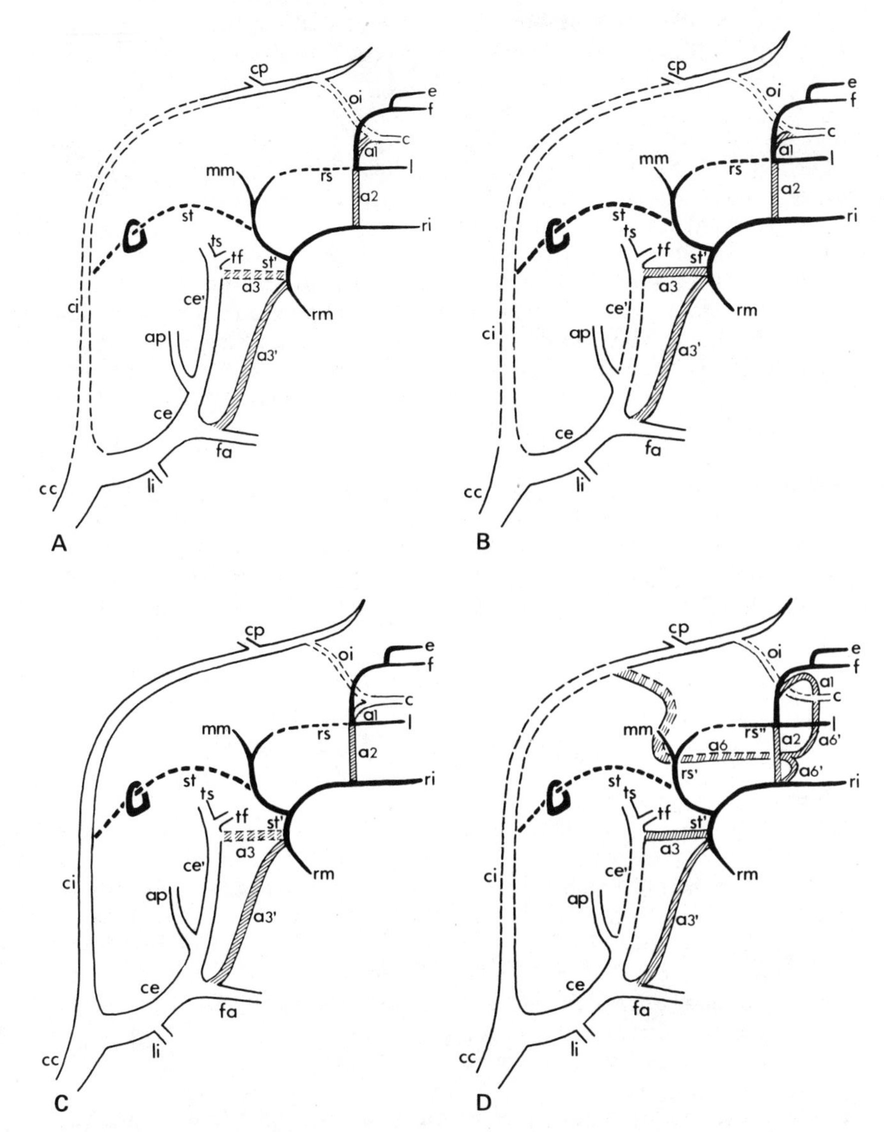

Fig. 7. The cephalic arterial pattern in certain caviomorphs, viz. chinchilloids (A); and octodontoids and certain Old World hystricognaths, viz. thryonomyoids and bathyergoids (B); in erethizontoids (C); and in certain Old World hystricognaths, viz. hystricoids (D). (A) Chinchillidae, e.g. Chinchilla laniger. (B) Octodontoidea, e.g. Octodon degus; Thryonomyidae, viz. Thryonomys swinderianus; and Bathyergidae, e.g. Bathyergus suillus. (C). Erethizontidae, viz. Coendou prehensilis. (D) Hystricidae, e.g. Hystrix cristata. (Reproduced from Bugge, 1974b.)

stapedial area of supply is assumed by the external carotid artery system, by means of five anastomoses (Fig. 7D). The upper and lower jaws (Fig. 7D, ri and rm) and the dura (mm) are supplied by the external carotid artery via anastomosis a 3' from the proximal part of the facial artery (fa), the anastomosis a 3 forming the common trunk for the superficial temporal artery (ts) and the transverse facial artery (tf) only, as in most caviomorphs. Most of the orbit (Fig. 7D, l, f, e, and c) is also supplied by the external carotid artery system, as in most caviomorphs, but in a rather specific way. The infraorbital branch (ri) not only assumes the supply of the extrabulbar part of the orbit (l, f, e) by means of anastomosis a 2, lateral to the 1st and 2nd trigeminal branches, but it also annexes the supply of the bulbar part (c) by the distal part of anastomosis a 6 (= a 6') medial to the 1st trigeminal branch, the anastomosis a 1 only forming a weak connection across the muscle cone between the bulbar (c) and the extrabulbar artery system (l, f, e).

Superfamily Thryonomyoidea: Family Thryonomyidae; and Superfamily Bathyergoidea: Family Bathyergidae. In the thryonomyoids (*Thryonomys swinderianus*) and bathyergoids (*Bathyergus suillus*, *Cryptomys natalensis*, *Heterocephalus glaber*) studied, the cephalic arterial pattern is nearly identical to that of octodontoids (Fig. 7B).

Superfamily Ctenodactyloidea

Family Ctenodactylidae. The results of a preliminary study of the carotid circulation in *Ctenodactylus vali* are included in Table 1.

DISCUSSION

Some of the most important of the characters used to classify the rodents at a higher taxonomic level are the relationship of the masseter muscle to the zygomatic arch and infraorbital foramen, the number and pattern of the molar dentition in fossil and Recent rodents, and the position of the origin of the angle of the mandible in relation to the incisive alveolus. Also, the histology of the incisor enamel has been used taxonomically.

It is a matter of fact, however, that using these criteria, it is often extremely difficult to distinguish between parallel or convergent changes and those indicating phylogenetic relationship. Therefore, there has been a need in recent years for a revision of the systematic classification of rodents, and especially of the classical subordinal division into Myomorpha, Sciuromorpha and Hystricomorpha, originally proposed by Brandt (1855), and followed by most zoologists in the first half of this century, for instance Tullberg (1899), Miller and Gidley (1918), Ellerman (1940-41), and Simpson (1945). The most convincing attempt to revise the classical rodent taxonomy is that of Wood (1955, 1959), mainly based on paleor tological evidence, with special reference to cranial morphology. In the following, the systematic value of the carotid arterial pattern will be discussed, focusing especially on taxonomically ambiguous groups, where earlier investigations of different organ systems have resulted in a varying taxonomic allocation.

Muroidea and Dipodoidea

The central groups within the classical myomorphs are the muroids (cricetids and murids) and dipodoids (zapodids and dipodids). The uniform cephalic arterial pattern is consistent with a close relationship between muroids and dipodoids. The cephalic supply pattern is generally very primitive, with well developed internal carotid and stapedial arteries, although certain New World mice (Fig. 2B-C) present some important exceptions. However, although the cephalic pattern of supply is radically altered in this group by the presence of some extra anastomoses, the general arterial pattern is still quite primitive, with most of the original system preserved. Table 1 summarizes the distribution of carotid traits in Recent rodent taxa.

The generally primitive pattern of cephalic arterial supply in muroids and dipodoids is a symplesiomorphic character, which cannot normally be accorded special taxonomic value. The cephalic arterial pattern in the other classical rodent suborders, however, is very specific and radically different from the primitive pattern found in muroids and dipodoids. Paleontological evidence seems to show that the loss in the other rodents groups of important parts of the original system, i.e., internal-external carotid arteries and the stapedial artery, in connection with development of specific anastomoses, can be followed back to the close of the Eocene (see Lavocat and Parent, this volume).

Geomyoidea

The geomyoids have in the first half of the present century been placed among the Sciuromorpha (cf. Simpson, 1945), although earlier zoologists regarded them as Myomorpha. However, Wilson (1949) and Wood (1955, 1959) in particular have more recently emphasized the myomorph affinities of the geomyoids, in spite of the sciuromorphic development of the zygomatic arch and origin of the masseter muscle. Wilson (1949), in particular, has pointed out that a myomorph position of the geomyoids would account for many features that are difficult to explain as parallelism across a subordinal boundary.

The development of the cephalic arterial system endorses the non-sciurid character of the geomyoids. In this respect, it is important that especially the more advanced cephalic supply system in the geomyids is radically different from that in all sciurids. The internal carotid artery is well developed in the geomyids (Fig. 3D) but is obliterated in the sciurids (Fig. 5B), while the stapedial

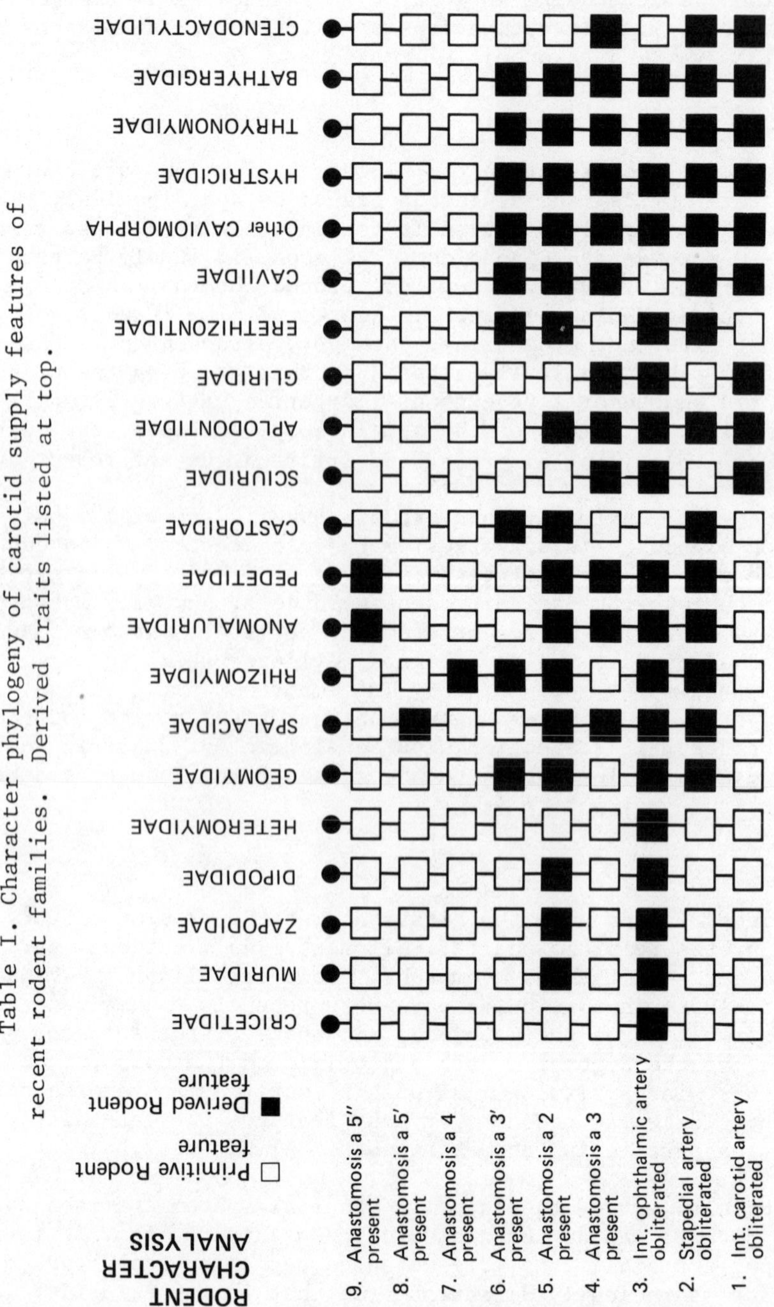

Table I. Character phylogeny of carotid supply features of recent rodent families. Derived traits listed at top.

artery system is lacking in the geomyids (Fig. 3D) but well developed in sciurids (Fig. 5B). On the other hand, the carotid supply pattern (Fig. 3D) in the most primitive geomyoids (e.g. heteromyids) is consistent with their affinities with the muroids (Table 1).

Spalacidae and Rhizomyidae

The taxonomic position of the spalacids and rhizomyids has been more problematic (Bugge, 1971a, 1974b). Most zoologists have underlined their myomorphic character (Gorgas, 1967) and have regarded mole-rats (Spalacidae) and bamboo-rats (Rhizomyidae) as specialized muroids (Simpson, 1945; Thenius, 1960, 1969). Wood (1955) assigned the two families either to the Muroidea *inc. sed.* or, as a special superfamily, Rhizomyoidea, to a revised suborder Myomorpha (Wood, 1959), while Stehlin and Schaub (1951) and Schaub (1953), on the basis of detailed studies of the molar pattern in fossil and Recent rodents, classified spalacids and rhizomyids as hystricomorphs.

The cephalic arterial pattern of supply seems to verify that spalacids and rhizomyids are related, the annexation by the internal-external carotid artery system of the stapedial area of supply having occurred fundamentally in the same rather specific manner (cf. Figs. 4A, B). Also, Lavocat and Parent (see this volume) found several common derived features in the middle ear region of spalacids and rhizomyids.

Anomaluridae and Pedetidae

The rather specific, but principally uniform, pattern of cephalic arterial supply (Figs. 4C, D) in the scaly-tailed "flying" squirrel (Anomaluridae) and spring hare (Pedetidae) seems to support Simpson's (1945) much discussed allocation of the two families to one superfamily, Anomaluroidea. The only difference with respect to the pattern of cephalic arterial supply is that the dura in the anomalurids (Fig. 4C) is supplied from the external carotid artery, whereas in the pedetids (Fig. 4D) it is supplied from the internal carotid artery. Simpson (1945), however, allocated the anomaluroids to the sciuromorphs, while earlier students have successively placed anomalurids and pedetids, singly or together, in all three classical rodent suborders (cf. Alston, 1875, 1876; Tullberg, 1899; Weber, 1928; Ellerman, 1940-41). The carotid supply pattern, however, in anomalurids and pedetids (Fig. 4C, D) is markedly different from the morphotypic condition of sciurids (Fig. 5B, C) and all other major groups of rodents (Table 1).

Castoridae

In the classical taxonomy the castorids are also placed among the sciuromorphs (cf. Simpson, 1945), whereas both Schaub (1953) and Wood (1955) considered a closer relationship with the sciurids as most improbable.

When the development of the cephalic arterial system is taken into consideration as a basis of classification, a position for the castorids in the neighbourhood of the sciurids seems to be most improbable, the cephalic arterial pattern in the castorids (Fig. 5A) being quite different from that of the sciurids (Fig. 5B, C). The stapedial artery, for instance, is completely lacking in the castorids (Fig. 5A), but well developed in the sciurids (Fig. 5B, C), whilst the internal carotid artery is well developed in the castorids (Fig. 5A), but is absent in the sciurids distal to the departure of the stapedial artery (Fig. 5B, C).

The well developed internal carotid artery in castorids (Fig. 5A) seems also to rule out the closer relationship with the Old World hystricognaths (Fig. 7B, D) assumed by Schaub (1958). The cephalic arterial pattern in general seems therefore to confirm the view of Wood (1959) that the castorids do not show any signs of being more closely related to any other Recent rodent group. The isolated position of the castorids has recently been emphasized by Wahlert (1977).

Sciuridae and Aplodontidae

In the classical rodent taxonomy (cf. Simpson, 1945), the sciuromorphs comprise aplodontoids, sciuroids, geomyoids and castoroids. Wood (1937, 1955, 1959), Stehlin and Schaub (1951), Schaub (1953), and Lavocat (1956), however, pointed out that these four superfamilies could hardly be closely related.

The pattern of cephalic arterial supply endorses the non-sciurid character of the geomyoids and the castorids (see above) and a rather isolated position of the aplodontids (see below), as pointed out by Wood (1959).

In the aplodontids, the cephalic arterial pattern (Fig. 5D) is rather advanced, both the internal carotid artery and the stapedial artery being obliterated and, like the dentition, fairly far removed from the primitive pattern one would expect from its systematic position (Wood, 1937; Simpson, 1945). The pattern of the carotid circulation is also considerably more advanced in the aplodontids then in the sciurids (Fig. 5B, C), where only the internal carotid artery is obliterated (distal to the departure of the stapedial artery).

The cephalic arterial pattern in aplodontids and sciurids and their presumptive ancestors makes it likely that the two families have developed independently of each other since the late Eocene (cf. also Vianey-Liaud, this volume). With the aplodontids, geomyoids and castorids removed from the classical sciuromorphs, the sciurids (possibly with the glirids (see below)) therefore remain as the only true representatives of this group.

Gliridae

In the classical rodent taxonomy, the glirids are assigned to the myomorphs (cf. Simpson, 1945), a position partly accepted by Wood (1955, 1959), who placed them in the vicinity of the myomorphs under Myomorpha *incertae sedis*. Grassé and Dekeyser (1955), on the other hand, regarded them as a sciuromorph superfamily, while Stehlin and Schaub (1951) emphasized their isolated position among the rodents.

The pattern of cephalic arterial supply in the glirids (Fig. 6A-C), however, is markedly different from that of typical muroids (Fig. 2A), whereas the cephalic pattern of supply in the more primitive glirids (*Eliomys* and *Graphiurus*, Fig. 6A) and that in typical sciurids (Fig. 5B) are nearly alike. The more advanced pattern of supply in *Muscardinus* (Fig. 6C) is probably connected with the marked specialization of the hazel mice in other respects too, and with their probable position as a strongly divergent side branch from the main glirid stem (cf. Thenius, 1969).

The striking similarity between the cephalic pattern of supply in typical sciurids (Fig. 5B) and that of primitive glirids (Fig. 6A), in conjunction with other common morphological features, especially concerning the dentition, therefore seems to confirm the sciurid character of the glirids as emphasized by Grassé and Dekeyser (1955). Recently Vianey-Liaud (this volume) has pointed out that the "myomorphy" of the glirids is certainly based on parallel development. Also, the investigation of the middle ear anatomy by Lavocat and Parent (this volume) indicates that the glirids are not closely related to muroids, but instead might be closer to the sciurids.

Hystricognathi

If the paleo-geographical situation is taken into account, it seems fair to divide the rodent suborder Hystricognathi into several relatively independent lines, whereas the New and Old World hystricognaths morphologically seem to form a natural entity, characterized by many common features, in particular with respect to the cephalic arterial pattern and masticatory apparatus, which are difficult to explain as pure parallelism. The cephalic arterial patterns of the four South American octodontoid families investigated, for instance, and those of the African thryonomyoids and bathyergoids, are remarkably alike, and all these families were also placed in the same superfamily by Simpson (1945). Apart from the presence of anastomosis a 6', the hystricids (Fig. 7D) attach to the same group.

Although the contribution from the external carotid artery system to the supply of the brain in cavioids (Fig. 6D), via a persistent internal ophthalmic artery, is rather specific, the only important departure from the general cephalic arterial pattern in New and Old World hystricognaths (see Table 1) is the persistent internal carotid artery in New World porcupines (Erethizontoidea). This is doubtlessly the primitive condition, retained from the early Tertiary common ancestor of all hystricognaths. Recent investigation has shown that some African Miocene thryonomyoids apparently still retained a carotid canal, and presumably, an internal carotid artery (Lavocat, 1973; Lavocat and Parent, this volume). Thus, it is possible that the internal carotid artery was lost independently within Old and New World hystricognaths (except erethizontoids).

The study of the pattern of the cephalic arterial supply in the hystricognaths has first and foremost called attention to the isolated position of the erethizontoids. They undoubtedly separated from the common caviomorph stem before the obliteration of the internal carotid artery took place, and they have evolved independently since that time. This wide separation of erethizontoids from other caviomorphs is also supported by Woods' (this volume) analysis of a variety of myological, dental, and chromosomal features. Apart from this difference, the pattern of the cephalic arterial supply, together with numerous other morphological and molecular data, seems to indicate a close phyletic relationship between New World and Old World hystricognaths. In this respect, it is important to note that research of recent years into continental drift and rotation has revealed that, although the South Atlantic existed during the Eocene, it was not then an impassable barrier to east-west migration (see Thenius, 1971; Tarling, 1980).

CONCLUSION

The carotid arterial pattern in rodents (and all other mammals) is a very valuable taxonomic tool at the subordinal and superfamilial levels, but it does not carry the same weight taxonomically at the familial and subfamilial levels. This carotid pattern can be studied in detail by special plastic injection and corrosion techniques in living taxa, and these findings can be compared with fossil basicranial remains, which often retain traces of this vascular pattern (see Lavocat and Parent, this volume). In this survey, an analysis of the primitive and derived character states of the internal and external carotid arteries, with their branches and anastomoses, has been shown to be valuable for testing various hypotheses of relationships among higher taxa of rodents.

REFERENCES

Alston, E. R. 1875. On Anomalurus, its structure and position. Proc. Zool. Soc. Lond. 1875: 88-97.

Alston, E. R. 1876. On the classification of the order Glires. Proc. Zool. Soc. Lond. 1876: 61-98.

Brandt, J. F. 1855. Untersuchungen über die craniologischen Entwickelungsstufen und die davon herzuleitenden Verwandtschaften und Classificationen der Nager der Jetztwelt, mit besonderer Beziehung auf die Gattung Castor. Mém. Acad. imp. Sci. St.-Pétersbourg, Ser. 6, 9: 125-336.

Bugge, J. 1963. A standardized plastic injection technique for anatomical purposes. Acta Anat. 54: 177-192.

Bugge, J. 1970. The contribution of the stapedial artery to the cephalic arterial supply in muroid rodents. Acta Anat. 76: 313-336.

Bugge, J. 1971a. The cephalic arterial system in mole-rats (Spalacidae), bamboo rats (Rhizomyidae), jumping mice and jerboas (Dipodoidea) and dormice (Gliroidea) with special reference to the systematic classification of rodents. Acta Anat. 79: 165-180.

Bugge, J. 1971b. The cephalic arterial system in sciuromorphs with special reference to the systematic classification of rodents. Acta Anat. 80: 336-361.

Bugge, J. 1971c. The cephalic arterial system in New and Old World hystricomorphs, and in bathyergoids, with special reference to the systematic classification of rodents. Acta Anat. 80: 516-536.

Bugge, J. 1972. The cephalic arterial system in the insectivores and the primates with special reference to the Macroscelidoidea and Tupaioidea and the insectivore-primate boundary. Z. Anat. Entwg. 135: 279-300.

Bugge, J. 1974a. The cephalic arteries of hystricomorph rodents. Symp. Zool. Soc. Lond. 34: 61-78.

Bugge, J. 1974b. The cephalic arterial system in insectivores, primates, rodents and lagomorphs, with special reference to the systematic classification. Acta Anat. 87 (Suppl. 62): 1-160.

Bugge, J. 1975. Corrosion casts of the cephalic arterial system, prepared by means of a standardized plastic injection and corrosion technique. 10th Int. Congr. Anat. Tokyo, p. 528.

Bugge, J. 1978. The cephalic arterial system in carnivores, with special reference to the systematic classification. Acta Anat. 101: 45-61.

Bugge, J. 1979. Cephalic arterial pattern in New World edentates and Old World pangolins with special reference to their phylogenetic relationships and taxonomy. Acta Anat. 105: 37-46.

De la Torre, E. and Netsky, M. G. 1960. Study of persistent primitive maxillary artery in human fetus: Some homologies of cranial arteries in man and dog. Amer. J. Anat. 106: 185-195.

Ellerman, J. R. 1940-41. The Families and Genera of Living Rodents, Vol. 1-2. British Museum (Natural History), London.

Gorgas, M. 1967. Vergleichend-anatomische Untersuchungen am Magen-Darm-Kanal der Sciuromorpha, Hystricomorpha und Caviomorpha (Rodentia). Eine Studie über den Einfluss von Phylogenie,

Spezialisation und funktioneller Adaptation auf den Säugetierdarm. Z. wiss. Zool. 175: 237-404.
Grassé, P.-P. and Dekeyser, P. L. 1955. Ordre des rongeurs. In: Traité de Zoologie, P.-P. Grassé, ed., Vol. 17, Part 2, pp. 1321-1525, Masson et Cie, Paris.
Grosser, O. 1901. Zur Anatomie und Entwickelungsgeschichte des Gefässsystemes der Chiropteren. Arb. Anat. Inst., Wiesbaden 17: 203-424.
Guthrie, D. A. 1963. The carotid circulation in the Rodentia. Bull. Mus. Comp. Zool. Harv. 128: 455-481.
Hennig, W. 1950. Grundzüge einer Theorie der phylogenetischen Systematik. Deutscher Zentralverlag, Berlin.
Hennig, W. 1966. Phylogenetic Systematics. Univ. of Illinois Press, Urbana.
Hofmann, L. von 1914. Die Entwicklung der Kopfarterien bei Sus scrofa domesticus. Morph. Jb. 48: 645-671.
Lavocat, R. 1956. Réflexions sur la classification des rongeurs. Mammalia 20: 49-56.
Lavocat, R. 1973. Les Rongeurs du Miocène d'Afrique orientale. I. Miocène inférieur. Mém. Trav. E. P. H. E., Inst. Montpellier 1: 1-284.
Le Gros Clark, W. E. 1959. The Antecedents of Man. Univ. of Edinburgh Press, Edinburgh.
Martin, R. D. 1968. Towards a new definition of primates. Man 3: 377-401.
Matthew, W. D. 1909. The Carnivora and Insectivora of the Bridger Basin, Middle Eocene. Mem. Amer. Mus. Nat. Hist. 9: 291-567.
McKenna, M. C. 1966. Paleontology and the origin of the primates. Folia Primatol. 4: 1-25.
Miller, G. S. and Gidley, J. W. 1918. Synopsis of the supergeneric groups of rodents. J. Wash. Acad. Sci. 8: 431-448.
Padget, D. H. 1948. The development of the cranial arteries in the human embryo. Contrib. Embryol. Carneg. Inst. Wash. 32: 205-261.
Presley, R. 1979. The primitive course of the internal carotid artery in mammals. Acta Anat. 103: 238-244.
Schaub, S. 1953. Remarks on the distribution and classification of the "Hystricomorpha." Verh. naturf. Ges., Basel 64: 389-400.
Schaub, S. 1958. Simplicidentata (= Rodentia). In: Traité de Paléontologie, J. Piveteau, ed., Vol. 6, Part 2, pp. 659-818, Masson et Cie, Paris.
Simpson, G. G. 1945. The principles of classification and a classification of mammals. Bull. Amer. Mus. Nat. Hist. 85: 1-350.
Stehlin, H. G. and Schaub, S. 1951. Die Trigonodontie der simplicidentaten Nager. Schweiz. Paleont. Abh. 67: 1-385.
Struthers, P. H. 1930. The aortic arches and their derivatives in the embryo porcupine (Erethizon dorsatum). J. Morph. 50: 361-392.
Tandler, J. 1899. Zur vergleichenden Anatomie der Kopfarterien bei den Mammalia. Denksch. Acad. Wiss., Wien 67: 677-784.

Tandler, J. 1902. Zur Entwicklungsgeschichte der Kopfarterien bei den Mammalia. Morph. Jahrb. 30: 275-373.

Tarling, D. H. 1980. The geologic evolution of South America with special reference to the last 200 million years. In: Evolutionary Biology of the New World Monkeys and Continental Drift, R. L. Ciochon and A. B. Chiarelli, eds., pp. 1-41, Plenum Press, New York.

Thenius, E. 1960. Nagetiere (Rodentia oder Simplicidentata). In: Stammesgeschichte der Säugetiere, E. Thenius and H. Hofer, eds., pp. 128-147, Springer, Berlin.

Thenius, E. 1969. Phylogenie der Mammalia. Stammesgeschichte der Säugetiere (einschliesslich der Hominiden). Handb. Zool. 8: 1-722.

Thenius, E. 1971. Zum gegenwärtigen Verbreitungsbild der Säugetiere und seiner Deutung in erdgeschichtlicher Sicht. Natur. Mus., Frankfurt 101: 185-196.

Tullberg, T. 1899. Ueber das System der Nagethiere: eine phylogenetische Studie. Nova Acta Reg. Soc. Scient. Upsala, Ser. 3, 18: 1-514.

Van Valen, L. 1966. Deltatheridia, a new order of mammals. Bull. Amer. Mus. Nat. Hist. 132: 1-126.

Wahlert, J. H. 1977. Cranial foramina and relationships of _Eutypomys_ (Rodentia, Eutypomyidae). Amer. Mus. Nov. 2626: 1-8.

Weber, M. 1928. Die Säugetiere, 2nd ed., Vol. 2. Systematischer Teil. G. Fischer, Jena.

Wilson, R. W. 1949. Early Tertiary rodents of North America. Publ. Carneg. Inst. Wash. 584: 67-164.

Wood, A. E. 1937. The mammalian fauna of the White River Oligocene. Part II. Rodentia. Trans. Amer. Phil. Soc. 28: 155-269.

Wood, A. E. 1955. A revised classification of the rodents. J. Mammal. 36: 165-187.

Wood, A. E. 1959. Eocene radiation and phylogeny of the rodents. Evolution 13: 354-361.

HOMOLOGIES OF MOLAR CUSPS AND CRESTS, AND THEIR BEARING ON ASSESSMENTS OF RODENT PHYLOGENY

P.M. Butler

Department of Zoology, Royal Holloway College
Egham, Surrey, TW20 OEX, England

INTRODUCTION

Evidence for phylogeny derives basically from morphological comparisons, whether of the structure of protein molecules, chromosomes, skulls, or, in the case of paleontological material, mainly teeth. The concept of homology arises from the recognition that some resemblances are more significant than others that are superficial or accidental. It is necessary to decide, when comparing two species, which structural element in one species should be compared with an element in the other. Such corresponding structures in different animals are given the same name (Gr. homos, same; logos, word), and resemblances and differences between them form the raw data on which phylogenetic hypotheses are based. Whether resemblances are due to inheritance from a common ancestor, or whether they have been produced by parallel evolution, is a question that can be decided only after considerable investigation, if at all. Therefore, in this paper I will use the term homology in its etymological sense, to mean morphologically comparable structures, whether or not their resemblance is due to common ancestry.

Parallelism is a widespread feature of evolution, and it is not confined to the teeth: we meet it again in the problem of hystricomorphy. It arises from the principle of functional continuity; a complex functional-morphological system can change only in such ways that it does not pass through an nonviable intermediate stage. Because of this limit on the possible directions of advance, similar animals, exposed to similar selective forces, are likely to respond in the same manner (Butler, 1982). Parallelism is not the same as convergence, which implies the evolution of dissimilar animals in different directions to reach a similar final result.

Recognition of resemblances is an intuitive, subjective process, and it is possible to be misled. This is especially so in the case of teeth, which are comparatively simple structures with few landmarks. The rodent literature contains many propositions of cusp homology that have subsequently been abandoned. Thus Wortman (1902) was deceived by the high metaconid of lower P4 of Paramys into identifying it with the primary cusp of the tooth. Winge (1941) thought that Aplodontidae were primitive because their selenodont molars resembled those of certain insectivores. Stehlin and Schaub (1951) homologized the anteroconid of Sciuridae with the paraconid of primitive mammals, but it is now believed to be a secondary development. The five-crested pattern that characterized their suborder Pentalophodonta has evolved several times independently, and according to Wood (1974) the metacone relates to the third crest in Caviomorpha and the fourth crest in Theridomyidae.

Early stages of development often reveal similarities that are less apparent in completed structures. In mammalian molars the first cusp to develop is generally the paracone on the upper teeth and the protoconid on the lower teeth; the other cusps arise from a marginal zone around these primary cusps (Butler, 1956). Few rodents have been investigated from this point of view, and the results from these are not clear. Gaunt (1955, 1961b), who worked on the mouse and the hamster (Mesocricetus auratus), was able to identify the protoconid on lower molars, but in early stages of the mouse the protocone is more prominent than the paracone; the tooth appears to be tilted buccally so that the paracone is situated low down on the buccal side. In Cavia (Tims, 1901; Santoné, 1935; Harman and Smith, 1936) cusps are very indistinct, and development is dominated by the in-pushing of folds from the sides of the tooth, a precocious formation of the ridge pattern. Tims (1901) described a transient early stage with a single cusp that became anterobuccal in both jaws, i.e. it appeared to represent the paracone and protoconid. Tooth development is direct: teeth of different patterns follow divergent paths from the beginning, even adjacent teeth in the same jaw, as in the mouse. Selection has operated on developmental processes so that functional teeth are produced in the most efficient manner. Early stages therefore do not necessarily reflect ancestral states.

Unworn teeth may of course help with the interpretation of worn teeth, but their characters are not necessarily primitive. The enamel-free areas on the cusps of muroids (Gaunt 1961a; Monmignaut, 1963), for example, are clearly adaptive in that they influence the shape of the functional, worn teeth. Also, deciduous teeth are not necessarily more primitive than permanent teeth. Being less hypsodont, they may help to interpret the molars, but they frequently possess additional anterior elements.

The range of variation of tooth patterns within populations is limited, except for occasional major abnormalities. It seems that the developmental processes are rather closely controlled, and departures from the norm are eliminated by natural selection. If selection is involved, it follows that the patterns are of adaptive significance. Functional aspects of molar evolution have received much attention in recent years (reviewed by Butler, 1983). Much evidence has been obtained from a study of the wear facets produced during mastication. These facets, formed where opposing teeth come into close contact, are scored with parallel scratches that reveal the direction of relative movement. Functional studies of rodent molars have been made by Rensberger (1973, 1975, 1978, 1982), Woods and Howland (1979), and Wilkins and Woods (1983), and a preliminary general survey of rodent wear facets has been published (Butler, 1980). In mammals generally, the interrelations of opposing cusps are very conservative, so that it is possible to trace homologous facets on homologous cusps through a wide range of mammals. Comparison of the wear facets of rodents with those of other mammals provides us with an additional criterion of cusp homology. This theme is developed further in the present paper.

MOLAR OCCLUSION IN PRIMITIVE RODENTS

It is generally accepted that the molars of primitive rodents were cuspidate, rather than crested or lophate (Fig. 1), and that their masticatory movements were not propalinal but more or less transverse. The facets show that chewing in paramyids involved a two-phase movement, resembling that of primates (Fig. 1) and some condylarths (Butler, 1973; Kay and Hiiemae, 1974). The lower molar passes medially across the upper, at first in a relatively transverse direction (buccal phase), and then obliquely forward (lingual phase). This mode of chewing was retained in aplodontoids (Rensberger, 1982) and sciurids; it can be illustrated by reference to a living sciurid, *Sciurus carolinensis* (Fig. 2).

In *Sciurus*, the teeth first make contact when the buccal lower cusps (protoconid and hypoconid) are in line with the buccal upper cusps (paracone and metacone), and alternate with them: the hypoconid between the paracone and metacone, and the protoconid between the metacone of one tooth and the paracone of the next. The tips of the protoconid and hypoconid oppose the parastyle and mesostyle, and the mesoconid opposes the top of the paracone. Each numbered wear facet on the upper molar (Fig. 2) makes contact with the similarly numbered facet on the opposing lower molar during mastication (for further details on rodent wear facets, see Butler, 1980).

As the upper molar travels medially, the protoconid follows the anterior cingulum of the upper molar, passing along the valley bounded by the metaloph of one tooth and the protoloph of the next; at the same time, the hypoconid moves into the trigon basin. The same medial movement produces contacts between upper and lower lingual cusps. The metaconid passes in front of the protocone, and the entoconid behind it. There is no distinct hypocone in Sciurus, but its potential position is shown by two facets on the posterior crest of the protocone, one (facet 8) due to the entoconid, and the other (facet 4) to the metaconid of the following tooth. In other rodents in which the hypocone is differentiated, it carries these facets on its anterior and posterior surfaces.

When the lower molar has traveled about half-way across the upper molar it begins to move more anteriorly, so that the tip of the hypoconid passes over the lingual part of the protoloph, between the protocone and the paraconule, and the posterior arm of the hypoconid crosses the top of the protocone; the mesoconid crosses the

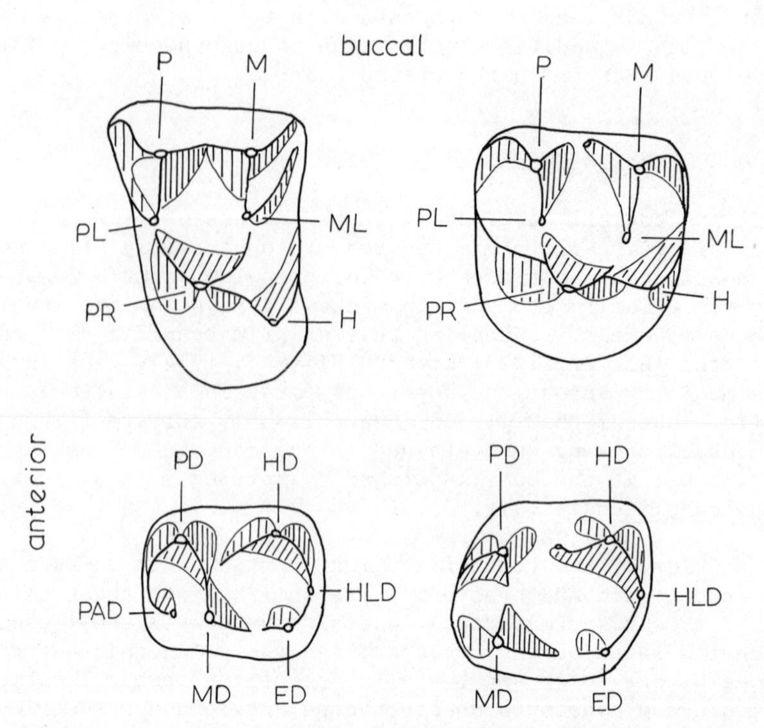

Fig. 1. Molar wear facets (hatched) and cusps of a primitive primat(left) and a primitive rodent (right). In this and all other figures, crown views of teeth are drawn with the mesial (anterior) side to the left, and the buccal (lateral) side above. Upper molar cusps: P, paracone; M, metacone; PR, protocone; H, hypocone; ML, metaconule; PL, paraconule. Trigonid cusps of lower molar: MD, metaconid; PAD, paraconi PD, protoconid. Talonid cusps of lower molar: ED, entoconi HD, hypoconid; HLD, hypoconulid.

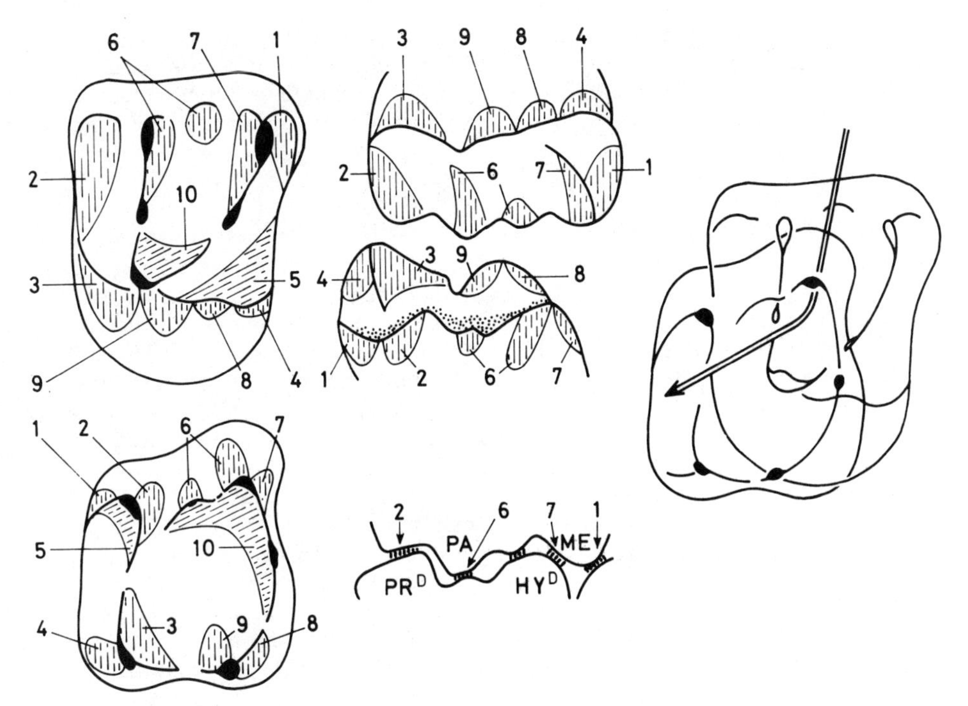

Fig. 2. Wear facets of *Sciurus* molars. On left, crown views of left upper and right lower molars. Center, above, oblique lingual views of right upper and left lower molars. Center, below, buccal cusps of left molars in occlusal contact. On right, the teeth in occlusion; arrow indicates the path of the hypoconid. Cusps are shown in black, and wear facets are numbered and hatched.

paraconule. During the same movement, the protoconid travels anterolingually across the lingual part of the posterior cingulum of the more anterior upper molar and crosses the virtual hypocone.

Considering the movement of the upper molar relative to the lower molar, the protocone passes between the metaconid and entoconid into the talonid basin, and then in the second phase it travels posterobuccally to cross the posteroloph between the hypoconid and the hypoconulid. The hypocone, if present, would pass in front of the metaconid and then posterobuccally to cross the posterior arm of the protoconid. The paracone and metacone function against the protoconid and hypoconid only in the first phase.

These occlusal relations of cusps and crests can be homologized readily with those of other eutherians, such as primates (Fig. 1). They show that rodent molars have passed through a tribosphenic stage in their evolution, and the Osbornian names can be applied confidently to their cusps.

COMPARISON WITH PRIMATES

If primitive rodents are compared with primitive primates, probably the most significant differences lie in the more horizontal chewing of the rodents (Fig. 1). In primitive primates, the movement of the lower teeth during the buccal phase of occlusion was directed upward at an angle of about 40° to the horizontal. The paracone and metacone stood higher above the trigon basin, and they had anterior and posterior crests with a cutting function against the crests of the protoconid and hypoconid. In this respect, the primates are intermediate between rodents and primitive Cretaceous eutherians such as Kennalestes and Zalambdalestes (Fig. 3). In these an important element of mastication was a semi-vertical shear by the high crests of the paracone and metacone, particularly against the crests of the trigonid, which was greatly elevated above the talonid. At the end of the stroke, food was crushed between the protocone and talonid, but there was no grinding function. The lingual phase of occlusion, with the hypoconid moving across the protocone, developed and became of increasing importance in the primates as the shearing function of the buccal cusps was reduced. The trigonid and talonid became more equal in height, and the talonid increased in area at the expense of the trigonid, which was compressed so that the paraconid was closely applied to the anterior face of the metaconid.

Even by the early Eocene, rodents had carried this trend further than the contemporary primates. The paracone and metacone were quite low and no longer possessed shearing crests. The buccal cingulum had disappeared except for the mesostyle. On the lower molars, the trigonid was still more compressed, and the paraconid had disappeared altogether. The metaconid, at the anterior margin of the tooth, developed an occlusal contact with the posterior cingulum of the preceding upper molar, a contact which developed within primates only in Notharctus and Anthropoidea. The protoconid-metaconid crest, primitively of major importance as a shearing crest, was obsolete except at its ends. The talonid made up most of the tooth. The oblique crest, from the hypoconid to the trigonid, which primitively sheared against the posterior side of the paracone, was represented only by the mesoconid. Thus, with the horizontal chewing movement, shearing had been eliminated in favor of grinding.

It is tempting to think, following Wood (1962), that the origin of rodents is somehow tied up with the plesiadapiform primates, many of which had enlarged incisors. Without succumbing to this tempta-

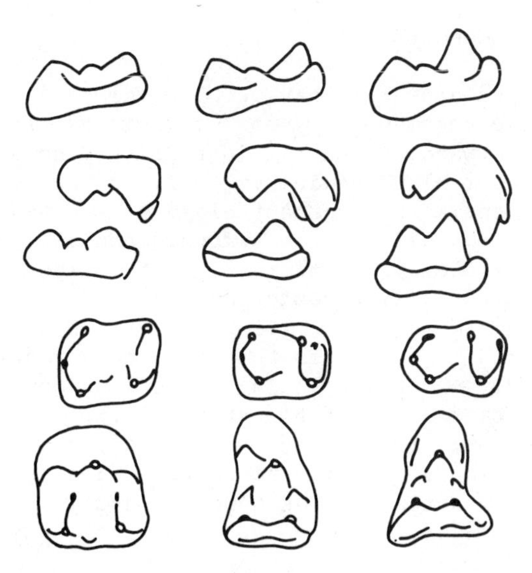

Fig. 3. Comparison of molars of (left) a Cretaceous placental, Kennalestes, (center) a Paleocene primate, Palaechthon, and (right) a primitive rodent, Paramys. Upper and lower molars in crown and buccal views, and upper molar in anterior view.

tion, I will go so far as to say that rodents and primates, though not necessarily related, advanced in the same direction, and the primates illustrate the way in which rodent molars probably evolved. If one wanted to transform a potential primate into a potential rodent, it would be necessary to consider three changes in the dentition: (1) A pair of incisors enlarged in each jaw, with loss of the remaining anterior teeth. Such a change has taken place many times in mammalian evolution. (2) To increase the force exerted by the incisors, the jaw muscles shifted forward, so that the molars came to stand behind the zygoma root and close to the coronoid process. Though this has happened to some extent in many mammals, it is particularly far advanced in rodents. (3) The relatively posterior position of the molars inhibited vertical movements of the jaw in chewing, thus preventing shearing actions, but at the same time it resulted in increased pressure between the teeth for grinding. In primates, the last molars, which stand nearest to the muscles, have the lowest cusps, the most abbreviated trigonids, and often the largest talonids. One might also add (4) the ability to protrude the mandible to bring the incisors into operation, and retrude it when the molars are in use. This ability is possessed by primates and the wombat, and probably some other mammals. It would be preadaptive in rodents to the forwardly directed chewing movements that characterize most members of the order.

COMPARISON WITH LAGOMORPHA

The Lagomorpha appear to have evolved along a different path. Although they have continuously growing incisors, their jaw muscles have not spread forward as much as in rodents, the zygomatic root being opposite upper M1. Chewing is fully transverse, without a propalinal component. An important element is a transverse shear between the posterior wall of the trigonid and a facet along the anterior margin of the upper molar (Fig. 4F). The top of the trigonid is worn against a wide posterior cingulum on the upper molar. The trigonid leans forward, and its worn surface is continuous with that of the talonid in front, so that the cheek dentition forms a series of forwardly sloping grinding areas separated by transverse ridges, the posterior edges of the trigonids. This arrangement restricts masticatory jaw movement to a transverse direction. The lower incisors are only slightly retracted during chewing; their tips pass through a notch between upper I1 and I2 as the jaw moves transversely.

These characters are shared with the Paleocene Eurymyloidea (Fig. 4), except that some of the latter agree with rodents in having lost the second upper incisors. All eurymyloids have a conspicuous transverse facet, extending from parastyle to protocone, which shears against the posterior side of the trigonid. Compared with rodents, a trigonid is more clearly differentiated from the talonid, and it occupies nearly half of the crown area. There is a strong protoconid-metaconid crest, obsolete in rodents. The talonid has a high hypoconid, comparable in height to the metaconid, and there is a strong oblique crest joining the hypoconid with the middle of the trigonid wall. An elevation on the crest represents the mesoconid. The hypoconulid is large, and it is continued as a lingual ridge or cingulum behind the entoconid. A transverse crest from the entoconid runs towards the hypoconid. There is no evidence of an oblique lingual phase of occlusion. At the end of the chewing stroke food would be compressed between the protocone and the steeply inclined lingual face of the hypoconid and oblique crest, but there is no indication that grinding occurred by passage of the hypoconid across the top of the protocone, such as characterizes rodents and primates.

Mimotona is regarded by McKenna (1982) as a primitive lagomorph, and he may well be right. However, I cannot understand his interpretation of the lagomorph lower molar. He says (p. 220) "The talonid basin of the lower molar has...shifted buccally to become a mere hypoflexid". The lagomorph talonid is very much like that of *Mimotona* (Fig. 4), the main difference being that the entoconid-hypoconid crest is complete in lagomorphs (also, see Tobien, 1974, Figs. 15,77). With regard to the upper molar pattern, McKenna has in my opinion been led astray by comparing lagomorphs with Anagalidae such as *Hsiuannania* rather than with Eurymylidae. The central V is not the protocone but the metacone, as Bohlin (1942) and Wood (1957)

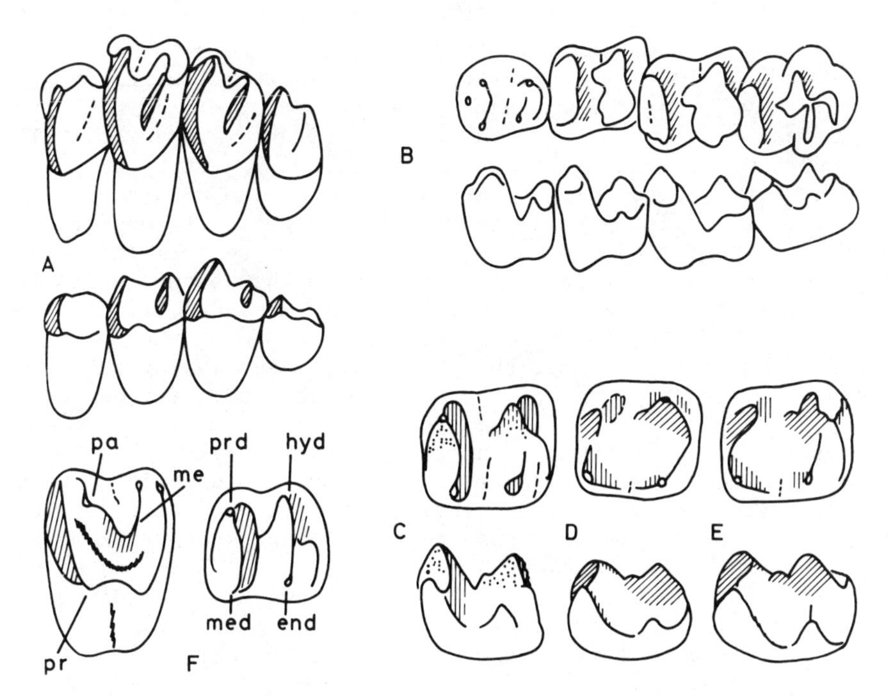

Fig. 4. A, B, left upper and right lower P4-M3 of Eurymylus, in crown and semi-lingual views. C, D, E, right lower molars of (C) Mimotona (eurymyloid), (D) Cocomys, and (E) Tamquammys (ctenodactylid). F, interpretation of cusp homologies of Lagomorpha.

postulated, and Lopez Martinez (this volume) has confirmed from the occlusal relations. The V-shaped synclinal fold is the trigon basin. On upper P3 the metacone has not differentiated from the paracone, as in upper P4 of Eurymylus. The least worn teeth of Eurymylus show the presence of a stylar cusp posterobuccally to the metacone, which also occurs in lagomorphs.

Heomys has been interpreted by Li (1977, this volume) as transitional between eurymyloids and rodents. Its upper molars are less transverse than in Eurymylus, they are lower crowned, and the metaconule is more distinct. These are primitive characters, not approaches to rodents. In other respects Heomys seems to be typically eurymyloid. It has a strong posterior trigonid wall, its talonid structure is eurymyloid, its zygomatic arch arises above upper M1, and there is no indication of oblique jaw movement.

The oldest true rodent in Asia seems to be Cocomys linchaensis (Dawson et al., 1984). Its molars (Fig. 4D) agree very closely with Paramys; they have the same facets, including large ones produced in an oblique lingual phase. It shows that rodents with paramyid-like molars were present in the Early Eocene of Asia, just as they were in North American and Europe. To get a Middle Eocene ctenodactylid from such a form would require the forward displacement of the entoconid, and the development of a transverse ridge, running from the entoconid towards the hypoconid (Fig. 4E). This change has occurred in several rodent families, e.g. Sciuravidae, Theridomyidae, Cricetidae, where it is associated with enlargement of the hypocone on the upper molar. The result has a superficial resemblance to a eurymylid, though the structure of the trigonid and the mode of wear are quite different.

Korth's (1984) argument that the Reithroparamyinae are the most primitive North American rodents is based on the supposition of an ancestry near to Heomys. In several characters, however, Paramyinae are nearer than Reithroparamyinae to the probable eutherian morphotype: hypocone weakly developed on the molars and absent on upper P4; entoconid connected with hypoconulid; hypoconulid relatively small; zygoma root less anterior in relation to upper P4; premaxilla proportionately smaller, not extending posteriorly as far as the nasals; bulla not co-ossified to the skull.

EVOLUTIONARY TRENDS IN RODENT MOLARS

Molar evolution within the rodents shows a high degree of parallelism, as has long been recognized (Wood, 1937). Changes such as hypocone enlargement, separation of the entoconid from the posterior cingulum, and development of the lingual sinus, the mesoloph and the mesolophid, have undoubtedly taken place several times independently in different phyletic lines, reducing the value of such characters in phylogenetic reconstruction. Rodent molar evolution is largely a matter of trends and grades, rather than of the sharp alternatives that are amenable to cladistic analysis. There are indeed some groups that have advanced along distinctive lines. Thus, the Aplodontoidea emphasized the transverse component of primitive rodent chewing, and developed longitudinal crests on the molars; the Gliridae have transverse crests which function as files in propalinal chewing, and which may not be homologous with the crests of other rodents. The Geomyoidea, also with propalinal chewing, developed an additional series of cusps. However, the majority of rodent groups seem to have followed a common trend of evolution, in which four grades can be distinguished (Fig. 5).

Primitively (grade A) the molar crowns are basined, with marginal cusps, as in paramyids. Chewing is of the two-phase type. In the buccal phase the movement is medial and upwards, so that the hypoconid passes into the trigon basin, and the talonid basin

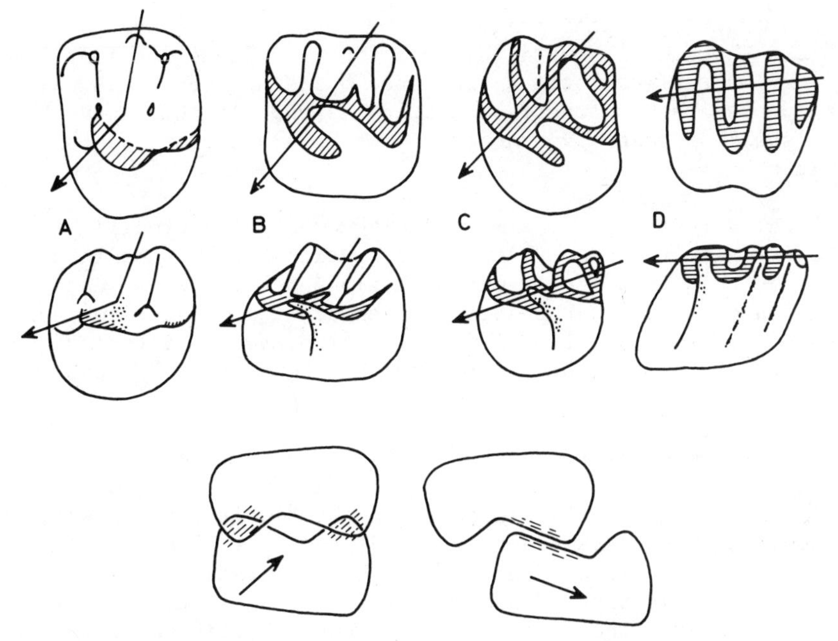

Fig. 5. Grades in the evolution of rodent upper molars. Left molars in crown and semi-lingual views. Arrows show the path of the hypoconid. A, sciurid; B, cricetodontid; C, *Erethizon*; D, *Dinomys*. Below, posterior view of teeth at grade B in buccal and lingual phases of occlusion, to show the formation of wear facets.

receives the protocone. In the lingual phase the movement turns obliquely forward and downward, and the hypoconid passes across the protocone. Lingual-phase wear is on the tops of the lower buccal cusps and upper lingual cusps. Sciuridae remain at this grade, but most families had advanced beyond it by the Oligocene.

In grade B, the basin is occupied by folds which wear at the top in the lingual phase to form a flat grinding surface. The paracone and metacone on the upper teeth, and the metaconid and entoconid on the lower teeth, stand up from the plane of the grinding area, and bear inclined facets on their anterior and posterior surfaces, produced in the buccal phase. The buccal phase movement is more oblique than in grade A, tending to line up with the lingual phase movement. There is, however, still a change of direction in the vertical plane; the lower teeth rise in the buccal phase and fall in the lingual phase. *Peromyscus* and *Sicista* are examples of living rodents in this grade.

By reduction in the height of the upstanding cusps, grade C is reached. Chewing is simplified to a single oblique movement, and the entire crown surface forms a flattened grinding area. Lingual and buccal phases can no longer be distinguished. At this grade the teeth usually become hypsodont. Examples are Castor, Hystrix and Erethizon.

Finally (grade D), the movement becomes fully propalinal, and the crests are aligned in a transverse direction, as in Anomalurus and Cavia. At this grade individual relations between upper and lower elements are lost, and the entire cheek dentition functions as a unit. Consequently, when additional folds develop, as in Microtinae and Hydrochoerus, they do not form on corresponding teeth of the two jaws.

The existence of rodents at each of these grades at the present day shows that the rate of advance differed from group to group. This is confirmed by paleontology. In the Middle Eocene, sciuravids and pseudosciurids advanced to grade B, and by the end of the Eocene, theridomyids and cylindrodontids had reached grade C. In the Oligocene eomyids, cricetids, dipodids, ctenodactylids and phiomyids were still at grade B. The Oligocene Caviomorpha, unlike African hystricognaths, were already at grade C. One might guess that the trend was toward increased efficiency in grinding hard or abrasive foods, and that selection pressure in this direction depended upon the nature of the diet.

HOMOLOGY OF CRESTS

Crests (also called ridges or lophs) develop as folds of epithelium, and are easily affected by small differences in the distribution of growth within the tooth. They therefore vary individually and may differ from one tooth to another within a dentition. Yet many crests have characteristic functions in occlusion, and it is possible to trace their evolution.

Instead of thinking of molars as an array of cusps, they can be considered as a series of basins and grooves that occlude with the cusps of the opposing teeth (Fig. 6). Thus in a primitive rodent the talonid basin might be called an anti-protocone. Between the anti-protocones are anti-hypocones, formed partly from one tooth and partly from the next. They are small in primitive rodents, but later they become similar in size and structure to the anti-protocones. The anti-paracone and anti-metacone are grooves on the buccal side of the lower teeth. In the same way the upper teeth have anti-hypoconid (trigon basin), anti-protoconid (formed from the anterior and posterior cingula of adjacent teeth), and lingual grooves for the metaconid and entoconid.

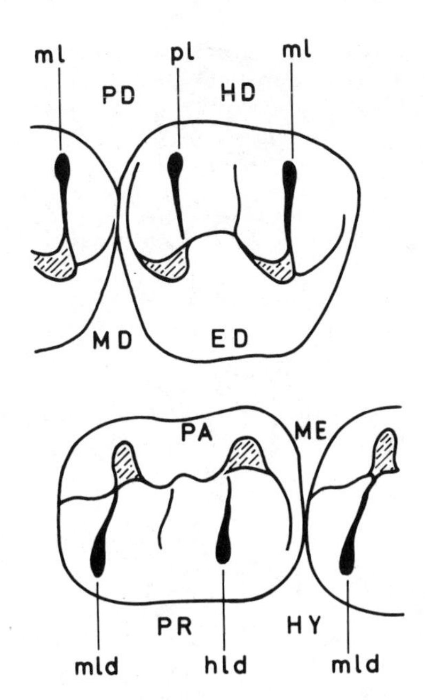

Fig. 6. Upper and lower molars to show ridges and basins. Upper molar: ED, anti-entoconid; HD, anti-hypoconid; MD, anti-metaconid; ml, metaloph; PD, anti-protoconid; pl, protoloph. Lower molar: hld, hypolophid; HY, anti-hypocone; ME, anti-metacone; mld, metalophid; PA, anti-paracone; PR, anti-protocone.

These basins and grooves can be recognized in all teeth that retain cusps, i.e., those at grades A and B (Fig. 7). Their homology derives from the homology of the cusps that work in them. It follows that the crests that separate homologous basins are themselves homologous.

On the trigonid of a primitive eutherian mammal (Fig. 2), there is a transverse crest from protoconid to metaconid, called the protolophid, and a more anterior crest from protoconid to paraconid, called the paralophid. The paralophid forms the posterior edge of the groove in which the metacone works (anti-metacone). It is equivalent to the buccal part of the anterolophid of rodents. The protolophid of primitive eutherians has a dual function. Its buccal portion, on the protoconid, forms the anterior edge of the anti-paracone groove, and its lingual portion, on the metaconid, is the anterior edge of the anti-protocone basin. In rodents these two portions are partly or completely separated, and the lingual portion

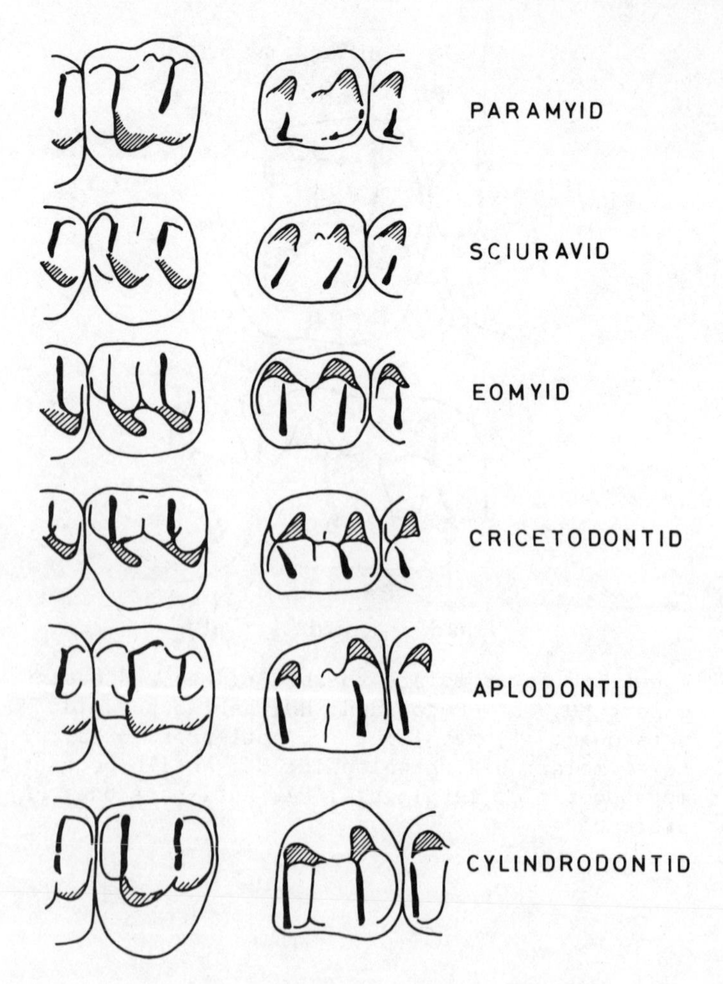

Fig. 7. Crest morphology of a number of typical rodents.

is known as the metalophid. It separates the anti-protocone basin from the anti-hypocone basin.

In some rodents the metalophid approaches the anterior margin of the tooth, where it has joined up with the anterolophid. This is the situation in Cylindrodontidae (Fig. 7), Aplodontoidea, Ctenodactylidae, Thryonomyoidea and Caviomorpha. The resulting crest from protoconid to metaconid has been mistaken for the anterior crest of the primitive trigonid, but this is not correct. The metalophid of rodents has no constant connection with the protoconid, and its varying relations to that cusp do not affect its homology. The so-called metalophulid-II of Cylindrodontidae and Caviomorpha is not homologous with the posterior trigonid crest, as it forms within the anti-protocone basin on the posterior side of the metalophid.

The posterior edge of the anti-protocone basin is formed by the hypolophid, which extends buccally from the entoconid. It divides the anti-protocone from the more posterior anti-hypocone, and its development is associated with the differentiation of the hypocone. Its relation to the crests of the hypoconid tends to parallel the relation of the metalophid to the crests of the protoconid, and again variation in this respect does not affect its homology.

The lingual sinus of the upper molar accommodates an enlarged entoconid. The primitive posterior crest of the protocone is displaced toward the center of the tooth to form the mure. Similarly, in the lower molar the buccal sinus is an anti-paracone, and the mure is the lingually displaced posterior crest of the protoconid.

Crests that develop within basins are less constant than those that form the edges of basins, and their homologies can seldom be traced above low taxonomic levels. Their function seems to be to improve the contact between the cusp and the basin in which it works. This function was performed in many primitive rodents by irregular wrinkles or crenulations, and the more definite ridges of other rodents may have developed by a process of simplification. Short ridges, running across the direction of motion, are common; for example, a link between the metaloph and the posteroloph in theridomyids, and the anterior extension of the paraconule of aplodontids. Other ridges run along the bottom of basins in the direction of motion; such are the mesoloph and mesolophid in the anti-hypoconid and anti-protocone basins respectively. These seem more likely to develop when the basin is broad and shallow. When the basin gets deeper and the top of the cusp does not penetrate to its floor, the ridges either disappear or they become more elevated to maintain contact with the cusp. They may be present only at the shallower end of the basin, near the mure. Some groups, such as Cylindrodontidae and Ctenodactylidae, did not develop a mesoloph; perhaps their anti-hypoconid basin deepened precociously.

Once stage C has been reached, and the crown is flat, it is difficult to know whether a loph was originally in the floor of a basin or at its edge. This is the reason for the controversy between Lavocat (1974) and Wood (1974) over the presence of a mesoloph in Caviomorpha (Fig. 8). Patterson & Wood (1982) have presented a strong case that caviomorphs primitively had only four lophs, but difficulties remain. Why, for instance, are the paracone and metacone closer together than the protoconid and hypoconid, which in other rodents alternate with them? Also, in *Branisamys*, which has five lophs, Patterson and Wood think that the "neoloph" is the fourth, but on upper P4 the third loph is the last to come into wear. The problem will be resolved only when the position of the metacone is

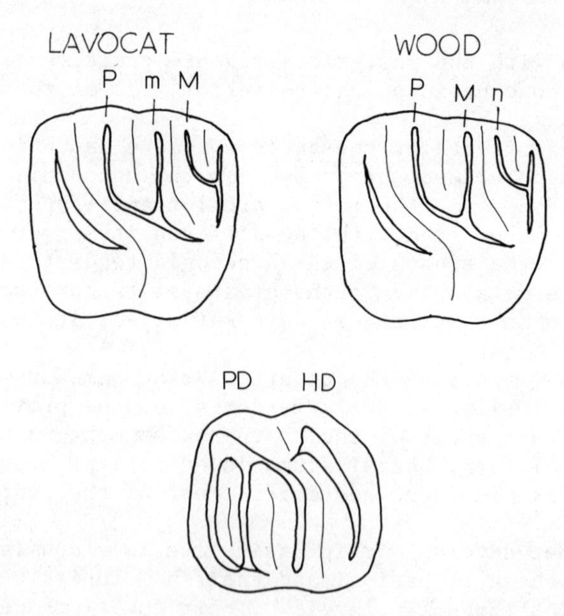

Fig. 8. Above, upper first molars of an Oligocene caviomorph (Branisamys) to illustrate the different interpretations of the lophs by Lavocat and Wood. P, protoloph; M, metaloph; m, mesoloph; n, neoloph. Below, lower first molar, to show the positions of the protoconid (PD) and hypoconid (HD). Redrawn after Patterson and Wood (1982).

established by the discovery of a primitive form that still has traces of cusps.

CUSPS HOMOLOGIES IN MURIDAE

The Muridae are anomalous in that they combine cuspidate molars with propalinal chewing. Their cusp pattern departs markedly from those of other rodents, requiring a special nomenclature (Miller, 1912). The upper molars have three rows of cusps arranged in transverse lophs or chevrons. It is generally agreed that the two outer (buccal) rows are homologous with the buccal and lingual cusps of Cricetidae, and the inner (lingual) row is new development. Comparison with Cricetidae shows that cusp 6 is the paracone (Fig. 9) and cusp 5 is the protocone (Schaub, 1938). In the lower molars there are two rows of cusps as in Cricetidae, but many murids have an additional buccal row.

The occlusal relations are in accordance with this interpretation. The basic difference lies in the direction of the chewing movement: in Cricetidae it is oblique, having a significant transverse component; in Muridae it is fully propalinal, the lower cusps

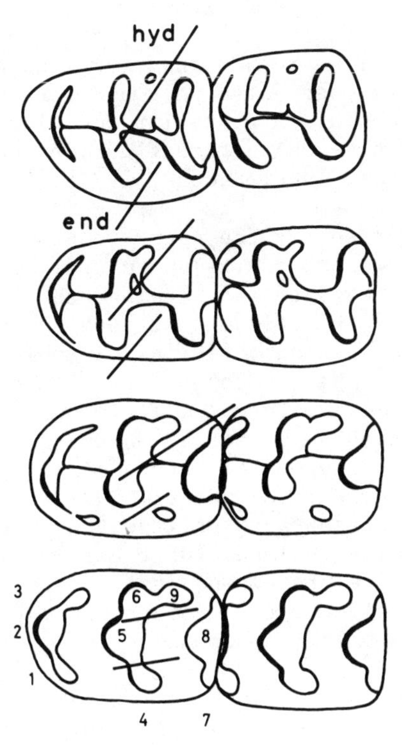

Fig. 9. Hypothetical stages in the transformation of upper M 1-2 from a cricetodontid (above) to a murid (below). Straight lines indicate paths of hypoconid and entoconid.

moving in longitudinal grooves between the upper cusps (Fig. 9).

One could envisage the change in the following manner: In the cricetid ancestor of murids the hypoconid passed anterolingually between the paracone and the metacone, but owing to the obliquity of its movement, its contact with the paracone faced partly lingually, instead of more directly posteriorly as in primitive rodents. At the same time the entoconid occluded with the posterolingual surface of the protocone. Continuing its movement, the hypoconid would pass obliquely across the protocone. If the movement now became more longitudinal, the facet on the paracone would move round to the lingual side of that cusp, and eventually the hypoconid would be travelling directly forward between the paracone and the protocone. Likewise, the entoconid would pass along the lingual side of the protocone.

In murids the entoconid moves along a groove between the protocone and the additional cusp 4 (Fig. 9). If cusp 4 developed before chewing became fully propalinal, the groove for the entoconid would at first be oblique rather than longitudinal, and cusps 4 would stand more posteriorly, so as to wear against the posterolingual surface of the entoconid. A posterior position of cusp 4 would therefore be a primitive character of murids, and this indeed seems to be the case (Petter, 1966; Jacobs, 1977).

At the posterior end of the upper molar, cusp 8 occludes with the protoconid of lower M2 and is undoubtedly the hypocone. However, cusp 9 is too far forward to wear against the protoconid; instead it supplements cusp 6 in occluding with the hypoconid of lower M1. This is obvious in stephanodont murids, in which cusps 6 and 9 are joined by a crest and share a common wear facet. Cusp 9 therefore does not function like a metacone. It is much like the mesostyle of some Hesperomyinae such as _Oryzomys_, which supplements the paracone in wearing against the hypoconid, and a similar structure occurs in the myocricetodontine, _Zramys_ (Jaeger, 1977). If cusp 9 is a mesostyle, it would be necessary to believe that the metacone of murids has been reduced, being replaced functionally by cusp 3 of upper M2. The chevron would therefore extend partly onto upper M2. In the same way, at the lingual side of the tooth, cusp 7 tends to disappear and to be replaced by cusp 1 of upper M2.

Antemus (Jacobs, 1977) does not fit this interpretation. From the position of cusp 4, chewing was probably oblique, but the wear facets have not been studied. Cusp 9 seems to be in a position to wear on its postero-lingual surface against the protoconid of the more posterior lower molar, like a metacone, and there is no mesostyle. If stephanodont murids were derived from a form like _Antemus_ the occlusal relations of cusp 9 must have changed. The origin and functional significance of stephanodonty need further investigation.

The evolution of propalinal chewing before the loss of cusps also occurred in the Zapodidae, where the Sicistinae chew obliquely like cricetids and the Zapodinae chew longitudinally, but in this case the third row of cusps did not develop. The Geomyoidea provide a third case, but their mode of origin, perhaps from Eomyidae (Harri and Wood, 1969; Fahlbusch, this volume) has not been investigated from a functional point of view.

CONCLUSION

Cusps and crests are not merely useful recognition marks, but structures that function in the living animal. When this is taken into account, some evolutionary modifications of rodent molars become more understandable. The primitive contacts between cusps in occlusion tend to persist through changes of molar pattern, and thus they provide additional data for the determination of homologies. Changes in the mode of chewing, from the primitive two-phase method to oblique and finally propalinal movement, have been important factors in rodent evolution. These changes involve modifications of the jaw muscles, and it should be possible to link the evolution of these muscles with that of the teeth.

Owing to the prevalence of parallel evolution, it is necessary to exercise caution when deducing phylogeny from the teeth alone. Comparative functional morphology tests the feasibility of phylogenetic hypotheses, but it does not indicate whether the postulated changes have occurred more than once. Other characters, preferably those unrelated to the feeding mechanism, need to be taken into account.

REFERENCES

Bohlin, B. 1942. The fossil mammals from the Tertiary deposit of Tuben-buluk, Western Kansu. Part I. Insectivora and Lagomorpha. Palaeont. sinica (n.s.) C, 8a: 1-110.

Butler, P. M. 1956. The ontogeny of molar pattern. Biol. Rev. 31: 30-70.

Butler, P. M. 1973. Molar wear facets of early Tertiary North American primates. Symp. 4th. int. Cong. Primatol. 3: 1-27.

Butler, P. M. 1980. Functional aspects of the evolution of rodent molars. Palaeovertebrata Mém. Jubil. R. Lavocat: 249-262.

Butler, P. M. 1982. Directions of evolution in the mammalian dentition. In: Problems of Phylogenetic Reconstruction, K. A. Joysey and A. E. Friday, eds., pp. 235-244, Academic Press. London.

Butler, P. M. 1983. Evolution and mammalian dental morphology. J. Biol. buccale 11: 285-302.

Gaunt, W. A. 1955. The development of the molar pattern of the mouse, *Mus musculus*. Acta anat. 24: 249-268.

Gaunt, W. A. 1961a. The presence of apical pits on the lower cheek teeth of the mouse. Acta anat. 44: 146-158.

Gaunt, W. A. 1961b. The development of the molar pattern of the golden hamster (*Mesocricetus auratus* W.), together with a re-assessment of the molar pattern of the mouse (*Mus musculus*). Acta anat. 45: 219-251.

Harman, M. T. and Smith, A. 1936. Some observations on the development of the teeth of *Cavia cobaya*. Anat. Rec. 66: 99-106.

Harris, M. T. and Wood, A. E. 1969. A new genus of eomyid rodent from the Oligocene Ash Spring local fauna of Trans-Pecos Texas. Pearce-Sellards Ser. Tex. mem. Mus. 14: 1-7.

Jacobs, L. L. 1977. A new genus of murid rodent from the Miocene of Pakistan and comments on the origin of the Muridae. Paleobios 25: 1-11.

Jaeger, J. J. 1977. Les rongeurs du Miocène moyen et supérieur du Maghreb. Palaeovertebrata 8: 1-166.

Kay, R. F. and Hiiemae, K. M. 1974. Jaw movement and tooth use in recent and fossil primates. Am. J. Phys. Anthrop. 40: 227-256.

Korth, W. W. 1984. Earliest Tertiary evolution and radiation of rodents in North America. Bull. Carneg. Mus. Nat. Hist. 24: 1-71.

Lavocat, R. 1974. What is an hystricomorph? Symp. Zool. Soc. Lond. 34: 7-20.

Li, C. K. 1977. Paleocene eurymyloids (Anagalida, Mammalia) of Qianshan, Anhui. Vert. Palasiat. 15: 103-118.

McKenna, M. C. 1982. Lagomorph interrelationships. Geobios, mém. spec. 6: 213-223.

Miller, E. S. 1912. Catalogue of the Mammals of Western Europe in the Collection of the British Museum. British Museum, London.

Monmignaut, C. 1963. Étude d'une anomalie de l'email des dents jugales chez des rongeurs nouveau-nés. Mammalia 27: 218-237.

Patterson, B. and Wood, A. E. 1982. Rodents from the Deseadean Oligocene of Bolivia and the relationships of the Caviomorpha. Bull. Mus. Comp. Zool. Harv. 149: 371-543.

Petter, F. 1966. L'origine des muridés. Plan cricétin et plans murins. Mammalia 30: 205-225.

Rensberger, J. M. 1973. An occlusion model for mastication and dental wear in herbivorous mammals. J. Paleont. 47: 515-528.

Rensberger, J. M. 1975. Function in the cheek tooth evolution of some hypsodont geomyoid rodents. J. Paleont. 49: 10-22.

Rensberger, J. M. 1978. Scanning electron microscopy of wear occlusal events in some small herbivores. In: Development, Function and Evolution of Teeth, P. M. Butler and K. A. Joysey, eds., pp. 415-438, Academic Press, London.

Rensberger, J. M. 1982. Patterns of dental change in two locally persistent successions of fossil aplodontid rodents. In: Teeth Form, Function and Evolution, B. Kurtén, ed., pp. 333-349, Columbia University Press, New York.

Santoné, P. 1935. Studien über den Aufbau, die Struktur und die Histogenese der Molaren der Saügetiere. I. Molaren von _Cavia cobaya_. Z. mikr. anat. Forsch. 37: 49-100.

Schaub, S. 1938. Tertiäre und quartäre Murinae. Abh. Schweiz. Pal. Ges. 61: 1-38.

Stehlin, H. G. and Schaub, S. 1951. Die Trigonodontie der simplicidentaten Nager. Schweiz. Pal. Abh. 67: 1-385.

Tims, H. W. M. 1901. Tooth genesis in the Caviidae. J. Linn. Soc. (Zool.) 28: 261-290.

Tobien, H. 1974. Zur Gebissstruktur, Systematik und Evolution der Genera _Amphilagus_ und _Titanomys_ (Lagomorpha, Mammalia) aus einigen Vorkommen im jüngeren Tertiär Mittel- und Westeuropas. Mainzer geomiss. Mitt. 3: 95-214.

Wilkins, K. T. and Woods, C. A. 1983. Modes of mastication in pocket gophers. J. Mammal. 64: 636-641.

Winge, H. 1941. The Interrelationships of the Mammalian Genera. Vol. 2. C. A. Reitzels Forlag, Copenhagen.

Wood, A. E. 1937. Parallel radiation among the geomyoid rodents. J. Mammal. 18: 171-176.

Wood, A. E. 1957. What, if anything, is a rabbit? Evolution 11: 417-425.

Wood, A. E. 1962. The early Tertiary rodents of the family Paramyidae. Trans. Am. Phil. Soc. 62: 1-261.

Wood, A. E. 1974. The evolution of the Old World and New World hystricomorphs. Symp. zool. Soc. London. 34: 21-60.

Woods, C. A. and Howland, E. B. 1979. Adaptive radiation of capromyid rodents: anatomy of the masticatory apparatus. J. Mammal. 60: 95-116.

Wortman, J. L. 1902. Studies of Eocene Mammalia in the Marsh Collection, Peabody Museum. Am. J. Sci. 13: 39-46.

EVOLUTIONARY TRENDS IN THE ENAMEL OF RODENT INCISORS

Wighart von Koenigswald

Hessisches Landesmuseum
Friedensplatz 1
D-6100 Darmstadt, West Germany

INTRODUCTION

Mammalian teeth are preferred objects of study in vertebrate paleontology. Their complicated morphology in most cases offers a specific determination, as well as many arguments to distinguish phylogenetic relationships. In this respect, rodent incisors are less interesting, because their morphology is fairly simple, due to their continuous growth. Only a few taxa can be identified by the characteristic cross-section of their incisors, or by grooves or crests on the enamel surface. However, the thin enamel, which covers only the buccal side of the tooth, contains quite a number of characters in its internal structure. Studies of these structures extend back to Tomes (1850), and were continued by Korvenkontio (1934) and Wahlert (1968). Several different structures within the enamel were observed and used for systematics, but very little was said about phylogenetic interrelationships, and the trends of the enamel evolution remained unknown.

For rodents, the incisors are important tools, with uses ranging from processing food to digging dens. Therefore, the approach of this study is based on functional aspects of the enamel structure. This point of view facilitates a comparison with enlarged or gliriform incisors that have developed in several other mammalian groups besides rodents. Some of these are classical examples of parallel or convergent evolution, such as the Malagasy aye-aye, *Daubentonia*, or the widespread Lagomorpha. This comparison of rodent incisors with similar structures in other taxa allows one to distinguish which level of enamel differentiation can be obtained by most mammals, and in which direction rodents could reach higher specializations.

Furthermore, this broad comparison helps to determine the primitive, as well as the derived, stages of a character sequence. This is of great importance, because rodent incisors are already highly differentiated when this group first occurs in the fossil record, and no primitive ancestors are known. The determination of primitive and derived characters in the enamel indicates phylogenetic relationships which, to some extent, contradict the classical systematic scheme, especially for the Hystricognathi.

MATERIALS AND METHODS

Enamel can be studied at various levels of magnification. Very high magnification is needed to determine the shape of the cross section of enamel prisms. With lower magnifications of about 300 to 1000 times, the interlacing of the prisms becomes observable, and the enamel types that form a compound structure, the enamel pattern (Schmelzmuster), can be observed. In this paper, the enamel types and their arrangement into enamel patterns are of most relevance.

These structures can be examined under the scanning electron microscope (SEM), or with the reflecting light microscope. The specimens normally were embedded in artificial resin and sectioned in various planes. After grinding, a brief etching with 2 N hydrochloric acid for two to three seconds makes the internal structure of the enamel visible. For samples of Recent teeth, a brief treatment with hydrogen peroxide improves the quality of observation.

This study is part of an extensive survey of mammalian tooth enamel. Of special interest here are gliriform incisors, which can be defined as evergrowing teeth, covered with enamel only on their buccal surface. The number of incisors is greatly reduced. In most cases, upper and lower incisors work as antagonists and form a self-sharpening device, particularly because the softer dentin is removed more easily than the more resistant enamel.

Moeller (1974) described rodent-like incisors in the extant mammalian genera *Vombatus* and *Lasiorhinus* (Marsupialia), *Procavia* (Hyracoidea), *Daubentonia* (Primates), and in Lagomorpha. This list can be extended by several fossil taxa: *Diprotodon* (Marsupialia), *Myotragus* (Bovidae), *Eurymylus* and *Rhombomylus* (Mixodontia). The Pleistocene bovid from the Baleares, *Myotragus*, is somewhat exceptional, because its lower incisors are evergrowing, even though there are no upper antagonists, as is usual in ruminants. However, the teeth are abraded and resharpened by the horny plate of the premaxilla. Incisors of all these genera, except *Eurymylus*, were studied for comparison with many Recent and fossil rodents from all three classic suborders.

THE NUMBER OF LAYERS IN THE INCISOR ENAMEL

The number of layers in the incisor enamel has been used to distinguish Lagomorpha from Rodentia. Tomes (1850) observed that only one layer can be seen in lagomorph incisors, whereas rodents have two. Korvenkontio (1934) called the inner layer of rodent incisor enamel, which contains decussating Hunter-Schreger bands (HSB), the portion interna, while the portio externa is formed by radial enamel. Some students have assumed that a one-layered stage should be less derived than a two-layered stage, but a general survey of mammalian enamel indicates that simply determining the number of layers is insufficient to characterize the enamel.

The most primitive type of enamel in mammalian teeth is radial enamel (Koenigswald, 1980), in which all prisms run parallel and radially toward the outer surface of the enamel. The prisms may be straight or slightly curved, but they never cross over each other. When radial enamel occurs exclusively, the enamel pattern (Schmelzmuster) is single-layered and can be regarded as primitive. In late Mesozoic mammals, radial enamel formed the entire enamel thickness, and radial enamel is still dominant in insectivorans and chiropterans.

Most mammals with gliriform incisors have developed a much more differentiated enamel pattern, in which radial enamel may have disappeared completely, as in lagomorphs, or else it forms only a thin layer, as in many rodents. Only in _Procavia_ was radial enamel found throughout the thickness of the incisor enamel in the present study.

Radial enamel is found only in small teeth usually. When tooth size is increased during evolution, a more complex internal structure is needed for strengthening the enamel. In mammalian teeth, there is only one basic pattern to fulfill this need, the Hunter-Schreger bands (HSB). Hunter-Schreger bands are layers of prisms which decussate, so that in a longitudinal section bands are visible, where prisms are truncated under different angles. The angle between prisms of adjacent layers may rise to 90°.

Within a homogeneous layer of radial enamel, HSB tend to occur first in the center, from where the HSB expand to either or both sides. This can be shown in the early Paleocene arctocyonids (Koenigswald and Rensberger, in prep.), as well as in molars of some arvicolid rodents (Koenigswald, 1980). Therefore, this sequence of appearance is accepted as a general rule. The occurrence of HSB increases the number of layers visible in a longitudinal section abruptly from one to three (Fig. 1). In the gliriform incisors of _Vombatus_, the HSB occupy most of the thickness of the enamel, but an inner radial enamel is clearly preserved, as well as an outer layer (Beier, 1981).

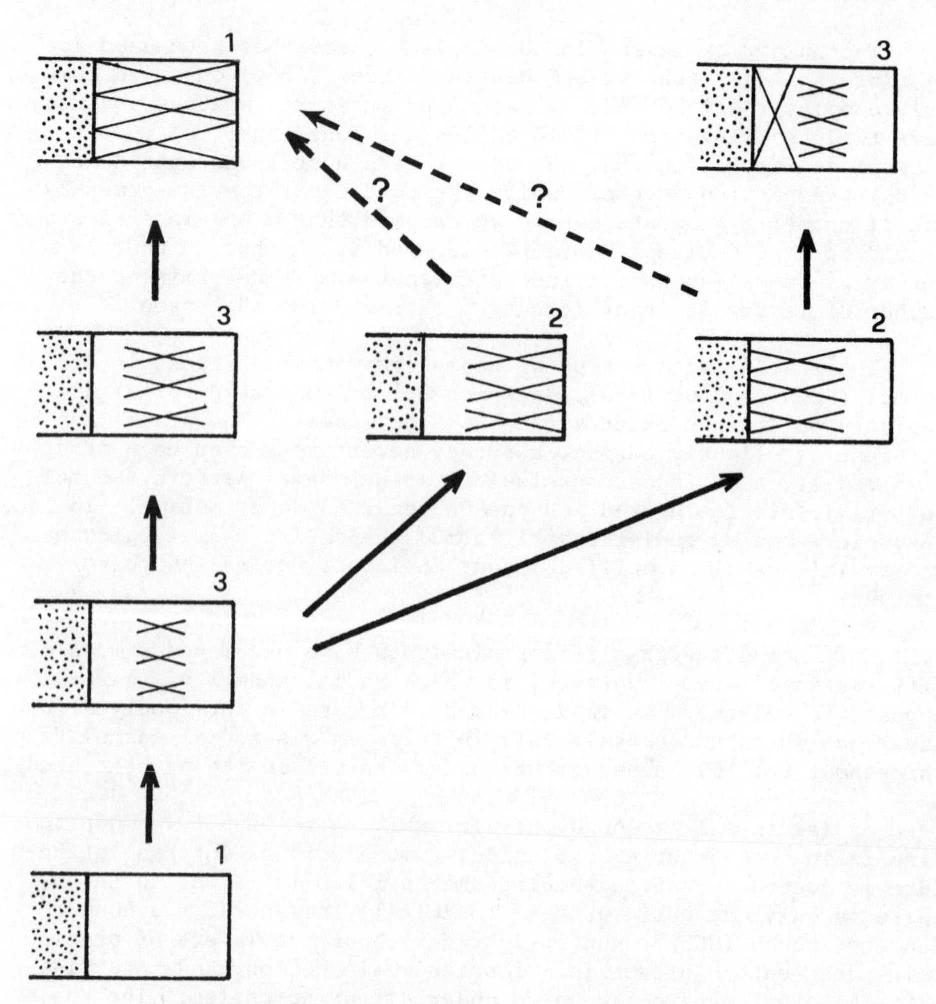

Fig. 1. Hypothesis of evolutionary expansion of Hunter-Schreger bands and its role in determining the number of enamel layers in gliriform incisors in mammals. Dentin, stippled; radial enamel, white; Hunter-Schreger bands, crossed lines. Numbers indicate number of enamel layers present.

During further evolution, three different pathways may be followed (Fig. 1). (1) The HSB expand toward the dentin and suppress the inner radial enamel. (2) The HSB expand toward the outer surface of the enamel and eliminate the outer radial enamel. (3) The HSB expand to both sides and finally eliminate all radial enamel. Therefore, a single layer of radial enamel may represent the primitive mammalian condition, whereas a single enamel layer containing

HSB, such as occurs in lagomorphs, is a highly derived pattern or Schmelzmuster.

Rodent incisors are generally regarded as having two layers of enamel. According to the first pathway described above, the inner radial enamel was most probably reduced by the HSB forming the portio interna. The outer radial enamel, the portio externa, varies in thickness. That this stage is most probably preceded by the three-layered stage, as indicated in the above model, can be supported directly by the fact that in some taxa, especially in the caviomorph *Chinchilla* (Fig. 2) and in *Ctenodactylus*, some inner radial enamel can be observed close to the dentin-enamel junction.

Further support for this evolutionary model is found in the enamel pattern of arvicolid molars, which can be regarded as a classical example of parallel evolution. Their molars are composed of dentin triangles surrounded by enamel, the leading edge of which displays a Schmelzmuster or enamel pattern nearly identical to that of muroid incisors. The inner part of the enamel is occupied by HSB, while the outer layer is radial enamel. This enamel pattern is widely distributed, for instance, in recent and fossil *Microtus*, *Arvicola*, *Lemmus*, and *Dicrostonyx*. However, in contrast to rodent incisors, where the HSB Schmelzmuster was derived early, in arvicolid molars the fossil record shows a step by step development of the enamel pattern. In early species of *Mimomys*, the molar enamel band consists only of radial enamel. At the evolutionary level of *Mimomys stehlini*, primitive HSB occur as discrete lamellar enamel, intercalated between a thick outer layer and a thin inner layer of radial enamel. In the extant arvicolids a similar stage is preserved in *Dinaromys bogdanovi*, whereas in the more recent species of *Mimomys*, such as *M. reidi*, *M. polonicus* and *M. savini*, the inner radial enamel has disappeared totally (Koenigswald, 1980). According to this model, the typical rodent incisor was preceded by an early three-layered stage.

The Schmelzmuster containing two layers is not the ultimate stage of evolution, however. Wahlert and Koenigswald (in prep.) show an unique differentiation in eomyid rodents, which increases the number of layers anew. In these, the portio interna is separated into two layers, so that the number of layers increases to three again. It is surprising that the outer radial enamel is retained in most rodents, even though it sometimes becomes very thin. The functional significance of the retention of this layer will be discussed below.

From the proposed three-layered stage with HSB in the middle, a second independent pathway of development occurred when the HSB expanded toward the outer surface of the enamel, so that the outer radial enamel disappeared. In this condition, the Schmelzmuster remains two-layered, but the radial enamel is on the inside. The

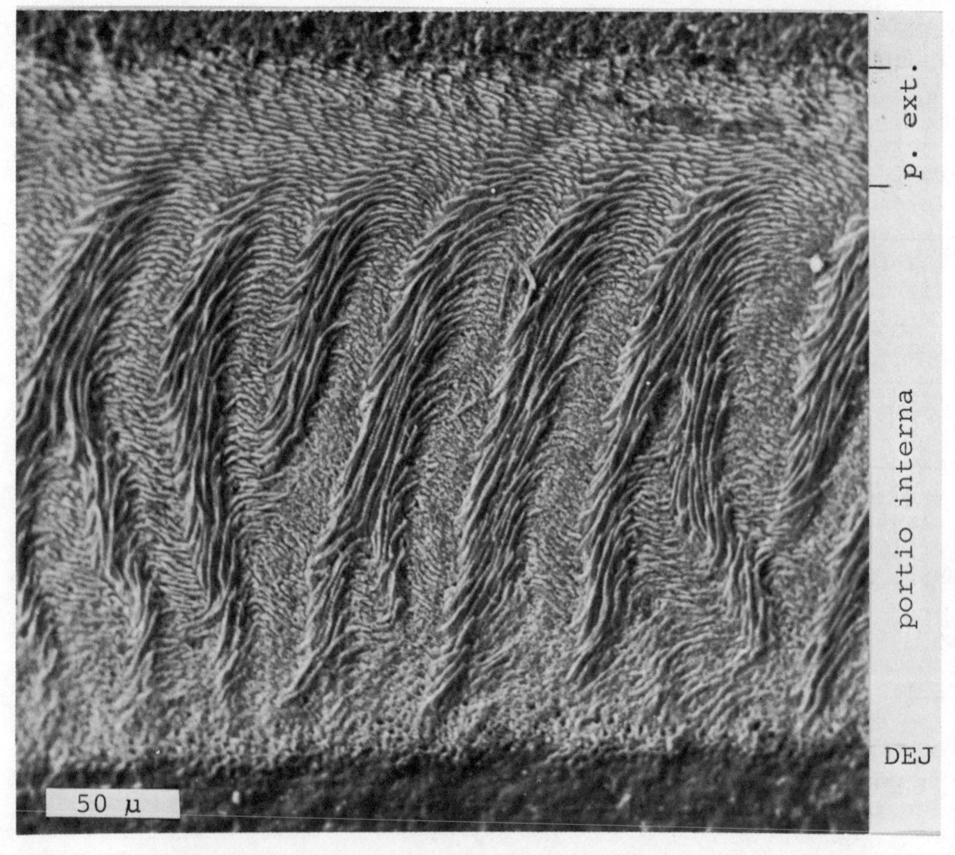

Fig. 2. Longitudinal section of incisor enamel in the caviomorph Chinchilla laniger. Multiserial Hunter-Schreger bands show their characteristic bifurcation in the portio interna. A thin layer of radial enamel is found close to the dentin-enamel junction (DEJ), as well as in the portio externa.

lower incisor of the bovid Myotragus has such an inner radial enamel, which occupies about one-fourth of the enamel thickness. The rest is filled by HSB, with no outer radial enamel. In molars of several rodents as well, such as Aplodontia, Pedetes, and some geomyids, this Schmelzmuster is present, although the inner radial enamel may vary in thickness.

A third pathway is developed when HSB expand in both directions and penetrate throughout the enamel, eliminating radial enamel from both the inner and outer sides (Fig. 1). In vombatid marsupials, Lasiorhinus HSB expand more extensively in both directions than in Vombatus, reducing the inner and outer radial enamel. Thus, this third pathway is also preceded by the three-layered stage. It is possible that expansion of the HSB is asymmetrical, so that a pattern like that in *Myotragus*, where the inner radial enamel is fairly thin, may be an evolutionary stage toward a Schmelzmuster in which HSB penetrate throughout the enamel.

In Lagomorpha the trend of expanding HSB has reached its final limits, where multiserial HSB begin immediately at the dentin-enamel junction and continue to the outer surface (Fig. 1, top). Because data from early fossil lagomorphs are lacking, it can not be proved whether both layers of radial enamel were reduced simultaneously, which seems to be most likely, or whether one layer of radial enamel was retained somewhat longer. It is unlikely that the lagomorph condition was derived from the two-layered stage as in rodents, because the borders of the HSB are more distinct in rodents, even in those with multiserial HSB. A common ancestor would have occurred much earlier.

In this context, it is very interesting that Li and Ting (this volume) indicate that the one-layered incisor enamel stage of the eurymyloid *Rhombomylus*, which closely resembles lagomorphs, is preceded by a two-layered stage in *Eurymylus*. Unfortunately, no information is yet available about which enamel types form the Schmelzmuster of *Eurymylus*. Thus, the eurymyloid enamel can not be interpreted in terms of the model presented here. It seems evident, however, that the one-layered enamel of lagomorphs is the result of a long evolutionary process and is very distinct from the initial one-layered stage formed solely by radial enamel.

In summary, it can be stated that simply identifying the number of layers in enamel is insufficient for distinguishing evolutionary relationships. The number may increase, as well as decrease, and only a determination of the enamel types justifies an evolutionary interpretation. The initial question of whether lagomorph enamel is less derived than that of the rodent incisor pattern can not be answered, because the two groups belong to different evolutionary pathways.

THE THICKNESS OF HUNTER-SCHREGER BANDS

The portio interna of rodent incisors is characterized by Hunter-Schreger bands (HSB), of which three types have been distinguished: multiserial, pauciserial, and uniserial HSB (Korvenkontio, 1934; Wahlert, 1968). The definition is based on the thickness of a single band measured in prism cross-section, but for phylogenetic interpretation further characters, such as the inclination or the orientation of the interprismatic matrix, should be considered (see below).

HSB are not restricted to rodent incisors, but are common in enlarged teeth of most mammalian orders. The first occurrence of this feature may be related to the increase in body size in Paleocene mammals (Koenigswald and Rensberger, in prep.).

The identity of HSB in mammals in general, and in rodent incisors in particular, can be demonstrated by several common features. HSB were described originally as light and dark bands, but the main feature is a decussation or crossing of layers of enamel prisms. The relationship between the occurrence of light and dark bands and the prism orientation was demonstrated by Rensberger and Koenigswald (1980). Basically, the light and dark bands can be seen in rodents as well as in other mammalian teeth, but outer radial enamel may hide this pattern. On the other hand, the reduced thickness of the HSB requires higher magnification for observation, and this makes it more difficult to illuminate the enamel tangentially.

In addition to the basic feature of decussating prisms, there is a characteristic pattern of bifurcation of the HSB. In longitudinal or tangential sections, one set of bands with identical orientation of prisms separates in one direction, while the other set of bands bifurcates exclusively in the opposite direction. This pattern can be observed in the various types of HSB in rodent incisors (Fig. 2), as well as in most other mammalian teeth.

Between adjacent HSB, zones of an intermediate prism orientation can be observed, especially in thick HSB. These zones are all related to the fact that prisms normally rise somewhat steeper than the HSB are inclined. Thus, the prisms within a band reach the upper border of a band and have to be incorporated into the next higher band by a more or less sharp turn of the prism direction (Fig. 3). In thinner HSB, such as in pauciserial and especially uniserial HSB (Fig. 4), the prisms are almost parallel to the bands, so that a change from one band to the next usually occurs only in case a band bifurcates. The character is still present regardless of the thickness of the bands, however.

In uniserial HSB (Fig. 4), the prisms frequently possess an oval cross-section, and the orientation of the long axis is somewhat inclined. In adjacent bands this inclination is in the opposite direction, so that the enamel shows a zigzag structure in tangential sections. In pauciserial, as well as in uniserial, HSB of rodents, the same undulating pattern can be observed, but at a somewhat larger scale. This kind of undulating arrangement was described first in lagomorphs (Korvenkontio, 1934), and it was thought to characterize only this taxon. However, this arrangement can occur in any HSB, but it is more obvious when the interprismatic matrix is arranged mostly as inter-row sheets. The aye-aye *Daubentonia* shows a very similar pattern, but here the HSB are separated more precisely, so that the intermediate zone between the bands is thinner.

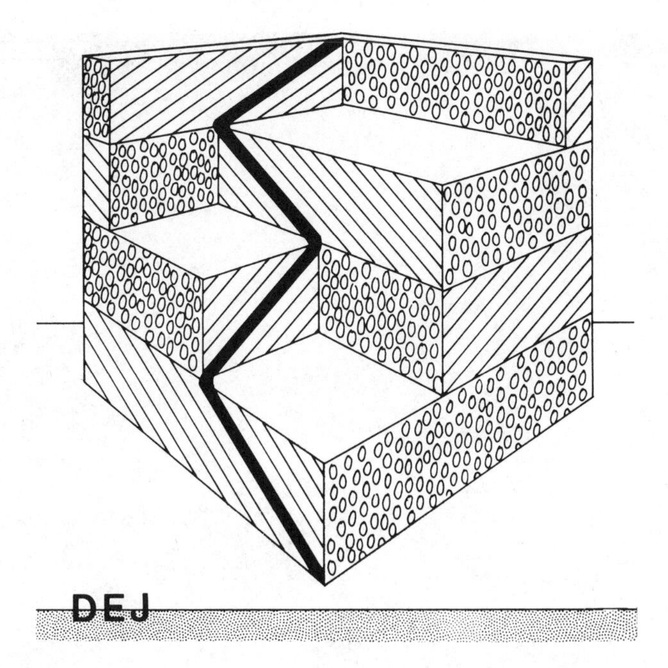

Fig. 3. Schematic course of an enamel prism passing through several HSB. Because prisms rise somewhat steeper normally than the HSB are inclined, prisms have to change direction when they pass from one band to the next higher one. When prisms change their direction gently, a zone of intermediate orientation is developed. DEJ, dentin-enamel junction.

This set of characters in common demonstrates the identity of HSB in mammalian teeth, including rodent incisors, even when HSB in the latter tend to become thinner. Despite the wide distribution of HSB in mammalian teeth, they have evolved in parallel several times (Koenigswald and Rensberger, in prep.), but obviously there is only a limited number of possibilities for strengthening the enamel. Enamel has to be strengthened whenever teeth become larger or more hypsodont. In contrast to human teeth, for instance, where the borders of the bands are not very distinct, the transitional zones tend to become narrower in hypsodont teeth.

Thus, HSB are very evident in hypsodont teeth and especially in evergrowing gliriform incisors, such as in _Vombatus_, _Lasiorhinus_, _Daubentonia_, and _Myotragus_, as well as in rodents and lagomorphs. The only exception so far known is the enlarged incisors of the hyracoid _Procavia capensis_, where I could not find any HSB. Shellis and Poole (1979) described an unusual helicoid orientation of prisms

Fig. 4. Longitudinal section of incisor enamel in the muskrat Ondatra zibethicus (family Muridae). Uniserial HSB are covered by a thin portio externa (p. ext.) with radial enamel. DEJ, dentin-enamel junction.

for Daubentonia, but my own observations with SEM and light microscopy showed that the transverse HSB do not differ basically from any other HSB, such as those in Chinchilla, for instance. Kawai (1955) surveyed the thickness of HSB in various mammals. According to his results, the average number of prisms for each band is about 10 or somewhat lower. In very thick enamels, however, such as in the rhinoceros, the number may be about three times as high.

The biomechanical advantage of HSB has been suggested by Lehner and Plenk (1936) as a protection against cracks penetrating the enamel. Cracks, even hairline cracks, which might do severe damage to the enamel, can best be prevented by a multiple crossover of fibrous elements.

In gliriform incisors of rodents and other mammals, the enamel tends to be somewhat thinner than in most mammalian molars, in order to form an extremely sharp cutting edge. Normally, HSB are well developed, and the thickness is about six to nine prisms per band in non-rodent incisors (a multiserial pattern). These numbers are also found in some hystricognath rodent incisors as well, such as *Cuniculus paca* with eight prisms per band, and *Hystrix cristata* with six. However, in other hystricognaths (Fig. 2), the thickness of the HSB is reduced even further, such as in *Myocastor coypus* (five prisms per band) and *Chinchilla laniger* (four to five prisms). This coincidence will be discussed below.

Korvenkontio (1934) and Wahlert (1968) have given a broad survey of the thickness of HSB in rodent incisors. The earliest rodents known from the fossil record, the Paramyidae, are described as having pauciserial HSB, that is, bands with a width of two to four prisms. From this pauciserial stage, the uniserial condition of myomorph and sciuromorph rodents may have evolved. In enamel that shows uniserial HSB, there is generally only one prism per band. The reduction in width of the HSB can be observed in the theridomyoid families Pseudosciuridae and Theridomyidae of Europe, in which pauciserial HSB change into uniserial ones. In molars of the extant geomyid *Thomomys*, the sequence from multiserial through pauciserial to uniserial HSB can be observed (Koenigswald, 1980). Thus, the reduction in thickness of HSB is a general evolutionary phenomenon (Fig. 5).

The physical advantages of pauciserial, or, even better, uniserial enamel, as compared to multiserial HSB, are many. Certainly, areas with HSB are generally not more resistant to wear than the more primitive radial enamel in the portio externa. Observations made on rhinoceros molars (Rensberger and Koenigswald, 1980) suggest that, in general, the resistance against wear is greatest when prisms penetrate the occlusal surface at an angle close to the acting forces, as in the outer radial enamel. However, it is likely that the enamel will split, due to the parallel arrangement of prisms. The functional advantage of Hunter-Schreger bands, therefore, is not to reduce wear but to prevent cracks in the enamel. The more often the direction of the prisms is changed, the less likely the enamel will be effected by cracks. From this point of view, a reduction in thickness of the HSB, as in uniserial enamel, provides the greatest protection. This hypothesis is supported by observations made on enamel cracks that develop during the process of specimen preparation. These cracks passed through the outer radial enamel, but they were stopped at the outer surface of the uniserial HSB.

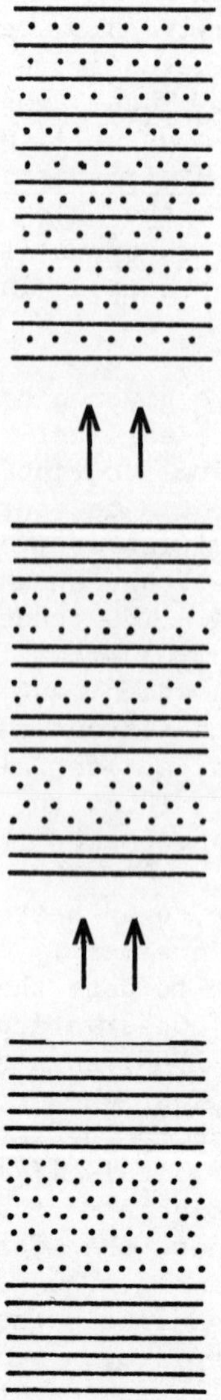

Hunter-Schreger bands (HSB)	Distribution in gliriform incisors	Comments
uniserial 1 prism	Muroidea Dipodoidea Geomyoidea Gliridae Sciuridae Castoridae Anomaluridae Aplodontidae	Most derived HSB, certainly developed in parallel several times
↑↑		
pauciserial 4 – 2 prisms	Paramyidae Pseudosciuridae Sciuravidae (Most ?) Ctenodactylidae	Derived stage of HSB, probably developed in parallel more than once
↑↑		
multiserial >4 prisms	Hystricognathi Pedetidae Lagomorpha Rhombomylus Daubentonia Myotragus Lasiorhinus Vombatus	Symplesiomorph character of HSB, in larger teeth of most mammalian orders

Fig. 5. Progressive reduction in thickness of HSB in mammalian gliriform incisors.

Shearing forces, which may occur in different ways depending on the various uses of the incisors, are neutralized better when HSB with decussating prisms are thinner. Because of these functional reasons, it may be proposed as an evolutionary model that the gradual reduction in thickness of HSB increases the physical properties important for incisor functions. The three enamel stages described as multiserial, pauciserial, and uniserial HSB form a continuum. Multiserial HSB, which occur widely in mammalian teeth, especially in gliriform incisors of non-rodents, are the least derived stage. The occurrence of multiserial HSB in rodent incisors therefore should be regarded as primitive, but the problem related to this proposition will be discussed below. Pauciserial HSB, although found only in Eocene-Oligocene fossil rodents, have to be regarded as more derived, and finally the uniserial enamel of myomorph and sciuromorph rodents forms the most highly derived condition. No functional reason can be detected to explain why HSB should become thicker secondarily and lose valuable physical properties.

As indicated above, however, this model (Fig. 5) creates problems, because it contradicts, to some extent, some general ideas of rodent phylogeny based on the fossil record (see Wood, this volume). Because the hystricognathous rodents, both the South American caviomorphs and the various Old World families, have multiserial HSB, these taxa can be regarded as very early offshoots from the basal rodent stock (also see Hartenberger, this volume). The occurrence of multiserial HSB in all hystricognaths does not contribute to the discussion about monophyly of this group (Landry, 1957; Luckett, this volume; Wood, this volume), because this is a symplesiomorphic character, which Hystricognathi share with other non-rodent mammals with gliriform incisors. On the other hand, this character indicates that the earliest known rodents from North America, the Paramyidae and Reithroparamyinae, which have pauciserial enamel, are already too derived to have given rise to Hystricognathi, contrary to the proposal by Wood (1959, this volume).

When Wahlert (1968) surveyed the distribution of the three types of HSB in rodent incisors, he assumed that the pauciserial HSB of Eocene paramyids should be ancestral to both the multiserial and the uniserial HSB of later rodents. Following publication of the model described above (Koenigswald, 1980), which proposed a continuous reduction in thickness of HSB during rodent evolution, Wahlert reviewed this topic and came to the same conclusion, i. e., that multiserial HSB is the primitive rodent condition (Wahlert, 1983).

Landry (1957) compared the osteology of hystricognaths with that of other rodents, and he found quite a number of primitive characters that led him to conclude that hystricognaths could not have been derived from paramyids. It seems possible that the history of the Hystricognathi goes back much further than indicated by the available fossil record. Perhaps even the very peculiar paleogeographic distribution of the Hystricognathi can be reconstructed more easily when this group is regarded as a very ancient radiation of rodents.

OTHER CHARACTERS OF RODENT INCISOR ENAMEL

Inclination of the Hunter-Schreger Bands

In rodent incisors, the transverse HSB are normally somewhat inclined toward the occlusal surface, but in sciurids, as well as in some early rodents such as paramyids and Eocene ctenodactylids, the HSB are almost vertical to the dentin-enamel junction (Sahni, 1980). For gliriform incisors it can be proposed that a stronger inclination indicates a more derived condition. This is supported by observations on the marsupials *Vombatus* and *Lasiorhinus*. In the gliriform incisors of *Vombatus*, the HSB are less pronounced and less inclined than in *Lasiorhinus* (Beier, 1981). Therefore, *Lasiorhinus* can be regarded as more derived in its enamel.

Functional interpretations, which add further arguments to the belief that inclined HSB represent a more derived stage in rodent incisors, are presented in a later section. They demonstrate that the inclination is not derived in itself, but that the direction of biting forces must also be considered. Thus, arvicolid molars developed a decrease in the inclination of the HSB due to different functional conditions.

Interprismatic Matrix

Boyde (1978) drew attention to some differences in the interprismatic matrix of the portio interna of rodent incisors. This matrix is clearly evident in muroids, crossing over prisms at a high angle, whereas it was seemingly absent in sciurids. However, my own observations have shown that in sciurids the interprismatic matrix is present as well, although it runs almost parallel to the prisms. Further investigation, however, reveals such a complicated distribution of this feature that it cannot be used for systematics as yet; at present, its relatively primitive and derived states remain to be defined.

External Index

It is obvious that the thickness of the outer radial enamel in rodent incisors varies greatly. The hypothesis of expanding HSB, which would reduce the thickness of inner and outer radial enamel, could mislead one to assume that the phylogenetic position might be estimated from the thickness of the persisting outer radial enamel. The comparatively thick layer of radial enamel in *Sciurus* may represent a less derived stage, especially because the inclination of the HSB is low. However, it is uncertain whether the thick layer of radial enamel in Castor can be interpreted in the same way. On the one hand, its HSB are more inclined, and, in addition, ecological factors may require a thicker portio externa as well. Korvenkontio (1934), who surveyed the relative thickness of the outer radial enamel in many rodents and calculated an "external index," argued also that burrowing rodents might have a thicker layer of radial enamel than other rodents.

FUNCTIONAL ASPECTS OF INCISOR ENAMEL

For arvicolid molars (Fig. 6) it has been demonstrated that the Schmelzmuster or enamel pattern is closely related to the pattern of stress during mastication (Koenigswald, 1977, 1980, 1982). All cutting edges are within a horizontal occlusal plane and, due to propalinal movement of the jaw, the leading and trailing edges in each dentin triangle can be determined. The Schmelzmuster obviously differentiates exactly between leading and trailing edges, in both upper and lower molars.

In comparison to this condition, the stress pattern of gliriform incisors is much more complex, even when the morphology is simpler. Incisors are used for gnawing, cutting, pulling, and even burrowing,

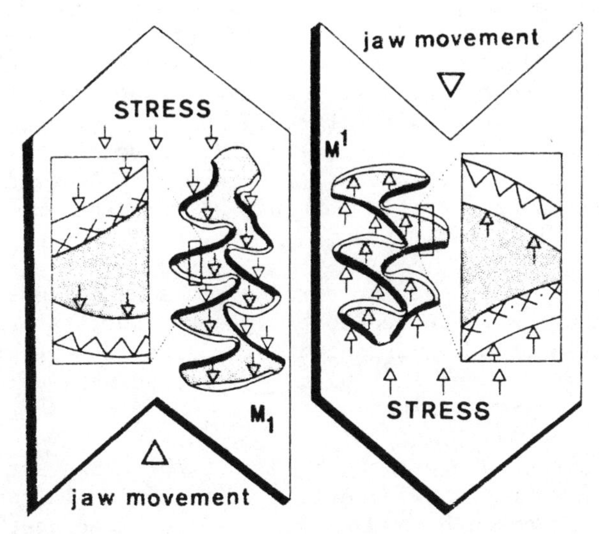

Fig. 6. In arvicolid molars (here Mimomys savini, from middle Pleistocene, Europe), the leading and trailing edges are clearly differentiated, because chewing forces act always in the same direction (adapted from Koenigswald, 1980).

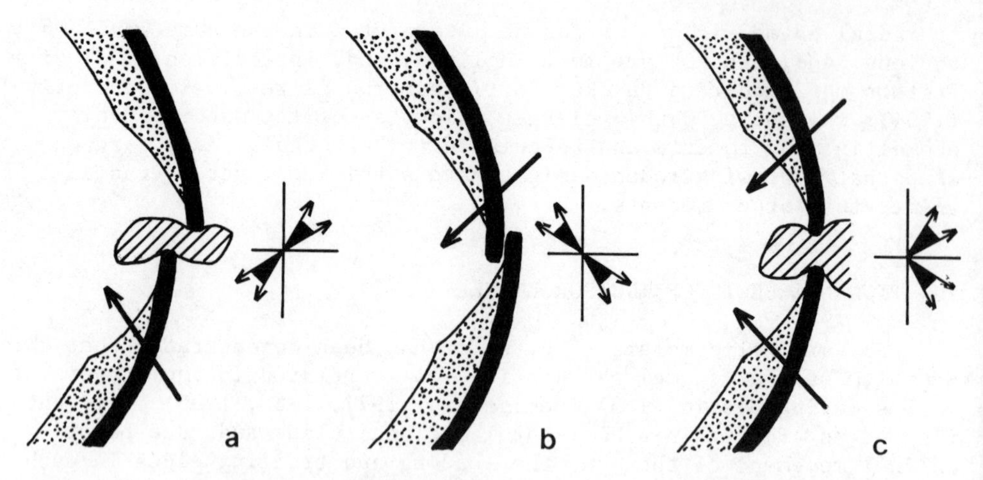

Fig. 7. In rodent incisors, stress on the enamel is produced in different directions during various activities, such as (a) cutting; (b) sharpening of the lower incisor; and (c) tearing of hard material. The Schmelzmuster developed has to serve all these functions.

to select only a few possible activities, and this creates a very different stress pattern in the enamel region of the tooth. Because upper and lower incisors work as antagonists, the direction of stress in both teeth is often reversed, even when the sequence of layers within the enamel is similar (Fig. 7). Compressive forces, as well as tension and shearing forces, occur reciprocally on the inner or outer side of the enamel, but in the upper and lower incisors the HSB always lie in the portio interna, whereas the portio externa is formed by radial enamel.

Because of these varying stresses, the Schmelzmuster in rodent incisors can not be expected to be selected for a single stress pattern; instead, it was selected for a general strengthening of the enamel. It has to cope with high tension, as well as with great stresses from various directions.

It was argued above that the HSB of rodents are reduced in their thickness from a multiserial stage, common in many mammalian orders, passing through an intermediately derived pauciserial stage, and culminating in an uniserial stage. By this development, the quality of the enamel has improved continuously to prevent penetration of cracks. Therefore, the continuous reduction in thickness of HSB offers a general advantage and can not be related to an adaptation for some specific diet. Multiserial enamel in Hystricognathi does not appear to be related to any specific function or dietary specialization.

In rodent incisors, the thinner the HSB, the greater the functional advantage; therefore, why does not this enamel type penetrate throughout the entire enamel? Instead, it is overlain by an outer radial enamel. This question can not be answered with certainty, but observations by Rensberger and Koenigswald (1980) showed that the resistance against wear of enamel varies with the direction in which the prisms are oriented to the chewing forces. As shown above, the direction of stress on the tip of the incisors may vary, but it can be assumed that it meets the incisor mostly from the tip, close to the long axis of the tooth. Generally, the resistance is greater when the prisms are more oriented within the direction of forces, whereas the enamel is abraded more easily when prisms are set at a large angle to these forces. In terms of incisor enamel, this means that, despite their other valuable properties, the HSB are worn out more easily than the outer radial enamel, in which prisms are oriented in a much smaller angle, with regard to the main forces. This correlation may be one reason why the outer radial enamel forms the tip of the cutting edge. In this regard, Korvenkontio's (1934) assumption that burrowing rodents may have an extremely thick outer radial enamel is of interest. A stronger inclination of the transverse HSB toward the tip of the incisors might be regarded as a more derived stage. In strongly inclined HSB, the prisms of the bands become oriented much more closely to the direction of the chewing forces (Wahlert and Koenigswald, in prép.). However, according to the measurements of Korvenkontio (1934), inclined HSB never reach the same inclination of prisms as do those in the portio externa.

Support for this interpretation can be found in the evolution of arvicolid molars, even though some aspects are reversed. In primitive species, such as *Mimomys stehlini*, HSB penetrate the occlusal surface at some angle (discrete lamellar enamel), while in more derived species, such as *Mimomys polonicus* and *M. savini* and their descendants, the HSB are almost perfectly parallel to the occlusal surface (typical lamellar pattern; Koenigswald, 1980). Here, the inclination of the HSB, in contrast to the pattern in the incisors, is obviously reduced, in agreement with the almost horizontal chewing forces acting on the occlusal surface of the molars. Even when the angle of the forces in relation to the occlusal surface is different, the HSB in molars, as well as in incisors, tend to meet the direction of these forces.

Another possibility to have prisms oriented close to the direction of the chewing forces is attained by having a more oblique, or even longitudinal, orientation of the HSB, instead of the normal transverse one. In the survey by Korvenkontio (1934), such a change in orientation of the HSB was noted in the genera *Tachyoryctes*, *Sicista*, and *Alactaga*. This change obviously occurred in several groups by parallelism. Vertical HSB are present in rhinocerotids as well, and their functional aspects have been discussed by Rensberger and Koenigswald (1980).

SUMMARY

The various enamel structures in rodent incisors are interpreted on the basis of comparison with the enamel of other mammalian orders. In particular, cases of parallel evolution, where evergrowing, large incisors are developed, allow us to distinguish between primitive and derived conditions. The number of layers in the enamel is not sufficient to distinguish between a more or less derived stage. As soon as the various evolved enamel types are determined, however, hypotheses on character state polarity can be made. The one-layered stage present in lagomorphs can not be regarded as ancestral to the two-layered stage in rodents, because both taxa have followed different pathways for the expansion of the Hunter-Schreger bands. The two-layered stage in rodent incisors was preceded by a three-layered stage, as evident in some hystricomorphous rodents.

Multiserial Hunter-Schreger bands, distributed widely among mammals with gliriform incisors, are regarded as less derived than either pauciserial or uniserial Hunter-Schreger bands, which appear to be restricted to rodents. Functional considerations also support this idea. Because the Hystricognathi has preserved multiserial HSB, this suborder could not have been derived from taxa with pauciserial Hunter-Schreger bands, such as North American Paramyidae.

A stronger inclination of the HSB in gliriform incisors can be regarded as a more derived condition, because of functional reasons. In contrast, our knowledge of the orientation of interprismatic matrix in the portio interna of rodent incisors is still insufficient, and, at present, it can not be used for systematic interpretation. The thickness of the outer radial enamel in rodent incisors reflects ecological adaptations, and caution should be adopted in interpreting its evolutionary significance.

We are only beginning to understand the evolutionary trends within enamel structures. Even though some rodent groups can be characterized already by their enamel, much further work remains to be done before a detailed hypothesis of rodent phylogeny can be based mainly on the enamel, and then be compared to those based on other characters.

ACKNOWLEDGMENT

During recent years, our research on enamel was supported generously in various aspects by the Deutsche Forschungsgemeinschaft, Bonn (Ko 627/7-1). I am also very much indebted to all my colleagues who contributed fossil and Recent material for this study, even when conscious that the study of enamel structure requires partial destruction of the specimen. Ms. I. Lehnen prepared the drawings. Cordial thanks are due to W. Patrick Luckett for his careful comments on this manuscript. He and Ann Forsten helped me with the English version, for which I am very grateful.

REFERENCES

Beier, K. 1981. Vergleichende Zahnuntersuchungen an Lasiorhinus latifrons Owen, 1845 und Vombatus ursinus Shaw, 1800. Zool. Anz. 207: 288-299.

Boyde, A. 1978. Development of the structure of the enamel of the incisor teeth in the three classical subordinal groups of the Rodentia. In: Development, Function, and Evolution of Teeth, Butler, P. M. and Joysey, K. A., eds., pp. 43-58, Academic Press, London.

Kawai, N. 1955. Comparative anatomy of bands of Schreger. Okijama Folia Anat. Jap. 27: 115-131.

Koenigswald, W. von. 1977. Mimomys cf. reidi aus der villafranchischen Spaltenfüllung Schambach bei Treuchtlingen. Mitt. Bayer. Staatsslg. Paläont. hist. Geol. 17: 197-212.

Koenigswald, W. von. 1980. Schmelzstruktur und Morphologie in den Molaren der Arvicolidae (Rodentia). Abh. Senckenb. naturforsch. Ges. 539: 1-129.

Koenigswald, W. von. 1982. Enamel structure in the molars of Arvicolidae (Rodentia, Mammalia), a key to functional morphology and phylogeny. In: Teeth: Form, Function and Evolution, Kurtén, B., ed., pp. 109-122, Columbia Univ. Press, New York.

Korvenkontio, V. A. 1934. Mikroskopische Untersuchungen an Nagerincisiven unter Hinweis auf die Schmelzstruktur der Backenzähne. Ann. Zool. Soc. Zool.-Bot. Fennicae Vanamo 2: 1-274.

Landry, S. O. 1957. The interrelationships of the New and Old World hystricomorph rodents. Univ. Calif. Publ. Zool. 56: 1-118.

Lehner, J. and Plenk, H. 1936. Die Zähne. In: Handbuch der mikroskopischen Anatomie des Menschen, Möllendorff, W. von, ed., vol. 5/3, pp. 447-708, Springer-Verlag, Berlin.

Rensberger, J. M. and Koenigswald, W. von. 1980. Functional and phylogenetic interpretation of enamel structure in rhinoceroses. Paleobiol. 6: 477-495.

Sahni, A. 1980. SEM studies of Eocene and Sivalic rodent enamel. Geosci. J. 1/2: 21-30.

Shellis, P. and Poole, D. F. G. 1979. The arrangement of prisms in the enamel of the anterior teeth of aye-aye. Scan. electron microsc. 1979/11: 497-506.

Tomes, J. 1850. On the structure of the dental tissues of the order Rodentia. Phil. Trans. Roy. Soc. Lond. 1850: 529-567.

Wahlert, J. H. 1968. Variability of rodent incisor enamel as viewed in thin section, and the microstructure of the enamel in fossil and Recent rodent groups. Breviora Mus. Comp. Zool. 103: 1-18.

Wahlert, J. H. 1983. Multiserial enamel and antiquity of hystricognath rodents. Amer. Soc. Mamm. 63rd Ann. Meeting, Gainesville, Fla. Abstract 152.

Wood, A. E. 1959. Eocene radiation and phylogeny of the rodents. Evolution 13: 354-361.

RODENT MACROMOLECULAR SYSTEMATICS

Vincent M. Sarich

Departments of Anthropology and Biochemistry
University of California
Berkeley, California 94720 USA

INTRODUCTION

Protein and nucleic acid data have a limited though vital role in the development of an understanding of the systematics of a group. They have the potential of providing, independent of any other information, the cladistic and temporal dimensions of the phylogeny of that group. I do not think it overstates the case to argue that until we have the cladistics we, for all practical purposes, really have nothing at all. The branching pattern forms the necessary framework upon which to properly assess the meaning of the evidence gleaned from other areas of research. Once the branching order has been worked out, and if there is a significant body of fossil data to integrate into our understanding of the group, we can attempt to provide a time estimate for each node or branch point.

On the lineages so defined are then placed such anatomical, behavioral, physiological, and molecular events as we can infer to have occurred along them. Finally, we attempt to answer the basic evolutionary questions of the selective "whys" and "hows" for each of these events. Thus the cladistic and temporal dimensions of a phylogeny become the framework upon which an understanding of what happened within it is to be structured.

I have, and will here continue to argue that the potential of proteins and, increasingly, nucleic acids, to delineate this framework is inherently greater than those of more traditional lines of evidence. The claim, of course, is empirically testable, but derived originally from three observations. First, the vast majority of changes at the molecular level are seen as single nucleotide or amino acid substitutions. They are thus countable in the same units across molecules and species. Second, we can, though here more problems can arise, work with homologous molecular structures over vast taxonomic ranges, thus using, for example, the same standards of comparison between *Homo* and *Pan*, *Rattus* and *Mus*, and *Tursiops* and *Bos*. Third, the range of amino acid sequences compatible with a given biochemical function would appear to be so large as to ensure a sufficiently useful approach to the ideal of continual differentiation after the achievement of genetic separation between two populations. The differences among extant forms are then measurable in the same units along a common scale and are patently derived characters. Thus one should be able to simply count these amino acid or nucleotide sequence differences and apportion them along a unique, derived phylogeny. Conversely, one measure of the phylogenetic quality of our molecular data sets will be the ease with which they are indeed so apportionable. Proteins and nucleic acids are, however, finite structures, and thus varying amounts of homoplasy will be present. The limitations of the apportionment processs relative to producing a cladogram congruent with the actual history of the taxa being studied will then stem from imperfections in the procedures used to ascertain the measure differences, the levels of homoplasy present, and, finally, any non-random distribution of the homoplasies.

The true saving grace in all of this effort in evolutionary reconstruction is the immutable fact that there can be only a single history for any group of organisms; thus ultimately all the anatomy, behavior, and physiology (past and present) must be fitted into it through some sort of successive approximation procedure. It would then appear that a cladogram, however derived, simply represents a working hypothesis to be tested in terms of the degree to which it facilitates an evolutionary understanding of the other available comparative data. Disagreements and inconsistencies among interpreparations will be inevitable, and it will prove useful to remember that no comparative technique can produce the ultimate, actual history.

The actual history or phylogeny does exist and is in that sense real. Any comparative data set and the resulting cladogram are also real enough, but can, for various reasons, only approximate the actual history. They are, nonetheless, all that we have to work with today. This same logic applies to any body of contemporary comparative data and, in a slightly altered form, to such information as is available from the fossil record. The fossil record, it should be noted here, is invaluable in telling us what happened and in what sequence; but it has certainly been much less important that contemporary comparative anatomy in providing phylogenies and taxonomies. Thus the desired actual history can have no operational reality save that give it through our efforts at reconstruction. Nor, obviously, can its reality be tested except in terms of the facilitation of our understanding of all the available comparative data.

SOURCES AND QUALITY OF MOLECULAR DATA

Comparative data concerning organisms necessarily come in two versions: continuous and discontinuous. Macromolecular data are no different in this regard -we have the discontinuous (amino acid or nucleotide sequences, electrophoretically assessed charge states), and continuous (distance data from immunological and DNA annealing comparisons). My concern here will be in the main with distance data. It will perhaps be best if we recognize at the outset that each measured distance is most usefully considered as a probability statement; that is, as a point somewhere along a distribution. For immunological distances, the distribution can itself be approximated through considerations of non-reciprocity and fit to additive phylogenies; for DNA annealing measurements, it would appear more realistic to attribute uncertainty more to the process of measurement itself. In other words, in immunology the comparison between protein species A and B is effected through a medium external to both of them (the antisera produced in another organism) and the uncertainty in the measurement derives mainly from variation in that medium; in DNA annealing, the uncertainty is intrinsic to the DNA structures themselves and to actual experimental errors. In either case, we are now in a very good position to provide precise assessments of the uncertainties of measurements and the approaches are mature enough to accept those irreducible uncertainties intrinsic to them. It is however important to emphasize here that the molecular worker is generally in a much better position than any other to actually provide such estimates of uncertainty, and we should be careful not to allow this strength to be seen as a weakness; that is, we should not fall into the position of assuming we can ever be free of uncertainty concerning the past. Accepting this limitation, being able to measure the uncertainty - to, in effect, book odds on each statement made and conclusion drawn - must be seen as an enormous advance over more traditional approaches to the evolutionary history of organisms.

Our primary goal is the branching order among the group of organisms under study. Assessing the validity of any conclusions drawn concerning branching order will be illustrated with specific examples, but some general comments are in order here. As already emphasized, each conclusion must be seen as a probability statement, and the probabilities involved are to be assessed in two quite distinct ways. First, the inferred distance or amount of change allocated to a lineage needs to be judged against the uncertainties implicit in the measurements from which the distance was inferred. Then the congruence of branching orders derived from different data sets must be tested, always keeping in mind that each conclusion about a lineage or branching order is a probability statement. We want to avoid at all costs imputing an undeserved power or precision to any particular technique, approach, system, method of data analysis, or set of theoretical considerations.

Thus we come to recognize that the nature and dimensions of the problem at hand need to be properly assessed so that we develop a feeling for what is possible and how that possible is to be approached. In general this is going to depend on at least three factors, two of which we have no control over at all. What we do have control over is the choice of an approach most appropriate to the nature of the problem at hand, and, of course, the informed application of the approach. Our resolving power (temporal length of the minimum definable lineage) is, at the recent, lower taxon end, basically a matter of the technique used. It is probably fair to say that the molecular worker can answer any useful phylogenetic question put to him to any desired level of resolution over the last few million years. As the questions begin to involve events of the more distant past, however, absolute resolving power must decrease. This is simply because a lineage is invisible to us unless something happened along it that has consequences still visible to us today, and the shorter the lineage in relation to its distance in time, the more likely it is that this something is no longer retrievable today. This logic tells us that many questions will have answers irretrievably lost in the mists of time, and many others will have answers too expensive to obtain at current levels of technology and understanding. We will have to learn not only what questions to ask, but also what questions not to ask, and to develop ways of discriminating between the two classes.

EXEMPLARS

Because so much of the current molecular data base bearing on rodent systematics is immunological, it is perhaps best to demonstrate both the strengths and the weaknesses of the approach with a specific group -the hystricognaths. It should be noted at the outset that the possibility that a group that you choose to work on may turn out to not be a monophyletic unit is always there, though generally one will be able to test the question with the molecular data themselves. I will also use throughout this paper a scale derived from quantitative precipitin (QP) rather than micro-complement fixation (MC'F) data, as in fact most of the rodent data derive from the former and the quantitative precipitin scale is conceptually more realistic and easier to deal with. Thus we will have a closed (0 - 100%) scale of reaction strengths rather than an open (0 to $>$ 200) scale of differences. The two scales are mutually interconvertible, however, where MC'F distance $\cong$ 2(100 - %QP reaction), and apparently linear with respect to the actual number of surface sequence differences in the protein species being compared.

The hystricognath data were originally presented in Sarich and Cronin (1976) and are given in Table 1.

These data then allow some preliminary groupings so as to both

Table 1. Quantitative precipitin cross-reactions among the various hystricognath albumins[a]

	Antisera									
	0.75	0.75	1.26	0.92	0.98	1.07	1.03	1.07	1.07	Rate
Antigens	H	Hy	Ba	Er	Da	Ho	Cav	Ch	Cap	tests
Hydrochoerus	100	33	28	36	38	26	32	45	32	+5
Hystrix	29	100	57	42	36	25	37	56	31	-6
Bathyergus	24	44	100	30	31	39	35	32	24	+2
Erethizon	34	39	29	100	48	32	42	52	33	-10
Dasyprocta	37	38	32	45	100	28	43	55	34	-1
Hoplomys	27	25	32	30	28	100	25	34	38	+7
Cavia	37	40	25	29	36	23	100	39	32	0
Chinchilla	38	52	40	54	47	34	49	100	38	-7
Capromys	32	29	41	40	30	29	36	35	100	-3
Myocastor	32	37	38	44	36	62	37	43	45	-2
Petromus	16	27	47	16	25	30	15	21	20	+7
Octodon	40	43	39	48	40	51	39	41	39	-1

Anti-Capromys		Anti-Hoplomys		Anti-Dasyprocta	
Geocapromys	70	Proechimys	86	Cuniculus	51
Plagiodontia	60	Diplomys	54	Anti-Erethizon	
		Cercomys	49	Coendu	95

[a] All cross-reactions are reported as the percentage of the amount of precipitate given by the homologous antigen. The number above each of the column headings is the factor by which each of the measured cross-reactions given by that antiserum was multiplied to give the reported values. This is the correction for non-reciprocity. The numbers in the last column are the amounts by which the albumins reacted more or less well with a series of nonhystricognath albumin antisera (carnivores and primates) relative to the reaction given by Cavia. Nonreciprocity is 3.6%.

simplify the exposition and help average out immunological discrepancies: Hoplomys-Myocastor-Octodon, Dasyprocta-Erethizon-Chinchilla, Bathyergus-Petromus, Hydrochoerus-Cavia. The data in Table 1 then collapse into the more manageable Table 2 (top section). The distance between any pair can then be allocated along each of the two lineages involved using the data in the rate test column. For example, the HMO - DEC distance is 62, while HMO is +7 relative to DEC. Thus we allocate 7 more of the 62 to the HMO lineage than to the DEC lineage (i.e., HMO, 34.5; DEC, 27.5). This then gives us the middle portion of Table 2. Even at this early level of analysis certain details of hystricognath phylogeny are evident. To a close approxi-

Table 2.

	HMO	DEC	BP	HC	Cap	Hys	Rate
HMO	-						+1
DEC	38	-					-6
BP	36	29	-				+5
HC	31	39	24	-			+2
Cap	38	35	28	33	-		-3
Hys	35	44	35	35	30	-	-6

Allocation of distance along the two lineages involved:

HMO	34.5	HMO	30	HMO	34.5	HMO	33	HMO	36
DEC	27.5	BP	34	HC	34.5	Cap	29	Hys	29
DEC	30	DEC	26.5	DEC	31	DEC	28	Cap	36.5
BP	41	HC	34.5	Cap	34	Hys	28	Hys	33.5
BP	39.5	BP	40	BP	38	HC	36	HC	36.5
HC	36.5	Cap	32	Hys	27	Cap	31	Hys	28.5

Amounts of change along the column lineage from its divergence from the row lineage:

	HMO	DEC	BP	HC	Cap	Hys
HMO	-	27.5	34	34.5	29	29
DEC	34.5	-	41	34.5	34	28
BP	30	30	-	36.5	32	27
HC	34.5	26.5	39.5	-	31	29.5
Cap	33	31	40	36	-	33.5
Hys	36	28	38	35.5	36.5	

mation one could envision our six lineages as diverging from one another, within the limits of resolution provided by this data set, at very nearly the same time. For example, along the HC lineage fro its separation from each of the other five we get 34.5, 34.5, 36.5, 35.5, and 36 percent reaction units of change for a mean of 35.4 $\pm$ 0.9. Thus the HC lineage does not specifically associate with any one of the other five. For DEC a similar calculation gives 28.6 $\pm$ 1.8, and again no specific associations are indicated. Indeed, the only significant paired deviations in the entire data set would appe to be HMO-BP, where the fact that the values involved are in each ca at least 4 percentage units lower than for any of the other comparisons, and Cap-Hys, where both values are the highest. The former suggests, given that we are not dealing with monotypic units, a weak phylogenetic association (for perhaps 10% of the history of the hystricognaths); the latter is more probably an immunological artifact, as Cap and Hys cannot both be further removed in time from one another than from the other four lineages.

Now one might be tempted to argue from this analysis that we are not being told very much; that five or six-way splits are not a very realistic representation of what "actually happened." Or put another way, one might argue that a technique with such "poor" resolving power is not worth very much relative to answering any question. Now that could be true, but not for the above reasons. It is not as if we do not get a graded set of reaction strengths which make taxonomic sense: Erethizon-Coendu, 95%; Hoplomys-Proechimys, 86%; Capromys-Geocapromys, 70%; Hoplomys-Myocastor, 62%; Bathyergus-Petromus, 47%; and so on. Perhaps we get this apparent "lack of resolution" simply because the lengths of time involved in getting this adaptive radiation underway were not very great, and that the combination of this possibility along with the antiquity of the radiation and the implicit limitations of the immunological approach then does not allow us to do any "better." I am much more impressed with the overall low level of uncertainty in measuring the amounts of albumin change along the six lineages (mean SD = 1.3% where the experimental error is ±1% and the known non-reciprocity in the entire data set is 3.6%). One could not do any better within the limitations of the technique, and clearly those limitations are not very great if one considers that the lineages average 33.3 percent reaction units of change from the origin of the exant hystricognaths. 33.3 ± 1.3 does not look at all bad; indeed, most immunological analyses should not, and do not, produce such high levels of internal consistency. Indeed, this data set probably provides as much resolving power as immunological comparisons of one protein (in this case, albumin) can provide. The overall phylogeny implied by this data set is given in Figure 1. It should be noted that this analysis suggests that a previous effort with this same data set (Sarich and Cronin, 1980) provided two associations not uniquely implicit in the data. It is not that the associations (for example, between the HC and DEC units) were "wrong;" it is just that they cannot now be defended as statistically significant. This is a most useful lesson to learn - to not allow the temptations induced by the fact that dichotomies are seen as evolutionarily more "real" than polychotomies to induce the belief that even the best data sets contain more resolving power than they actually do. If a series of divergences occur within a period of time not reliably resolvable by the given technique used, then a polychotomy is the reality and a series of dichotomies imaginary - though of course better techniques or new sources of data may provide the desired resolution in the future.

MOLECULAR CLOCKS AND THEIR CALIBRATION

Here one gets into more controversial and slightly embarassing matters - embarassing because of some miscalibrations of ours which are only now becoming clear. One first notes that molecular clocks, if they exist at all, are stochastic in character and data-bound. One does not assume a clock and impose it, willy-nilly, on any given

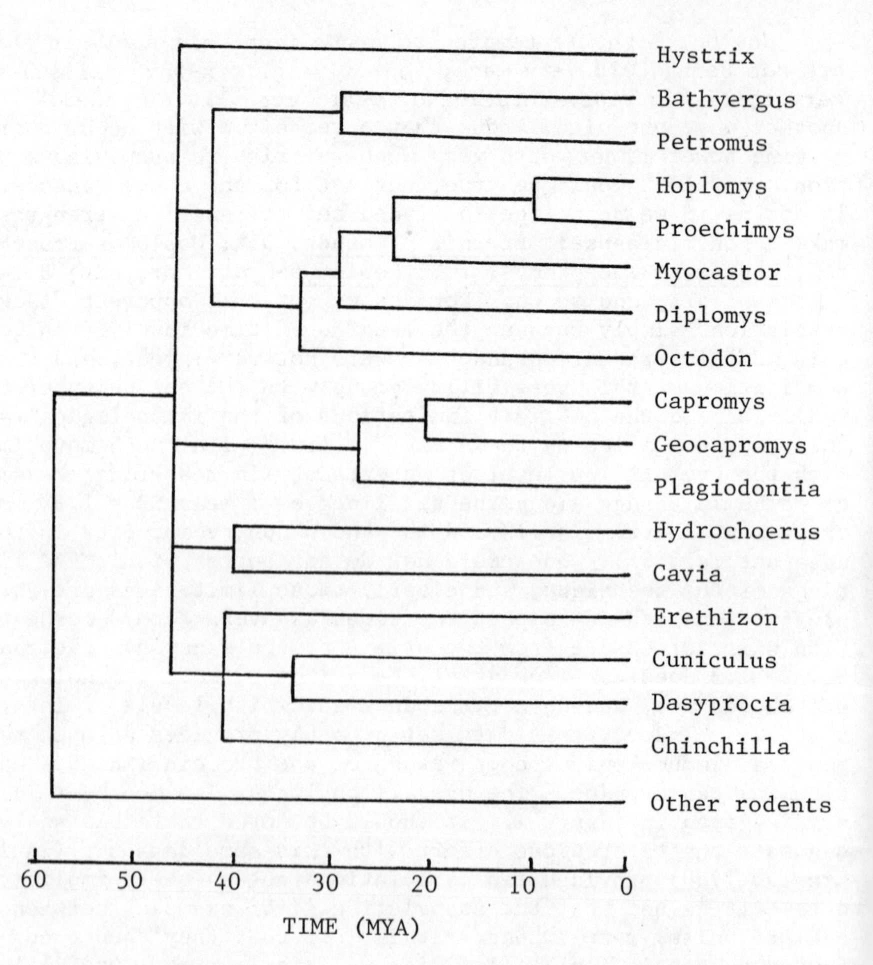

Fig. 1. Albumin immunological phylogeny of the Hystricognathi. This phylogeny was generated using the iterative procedure outlined in Table 2. It is to be understood that each posited divergence is a probability statement, even though technical limitations prevent representation of this. Thus the vertical lines are drawn at the most probable time of divergence, and it is entirely possible (indeed, to be expected) that some of the indicated associations will not stand the test of time. At this point, the non-association about which I am most dubious is that between Hystrix and Bathyergus/Petromus, where the strength of the reaction using anti-Bathyergus, along with the relatively weak reactions shown by that antiserum with the South American forms and the strength of the Thryonomys reaction, make me wonder as to whether the nature of the anti-Hystrix reactions might be misleading me. This, however, can not be clearly resolved with the material on hand. One needs more antisera

data set; one has to demonstrate that the data set shows sufficient regularity of change to justify analysis within a molecular clock context. We test this regularity by using out-group comparisons. In Table 1, for example, a set of antisera to various carnivore albumins react more or less strongly with those of the hystricognaths tested. (Anti-carnivore albumin sera are used because they tend to react more strongly than most other anti-placental albumin sera. This is almost certainly due to carnivore albumin evolution having proceeded at a somewhat slower rate than that modal for placentals. Obviously one would have to calibrate a carnivore albumin-specific clock if one were working within that group.) Erethizon albumin reacts most strongly, indicating it to have changed least; those of Hoplomys and Petromus react most weakly, indicating more change along those lineages. But of course variation of this sort is to be expected with a probabilistic clock; the useful question to answer has to do with the nature of the distribution of that variation. One might, for example, ask as to what the result would be if a lineage had been changing at twice or one-half the hystricognath mean over the history of the group. The mean, as we have seen is 33.3; doubling the rate would give 66.6; halving it would give 16.6. A lineage envolving at twice the hystricognath rate would then give (+33) in our rate test column - nothing extant and tested comes close, and this is not because the antisera used could not show it. Any number of placental albumin reactions, including some rodents (see Table 3), give very weak reactions. One evolving at one-half the hystricognath rate would give (-16.6) in the rate test column - again no hystricognath extant and tested reacts that strongly. Now albumins which appear to have changed at twice the rate modal for their taxon do exist (e.g., Phaner among the prosimians), and so do some which evolve at one-half or less the rate of others (e.g., birds as a group relative to mammals), but those are "problems" to be isolated experimentally and dealt with when necessary. Clearly no such cases exist for the hystricognaths. For a perfect stochastic clock, the variance would equal the mean, and the hystricognaths come close to this if one assumes that 1% reaction difference is equal to about 2 amino acid differences (Benjamin et al., 1984). In any case, we certainly have a system that seems to be accumulating change in statistically predictable fashion. So we then ask, if

$$\text{(Time of separation between two lineages)} = k(100 - \%\ \text{albumin cross- reaction})$$

then what is k? This is where a combination of circumstances has led us astray in the past and this is an excellent opportunity to put the whole matter in a more realistic perspective.

Our problems in this area stem from a combination of circumstances. First and foremost is the fact that our first two major albumin immunological investigations involved the primates and carnivores, and it turns out that most of the prosimian primates and carnivores in general have albumins that have changed less than those of other mammals. This was evident for the prosimians as long ago as 1970 when our first interordinal tests involving primates showed that various anti-carnivore albumins consistently reacted better with prosimian than with anthropoid albumins (Sarich, 1970). Because I did not see a prosimian clade in our data, I tended, when thinking about it at all, to attribute this to a singular, conformation-altering change early in anthropoid history, and to let it go at that That carnivore albumins as a group had changed less than the average was indicated by the generally greater cross-reactivities shown by antisera made against them (e.g., interordinal cross-reactions using antisera to carnivore albumins tend to average about 30%; using antisera to the albumins of other orders gives values averaging 15-20%). Added to this was our relative lack of concern with attempting to date relationships back beyond the Oligocene and the indication from the globin sequence data of lineages with substantial numbers of amino acid substitutions along them which were ancestral to some placental orders and not to others. We took these data as indicating that a number of interordinal separations within the placental mammals took place long after others. To this was added the increasing tendency of at least some paleontologists in the 1960s and 1970s to extend the classic beginning Tertiary explosive radiation of the placentals much further back in time [for example, see Simons (1976) Thus we came to think in terms of an antiquity for this radiation extending well back into the Cretaceous some 80-90 MYA, and from thi came the calibration that one accumulated a 50% difference in albumi cross-reactivity between two species last sharing a common ancestor about 55 MYA. The one clearly awkward datum here involved the fossi rich artiodactyls and perissodactyls, where application of the above time scale suggested separations within the orders well back into th Cretaceous, even though neither is seen in the fossil record until the early Eocene. This we tended to treat as an artifact of the use of rabbits to produce the antisera, arguing that there might be phylogenetic affinities between rabbits and condylarth-derived forms, thus biasing the immunological distances obtained within the latter upward. As can be seen from Sarich and Cronin (1980), this was stil our view as late as 1979. It no longer is.

What finally pushed me into a more realistic temporal framework began with an investigation into lagomorph origins encouraged and aided by a graduate student in paleontology named John Chiment. Th two of us looked into the question of lagomorph affinities by makin antisera against *Oryctolagus* albumin in both guinea pigs and chickens. These showed clearly that there was no special affinity betwe lagomorphs and either rodents or artiodactyls, or, for all that goe any other major group of mammals tested. Although the lagomorph pr ject was never completed, it did force the recognition that the rab bit was probably not biasing the albumin immunological studies in a group of mammals significantly more than in any other. That being the case, it was time to reset the albumin clock for placental mammals.

After one allowed for measured differences in amounts of change along various lineages, it became clear that the mean maximum immunological distances for albumin comparisons within placental orders involved QP cross-reactions in the 20-25% range. The only exceptions are the carnivores, as already indicated; cetaceans, which have a much more recent origin and are in any case clearly artiodactyl-derived; and the insectivores, which almost certainly do not form a monophyletic unit so that many within-insectivore comparisons are really interordinal and not intraordinal. The rodents fall comfortably into this general pattern.

Now, as Novacek (1982) has so nicely summarized, only three extant orders can be shown in the Paleocene, and it is not clear that for any of these (primates, carnivores, rodents) can the Paleocene lineages be associated with extant ones. The real beginnings of diversification leading to extant intra-ordinal lineages would then appear to be no earlier than late Paleocene and early Eocene, and indeed most orders do not even appear in the fossil record until the Eocene or later. One might then envision that the 20-25% intra-ordinal cross-reactions observed (75-80% difference) took some 50-55 MY of separation to produce, giving us a calibration of 1.4-1.5% difference in the albumin cross-reaction for every 1 MY of separation between two lineages. This compares to a 0.9%/MY figure assuming 80-90 MY for the beginnings of the placental radiation. It should of course also be noted here that this new picture can be seen as lending support to, or being supported by, the asteroid impact theory of the Cretaceous-Tertiary boundary -a theory which would then suggest that perhaps only a small number of placental lineages made it through the darkness, and that of those only one is the progenitor of all extant placentals.

INTERORDINAL RELATIONSHIPS

If the picture just sketched is in fact realistic, it also goes a long way toward answering the question of why it has proved so difficult to determine interordinal affinities among the placentals - to, for example, determine the sister group to rodents. If we are now only allowing some 10 MY to proceed from the adaptive radiation which gave rise to placental orders to the late Paleocene to early Eocene intraordinal radiations, then any periods of common ancestry among extant orders are going to be rather brief indeed, and one might argue that we are fortunate to be able to see any, and not worry about our inability to see more. This may be one of those questions that doesn't have accessible answers; the answers (shared, derived characters linking orders) having been destroyed or distorted beyond recognition over time.

This is not to say we have no evidence bearing on the question. The following interordinal associations are reasonably secure in term of our own data.

1. Primates, Dermoptera, Tupaiidae (PDT) (Cronin and Sarich, 1980).
2. PDT, Chiroptera, Rodentia, Carnivora, and probably Talpidae and Soricidae. The rodent inclusion in this unit is discussed below and that of the carnivores in Cronin and Sarich, 1980. The mole and shrew inferences are drawn from recent collaborative work with Terry Yates and Dwight Moore of the University of New Mexico (manuscript in preparation), and the bat evidence will be discussed in forthcoming manuscripts co-authored with Rodney Honeycutt and Elizabeth Pierson.
3. Proboscidea, Hyracoidea, Sirenia, Tubulidentata (Rainey, Sarich, Lowenstein, and Magor, manuscript submitted).
4. Artiodactyla, Cetacea (AC). This association is strongly supported by immunological comparisons using antisera to numerous artiodactyl and cetacean albumins which show, first, that the extant cetaceans share a relatively recent common ancestry - they are surely as closely related in time as, fo example, *Rattus* and *Mesocricetus* - and, second, that the cetacean lineage shares more recent common ancestry with the non-suid artiodactyls and hippos than those two groups do with the suids (*Sus and Tayassu*). The more likely association is with the hippos, but has not been verified using antisera to *Hippopotamus* proteins.
5. AC, Perissodactyla. This association is strongly supported by the globin sequence data discussed by Shoshani et al. in this volume.

If we now go back to the rodents, it would be fair to say that we can, at present, shed no real light on the details of relationshi within Group 2. As just noted, it may be that short of a great deal of DNA sequence work, this matter cannot be addressed productively at the molecular level. The question of course is whether, once having recognized Group 2, enough alternatives have been excluded t be able to make some morphological progress relative to rodent origins. I will say that I know of no molecular data which would suggest any rodent-lagomorph affinities, and there are appreciable, if not sufficient, data to in fact reject the rodent-lagomorph hypothesis.

The first of these has already been mentioned - antisera made i chickens against *Oryctolagus* and *Sylvilagus* albumins do not react especially strongly with rodent albumins. Second, the globin seque data, though admittedly somewhat suspect because of the numerous ge duplications at those loci, tend to align lagomorphs with the artio dactyl-perissodactyl clade (Sarich and Cronin, 1980; Cronin and Sarich, 1980). Third, we have noted that antisera made in rabbits against the non-primate Group 2 albumins (including those of rodents) will consistently assess Aotus albumin as much less changed than that of Cebus, or that of Lemur as much less changed than Homo albumin. Antisera made in rabbits or chickens to various artiodactyl and perissodactyl albumins, tend not to so discriminate, and neither do the chicken anti-lagomorph albumin sera. Finally, to mention one piece of exotic morphology, only Group 2 forms have a baculum, which is certainly a synapomorphy for them; lagomorphs do not.

This was the state of affairs before preparations for this conference were begun; since then more data have accumulated - in particular bearing on the suggestion made by Hartenberger and Luckett, among others, of a sister-group relationship between lagomorphs and rodents. The next section is then aptly entitled, echoing Wood's famous paper of 1957,

WHAT, IF ANYTHING, IS A RABBIT?

I will answer, "certainly not a rodent." The hypothesis of a sister-group relationship between rodents and lagomorphs requires the demonstration of derived features linking the two orders and excluding all others. At the molecular level, of course, "all others" is limited to extant lineages. Now it should be noted at the outset that one is in no position to effectively tackle the question of interordinal relationships among mammals in general at the molecular level, as we lack access to a molecular data set which fulfills the combined requirements of broad coverage, adequate resolving power, and, most critically, the availability of appropriate out-groups. Perhaps the globin data come closest at this time, but there are serious problems with them when it comes to questions of homology and resolving power. It is, on the other hand, much easier to structure a stringent test of a proposed association between two lineages relative to a third among eutherian mammals, as the proposal itself guarantees that the necessary out-group will be available.

This can be illustrated starting with a data set obtained using a pool of antisera made in chickens to the albumins of Oryctolagus and Sylvilagus (Table 3). Note that the pika (Ochotona princeps) does not give the expected strong reaction relative to any other albumin and that the albumin of an elephant shrew (Rhynchocyon) reacts just as well as that of the pika. The rodent reactions are appreciably weaker. Thus we can argue that if rodents and lagomorphs are to form a clade relative to Rhynchocyon (or any other lineage in the data set), much more of the lagomorph to rodent distances must be allocated to the rodent than to the lagomorph lineages. Otherwise the lagomorph -Rhynchocyon distance could not be less than the lagomorph - rodent distance. Now a second out-group can be introduced to test this conclusion independently, specifically an out-group composed of antisera to two primate albumins (Aotus and Nycticebus) (Table 3). Note that

Table 3.

Quantitative precipitin cross-reactions
(Antisera made in chickens)

anti-(*Sylvilagus*/*Oryctolagus*)		anti-(*Aotus*/*Nycticebus*)	
Rhynchocyon	45	Ursus	51
Ochotona	44	Castor	50
Ursus	41	Scapanus	48
Scapanus	38	Hystrix	46
Elephantulus	33	Peromyscus	43
Elephas	32	Elephas	43
Hystrix	30	Cavia	41
Erethizon	30	Bos	40
Procavia	29	Procavia	40
Castor	25	Rattus	40
Nycticebus	25	Acomys	40
Aotus	24	Rhynchocyon	39
Acomys	24	Ochotona	37
Dipodomys	22	Aplodontia	32
Sciurus	20	Oryctolagus	31
Cavia	18	Sciurus	26
Rattus	17	Phalanger	24
Aplodontia	17	Didelphis	23
		Echymipera	21
		Tachyglossus	19

anti-*Rhynchocyon*

Elephantulus	30
Scapanus	27
Oryctolagus	23
Ursus	24
Elephas	25

No other albumin tested gives a reaction of more than 8%, and almost all are less than 5% (*Ochotona*, *Rattus*, *Bos*, *Homo*, *Dipodomys*, *Hystrix*, *Cavia*, *Aotus*, *Nycticebus*).

the requirement is not fulfilled, as only the most changed rodent albumins (*Aplodontia*, *Sciurus*) are as different from those of the primates as is that of the rabbit. Rodent albumins with amounts of change modal for the order, and for eutherians in general, such as those of *Cavia* and *Rattus*, react about as well as those of *Rhynchocyon* and *Ochotona*, indicating no acceleration of albumin change along these lineages, and, therefore, no phylogenetic association with lagomorphs. The appearance of *Rhynchocyon* in this context is perhaps not totally surprising (McKenna, 1975) but nonetheless of some appreciable interest. Data obtained with chicken antisera to *Rhynchocyon* albumin, while leaving a good deal to be desired in terms of reciprocity and specificity, are consistent with the above (Table 3). Note that by far the strongest reactions are given by *Elephantulus*, *Scapanus*,

Oryctolagus, Ursus, and Elephas. Of these, Ursus, Elephas and Scapanus would appear to have conservative albumins which generally react more strongly no matter what out-group antiserum is used and Elephantulus is, after all, another elephant shrew, albeit an obviously distantly related one (see also Goodman, 1965). The appearance of Oryctolagus in the reactive group is then intriguing, though at this point little more than that. Clearly a great deal more work needs to be done, especially in looking at the questions of the association between the two groups of elephant shrews on the one hand, and between leporids and ochotonids on the other. Then the four lines can be more definitively related to one another and to other eutherians.

Finally, a recent article on several α-lactalbumin sequences (Shewale et al., 1984) provides a phylogenetic statement which clearly aligns rodents (Cavia and Rattus) with primates (Homo) and leaves the rabbit (Oryctolagus) an out-group to both artiodactyls (Bos, Capra) and the rodent-primate unit (Fig. 2). All of this is quite consistent with the immunological data just summarized and discussed.

Thus, although I have argued above that the problem of detailing the pattern of interordinal relationships among eutherian mammals should be relatively intractable, in fact the relative wealth of available molecular data (little of it gathered as a specific attack on the problems and more of it discussed in the contributions by Shoshani and DeJong in this volume) provides a number of rather better-defined associations than might have been expected. How many of these will survive the rigors of being tested by better data remains to be seen, as does the indication that we may still retain a fair number of lineages representing the basal eutherian adaptive radiation, but it is clear that molecular progress is being made and may even in this tricky area soon provide a number of definitive answers.

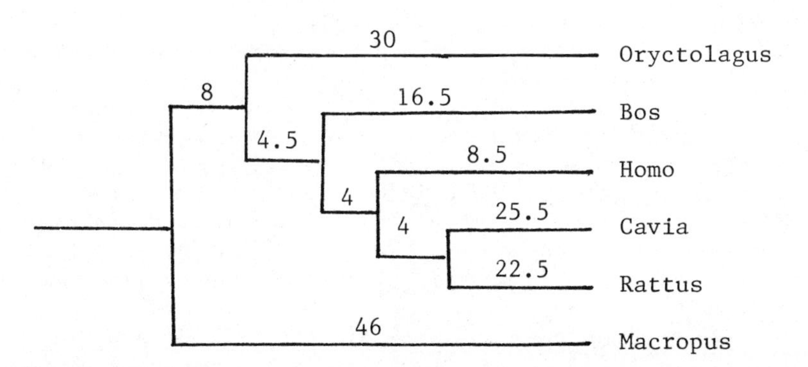

Fig. 2. Phylogeny of some α-lactalbumins. Redrawn from Shewale, Sinha. and Brew (1984, Fig. 14). The number of estimated amino acid substitutions is given above each lineage.

RELATIONSHIPS AMONG RODENTS

Here I open with a plea for cooperation and collaboration. It is far easier to generate molecular data than to know what to do with them. Unless they are applied to the resolution of problems concerning the organisms themselves they remain sterile, having very little intrinsic interest in other than exceptional circumstances. Thus there is no longer any point in doing molecular comparisons just because, as with Mallory's Everest, they are there to be done. We have learned to do the comparisons effectively and efficiently, now we need to get more workers dealing with the organisms themselves (fossil as well as living) to make use of the available molecular data, and, more important, to initiate molecular investigations designed to resolve issues which appear as intractable using more conventional approaches.

Hystricognathi

In addition to the data discussed above, I have recently carried out comparisons of Thryonomys, Lagidium, and Lagostomus albumin. The preliminary indications are of the former associating with the Bathyergus-Petromus clade, and Lagidium showing quite close affinities with Chinchilla (80% cross-reaction). The plains viscacha (Lagostomus), on the other hand, has an albumin that reacts no more strongly with anti-Chinchilla (43%) than is average for other hystricognath lineages. Lagostomus then, as well as a number of other forms (e.g., Dinomys, Ctenomys, Abrocoma), need to be investigated in much more detail, and it is likely that several more ancient hystricognath lineages will be so defined. As already noted, and in contrast to the unrealistically precise phylogenetic analysis in Sarich and Cronin (1980), it would now appear that a single, complex, and ancient adaptive radiation characterizes the group. I would date it, according to the new calibration, at about 45-48 MYA (67 divided by 1.4-1.5), some 10-12 MY after the basal adaptive radiation of the order Rodentia as we know it today. This estimate is entirely congruent with that made by Wood (this volume) using entirely independent data and logic. Though clearly a good deal more in the way of albumin comparisons, especially among the Old World hystricognaths, needs to be done, to say nothing of work involving other molecules, it would appear that we still require at least two trans-Atlantic crossings. Neither the Old World nor the New World forms seem to be forming clades relative to one another, though there is no question of the monophyly of the group relative to other rodents. I have looked at Pedetes and it certainly does not appear hystricognath-related; as yet no ctenodactylid or anomalurid material has been made available to me.

A Possible Synapomorphy Defining a Non-Hystricognath Clade

At present it is not possible to provide much detail from the available molecular perspective (protein immunology, DNA annealing) on the question of relationships at the highest taxon levels within rodents. Pedetes, for example, is not a hystricognath but what is it besides a monotypic lineage? As I have already pointed out, this may prove to be a non-question; that is, one that cannot be answered at the desired level. It may indeed turn out that a very distant adaptive radiation has left few details today as to its precise phylogenetic path; and this could be just as true for rodents as for the placentals in general. I would argue that we need first to define a problem whose solution would be aided through knowing the phylogenetic details; the details should no longer be an end in themselves. Nonetheless one intriguing indication has come out of some long-term electrophoretic studies we have been involved in. Typically such lower-taxon comparisons will include a stain for non-specific esterases present in liver and kidney tissue. I have noted that the intensity of staining and band multiplicity is consistently much greater for the various muroids, sciurids, and geomyoids sampled than for any other forms, including the hystricognaths. Thus at present we seem to have a clearly synapomorphic marker present only in myomorph and sciuromorph rodents, suggesting that various lineages within the two suborders share some period of common ancestry subsequent to the divergence of the hystricognath line. I note that it is not at all clear that either the Myomorpha or Sciuromorpha represent monophyletic units themselves.

Myomorpha - General Considerations

I have given much more attention to the myomorphs in general, and the murids and cricetids specifically, for several reasons. First, there was the early and continuing claim that short generation times for forms such as Rattus and Mus accelerated their genomic evolutionary rates relative to, for example, the primates. Then there is the enormous emphasis on studies of the New World cricetids among North American mammalogists. One aspect of that emphasis has focused on the "hesperomyines" with suggestions that much of the radiation of this enormously complex and successful group had taken place in South America following the rise of the Isthmus of Panama some 4-5 MYA. I found this very difficult to accept. Finally there was the opportunity in 1969 to collect mammalian material in New Guinea, and this led to a fairly detailed look at the Australasian Muridae. All of these investigations are continuing slowly.

I will begin with the Rattus-Mus matter and its impact on questions of rodent systematics, the generation length question as it applies to the molecular clock concept, and ultimately to the whole question of the general cross-taxon utility of the molecular approach. Though possibly I have been too close to this matter and may therefore overstate the case, it still seems to me that the Rattus-Mus situation has caused problems all out of proportion to the actual issues involved. I sometimes wonder if perhaps one reason for this is that these are the rodents most familiar to us, in and out of the laboratory. One is thus conditioned to thinking of mice as just little rats or rats as just big mice. Indeed I once gave a paper to the American Society of Mammalogists (1971) entitled: "Rats are not just big mice: a lesson in molecular evolution" as a reaction to this problem. Yet when the first molecular comparisons between the two genera were carried out around 1970, both the albumin immunological and DNA annealing differences seemed very large for two such apparently similar forms. Reinforcing this surprise was the fact that Simpson had some years before (1959) placed the murid-cricetid separation at around 10 MYA, so that clearly then a within-murine separation would have to be much more recent, and a figure of 5 MY or so was usually quoted. Yet the albumin difference (on the scale used in this paper) is 33%, and the DNA difference was first measured at 20°C (Kohne, 1970), then went to 15°C (Rice, 1974), and remains in the 10-15°C area (Brownell, 1983). These are large differences, particularly if we compare them to the corresponding *Homo*-*Pan* figures of 5% and 1.9°C (Sibley and Ahlquist, 1984). As the separation for the latter pair was by consensus at least 20 MYA in 1970, *Rattus* and *Mus* had accumulated some 6 times more genomic change in one-fourth the time. So, what factor could account for this? Obviously, generation length, as any geneticist could tell you. So the ideas of *Rattus* and *Mus* as closely similar in morphology, closely related in time, and yet changing very rapidly genomically all conspired to "make nonsense" of our observations. It should also be noted that this argument was also available for "refuting" the notion of taxon-independent molecular clocks which suggested "impossible" separation times of 4-5 MYA for *Homo* and *Pan* and 6-10 times that for *Rattus* and *Mus*.

Well, we are some 15 years from that time and it is clear that we did not know as much as we thought we did. First, it was shown that the albumins of *Rattus* and *Mus* are no more different from those of several carnivores than are those of *Homo* and *Pan* (Sarich, 1972). Thus by direct test there was no *demonstrable* generation length effect. Second, further consideration of the hominid fossil record, particularly after the discovery of Lucy and her *Australopithecus afarensis* brethren, indicated just how African ape-like the earliest hominids were. If one was not all that far removed from a chimp at 3 MYA, then why go all the way to 20 MYA to find a common ancestor? Finally there came the recognition that the alleged fossil hominid *Ramapithecus* was not (either *Ramapithecus* or a hominid) - in most cur rent thought (though certainly not mine), it is seen as a proto-orang (Andrews, 1983; Ward and Pilbeam, 1983). I can not resist calling attention to the fact that large hominids simply do not get any more different than *Pongo* and *Homo*; and having a fossil taxon shifted from affinity with the *Homo* line to affinity with the *Pongo* line says some very questioning things about the paleo-anthropological enterprise as recently practiced. Neither can I resist suggesting that in another few years *Ramapithecus*/*Sivapithecus* will be no more *Pongo*-related than *Homo*-related and seen finally for what it is - an extinct early bran of the large hominoid stock separating from it before the *Pongo* -

African ape divergence and after the hylobatids (thus about 11-13 MYA.

So most people no longer talk of a Homo-Pan separation in the early Miocene some 20 MYA, and this begins to get the primate half of our problems into a more realistic framework for discussion - though the "Ramapithecus/Sivapithecus as orang-related" movement is going to provide a brief diversion in that progression. Now what about the rodent half?

One begins to sense progress there too, though we have much further to go. Perhaps the most important event here is the discovery of a more ancient murid fauna, so that Rattus-Mus divergence times of more than 10 MYA are being seriously envisaged (see, for example, Jacobs and Pilbeam, 1980). This at least opens the way for a meaningful dialogue to develop between the paleontologists and molecular workers, which certainly was not the case with 20 MY old hominids and no Rattus or Mus lineage older than 5 MY. That mistakes have been made in reading the evolutionary record by biochemists as well as paleontologists and comparative neoanatomists is one of the major messages of this effort, and what may prove a particularly telling example has recently surfaced as part of our continued effort in murid systematics.

A visiting Israeli scholar, Uzi Ritte, provided us with samples of Acomys tissue and, as Jacobs (1978) had suggested that it was the living genus most closely related to Mus, a doctoral student working on Mus lysozymes, Michael Hammer, asked to include Acomys to provide an out-group perspective on within-Mus relationships using albumin immunological comparisons. That comparison provided perhaps the greatest surprise in the 20 years of doing molecular systematics, as it turned out that not only did Acomys albumin not behave as if it were closely related to those of Mus, it did not even appear to be murine or even clearly murid (Table 4). That this was not an albumin idiosyncrasy was indicated by subsequent transferrin and lysozyme comparisons, and we thus have an example of a relatively well-known living genus (not a fossil or some obscure form that practically no one has seen or worked with) whose murine status has, as far as I am aware, never been questioned, apparently having its affinities seriously misread until a serendipitous series of events in a biochemistry laboratory. Now I do not wish to be seen as suggesting that this sort of situation is the rule - it surely is not - nor that the Acomys story is completed - actually it is barely begun - but simply that mistakes can be made and that one ought to be careful about hanging too much in the way of conclusions on too little in the way of evidence. It is no doubt true that my experiences with paleontologists working on primates have left me with an unduly jaundiced view of their competence to draw valid phylogenetic inferences from the fossil record in general and from dental evidence in particular, but Acomys, to repeat, is not a fossil. Neither, to repeat, is the

Table 4.

Immunological Comparisons Involving Acomys

ALBUMIN

anti-Mus		anti-[Peromyscus, Onychomys] pool		anti-[Praomys, Rattus, Notomys, Pseudomys] pool	
Rattus	67	Mus	55	Mesocricetus	50
Mesocricetus	50	Rattus	53	Peromyscus	48
Acomys	45	Acomys	53	Sigmodon	42
				Tylomys	42
				Acomys	41
				Oryzomys	31

LYSOZYME* anti-Rattus		TRANSFERRIN* anti- Praomys	
Mus	67	Rattus	56
Acomys	17	Acomys	37
		Mesocricetus	35

* The lysozyme and transferrins were compared by Dr. Ellen Prager using MC'F.

biochemical investigation completed. Nonetheless, this ongoing story ought to give one pause, and that pause leads to reflection as to what we can learn from whole, live, behaving animals; from the fragmentary bits and pieces of them that survive so haphazardly in the fossil record; and from their proteins and nucleic acids. These considerations lead finally to a mature collaboration, which recognizes that all lines of evidence have something to contribute to evolutionary analysis, although there is, ultimately, only one story to tell. This last singular fact is the saving grace in our efforts, and we ought to take greater cognizance of it than we do.

So where are we on the Rattus-Mus divergence time question specifically at the present time? The albumin data would suggest a date of 22-24 MYA (33/1.4-1.5), and the latest DNA annealing data of Brownell (1983) are consistent with this figure. It is difficult to see, from a molecular perspective, how that figure could be significantly reduced (say to 15 MYA); at the same time, it needs to be asked whether the fossil record, read in the context of the anatomy of extant forms (such as Acomys), can be seen as being consistent with such antiquity.

Myomorpha - Specifics

The surprising implications of, among other things, the data, have prompted a more intensive survey of the entire myomorph picture and led to a much more realistic assessment of how little we actually know and how much remains to be done. We first need to recognize that the molecular effort in this area is mainly confined to our albumin and transferrin immunological comparisons in Berkeley and the DNA annealing efforts of Rice (1974) and Brownell (1983). Though the results are pleasingly congruent, both qualitatively and quantitatively, they are hardly extensive enough to serve as anything more than a beginning attack on a most complex situation - complex enough, it should be noted, before any molecular effort was involved. As the *Acomys* situation, among others to be discussed, indicates, it is rather likely that the complexity is likely to increase, at least to begin with. I will also note that our own recent efforts in this area have used only antisera on hand from previous more circumscribed investigations, and thus have not been able to look into any number of relationships. Similarly, the DNA annealing efforts would appear to be limited technically, as there seems to be something about the organization of the myomorph genome which gives greater differences than one might expect from consideration of the immunological data, fossil record, and evidence from other groups such as carnivores and primates from which we have DNA annealing data. It would not appear that these larger measured DNA differences are readily explicable as a modest increase in the rate of nucleotide substitution among myomorphs (though that would not be an especially surprising finding); it is more that the data suggest that any given DNA annealing comparison between two myomorph taxa starts with a few degrees of ΔTm built in to the system. More specifically one notes (Brownell, 1983, Fig. 5) that 100% hybridization between two rodent DNAs can give ΔTm values up to 6°C. This gives one empirical pause, and suggests that whatever the underlying structural cause we need to automatically subtract some roughly constant amount from any reported myomorph DNA annealing ΔTm value, and that DNA annealing comparisons within the group will more rapidly approach the technical limitations of the approach. These conclusions emerge from lengthy correspondence and direct discussion with Dr. Brownell over the past 3 years or so. In any case, there is no question that immunological comparisons can at present see much further within the rodents than can the DNA annealing approach, and that the difference is greater for the rodents than for other groups where both approaches have been used. Again, however, it is not that this makes the one approach or the other better or worse; it is simply that their strengths and limitations need to be recognized, assessed, and taken into account in designing research strategies and drawing conclusions. Thus, the conclusions in the rest of this effort may be somewhat more circumspect than they might have been a few years ago. The complexity and magnitude of the matter of myomorph systematics are only now beginning to become clear.

The relevant data are presented in Tables 5 - 7. In surveying them, one is struck with the thought that the myomorph adaptive radiation must have begun not long after the original rodent radiation, as the intra-myomorph albumin differences, after correction for observed rate variation (Table 5), are rather larger than those seen among the hystricognaths. I have already suggested the period of common ancestry for the latter to be on the order of 10 MY; if this and the recognition of a sciurid/geomyoid/myomorph clade are valid conclusions, then the time span within which to resolve divergences leading to dipodids, rhizomyids, spalacids, and gliroids, to say nothing of more enigmatic forms such as Castor, Pedetes, ctenodactylids, and anomalurids, is bound to be relatively brief. All of this makes it very unlikely, as has already been noted for other adaptive radiations at issue here, that any single approach is going to be sufficiently powerful to provide the necessary resolving power. This situation is compounded by the lack of antisera for any of these non-cricetid/murid forms except Spalax and of any other relevant molecular data from any source on them. Thus logic demands that we start with a unit that is more or less well-defined by the available data, say what we can within it, and only then try to deal with less tractable taxa.

Dimensions of the Murid/Cricetid Unit

The albumin immunological and DNA annealing data define a unit which includes the New World cricetids, microtines, gerbillines, Old World cricetines, and murids. Within that unit the microtines form, as would be expected, a compact, though speciose, unit, with albumin cross-reactions greater than 85% and implied separation times within the last 10 MY. I very much doubt, however, that the very recent dates for within-microtine radiations posited by paleontologists (for example, Chaline, 1977; Chaline and Mein, 1979; Chaline, this volume) are realistic. This situation, where the fossil record is allegedly excellent, will provide another very interesting test of molecular dating techniques.

Except for the single Meriones-Gerbillurus comparison in Brownell (1983) suggesting a distance comparable to Rattus-Mus, no within-gerbilline data are available, and the within-Old World cricetine comparisons include only Mesocricetus, Cricetulus, and Mystromys. The DNA data of Rice (1974) here suggest, and our albumin immunological data concur, a Mesocricetus-Cricetulus distance somewhat less than that for Rattus-Mus and thus a separation time within the last 20 MY. Mystromys, on the other hand, is simply seen as part of an original cricetid radiation having no special affinities with the Old World cricetines proper. That leaves the much more complex questions raised by the data on murids and New World cricetids.

Among the murids the Acomys problem has already been raised and

Table 5.

Outgroup assessment of albumin evolution among rodents (QP cross reaction %)

anti - (Canis, Ursus, Felis, Hyaena) pool

Castor	45	Rattus	30
Clethrionomys	40	Cavia	30
Zapus	39	Sigmodon	30
Peromyscus	36	Acomys	30
Mesocricetus	35	Rhizomys	27
Eliomys	33	Pedetes	24
Spalax	33	Gerbillus	19
Mus	31	Aplodontia	12
Calomys	31		

discussed. In addition, the exclusion of a cricetomyine unit (Cricetomys, Steatomys, Saccostomus) from close affinity with other murids (Flynn, Jacobs, and Lindsay, this volume) is entirely consistent with the magnitude of the albumin differences between these forms and the murid unit which might be bounded molecularly by the Rattus and Mus lineages. Of the taxa in our collection only Praomys and Apodemus seem to definitely associate with the Mus lineage; all the others (Aethomys, Thallomys, Rhabdomys, Hydromys, Pogonomys, Melomys, Pseudomys, Conilurus, Mastacomys, Zyzomys, Mesembriomys, Notomys, Uromys, Xeromys, Leggadina, Leporillus) would appear to associate with the Rattus line or else to be part of the basal murine radiation. Resolving this very complex unit is well beyond the capacity of the range of antisera currently available to me. The one other group which does appear to associate more closely with this unit than do the cricetids proper (cricetines, New World cricetids, microtines, gerbillines) is Otomyinae, represented in our collection by Otomys. Again the beginnings of the murid adaptive radiation, leaving the Otomyinae out for the time being, appear to be dated by the Rattus-Mus separation at 22-24 MYA. Going back to the cricetid-murid radiation increases the observed albumin immunological distances by a factor of at least 1.5 and perhaps as much as 1.8, and the DNA data of Brownell and Rice are consistent with this estimate. Thus I would date the beginning of this radiation towards the end of the Eocene some 35-40 MYA. Our data are not sufficient to definitively deal with the phylogenetic placement of Spalax, the various cricetomyines and dendromurines (Saccostomus, Cricetomys, Steatomys), gliroids (Eliomys), zapodids (Zapus), and rhizomyids (Rhizomys) relative to one another and the murid/cricetid unit, though there is little molecular doubt as to their affinities with this unit. It is likely that the zapodids

Table 6.
Higher taxon molecular comparisons within Myomorpha

DNA annealing mean % hybridization

	NWC	OWC	Mi	Mu	G	N
New World Cricetidae	100					
Old World Cricetidae	40	100				
Microtinae	44	42	100			
Muridae	30	33	29	100		
Gerbillinae	30	21	26	30	100	
Napaeozapus	18	22	18	19	22	100
non-Myomorpha	19	17	17	18	19	19

from Brownell (1983)

Albumin cross reactions (QP %)

Reaction mean using antisera to various Cricetidae

Muridae	43
Spalax	38
Gerbillus	33
Eliomys	32
Zapus	28
Rhizomys	18
Hystricognath mean	20

(on the basis of both the albumin and DNA data) and rhizomyids (only albumin data) represent the earliest divergences among extant myomorph lineages, probably dating back to close to the origin of the suborder some 55 MYA. Thus I have to wonder about the assignment by Flynn, Jacobs and Lindsay of the rhizomyids to the "muroids of modern aspect" clade, and hesitate in making a stronger statement only because I have only the one *Rhizomys* sample and in such situations feel uneasy as to possible errors in identification or labeling The sample was collected by Dr. Murray Johnson many years ago and obtained from him along with the rest of his plasma collection in 1972. I should note that no error of identification or labeling has yet evidenced itself in the intervening years so that I think it quite likely that the *Rhizomys* lineage is indeed a good deal older than Flynn, Jacobs and Lindsay suggest. An albumin phylogeny of muroid and other non-hystricognath rodents is presented in Figure 3.

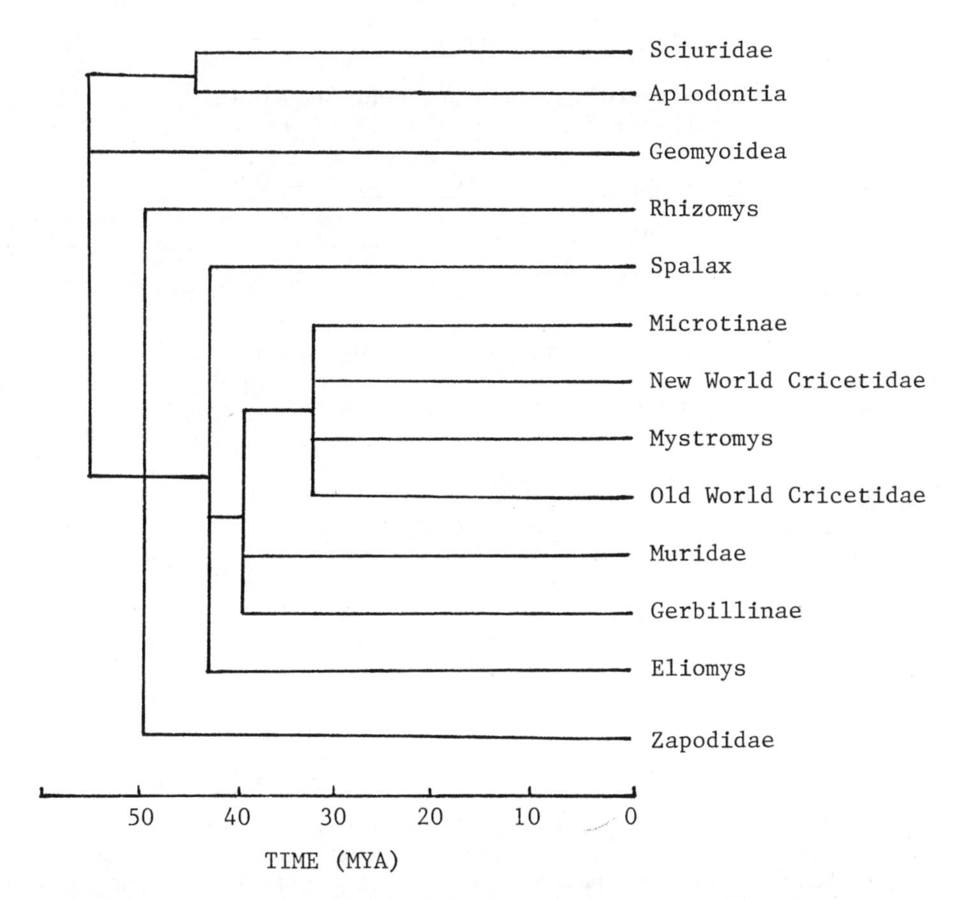

Fig. 3. A higher taxon phylogeny of non-hystricognath rodents. The data and derivation of this tree are given and discussed in the text with the exception of the placement of Aplodontia. Both Aplodontia and the sciurids seem to have albumins appreciably more changed than is modal for rodents; thus the Aplodontia - Spermophilus albumin cross-reaction of 25% would appear to indicate a significant association between the two lineages. The removal of Mystromys from the Cricetinae is based on the data of Rice (1974). The recognition of a cricetid unit (microtines, hesperomyines, Mystromys, cricetines) is strongly indicated by the DNA data of Brownell (1983) and consistent with my albumin data. The gerbils are pretty clearly not associated with the cricetid unit by either the DNA or albumin data, and could be slightly outside the murid/cricetid unit given the data available. The content of Muridae is given in the text. It is quite likely that Cricetomys, Saccostomus, and Steatomys are at least as distant from the murid/cricetid unit as the gerbils, but at present I lack appropriate antisera to test this matter in the requisite detail.

Table 7.

Albumin cross reactions among New World Cricetidae + Mesocricetus and Rattus (QP %)

	P	S	C	T	O	M	R
Peromyscus	100						
Sigmodon	63	100					
Calomys	64	57	100				
Tylomys	58	37	49	100			
Oryzomys	55	55	72	35	100		
Mesocricetus	57	51	49	51	40	100	
Rattus	50	43	40	39	31	49	100
Phyllotis	70	66	84	38	60	63	45
Ichthyomys	58	58	63	36	55	52	37
Holochilus	55	44	64	28	65	59	37
Nectomys	57	45	70	28	78	57	35
Zygodontomys	40	34	54	23	60	47	31
Akodon	68	44	69	35	58	47	37

New World Cricetids

Perhaps the most enigmatic or difficult situation in terms of the desired reconciliation of the molecular, paleontological, and neoanatomical data is that of the New World cricetids. The problem is simple to state but not so simple to solve. As Flynn, Jacobs, and Lindsay note in this volume, the oldest fossil South American cricetids date from only 3.5 MYA, and the oldest fossil which can reasonably be assigned to the group which includes South American cricetids is about 6 MY old, and is assigned to the extant genus Calomys. Yet we look at the distribution at the generic level and find (approximately) 31 genera restricted to South America, another 12 in South and Central America (Sigmodon and Oryzomys also in the Southern U.S.), and 10 belonging to the peromyscine/neotomine unit. About the reality of the last there is little doubt from a morphological perspective (e.g. Carleton, 1980), provided Tylomys and Ototylomys are excluded, and our albumin immunological distances obtained using antisera to two Peromyscus species and Onychomys are entirely consistent with this formulation. The cross-reaction between Peromyscus and Scotinomys, the two outliers on Carleton's phylogeny, is 86% and Neotoma is at 84%, while the next closest reaction is with Phyllotis at 70%. Peromyscus-Tylomys is at 58%, while the Tylomys-Ototylomys cross-reaction is 82%. These data suggest that the peromyscine/neotomine radiation began within the last 10 MY or so and we can, for the time being, leave it at that.

The other some 40 genera are the real problem. If one were to look simply at the biogeography, then the only possible conclusion would be that this is a South American radiation already well differentiated at the generic level (at least) by the time the Isthmus of Panama appeared and allowed its extension into Central America. If it were not for the lack of a South American fossil record for the group prior to the emergence of the Isthmian connection - a very large if - no one would have ever given the matter a second thought. But the fossil record is not there (is absence of evidence evidence of absence here?), and this fact necessarily leads to the consideration of other scenarios, however unlikely they may appear at first glance.

This possibility that the large observed albumin differences are due to a much more rapidly running albumin clock is easily excluded by the rate data. There are two lineages with rather more than the average amount of change (*Oryzomys* and *Zygodontomys*), but all the others cluster quite tightly around the general rodent mode. Thus by direct test there is nothing unusual about the pattern of albumin differentiation in the group. About the only other rational scenario was sketched by Wood at this conference. He argued for a Central America/Mexico staging area within which much of the diversity of the group accumulated with a post-Isthmian migration of a number of these lineages into South America. This model has the virtue of being readily testable, though the critical tests have not yet been done. What we do have is consistent with the model but also with the alternative of having the basic radiation being South American and old. We find that several of the Central American genera are speciose, most extend into South America, and that a number of them are at substantial albumin immunological distances from one another (e.g. *Sigmodon*, *Oryzomys*, *Zygodontomys*, *Tylomys*, *Nyctomys*, *Nectomys*, *Rhipidomys*, *Ichthyomys*). Thus the critical test comes with the endemic South American forms, for there the Wood scenario requires that the vast majority of those lineages be within less than 4 MY worth of genetic distance of one another; that is, the sorts of distances one finds within, for example, the *Thomomys bottae-umbrinus* complex of Western North America (Patton and Feder, 1978). This should be true both intra- and intergenerically and most expeditiously sampled using standard electrophoretic procedures. If the Wood scenario turns out to be the correct one, it will be the most remarkable example of rapid morphological differentiation documented in the recent history of mammals.

The available data for the group (Table 7) indicate great antiquity for a number of lineages relative to one another, including two South American endemics, *Akodon* and *Phyllotis*. Figure 4 attempts to place these data into the beginnings of a temporal perspective. Note that for the Wood scenario to be correct, we will need to find a number of South American clades where the intra-clade divergence times are less than for any pair of genera in Figure 4.

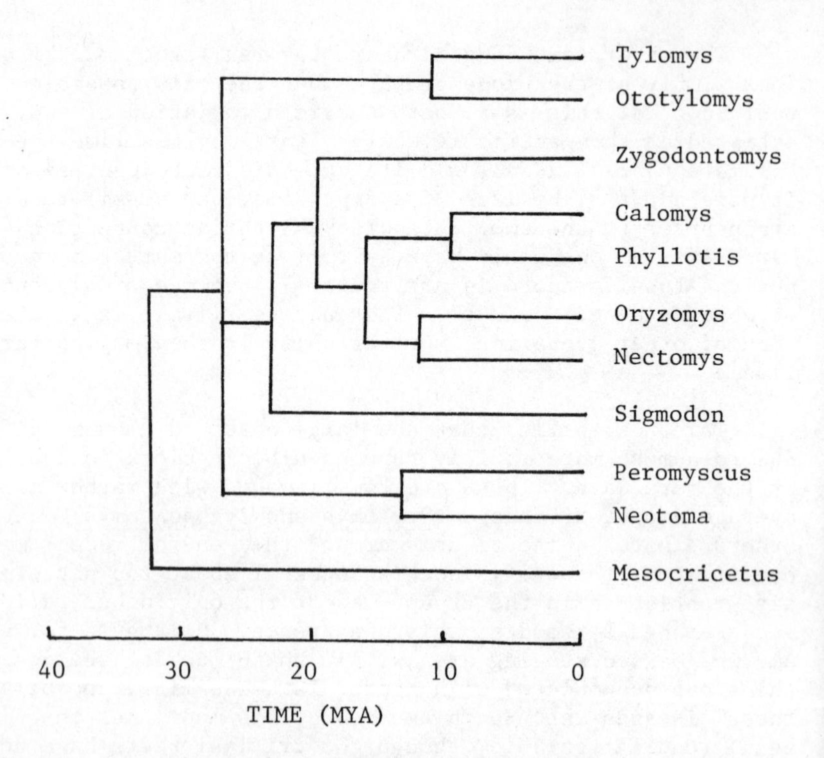

Fig. 4. Albumin immunological phylogeny of New World Cricetidae. This tree was developed from the data in Table 7. The genera not in the figure (Ichthyomys, Holochilus, Akodon) also would appear to associate with the clade of five genera which includes Oryzomys, but in the absence of any other antisera to taxa in that clade one cannot place these three relative to the other five.

CONCLUSIONS

I was much encouraged by this conference. The genuinely collegial atmosphere throughout the whole of the five days indicated that virtually everyone there - and it was a very diverse group indeed - wanted to be part of an ongoing give and take dialogue that would lead to major clarifications of a number of rather vague evolutionary scenarios. Some major problems were brought much closer to final solution; other previously unsuspected problems took their place. But all in all progress was clear. Things that seemed impossible began to make sense. I offer this article in that spirit.

REFERENCES

Andrews, P. J. 1983. The natural history of Sivapithecus. In: New Intrepretations of Ape and Human Ancestry, R. L. Ciochon and R. S. Corruccini, eds., pp. 441-463, Plenum Press, New York.

Benjamin, D. C., Berzossky, J. A., East, I. J., Gurd, F. R. N., Hannum, C., Leach, S. J., Margoliash, E., Michael, J. G., Miller, A., Prager, E. M., Reichlin, M., Sercarz, E. E., Smith-Gill, S. J., Todd, P. E. and Wilson, A. C. 1984. The antigenic structure of proteins: a reappraisal. Ann. Rev. Immun. 2: 67-101.

Brownell, E. 1983. DNA/DNA hybridization studies of muroid rodents: symmetry and rates of molecular evolution. Evolution 37: 1034-1051.

Carleton, M. D. 1980. Phylogenetic relationships in neotomine-peromyscine rodents (Muroidea) and a reappraisal of the dichotomy within New World Cricetinae. Misc. Publ. Mus. Zool. Univ. Mich. 157: 1-146.

Chaline, J. 1977. Rodents, evolution, and prehistory. Endeavour 1: 44-51.

Chaline, J. and Mein, P. 1979. Les Rongeurs et l'Evolution. Doin, Paris.

Cronin, J. E. and Sarich, V. M. 1980. Tupaiid and Archonta phylogeny: the macromolecular evidence. In: Comparative Biology and Evolutionary Relationships of Tree Shrews, W. P. Luckett, ed., pp. 293-312, Plenum Press, New York.

Goodman, M. 1965. The specificity of proteins and the process of primate evolution. In: Protides of the Biological Fluids, H. Peters, ed., pp. 70-86, Elsevier, Amsterdam.

Hafner, M. S. 1982. A biochemical investigation of geomyoid systematics. Z. Zool. Syst. Evol. Forsch. 20: 118-130.

Jacobs, L. L. 1978. Fossil rodents (Rhizomyidae and Muridae) from Neogene Siwalik deposits, Pakistan. Mus. North. Arizona Bull. Ser. 52: 1-103.

Jacobs, L. L. and Pilbeam, D. 1980. Of mice and men: fossil-based divergence dates and molecular "clocks". J. Human Evol. 9: 551-555.

Kohne, D. 1970. Evolution of higher organism DNA. Quart. Rev. Biophysics 3: 327-375.

McKenna, M. C. 1975. Toward a phylogenetic classification of the Mammalia. In: Phylogeny of the Primates, W. P. Luckett and F. S. Szalay, eds., pp. 21-40, Plenum Press, New York.

Novacek, M. J. 1982. Information for molecular studies from anatomical and fossil evidence on higher eutherian phylogeny. In: Macromolecular Sequences in Systematic and Evolutionary Biology, M. Goodman, ed., pp. 3-41, Plenum Press, New York.

Patton, J. L. and Feder, J. H. 1978. Genetic divergence between populations of the pocket gopher Thomomys umbrinus. Z. Saugetierkunde 43: 17-30.

Rice, N. R. 1974. Single copy DNA relatedness among several species of the Cricetidae (Rodentia). Carneg. Inst. Wash. Yrbk 73: 1098-1102.

Sarich, V. M. 1970. Primate systematics with special reference to Old World monkeys. In: Old World Monkeys: Evolution, Systematics, and Behavior, J. R. Napier and P. H. Napier, eds., pp. 175-226, Plenum Press, New York.

Sarich, V. M. 1972. Generation time and albumin evolution. Biochem. Genet. 7: 205-212.

Sarich, V. M. and Cronin, J. E. 1976. Molecular systematics of the primates. In: Molecular Anthropology, M. Goodman and R. E. Tashian, eds., pp. 141-170, Plenum Press, New York.

Sarich, V. M. and Cronin, J. E. 1980. South American mammal molecular systematics, evolutionary clocks, and continental drift. In: Evolutionary Biology of New World Monkeys and Continental Drift, R. L. Ciochon and A. B. Chiarelli, eds., pp. 339-422, Plenum Press, New York.

Schewale, J. G., Sinha, S. K. and Brew, K. 1984. Evolution of α-lactalbumins. J. Biol. Chem. 259: 4947-4956.

Sibley, C. G. and Ahlquist, J. E. 1984. The phylogeny of hominoid primates, as indicated by DNA-DNA hybridization. J. Mol. Evol. 20: 2-16.

Simons, E. L. 1976. The fossil record of primate phylogeny. In: Molecular Anthropology, M. Goodman and R. E. Tashian, eds., pp. 35-62, Plenum Press, New York.

Simpson, G. G. 1959. The nature and origin of supraspecific taxa. Cold Spring Harbor Symp. Quant. Biol. 24: 255-271.

Ward, S. C. and Pilbeam, D. R. 1983. Maxillofacial morphology of Miocene hominoids from Africa and Indo-Pakistan. In: New Interpretations of Ape and Human Ancestry, R. L. Ciochon and R. S. Corruccini, eds., pp. 211-238, Plenum Press, New York.

Wood, A. E. 1957. What, if anything, is a rabbit? Evolution 11: 417-425.

REPRODUCTIVE AND CHROMOSOMAL CHARACTERS OF CTENODACTYLIDS AS A KEY TO THEIR EVOLUTIONARY RELATIONSHIPS

Wilma George

Department of Zoology
University of Oxford
Oxford, England

INTRODUCTION

When Zittel named the family Ctenodactylidae in 1893 to include *Ctenodactylus*, *Pectinator*, *Petromys* and the fossil *Pelligrinia*, the genera of gundis had had a chequered career in the classification systems. Originally, *Ctenodactylus* had been classified with *Petromus* as a dipodid by Gervais (1848), but gundis were octodontids for Brandt (1855), Flower and Lydekker (1891). Blyth (1855) assigned *Pectinator* to the chinchillids, and the two genera became murids in Alston's classification of 1876. With the addition of two more genera, the gundis achieved family status, but this did not make their position any more certain, and they wandered round the sub-orders for the next 90 years. Tullberg (1899) and Ellerman (1940) claimed them for the Myomorpha; Wood (1955) for the Sciuromorpha; Thomas (1896), Winge (1924), Weber (1928) and Landry (1957) for the Hystricomorpha; and the majority were unable to assign them to any of the three conventional suborders (Miller and Gidley, 1918; Bohlin, 1946; Simpson, 1945; Grassé and Dekeyser, 1955; Wood, 1965, 1974, 1977; Shevyreva, 1971; Chaline and Mein, 1979; Patterson and Wood, 1982). They gave rise to feelings of despair in some authors: "this group vies with or exceeds the bathyergids in uncertainty" (Simpson, 1945) - and "les rapports avec les autres formes restent de plus en plus énigmatiques" (Grassé and Dekeyser, 1955).

The taxonomists relied mainly on skeletal and muscle morphology for their criteria, and ambiguities arose over the apparent mixture of hystricomorphous, sciuromorphous and myomorphous characters in the family. Thus, they are hystricomorphous and sciurognathous. They have multiserial enamel on the incisors but lack the posterior part of the *M. masseter lateralis profundus*.

CHARACTER ANALYSIS OF CTENODACTYLIDS AND OTHER RODENTS

Increasing knowledge of rodent soft-part anatomy, karotypes, ecology and behaviour has made it possible to approach the classification of the ctenodactylids phenetically.

Two methods of phenetic classification show the affinities of modern ctenodactylids with other modern rodent families: a graph-linkage variant of single-link cluster analysis (Estabrook, 1966; Bisby, 1973) and cladistic analysis of shared characters (Wiley, 1981).

In cluster analysis, characters are not weighted for phylogenetic significance. It is, therefore, important to use large numbers of characters to reduce the effect of convergent characters. Thus, the analysis was based on 69 characters ranging from skeletal structures through soft-part anatomy to hair patterns and chromosome characters. Most of the information on soft-part anatomy and karyotypes comes from personal observation. The rest is from published sources and is summarised in an article in preparation. Sufficient information was obtained from 29 rodent families and included ten South American and seven African hystricognath families.

The graph-linkage cluster analysis joins families into larger groupings at different levels in a hierarchy. As each family joins the cluster it is linked to the family most like itself. The value of the similarity coefficient indicates relative levels in the hierarchy though, in itself, it cannot be tested for statistical significance. It is a mixed data coefficient that allows mixing different types of characters, including multi-state characters. Thus, protrogomorphy, sciuromorphy, hystricomorphy and myomorphy are treated as four states of a single character. It allows for the use of data from which some values are missing; but in the present analysis, more than ten gaps in a family or in a character eliminated it from further consideration. For this reason, several families are omitted, and all biochemical information is omitted.

Figure 1 shows the links between the 29 modern families. Finer detail, such as the exact level and point of linkage of caviids with the other hystricognaths, is not illustrated although it emerged in the analysis. The similarity coefficient at which precise links are illustrated starts at 0.83 in Figure 1, by which time most of the South American caviomorphs and the thryonomyids have joined their cluster, while murids, cricetids, gerbillids and dipodids have made another cluster. The analysis shows a close-knit hystricognath block linking in hystricids and petromurids less closely. Erethizontids link at the next level to a capromyid group and are quickly followed by bathyergids and ctenodactylids. Ctenodactylids link equally to petromurids and thryonomyids. They have no links with any myomorph or sciuromorph families, nor with pedetids or anomalurids. These last two families make a tenuous link between the two major blocks and notably resemble one another (possibly owing to the retention of an array of generalised rodent characters).

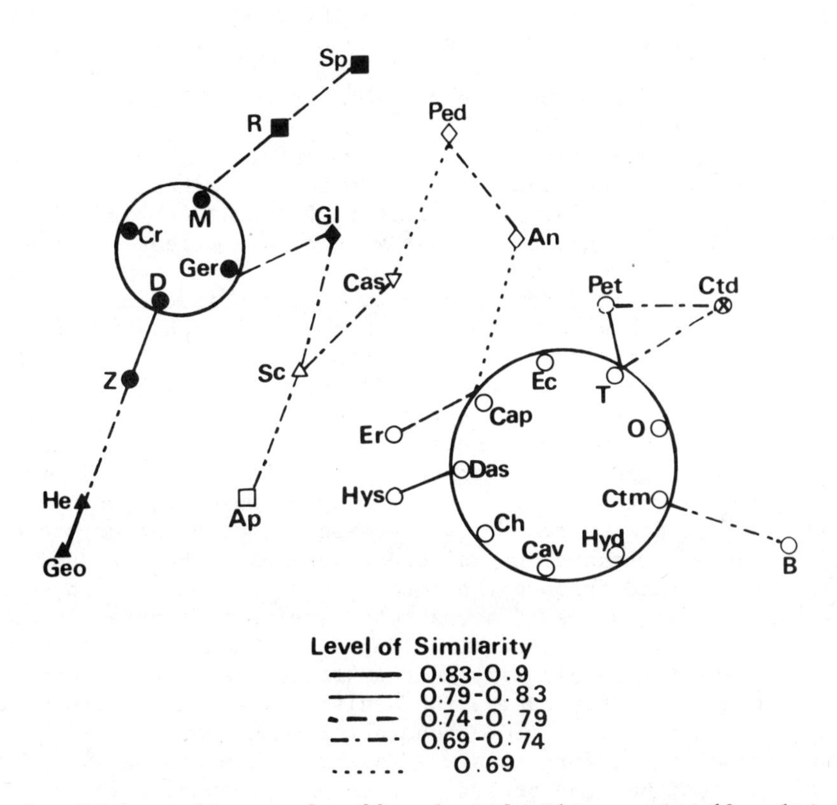

Fig. 1. Linkage diagram for 29 rodent families using 69 multi-state characters. An = Anomaluridae, Ap = Aplodontidae, B = Bathyergidae, Cap = Capromyidae (including Myocastor), Cas = Castoridae, Cav = Caviidae, Ch = Chinchillidae, Cr = Cricetidae, Ctd = Ctenodactylidae, Ctm = Ctenomyidae, D = Dipodidae, Das = Dasyproctidae, Ec = Echimyidae, Er = Erethizontidae, Geo = Geomyidae, Ger = Gerbillidae, Gl = Gliridae, He = Heteromyidae, Hyd = Hydrochoeridae, Hys = Hystricidae, M = Muridae, O = Octodontidae, Ped = Pedetidae, Pet = Petromuridae, R = Rhizomyidae, Sc = Sciuridae, Sp = Spalacidae, T = Thryonomyidae, Z = Zapodidae.

Cluster analysis groups like with like, but it does not distinguish between so-called primitive characters, characters derived through a common ancestor and those acquired in parallel or convergently. It does not indicate which characters are shared by families within the main blocks, which are spurious and which are positively correlated with one another. Interestingly, it links bathyergids to ctenomyids, owing to the convergence of many characters associated with burrowing, but bathyergids are almost equally linked with the other octodontids. Ctenodactylids may owe some of their linkage with petromurids to the similarity of their habitats, but their linkage to thryonomyids - and, at a slightly lower level, to chinchillids - makes their hystricognath connections fairly certain.

The next analysis was a search for shared characters on the basis of the clustering results. The criterion for the choice of characters was the generality of the characters within the rodents, the less general being regarded as of more significance than the more general (Nelson, 1978). Characters that varied significantly within families and characters that were spread across families within all the suborders were eliminated on this criterion. The scale pattern of hairs, for example, was eliminated because the same pattern occurred in some families of myomorphs and hystricognaths, in geomyoids, glirids and protrogomorphs. There remained 31 characters, shown in Table 1. The characters and their alternatives represent states that have at various times been identified as rodent plesiomorphs and apomorphs. However, in many cases it is impossible to be certain of the derived or generalised state of the character for the rodents without assuming a phylogeny which it was the aim of the analysis to discover. For this reason, the character states in Table 1 should be regarded as alternatives only, the one showing a more general spread through the rodents than the other. In some cases, the embryogenesis of a character was used to determine the likelihood of its being general or derived. Such a criterion was used, for example, in classifying the attributes of the blood vascular system. In other cases, the decision was made according to the distribution pattern of the character among mammals in general; the nipple position, for example. Fossil evidence usually determined skeletal states, but the presence of an entepicondylar foramen in the humerus proved a puzzle. It occurs in insectivores and tree shrews (Novacek, 1980) and is reported in a eurymyloid (Li and Ting, this volume), but Davies (1982) argued a convincing case for its derived state in pedetids.

The distribution of the 31 characters of the shared-character analysis produced the same two blocks of families as the cluster analysis: the myomorph-sciuromorph and the hystricognath, with anomalurids and pedetids in between. Ctenodactylids were associated most closely with the hystricognath block. Of the 31 characters, 24 (77%) are shared, though not exclusively, by ctenodactylids and some hystricognaths; 3 (10%) are shared between hystricognaths, ctenodactylids and either the pedetids or anomalurids or both. Sciuromorphs and myomorphs had no characters shared exclusively with ctenodactylids but sciuromorphs shared another 3 characters with ctenodactylids, aplodontids, anomalurids and pedetids (all characters of placentation).

Table 1. Alternative character states used in grouping rodent families into major blocks.

	Character state (O in Fig. 2)	Alternative character state (● in Fig. 2)
1.	Malleus and incus free	Malleus and incus fused
2.	Uniserial enamel	Multiserial enamel
3.	Vagina patent	Closure membrane
4.	Penis simple	Sacculus urethralis
5.	Parietal endoderm of blastocyst complete	Parietal endoderm incomplete
6.	No *M. scapuloclavicularis*	*M. scapuloclavicularis* present
7.	Ventral nipples	Lateral nipples
8.	Chromosomes uniform	Chromosome with big n.o.r.
9.	Subplacenta absent	Subplacenta present
10.	*M. masseter lateralis profundus* undivided	*M. masseter lateralis profundus* divided
11.	Implantation superficial	Implantation interstitial
12.	Sciurognathy	Hystricognathy
13.	Acromial process simple	Acromial process bifurcate
14.	Ovary simple	Ovary with accessory corpora lutea
15.	Sciuromorphy	Hystricomorphy
16.	Straight cochlea	Elbowed cochlea
17.	Stapedial muscle absent	Stapedial muscle present
18.	Cricetid systemic arches	Sciurid systemic arches
19.	Pterygoid fossa not opening into orbit	Pterygoid fossa opens into orbit
20.	Amniogenesis by folding	Amniogenesis by cavitation
21.	Allantoic vesicle permanent	Allantoic vesicle absent
22.	Teeth more than 20	Teeth fewer than 20
23.	Cervical vertebrae free	Cervical vertebrae fused
24.	*M. stylohyoideus* absent	*M. stylohyoideus* present
25.	Lung lobes 2-4:3	Lung lobes 1:3
26.	Tibia-fibula free	Tibia-fibula fused
27.	No entepicondylar foramen	Entepicondylar foramen present
28.	Lachrymal duct in orbit	Lachrymal duct outside orbit
29.	Stapedial artery present	Stapedial artery absent
30.	Sciuromorphy	Myomorphy
31.	Bilateral or right azygous veins	Left azygous vein

Following the 31-character analysis all characters that were present in both the myomorph-sciuromorph block and the hystricognath block were eliminated in spite of the fact that some characters have been shown to be convergent. This elimination removed hystricomorphy and the bifurcate acromium, for example. The characters were reduced to 20, shown in Figure 2. The two blocks remain and ctenodactylids, though on the edge, are firmly associated with hystricognaths. The characters they share with the myomorph-sciuromorph block, like sciurognathy and superficial implantation, are probably plesiomorphs. In fact, in this figure, ctenodactylids have more seemingly derived characters in common with any caviomorph family than a geomyid has, for example, with a myomorph or a sciuromorph. The anomalurids and pedetids remain anomalous, in between the two main blocks.

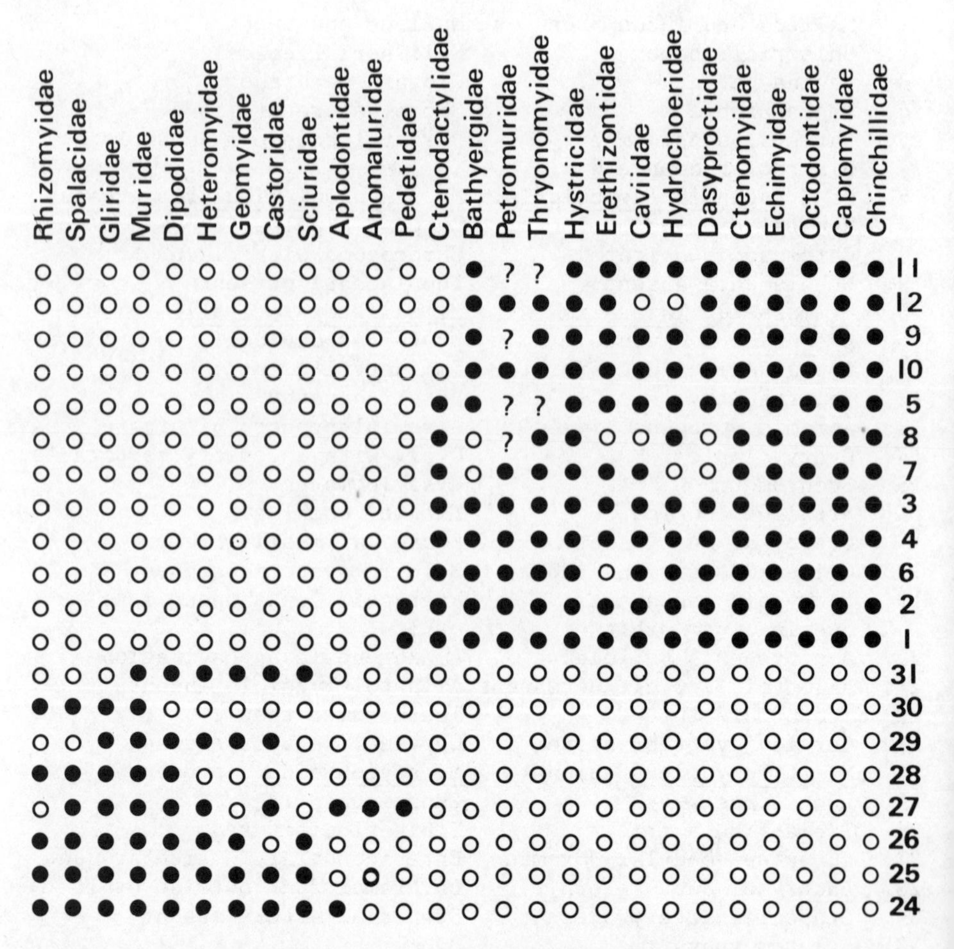

Fig. 2. Character phylogeny of rodent families using 20 of the characters listed in Table 1.

Figure 3 was constructed from 12 characters that unite all hystricognath families. It shows the significant links ctenodactylids have with hystricognaths. Bathyergids are more closely linked than ctenodactylids but anomalurids have no connection and pedetids little. The characters that unite ctenodactylids with hystricognaths come from a variety of systems: masticatory, reproductive, locomotory, auditory and karyotypic.

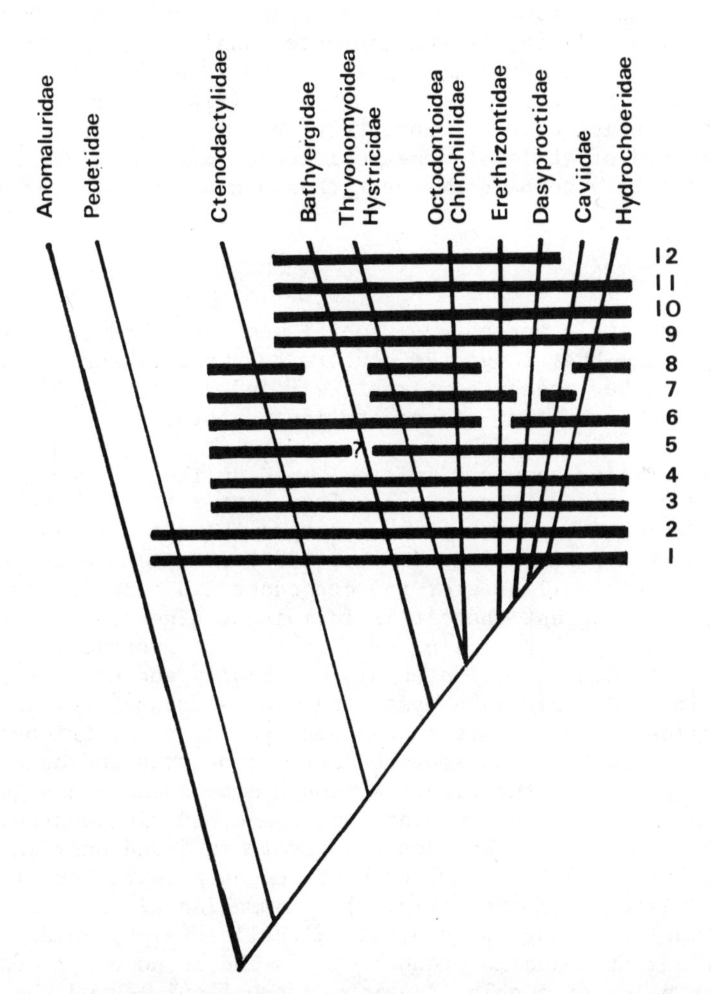

Fig. 3. Cladogram of hystricognath rodents and the relationship to them of Ctenodactylidae, Pedetidae and Anomaluridae. Characters numbered from Table 1.

AFFINITY OR CONVERGENCE?

From the phenetic analyses of modern rodent families it looks likely that ctenodactylids have a connection, phylogenetically, with hystricognaths. Their closest links, demonstrated in the cluster analysis (Fig. 1), are with African thryonomyids and petromurids.

However, there is a problem that characters that emerge as apparently significant phylogenetically could equally well be ascribed to convergence. The use of isolated single characters is particularly suspect. Petromurids and ctenodactylids are arid area rock-dwellers and could owe their similarities to similarity of habitat. They both have combs on their hind toes (though this is not a character that has been included in the shared-character analysis). A whole masticatory system may be more important than the single attribute of hystricomorphy or hystricognathy (Wahlert, 1974). For this reason, patterns or mosaics of some characters were investigated. A pattern though composed of at least some positively correlated characters may be less likely to be convergent than a single character state.

REPRODUCTIVE PATTERN

Male and female morphology, foetal membranes and reproductive physiology among hystricognaths conform to a pattern which differs from that of other rodent families (Weir and Rowlands, 1973; Weir, 1974). To what extent do ctenodactylids conform?

Ctenodactylids have a simple os penis unlike, for example, many muroids and sciurids (Burt, 1960). They have a small sacculus urethralis which is a dorsal pocket-outgrowth on the penis. It occurs in most hystricognaths (Dathe, 1937) and in no other group of rodents. The small size of the ctenodactylid sacculus has led some authors to suggest that it is incipient, degenerate (Wood and Patterson, 1959) or not homologous with that of hystricognaths. It has not yet been investigated histologically and until that has been done it is defined as a hystricognath sacculus. Female ctenodactylids too have some characters typical of hystricognaths. The vagina is closed by a membrane during anoestrus and opens only at oestrus and for parturition. Although many rodents have a seasonally sealed vagina the regular appearance and disappearance of the closure membrane at each oestrous cycle is found only in hystricognaths (Weir, 1974). A further hystricognath character is the presence of lateral nipples (Fig. 4). Mammalian nipples normally develop at points along two ventral epithelial thickenings. In most hystricognaths and in ctenodactylids and in no other rodents one or more pairs of nipples develop on the flank behind the armpit. Hydrochoerids, dasyproctids, bathyergids and <u>Cavia</u> are exceptions bu hydrochoerids and dasyproctids are rodents that stand high off the ground suckling their young and bathyergids are subterranean.

Fig. 4. Young Ctenodactylus vali at a lateral nipple.

Thus, lateral nipples can be assumed to have been secondarily lost in tall rodents and rodents in cramped burrows. Cavia is the more exceptional in that it has fewer nipples than offspring and none is lateral. This may be accounted for by its habit, shared with hydrochoerids, of communal suckling.

Finally, the ovary of ctenodactylids resembles closely the ovary of Thryonomys (Oduor-Okelo, 1978) in having accessory corpora lutea formed from atretic follicles and numbering more than the number of embryos. Whether these ovaries are strictly "hystricognathous" is uncertain.

Thus, ctenodactylids share four to five structural reproductive characters with the hystricognath block. In addition, ctenodactylids have a comparatively long oestrous cycle (23-29 days), long gestation (56 days), and a small number of young born furred and open-eyed (George, 1978). A good deal has been made of this aspect of rodent reproduction, though it is difficult to assess how much it really sets hystricognaths and ctenodactylids apart from the other rodents. Castorids have a long gestation and well-formed neonates, so that the significant factor could be size. Various attempts have been made to standardise these characters: relating gestation length to maternal weight; litter weight to maternal weight and gestation length (Eisenberg, 1981); and even neonate brain size to gestation length (Sacher and Staffeldt, 1974).

If the gestation length is related to maternal weight, there is no difference between the sciurid Cynomys ludovicianus and the hystricognaths Thyronomys swinderianus and Myocastor coypus; both have a relatively short gestation for their weight. Hydrochoerus and Hystrix are heavyweights and have gestation lengths of 150 and 112 days, but this is relatively short compared with their weights. Ctenodactylids, at an average weight of 180g and with a gestation of 56 days, fall into the same category as Lagostomus maximus but also as Gerbillus pyramidum. The relatively longest gestation for a hystricognath is found for Octodontomys gliroides which, with a weight of 100g, gestates for 99 days. In contrast, the rodents with the relatively longest gestations are Reithrodontomys montanus and Baiomys taylori which weigh under 10g but gestate for around 20 days. The seemingly long gestation period of hystricognaths is not relatively long compared with adult size.

Another method of standardisation is to relate litter weight to gestation length. Here, again, there is considerable scatter across the families. On the whole, big rodents produce a heavier litter per unit time than small rodents. Thus, Castor canadensis and Ondatra zibethicus produce as heavy a litter per unit time as Agouti paca. Ctenomys talarum and ctenodactylids are the equivalents of the chipmunk Tamias striatus and Mus musculus.

Eisenberg (1981) suggested relating the litter weight divided by the female weight to the gestation length plus days from birth to eye opening. He worked this through for a few cricetids. I have extended this to other rodent families. A plot of 55 species representing 19 families (often not more than one species in a family) shows considerable scatter (Fig. 5). The myomorph boundaries do not enclose any hystricognaths and only one sciuromorph. However, the sciuromorph boundaries contain echimyids, caviids, pedetids and ctenodactylids (Fig. 5). Many hystricognaths are well outside the boundaries of the myomorphs and sciuromorphs, and the myomorphs are self-contained at the other extreme. Thus, two distinct patterns of gestation emerge by this method: a myomorph pattern with short gestation for a relatively heavy litter, and a hystricognath pattern with greater scatter but, on the whole, a longer gestation to produce a relatively smaller litter weight. Ctenodactylids are on the borders of the hystricognath group and just outside the myomorph. Sciuromorphs have an intermediate pattern, overlapping both myomorph and hystricognath groups.

All hystricognath young are born with eyes open, a character shared only with castorids and exceptional species like Sigmodon hispidus (Eisenberg, 1981). Ctenodactylid young are fully-furred and open-eyed at birth.

The foetal membranes of ctenodactylids are ambiguous and, in many ways, resemble sciuromorphs. As Luckett (1971 , 1980a, 1980b) has demonstrated, ctenodactylids fall into an intermediate position. Of his 11 characters, only four are shared with the hystricognath

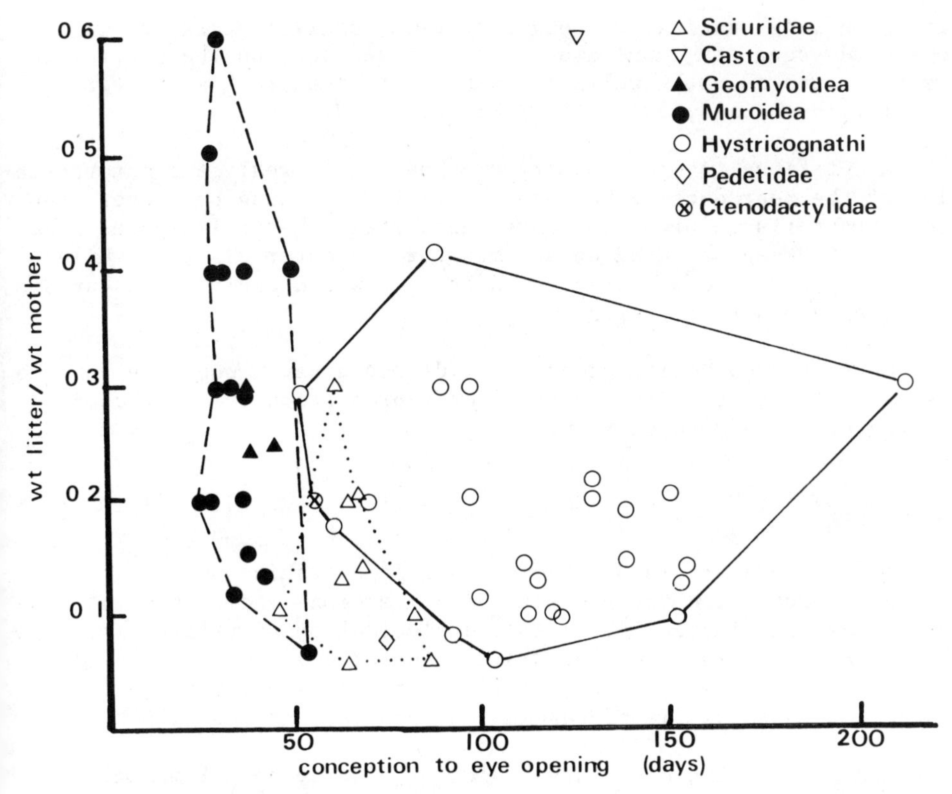

Fig. 5. Ratio of litter weight divided by weight of mother plotted against gestation to eye opening (in days) in 55 species of rodent.

block and three of the four are found among myomorphs. Thus, ctenodactylids, unlike hystricognaths, have superficial implantation and a plesiomorphic placenta but like hystricognaths have a reduced endodermal lining to the blastocyst.

Taking all the reproductive characters together and comparing them with other patterns, ctenodactylids differ least from pedetids, petromurids and thryonomyids. They differ most from myomorphs, protrogomorphs and sciuromorphs (but it must be noted that they differ, to a lesser degree, from most of the caviomorphs).

To what extent can this mosaic of reproductive characters be superficially adaptive? It has been argued (George, 1978) that many of the ctenodactylid reproductive features are typical of mammals living in rocky habitats. Long gestation, small litters and well-developed young are found among rock hyraxes but, on the contrary, are not found in rock-dwelling pikas and Mexican rock squirrels (Asdell, 1964; Sale, 1965, 1969; Millar, 1972).

Adaptations to open country provide a more realistic interpretation of the reproductive pattern. Well-furred young born eyes open are characteristic of open-country ungulates. Lateral nipples allow a rodent to keep her head up and monitor a greater distance while suckling young. Such character adaptation would apply where burrows or trees were not an option.

On this hypothesis, ctenodactylids would have gone some way from a supposedly squirrel-type pattern of reproduction toward the open-country hystricognath pattern.

KARYOTYPES

Karyotype analysis has proved useful in tracing speciation through Robertsonian transformations, inversions and translocations, but it has contributed little to the elucidation of relationships between higher taxa. Neither diploid number nor nombre fondamental (nf) is a useful measure of affinity, and following more subtle changes by the analysis of banding patterns has proved difficult.

Rodents are the most karyotypically diverse of all mammals, with diploid numbers ranging widely from family to family and from species to species. The hystricognath block has a range of diploid numbers from 2n=22 in *Ctenomys occultus* to 2n=88 in *Geocapromys brownii*, with no perceptible peak in the distribution curve (George and Weir, 1974)

The five ctenodactylid species have 2n=40 and 2n=36 (Figs. 6,8) In addition, all ctenodactylids have a metacentric chromosome with a strikingly big n.o.r. or nucleolar organiser region (George, 1979a Permanent chromosomes of this type are found in a few insectivores (Borgoankar, 1969; Gropp, 1969), in the giraffe (Koulischer, 1973), and, consistently among the Carnivora (Wurster and Benirschke, 1968; Wurster, 1969). In this last group n.o.r. chromosomes have been used to establish phylogenetic relationships, between Ursidae and Canidae, for example (Wurster and Benirschke, 1968). Among rodents, only members of the hystricognaths have such a chromosome as a permanent feature of the karyotype (Figs. 6,7). It is present in the majority of hystricognaths though absent from most caviids, dasyproctids and erethizontids. It is found in thryonomyids (George, 1980) and some hystricids (Renzoni, 1967), though it appears not to be present in bathyergids (George, 1979b). There seems, therefore, to be some evidence for regarding the n.o.r. chromosome in rodents as a derived character.

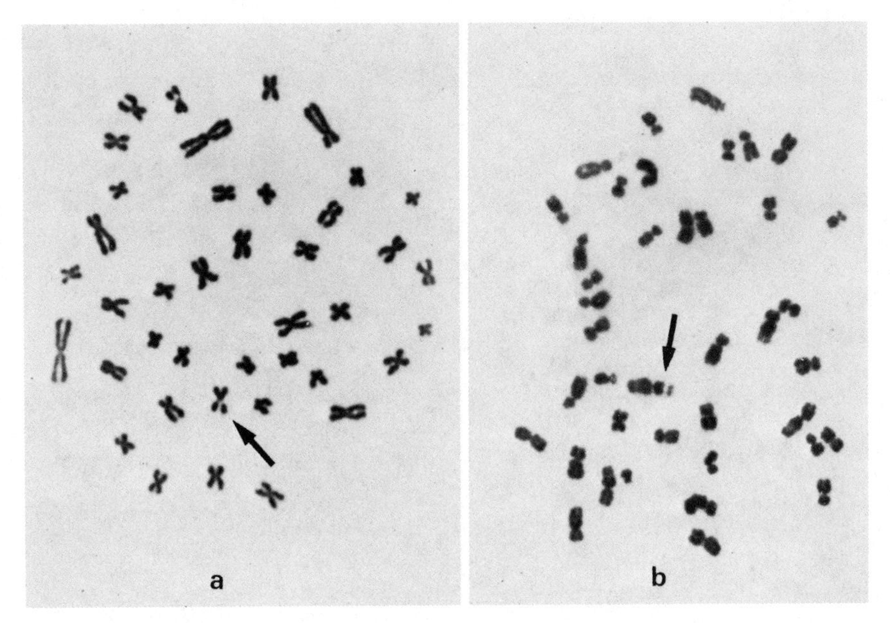

Fig. 6. Metaphase plates of (a) female Ctenodactylus gundi and (b) male Thryonomys gregorianus. Nucleolar organiser chromosomes (n.o.r.) arrowed.

Banding techniques have revealed a close similarity of the x-chromosomes of those hystricognaths that have been treated in this way (George, 1980). Ctenodactylids again fit the hystricognath pattern (Figs. 8,9). They resemble pedetids and hystricognaths but differ markedly from sciuromorphs (Nadler et al., 1975) and myomorphs (Gallimore and Richardson, 1973).

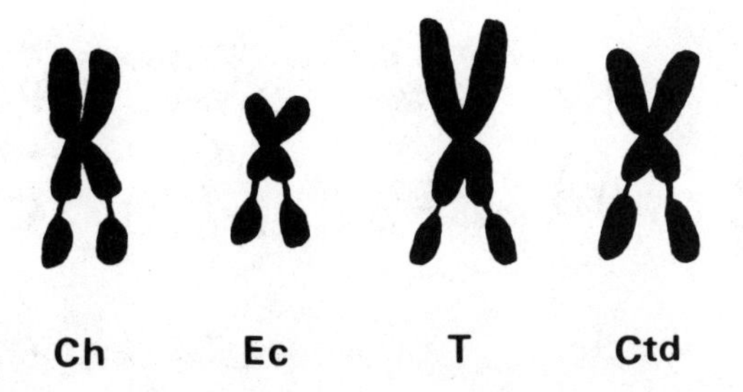

Fig. 7. Nucleolar organiser chromosomes (n.o.r.). Ch, chinchillid; Ec, echimyid; T, thryonomyid; Ctd, ctenodactylid.

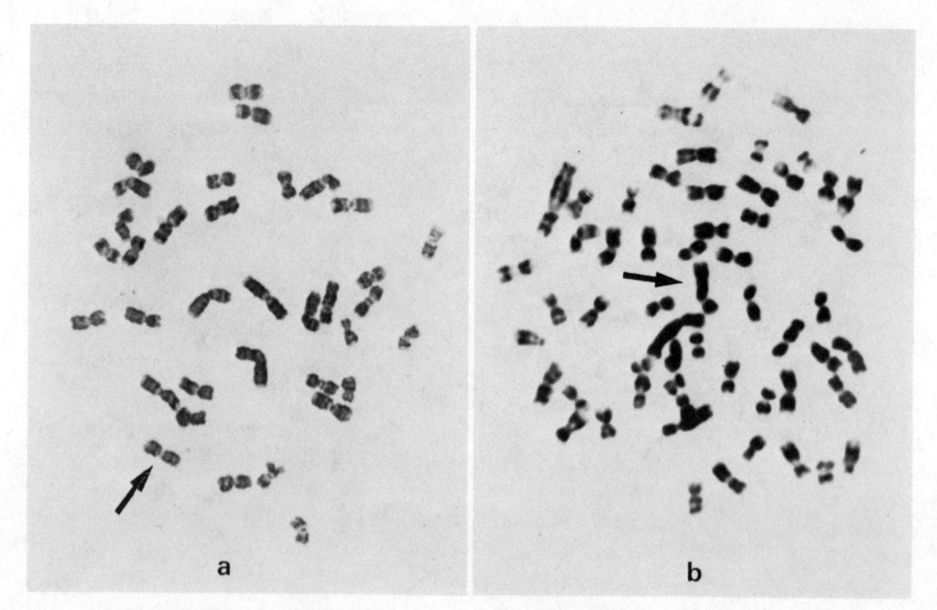

Fig. 8. Metaphase plates of (a) female Massoutiera mzabi and (b) female Lagostomus maximus. G-banded X chromosomes arrowed.

If there is any significance in similarities of karyotype, then the ctenodactylids are like the African and South American hystricognaths in high NF, possession of an n.o.r. chromosome, and in the banding pattern of the X-chromosome. They do not resemble myomorphs or sciuromorphs. There seems no reason to suppose that the big n.o.r. and X-chromosome pattern of the ctenodactylids are convergent characters, but the possibility cannot be excluded. Karyotypes are characteristically variable.

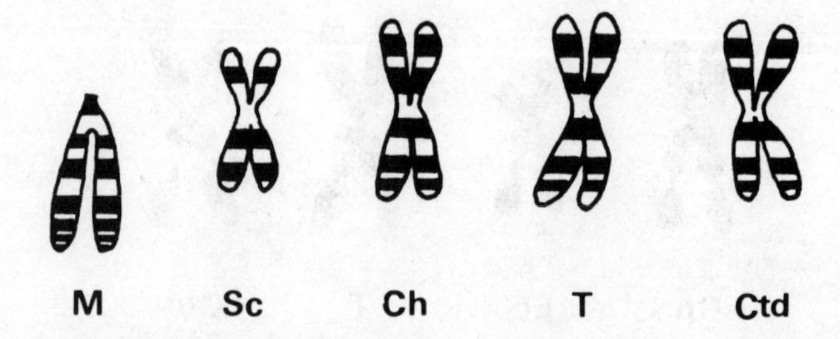

Fig. 9. G-bands of X-chromosomes. M, murid; Sc, sciurid; Ch, chinchillid; T, thryonomyid; Ctd, ctenodactylid.

BLOOD VASCULAR SYSTEM

From a study of seven characters of the blood vascular system, it was concluded that ctenodactylids were more like African hystricognaths and some caviomorphs than either of the other suborders (George, 1981). There was no single feature that marked the suborders from one another, but the combination of the patterns of veins of the neck, thorax and gonads and the arrangement of the arterial arches and cephalic arteries revealed a muroid group (which included geomyids, rhizomyids and spalacids), a loose aplodontid-sciurid-glirid group, and the hystricognath block (Fig. 10). Erethizontids were out on their own, and pedetids and anomalurids more like one another than like anything else.

The hystricognath block and the ctenodactylids tend to have either a right azygous vein draining the thorax or none at all, but the myomorphs have a left azygous. There is some controversy over the status of the azygous in evolutionary terms and its usefulness as a taxonomic character. From embryological studies, Arey (1952) concluded that the most generalised condition for a mammal would be bilateral azygous veins. Beddard (1907), too, regarded the bilateral condition as primitive but noted that the commonest mammalian pattern

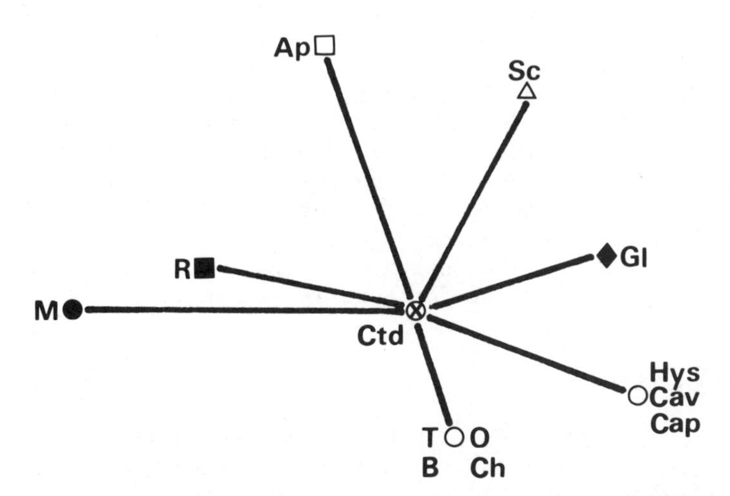

Fig. 10. Differences in circulatory system patterns between some rodent families and superfamilies. M, muroid; R, rhizomyoid; Ap, aplodontid; Sc, sciurid; Gl, glirid; Hys, hystricid; Cav, caviid; Cap, capromyid; T, thryonomyid; B, bathyergid; Ch, chinchillid; Ctd, ctenodactylid. The length of the lines is proportional to the differences between the families.

was to have a right azygous. In this respect, then, myomorphs form a homogeneous group (with geomyoids) with the specialised character of a left azygous vein. No hystricognath has a left azygous vein. Ctenodactylids share with most hystricognaths, the rhizomyids and spalacids the alternative specialisation of having lost both azygous veins.

Ctenodactylids share with hystricognaths, sciurids and glirids a pattern of systemic arches where an innominate (brachiocephalic) artery leaves the aorta before dividing into the common carotids. The right common carotid gives off the right subclavian artery but the left subclavian leaves the aortic arch independently. Whether this "sciurid" pattern is a derived character is difficult to determine. Embryological evidence does not solve the problem unequivocally. The "sciurid" pattern is not an obviously convergent character associated with long necks, although the short-necked bathyergids, alone among hystricognaths, do have the "cricetid" pattern, in which the two common carotids leave the aortic arch independently. Monotremes and most marsupials have the "cricetid" pattern. It is clearly a variable character which may be associated with any one of a number of anatomical features, but the pattern is common to hystricognaths (except bathyergids) and otherwise found only in sciurids and glirids among rodents.

The blood drainage from the gonads (Barnett et al., 1958) is in a specialised condition in ctenodactylids, with both gonadal veins going direct to the posterior vena cava instead of joining the renal veins on one or both sides. The ctenodactylid pattern occurs in erethizontids, pedetids, anomalurids and heteromyids but is never found in myomorphs. Finally, both the stapedial artery and the internal carotid arteries are much reduced or lost in hystricognaths and ctenodactylids, which distinguishes them from myomorphs, considered by Bugge (1974) to be the most primitive. Thus, the pattern of the circulatory system indicates a close similarity of ctenodactylids with hystricognaths. The shared characters seem to be derived, though definitive evidence for this is lacking.

SHOULDER PATTERN

Ctenodactylids share with hystricognaths a scapuloclavicularis muscle, bifurcate acromial process on the scapula and the absence of an entepicondylar foramen in the humerus. Whether the lack of an entepicondylar foramen is a specialised or generalised character of rodents is uncertain, although it is present in Paramys (Wood, 1962) Most myomorphs have an entepicondylar foramen, but sciurids, aplodontids, hystricognaths and ctenodactylids do not.

Aplodontids have a bifurcate acromial process but do not have a scapuloclavicularis muscle. Lagomorphs and equids have a scapuloclavicularis muscle but only leporids have a bifurcate acromium. Leporids complete their similarity to hystricognaths in lacking the entepicondylar foramen.

There would seem to be a positive correlation between these structures. It has been suggested that the scapuloclavicularis might hold a weak and loosely attached clavicle in position and might, phylogenetically speaking, eventually substitute for it. Equids have no clavicle and leporids have a small one. Some caviomorphs have small clavicles, loosely attached to the sternum (caviids), but others have good firm clavicles. The scapuloclavicularis muscle is part of the flexor system, while the muscles associated with the acromium and metacromium are extensors (Woods, 1972). Could it be that the combination of the flexor scapuloclavicularis and extensor muscles like acromodeltoideus associated with the acromial process serves to keep the humerus securely at right angles to the trunk in animals that were increasing their stride and, typically, running?

If this does indicate an adaptation to cursorial locomotion, the retention of the clavicle would have avoided the specialised restrictions on limb movement of the equids. Hystricognaths and rabbits use their forelimbs in eating, and some hystricognaths and at least one rabbit climb trees (Verts et al., 1984).

If this combination of characters is synapomorphic, then the ctenodactylids were part of the early hystricognath stock which was already becoming the "marcheurs" of Blainville (1816).

CONCLUSIONS

It has been suggested that ctenodactylids are an evolutionary line from some paramyid-like ancestor, with a cluster of characters converging on the hystricognaths (Wood, 1977; Patterson and Wood, 1982). Ctenodactylids could, however, be offshoots or even relicts of an ancestral proto-hystricognath line which had already acquired some of the characteristic hystricognath adaptations. If affinity rather than convergence seems a likely explanation for the many similarities ctenodactylids undoubtedly have with hystricognaths, then ctenodactylids could be an important early radiation of the group.

Phenetic cluster analysis, the construction of cladograms of shared (though not necessarily derived) characters, and the analyis of patterns of reproduction, circulatory system and karyotypes have allied ctenodactylids more closely with African hystricognaths than with any other group of rodents. Ctenodactylids are linked as closely as bathyergids to hystricognaths and only slightly less than erethizontids to hystricognaths. Ctenodactylids have no unique linkages with myomorphs or sciuromorphs.

Ctenodactylids were abundant in Eocene Asia when there were few other rodents around (Shevyreva, 1972; Dawson, 1977; Wood, 1977, Li and Ting, this volume). Unfortunately, flora of Eocene Asia have not been investigated and there is very little information on climatic conditions. The northernly part of the northern hemisphere seems to have enjoyed an equable climate with broad-leaved evergreen rain forests; but, further south (in North America at latitude 35°N, for example), there is evidence of much drier tropical vegetation dominated by legumes (Wolfe, 1978). If these latitudinal bands were repeated in Asia it is possible that early hystricognaths-ctenodactylids among them - were evolving in dry warm conditions unfavourable to forest. The running hyracodont rhinoceroses were around at the time (Savage and Russell, 1983). In open country dominated by scrub legumes, small rodents burrow, but ctenodactylids neither burrow nor nest. The terrain was either periodically baked or flooded thus loading the odds against burrowers and favouring the "ungulate" reproductive pattern. Assuming the terrain selected ctenodactylids and their relatives for open-country adaptation, a locomotory pattern developed which increased the length of the stride and improved running ability but which did not eliminate limb flexibility.

This pattern could have formed the base from which later hystricognaths evolved, adapted to the varying conditions of climate, vegetation, predation and competition. Meanwhile, ctenodactylids continued to flourish in Asia with little morphological change (Hussain et al., 1978) until they were pushed into marginal habitats in the Miocene and, eventually, into today's typically relict distribution pattern.

REFERENCES

Alston, E. R. 1876. On the classification of the order Glires. Proc. Zool. Soc. Lond. 1876: 61-98.

Arey, L. B. 1954. Developmental Anatomy. W. B. Saunders, Philadelphia.

Asdell, S. A. 1964. Patterns of Mammalian Reproduction. Cornell University Press, Ithaca.

Barnett, C. H., Harrison, R. J., and Tomlinson, J. D. W. 1958. Variations in the venous systems of mammals. Biol. Rev. 33: 442-487.

Beddard, F. E. 1907. On the azygous veins in the Mammalia. Proc. Zool. Soc. Lond. 1907: 181-223.

Bisby, F. A. 1973. The role of taximetrics in angiosperm taxonomy. New Phytol. 72: 699-726.

Blainville, H. M. D. de 1816. Prodrome d'une nouvelle distribution systématique du règne animal. Bull. Sci. Soc. Philom. Paris sér. 3, 3: 105-124.

Blyth, E. 1855. Report on a zoological collection from the Somali country. J. Asiatic Soc. Bengal 24: 291-306.

Bohlin, B. 1946. The fossil mammals from the Tertiary deposit of Taben-Buluk, western Kansu part 2. Paleont. Sinica ser. C, 8b: 1-259.

Borgaonkar, D. S. 1969. Insectivora cytogenetics. In: Comparative Mammalian Cytogenetics, K. Benirschke, ed., pp. 218-246, Springer-Verlag, Berlin.

Brandt, J. F. 1855. Beiträge zur nähern Kenntniss der Säugethiere Russlands. Mem. Acad. Imp. Sci. St. Petersbourg, ser. 6, 9: 1-365.

Bugge, J. 1974. The cephalic arterial system in insectivores, primates, rodents and lagomorphs with special reference to the systematic classification. Acta. Anat. 87 (suppl. 62): 1-160.

Burt, W. H. 1960. Bacula of North American mammals. Misc. Publ. Mus. Zool. Univ. Mich. 113: 1-175.

Chaline, J., and Mein, P. 1979. Les Rongeurs et l'Evolution. Doin, Paris.

Dathe, H. 1937. Uber den Bau des männlichen Kopulationsorganes beim Meerschweinchen und anderen hystricomorphen Nagetieren. Morphol. Jahrb. 80: 1-65.

Davies, C. 1982. The recent and fossil affinities of the genus _Pedetes_ (Mammalia, Rodentia). D. Phil. thesis, Oxford University.

Dawson, M. R. 1977. Late Eocene rodent radiations in North America, Europe and Asia. Géobios Mém. Spéc. 1: 195-209.

Eisenberg, J. F. 1981. The Mammalian Radiations. Athlone Press, London.

Ellerman, J. R. 1940. The Families and Genera of Living Rodents. British Museum (Natural History), London, 2 vols.

Estabrook, G. F. 1966. A mathematical model in graph theory for biological classification. J. Theor. Biol. 12: 297-310.

Flower, W. H., and Lydekker, R. 1891. An Introduction to the Study of Mammals Living and Extinct. Black, London.

Gallimore, P. H., and Richardson, C. R. 1973. An improved banding technique exemplified in the karyotype analysis of two strains of rat. Chromosoma 41: 259-263.

George, W. 1978. Reproduction in female gundis (Rodentia: Ctenodactylidae). J. Zool. 185: 57-71.

George, W. 1979a. The chromosomes of the hystricomorphous family Ctenodactylidae (Rodentia: ? Sciuromorpha) and their bearing on the relationships of the four living genera. Zool. J. Linn. Soc. 65: 261-280.

George, W. 1979b. Conservatism in the karyotypes of two African mole rats (Rodentia: Bathyergidae). Z. Säugetierk. 44: 278-285.

George, W. 1980. A study in hystricomorph rodent relationships: The karyotypes of _Thryonomys gregorianus_, _Pedetes capensis_, and _Hystrix cristata_. Zool. J. Linn. Soc. 68: 361-372.

George, W. 1981. Blood vascular patterns in rodents: Contributions to an analysis of rodent family relationships. Zool. J. Linn. Soc. 73: 287-306.

George, W. and Weir, B. 1974. Hystricomorph chromosomes. In: The Biology of Hystricomorph Rodents, I. W. Rowlands and B. J. Weir, eds., pp. 143-160, Academic Press, London.

Gervais, P. 1848. Rongeurs. In: Dictionnaire Universel d'Histoire Naturelle, C. D'Orbigny, ed., vol. 11, pp. 198-204.

Grassé, P.-P. and Dekeyser, P. L. 1955. Traité de Zoologie. Masson, Paris, vol 17, 2: 1542.

Gropp, A. 1969. Cytologic mechanisms of karyotype evolution in insectivores. In: Comparative Mammalian Cytogenetics, K. Benirschke, ed., pp. 247-266, Springer-Verlag, Berlin.

Hussain, S. T., Bruijn, H. de and Leinders, J. M. 1978. Middle Eocene rodents from the Kala Chitta Range (Punjab, Pakistan). Proc. K. Ned. Akad. Wet. 81B: 74-112.

Koulischer, L. 1973. Common patterns of chromosome evolution in mammalian cell cultures or malignant tumours and mammalian speciation. In: Cytotaxonomy and Vertebrate Evolution, A. B. Chiarelli and E. Capanna, eds., pp. 129-164, Academic Press, New York.

Landry, S. O. 1957. The interrelationships of the New and Old World hystricomorph rodents. Univ. Calif. Publ. Zool. 56: 1-118.

Lindsay, E. H. 1977. Simimys and the origin of the Cricetidae (Rodentia: Muroidea). Géobios 10: 597-623.

Luckett, W. P. 1971. The development of the chorio-allantoic placenta of the African scaly-tailed squirrels (family Anomaluridae). Amer. J. Anat. 130: 159-178.

Luckett, W. P. 1980a. Fetal membrane and placenta development in the African hystricomorphous rodent Ctenodactylus. Anat. Rec. 196: 116A.

Luckett, W. P. 1980b. Monophyletic or diphyletic origins of Anthropoidea and Hystricognathi: Evidence of the fetal membranes. In: Evolutionary Biology of the New World Monkeys and Continental Drift, R. L. Ciochon and A. B. Chiarelli, eds., pp. 347-368, Plenum Press, New York.

Millar, J. S. 1972. Timing of breeding of pikas in southwestern Alberta. Canad. J. Zool. 50: 665-669.

Miller, G. S. and Gidley, J. W. 1918. Synopsis of the super-generic groups of rodents. J. Wash. Acad. Sci. 8: 431-448.

Nadler, C. F., Lyapunova, E. A., Hoffman, R. S., Vorontsov, N. N., and Malygina, N. A. 1975. Chromosomal evolution in holarctic ground squirrels (Spermophilus) I. Giemsa band homologies in Spermophilus columbianus and S. undulatus. Z. Säugetierk. 40: 1-7.

Nelson, G. 1978. Ontogeny, phylogeny, paleontology and the biogenetic law. Syst. Zool. 27: 324-345.

Novacek, M. J. 1980. Cranioskeletal features in tupaiids and selected Eutheria as phylogenetic evidence. In: Comparative Biology and Evolutionary Relationships of Tree Shrews, W. P. Luckett, ed., pp. 35-93, Plenum Press, New York.

Oduor-Okelo, D. 1978. A histological study on the ovary of the

African cane rat Thryonomys swinderianus. East Africa Wildl. J. 16: 257-264.

Patterson, B. and Wood, A. E. 1982. Rodents from the Deseadan Oligocene of Bolivia and the relationships of the Caviomorpha. Bull. Mus. Comp. Zool. Harv. 149: 371-453.

Renzoni, A. 1967. Chromosome studies in two species of rodents. Mammal Chromosome Newsletter 8: 111-112.

Sacher, G. A. and Staffeldt, E. F. 1974. Relation of gestation time to brain weight for placental mammals. Amer. Nat. 108: 593-615.

Sale, J. 1965. Gestation period and neonatal weight of the hyrax. Nature 205: 1240-1241.

Sale, J. 1969. Breeding season and litter size in Hyracoidea. J. Reprod. Fert. Suppl. 6: 249-263.

Savage, D. E. and Russell, D. E. 1983. Mammalian Paleofaunas of the World. Addison-Wesley, Reading, Mass.

Shevyreva, N. S. 1971. New middle Oligocene rodents of Kazakhstan and Mongolia. Acad. Sci. USSR. Paleont. Inst. Proc. 130: 70-86.

Shevyreva, N. S. 1972. New rodents from the Palaeogene of Mongolia and Kazakhstan. Akad. Nauk. SSSR, Paleontol. Zh. 3: 134-145.

Simpson, G. G. 1945. The principles of classification and a classification of mammals. Bull. Amer. Mus. Nat. Hist. 85: 1-350.

Thomas, O. 1896. On the genera of the rodents: An attempt to bring up to date the current arrangement of the order. Proc. Zool. Soc. Lond. 1896: 1012-1028.

Tullberg, T. 1899. Ueber das System der Nagethiere, eine phylogenetische Studie. Nova Acta Reg. Soc. Sci. Upsala, ser. 3, 18: 1-514.

Verts, B. J., Gehman, S. D., and Hundertmark, K. J. 1984. Sylvilagus nuttalii a semiarboreal lagomorph. J. Mammal. 65: 131-135.

Wahlert, J. H. 1974. The cranial foramina of protrogomorphous rodents, an anatomical and phylogenetic study. Bull. Mus. Comp. Zool. Harv. 146: 363-410.

Weber, M. 1928. Die Säugetiere, Vol. 2. Fischer, Jena.

Weir, B. J. 1974. Reproductive characteristics of hystricomorph rodents. In: The Biology of Hystricomorph Rodents, I. W. Rowlands and B. J. Weir, eds., pp. 265-301, Academic Press, London.

Weir, B. J. and Rowlands, I. W. 1973. Reproductive strategies of mammals. Ann. Rev. Ecol. Syst. 4: 139-163.

Wiley, E. O. 1981. Phylogenetics. Wiley, New York.

Winge, H. 1924. Pattedyr-Slaegter, Vol. 2. Hagerups, Copenhagen.

Wolfe, J. A. 1978. A palaeobotanical interpretation of Tertiary climates in the northern hemisphere. Amer. Sci. 66: 694-703.

Wood, A. E. 1955. A revised classification of the rodents. J. Mammal. 36: 165-187.

Wood, A. E. 1962. The early Tertiary rodents of the family Paramyidae. Trans. Amer. Phil. Soc. 52: 1-261.

Wood, A. E. 1965. Grades and clades among rodents. Evolution 19:

115-130.

Wood, A. E. 1974. The evolution of the Old World and New World hystricomorphs. In: The Biology of Hystricomorph Rodents, I. W. Rowlands and B. J. Weir, eds., pp. 21-54, Academic Press, London.

Wood, A. E. 1977. The evolution of the rodent family Ctenodactylidae. J. Paleont. Soc. India 20: 120-137.

Wood, A. E. and Patterson, B. 1959. The rodents of the Deseadan Oligocene of Patagonia and the beginnings of South American rodent evolution. Bull. Mus. Comp. Zool. Harv. 120: 281-428.

Woods, C. A. 1972. Comparative myology of jaw, hyoid and pectoral appendicular regions of New and Old World hystricomorph rodents. Bull. Amer. Mus. Nat. Hist. 147: 115-198.

Wurster, D. H. 1969. Cytogenetic and phylogenetic studies in Carnivora. In: Comparative Mammalian Cytogenetics, K. Benirschke, ed., pp. 310-329, Springer-Verlag, Berlin.

Wurster, D. H. and Benirschke, K. 1968. Comparative cytogenetic studies in the order Carnivora. Chromosoma 24: 336-382.

Zittel, K. A. von, 1893. Handbuch der Paleontologie Abt. 1. Palaeozoologie Band 4. Vertebrata (Mammalia). Oldenbourg, Munich.

THE RELATIONSHIPS, ORIGIN AND DISPERSAL OF THE HYSTRICOGNATHOUS RODENTS

Albert E. Wood[1]

Professor of Biology
Emeritus, Amherst College
Amherst, Mass., U.S.A.

INTRODUCTION

General

My topic - the relationships, origin and dispersal of hystricognathous rodents - is complicated by a very considerable amount of confusion among authors as to what should be included in the Suborder Hystricognathi. This confusion results, at least in part, from the fact that simplicity of arrangment of taxa has a great appeal, but that simplicity of arrangement does not necessarily reflect either what must have been the facts of the evolution of rodents, or our less than perfect knowledge of these facts.

As a background, it is necessary to review some aspects of the history of rodent classification. It is also necessary to emphasize, as I have been doing for fifty years, that the presence of identical features in two lines of rodents does <u>not</u> necessarily mean that these features were inherited from a common ancestor. Common features have evolved, over and over again, in different lines of rodents, <u>after</u> their ancestral lines had diverged from a common ancestor in which these features had not been acquired. This is the process of parallelism. It implies, of course, that the genetic potentiality for evolving these structures existed in the common ancestor. Parallelism has been rampant in dental, osteological and myological features. I do not know any reason why the same should not have been true for

---

1. Present address: 20 Hereford Avenue, Cape May Court House, New Jersey 08210 U.S.A.

embryology or biochemistry. Certainly there is no *a priori* reason for assuming that it did not occur.

History of Rodent Classification

In 1839, Waterhouse published on the classification of rodents, and pointed out that the animals he included in the order could be divided into two major groups, the rabbits and all others. The latter, in turn, were divisible into three groups, in which, respectively, (1) a portion of the masseter muscle arose in front of the orbit, descending to insert on the lower jaw; (2) a deeper portion of the masseter, likewise arising from the face, passed through an enlarged infraorbital foramen before descending to insert on the mandible; and (3) the first two conditions were combined. Later, Brandt (1855) named subdivisions for these three conditions and for the quite different condition in the rabbits, the Sciuromorpha, Hystricomorpha, Myomorpha and Lagomorpha, respectively. Tullberg (1899) brought a second structural condition into his classification, namely the condition of the angle of the jaw. He divided what are now called rodents into the Sciurognathi, in which the angle arises from the ventral side of the horizontal ramus in the plane of the incisor, and the Hystricognathi, in which it arises from the lateral side of the mandible. Whereas all living rodents (except *Aplodontia* and the Bathyergidae) fit readily into one or another of the subdivisions proposed by Waterhouse and Brandt, Tullberg's (1899) arrangement unfortunately introduced confusion into rodent classification, which has still not been removed, because the zygomasseteric pattern and the type of angular process have evolved independently. As a result, there have been major disagreements concerning the numerous rodents that are hystricomorphous but not hystricognathous, or hystricognathous but not hystricomorphous.

One tendency that must be guarded against is the use of a familiar term but with highly diverse meanings. This has been particularly true of "Hystricomorpha". As used by Brandt (1855), and by many authors since, it is synonymous with Tullberg's Hystricognathi plus the first two "Sectios" (essentially superfamilies) of his Myomorphi, the Ctenodactyloidei and the Anomaluroidei, the latte including the Anomaluridae and the Pedetidae. On the other hand, some authors have restricted the scope of the term by eliminating not only the non-hystricognathous hystricomorphs but also the South American hystricognaths (Wood, 1955; Wood and Patterson, 1959; Bugge 1974a). Still others have used the term as a synonym of Tullberg's Hystricognathi. Conclusions as to the nature and relationships of the Hystricomorpha, using one definition, cannot automatically be extended to the Hystricomorpha as the term is used by someone else.

Hystricomorphy and Hystricognathy

Hystricomorphy involves the origin of *M. masseter medialis* fror the face, usually as far forward as the middle of the lateral surface of the premaxilla, and the passage of this muscle through the greatly hypertrophied infraorbital foramen, to descend lateral to the cheek teeth and insert on the mandible. In extreme cases, the infraorbital foramen is comparable in size to the orbit. Maier and Dierboch (this volume) have proposed that the embryology of this area indicates that the space occupied by the muscle is not an enlarged infraorbital foramen but a neomorph, because it arises embryologically, in the guinea pig, as a separate opening. It seems more probable that this is an embryonic adaptation, serving to increase the strength of this region in the growing embryo and to permit the muscle to develop before the onset of ossification. The large size of the infraorbital foramen in many Eocene protrogomorphs would suggest that migration of muscle fibers through the foramen should have been relatively easy.

In hystricognathy, the angular process of the mandible arises from the side of the incisive alveolus and spreads laterally as well as posteriorly. In fully developed hystricognathy, as seen in Recent rodents, there is a groove at the anterior end of the angle, carrying *M. masseter lateralis, pars reflexa* (Woods, 1972), which arises from the medial surface of the angular process of the mandible. Among Recent rodents, hystricognathy is always accompanied by a deepening of the pterygoid fossa, which has resulted in an opening from that fossa to the orbit, as occurs also in the geomyids (Fig. 1), *Spalax* and *Aplodontia*.

Both hystricomorphy and hystricognathy result in increases in the length of parts of the masseter muscle, and especially in increases of the anteroposterior component of pull of the muscle. Presumably these changes originated as adaptations increasing the efficiency of the gnawing mechanism, but subsequently, with the acquisition of propalinal chewing (Landry, 1957), they became important in mastication. The deepening of the pterygoid fossa and the lateral migration of the angular process resulted in lengthening *M. pterygoideus internus*, which assists the masseter in its functions. We do not know whether the changes in the masseter and the pterygoid occurred simultaneously or sequentially.

Classificatory Problems

In all studies of classification, a major problem is separating characters that arose by convergence or parallelism from those that were inherited from common ancestors. Among rodents, parallelism is especially frequent, due to the common basic adaptations of all rodents and the large number of separate but closely related lineages. Some authors claim that it is easily possible to distinguish between parallelism and common inheritance in features that they have studied, but this seems improbable to me, an opinion in which I am supported by the studies of Beintema (this volume) and other molecular biologists. They indicate that, in studies of protein evolution, they have found as much demonstrable parallelism as has been reported for any other types of characters.

The sciurognathous hystricomorphs include the living Anomaluridae, Ctenodactylidae and Pedetidae, from Africa, and the Dipodoidea, as well as the fossil Theridomyoidea from the Eocene and Oligocene of Europe, and the genus *Protoptychus* from the middle Eocene of North America. There is no suggestion in any of these animals of any changes in the angular region that might represent initial changes toward hystricognathy.

Ever since Tullberg (1899), these sciurognathous but hystricomorphous rodents have caused major disagreements. A wide variety of solutions to their relationships to each other or to hystricognathous rodents have been proposed, but no solution seems possible that does not require extensive parallelism.

One group of living hystricognaths, the Bathyergidae, is either not hystricomorphous (= protrogomorphous) or at most show only very slight indications of hystricomorphy (Patterson and Wood, 1982; Tullberg, 1899). As indicated below, there are likewise questions as to where these rodents fit into the classification.

Many investigators present a "most parsimonious tree" as being an indicator of the true relationships, even when alternative proposals are almost as "parsimonious". I have never seen a report of a probability study to determine the significance, if any, of the differences between the different possible cladograms. As an example, Shoshani et al. (this volume) present cladograms showing the relationships among *Rattus*, *Mus*, *Ondatra*, *Spalax* and *Cavia* according to four schemes. The minimum number of nucleotide replacements for the four are given as 2295, 2298, 2304 and 2305, respectively, the first being preferred as the most parsimonious and the second as the best according to conventional data. I very much doubt that there is any mathematical significance to the differences in numbers of replacements among these cladograms. Furthermore, such diagrams neglect the possibility of parallelism which, it should by now be evident, is a major feature of rodent evolution.

SUBORDER HYSTRICOGNATHI

Definition

The taxon that I call the Suborder Hystricognathi was defined by Tullberg (1899, p. 69) as follows: "Sämtliche *Hystricognathen* zeichen sich dadurch aus, dass der Angularfortsatz des Unterkiefers von seitwärts verschoben ist, so dass sein Vorderteil von der äusseren Seite des Corpus ausgeht, wie auch dadurch dass seine Margo inferior nahezu mit dem Jochbogen parallel und in seitlicher Aussicht, wenigstens im grössten Teile seiner Länge, ganz horizontal verlaüft." There are several advanced fetal membrane features (Luckett, this volume) and some biochemical ones that unite all modern hystricognaths and that are found nowhere else in the order. I know of no other features where this is the case, except for hystricognathy.

Using Tullberg's definition, the Suborder Hystricognathi includes the Hystricidae, the Thryonomyoidea, the Bathyergoidea, the Caviomorpha, and the Eocene-Oligocene rodents that I have called the Franimorpha.

Papers presented at this symposium have demonstrated a number of other indications of affinity among the hystricognathous rodents. I have not included some of these on the chart (Fig. 1) because of inability to express them diagrammatically. Another problem is uncertainty as to what is a character. For example, are the fetal membrane resemblances that Luckett (this volume) has described a series of independent developments, or do they form an evolutionarily interrelated complex? In living hystricognaths, _M. masseter lateralis, pars reflexa_, passes around the ventral surface of the mandible to insert on the medial side of the angular process; a groove for this muscle develops along the anteroventral border of the angle; and the pterygoid fossa opens dorsally into the orbit. I have treated the first two of these developments as part of hystricognathy, but the third as a separate character (Fig. 1). All of them, however, appear to have evolved independently of each other and later than hystricognathy _sensu stricto_. What I call "fully developed hystricognathy" is found in the Hystricidae, Thryonomyoidea and Caviomorpha, and is exaggerated in the Tsaganomyidae and Bathyergoidea, where the angle is much more everted than in other hystricognaths.

Other Features Supporting Hystricognath Interrelationships

Numerous other features have been proposed as showing that the hystricognaths are a related stock. These vary as to their exclusive nature or as to the ease of interpreting their significance.

Hystricomorphy is often considered almost as a diagnostic feature of the Hystricognathi, although it has frequently been pointed out, beginning with Tullberg (1899), that many hystricomorphous rodents are not hystricognathous. Hystricomorphy has characterized caviomorphs, thryonomyoids and hystricids since their first known occurrences; it is reported in the Dipodoidea in the late Eocene; it may have characterized Eocene muroids (see Vianey-Liaud, this volume); it developed in the Theridomyoidea in the mid-Eocene (Wood, 1974b); it characterizes ctenodactyloids, probably having been acquired in the superfamily in early to middle Eocene; it occurs in all known anomalurid and pedetid skulls (early Miocene to Recent); a condition that has been called hystricomorphy (although perhaps a different term would be preferable) is found in the early Miocene bathyergid *Bathyergoides* (Lavocat, 1973); and the mid-Eocene *Protoptychus* is also hystricomorphous (Fig. 1). I believe that hystricomorphy, among these groups, developed independently at least seven and possibly as many as ten or twelve times, as parallelisms.

Multiserial incisor enamel is found, so far as we know, in all living and fossil members of the Caviomorpha, Thryonomyoidea, Hystricidae and Bathyergidae, as well as in *Pedetes* and the Oligocene and later, but not Eocene, Ctenodactyloidea (i.e., merely in the Family Ctenodactylidae). It has not yet been reported elsewhere in the order except in the north Asiatic Oligocene Tsaganomyidae, where the enamel of both *Cyclomylus* and *Tsaganomys* is intermediate between pauciserial and multiserial, that of *Cyclomylus* being closer to pauciserial and that of *Tsaganomys* to multiserial. The incisor enamel of the mid-Eocene cylindrodont *Mysops* (perhaps ancestral to the tsaganomyids) is pauciserial with a slight inclination and is very much like that of *Cyclomylus* (J. W. Wahlert, personal communication). Therefore, since multiserial enamel was being evolved among the tsaganomyids during the Oligocene, either this family was ancestral to all latter hystricognaths, or multiserial enamel evolved independently at least three and probably at least four times (main group of Hystricognathi, Tsaganomyidae, Ctenodactylidae and *Pedetes*, unless the last was a ctenodactylid derivative, which seems unlikely).

Among living Old World hystricognaths, the malleus and incus seem universally to be fused. The fusion is much more extensive in the bathyergids than in the others, extending to the crus longus of the malleus. Fusion is usual but by no means universal among caviomorphs, and seems to develop at a later ontogenetic stage than in the Old World forms. Examples of non-fusion, though relatively rare, are widespread (Patterson and Wood, 1982). Auditory ossicles are rarely reported among fossils. Fields (1957) reported fusion of the ossicles in the Miocene dinomyid *Drytomomys aequatorialis*. His figures show a marked groove between the bones. Subsequently, the specimen was badly damaged. Further observation indicated that a ... "thin film of matrix extended over the entire articular surface of the malleus separating it from the incus; fusion had not occurred" (Patterson and Wood, 1982, p. 483). In African fossil hystricognaths "Lavocat reported that the malleus and incus of *Paraphiomys pigotti* were fused, presumably because they were lying within the bulla in an articulated position. His figure (1973: Pl. 13, Fig. 4) suggests that there is a matrix-filled groove at their junction. This raises the possibility that, as in the case of *Drytomomys*, they may have been separable in life" (Patterson and Wood, 1982, p. 483), and that in the Miocene the hystricognath ossicles, like those of *Pedetes* (Wood and Patterson, 1959), were closely appressed but not fused. Therefore, it seems probable that actual fusion of the ossicles occurred at least twice (and probably several more times) among

	Caviomorpha	Thryonomyoidea	Hystricidae	*Reithroparamys and Franimys*	*Prolapsus*	*Cylindrodon*	Bathyergidae	Tsaganomyidae	Ctenodactyloidea	Anomaluridae	Pedetidae	Theridomyoidea	*Protoptychus*	Dipodoidea
Hystricognathy	X	X	X	X^2	X^2	X^2	X^3	X^3	–	–	–	–	?	–
Hystricomorphy	X	X	X	–	?	–	$–^1$	–	X	X	X	X^8	X	X
Incisor enamel	M	M	M	P	P:M	U	M	P/M	P/M	U	M	P/U	P?	U
Fused malleus and incus	$\pm$	X	X	?	?	?	X^7	?	X	–	$\pm$	–	?	–
Pterygoid fossa opens into orbit	X	X	X	?	?	?	X	?	–	–	–	–	?	–
Strong pars reflexa	X	X	X	?	?	X	X	X	X	–	–	–	–	–
Ventral margin of angular process parallel to zygoma	X	X	X	X	X	X	X	X	X	–	–	–	?	–
Internal carotid lost	X^4	X	X	?	?	?	X	?	X	–	–	–	?	–
Retention of dP4/4	X^5	X^6	–	–	–	–	?	–	–	–	X^9	–	–	–

Fig. 1. Distribution of a variety of characters among hystricognathous and other rodents. (Continued on following page.)

hystricognaths, although close appression and elimination of interossicular movement might have evolved only once in this group. Ossicular fusion in ctenodactylids and the close appression in Pedetes must have been parallelisms to what happened in the hystricognaths.

Parent (1976, 1980) and Lavocat and Parent (this volume) have stressed similarities between the middle ears of various hystricognaths. Some of these features, according to their reports, are found also among the sciurognathous Theridomyoidea, Ctenodactylidae, Anomaluridae and Pedetidae. Similarities among the various groups are clear, but when or at what stage of evolution these features were acquired are unknown. Lavocat and Parent indicate that there are both abundant resemblances and differences between ctenodactylids and thryonomyids, and they believe that the resemblances are largely parallelisms. Given what we know of the evolutionary history of these animals, it seems exceedingly difficult to assume special relationships between the hystricognaths and both the Theridomyoidea and Ctenodactyloidea.

All known members of the Thryonomyoidea, later than the early Oligocene, retain their deciduous teeth (dP4/4) throughout life, and the permanent premolars are suppressed. Even in the early Oligocene, replacement was greatly delayed (Wood, 1968; Patterson and Wood, 1982). This same peculiarity occurs in all known members of the caviomorph families Echimyidae, beginning with the Miocene Colhuéhuapian, and Capromyidae. There is no trace of any delay in the eruption of the permanent premolars in other caviomorphs, in the Deseadan echimyids, or in the Hystricidae. The suppression of the permanent premolars clearly occurred independently in echimyids

---

Fig. 1. (continued). X, character present; -, character absent; ?, situation unknown; ±, ossicles not fused in all; M, multiserial incisor enamel; P, pauciserial incisor enamel; U, uniserial incisor enamel; P:M, condition intermediate between pauciserial and multiserial; P/M, primitive members pauciserial, advanced members multiserial; P/U, primitive members pauciserial, advanced members uniserial. Notes: 1, the hystricomorphy of Bathyergoides is completely sui generis; 2, hystricognathy according to Tullberg's (1899) definition; this is incipient hystricognathy according to Wood; 3, ultra-hystricognathy, in which eversion of the angle is highly exaggerated; 4, lost except in Erethizontidae; 5, retained in all known post-Deseadan Echimyidae (including capromyids); 6, retained in all after early Oligocene, and in most during early Oligocene; 7, ultrafusion, including the crus longus of the malleus; 8, hystricomorphy evolved within the group; 9, dP4/4 replaced in the Miocene Parapedetes; no replacement in Pedetes; condition unknown in the Miocene Megapedetes.

and the thryonomyoids. Retention of deciduous teeth has been reported in the Bathyergidae, but the evidence for it is equivocal, and I suspect that in *Heliophobius*, at least, the extra teeth are neomorphic splits of molar tooth germs. It seems probable that the anterior teeth in the Miocene *Bathyergoides* (Lavocat, 1973) were dP4/4, whether or not they were replaced by P4/4.

The internal carotid artery and the carotid canal in the basicranium have been lost in all hystricognaths that Bugge (1971, 1974a, b) studied except *Coendou prehensilis*, the only erethizontid he described. Lavocat (1973) stated that, in the African Miocene thryonomyoid *Diamantomys luederitzi*, "Le foramen carotidien s'ouvre en avant de la gouttière ... qui se creuse entre la région postérieure du basi-occipital et la bulle. C'est un trou de toute la hauteur de la gouttière." Unfortunately, we do not know the condition of the internal carotid in Oligocene thryonomyoids or caviomorphs, because the basicranium is not preserved in any of these except in *Platypittamys*, where it is so damaged that it "... is impossible to interpret the ventral surface of the skull" (Wood, 1949).

These data strongly suggest that the loss of the internal carotid artery occurred independently in the Caviomorpha and the Thryonomyoidea, and this would argue against using the presence of this blood vessel in *Coendou* as the basis for erecting an infraorder Erethizontomorpha, *contra* Bugge (1974a, b). In addition, the internal carotid artery has also been lost in *Marmota monax* (Wahlert, 1974) and *M. marmota* (Bugge, 1974a), as well as in several glirids and *Aplodontia* (Bugge, 1974a, this volume), so that this loss occurred at least three times independently among sciurognaths, as well as at least twice in hystricognaths.

George (1981) has studied the branches of the thoracic aorta, the vena cava, the azygous and gonadal veins, and the presence or absence of a ring of the external jugular vein around the clavicle. The patterns of these vessels give additional support for the separation of the Erethizontidae from the other Caviomorpha. The distinctness of the bathyergids in these features from the thryonomyoids and hystricids is rather notable, as is the suggestion of their possible relationships to the Rhizomyidae and Spalacidae (George, 1981). It is worth pointing out that the cricetid and bathyergid types of aortic branches, as described by George, are probably more primitive than the sciurid and thryonomyid types. Bilateral anterior venae cavae would appear to be the primitive condition, and the union of the vessels would seem to be a secondary feature, occurring at random among hystricognaths as well as in *Ctenodactylus*.

A peculiar structure, of unknown function, the sacculus urethralis, is present in all living hystricognaths except *Lagostomus* (Wood and Patterson, 1959). It is either unknown outside the suborder, or present in the Ctenodactylidae, where there is a structure that may or may not be a sacculus urethralis. It is entirely possible that this structure has been inherited from the Eocene ancestors of the modern hystricognaths.

Luckett (1980) has supported the unity of the hystricognaths on the basis of similarities of their fetal membranes, in which they share a number of advanced features not found elsewhere in the order. In the hystricognaths and Ctenodactylidae a parietal endodermal layer fails to develop in the early blastocyst (Luckett, 1980, this volume). Except for this, however, the fetal membranes do not tend to unite any sciurognaths with the hystricognaths.

A number of biochemical approaches to the subordinal relationships of rodents have been formulated, agreeing with the morphological ones in yielding a diversity of interpretations. Sarich (this volume), using albumin immunological comparisons, has found the hystricognaths to be a natural group, although the hystricids were rather distinct. He reported that the hystricognaths studied fell into six groups, which he believed had separated, close to the same time, about 45-48 MY ago. One of these groups united *Petromus*, *Thryonomys* and *Bathyergus*, whereas *Hystrix* was alone in a second group. The third group contained *Hoplomys*, *Myocastor* and *Octodon*, rather widely separated by Woods (1984), but united by Patterson and Wood (1982) in the superfamily Octodontoidea. Sarich's other groups were *Cavia* and *Hydrochaeris*, an eminently reasonable combination, and *Capromys*. The distinctness of this last genus supports the views of Woods (1984), rather than those of Patterson and Wood (1982). In Sarich's data, the pedetids do not fit with the hystricognaths, and he had no evidence on ctenodactylids or anomalurids.

Beintema (this volume) concluded that the amino acid sequences of ribonuclease and insulin that he had studied in hystricognaths (limited to *Hystrix* and six caviomorphs) indicated that they were related. To date, however, these hystricognath sequences have been compared only to those of muroids among the other Rodentia.

The studies by Durette-Desset (1971) and Quentin (1973a, b, c) on the nematodes of rodents, and those of Traub (1980) on the sucking lice and fleas of rodents, together with the conclusions of these authors on the implications of their studies for hystricognath relationships, are very difficult to interpret. They all involve apparently related parasites in hosts widely separated geographically and often taxonomically. The data can be satisfactorily explained in a variety of manners, either demonstrating or refuting close trans-Atlantic relationships between the New World and Old World hystricognaths. Any interpretation raises at least as many unanswered questions as it solves. The complete lack of a paleontologic record for nematodes and sucking lice, and the almost complete lack of one for fleas (the modern genus *Palaeopsylla* is present in the Oligocene

Baltic amber), means that we have no idea of the rates of their evolution and hence of how to interpret their distribution, both geographic and among hosts. For a more extensive discussion, see Patterson and Wood (1982).

Lavocat (1971) discussed a number of features that he believed indicated close relationship between caviomorphs and Old World hystricognaths. He believed that these demonstrated that the caviomorphs were derived from thryonomyoids, an opinion that he reiterated in 1973, when he listed these as "principal common characters" of the Old World hystricognaths (his Phiomorpha) and the Caviomorpha, and again in 1976. Several of these features are discussed above. Others were demonstrated by Patterson and Wood (1982) either to be generally absent in hystricognaths, or to be found in a wide variety of sciurognaths as well.

Even the features given above as characteristic of fully developed hystricognathy are not limited to the Hystricognathi. Hystricomorphy is found in a considerable variety of sciurognaths (Fig. 1). The deepening and perforation of the pterygoid fossa "occurs elsewhere in the order among geomyoids and in _Spalax_; it is also present in _Aplodontia_, in which the perforation is large .... In all forms with a perforated fossa, the angular region of the mandible, either the whole, as in hystricognaths, or the posterior part, as in the sciurognathous forms, is everted" (Patterson and Wood, 1982, p. 479).

In trying to analyze these features in rodents, there are three very distinct questions that must be answered. First, are the Hystricognathi, as I define them, a natural group? I believe that the answer is "yes," provided that one accepts the possibility that many of the features that characterize the Recent forms may have evolved independently, by parallelism. Second, how and when did the hystricognaths achieve their present distribution? This question is very much _sub judice_ at present. And finally, from what ancestral stock were the hystricognaths derived? I shall attempt to answer this question a little later.

Lavocat and Parent (this volume) support the sciurognathous Theridomyoidea as being the ancestral stock of the Hystricognathi. Because hystricomorphy (Wood, 1974b) and uniserial incisor enamel (Korvenkontio, 1934) were evolved within the Theridomyoidea, their conclusion requires the assumptions that hystricomorphy evolved, in the hystricognath group, before hystricognathy, and that the theridomyoids (alone among rodents, except possibly for Eocene cylindrodonts - see below) were able to evolve _both_ uniserial and multiserial incisor enamel from their ancestral pauciserial type. Neither of these assumptions have been proven.

It seems probable to me that hystricognathy, hystricomorphy, the shift of the insertion of _M. masseter lateralis, pars reflexa_, and the deepening of the pterygoid fossa, are all reactions to a selection for increased strength of the anteroposterior component of muscular pull on the lower jaw, which probably evolved, initially, to strengthen the action of the incisors. As such, it is possible that these all evolved as a single unit, produced by a single set of genetic actions. On the other hand, it is perhaps more probable that they did not, and that they evolved at slightly different times, perhaps in different sequences in different hystricognath lines. The only evidence at present that bears on this question is the existence of the rodents that Wood (1975) called the Franimorpha, which had already acquired hystricognathy and, at least in some instances, the shift of the insertion of the *pars reflexa*, but none of the other associated changes. Certainly, according to Tullberg's (1899) definition of the Hystricognathi, they belong in the suborder.

I find it difficult to believe that the development of multiserial incisor enamel was part of the same hystricognath-hystricomorph complex, although its evolution might have followed that of the other features by a relatively short time. The fact that the tsaganomyids show the initial stages in the acquisition of multiserial incisor enamel, together with an ultra-hystricognathous angle, very possibly a perforated pterygoid fossa, but no trace of hystricomorphy, clearly demonstrates the lack of necessary correlation among these features. I can see no basis for assuming that the development of hystricognath similarities in fetal membranes, or the growth of a sacculus urethralis, or the fusion of the ear ossicles, had anything to do with the functioning of the jaw muscles or incisors. Unfortunately, we have no evidence as to the dates at which any of this last group of characters, all advanced, may have evolved.

CONTENT OF THE SUBORDER HYSTRICOGNATHI

Superfamily Thryonomyoidea

The thryonomyoids are a group of African rodents, very abundant and diversified in the early Miocene (Stromer, 1926; Hopwood, 1929; Lavocat, 1973; Patterson and Wood, 1982). After the invasion of Africa by other rodents, apparently early in the Miocene and presumably through Asia Minor (an invasion that included, at the minimum, the sciurids, cricetids, and probably the ctenodactylids, but also, perhaps, the bathyergids and pedetids), the thryonomyoids were outcompeted, and were rapidly reduced in numbers and variety, so that today there are but two living genera, *Thryonomys* and *Petromus*, representing two distinct lines that were already present in the Oligocene (Patterson and Wood, 1982). No rodents other than thryonomyoids have been described from the Oligocene of Africa, and at that time they do not appear yet to have become highly diverse, nor to have become the dominant factor in the African mammalian fauna that they had become by the Burdigalian. At least by the early

Miocene, the thryonomyoids had already acquired all the osteological and dental features of the Hystricognathi, with the possible exception of the fusion of the malleus and incus (see above).

By the first record of the thryonomyoids, near the beginning of the Oligocene, the lower jaw was already fully hystricognathous and the infraorbital foramen seems to have been large (Wood, 1968). These animals had already developed the peculiarity that characterizes all later members of the superfamily, the delay in or elimination of the replacement of the deciduous teeth. As a result, permanent premolars are unknown for any member of the superfamily after the early Oligocene Lower Fossil Wood Zone of the Jebel el Qatrani Formation of Egypt. In one genus, Phiomys, two lower jaws are known with the permanent premolars in place, whereas 16 other specimens, some of more advanced age, still retained lower dP4 (Wood, 1968). In another genus, Gaudeamus, replacement of the deciduous teeth is not known to have occurred, but a permanent premolar was excavated from beneath lower dP4 in one specimen (Wood, 1968), so that the tooth had not been completely lost. No other permanent premolars are known for the superfamily. Such retention of deciduous teeth is unknown in any other group of Oligocene hystricognaths - the South American caviomorphs, the north Asiatic tsaganomyids, or the North American cylindrodonts.

Family Hystricidae

The Old World porcupines are found at present over much of Africa, and from Italy east to southern China, Indonesia and the Philippines (Walker, 1975). Their fossil distribution is similar. They share many features with the thryonomyoids, but no suggestion of the retention of deciduous premolars has ever been reported. No trace of hystricids has been reported in the rich early Miocene of Kenya (Lavocat, 1973), and their large size should have rendered them readily collectable, had they been present at that time. The family is first known in the late Miocene of Egypt, Hungary and India (Chaline and Mein, 1979), in which areas they appear suddenly, apparently as recent immigrants. A number of fossils from the Oligocene and earlier Miocene of Europe were originally referred to Hystrix (reviewed by Schlosser, 1884). Some of these have now been referred elsewhere, and the rest are completely indeterminate, but memories of these citations still creep into later reference works (Wood, 1955; Walker, 1975).

Infraorder Caviomorpha

The caviomorphs are first known from the Oligocene (Deseadan) of South America, at which time they were fully hystricognathous and hystricomorphous. None of the 14 Deseadan genera in which the premolar region is known shows any suggestion of retention of the deciduous teeth (Lavocat, 1976; Patterson and Wood, 1982; Wood and

Patterson, 1959). This is true of the Echimyidae, as well as of the other families. From the early Miocene Cohuéhuapian on, all known echimyids retain their deciduous teeth throughout life (Patterson and Wood, 1982). If the Erethizontidae are included in this infraorder, as is done here, some caviomorphs are more primitive than any living thryonomyoid or hystricid in that they retain the internal carotid artery and the carotid canal throughout life, although these structures seem to have been present also in the thryonomyoid Diamantomys from the Miocene of East Africa (Lavocat, 1973). The presence of these structures in erethizontids led Bugge (1974a) to propose that the New World porcupines deserve a separate infraorder of their own. Up to the present, no Deseadan caviomorph skull or skull fragment has been reported that is adequately preserved to permit determination of the presence or absence of the carotid canal. Since, at one time or another, the ancestors of the other caviomorphs must have possessed an internal carotid artery and a carotid canal, I find it simpler to believe that the erethizontids are merely caviomorphs that have retained this primitive condition, lost at some unknown time or times by other members of the infraorder. Such a situation should cause no mental problems to anyone but strict cladists. The loss of the artery and canal presumably occurred independently, as parallelisms, in the New and Old World hystricognaths. The pattern of the gonadal veins of the erethizontids and the presence of a "squirrel ring" of the jugular vein (George, 1981) support the wide separation of the erethizontids from the other caviomorphs, as proposed by Bugge (1974a), and suggest a closer relationship to the Sciuridae, which last postulate I would find difficult to accept. These caviomorph-sciurid resemblances seem clearly to be an example of parallelism or convergence.

The presence of the internal carotid artery and carotid canal in erethizontids and the universal delay or absence of replacement of dP4/4 in thryonomyoids would seem to demonstrate that the earliest known caviomorphs are significantly more primitive than the earliest known thryonomyoids, and to be strong arguments against the derivation of the Caviomorpha from the Thryonomyoidea, though they do not oppose a reverse derivation.

Superfamily Bathyergoidea

I place two families here, the Bathyergidae and the Tsaganomyidae.

The Bathyergidae are exclusively African, and unknown before the early Miocene of Southwest Africa (Stromer, 1926) and Kenya (Lavocat, 1973), at which time they were fully adapted to a burrowing life. They show an exaggerated type of hystricognathy, with a highly flaring angle, developed elsewhere only in the Tsaganomyidae (Fig. 1).

Most participants in the present symposium believe that their data indicate that the Bathyergidae are closely related to the Thryonomyoidea, Hystricidae, and Caviomorpha, as part of a closely knit group, the Hystricognathi, to which other families might or might not be added. Some authors, however, have presented data suggesting alternative possibilities. George (1981) indicated strong similarities of the bathyergids to rhizomyids and spalacids, and to anomalurids, in their blood vascular patterns, although she concluded that the best representation of their relationships might be to recognize three independent superfamilies, Thryonomyoidea, Bathyergoidea and Ctenodactyloidea.

There are fundamental unanswered questions as to the dental formula of the bathyergids. Among living genera, Grassé and Dekeyser (1955) give the dental formulae of _Georychus_ and _Heterocephalus_ as M3/3; of _Cryptomys_ and _Bathyergus_ as P1/1, M3/3; and of _Heliophobius_ as P3/3, M3/3. In this last genus, the six teeth of each series are never present at the same time, the usual number in service being three. Other authors have suggested different dental formulae. In _Proheliophobius_ from the early Miocene of East Africa (Lavocat, 1973), the anterior tooth is less worn than the second of the four in the skull, so in this specimen the upper teeth are clearly P4/, M1/-M3/. _Bathyergoides_, according to Lavocat (1973), possessed 5 lower and 4 upper cheek teeth, a most unorthodox arrangement. Apparently only one specimen (SO 736) indicated the presence of 5 lower teeth, showing what seems to be an alveolus, followed by 3 teeth in use and a fourth erupting (Lavocat, 1973, Pl. 30, Fig. 4). All three of the worn teeth have patterns of molars, not premolars, and the photograph looks as though there is a wear surface on the anterior face of the anterior tooth. It is possible that this specimen has duplicated one of the molars, forecasting the condition that I believe exists in _Heliophobius_ (see below). Among other specimens of _Bathyergoides_, some show an anterior tooth more worn than the second, and hence dP4/ or dP/4 (Lavocat, 1973, Pl. 23, Fig. 4; Pl. 29, Fig. 6; Pl. 30, Fig. 5), whereas, in others, the anterior tooth is less worn than the second, and hence P4/ or P/4 (Lavocat, 1973, Pl. 7, Fig. 5). It would seem, then, that _Bathyergoides_ normally had P4/4, M1-3/1-3, with P4/4 replacing dP4/4, but that one specimen had acquired an extra tooth of some sort. In _Heliophobius_, with a cheektooth formula of 6/6, even if one assumes that these include P3-4/4, M1-3/1-3 and dP4/4, not enough lower teeth are available anywhere among the Rodentia to account for the formula without the secondary duplication of at least one tooth. Lavocat (1973) suggested that the extra tooth of _Bathyergoides_ might be dP/3 or P/3. It seems highly improbable to me that either of these teeth were present in the Bathyergidae, as they are otherwise completely unknown in the Rodentia. If either P/3 or dP/3 were present in the Bathyergidae, it would mean that the family must have had an ancestry independent of that of all other known rodents since some unknown date in the Paleocene. It seems much more likely to me that what is involved in _Heliophobius_ and _Bathyergoides_ is a secondary duplication of teeth (probably molars) in both the upper and lower jaws. We know that such duplication occasionally occurs in mammals, producing the accessory teeth of odontocetes, armadillos and Myrmecobius, as well as the clearly duplicated teeth in Otocyon (Wood and Wood, 1933).

The condition of the masseter muscle sets the bathyergids apart from other hystricognaths. In most bathyergids, the infraorbital foramen is small (protrogomorphous), with no indication of penetration by M. masseter medialis. In Cryptomys coecutiens (Tullberg, 1899; cited as Georychus coecutiens), the muscle barely penetrates the foramen, and seems to leave no scar on the snout in front of the foramen. In Bathyergoides (Lavocat, 1973, Pl. 7, Fig. 5), the muscle obviously penetrated the infraorbital foramen, but, instead of having an anteroposterior alignment, as in all other hystricomorphous forms, the foramen faced upward and the fibers ran vertically. Lavocat interpreted these conditions to mean that Bathyergoides was intermediate between the Thryonomyoidea and the other bathyergids, and was in the process of retrogressing toward the protrogomorphous condition, present in other members of the family, and that Cryptomys is the final stage in such a retrogression.

It seems much more probable to me that Bathyergoides was developing hystricomorphy, independently of all other hystricognaths. If this is correct, Cryptomys may currently be about to do the same thing, or this much penetration of the foramen may be a normal variant in protrogomorphous forms. I find it very difficult to envisage any selective pressures that could result in the drastic reduction in size of the masseter muscle, once it had evolved to the hystricomorph stage, and this would be especially true of rodents that used their lower incisors for tunnel-digging. It is even more difficult for me to believe that, should such reverse evolution of the muscle have occurred, the bony structure could revert to a condition indistinguishable from that of the protrogomorphous ancestors of the bathyergids. I can easily imagine animals in which it might have been selectively advantageous to have acquired hystricomorphy, but which had not done so until the appropriate mutations came along, which could well have been at different times in different lines. Therefore, I believe that the ancestors of the Bathyergidae have never been hystricomorphous.

The structures of the hyoid, laryngeal and pharyngeal regions unite the bathyergids as a group, and set them apart from all other rodents, even those that also use their lower incisors for digging (Woods, 1975). There are basic similarities that unite them with other hystricognaths, but the differences seem to me (especially since they are not necessarily associated with incisor digging) to support a long separation between the Bathyergoidea and the other hystricognaths (Fig. 2).

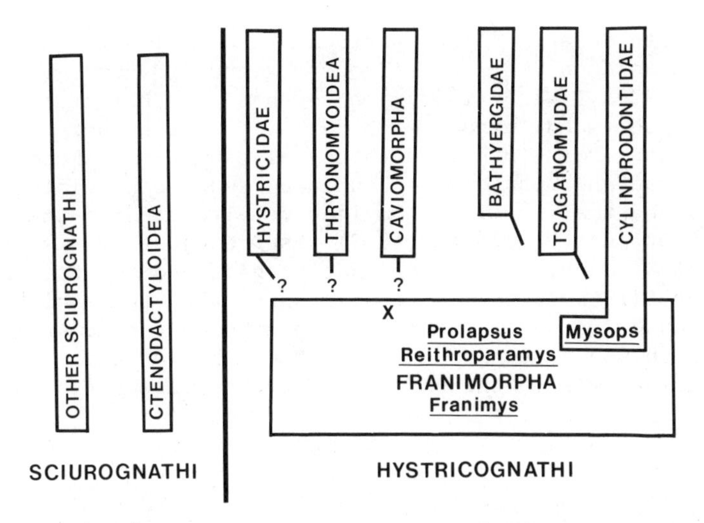

Fig. 2. Suggested interrelationships of the members of the Suborder Hystricognathi. If Protoptychus should prove to have been hystricognathous, it would probably fit in at about X.

Apparently, the most primitive patterns of branches from the aortic arch occur in what George (1981) identified as the cricetid and bathyergid types. These are very similar, but distinguishable. The bathyergid type, found also in Anomalurus and in some specimens of Ctenodactylus, is more primitive than the sciurid pattern, which "occurs in sciurids, glirids, caviomorphs [including erethizontids], hystricids, thryonomyids and in the majority of the ctenodactylids ..." (George, 1981, p. 289). The bathyergid pattern is also like that of spalacids and rhizomyids. She further suggested that the karyotypes separate the bathyergids from both ctenodactylids and thryonomyoids. These results would seem to support the postulate that the bathyergids were not derived from, although related to, the thryonomyoids. It is interesting to note that, in the pattern of the aortic arches, the bathyergids are among the more primitive rodents.

Thus, although most features point toward a relationship between the Bathyergidae and the other Recent hystricognaths, their protrogomorphy, the hyoid, laryngeal and pharyngeal construction, and the aortic arch pattern, suggest that they may represent an early offshoot from the basic hystricognath stock (Figs. 1, 2).

The Tsaganomyidae are exclusively north Asiatic, ranging through the Oligocene. They were originally (Matthew and Granger, 1923) described as members of the Bathyergidae. Burke (1935) believed them to have been related to the North American cylindrodonts, then represented in Asia only by the Oligocene genus *Ardynomys* (Matthew and Granger, 1925). Wood (1937), on the basis of his studies then (and still) unpublished, agreed with Burke. Other authors have generally accepted one or the other of these relationships. Wood (1980, and *in* Patterson and Wood, 1982) concluded that both points of view were correct, and that the Tsaganomyidae were related to the ancestry of the Bathyergidae. Lavocat (1973) reserved judgement as to the relationships, if any, between the bathyergids and the tsaganomyids, but felt that, if there were a relationship, the best explanation was that the tsaganomyids were descended from unknown Eocene African thryonomyoids, which would seem to contradict his belief that the Miocene *Bathyergoides* was in the process of giving rise to the Bathyergidae.

The tsaganomyids are hystricognathous but protrogomorphous, with no suggestion of penetration of the infraorbital foramen by *M. masseter medialis*. The angular process is ultra-hystricognathous as in the bathyergids. The pterygoid fossa is apparently deep and perforated (Patterson and Wood, 1982). The *pars reflexa* of *M. masseter lateralis* seems to have developed a large fossa for its insertion on the median side of the angular process, but a groove had not yet evolved for its passage around the front of the angle. As in the bathyergids, the growing base of the upper incisor extends back into the orbit, sometimes reaching the alveolar border by M2/. The permanent teeth of *Tsaganomys* were hypselodont, whereas dP4/4 were quite brachydont. The pattern-bearing portion of the permanent teeth of *Tsaganomys* had not taken part in the increase in height of the crown. Presumably related to the highly advanced burrowing mode of life (Vinogradov and Gambarian, 1952), there was extremely rapid wear of the cheek teeth. Therefore, *Tsaganomys* never retained identifiable patterns of more than two adjacent teeth at the same time. The pattern of unworn teeth can, however, be reconstructed from a series of juvenile specimens of slightly different ages. *Tsaganomys* was presumably too specialized in the hypselodonty of its cheek teeth to have given rise to the Bathyergidae, but the same does not follow for other members of the family, such as *Pseudotsaganomys* which had much lower-crowned and rooted cheek teeth (Vinogradov and Gambarian, 1952).

The Tsaganomyidae and Bathyergidae agree in being protrogomorphous, hystricognathous with hypertrophy of the angle, having the upper incisor arising well back in the orbit, and either possessing or developing multiserial incisor enamel. The two families are likewise very similar in structure of the skull (including the backward migration of the base of the upper incisor) and lower jaw to the cylindrodonts, and the tsaganomyids, at least, agree with the cylindrodonts in cheek-tooth pattern. The Oligocene cylindrodonts cannot be ancestral to the other two families because they possessed uniserial incisor enamel, but the mid-Eocene cylindrodont *Mysops* still retained pauciserial enamel. The backward extension of the upper incisor, into the orbit, as in *Tsaganomys* and the Bathyergidae, has been described in the American species of Ardynomys. The same situation possibly existed in the early Chadronian Pseudocylindrodon texanus. I do not know of the occurrence of this development elsewhere in the order except in Spalacopus (Wood, 1974a).

The union of the Tsaganomyidae with the Bathyergidae in the Superfamily Bathyergoidea seems warranted. The derivation of both from primitive, mid-Eocene, pauciserial cylindrodonts, seems likely to me (Fig. 2). Certainly this is the only case where a potential ancestor for one of the modern groups of hystricognaths can be pointed out. If the Bathyergidae were recent immigrants to Africa in the early Miocene, they could well have had a tsaganomyid ancestry.

Infraorder Franimorpha

There are a number of fossils from the Eocene and early Oligocene of North and Middle America that, according to Tullberg's (1899) definition, clearly belong to the Hystricognathi. These were united by Wood (1975) as the Infraorder Franimorpha, although they obviously had not acquired what I identified above as "fully developed hystricognathy". This allocation has been rejected by some authors, who apparently do not accept animals as being hystricognathous unless they had developed all the characteristics of hystricognathy, or even, perhaps, all features ever associated with hystricognathy. Dawson (1977) commented: "The 'very incipient' hystricognathous condition of Reithroparamys ... seems to me to be still a sciurognathous jaw, 'incipient' only a posteriori." This could be interpreted either as agreeing or disagreeing with Wood's point of view. Apparently, from Korth's comment (1984), it was meant as a disagreement. Korth specifically denied that any of the specimens Wood had referred to the Franimorpha were hystricognathous, as defined by Tullberg (1899). However, the angle did arise from the lateral surface of the mandible in Prolapsus from the middle Eocene of southern Texas (Wood, 1972); in the early Oligocene Cylindrodon (Wood, 1984); in the mid-Eocene Reithroparamys (Landry, 1957; Wood, 1962), especially in R. huerfanensis; and I believe that it did the same, though in a less pronounced fashion, in Franimys from the earliest Eocene (Wood, 1962).

Korth's (1984) only documentation that the angle does not arise from the lateral side of the mandible in these forms is a stereoscopic photograph of the lower jaw of Franimys, which, unfortunately, is not clear enough as reproduced to provide any useful information. Korth stated that the angle of Prolapsus looks as though it is hystricognathous, but is not, because the jaw is robust. The robustness does not seem to me to have any bearing on where the angle arises from the ramus. Black and Sutton (1984) state that both Prolapsus and Guanajuatomys are hystricognathous, but that the hystricognathy is of two different patterns. They also indicate that the lower jaw of Protoptychus appears to be hystricognathous, but deny any suggestion of hystricognathy in Reithroparamys.

Moreover, in all of these forms, the ventral margin of the angular process, as indicated in Tullberg's definition, is more nearly horizontal than in most other contemporaneous rodents (Wood, 1962), although it is also nearly horizontal in ctendodactylids. In *Cylindrodon* (Wood, 1984) *M. masseter lateralis, pars reflexa* had extended its insertion onto the medial side of the mandible. In none of the other franimorphs has this area been adequately studied. Unfortunately, at present nothing can be told of the nature of the pterygoid fossa in any of these genera except *Cylindrodon*, where it apparently was not unusually deep. In all the others, the skulls are broken through the pterygoid region, which might suggest that deepening of the fossa had weakened the skull in this area.

Korth (1984) concluded that what Wood has called incipient hystricognathy is quite common in early Eocene rodents, and might be the pattern that could have given rise to both hystricognathy and sciurognathy.

FAMILIES EXCLUDED FROM THE HYSTRICOGNATHI

General

Six groups of sciurognathous hystricomorphs are known - the Dipodoidea, Ctenodactyloidea, Anomaluridae, Pedetidae, Protoptychidae and Theridomyoidea, to which perhaps the Muroidea should be added, assuming that they were originally hystricomorphous (see Vianey-Liaud, this volume). In most of these groups (except the Theridomyoidea and perhaps the Ctenodactyloidea), we have no indication of the stages through which the animals passed in the evolution of hystricomorphy, so that we cannot be certain that the hystricomorphy is strictly homologous in all forms. In all of the above groups, the angular process is completely sciurognathous, but the infraorbital foramen is (or becomes) as large as in typical hystricognaths.

The ctenodactylids, anomalurids and pedetids have often been associated with each other and with the hystricognaths, perhaps on the basis of their present restriction to Africa, to their very limited (until recently) known fossil record, or to the fact that there have been no clear indications of any other relationships.

All of the hystricomorphous sciurognaths (except the Dipodoidea and Muroidea, if originally hystricomorphous) have been associated with the hystricognaths by some of the participants in the present symposium, and all of them have been dissociated from the hystricognaths by others.

Luckett (1971) pointed out that there is great similarity of th fetal membranes of the Anomaluridae, Ctenodactylidae, Pedetidae, Aplodontidae and Sciuridae, which he interpreted as independent retentions of the primitive rodent condition, and not as indicative of special relationships. George (1981) has suggested that the "overall blood vascular pattern is not inconsistent with a subordinal rank" for the Anomaluridae and Pedetidae as Anomaluromorpha.

I see very little hard evidence that any of the groups listed at the beginning of this section have significant relationships to any other, although there is little good evidence that the anomalurids and pedetids are not related.

Superfamily Dipodoidea

No one currently seems to believe that the dipodoids have any close relationship to the hystricognaths, but rather they are usually associated with the muroids. They are included here only for completeness and because, if they are not hystricognaths, they demonstrate that hystricomorphy evolved at least twice, and that, therefore, caution must be used in accepting it as evidence of relationships among forms that possess this feature.

Superfamily Ctenodactyloidea

In this superfamily, first known from the early Eocene of northern Asia, the cheek tooth pattern, even in the earliest members, is quite distinctive (Wood, 1977; Hartenberger, 1980; Dawson et al., 1984), not suggesting relationships with any typical hystricognath. It has been suggested that the earliest members of the superfamily may be close to being ancestral to all, or most, other rodents (Hartenberger, this volume), but this involves acceptance of numerous still unproven hypotheses. Of course, such a statement might be valid for all early Eocene rodents.

Recent ctenodactylids agree with hystricognaths in lacking a parietal endodermal layer in the early blastocyst (Luckett, this volume); in fusion of the malleus and incus; in possession of a structure that may be a sacculus urethralis, and, if so, may be either a primitive or a degenerate one; and in having multiserial incisor enamel (Fig. 1). The evidence is very good that several of these features were acquired independently of their development in the hystricognaths. "Although there is ... resemblance [in the circulatory patterns] between ctenodactylids and thryonomyids on the one hand and between bathyergids and thryonomyids on the other ... bathyergids and ctenodactylids show slightly less similarity in their blood vascular patterns .... While the karyotypes unite the ctenodactylids with thryonomyids, they do not suggest an association of either with the bathyergids ..." (George, 1981). She further suggests (this volume) that some or all of the reproductive resemblances of ctenodactylids and hystricognaths may be adaptations for life in open country. Bugge (this volume) reports the internal carotid artery to have been lost in *Ctenodactylus*.

Ctenodactylids have long been recognized from the Oligocene of north Asia (Bohlin, 1946), but it has recently been demonstrated (Wood, 1977; Hartenberger, 1980) that they were quite well represented there in the early Eocene Tamquammys (Shevyreva, 1971) and Saykanomys (Shevyreva, 1972a, b). Dawson et al. (1984) named two new families of Eocene ctenodactyloids from north Asia. The superfamily had reached the Indian subcontinent by the middle Eocene, where they were variously considered paramyids (Sahni and Khare, 1973; Sahni and Srivastava, 1976), thryonomyoids (Lavocat, 1973) or referred to a new family, "probably ancestral to the Ctenodactylidae, the Cylindrodontidae and the Phiomorpha" (Hussain et al., 1978). These last authors published illustrations of the incisor enamel of Chapattimys, clearly showing it to be of the typical pauciserial pattern (Sahni, 1980; Patterson and Wood, 1982), although Hussain et al. considered it to be multiserial. The incisor enamel of ctenodactylids, however, had become multiserial by the late Oligocene (Bohlin, 1946).

In Africa, the family is first known from the later Miocene of Beni Mellal (Lavocat, 1961), where it presumably represents a newly arrived immigrant stock, showing close relationships to contemporary Indian forms. It is not known from the Burdigalian of East Africa. Ctenodactylids have, however, been reported from Majorca, from rocks identified as middle to late Oligocene (Adrover and Hugueney, 1976; Adrover et al., 1977). They are also present in Sardinia, perhaps contemporaneous with their occurrence at Beni Mellal (Bruijn and Rümke, 1974).

In a very clear-sighted comment, Bohlin (1946, p. 143) remarked that, on the basis of the incisor enamel, "Ctenodactylus does not belong to the Sciuromorpha; this does not however necessarily entail that the genus must be referred to the Hystricomorpha." Hartenberger (1980) has proposed that the ctenodactylids were one of the initial subdivisions of the Rodentia. He also indicated that they have had a long history independent of all other groups of rodents, a conclusion with which I find myself fully in accord, and one that is fully supported by Dawson et al. (1984).

The other contributors to the present volume are divided as to whether the features of ctenodactyloids do or do not indicate relationships to the hystricognaths. Perhaps the best solution to the conflicting data (Fig. 1) is that they occupy a position somewhat as indicated in Figure 2, members of the Sciurognathi that have retained many ancestral rodent characteristics from a time prior to the hystricognath-sciurognath split, including the potential for evolving a number of hystricognath-like features by parallelism, and that, in many respects, they have evolved less than have other members of the Sciurognathi.

Family Anomaluridae

The anomalurids are first known from the Burdigalian of Kenya, although they may have been present earlier (Lavocat, 1973; also see Jaeger et al., this volume). The anomalurids have separate malleus and incus, uniserial enamel, and no sacculus urethralis (Fig. 1). The fetal membranes (Luckett, 1980, this volume) are more like those of the geomyoids, dipodoids and muroids than like those of any other of the modern families here under discussion. The cheek-tooth pattern of the anomalurids is very reminiscent of that of the Theridomyoidea, from which it is theoretically possible that they might have been derived. Bugge (1974b, this volume) has reported that the cephalic arteries of *Anomalurus* and *Pedetes* are almost identical, which he feels warrants uniting them in a Superfamily Anomaluroidea. George's data (this volume) suggest that there is a hystricognath group of rodents and a sciuromorph-myomorph one, with the anomalurids and pedetids intermediate. She stated earlier that the "pedetids and anomalurids have similarities in blood vascular pattern, differing only in the presence of a 'squirrel ring' in the pedetids and in the anomaly of the gonadal vessels" (George, 1981). The cheek-tooth patterns of anomalurids and pedetids, however, have very little or nothing in common.

Family Pedetidae

This family is, to me, very enigmatic. Again, they are hystricomorphous sciurognaths. The malleus and incus are not fused in *Pedetes*, although so closely appressed as apparently to prevent motion between them (Wood and Patterson, 1959). The incisor enamel of *Pedetes* (the only genus so far reported on) is multiserial (Korvenkontio, 1934), but there is no suggestion of a sacculus urethralis. Their fetal membranes show no apparent similarities to those of hystricognaths (Luckett, 1980, this volume), except as indicated by Luckett (1971).

Three described genera are included in the family, very distinct from one another but all clearly pedetids. In *Pedetes*, the cheek teeth are all hypselodont, with traces of pattern on the unworn crowns, insufficient to permit the determination of the ancestral pattern (Wood, 1965). The anterior cheek tooth, in front of the three molars, erupts before the first molar and is retained throughout life. Hence these teeth are presumably dP4/4; certainly there is no deciduous predecessor that ever became calcified (Wood, 1965). In the early Miocene *Parapedetes*, from Southwest Africa, the cheek teeth had already become nearly or quite hypselodont. Here, however, there was clearly a low-crowned deciduous tooth in front of the molars (Stromer, 1926), and there was normal mammalian replacement. Finally, in *Megapedetes*, best known from the early Miocene of Kenya (MacInnes, 1957), but also reported from the Miocene of Chios (Tobien, 1968) and Anatolia (Sen, 1977), the animal was much larger, with a skull length of 120mm, as contrasted with 85mm in *Pedetes* (MacInnes, 1957) and about 55mm in *Parapedetes* (Stromer, 1926). In *Megapedetes*, the cheek teeth are brachydont, but details of the pattern did not persist very long (MacInnes, 1957). Lavocat and Michaux (1966) illustrated one juvenile specimen in which the first molar and the tooth in front of it are present. The anterior tooth is slightly worn and the second is not, which should mean that these are dP/4 and M/1, although Lavocat and Michaux described the anterior tooth as P/4. No data have been reported for this genus, as to whether normal replacement did or did not occur. I would guess that there was normal replacement.

Either these three genera represent three very divergent evolutionary lines, or, if Parapedetes is related to Pedetes, there must have been progressive reduction of dP4/4, so that P4/4 erupt before M1/1, a most unusual condition. The tooth pattern of the pedetids is unclear, but it has little resemblance to that of any other sciurognathous hystricomorph. I see no liklihood that the tooth pattern of pedetids could have been derived from that of theridomyoids.

I believe that there is no valid basis for assuming any relationships between the Anomaluridae and Pedetidae on the one hand and the Hystricognathi on the other. The relationships of these two families must remain in doubt until more data are available (Fig. 1). In my opinion, the only presently viable hypotheses are that the Anomaluridae and Pedetidae are related to each other but are not close to anything else, or that neither family has any close relatives.

Family Protoptychidae

This family is known only from the genus Protoptychus from the middle Eocene of North America. Two skulls and one maxilla have been described; considerable additional material, including the lowe jaw, has been collected from the Washakie of Wyoming, but has not ye been studied. The animal had a hystricomorphous infraorbital foramen, but M. masseter medialis only penetrated the dorsal part of the foramen (Wahlert, 1973). The incisive foramina are long, extending back to the middle of P4/. The carotid canal appears to be absent. These, as indicated by Wahlert (1973), are hystricognath characteristics. The incisor enamel seems to be pauciserial. Until the undescribed material of Protoptychus is reported on, the most one can do is to say, with Wahlert, that it might be a franimorph, it might have something to do with the origin of the caviomorphs, or it migh be something quite different.

Superfamily Theridomyoidea

The European Pseudosciuridae and Theridomyidae have generally been considered to have arisen in Europe from early Eocene paramyid (Wood, 1962; Thaler, 1966). Recently, Hartenberger (1980), citing the absence of a known transition, has suggested that they might ha been derived from some equally unknown Asiatic group that managed to reach Europe in the middle Eocene by waif dispersal. Certainly migration from Asia to Europe was not easy at that time. Theridomyoids have never been reported from outside Europe, and there is no evidence that they were ever able to spread beyond the limits of that continent. They became extinct toward the end of the Oligocene.

Hystricomorphy developed within the family Pseudosciuridae, as illustrated by the skulls of *Protadelomys*, *Sciuroides* and *Adelomys* from the middle and late Eocene (Wood, 1974b). Auditory ossicles have been described in two theridomyid genera, not identified below subfamily level (Lavocat, 1967). No evidence was reported in these specimens that there had been any fusion. There is never any suggestion of the retention of deciduous teeth in any member of the superfamily. Korvenkontio (1934) listed the incisor enamel of pseudosciurids as being pauciserial, that of the theridomyids *Theridomys* and *Protechimys* also as pauciserial, whereas that of *Nesokerodon* was termed "pauci-uniserial" and that of *Archaeomys* was uniserial. This is the only case of transition from pauciserial to either uniserial or multiserial that Korvenkontio found.

The cheek teeth of theridomyids have a pattern very common among rodents, that Stehlin and Schaub (1951) termed the "Theridomys-Trechomysplan," originally apparently intended purely as a comparative-anatomical concept, but which quickly became converted to a phylogenetic one (Wood and Patterson, 1959). Because of a belief that the theridomyoid tooth pattern was a structure of overriding importance, and because the later theridomyoids were fully hystricomorphous, they have often been considered to have been ancestral to some or all of a wide variety of rodents. Schlosser (1884) spelled out this idea in detail, believing them to have been ancestral to the Hystricidae, Caviomorpha and Castoridae. Schaub (1953) proposed a Suborder Pentalophodonta, with two infraorders, the Palaeotrogomorpha and the Neotrogomorpha. The latter is identical to my use of the term Caviomorpha. Schaub's Palaeotrogomorpha included the Theridomyoidea, Hystricidae, Castoridae (but not the Eutypomyidae), Thryonomyidae, Eupetauridae (generally considered to be sciuroids), Petromuridae, Bathyergidae, Spalacidae and Rhizomyidae. The Anomaluridae were not mentioned. Later Schaub (1958) repeated this classification, adding a Family Pellegriniidae (now known to be ctenodactylids) to the Palaeotrogomorpha. At this time, he tentatively included the Anomaluridae in a superfamily Eomyoidea, saying (p. 768) "Deux familles distinctes, les Eomyidae et les Anomaluridae sont réunis à titre d'essai, eu égard à la structure de ses molaires." The Eomyidae are now generally considered to be geomyoids (Fahlbusch, this volume). Lavocat (1978) thought that the Thryonomyoidea, Bathyergidae, Anomaluridae and Pedetidae were present in the African Eocene and that they had originated from the theridomyoids. I know of no evidence in favor of this last idea.

I continue to believe that the Theridomyoidea were a European phenomenon, members of the Suborder Sciurognathi, that were not ancestral to any other families of rodents. They paralleled many other rodents in developing hystricomorphy, and were one of the early groups to develop the "Theridomys-Trechomysplan" in their dentition. This again was a case of parallelism.

DEVELOPMENT OF HYSTRICOGNATH FEATURES

Origin of Hystricognathy

Presumably hystricognathy developed from sciurognathy. No one has yet suggested the reverse except by implication. Deriving sciurognathy from hystricognathy seems highly improbable, because it would involve the reverse evolution of too many characteristics that must have been of great functional importance. Either hystricognathy developed gradually, or it developed rapidly. In the latter case, there is probably no use looking for ancestral groups, as we probably wouldn't recognize them if we saw them. I prefer the first alternative, until the evidence to the contrary becomes very compelling, because I believe that the available evidence documents slow rather than rapid evolution throughout the known parts of rodent history.

Either hystricognathy developed only once, or it arose several times, by parallelism. To a certain extent, this may be a matter of semantics, involving the definition of animals that had begun to make the changes involved in hystricognathy, but that had not yet reached the level of complete hystricognathy. Knowing how extensive parallelism is among rodents, I see no reason for not believing that fully-developed hystricognath features evolved several times, independently, from an ancestral stock that had become part of the Hystricognathi by Tullberg's (1899) definition, but which was only in the first stage of the changes needed to achieve full hystricognathy and in which there was the genetic potential to permit such an evolution. If, as I believe, this early hystricognathous stock was widespread in the northern hemisphere, this postulate would solve all the paleogeographic problems involved in the present and past distribution of the Hystricognathi.

I believe that the first step in the development of hystricognathy was the increase in the anteroposterior component of the jaw muscles. Which came first - the elongation of *M. masseter medialis*, of *M. masseter superficialis, pars reflexa*, or of *M. pterygoideus internus*? Or did they all elongate simultaneously? It is impossible at this time to be certain in this matter, but, in my present opinion, the initial step was a shift of the angular process so that it arose lateral to the incisive alveolus, which would have increased the length of *M. pterygoideus internus*, and, to a lesser extent, of the superficial masseter. Such a condition is represented, among known fossils, by Prolapsus from the middle Eocene of southern Texas, Reithroparamys from rocks of the same age farther north, and, in a less advanced stage of development, by Franimys from the earliest Eocene of Wyoming. The same stage, though perhaps independently derived, is shown by Cylindrodon from the early Oligocene of the United States.

The next step, I believe, would have involved (either as independent changes or as an associated complex) the deepening of the pterygoid fossa, the movement of the insertion of M. pterygoideus internus backward to become more nearly horizontal (Woods, 1972), and the migration of the insertion of M. masseter superficialis, pars reflexa onto the medial surface of the angular process. This last condition is seen in Cylindrodon. A later development would be the formation of a channel at the anterior end of the angle, involving further lateral movement of the angle. All these changes seem to have occurred in the Tsaganomyidae. The lengthening of the pterygoid is not, however, a feature unique to the hystricognaths, having also occurred in the geomyoids, Spalax and Aplodontia, where it likewise resulted in the perforation of the pterygoid fossa (Fig. 1) and the shift of the origin of the muscle into the temporal fossa or the rear of the orbit (Patterson and Wood, 1982). In Prolapsus, I believe, the pars reflexa arose from the medial side of the angle, and the formation for the groove to carry it to the medial surface had begun (Wood, 1973).

Origin of Hystricomorphy

The third step, the forward growth of the origin of M. masseter medialis through the infraorbital foramen and onto the face, might have been associated with the first two stages listed above, but was not necessarily so associated, as indicated by the conditions in the Tsaganomyidae and Bathyergidae. It seems probable, to me, that the beginnings of hystricomorphy first occurred in a small rodent, because in such forms the diameter of the infraorbital foramen, transmitting only the infraorbital nerve and blood vessels, is a larger percentage of snout height than in a larger animal. For example, in Sciuravus and Prolapsus (in both of which there is some breakage of the margin of the foramen), the infraorbital foramina seem to be at least 25% of the distance from the top of the skull to the alveolar border by P4/, whereas in larger forms, such as Paramys and Reithroparamys, it is considerably less. The skull of Prolapsus shows no evidence of the penetration of the foramen by the medial masseter, but damage to the specimen has eliminated the possibility of detecting slight penetration, had any been present.

These observations support the conclusions that the forward movement of M. masseter medialis through the infraorbital foramen occurred after the changes in the angular region which produced hystricognathy; that hystricomorphy probably developed among smaller members of the Hystricognathi; and that it had been preceded and facilitated by an increase in the proportionate size of the infraorbital foramen, probably related to the size of the animal.

Origin of Multiserial Incisor Enamel

During the time period in which the above changes were taking place, there was also selective pressure for strengthening the incisor enamel (Wilson, 1972), a pressure that seems (from the universality of the change) to have been acting on all groups of rodents. By the Oligocene, the Caviomorpha and Thryonomyoidea had already acquired multiserial enamel. The Oligocene tsaganomyids, however, apparently show stages in the development of multiserial enamel, that of *Cyclomylus* being more like pauciserial whereas that of *Tsaganomys* was closer to being fully multiserial. The acquisition of multiserial incisor enamel in the ctenodactyloids occurred between the middle Eocene (Sahni, 1980) and the late Oligocene (Bohlin, 1946). It seems certainly to have been independent of what happened in the caviomorphs, thryonomyoids and tsaganomyids. There is no evidence, among the Caviomorpha, Thryonomyoidea and Hystricidae, as to what the temporal relationships may have been among the acquisition of hystricognathy, hystricomorphy and multiserial incisor enamel. I believe that, in the Bathyergoidea, the acquisition of hystricomorphy was the last of the three conditions to arise.

I think that it is clear that rodents with multiserial incisor enamel were derived from ancestors with pauciserial enamel. This change in the incisor enamel clearly was *not* a part of an association that involved the lengthening of the anteroposterior pull of muscles. It presumably evolved, after the muscle changes, to strengthen the incisor enamel to withstand the increased stresses (Wilson, 1972).

Structures of the Circulatory System

It seems to me, based on the presence of the internal carotid artery in the Erethizontidae, that Bugge (this volume) has demonstrated that this artery was lost independently in the Caviomorpha and in the Old World forms, unless the caviomorphs were able to migrate from South America to Africa. I have not the slightest idea when the loss occurred, but I see no reason to assume that it was genetically associated with any of the other features. Other aspects of the circulatory system, as reported by George (1981) do not seem to indicate any important distinction of the erethizontids from the other caviomorphs, except for the peculiar occurrence of the "squirrel ring," which must surely be a case of convergence with the sciurids and aplodontids.

Embryological and Biochemical Characters

It seems reasonable to assume that many of the fetal membrane structures that Luckett (1980, this volume) has listed have been inherited from the common ancestral hystricognath stock, although there has probably also been some parallelism. However, I do not know how to use these features to determine the date at which such an ancestral stock lived. On the available data, I believe that it is legitimate to assume that these involve inheritance from early or middle Eocene ancestral Hystricognathi.

Likewise, the biochemical data indicate hystricognath relationships, but some similarities could easily have been inherited from franimorphs, and others might represent parallelisms. Again, I do not believe that the data can distinguish a common origin 35-40 m.y. ago from one 50 m.y. ago.

Other Supposed Hystricognath Features

The structures of the middle ear are undoubtedly of great importance, but I feel that there is still much that is unknown on the subject, and that we still are a long way from being able to separate common ancestry from parallelism, particularly since the region has not been adequately studied in so many forms.

There is no way, so far as I can see, that anyone can draw conclusions at present as to the function of the sacculus urethralis, which does not mean that it may not be highly adaptive. Without some information on this point, it is hard to determine its significance.

The only suggestion that I have encountered as to the function of the fusion of the malleus and incus is that it is an adaptation to prevent the dislocation of heavier ossicles (Patterson and Wood, 1982). There is no evidence as to when fusion began to develop, except that close appression of the ossicles was present by the early Miocene in both Africa and South America (Lavocat, 1973; Patterson and Wood, 1982). The fairly frequent absence of fusion among caviomorphs (Patterson and Wood, 1982) indicates that universal fusion, as seems to be the case among thryonomyoids, hystricids and bathyergids, was probably not present in the latest common ancestors of the modern hystricognaths.

ORIGINS OF THE PRESENT DISTRIBUTION OF THE HYSTRICOGNATHI

Thryonomyoidea

The thryonomyoids have inhabited Africa at least since the latest Eocene (Jaeger et al., this volume). They have been reported, elsewhere, from the middle to late Oligocene of the Balearics (Adrover and Hugueney, 1976; Adrover et al., 1977) and from the Miocene of Chios (Tobien, 1968) and India (Hinton, 1933; Black,

1972). These are clearly marginal occurrences, and the fossils are of African types and presumably represent migrants from Africa. Thryonomyoids or potential thryonomyoid ancestors are unknown from the Eocene of India or Pakistan, and the Indian Eocene faunas show no close relationships to those of the African Eocene or Oligocene, but rather to the faunas of northern Asia, although there are a number of endemic elements (Sahni et al., 1981). I know of no evidence that the thryonomyoids had European ancestors, and it seems most probable to me that they reached Africa from northern Asia, through Asia Minor. Either the thryonomyoids had been in Africa a lot longer before the beginning of the Oligocene than I believe to have been the case (Wood, 1983), or their Asian ancestors should already show delayed replacement of the deciduous teeth, as well as the more usual suite of hystricognath features, at least partially developed. I have no idea what type of rodents these ancestors may have been, except that I believe that they would have been franimorphs (Fig. 2).

Hystricidae

The Old World porcupines are first known from the late Miocene of Egypt, Hungary and India. There have been reports of earlier hystricids from Europe (Schlosser, 1884), but none seems to have been a hystricid. The incisor enamel of "*Hystrix*" *lamandini* is uniserial (Korvenkontio, 1934), and it is now generally agreed that this animal was not a hystricid (Lavocat, 1962). There is no suggestion of hystricids in the early Miocene of Africa (East Africa - Lavocat, 1973; Southwest Africa - Stromer, 1926; Hopwood, 1929) or from the later Miocene of Morocco (Lavocat, 1961). No early hystricids or near-hystricids have been reported from northern Asia. Because there is always normal replacement of the deciduous teeth, hystricids cannot have been descended from any post-Eocene thryonomyoids, but might have been ancestral to them, although I do not believe this to have been the case. Presumably hystricids reached Africa after the Miocene of Southwest Africa, East Africa and Morocco, and presumably by dry land from southwest Asia. No potential source groups are known, and our general lack of information suggests that they may have originated in southeast Asia, where they may have had a long pre-Miocene history (Wood, in press).

Caviomorpha (including the Erethizontidae)

It would seem that it should have required about the same length of time for the Oligocene caviomorphs to have evolved from a single (or double, if Bugge is correct) invading ancestral stock into the diverse and dominant sector of the population that they were in the Deseadan, as for the African early Miocene thryonomyoids to have done the same, and much longer than it has taken the Australian murids to have reached their present diversity (Wood, 1983). No rodents are known from the middle Eocene Mustersan of Patagonia. The only continental Tertiary deposit intermediate between this and the Deseadan is the Divisadero Largo, of uncertain age, with a fauna perhaps largely arboreal (Simpson et al., 1962). Because of what we know of their history, we can assume that then, as now, the caviomorphs were largely terrestrial. Therefore, the caviomorphs could have reached South America any time after the Musters, and their extensive diversification in the Deseadan suggests that they arrived there very shortly after Mustersan time.

Except for Croizat (1979), who has suggested that the caviomorphs were in South America "since the Cretaceous or even earlier," everyone agrees that the caviomorph ancestors came to South America either from Africa or from Middle America. If (as I do not believe) the ancestral caviomorphs came to South America from Africa (Lavocat, 1973, 1976, 1980), the migration must have been long enough before the Deseadan to have allowed the differentiation of the caviomorphs that we see there, which, I believe, would mean that they separated from the thryonomyoids well over 40 m.y. ago. If the caviomorphs arose from Middle American ancestors, as seems much more probable to me, the common ancestors of all the Hystricognathi would have been Holarctic rodents that lived about 45-50 m.y. ago.

Bathyergidae

The bathyergids first appear in the early Miocene of Southwest Africa and East Africa, being very abundant in the latter area. Lavocat (1973) believed that *Bathyergoides*, the common early Miocene genus, was descended from thryonomyoids. He thought that the tsaganomyids perhaps represented an early off-shoot of the bathyergids that reached northern Asia from Africa by the early Oligocene, at a time when no bathyergids are known in Africa, when there was no land connection with Asia, and when it seems unlikely that the bathyergids would have been able to have evolved from the thryonomyoids. This postulate likewise ignores the strong evidence for tsaganomyid-cylindrodontid relationships. It seems much more probable to me that the bathyergids were derived from tsaganomyids, and that they reached Africa in the same early Miocene dry-land invasion that brought the cricetids, sciurids and presumably the ctenodactylids and perhaps the pedetids. The majority of bathyergids have normal replacement of the deciduous teeth. *Bathyergoides*, *Proheliophobius* and perhaps *Cryptomys* apparently represent independent evolution of the beginnings of hystricomorphy.

Tsaganomyidae

The northern Asiatic Oligocene tsaganomyids agree with the bathyergids in having a posteriorly extended upper incisor, arising posterior to at least some of the upper cheek teeth, procumbent upper incisors, a highly everted angle, the beginnings of the migration of the *pars reflexa* of *M. masseter lateralis* onto the median surface of the angle, an open pterygoid fossa, and they were protrogomorphous, as are most bathyergids. Like the latter, they were highly adapted for burrowing (Vinogradov and Gambarian, 1952). The incisor enamel was transitional between pauciserial and multiserial. The cheek teeth were hypselodont, four-crested, and there was normal replacement of dP4/4. They appear to me to be the ideal ancestral group for the bathyergids. In the other direction, I believe them to have been derived from the cylindrodonts (Fig. 2).

CONCLUSIONS

In summary, I would postulate that the relationships of the various groups that I have discussed (Fig. 2) might be as follows.

(1) The Hystricognathi are a natural group, but only if one includes the basic hystricognathous stock, which I have termed the Franimorpha, in which many of the features secondarily associated with hystricognathy had not yet developed.

(2) The Caviomorpha were descended from middle Eocene Middle American franimorphs, that were able to reach South America shortly after Mustersan time, either directly through Central America or, perhaps somewhat more probably, via the Nicaragua Plateau, the Greater Antilles, and either the Aves Ridge or the Lesser Antilles. The erethizontids should be included in the Caviomorpha, but, if separable, they represent a closely allied group, descended from the same general ancestral stock from Middle America.

(3) The Bathyergidae invaded Africa, probably by dry land, from Asia Minor at the time of the formation of the Burdigalian land bridge, which took proboscideans and some thryonomyoids out of Africa and brought in (at the minimum) ctenodactylids, cricetids and sciurids. The bathyergids were probably descended from some member or members of the north Asiatic Oligocene Family Tsaganomyidae, which should be included in the Bathyergoidea. The Tsaganomyidae were apparently derived from northern Asiatic members of the Cylindrodontidae. This latter family was apparently of North American origin, and the American species referred to Ardynomys are quite similar to the Asiatic one, whether or not they deserve to be congeneric.

(4) The Hystricidae, first known from the late Miocene, probably had a southeastern Asiatic origin, and, I believe, had inhabited that area for some time before the late Miocene. No possible ancestors are now recognized, either in southern or in northern Asia.

(5) The Thryonomyoidea originated outside of Africa, and invaded that continent, probably from the northeast, in late Eocene time, presumably by rafting. No potential ancestors are known. The ancestors could have been primitive hystricids, but I know of no particular evidence for this. Certainly there are no fossils that I

have heard of from Europe that could, in my opinion, have had anything to do with the thryonomyoid ancestry.

(6) If I am correct that Franimys, Reithroparamys, Prolapsus and the cylindrodonts are hystricognathous, and that this assemblage is a natural group, the Franimorpha, then the modern peculiarities in the flea and nematode associations of hystricognaths could have been inherited from Eocene northern hemisphere ancestors, retained independently in the various modern groups. This is especially worthy of consideration in view of the uncertainties as to evolutionary rates in fleas and nematodes.

(7) If the Franimorpha are not accepted as primitive members of the Hystricognathi (which would require a redefinition of the suborder), then there are no Eocene rodents known from any part of the world that can legitimately be considered to have anything to do with the origins of either hystricognathy or of the Suborder Hystricognathi.

(8) Obviously, these conclusions require that there has been extensive parallelism in the evolution of the Hystricognathi, but it should have become obvious to everyone by now that parallelism is an all-pervading mechanism among the Rodentia.

(9) The series of non-hystricognathous hystricomorphs (Anomaluridae, Ctenodactyloidea, Dipodoidea, Pedetidae, Protoptychidae, and Theridomyoidea) have nothing to do with the hystricognaths, even though some of them have paralleled the hystricognaths in one or more other features besides hystricomorphy. Any attempt to include these groups in the hystricognath family tree would have to invoke even more (and, I believe, less believable) parallelism than what I have proposed.

(10) The use of the term "Hystricomorpha" as a taxonomic entity should be avoided, at least in the near future, because it means all things to all people and only leads to confusion.

REFERENCES

Adrover, R. and Hugueney, M. 1976. Des Rongeurs (Mammalia) africains dans une faune de l'Oligocène élevé de Majorque (Baléares, Espagne). Nouv. Arch. Mus. Hist. nat. Lyon 13(suppl.): 11-13.

Adrover, R., Hugueney, M. and Mein, P. 1977. Fauna Africana Oligocena y nuevas formas endemicas entre los micromamiferos de Mallorca (Nota preliminar). Bol. Soc. Hist. Nat. Baleares 22: 137-149.

Black, C. C. 1972. Review of fossil rodents from the Neogene Siwalik beds of India and Pakistan. Palaeontology 15: 238-266.

Black, C. C. and Sutton, J. F. 1984. Paleocene and Eocene rodents of North America. In: Papers in Vertebrate Paleontology Honoring Robert Warren Wilson, R. M. Mengel, ed., pp. 67-84, Carnegie Mus. Nat. Hist., Spec. Publ. 9, Pittsburgh.

Bohlin, B. 1946. The fossil mammals from the Tertiary deposits of Taben-Buluk. Part II: Simplicidentata, Carnivora, Artiodactyla, Perissodactyla, and Primates. Paleont. Sinica n. s. C (8b): 1-259.

Brandt, J. F. 1855. Beiträge zur nähern Kenntniss der Säugethiere Russlands. Mém. Acad. Imp. St.-Pétersbourg (6) 9: 1-375.

Bruijn, H. de and Rümke, C. G. 1974. On a peculiar mammalian association from the Miocene of Oschiri (Sardinia). I, II. Proc. Kon. Ned. Akad. Wetensch., Ser. B, 77: 48-79.

Bugge, J. 1971. The cephalic arterial system in New and Old World hystricomorphs, and in bathyergoids, with special reference to the classification of rodents. Acta anat. 80: 516-536.

Bugge, J. 1974a. The cephalic arterial system in insectivores, primates, rodents and lagomorphs, with special reference to the systematic classification. Acta anat. 87 (Suppl. 62): 1-160.

Bugge, J. 1974b. The cephalic arteries of hystricomorph rodents. Symp. Zool. Soc. Lond. 34: 61-78.

Burke, J. J. 1935. Pseudocylindrodon, a new rodent genus from the Pipestone Springs Oligocene of Montana. Ann. Carneg. Mus. 25: 1-4.

Chaline, J. and Mein, P. 1979. Les Rongeurs et l'Evolution. Doin Editeurs, Paris.

Croizat, L. 1979. Review of: Biogeographie: Fauna und Flora der Erde und ihre geschichtliche Entwicklung, by P. Bănărescu and N. Boscaiu. Syst. Zool. 28: 250-252.

Dawson, M. R. 1977. Late Eocene rodent radiations: North America, Europe and Asia. Géobios Mém. Spéc. 1: 195-209.

Dawson, M. R., Li, C.-K. and Qi, T. 1984. Eocene ctenodactyloid rodents (Mammalia) of eastern and central Asia. In: Papers in Vertebrate Paleontology Honoring Robert Warren Wilson, R. M. Mengel, ed., pp. 138-150, Carnegie Mus. Nat. Hist., Spec. Publ. 9, Pittsburgh.

Durette-Desset, M.-C. 1971. Essai de classification des nématodes héligmosomes. Corrélations avec la paléobiogéographie des hôtes. Mém. Mus. Nat. Hist. Nat. (A) 69: 1-136.

Fields, R. W. 1957. Hystricomorph rodents from the late Miocene of Columbia, South America. Univ. Cal. Publ. Geol. Sci. 32: 273-404.

George, W. 1981. Blood vascular patterns in rodents: contributions to an analysis of rodent family relationships. Zool. J. Linn. Soc. 73: 287-306.

Grassé, P.-P. and Dekeyser, P. L. 1955. Ordre des Rongeurs. In: Traité de Zoologie. Anatomie, Systématique, Biologie, Vol. 17 (2), P.-P. Grassé, ed., pp. 1321-1525, Masson et Cie, Paris.

Hartenberger, J.-L. 1980. Données et hypothèses sur la radiation initiale des Rongeurs. Palaeovertebrata, Mém. Jubil. R. Lavocat: 285-301.

Hinton, M. A. C. 1933. Diagnoses of new genera and species of rodents from the Indian Tertiary deposits. Ann. Mag. Nat. Hist. (10) 12: 620-622.
Hopwood, A. T. 1929. New and little known mammals from the Miocene of Africa. Amer. Mus. Novit. 344: 1-9.
Hussain, S. T., Bruijn, H. de and Leinders, J. M. 1978. Middle Eocene rodents from the Kala Chitta Range (Punjab, Pakistan). Proc. Kon. Ned. Akad. Wetensch., Ser. B, 81: 74-112.
Korth, W. W. 1984. Earliest Tertiary evolution and radiation of rodents in North America. Bull. Carneg. Mus. 24: 1-71.
Korvenkontio, V. A. 1934. Mikroskopische Untersuchungen an Nagerincisiven unter Hinweis auf die Schmelzstruktur der Backenzähne. Histologisch-phyletische Studie. Ann. Zool. Soc. Zool.-Bot. Fennicae Vanamo 2: 1-274.
Landry, S. O., Jr. 1957. The interrelationships of the New and Old World hystricomorph rodents. Univ. Calif. Publ. Zool. 56: 1-118.
Lavocat, R. 1961. Etude systématique de la faune de mammifères. La gisement de vertébrés Miocènes de Beni Mellal (Maroc). Etude géologique. Notes et Méms. Serv. Géolog. Royaume de Maroc 155: 29-95.
Lavocat, R. 1962. Réflexions sur l'origine et la structure du groupe des Rongeurs. Colloque Internat. C.N.R.S. 163: 287-299.
Lavocat, R. 1967. Observations sur la région auditive des rongeurs théridomorphes. Problèmes actuels de Paléontologie, Colloque Internat. C.N.R.S. 611: 491-501.
Lavocat, R. 1971. Affinités systématiques des caviomorphes et des phiomorphes et origine Africaine des caviomorphes. An. Acad. Bras. Cienc. 43 (Supl.): 515-522.
Lavocat, R. 1973. Les rongeurs du Miocène d'Afrique Orientale. 1. Miocène inférieur. E.P.H.E., Inst. Montpellier, Mém. 1: 1-284.
Lavocat, R. 1976. Rongeurs caviomorphes de l'Oligocène de Bolivie. II. Rongeurs du Bassin Déséadien de Salla-Luribay. Palaeovertebrata 7: 15-90.
Lavocat, R. 1978. Rodentia and Lagomorpha. In: Evolution of African Mammals, V. S. Maglio and H. B. S. Cooke, eds., pp. 69-89, Harvard Univ. Press, Cambridge.
Lavocat, R. 1980. The implications of rodent paleontology and biogeography to the geographical sources and origin of platyrrhine primates. In: Evolutionary Biology of the New World Monkeys and Continental Drift, R. L. Ciochon and A. B. Chiarelli, eds., pp. 93-102, Plenum Press, New York.
Lavocat, R. and Michaux, J. 1966. Interprétation de la structure dentaire des rongeurs africains de la famille des Pédétidés. C. R. Acad. Sci. Paris 262: 1677-1679
Luckett, W. P. 1971. The development of the chorio-allantoic placenta of the African scaly-tailed squirrels (Family Anomaluridae). Amer. J. Anat. 130: 159-178.
Luckett, W. P. 1980. Monophyletic or diphyletic origins of Anthro-

poidea and Hystricognathi: evidence of the fetal membranes. In: Evolutionary Biology of the New World Monkeys and Continental Drift, R. L. Ciochon and A. B. Chiarelli, eds., pp. 347-368, Plenum Press, New York.

MacInnes, D. G. 1957. A new Miocene rodent from East Africa. Fossil Mammals of Africa, Brit. Mus. (Nat. Hist.) 12: 1-36.

Matthew, W. D. and Granger, W. 1923. New Bathyergidae from the Oligocene of Mongolia. Amer. Mus. Novit. 101: 1-5.

Matthew, W. D. and Granger, W. 1925. New creodonts and rodents from the Ardyn Obo Formation of Mongolia. Amer. Mus. Novit. 193: 1-7.

Parent, J.-P. 1976. Disposition fondamentale et variabilité de la region auditive des rongeurs hystricognathes. C. R. Acad. Sci. Paris, D 283: 243-245.

Parent, J.-P. 1980. Recherches sur l'oreille moyenne des rongeurs actuels et fossiles. Anatomie. Valeur systématique. E. P. H. E., Inst. Montpellier, Mém. 11: 1-286.

Patterson, B. and Wood, A. E. 1982. Rodents from the Deseadan Oligocene of Bolivia and the relationships of the Caviomorpha. Bull. Mus. Comp. Zool. 149: 371-543.

Quentin, J.-C. 1973a. Affinités entre les Oxyures parasites des rongeurs Hystricidés, Erethizontidés et Dinomyidés. Intérêt paléobiogéographique. C. R. Acad. Sci. Paris 276: 2015-2017.

Quentin, J.-C. 1973b. Morphologie et position systématique d'*Oxyuris stossichi* Setti, 1897. Intérêt paléobiogéographique de cette espèce. Bull. Mus. Nat. Hist. Nat. (3) 183: 1403-1408.

Quentin, J.-C. 1973c. Les Oxyurinae des Rongeurs. Bull. Mus. Nat. Hist. Nat. (3) 167: 1046-1096.

Sahni, A. 1980. SEM studies of Eocene and Siwalik rodent enamels. Geosci. J. 1: 21-30.

Sahni, A., Bhatia, S. B., Hartenberger, J.-L., Jaeger, J.-J., Kumar, K., Sudre, J. and Vianey-Liaud, M. 1981. Vertebrates from the Subathu formation and comments on the biogeography of the Indian subcontinent during the early Paleogene. Bull. Soc. géol. France 23: 689-695.

Sahni, A. and Khare, S. K. 1973. Additional Eocene mammals from the Subathu Formation of Jammu and Kashmir. J. Palaeont. Soc. India 17: 31-49.

Sahni, A. and Srivastava, V. C. 1976. Eocene rodents and associated reptiles from the Subathu Formation of northwestern India. J. Paleontol. 50: 922-928.

Schaub, S. 1953. Remarks on the distribution and classification of the "Hystricomorpha." Verh. naturf. Ges. Basel 64: 389-400.

Schaub, S. 1958. Simplicidentata (= Rodentia). In: Traité de Paléontologie, Vol. 6 (2), J. Piveteau, ed., pp. 659-818, Masson et Cie, Paris.

Schlosser, M. 1884. Die Nager des europäischer Tertiärs nebst Betrachtungen über die Organisation und die geschichtliche Entwicklung der Nager überhaupt. Palaeontographica 31: 1-140.

Sen, S. 1977. Megapedetes aegaeus n. sp. (Pedetidae) et à propos d'autres "rongeurs africains" dans le Miocène d'Anatolie. Géobios 10: 983-986.

Shevyreva, N. S. 1971. The first find of Eocene rodents in the U. S. S. R. Bull. Acad. Sci. Georgian S. S. R. 61: 745-747. [in Russian]

Shevyreva, N. S. 1972a. On the evolution of the rodent family Sciuravidae. Dokladi Akad. Nauk SSSR 206: 1453-1454. [in Russian]

Shevyreva, N. S. 1972b. New rodents from the Paleogene of Mongolia and Kazakhstan. Paleont. J. Moscow 3: 134-145. [in Russian]

Simpson, G. G., Minoprio, J. L. and Patterson, B. 1962. The mammalian fauna of the Divisadero Largo Formation, Mendoza, Argentine. Bull. Mus. Comp. Zool. 127: 237-293.

Stehlin, H. G. and Schaub, S. 1951. Die Trigonodontie der simplicidentaten Nager. Schweiz. paläontol. Abhandl. 67: 1-385.

Stromer, E. 1926. Reste Land- und Süsswasser-bewohnender Wirbeltiere aus den Diamantenfeldern Deutsch-Südwestafrikas. In: Die Diamantenwüste Südwestafrikas, by E. Kaiser, pp. 107-153, Vol. 2, Dietrich Reimer, Berlin.

Thaler, L. 1966. Les rongeurs fossiles du Bas-Languedoc dans leurs rapports avec l'histoire des faunes et la stratigraphie du Tertiaire d'Europe. Mém. Mus. Nat. Hist. Nat., n. s. (C) 17: 1-295.

Tobien, H. 1968. Paläontologische Ausgrabungen nach jungtertiären Wirbeltieren auf den Insel Chios (Griechenland) und bei Maragheh (NW-Iran). Jahr. Vereinigung "Freunde d. Universität Mainz" 1968: 51-58.

Traub, R. 1980. The zoogeography and evolution of some fleas, lice, and mammals. In: Fleas. Proc. Internat. Conference Fleas, R. Traub and H. Starcke, eds., pp. 93-172, A. A. Balkema, Rotterdam.

Tullberg, T. 1899. Ueber das System der Nagethiere: eine phylogenetische Studie. Nova Acta Reg. Soc. Scient. Upsala (3) 18: 1-514.

Vinogradov, B. C. and Gambarian, P. P. 1952. Oligocene cylindrodonts from Mongolia and Kazakhstan (Cylindrodontidae, Glires, Mammalia). Akad. Nauk SSSR, Trudi Paleontol. Inst. 41: 13-42. [in Russian]

Vucetich, M. G. 1975. La anatomía del oído medio como indicadora de relaciones sistemáticas y filogenéticas en algunos grupos de roedores Caviomorpha. Actas Primero Congr. Argent. Paleontol. Bioestrat., Tucumán, Arg. 2: 477-494.

Wahlert, J. H. 1973. Protoptychus, a hystricomorphous rodent from the late Eocene of North America. Mus. Comp. Zool., Breviora 419: 1-14.

Wahlert, J. H. 1974. The cranial foramina of protrogomorphous rodents, an anatomical and phylogenetic study. Bull. Mus. Comp. Zool. 146: 363-410.

Walker, E. P. 1975. Mammals of the World, 3rd. Ed., Vol. 1 and 2. Johns Hopkins Univ. Press, Baltimore.

Waterhouse, G. E. 1839. Observations on the Rodentia, with a view to point out the groups, as indicated by the structure of the crania, in this order of Mammals. Mag. Nat. Hist. n. s. 3: 90-96.

Wilson, R. W. 1972. Evolution and extinction in early Tertiary rodents. 24th Internat. Geol. Congr., Sec. 7: 217-224.

Wood, A. E. 1937. The mammalian fauna of the White River Oligocene. Part II. Rodentia. Trans. Amer. Phil. Soc. n. s. 28: 155-269.

Wood, A. E. 1949. A new Oligocene rodent genus from Patagonia. Amer. Mus. Novit. 1435: 1-54.

Wood, A. E. 1955. A revised classification of the rodents. J. Mammal. 36: 165-187.

Wood, A. E. 1962. The early Tertiary rodents of the Family Paramyidae. Trans. Amer. Phil. Soc. 52: 1-261.

Wood, A. E. 1965. Unworn teeth and relationships of the African rodent, Pedetes. J. Mammal. 46: 419-423.

Wood, A. E. 1968. Early Cenozoic mammalian faunas, Fayum Province, Egypt. Part II. The African Oligocene Rodentia. Bull. Peabody Mus. Nat. Hist. 28: 23-105.

Wood, A. E. 1972. An Eocene hystricognathous rodent from Texas: its significance in interpretations of continental drift. Science 175: 1250-1251.

Wood, A. E. 1973. Eocene rodents, Pruett Formation, southwest Texas; their pertinence to the origin of the South American Caviomorpha. Texas Mem. Mus., Pearce-Sellards Ser. 20: 1-40.

Wood, A. E. 1974a. Early Tertiary vertebrate faunas, Vieja Group, Trans-Pecos Texas: Rodentia. Texas Mem. Mus., Bull. 21: 1-112.

Wood, A. E. 1974b. The evolution of the Old World and New World hystricomorphs. Symp. Zool. Soc. London 34: 21-60.

Wood, A. E. 1975. The problem of the hystricognathous rodents. In: Studies on Cenozoic Paleontology and Stratigraphy in honor of Claude W. Hibbard, Claude W. Hibbard Memmorial Vol. 3, G. R. Smith and N. E. Friedland, eds., Univ. Mich. Papers Paleontol. no. 12, Ann Arbor, pp. 75-80.

Wood, A. E. 1977. The evolution of the rodent Family Ctenodactylidae. J. Palaeontol. Soc. India 20: 120-137.

Wood, A. E. 1980. The origin of the caviomorph rodents from a source in Middle America: a clue to the area of origin of the platyrrhine primates. In: Evolutionary Biology of the New World Monkeys and Continental Drift, R. L. Ciochon and A. B. Chiarelli, eds., pp. 79-91, Plenum Press, New York.

Wood, A. E. 1983. The radiation of the Order Rodentia in the southern continents; the dates, numbers and sources of the invasions. In: Wirbeltier-Evolution und Faunenwandel in Känozoicum, W.-D. Heinrich, ed., pp. 381-394, Schriftenr. geol. Wiss. 19/20, Berlin.

Wood, A. E. 1984. Hystricognathy in the North American Oligocene rodent Cylindrodon and the origin of the Caviomorpha. In: Papers in Vertebrate Paleontology Honoring Robert Warren Wilson R. M. Mengel, ed., pp. 151-160, Carnegie Mus. Nat. Hist., Spec.

Publ. 9, Pittsburgh.

Wood, A. E. in press. Northern waif primates and rodents. In: The Great American Biotic Interchange, S. D. Webb and F. S. Stehli, eds., Plenum Press, New York.

Wood, A. E. and Patterson, B. 1959. The rodents of the Deseadan Oligocene of Patagonia and the beginnings of South American rodent evolution. Bull. Mus. Comp. Zool. 120: 279-428.

Wood, A. E. and Wood, H. E., 2nd. 1933. The genetic and phylogenetic significance of the presence of a third upper molar in a modern dog. Amer. Midl. Nat. 14: 36-48.

Woods, C. A. 1972. Comparative myology of jaw, hyoid and pectoral appendicular regions of New and Old World hystricomorph rodents. Bull. Amer. Mus. Nat. Hist. 147: 117-198.

Woods, C. A. 1975. The hyoid, laryngeal and pharyngeal regions of bathyergid and other selected rodents. J. Morph. 147: 229-250.

Woods, C. A. 1984. Hystricognath rodents. In: Orders and Families of Recent Mammals of the World, S. Anderson and J. K. Jones, Jr., eds., pp. 389-446, John Wiley and Sons, New York.

MYOLOGY OF HYSTRICOGNATH RODENTS:

AN ANALYSIS OF FORM, FUNCTION, AND PHYLOGENY

Charles A. Woods and John W. Hermanson

Florida State Museum, Gainesville, Florida 32611
Emory University, Atlanta, Georgia 30322

INTRODUCTION

Muscles have been used as valuable morphological characters in a number of analyses of the phylogenetic relationships of rodents and other mammals (see literature cited in Rinker, 1954; Klingener, 1964; Woods, 1972). The location, innervation, size, shape, number of parts, relative position, presence or absence, and even function of muscles have been used to establish the phylogenetic relationships of various rodent taxa. While comparisons of the form of muscles go back to the time of Vesalius (1543), and Tyson (1699) used muscles in his comparison of various primates, most early works were mainly concerned with descriptive myology. It was not until the great flowering of comparative anatomy in the middle of the last century that careful comparative analyses of musculature were used to formulate hypotheses of phylogenetic relationships.

Many early investigations of rodent myology described and compared muscles, but rarely established whether myological conditions were primitive or derived, or compared large series of myological characters from a variety of anatomical regions. The most comprehensive treatments of this period are the works of Parsons (1894, 1896) and Tullberg (1899), who used a large number of muscles to establish the phylogenetic relationships of rodents. Parsons discussed various myological characters in detail and discounted the importance of some similarities of musculature. For example, while he considered dipodoids to be most closely related to hystricomorphs in 1894 based on the similarity of the condition of the medial masseter and infraorbital foramen, by 1896 he recognized that the jerboa lacked a scapulo-clavicularis and several other muscles that characterize hystricomorphs. Thus, Parsons concluded that jerboas, even though hystricomorphous, "are more nearly allied to the Myomorpha than to the Hystricomorpha." While Parsons did not determine if a character was primitive or derived, he was one of the first morphologists to question the validity of certain myological features in establishing heritage (= phylogenetic relationship), and to emphasize certain characters over others in making phylogenetic decisions. Parsons (1896:192) stated that "in rodents certain muscles are valuable for classificatory purposes and, if several are taken, are not likely to mislead." Parsons considered the muscles of the trunk, neck, and shoulder girdle to be the most reliable. Tullberg (1899), who was very cautious and complete in his descriptions and analyses, also negated the importance of the hystricomorphous condition in dipodoids because it was shared by so many apparently unrelated taxa. He understood the importance of primitive and advanced characters, and his phylogenetic tree (1899: 481), drawn on the basis of his analysis of a variety of morphological features, is remarkably close to current phylogenetic arrangements.

A number of important twentieth century works attempted to determine the relationships of rodents based on myological characters. These works included descriptive and experimental studies that provide information on the form and function of muscles in rodents. Examples of these works are Meinertz (1941, 1944) on facial muscles; Langworthy (1925), Enders (1934), and Woods and Howland (1977) on skin musculature; Adams (1919), Forster (1928-1929) Muller (1933), Rinker and Hooper (1950), Becht (1953), Van Vendeloo (1953), Turnbull (1970), Hiiemae (1971), Hiiemae and Houston (1971), Weijs (1973, 1975), Gorniak (1977), Woods and Howland (1979), Kesner (1980), and Byrd (1981) on masticatory muscles; Sprague (1941), House (1953), Sharma and Sivaram (1959), Sivaram and Sharma (1970), Kupper (1970), Doran and Baggett (1971), and Woods (1975) on hyoid, laryngeal, and pharyngeal muscles; Appleton (1928), Hill (1934), Howell (1936), Fry (1961), Dynowski (1974), Berman (1979), Aristov (1981), and McEvoy (1982) on appendicular muscles. Examples of important monographs on the myology of rodents are Merriam (1895) on pocket gophers, Howell (1926) on the wood rat, Howell (1932) on the kangaroo rat, Green (1935) on the rat, Hill (1937) on pocket gophers, Bryant (1945) on squirrels, Rinker (1954, 1963) on cricetines, Klingener (1964) on dipodoids, Woods (1972) on hystricognath rodents, and Cooper and Schiller (1975) on the guinea pig. The information provided in these (and other) monographs is useful in evaluating rodent phylogenies.

In theory, muscles should be excellent morphological features to use in evaluating phylogenetic relationships. It is possible to identify individual muscles and to establish homologies based on their innervation. Muscles are variable among taxa, and are associated with a wide range of functional conditions that may be unrelated to each other, (i.e., vocalizations, gnawing, chewing, facial expression, temperature regulation, posture, movement, and even reproduction). In addition, muscles come from several different embryonic sources (Cheng, 1955; Jones, 1979) and are subjected to different selective pressures during ontogeny as well as phylogeny.

USE OF MYOLOGICAL DATA IN PHYLOGENETIC STUDIES

Justification

The justification for using myological characters to seek phylogenetic relationships among rodents was summarized by Hill (1937: 159), who noted that even though variations in the origin and insertion of muscles occur, "heritage" exerts a great influence on the form of muscles. This is especially true below the family level, as noted by Hill (1937), Rinker (1954, 1963), and Klingener (1964), all of whom observed that the assumption of myological similarities being attributable to "heritage" above the subfamily level was probably invalid. This seemed true at that time because not enough data existed to provide detailed descriptions of the morphology and variation within various species (individual and intraspecific variation), genera, and subfamilies. With ca. 1749 species of known Recent rodents (Carleton, 1984), a large amount of morphological variation exists within various rodent lineages. A.E. Wood (in Patterson and Wood, 1982: 507) was unable to find a single feature in all members of New and Old World hystricognaths that is not also found somewhere else among rodents (both extinct and extant). A large data base is necessary before myological data can be used with confidence at the subordinal and superfamily level to group taxa into clades, and before it will be possible to identify cases of parallelism and convergence.

Methodology

The best way to insure the validity of myological information is to deal with data on as many different taxa within a taxonomic category as possible. In many cases, this will require extensive dissection work, although in some cases it would be possible to select phylogenetic lineages to study in which there are few individual extant taxa. For example, the New World Superfamily Erethizontoidea with one family, four Recent genera, and ten Recent species (Woods, 1984) is easier to compare with the Old World Superfamily Hystricoidea (= A.E. Wood's Infraorder Hystricomorpha) with one family, three Recent genera, and eleven Recent species (Woods, 1984) than it would be to make comparisons between the Family Dipodidae (with six subfamilies, 14 Recent genera, and 44 Recent species; Klingener, 1984) and the Superfamily Muroidea (with 15 Recent subfamilies, ca. 261 Recent genera, and ca. 1135 Recent species; Carleton and Musser, 1984). From these figures, the scope of the problem of working at higher taxonomic levels becomes apparent.

Myological data can best be used to formulate decisions on phylogenetic relationships at higher taxonomic levels in ancient lineages where there has been little diversification, such as in hystricognath rodents, ctenodactylids, or aplodontids (Woods, 1982). In young lineages where there has been a rapid radiation into a variety of morphotypes, the myological characters are more applicable at the generic and subfamilial level, as was pointed out by Hill (1937) and Rinker (1954, 1963). However, even in these lineages, certain morphological regions appear to have been extremely conservative and may provide characters that are useful in establishing phylogenetic relationships at higher taxonomic levels. As noted by Parsons (1896), the anatomy of the muscles of the trunk and neck vary less than the appendicular and masticatory muscles and are especially useful in "settling the position of animals."

The Search for Primitive Characters and Parallelism

The use of cladistics in establishing phylogenetic relationships remains controversial, and the method has been criticized as being directly circular, nondeductive, and probabilistic by Bock (1981). According to Bock synapomorphs are not identical to homologues, and the criteria used to test for synapomorphic characters are not the same as should be used in testing homologues. An important point of contention by Bock (1981: 18) is that "symplesiomorphs are perfectly good homologues," and therefore should not be rejected as important characters, as is required in cladistic analyses. We agree in part with Bock's main thesis that studies of phylogeny and classification should make use of functional analyses. However, functional data must conform to the principle of homology (Wiley, 1975), as do species-specific behaviors (or taxon-specific behaviors). The construction of cladograms and the establishment of plesiomorphic and apomorphic character states in phylogenetic systematics have been described extensively (see Cracraft, 1981, 1983, for recent reviews). It is not necessary, therefore, to discuss the philosophy of phylogenetic systematics or the mechanics of constructing cladograms in detail.

The methodology of cladistics has proven to be a valuable tool in interpreting data and in establishing primitive and derived character states in specific lineages, especially in studies of taxa where complete information on the form, function, and ontogeny of myological characters is not available. It is clear, however, that great care must be taken in choosing the taxa to be studied and in selecting an outgroup. For example, Parsons (1894), in his study of the myology of rodents, compared tree porcupines with ground porcupines. Parsons chose a member of the genus *Hystrix* (Old World porcupine) to compare with *Sphiggurus* (a New World porcupine). Parsons (1894: 295) concluded "it is difficult to point out many points which are characteristic of the porcupines as a group owing to the great differences between the muscles . . . ." He was biased in this study by his assumption that all porcupines are closely related. Thus, Parsons tried to mitigate the differences rather than accept their value as myological data to conclude that all "porcupines" are not closely related. An example of this problem can be found within New World porcupines (Erethizontidae), where it has recently become apparent that the genus *Chaetomys* is probably not a porcupine, but rather a distinct lineage of spiny rat of the Family Echimyidae (Patterson and Wood, 1982; Woods, 1982).

The inference as to whether the forms are closely related or whether they developed their resemblances independently (parallelism) can be strengthened by examining the functional significance of various characters to understand which characters are conservative (i.e., have not changed in spite of functional changes that should affect their form) or which are dynamic (have changed form in response to functional change). Woods and Howland (1979), for example, were able to show that a number of characters of the jaw, hyoid, and pharyngeal regions were similar in *Myocastor coypus* (the coypu) and various capromyid rodents in spite of dramatic differences in the mechanics of chewing. These conservative characters, many of which are used in this analysis and are listed in Table 1, were useful in establishing phylogenetic relationships at higher taxonomic levels as discussed above, because as synapomorphies they separate these taxa from other major groups of rodents. However, they were not useful for separating *Myocastor* from capromyids (family level analyses) or separating the various capromyids from each other (subfamily analyses), because all of these myological features were plesiomorphic within this clade.

Another set of osteological and myological characters were shown by Woods and Howland to be extremely responsive to chewing mechanics and were therefore dynamic characters. Some capromyid taxa, such as *Geocapromys brownii* and *Capromys pilorides*, chew propalinally, while others, such as *Plagiodontia aedium*, chew obliquely. *Myocastor coypus* chews obliquely. Capromyid rodents and *Myocastor coypus* can be shown by other means, such as serology and the morphology of other regions, to be related at the superfamily level (all members of the Octodontoidea) but distinct at the family level (Capromyidae and Myocastoridae). These characters therefore appeared to reflect the similar functional conditions of the masticatory region in *Myocastor coypus* and *Plagiodontia aedium* rather than a close phylogenetic relationship. It was possible to demonstrate this probable example of parallelism in characters associated with the masticatory system because of the presence of functional data. The myological characters that were most dynamic were associated with the temporal, internal pterygoid, external pterygoid, and posterior digastric muscles (Woods and Howland, 1979:111-113).

Functional data can also be of direct use in formulating phylogenetic hypotheses. Since propalinal chewing in most rodents is derived from oblique chewing (Percy Butler, personal communication), observations on chewing techniques in these rodents lead to the hypothesis that *Geocapromys* and *Capromys* are more derived than *Plagiodontia* and *Myocastor*. However, even though both *Plagiodontia* and Myocastor chew obliquely, there are significant differences in the mechanics of mastication between the two forms. Myocastor chews unilaterally with only one toothrow in contact during each anterolateral mandibular stroke. In Plagiodontia however, both toothrows are in contact during the anterolateral power stroke. These functional observations, based on the analysis of cineradiographic films, lead to the hypothesis that Myocastor is more primitive and on a separate lineage than is Plagiodontia. Even though mastication in Plagiodontia differs dramatically from Geocapromys and Capromys, it could easily have given rise to the condition in these two taxa which chew propalinally with both toothrows in contact at the same time. This hypothesis, which will be discussed and tested later in this paper, has been formulated totally on the basis of functional data, and points out the importance of doing functional analyses as well as descriptive morphology.

Extrapolation of function from form can often be misleading and lead to serious errors in establishing whether a character is conservative or dynamic (and therefore in establishing the probability of parallelism). A good example of this is in the classic paper on pocket gophers by Merriam (1895). He concluded that three groups of pocket gophers chewed in different ways based on the morphology of their crania. The analysis by Wilkins and Woods (1983), however, demonstrated a single propalinal mode of mastication in all pocket gophers, that masticatory mode did not account for the differences observed in jaw morphology. Other explanations for these differences must be sought. Therefore, in using myological data for phylogenetic analyses it is important to establish which characters are dynamic versus which are conservative, and to seek to establish the functional reasons for the morphological conditions observed (see Woods and Howland, 1979). Whenever possible, the functional explanations should come directly from primary analyses rather than via extrapolation from the condition in other taxa.

In summary, muscles appear to be useful in establishing phylogenetic lineages at various taxonomic levels, but not all muscles may be useful at the same level of analysis, nor equally effective at indicating phylogenetic relationships. Some idea of the influence of function on the form of the musculature should be determined, because some muscles are much more influenced by changed functions of an anatomical region than are others. The best myological analyses include a series of studies to: (1) establish the myological differences between less specialized and more specialized taxa; (2) compare the patterns for myological data from one region with myological data from another; (3) establish the relationships of various muscles to function; and (4) seek evidence from diverse sources, such as genetics, biochemistry, behavior, and ectoparasites, that might provide additional patterns of possible phylogenetic relationships.

RESULTS

Myological Analysis

The data from 26 muscles are presented in Table 1 in a format that is suitable for cladistic analysis. The status of each character is noted as either present or absent for each of 12 taxa. Taxa were selected from a cross-section of New and Old World hystricognath rodents to include forms that have been classified in previous works at levels ranging from the same genus to different suborders. This analysis investigates the phylogeny of the taxa based on myological data and compares the conclusions derived from a cladistic analysis of myological characters with similar analyses based on osteological, biochemical, chromosomal, arterial, and dental data, thereby evaluating the validity of myological information in determining heritage. In addition, an attempt is made to evaluate the taxonomic level at which myological analyses are most effective in predicting phylogenetic relationships.

Taxa Studied

The taxa chosen include three members of the Family Capromyidae *sensu stricto*: *Plagiodontia aedium*, the Hispaniolan hutia or zagouti; *Capromys pilorides*, the Cuban hutia; and *Geocapromys brownii*, the Jamaican hutia or Indian coney. *Plagiodontia aedium* is usually classified in a subfamily, the Plagiodontinae, separate from the remaining two genera which have been classified in the Subfamily Capromyinae (see Woods, 1984, 1985). While *Geocapromys* and *Capromys* are distinct enough from each other to be considered separate taxa by many authors (see Woods, 1984, 1985), they have been grouped together in the same genus by Varona (1974) and Corbet and Hill (1980). The nutria or coypu, *Myocastor coypus*, is frequently classified as a member of the Family Capromyidae (Corbet and Hill, 1980), but has also been considered a separate family (Woods and Howland, 1979; Woods, 1984, 1985) as well as a distinct subfamily of the Family Echimyidae (Patterson and Wood, 1982). Therefore, it is apparent that these four genera are related to each other at taxonomic levels ranging from generic to the family level (or above), and represent an ideal group to use for a myological investigation at lower taxonomic levels.

The South American spiny rat, *Echimys armatus* (considered as a distinct genus by Husson in 1978, whom we will follow in classifying as *Makalata armata*), is one of the more primitive members of the Echimyidae based on dental characteristics and habits. Patterson and Wood (1982) consider echimyids to be a primitive group of New World hystricognaths from which all of the above taxa evolved sometime after the Deseadan. They classify myocastorines, plagiodontines, and capromyines as separate subfamilies in the Family Echimyidae, all of which are grouped together by the shared derived dental character of retaining the deciduous premolar throughout life. While Deseadan echimyids replaced the deciduous premolar in the normal manner with a permanent premolar, all post-Deseadan members of the Echimyidae, as well as capromyids (sensu stricto) and myocastorids, retain the upper and lower deciduous premolars through their lives. While it is possible these teeth might represent permanent premolars and that the deciduous premolars are replaced during fetal development (W. Patrick Luckett, personal communication), it appears more likely that these taxa retain the deciduous premolar. The presence of a non-erupted permanent premolar in a Santacrucian (early Miocene) echimyid (Wood and Patterson, 1959) and the presence of a permanent premolar that replaced the deciduous tooth in a single specimen of Plagiodontia aedium (Woods, 1985) support this hypothesis.

The degu, Octodon degus, is a small semifossorial rodent from Chile that resembles many spiny rats in body form. The deciduous premolars in this taxon, and in other closely related taxa from the Andean region of South America, are replaced in the normal rodent manner, and the pattern of reentrant folds of the molariform teeth is simplified to one labial and one lingual reentrant, resembling the pattern of a figure-of-eight. These taxa are grouped together by most authors into the Family Octodontidae, which may be more closely related to the taxa discussed above than to any other group of New World hystricognaths. This relationship is usually indicated by grouping all forms together in the Superfamily Octodontoidea. George and Weir (1974), in their work on karyotypes, noted this relationship and united these taxa together in the "octocap" group. Both echimyids and octodontids have been known in the fossil record since the Deseadan.

The North American porcupine, Erethizon dorsatum, and the South American tree porcupines, Coendou and Sphiggurus, are members of the family Erethizontidae. This family has also been present in the fossil record of South America since the Deseadan, and many morphological features indicate that New World porcupines may be very distinct from the remaining New World hystricognaths (see Woods, 1982, 1984). The New World porcupines, therefore, appear to be distinct from octocaps at the highest taxonomic levels within the New World Hystricognathi (see Patterson and Wood, 1982: 522-523).

The New World taxa with chunky bodies and nearly vestigial tails appear to represent a distinct lineage in the radiation of South American hystricognaths. George and Weir (1974) distinguished this group from the octocap group discussed above by calling them the "cavyprocts." We have chosen the agouti, Dasyprocta punctata, from Costa Rica for analysis in this study.

Among Old World taxa of hystricognathous rodents we have selected specimens for analysis from two lineages: porcupines of the Family Hystricidae and cane and rock rats of the Superfamily Thryonomyoidea. These two lineages have recently been separated into different infraorders within the Suborder Hystricognathi by Patterson and Wood (1982), hystricids in the Infraorder Hystricomorpha and cane and rock rats in the Infraorder Phiomorpha. Old World porcupines are first known in the fossil record from the late Miocene of Egypt, Hungary, and India. Their recent distribution includes Africa, Italy eastward to Southern China, Indonesia, and the Philippines (see Wood, this volume). We have chosen *Atherurus macrourus* from Malaya for analysis in this study. Rock rats (*Petromus typicus*) are a monospecific group in the Family Petromuridae that is currently known only from southwestern Africa. Cane rats have a broader distribution and are found throughout Africa south of the Sahara where two species of the genus *Thryonomys* are known. They are classified in their own family, the Thryonomyidae. The two families appear to have been separate lineages in Africa since the Oligocene, and taxa from the superfamily are found in the Jebel el Qatrani Formation of Egypt, which is early Oligocene (Wood, 1983). Reports from French paleontologists working in Algeria (J.J. Jaeger, this volume) may push the history of this superfamily in Africa back into the Eocene. We have chosen the rock rat, *Petromus typicus*, from Namibia and the cane rat, *Thryonomys swinderianus*, from Zambia for analysis in this study.

Within the assemblage of 12 taxa discussed above, several taxonomic arrangements have been proposed in the literature and diversity exists at taxonomic levels ranging from the subgeneric to the subordinal. The diversity observed in this assemblage of taxa goes back as far as the Deseadan in South America and the Jebel el Qatrani Formation of the Fayum in Africa. The latter is dated as early Oligocene (36 million years old). The Deseadan has also been dated as early Oligocene (see Patterson and Wood, 1982) but recent data from the work of MacFadden in Bolivia (personal communication) indicates that some if not most of the earliest rodent material from South America is 25 million years old rather than 36.

Patterson and Wood (1982) and Wood (1983; this volume) think that within the Hystricognathi there have been five independent radiations that are unrelated to each other except at the most distant levels. These independent lineages are the Caviomorpha in the New World and the Hystricidae, the Thryonomyoidea, and the fossorial Bathyergidae (or mole rats) from the Old World. Patterson and Wood contend that most similarities between these groups are the result of parallel evolution. Sarich and Cronin (1980) and Bugge (1974; this volume) suggest that New World caviomorphs may not be monophyletic, and that the New World porcupines had a history distinct from other caviomorphs. Sarich and Cronin proposed that rodents entered South America by rafting there from Africa at two separate times. Patterson and Wood (1982) and Woods (1982) proposed that caviomorphs may have invaded South America via the Antilles (see Woods, 1985) and that the Capromyidae (*sensu stricto*) may be descendants from the ancestors of some (Woods) or all (Patterson and Wood) New World hystricognaths. This assemblage of 12 rodent taxa is an excellent group to investigate because the possible phylogenetic relationships between the various taxa range from low to high taxonomic levels. The establishment of the probable taxonomic affinities of these taxa based on data from a variety of sources would have far reaching significance to our understanding of the origin and evolution of South American hystricognath rodents. The taxa are also a good group to investigate because they satisfy many of the criteria discussed earlier in this paper. The taxa are not too diverse and earlier studies have made a wide range of data available on the morphology of a variety of systems in all taxa. In addition, some work has been completed on the functional morphology of all of these taxa (Woods and Howland, 1979; Woods, unpublished data).

Myology

The myological data presented in Table 1 are compiled from several previous studies as well as new information. The basic myology is from Woods (1972) with additional information from Woods and Howland (1977) on skin musculature, Woods and Howland (1979) on jaw musculature and functional aspects of mastication, as well as extensive notes and drawings made on the comparative and functional anatomy of hystricognath rodents during the past 15 years. Each of the above regions is part of the whole animal, but is also independent, and therefore influenced by different forces during ontogeny and phylogeny. The masticatory muscles (Fig 1; masseter, temporalis, pterygoid, digastric), facial muscles (platysma and derivatives), hyoid muscles, and pharyngeal muscles (Fig. 2) are all part of the branchiomeric musculature, and as such are innervated by cranial nerves V, VII, IX, and X. The tongue and deep hyoid muscles are part of the myotomic musculature and are innervated by branches of the hypoglossal nerve. The muscles of the neck and trunk are also part of the myotomic musculature. The various parts of the cutaneous maximus (skin muscles, Fig. 3) are derived from the pectoral muscles. They are therefore part of the appendicular musculature and innervated by branches of the anterior thoracic nerve.

The division of muscles into separate groups is based on the position and innervation of individual muscles. The myological information in Table 1 can either be used collectively by assembling an analysis based on all of the available myological data presented or by comparing the data in each separate group. The groups do not correspond exactly to groupings that would be based on embryonic derivation, since so few studies have been made of the embryology of rodent muscles (see Cheng, 1955; Jones, 1979). The groupings followed here were established by Hill (1937) and refined by Rinker (1954). They are based on innervation, comparison of position in a variety of forms, and, whenever possible, on developmental data. The major muscles listed in Table 1 are illustrated in Figures 1-3.

Table 1. Listing of 26 myological characters of 12 selected taxa of hystricognath rodents.

Character	Taxon (Genus) *											
	Ge	Ca	Pl	Ma	My	Oc	Er	Co	Da	At	Pe	Th
APPENDICULAR MUSCULATURE												
Pectoral Group												
Cutaneous maximus pars dorsalis												
1. interdigitates over thorax	+	+	+	+	+	0	+	+	+	+	+	+
2. humeral head present	0	0	0	0	0	0	+	+	+	+	+	0
Cutaneous maximus pars thoracoabdominalis												
3. perp. to *P. dorsalis*	0	0	0	0	0	0	+	+	0	+	0	0
4. sweep over knee	+	+	+	+	+	+	0	0	+	+	+	+
Cutaneous maximus pars femoralis												
5. Continues to tail	0	0	0	0	+	0	0	0	0	0	0	0
BRANCHIOMERIC MUSCULATURE												
Masticatory Group												
Masseter superficialis												
6. pars anterior present	+	+	+	+	+	+	0	0	0	0	+	0
7. pars anterior massive	0	0	+	+	+	0	0	0	0	0	0	0
Masseter posterior												
8. large and in jugal fossa	+	+	+	+	0	+	0	0	0	0	0	0
Masseter medialis												
9. pars posterior divided by masseteric nerve	0	0	0	0	+	+	+	+	+	+	+	+
Glossopharyngeal-vagus Group												
Glossopharyngeus												
10. large and two-parted insertion	+	+	+	0	0	0	0	0	0	0	0	0

(continued)

Table 1 (Continued).

Character	Taxon (Genus) *											
	Ge	Ca	Pl	Ma	My	Oc	Er	Co	Da	At	Pe	Th
Cricothyroideus												
11. long and thin	0	+	0	0	0	0	0	0	0	0	0	0
12. divided into more than one part	+	+	+	+	+	+	0	0	0	0	+	0
MYOTOMIC MUSCULATURE												
Lingual Group												
Hyoglossus												
13. part of O. free-floating	+	+	+	0	0	0	0	0	0	0	0	0
Styloglossus												
14. O. from pterygoid process	+	+	+	+	0	0	0	0	0	0	0	0
Medial Ventral Cervical Group												
Geniohyoideus												
15. I. free-floating	+	+	+	0	0	0	0	0	0	0	0	0
Sternohyoideus												
16. I. free-floating	+	+	+	0	+	0	0	0	0	0	0	0
17. O. distinct from sternothyroideus	+	+	0	0	0	0	0	0	0	0	0	0
Omohyoideus												
18. loss	+	+	+	0	0	0	0	0	+	0	0	0
Thyrohyoideus												
19. I. partially free-floating	+	+	+	0	0	0	0	0	0	0	0	0
Lateral Cervical Group												
Scalenus anticus												
20. present	+	+	+	+	+	+	0	0	0	0	+	+

Character	Ge	Ca	Pl	Ma	My	Oc	Er	Co	Da	At	Pe	Th
	Taxon (Genus) *											
APPENDICULAR MUSCULATURE												
Suprascapular Group												
Supraspinatus and Infraspinatus												
21. fused	+	+	+	+	+	+	0	0	0	0	+	0
Pectoral Group												
Pectoralis Major												
22. separated into two layers	0	0	0	0	0	0	0	0	+	0	0	0
Latissimus-Subscapular Group												
Latissimus dorsi and Teres major												
23. I. separate	0	0	0	0	0	0	+	+	0	0	0	0
Latissimus Achselbogen												
24. present	+	+	+	+	+	+	0	0	0	0	0	0
Flexor Group of Arm												
Coracobrachialis												
25. middle and long head divided by median nerve	0	0	0	0	0	0	+	+	0	+	0	0
Extensor Group of Forearm												
Brachioradialis												
26. present	0	0	0	0	0	0	+	+	0	0	0	0

* Ge = Geocapromys, Ca = Capromys, Pl = Plagiodontia. Ma = Makalata, My = Myocastor, Oc = Octodon, Er = Erethizon, Co = Coendou, Da = Dasyprocta, At = Atherurus, Pe = Petromus, Th = Thryonomys.

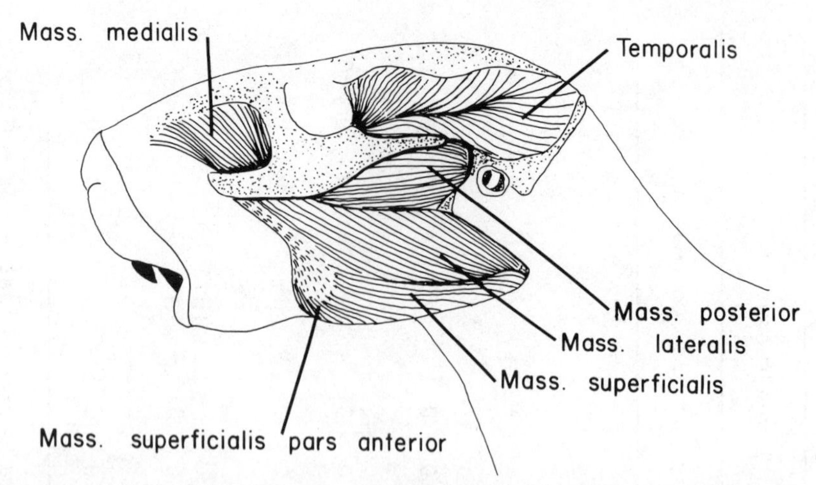

Fig. 1. Masticatory muscles of Plagiodontia aedium.

Phylogenetic Hypotheses Based on Myology

All data were analyzed by establishing an outgroup and forming a table listing derived character states. These data were then used to develop cladograms for the assemblage of taxa under consideration. The analyses were undertaken at various levels, and used different taxa as the outgroup to establish the polarity of characters.

The most parsimonious cladogram derived from myological data is presented in Figure 4. Capromys pilorides is the most derived of the taxa examined. The sister group for Capromys pilorides is Geocapromys brownii. Plagiodontia aedium is separable from both G. brownii and C. pilorides, and all three genera form a natural group that can be characterized by 5 to 13 synapomorphic characters (depending on the outgroup and number of genera included in the analysis). The pattern of synapomorphic myological characters is consistent in all of the cladograms constructed, and indicates that capromyid rodents share many derived characters. The myological data support the hypothesis that these rodents are members of the Family Capromyidae with Plagiodontia in the Subfamily Plagiodontinae and Geocapromys and Capromys as separate genera in the Subfamily Capromyinae.

When Makalata armata is included in the analysis, Makalata consistently groups more closely with the three capromyid genera than does Myocastor coypus. Five synapomorphic myological characters distinguish Makalata armata from capromyids. Myocastor coypus is separable from M. armata by three additional synapomorphic characters. These data indicate that Myocastor is less closely related

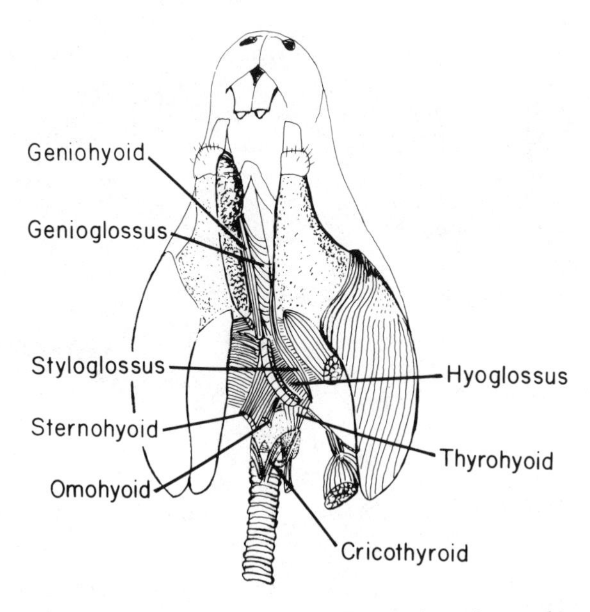

Fig. 2. Hyoid muscles of Plagiodontia aedium.

to capromyids than is Makalata, and that it is appropriate to consider all three groups as members of separate families (Capromyidae, Echimyidae, and Myocastoridae).

When Octodon degus is included, the taxon consistently groups below all of the above forms, from which it can be separated by one or two synapomorphic characters. This indicates that Octodon is distinct from capromyids, echimyids, and myocastorids, but part of the same clade. It seems reasonable to classify the taxon in a distinct family (Octodontidae) and to group all of the forms together in the same superfamily (Octodontoidea).

Erethizon dorsatum forms an appropriate outgroup, and is separable from all of the above forms (the Octodontoidea) by three synapomorphic characters. The resulting cladogram produces an hypothesis that is consistent with the results discussed above, and suggests that Erethizon is distinct from the other taxa at a high taxonomic level, but part of the same natural clade. Erethizon and Coendou are not separable from each other on the basis of the myological characters used in this study. Erethizon and Coendou are classified together in the Superfamily Erethizontoidea.

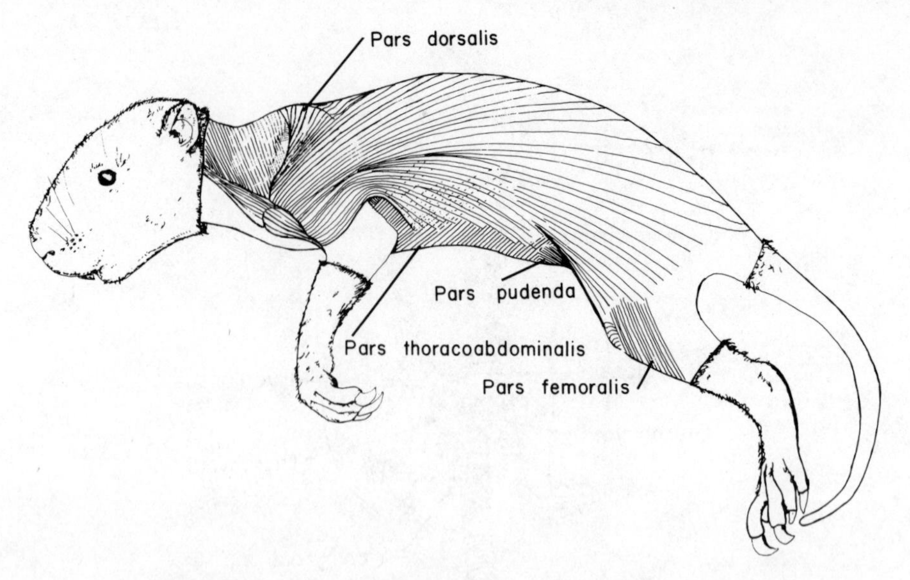

Fig. 3. Skin muscles of Plagiodontia aedium.

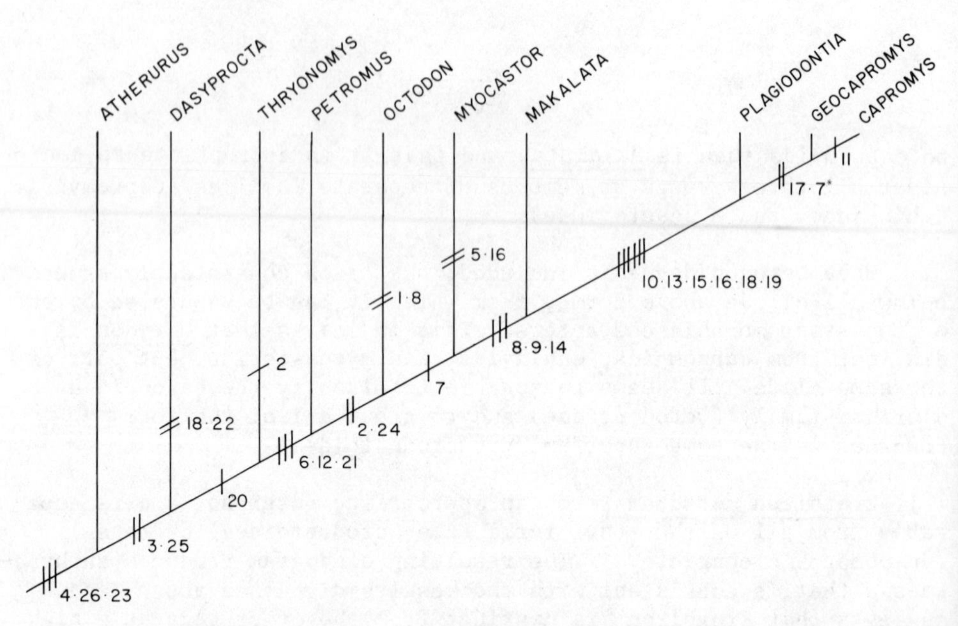

Fig. 4. Cladogram showing a hypothesis of the relationships of hystricognath rodents based on myological characters. Numbers correspond to characters in Table 1. * indicates secondary loss.

Cladograms are very difficult to construct and have conflicting character states when the outgroup used in a comparison is neither Erethizon nor Coendou. For example, when Makalata armata is considered as an outgroup, and Myocastor coypus is included in the analysis, then Myocastor is characterized by four autapomorphic myological characters. A similar complex cladogram results when Octodon degus is used as the ougroup and Erethizon is included in the analysis, and indicates that the hierarchial arrangement presented in Figure 4 is the most parsimonious cladogram based on the myological characters analyzed in this study.

When the Old World genera and the New World genus Dasyprocta are included in the analysis, the most parsimonious cladogram still results from the designation of Erethizon as the outgroup. Three characters separate erethizontids from all other hystricognaths considered. Two characters separate Atherurus macrourus from the remaining taxa, suggesting that New and Old World porcupines may be more closely related to each other than to any other group of New or Old World hystricognath. Several characters indicate that the Old World genera Thryonomys and Petromus are more closely related to octodontoids than is the New World cavyproct genus Dasyprocta. The analysis based on the myological data presented in Table 1 leads to the hypothesis that capromyids are the most derived New World taxa considered, and that octodontoids and thryonomyoids are more closely related to each other than to porcupines or cavyprocts.

Comparison with Other Data

In an effort to determine the validity of the taxonomic rankings based on myological characters (i.e. to test the hypothesis), and to evaluate the usefulness of myology in establishing phylogenetic relationships, data from analyses of postcranial material, blood serum proteins, chromosomes, teeth, and carotid arterial patterns can be compared with myological data. The blood data are based on an analysis by Woods et al. (manuscript in preparation) and on personal communication from Dr. C.W. Kilpatrick. These data were presented in a preliminary format by Woods (1982). Data from Sarich and Cronin (1980) and Sarich (this volume) on the relationships of hystricognath rodents based on quantitative precipitin cross-reactions of blood albumins are also compared with the results of the myological analysis of the present study. Chromosomal data for comparison are compiled from the study of Weir and George (1974). The data on carotid arterial patterns are compiled from the work of Jorgen Bugge (1974; this volume).

Analysis of Postcranial Skeletons

Osteological characters have frequently been employed as sources of data in systematic biology because of the availability and durability of skeletal remnants. Tarsal bones, for example, have been sensitive indicators of phylogeny and locomotor mode in primate studies (Szalay and Decker, 1974). Postcranial skeletons have been used less frequently than cranial material in systematics, in part because fossilized skeletons are often disarticulated when collected and as a result postcranial bones cannot usually be correlated with a particular cranium or dentition. Taxonomic designations are almost always based on cranial or dental material, and so the evolution of postcranial skeletons in the fossil record is poorly documented, especially in smaller taxa such as rodents.

Differences in postcranial morphology in recent forms are often too subtle to permit resolution among taxa. Hystricognaths, however, should lend themselves better than most other rodent groups to an analysis of postcranial elements because of their large body size and because of numerous locomotor modes (scansorial, cursorial, fossorial, arboreal and semiaquatic) that characterize these forms. We examined postcranial material from 13 species of New and Old World hystricognath rodents and compared our findings with the hypotheses formed from myological information. The taxa studied were the same as in our myological analysis with the exceptions of *Petromus typicus* and *Atherurus macrourus*. In addition, we studied *Elasmodontomys obliquus*, a large Pleistocene rodent from Puerto Rico that is now extinct, and *Dinomys branickii* from northern South America. *Elasmodontomys* and *Dinomys* were included because their systematic status is controversial and postcranial material was available (Woods 1984).

Several characters were not included in the construction of our cladograms because of their dynamic status. These included the height of the greater trochanter of the femur; orientation of the femoral head relative to the long axis of the femur; configuration of the patellar fossa; presence or absence of a peroneal tubercle on the fibula; condition of the iliac spine; configuration of the pubic bones; shape of supraspinous and infraspinous fossa of the scapula; condition of the metacromion process; shape of the distal articular surface of the humerus; and configuration of the ulna and radius. We included analysis of the calcaneus and talus bones, but were unable to include data on the other carpal and tarsal bones because of their absence in several specimens.

The presence of an elongate acromion process (at least twice the length of the solid portion of the scapular spine) is a significant character that is synapomorphous in all octodontoids (octodontids, echimyids, and capromyids). No functional explanation is available for this condition of the scapular spine, and we were unable to find this condition in any other New or Old World hystricognath.

The data on postcranial morphology support an hypothesis of common ancestry in the Capromyidae, Myocastoridae, Heptaxodontidae, and Octodontidae. All taxa are separable from *Erethizon* and *Coendou* by two synapomorphic characters: a narrow proximal epicondylar ridge of the humerus; a laterally deflected deltoid crest of the humerus. Two synapomorphies separate octodontids from *Dinomys*: an elongated acromion process; a perforated supratrochlear foramen at the distal end of the humerus. *Erethizon*, *Coendou*, and *Dinomys* exhibit fusion of the second and third cervical vertebrae (Ray, 1958). Grand and Eisenberg (1982) noted other similarities between *Dinomys* and erethizontids, such as proportions of their muscle masses, vocalizations, and behavioral characteristics, and suggested that *Dinomys* is closely related to erethizontids. The data from postcranial skeletons support the hypothesis that *Dinomys* (Family Dinomyidae *sensu stricto*) is in the Superfamily Erethizontoidea, rather than in the Superfamily Cavioidea as proposed by Patterson and Wood (1982).

Thryonomys swinderianus shares numerous synapomorphic characters with octodontoids with the exception of an elongated acromion process. The postcranial data support the hypothesis that *Thryonomys* is more closely related to octodontoids than it is to cavioids (*Dasyprocta*) or *Dinomys*. Among octodontoids, *Elasmodontomys obliquus* groups most closely with *Myocastor coypus* and is not very distinct from capromyids. *Elasmodontomys* stands clearly apart from *Dinomys*, with which it has been grouped by some authors in the past. *Elasmodontomys* is part of the enigmatic West Indian Family Heptaxodontidae, which was classified *insertae sedis* as to superfamily by Patterson and Wood (1982:523). We propose placing the heptaxodontids within the Superfamily Octodontoidea, based on the elongated acromion process of *Elasmodontomys*, and would associate this family adjacent to the West Indian Family Capromyidae with which it may share a common evolutionary history (Woods, 1984).

In conclusion, postcranial data are of limited value in formulating hypotheses about the phylogeny of hystricognath rodents, because few useful characters could be found. Based on the information available from the limited number of useful characters, *Myocastor*, *Elasmodontomys*, and the three genera of extant capromyid rodents constitute a cohesive group. *Octodon* and *Proechimys* are intermediate between erethizontids and capromyids, with *Proechimys* being more closely allied to capromyids than is *Octodon*. All of these taxa share the important synapomorphic character of an elongated acromion process allowing a close association of the infraspinatus and supraspinatus muscles. This character seems consistently to define the Superfamily Octodontoidea. *Dinomys branickii* groups with *Erethizon* and *Coendou* on the basis of the synapomorphic character of the fusion of the second and third cervical vertebrae. The deltoid crest is not deflected laterally in erethizontids, while it is in *Dinomys*, so we do not place these taxa in the same family. The fusion of the cervical vertebrae appears to be an important apomorphic character, and we propose grouping *Dinomys* and erethizontids in the same superfamily. There are not sufficient data to clarify the position of Dasyprocta, and so the status of cavioids relative to octodontoids cannot be resolved based on postcranial characters other than to say that cavioids appear to be distinct from erethizontoids (including Dinomys) and octodontoids (including Elasmodontomys).

Analysis of Electromorphs of Hemolysate and Serum Proteins

The bioserological data indicate that the genus Geocapromys is the most derived of all taxa considered, and can be defined by five apomorphic electromorphs (Fig. 5). The three genera of capromyid rodents stand apart from the other taxa analyzed and are characterized by four synapomorphic electromorphs. Capromys and Plagiodontia do not clearly separate from one another. Echimyids (as represented by the genus Proechimys rather than Makalata) group more closely with Myocastor than with Octodon. Either Proechimys or Octodon can be grouped with the capromyids to produce equally parsimonious trees.

According to Kilpatrick (personal communication), the biochemical data on Erethizon are very divergent, and the taxon does not make a very good outgroup, since only two electromorphs can be identified as truly plesiomorphic (shared with outgroup and group being analyzed). This suggests that Erethizon is very divergent from the other taxa under consideration and may have a different heritage or more distant relationship.

Some workers believe that electromorphs of blood proteins change at a more rapid rate than other systems, and that these data are of limited use in establishing phylogenetic relationships at the family level or above (James Patton, personal communication; discussion at this symposium). The rapid rate of change in many electromorphs might make it difficult to determine phylogenetic relationships that go back 20 to 40 million years. It should be noted, however, that the data gathered in association with C. W. Kilpatrick (Woods, 1982; Woods et al., manuscript) analyzed slowly evolving loci, many of which are normally associated with tissue proteins. We looked at these proteins from blood because tissue samples were not available.

The relationships of the various taxa within the Octodontoidea that were predicted on the basis of electromorphs are similar to hypotheses formed from myological data. At higher taxonomic levels the data from electromorphs predict that octodontoids and cavioids are very divergent, and that Atherurus groups more closely with cavioids than it does with octodontoids. These associations are based on data from slowly evolving loci. The relationships of the three capromyid genera are based on data from rapidly evolving loci. The differences in the arrangement of the capromyid genera from the hypothesis based on muscles may be because the radiation of Recent capromyids is more recent than can be revealed by the analysis of slowly evolving proteins, but more ancient than can be resolved by data from rapidly evolving loci. The insular nature of the distributions of capromyids may also be a factor in the evolutionary rates of these rodents (Woods, 1985). The most parsimonious cladogram constructed from the electromorph data is presented in Figure 5.

Analysis of Albumin Changes

The work of Sarich and Cronin (1980) indicates that, among capromyids, *Geocapromys* and *Capromys* are sister groups and are more derived than *Plagiodontia*. The three genera form a natural unit that is distinct from other New World hystricognaths. Sarich and Cronin suggest that *Plagiodontia* has been distinct from the *Geocapromys*/*Capromys* lineage for 40 units (the authors calculate 1.1 units of change per million years of time), which they estimate to be 38 million years. Their conclusions on the isolation of capromyids from all other octodontoids support our hypothesis based on myological data. In the remaining lineages, Sarich and Cronin (1980: 410) calculate that *Octodon* is the most primitive taxon with 37 units of change, followed by *Myocastor* with 28 units and the echimyids *Proechimys* and *Hoplomys* with 18 and 23 units respectively. While the authors are not able to locate a place where capromyids tie into the *Octodon*/*Myocastor*/echimyid lineage, the data from Sarich and Cronin do not refute the overall phylogenetic conclusions based on myological data for the relationships of capromyids, echimyids, *Myocastor*, and octodontids. They also conclude that *Thryonomys* and *Petromus* are more closely related to the above octodontoids than any of these taxa are to other New and Old World hystricognaths. As does Bugge in this volume, Sarich and Cronin (1980:fig. 3) group bathyergids with this same assemblage of New and Old World taxa that appear to form a natural clade. The molecular data also support the hypothesis that *Dasyprocta* and the allied cavyproct genera are distinct from octodontoids and are perhaps more ancient in their divergence.

There are two important features where the data from Sarich and Cronin are at variance with the analysis based on comparative myology. Quantitative precipitin cross-reactions indicate that extant capromyids originated as a taxon at least 40 units of change ago, whereas the Recent echimyids analyzed originated only 18-22 units ago. This would imply that echimyids should be more derived than capromyids, which is different from the conclusion based on myology. This difference could be the result of extant capromyids being limited in diversity and distributed on isolated islands of the Antilles, whereas echimyids, the most divergent and successful hystricognaths, are broadly distributed throughout Central and South America. Some species of *Proechimys* appear to be very recent in origin, and the genus is characterized by a high degree of intrageneric chromosome diversity (Reig and Useche, 1976). Another conclusion in which the two methods are at variance relates to *Erethi-*

zon dorsatum, which myological data indicate is the best outgroup. Sarich and Cronin conclude that erethizontids along with caviids, dasyproctids, and chinchillids form a natural group that is more primitive (ancient in origin) that the Octodontoidea, but that within this group Cavia and Hydrochaeris are more primitive than Erethizon. They also group Erethizon closer to cavyprocts than to Hystrix, while the myological data support the opposite hypothesis. They conclude that the distinction between octodontoids and cavioids and Erethizon is so great as to indicate the possibility of two separate invasions of hystricognath rodents into South America from Africa.

Analysis of Carotid Arterial Patterns

Jorgen Bugge (1974; this volume) has demonstrated differences in the pattern of the arterial supply to the brain that appear to be consistent within phylogenetic lineages. Coendou prehensilis has a cephalic arterial pattern that differs from all other hystricognaths, and is closest to the pattern observed in non-hystricognath rodents. Bugge did not study Erethizon dorsatum, but it appears to have the same pattern as Coendou based on cranial foramina. Bugge concludes (this volume) that New World porcupines diverged from the common stem of other New World hystricognaths before the obliteration of the internal carotid artery took place, which would mean the divergence was before the separation of the other New World hystricognath superfamilies in the Deseadan. This information indicates that New World porcupines are a good outgroup to use in establishing the polarities of myological characters. Since thryonomyoids and Old World porcupines also lack the internal carotid artery, erethizontids also appear to be a good outgroup to use in establishing morphological polarities in Old World hystricognaths. The evidence from carotid arterial patterns, therefore, is consistent with the hypothesis formulated by an analysis of myological data on the phylogenetic position of erethizontids relative to other hystricognaths.

Bugge (this volume) has also shown that there is a remarkable amount of similarity (="practically identical") between carotid arterial patterns in octodontoids and Thryonomys swinderianus. This observation supports the hypothesis developed on the basis of myological data that New World octodontoids and Old World thryonomyoids are more closely related to each other than octodontoids are to other New World hystricognaths. According to Bugge, the patterns he observed in the morphology of carotid arteries in hystricognath rodents could be explained as a result of a double invasion of these rodents into South America. One invasion gave rise to erethizontoids and a second gave rise to the remaining New World hystricognaths.

Analysis of Chromosome Data

The work by George and Weir (1974) on hystricognath rodent chromosomes presents two different phylogenetic schemes based on their synopsis of karyotypic features. One scheme is based on Robertsonian theory, while the other is based on Stebbins' hypothesis on the direction of changes in chromosome morphology and number. The final position of each taxon differs depending on which scheme is accepted. Following the conclusions in George and Weir (1974: 104) and Stebbin's hypothesis, the taxonomic organization is similar to the cladograms constructed from myological data. *Geocapromys brownii* is the most derived capromyid taxon and differs markedly from *Capromys pilorides*. Their scheme, however, can find no justification for separating *Myocastor* from *Geocapromys* and *Capromys* at the family level, even though they do consider *Myocastor* to be nearer the common ancestor than *Capromys*. They do consider it "possible that *Myocastor* is a separate octodontid line" (1974:102). The authors concluded that the three capromyid genera are divergent from one another but resemble each other more than they do any other hystricognath group. The lineages giving rise to *Octodon* and *Proechimys* are more primitive than the capromyid lineage, and octodontids and echimyids are closely related to each other. George and Weir (1974:92-93) group all of the above taxa together into what they consider to be a natural group (the octocaps). They did not study *Thryonomys swinderianus* or *Petromus typicus*. The octocaps were distinquishable in chromosome morphology from *Dasyprocta punctata*, which George and Weir grouped together with other taxa in the cavyproct group. This arrangement is consistent with our findings based on myological information. The chromosomal data for the genus *Hystrix* were incomplete, and the authors were unable to discuss the status of Old World porcupines conclusively. They did analyze chromosomal data for *Erethizon dorsatum* and *Coendou bicolor rothschildi* and concluded that the Erethizontidae are isolated from all other New World hystricognaths. Therefore, the greatest difference between the conclusions of George and Weir and the phylogenetic interpretations based on myological data is in the position of *Myocastor coypus*. Myological data indicate *Myocastor* is very distinct from capromyids and even echimyids, and is best considered its own separate and specialized lineage within the Superfamily Octodontoidea. We have not attempted to construct a cladogram based on chromosomal data because of the difficulty in establishing the polarity of the available data.

Analysis of Dental Characters

Data from dental characteristics have been discussed in detail by Woods (1985), who traced the evolution of various dental characteristics, such as the pattern of reentrant folds, the distribution of cement around molariform teeth, the degree of hypsodonty and the time of replacement of deciduous premolars in octodontoid rodents. The results of this analysis vary, depending on which synapomorphic characters are considered more important (i.e., loss of reentrant folds or extreme development of cement surrounding the molariform teeth). It is difficult to construct a parsimonious cladogram with the available data. The most parsimonious cladogram presented by Woods (1985), which can be modified by the addition of data from *Erethizon*, *Octodon*, and *Makalata*, indicates that capromyids, echimyids, and *Myocastor* are all closely related. The most derived taxon is *Plagiodontia*. *Capromys* and *Geocapromys* appear to be closely related but less derived than *Plagiodontia*. The extant Capromyidae are distinct in dental morphology from *Myocastor* and *Makalata*, which share several important features, such as a postero-lingual flexus. Dental data from the extinct West Indian capromyid *Hexolobodon phenax*, however, link capromyids with *Makalata* and *Myocastor* into a clade with diverse, but relatable dental morphology. *Octodon degus* appears to be specialized with autapomorphic characters that are difficult to relate to known echimyids, capromyids, or *Myocastor*. *Petromus typicus* is most similar in dental morphology to *Octodon degus*, while *Thryonomys swinderianus* resembles capromyids. *Erethizon dorsatum* has many primitive dental features, such as the replacement of deciduous premolars and low-crowned, rooted molariform teeth The pattern of the cheekteeth in *Dasyprocta punctata* is remarkably similar to the pattern in *Atherurus* and other Old World porcupines.

It was difficult to use the characters examined to produce a cladogram because so many conflicting character states existed. The reasons for this appear to be that there are at least three different functional trends present within the taxa examined. (1) A trend towards oblique chewing in which the reentrant folds are distinctly angled to the long axis of the body, and the length of the reentrant folds increases. (2) A trend towards propalinal chewing in which the reentrant folds are perpendicular to the long axis of the body, and cement surrounds the external surface of the molariform teeth. The upper premolar teeth become angled posteriorly in the alveolar cavity. (3) A trend towards the modification of the molariform teeth in response to a diet of coarse plant materials in which the high crowned molariform teeth are open-rooted throughout the life of the individual, and the pattern of the reentrant folds becomes modified by the addition of reentrant folds. These trends occur in different lineages, and therefore the structural modifications are imposed on dental patterns that are different to begin with (i.e. the trend toward oblique chewing in the genus *Isolobodon* vs the genus *Plagiodontia*). Function and phylogeny become intermixed in different clades, and the construction of a parsimonious cladogram based on dental characters in an assemblage of New World hystricognaths, even as closely related to each other as capromyid rodents, becomes difficult because of the large number of conflicting characters.

In summary, dental morphology indicates that capromyids are

difficult to separate from echimyids and *Myocastor* (Woods, 1985). These taxa appear to be members of the same clade, but it is not easy to find consistent characters that separate the groups into distinct lineages because of the presence of so many conflicting character states. *Octodon* is specialized and the dental morphology can be derived from a form with dental characteristics that could have given rise to echimyids, capromyids, and *Myocastor*. *Erethizon dorsatum* has dental features, such as an extra loph in the upper molariform teeth (neoflexus), that are very distinct from dental features in the above taxa and separate this taxon from all octodontids. The various patterns of dental morphology are difficult to relate to each other or rank into sequential schemes that appear to relate to phylogenetic patterns (Woods, 1985).

CONCLUSIONS

The myological data are consistent with biochemical and chromosomal data in indicating that *Geocapromys* is distinct from both *Capromys* and *Plagiodontia*, all of which form a natural group that is separate from other taxa considered in this study. With so many synapomorphic characters in common this group is easily defined as a family, the Capromyidae. The data from all sources also indicate that *Erethizon* and *Coendou* are distinct from the remaining ten genera studied, but are not separable from each other on the basis of myological data. *Erethizon* and *Coendou* have a common heritage with the other taxa studied and serve as a good outgroup in analyses of myology. The biochemical, chromosomal, and dental data discussed above, as well as data on cranial circulation (Bugge, 1974; this volume) and ectoparasites (Vanzolini and Guimaraes, 1955), however, indicate that erethizontids are very distinct from the other genera and may be only distantly related. The level of differentiation is at least at the level of a superfamily, and best represented by placing *Erethizon* and *Coendou* in the Superfamily Erethizontoidea. A relationship between *Makalata*, *Proechimys*, *Octodon*, *Myocastor*, and the capromyids is supported by all of the data, and this group appears to be monophyletic. The taxa are best grouped together in the Superfamily Octodontoidea. Within this clade, myological data indicate that *Makalata* is more closely related to capromyids than is *Octodon*, and that *Myocastor* stands apart from both *Makalata* and *Octodon*, while biochemical data group *Myocastor* more closely with *Makalata* and *Proechimys* than with *Octodon*. Chromosomal data are not able to separate *Myocastor* from the Capromyidae. Cladograms based on dental data have many incongruities, and it is difficult to construct a parsimonious cladogram that does not have conflicting characters between *Makalata*, *Myocastor*, and *Octodon*. All three taxa appear to be related, but dental characters are not able to separate them into distinct clades. Based on all of the above data, the most accurate phylogenetic scheme appears to be to treat each taxon in a distinct family, the Echimyidae (*Makalata* and *Proechimys*), Octodontidae (*Octodon*), and Myocastoridae (*Myocastor*) (Fig. 6).

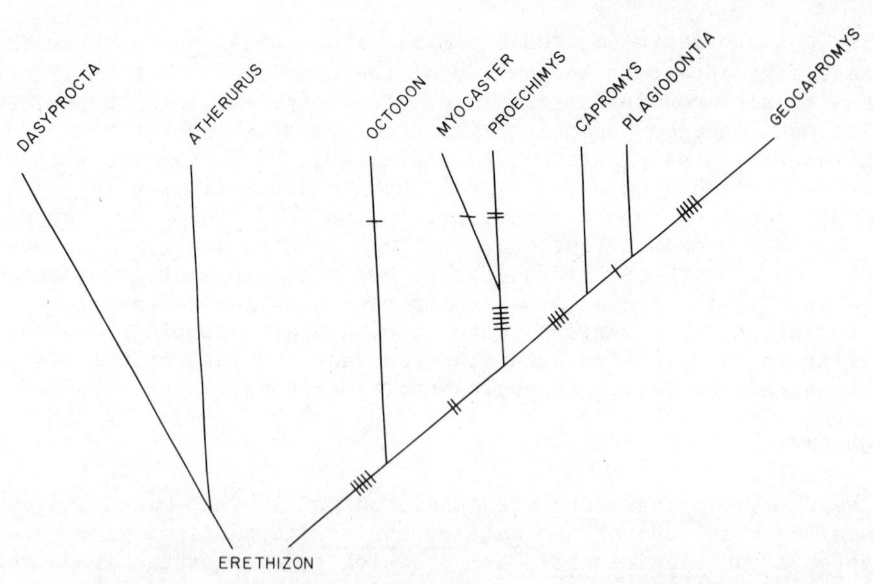

Fig. 5. Cladogram showing a hypothesis of the relationships of hystricognath rodents based on data from hemolysate and serum proteins. The units of similarity are electromorph substitutions.

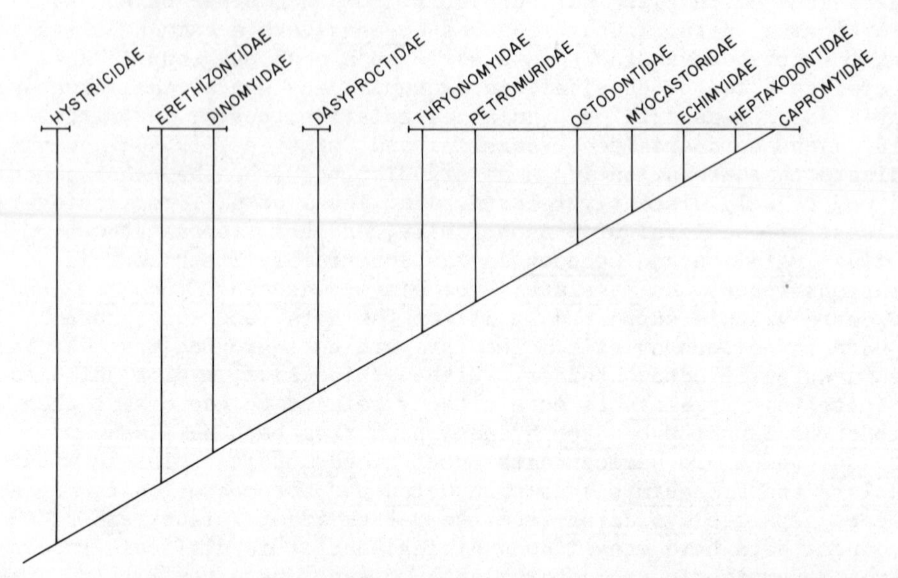

Fig. 6. Cladogram showing a hypothesis of the relationships of hystricognath families and superfamilies based on a synthesis of all the data discerned in this chapter. Major independent subradiations are suggested by the horizontal bars at the top of the figure. The outgroup is a non-spiny, subhystricognathous hypothetical ancestor.

A synthesis of all the above data groups these 12 taxa into four clades: the Erethizontoidea, the Octodontoidea plus Thryonomyoidea, the Cavioidea, and the Hystricidae. Within the Octodontoidea, the Capromyidae are a clearly defined separate clade, while the Octodontidae, Echimyidae, and Myocastoridae are less distinct from each other. All octodontoids appear to be about equally related to one another based on the data available from a variety of systems within the extant forms analyzed. The families Thryonomyidae and Petromuridae appear to be part of this same clade, but more distantly related than are the various New World taxa. Since it is known that both echimyids and octodontids were present during the Deseadan of South America and thryonomyoids first appear in the Jebel el Qatrani Formation of North Africa, both of which have been dated as early Oligocene, it is not clear how octodontoids and thryonomyoids are related. Recent evidence suggests that the radiation of octodontoids in South America may be more recent than previously thought (Bruce MacFadden, personal communication) and that thryonomyoids may go back into the Eocene in Africa (J.J. Jaeger, this volume). These findings support the hypothesis that thryonomyoids gave rise to octodontoids, and that there may have been more than one invasion and subsequent radiation of hystricognath rodents in South America. The Recent capromyid genera of the Antilles are more derived than are the other octodontoids, and so it does not appear likely that capromyids are ancestral to the remaining taxa. The results from the analyses presented here make it seem more likely that capromyids evolved at an early date from a common ancestor that was slightly closer to the Echimyidae than to the Octodontidae, and that myocastorids split off at an equally early date. These data, along with the known paleontological data for South America (Patterson and Wood, 1982; Woods, 1982), give the impression of an ancient radiation spreading northward from South America into the Caribbean, rather than the opposite, and tend to support a southern route (via Africa) for the origin and evolution of the Octodontoidea in South America (Fig. 6).

The Antillean rodents could have become established in the West Indies during the Oligocene when Jamaica and southern Hispaniola were much closer to Central America (see the geological/tectonic reconstruction of the Caribbean basin in Sykes et al., 1982). Jamaica appears to have been the center of evolution for heptaxodontids, while Hispaniola is the region of greatest diversity for capromyids. Geological evidence, however, indicates that Jamaica was submerged in the Upper Eocene and did not become emergent again until the early Miocene, and that southern Hispaniola was also submerged during the Eocene and Oligocene (Woods, 1985). This would make it unlikely that rodents could have dispersed to the Greater Antilles via Jamaica and southern Hispaniola. It is much more probable that Antillean rodents invaded the Greater Antilles directly from South America sometime after the Oligocene and maybe as late as the Upper Miocene. Osteological data support the hypothesis that heptaxodontids and capromyids are related, but that heptaxodontids split off from the Myocastor-echimyid-capromyid lineage early in the evolution of these groups. Therefore, if heptaxodontids and capromyids are closely related, their common origin goes back to near the time of origin of capromyids in the Antilles.

More data are necessary in order to resolve how heptaxodontids, echimyids, and capromyids are related and when the intrainsular exchanges between Hispaniola and Jamaica took place; data available at the present time support the hypothesis that both groups invaded the Antilles (either as the same group or closely related separate groups) directly from South America either via the Lesser Antilles or by direct dispersal to Hispaniola. Heptaxodontids and capromyids achieved their present level of diversity sometime after Jamaica and southern Hispaniola reached their present geographic locations approximately 9 million years ago (Sykes et al., 1982). The relatively late date of association of southern Hispaniola with northern Hispaniola, and therefore the rest of the Greater Antilles, may account for the very derived nature of capromyids relative to the rest of the New World hystricognaths, which appear to have been very conservative in their evolutionary histories.

Both erethizontoids and octodontoids were well developed in the fossil record of South America in the Deseadan (Patterson and Wood, 1982; Woods, 1982; Woods, 1984). Taking into account the great differences between Erethizon and octodontoids reported here and elsewhere, it appears likely that their common ancestor would have existed considerably farther back in time than the Deseadan. The next oldest known assemblage of fossil mammals in South America is the Mustersan, which is mid-Eocene. Since there are no rodent fossils in the Mustersan, the common ancestor of erethizontoids and octodontoids would have existed after that date if it were in South America. The new finding that the oldest known South American rodents may be only 26 million years old provides more time for the South American hystricognaths to have reached the remarkable diversity present in the Deseadan. However, our data and those of Bugge (this volume) and Sarich and Cronin (1980) on the probable relationship of thryonomyoids and octodontoids, as well as the new data on the presence of thryonomyoids in the upper Eocene of North Africa, support the hypothesis that hystricognath rodents in South America probably are not monophyletic, and that the old concept of the Caviomorpha should be abandoned. There were two and maybe three main radiations of hystricognath rodents in South America. One radiation is associated with the Superfamily Octodontoidea and is most closely related to the Old World Superfamily Thryonomyoidea. All members of these two groups possess a scalenus anticus muscle in the neck, which distinquishes them from other New and Old World hystricognaths. The two groups appear to have evolved from an early Old World thryonomyoid ancestor. The second radiation in the New World is most closely related to erethizontids. Erethizontids are probably members of the same overall clade as the remaining New World hystricognaths. Whether this is part of the same overall radiation, as Wood believes (1983:388), or represents a separate invasion of South America (see Bugge, this volume) is unresolved. Bugge classifies New World porcupines in their own separate infraorder, the Erethizontomorpha. The remaining New World taxa all appear to be based on the cavyproct body plan. The ease with which Erethizon serves as an outgroup in analyses of myological data for all of the taxa considered in this study indicates that it is likely they are all related at some level, and supports the hypothesis that hystricognaths are a monophyletic group (the Suborder Hystricognathi).

The conclusions based on myological data are generally consistent with conclusions from both biochemical and chromosomal data, and together these data appear to be more reliable than dental data in constructing cladograms and formulating phylogenetic conclusions. The above data suggest that dental morphology has been more dynamic in hystricognaths than we previously believed, and that patterns of convergence and parallelism in the evolution of rodent teeth are more common than in rodent musculature or biochemical characters. Teeth apparently are dynamic structures in all mammals. Riley (in press) investigated masticatory form and function in three mustelids, and found that the morphology of the cheekteeth varied greatly depending on the food habits of a species, while jaw musculature in the same species was much more conservative. Hiiemae (1978) has hypothesized that the greatest variation in the mammalian jaw apparatus is in the shape of teeth. Dental data should not be used exclusively in reconstructing rodent phylogenies. Jaw muscles, like teeth, contain many dynamic characters that must be carefully evaluated and can exhibit many examples of convergence and parallelism (Woods and Howland, 1979; Wilkins and Woods, 1983).

In conclusion, it is apparent that all muscles are not equally useful in constructing rodent phylogenies. The masticatory region has been the center of morphological change in rodent evolution, and there are many examples of convergence and parallelism in jaw musculature and dentition. The pharyngeal, laryngeal, and skin muscles appear to provide the most reliable data in constructing parsimonious cladograms. Myological and molecular data provide the best correlations of phylogenetic hypotheses. The data based on chromosomal analyses are limited in usefulness because of the lack of resolution of how karyotypes evolve, and therefore which are primitive and derived character states. Characters from blood proteins and dental morphology appear to be more dynamic (subject to change) than myological characters and of greater use in resolving phylogenetic questions at the generic and family levels rather than at higher taxonomic levels.

Muscles appear to be useful in separating phylogenetic lineages at upper as well as lower phylogenetic levels. They are very effective at separating Geocapromys from Capromys (generic level), as well as distinguishing taxa at the subfamily and family levels. The various capromyid genera may be more distinct from each other than are many other rodents because of the nature of their insular distributions in the Greater Antilles, and because these taxa may have been isolated for long periods of time (Woods, 1985). This may bias the present study by indicating that myology is more useful at lower taxonomic levels than is really the case. It is apparent in this study, however, that muscles are extremely useful characters to use in phylogenetic studies, and that data from myological analyses have the power to resolve phylogenetic lineages at a variety of taxonomic levels. The great diversity in location, kind, structure, and function of rodent musculature results in a system of characters that can be isolated to the level of phylogenetic analysis desired. Myological data provide a powerful and effective means of reconstructing the phylogenetic history of rodents.

ACKNOWLEDGEMENTS

We are indebted to Rhoda J. Bryant, David Forrestal, Elizabeth B. Howland, and Margaret K. Langworthy for technical assistance and to Dr. C. William Kilpatrick for his assistance in the collection, preparation, analysis, and interpretation of the hemolysate and serum proteins. We thank Dr. Farish A. Jenkins, Jr., of the Harvard University Museum of Comparative Zoology for allowing us to use the Museum's cineradiographic unit. Dr. David J. Klingener and Dr. Albert E. Wood and Gary Morgan read the manuscript and provided valuable commentary and criticisms. This paper is dedicated to David J. Klingener of the University of Massachusetts who has been an inspiration to us both. Thank you Dave!

REFERENCES

Adams, L.A. 1919. A memoir on the phylogeny of the jaw muscles in recent and fossil vertebrates. Annals N.Y. Acad. Sci. 28:51-166.

Appleton, A.B. 1928. The muscles and nerves of the post-axial region of the tetrapod thigh. J. Anat. 62:364-438.

Aristov, A.A. 1981. Short muscles of the hand in the Murinae (Rodentia, Muridae). Zool. Zhur 60(11):1675-1682.

Becht, G. 1953. Comparative biologic-anatomical researches on mastication in some mammals. Proc. Ned. Akad. Wet. 56:508-527.

Berman, S.L. 1979. Convergent evolution in the hind limb of bipedal rodents. Doctoral Dissertation, Univ. Pittsburgh. 302 pp.

Bock, W.J. 1981. Functional-adaptive analysis in evolutionary classification. Amer. Zool. 21:5-20.

Bryant, M.D. 1945. Phylogeny of the Nearctic Sciuridae. Amer. Midl. Nat. 33(2):257-390.

Bugge, J. 1974. The cephalic arteries of hystricomorph rodents. Zool. Soc. Lond. Symposium 34:61-78. Acta Anat. suppl. 62, 87:1-160.

Byrd, K.E. 1981. Mandibular movement and muscle activity during mastication in the guinea pig (<u>Cavia porcellus</u>). J. Morph. 170:147-169.

Carleton, M.D. 1984. Introduction to rodents. In: Orders and Families of Recent Mammals of the World, S. Anderson and J.K. Jones (eds.), pp. 255-265, Wiley and Sons, New York.

Carleton, M.D. and Musser, G.G. 1984. Muroid rodents. In: Orders and Families of Recent Mammals of the World, S. Anderson and J.K. Jones (eds.), pp. 289-379, Wiley and Sons, New York.

Cheng, C. 1955. The development of the shoulder region of the opossum, Didelphis virginiana, with special reference to the musculature. J. Morph. 97:415-471.

Cooper, G. and Schiller, A.L. 1975. Anatomy of the guinea pig. Harvard University Press, Cambridge. 417 pp.

Corbet, G.B. and Hill, J.E. 1980. A world list of mammalian species. British Museum (Natural History). 226 pp.

Cracraft, J. 1981. The use of functional and adaptive criteria in phylogenetic systematics. Amer. Zool. 21:21-36.

Cracraft, J. 1983. Cladistic analysis and vicariance biogeography. Amer. Scien. 71(3):273-281.

Doran, G.A. and Baggett, H. 1971. A structural and functional classification of mammalian tongues. J. Mamm. 52:427-429.

Dynowski, J. 1974. Comparative studies on the muscles of the limbs in some species of Rodentia. Acta Theriol. 19(8): 107-142.

Enders, R.K. 1934. The panniculus carnosus in an octodont rodent. Anat. Rec. 59(2):153-156.

Forster, A. 1928-1929. La crete en "s" du maxillaire inferieur chez certains rongeurs. Etude de specialisation particuliere du tissu osseux. Arch. Anat. Hist. Embr., Strasbourg 10: 327-346.

Fry, J.F. 1961. Musculature and innervation of the pelvis and hind limb of the mountain beaver. J. Morph. 109:173-197.

Gorniak, G.C. 1977. Feeding in golden hamsters, Mesocricetus auratus. J. Morph. 154:427-458.

Grand, T.I. and Eisenberg, J.F. 1982. On the affinities of the Dinomyidae. Sauget. Mitteilungen 30:151-157.

Greene, E.C. 1935. Anatomy of the rat. Trans. Amer. Phil. Soc. (N.S.) 27:1-370.

George, W. and Wier, B.J. 1974. Hystricomorph chromosomes. Zool. Soc. Lond. Symposium 34:79-108.

Hiiemae, K.M. 1971. The structure and function of the jaw muscles in the rat (Rattus norvegicus L.). III. The mechanics of the muscles. Zool. J. Linn. Soc. 50:111-132.

Hiiemae, K.M. 1978. Mammalian mastication: A review of the activity of the jaw muscles and the movements they produce in chewing. In: Development Function and Evolution of teeth, P.M. Butler and K.A. Joysey (eds.), pp. 359-398, Academic Press, New York.

Hiiemae, K. and Houston, W.J.B. 1971. The structure and function in the jaw muscles in the rat (Rattus norvegicus L.). I. Their anatomy and internal architecture. Zool. J. Linn. Soc. 50:75-99.

Hill, J.E. 1934. The homology of the presemimembranosus muscle in some rodents. Anat. Rec. 59:311-313.

Hill, J.E. 1937. Morphology of the pocket gopher mammalian genus Thomomys. Univ. Calif. Publ. Zool. 42(2):81-171.

House, E.L. 1953. A myology of the pharyngeal region of the albino rat. Anat. Rec. 116:363-378.

Howell, A.B. 1926. Anatomy of the Woodrat. Williams and Wilkins, Baltimore. 225 pp.

Howell, A.B. 1932. The saltatorial rodent Dipodomys: The functional and comparative anatomy of its muscular and osseous systems. Proc. Amer. Acad. Arts and Sciences 67(10):377-536.

Howell, A.B. 1936. The phylogenetic arrangement of the muscular system. Anat. Record 66:295-316.

Husson, A.M. 1978. The Mammals of Suriname. E.J. Brill, Leiden. 569 pp.

Jones, C.L. 1979. The morphogenesis of the thigh of the mouse with special reference to tetrapod muscle homologies. J. Morph. 162(2):275-309.

Kesner, M.H. 1980. Functional morphology of the masticatory musculature of the rodent subfamily Microtinae. J. Morph. 165:205-222.

Klingener, D.J. 1964. The comparative myology of four dipodoid rodents (Genus Zapus, Napeozapus, Sicista, and Jaculus). Misc. Publ. Mus. Zool. Univ. Mich. 124:1-100.

Klingener, D.J. 1984. Gliroid and dipodoid rodents. In: Orders and Families of Recent Mammals of the World, S. Anderson and J.K. Jones (eds.), pp. 381-388, Wiley and Sons, New York.

Kupper, W. 1970. Der Kehlkopf des afrikanischen Springhasen, Pedetes capensis. Z. Wiss. Zool. 181:140-178.

Langworthy, O.R. 1925. A morphological study of the panniculus carnosus and its genetical relationship to the pectoral musculature in rodents. Amer. J. Anat. 35:283-302.

McEvoy, J.S. 1982. Comparative myology of the pectoral and pelvic appendages of the North American porcupine (Erethizon dorsatum) and the prehensile-tailed porcupine (Coendou prehensilis). Bull. Amer. Mus. Nat. Hist. 173(4):337-421.

Meinertz, T. 1941. Das oberflachliche facialisgebiet der Nager. Zool. Jahrb., Abt. fur Anat. und Ontogenie der Tiere 67: 119-270.

Meinertz, T. 1944. Das superfizielle Facialisgebiet der Nager VII. Die hystricomorphen Nager. Zeit. fur Anat. und Entwick- lungs-geschichte 113:1-38.

Merriam, C.H. 1895. Monographic revision of the pocket gophers family Geomyidae (exclusive of the species of Thomomys). North Amer. Fauna 8:1-258.

Muller, A. 1933. Die Kaumuskulatur des Hydrochoerus capybara und ihre Bedeutung fur die Formgestaltung des Schadels. Morph. Jahrb. 72:1-59.

Patterson, B. and Wood, A.W. 1982. Rodents from the Deseadan Oligocene of Bolivia and the relationships of the Caviomorpha. Bull. Mus. Comp. Zool. Harvard Univ. 149(7):371-543.

Parsons, F.G. 1894. On the myology of the sciuromorphine and hystricomorphine rodents. Proc. Zool. Soc. Lond. 1894: 251-296.

Parsons, F.G. 1896. Myology of Rodents - Part II. Proc. Zool. Soc. Lond. 1896:159-192.

Ray, C.E. 1958. Fusion of cervical vertebrae in the Erethizontidae and Dinomyidea. Breviora Mus. Comp. Zool., Harvard Univ. 97:1-11 + 2 figs.

Reig, O.A. and Useche, M. 1976. Diversidad cariotipica y sistematica en poplaciones Venezolanas de Proechimys (Rodentia, Echimyidae), con datos adicionales sobre poplaciones de Peru y Colombia. Acta Cient. Venezolana 27:132-140.

Riley, M.A. (in press). An analysis of masticatory form and function in three mustelids (Martes americana, Lutra canadensis, Enhydra lutris). J. Mamm.

Rinker, G.C. 1954. The comparative myology of the mammalian genera Sigmodon, Oryzomys, Neotoma, and Peromyscus (Cricetinae), with remarks on their intergeneric relationships. Misc. Publ. Mus. Zool. Univ. Mich., 83:1-124.

Rinker, G.C. 1963. A comparative myological study of three subgenera of Peromyscus. Occ. Pap. Mus. Zool. Univ. Mich., 632:1-18.

Rinker, G.C. and Hooper, E.T. 1950. Notes on the cranial musculature of two subgenera of Reithrodontomys (harvest mice). Occ. Pap. Mus. Zool. Univ. Mich., 528:1-11.

Sarich, V.M. and Cronin, J.E. 1980. South American mammal molecular systematics, evolutionary clocks, and continental drift. In: Evolutionary Biology of the New World Monkeys and Continental Drift, R.L. Ciochon and A.B. Chiarelli (eds.), pp. 399-421, Plenum Press, New York.

Sharma, D.R. and Sivaram, S. 1959. On the hyoid region of the Indian gerbils. Mammalia, 23:149-167.

Sivaram, S. and Sharma, D.R. 1970. The hyoid complex of the porcupine Hystrix leucura. Saugetier. Mitteil., 18:52-61.

Sprague, J.M. 1941. A study of the hyoid apparatus of the Cricetinae. J. Mamm., 22:296-310.

Sykes, L.R., McCann, W.R. and Kafka, A.L. 1982. Motion of Caribbean plate during last 7 million years and implications for earlier Cenozoic movements. J. Geoph. Res., 87:10656-10676.

Szalay, F.S. and Decker, R.L. 1974. Origins, evolution, and function of the pes in the Eocene Adapidae (Lemuriformes, Primates). In: Primate Locomotion, F.A. Jenkins Jr. (ed.), pp. 239-259 Academic Press, New York.

Tullberg, T. 1899. Ueber das system der Nagethiere: eine phylogenetische studie. Nov. Act. Reg. Soc. Sci. Upsal, Ser. 3, 18:1-514.

Turnbull, W.D. 1970. Mammalian masticatory apparatus. Fieldiana: Geol., 18:148-356.

Tyson, E. 1699. Orang-outang; sive, Homo sylvestris; or, The anatomy of a pygmy compared with that of a monkey, an ape, and a man. London. pp. 12, 108, 55.

Van Vendeloo, N.H. 1953. On the correlation between the masticatory muscles and the skull structure in the muskrat, Ondatra zibethica. Koninkl. Nederl. Akad. Wetenschapen, Proc. Series C., 56:116-127, 265-277.

Vanzolini, P.E. and Guimaraes, L.R. 1955. South American land mammals and their lice. Evol., 9:345-347.

Varona, L.S. 1974. Catalogo de los mamiferos vivientes y extinguidos de las Antillas. Instit. Zool., Acad. Cienc. Cuba, 139 pp.

Vesalius, A. 1543. De Humani Corporis Fabrica. Basel.

Weijs, W.A. 1973. Morphology of the muscles of mastication in the albino rat, Rattus norvegicus (Berkenhout, 1769). Acta Morphol. Neerl. Scand., 11:321-340.

Weijs, W.A. 1975. Mandibular movements of the albino rat during feeding. J. Morph., 145:107-124.

Wiley, E.O. 1975. Karl R. Popper, systematics and classification: A reply to Walter Bock and other evolutionary taxonomists. Syst. Zool. 24-233-242.

Wilkins, K.T. and Woods, C.A. 1983. Modes of mastication in pocket gophers. J. Mamm. 64(4):636-641.

Wood, A.E. 1983. The radiation of the Order Rodentia in the southern continents: The dates, numbers and sources of the invasions. Schriftenr, Geol. Wiss. Berlin, 19/20:381-394.

Wood, A.E. and Patterson, B. 1959. The rodents of the Deseadan Oligocene of Patagonia and the beginnings of South American rodent evolution. Bull. Mus. Comp. Zool., Harvard Univ., 120(3):282-428.

Woods, C.A. 1972. Comparative myology of jaw, hyoid, and pectoral appendicular regions of New and Old World hystricomorph rodents. Bull. Amer. Mus. Nat. Hist., 147:117-198.

Woods, C.A. 1975. The hyoid, laryngeal and pharyngeal regions of bathyergid and other selected rodents. J. Morph., 147:229-250.

Woods, C.A. 1982. The history and classification of South American hystricognath rodents: Reflections on the far away and long ago. In: Mammalian Biology in South America, M.A. Mares and H.H. Genoways (eds.), pp. 377-392, Univ. Pittsburgh.

Woods, C.A. 1984. Hystricognath rodents. In: Orders and Families of Recent Mammals of the World, S. Anderson and J.K. Jones (eds.), pp. 389-446, Wiley and Sons, New York.

Woods, C.A. 1985. Adaptive radiation of capromyid rodents II: New taxa from Hispaniola, and the evolution and systematics of Antillean capromyids (Mammalia: Capromyidae). Bull. Florida State Mus., Biol. Sci.

Woods, C.A. and Howland, E.B. 1977. The skin musculature of hystricognath and other selected rodents. Zbl. Vet. Med. Comp. Anat. Hist. Embryol., 6:240-264.

Woods, C.A. and Howland, E.B. 1979. Adaptive radiation of capromyid rodents: Anatomy of the masticatory apparatus. J. Mamm., 60(1):95-116.

AMINO ACID SEQUENCE DATA AND EVOLUTIONARY RELATIONSHIPS AMONG HYSTRICOGNATHS AND OTHER RODENTS

Jaap J. Beintema

Biochemisch Laboratorium, Rijksuniversiteit
Nijenborgh 16
9747 AG Groningen, The Netherlands

INTRODUCTION

The study of metabolic pathways and of biomacromolecules like proteins and nucleic acids supplies molecular data useful for deriving evolutionary relationships between taxa.

The classical example of the use of metabolic pathways are the different nitrogen excretion products in vertebrates (Baldwin, 1949). Recent examples in the field of plant taxonomy are studies by Richardson (1983) and Mahlberg and Pleszczynska (1983). Differences in metabolism originate from differences in enzymic activities determined by the presence and level of expression of the genes coding for these enzymes. It has been suggested that evolution is more dependent on major metabolic shifts, which usually are achieved by altering the expression of genes coding for enzymes, than on changes in the structures of the enzymes (Wilson et al., 1977).

Studies of biomacromolecules like proteins and nucleic acids also may be used for the analysis of evolutionary relationships between organisms. Properties like charge (electrophoretic mobilities), molecular size or weight, the presence of post-translational modifications like carbohydrate moieties in glycoproteins, as well as the amino acid composition and enzymic properties, have little relationship with the degree of sequence (dis)similarity and do not supply reliable information about evolutionary relationships. Our studies on the digestive enzyme ribonuclease in different mammalian taxa showed that both charge and glycosylation state show similarities in non-related taxa, as a consequence of common adaptive requirements (Beintema and Lenstra, 1982).

As the evolutionary heritage of living species is encoded in the nucleotide sequence of their DNA, part of which is translated into the amino acid sequence of proteins, properties that have a direct relationship with the degree of sequence (dis)similarity supply more reliable information to analyze phylogenetic relationships. Quantitative data on sequence (dis)similarity are derived either directly from experimentally determined sequences, or indirectly from properties which are a reliable measure of sequence (dis)similarity, such as hybridization of nucleic acids and immunological analysis of proteins (Sarich, this volume). These latter approaches have the advantage that the actual sequences need not be known and that they can be performed relatively rapidly with crude preparations of biological material.

There are two different procedures used to derive phylogenetic trees from quantitative data on sequence (dis)similarity of homologous proteins and nucleic acids. The first method is a phenetic one and uses the degree of difference in a matrix form for analysis with cluster methods. Another approach is a cladistic one and can only be used with the actual sequences. In this latter approach, each nucleotide or amino acid residue position is treated as a separate character state and the minimum number of changes at each position in a given network (tree) is derived according to parsimony rules. The following step is to repeat this for other networks to find a network with a minimum of changes summed over all positions; this is the most parsimonious network or tree.

Both procedures have their advantages and disadvantages. Most parsimonious methods give better results than the matrix method if unequal evolutionary rates have occurred (Fitch, 1977). On the other hand, generally many parallel and back substitutions occur in most parsimonious trees and often quite a number of trees are found that require only slightly larger numbers of substitutions than the most parsimonious tree or trees. These trees often have widely different topologies. By using computer programs that produce model sequences by simulating the process of protein evolution, it was found that the ancestral sequence method, which uses the parsimony procedure, is capable of greater accuracy than the matrix method when sequences are less than about 50% different, but that for sequences more different, the matrix method is superior (Peacock and Boulter, 1975; Dayhoff, 1976a). When we analyzed our ribonuclease data (see below), we found that parallel and back substitutions of a certain type often occur in some related taxa. Such a situation impairs results obtained with the most parsimonious method more than those obtained with matrix methods (Beintema, 1983a).

Figure 1 shows in a schematic way the conversion of the one-dimensional information present in DNA into a functional protein molecule with a characteristic unique three-dimensional structure. As already mentioned, phylogenetic analyses of sequences use every

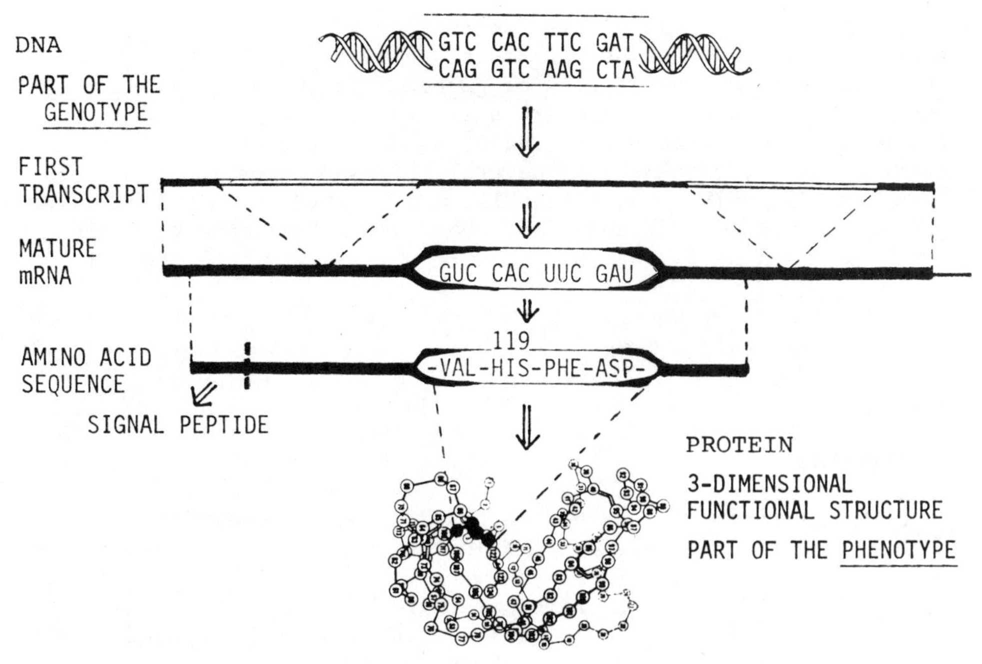

Fig. 1. Conversion of the one-dimensional information present in DNA into a functional protein molecule with a three-dimensional structure. The protein structure represents bovine pancreatic ribonuclease, and the sequence fragment is the region surrounding one of the active-site histidines. The messenger RNA sequence of rat ribonuclease has been determined recently (MacDonald et al., 1982). The structure of the first transcript of the ribonuclease gene (with exon/intron structure; see De Jong, this volume) remains unknown.

sequence position as a separate character state. Amino acid sequence data can be used directly or after translation into the nucleotide sequence (with uncertainties about the correct codon choices). There are 20 different character states for each amino acid position and four different states for each nucleotide position. A consequence of the limitation in the number of character states is the frequent occurrence of parallel and back substitutions. Another point is that no distinction can be made between primitive and advanced character states. Therefore, no direction in substitution patterns can be derived.

Rodents have been used rather frequently for sequence studies of nucleic acids and proteins, because the mouse and rat, the most frequently used laboratory animals, belong to this mammalian order. However, species other than mouse, rat, hamster, and guinea pig have been investigated only rarely, and the number of studies that are useful for a general study of rodent phylogeny is limited. Elsewhere in this volume, studies of globins (Shoshani et al.) and of eye lens proteins (De Jong) are summarized. Here, our studies on the pancreatic enzyme ribonuclease, as well as studies on insulin carried out predominantly by several British groups (Smith, 1966; Horuk et al., 1979, 1980; Blundell et al., 1982; Bajaj et al., 1984), will be presented. In addition, some comparative data on evolutionary rates will be discussed.

As is usually done in biochemical literature, species will be indicated by their common (English) names. The species studied are listed in Table 1.

RIBONUCLEASE

Mammalian species with a herbivorous diet generally have a rather high ribonuclease content in their pancreas (Barnard, 1969; Beintema et al., 1973). The enzyme probably has a function in the digestion of RNA from the symbiotic microorganisms occurring in the stomach or cecum of these species. Amino acid sequences of nearly 40 mammalian ribonucleases have been determined. The three-dimensional structure, which has been derived for the bovine enzyme (Fig. 1), is probably very similar for all pancreatic ribonucleases (Lenstra et al., 1977). Ribonucleases from four muroid and seven hystricognath rodent species have been investigated (Table 1, Fig. 2). Evolutionary trees of the sequences have been constructed using the maximum parsimony procedure (Fitch, 1971). Both a most parsimonious tree and a tree which is more in agreement with current

Table 1. List of rodent species studied.

	Common Name	Species
Muroidea	rat	*Rattus norvegicus*
	mouse	*Mus musculus*
	spiny mouse	*Acomys cahirinus*
	hamster	*Mesocricetus auratus*
	muskrat	*Ondatra zibethica*
Hystricognathi	porcupine	*Hystrix cristata*
	capybara	*Hydrochoerus hydrochoeris*
	guinea pig	*Cavia porcellus*
	cuis	*Galea musteloides*
	chinchilla	*Chinchilla brevicaudata*
	casiragua	*Proechimys guairae*
	coypu	*Myocastor coypus*

```
                            1         2         3         4         5         6
RIBONUCLEASES      1   5    0    5    0    5    0    5    0    5    0    5    0
RAT             GESR S  D  K     TE P K            QG  K S            E    I   GQ
MOUSE              R SA    Q     PD   IN            D  N S
HAMSTER              SA M          VAT              N               S    H
MUSKRAT         ---KETSAQKFERQHMDSTGSSSSPTYCNQMMKRREMTQGYCKPVNTFVHEPLADVQAVCSQEN
                                         XX              x
                                     x        x                              x
PORCUPINE       ---KESSAMKFERQHMDSSGSPSSNSNYCNEMMRRRNMTQDRCKPVNTFVHEPLADVRAVCFQKN
CAPYBARA           A       Q   V  E  S   A       V  K                    Q
GUINEA-PIG A       A           V  G  S   A       KK E  K               E Q   S R
GUINEA-PIG B       A       Q     PE    NS     V  I      G           S    Q
CUIS               A       Q      D H DT T       V  S   G             EA Q   S
CHINCHILLA                 Q           T A       KG     GY               Q
CASIRAGUA                K  Q   I       T P    A  KS     E                Q
COYPU              S    K         R    T P       KS     G                Q

                       7         8         9        10        11        12
                  5    0    5    0    5    0    5    0    5    0    5    0    5
RAT                   RN  H  S T R        S        T TDS  HI I  D N Y
MOUSE                RK      S        H            K T Y  HI      N Y       T
HAMSTER          A    K      H             A               I      N
MUSKRAT         VTCKNGNSNCYKSRSALHITDCRLKGNSKYPNCDYQTSQLQKQVIVACEGSPFVPVHFDASV----
                           X    X      XX        X
                                               x X Xx     X
PORCUPINE       VACKNGQTNCYQSNSLMHITDCRVTGSSKYPDCSYGMSQLERSIVVACEGSPYVPVHFDASVGPST
CAPYBARA         P           Y S         SN  F     RTT AQK        NL          E
GUINEA-PIG A     S           Y S    E  L SG  F N   RT  AQK  I     K        N  ----
GUINEA-PIG B     L           Y R R       S   F N   R   AQK  I     D           E
CUIS             P           H S R       S     N   R T AQK  I     T S       T ----
CHINCHILLA       P     S       N       L SN    N   RT RENKG I     N           ----
CASIRAGUA        P     S   E T N       L SN  F   L RT  E K  I     N           AA A
COYPU            L             N         SN D  N   RT  E K        N           AA A
```

Fig. 2. The amino acid sequences of rodent pancreatic ribonucleases in the IUB one-letter code (Stryer, 1981). Only the differences with the muskrat sequence for the other muroid sequences and with the African porcupine sequence for the other hystricognath rodent sequences are given. Deletions are indicated with (-). Below the muskrat sequence and above the porcupine sequence are indicated derived replacements, which are more (X) or less (x) typical for the muroid and the hystricognath sequences, respectively. References to sequences: muroids (Beintema and Gruber, 1973; Van Dijk et al., 1976; Lenstra and Beintema, 1979; Jekel et al., 1979; MacDonald et al., 1982; Beintema, 1983b); and hystricognaths (Van den Berg et al., 1976, 1977; Beintema et al., 1982; Beintema and Neuteboom, 1983).

biological opinion have been derived (Beintema and Lenstra, 1982).

Before discussing the results of these studies, several features of a set of character states such as is presented in Figure 2 should be mentioned. Derived replacements that are more or less typical for the muroid or hystricognath ribonuclease sequences are indicated below the muskrat and above the African porcupine sequences, respectively. Usual procedures for deriving trees from sequences do not give greater weight to such replacements than to others. However, we have made an approach in this direction in our investigation of the relation of the amino acid sequence of African porcupine ribonuclease to those of other rodent ribonucleases by making a separate character analysis at a large number of amino acid residue positions. This will be discussed later. Several, but not all, hystricognath ribonucleases have four additional amino acids at the C-terminus, which can be explained by substitutions of the stop codon at position 125 and by the presence of an additional stop codon at position 129 (Beintema, 1983b). During the analysis of sequences with the parsimony procedure, we did not take into account changes between coding and non-coding positions. Fitch and Yasunobu (1975) have discussed the problem of gap weighting in deriving phylogenies from amino acid sequences aligned with gaps, and they concluded that there is no unique solution.

Because ribonuclease is a rather rapidly evolving enzyme, in which many parallel and back replacements of amino acids occur, the ribonuclease data do not allow evolutionary relationships dating back to the origin of the eutherian mammalian orders and suborders in the distant past to be determined. Thus, the most parsimonious tree of ribonuclease has the hystricognath rodents grouping with other eutherian taxa, separate from the muroid rodents. However, within both rodent taxa the most parsimonious grouping of sequences agrees quite well with that derived from other biological data.

Cluster analysis of difference matrices of ribonuclease sequences gave results which, because of very unequal evolutionary rates in different lineages, differ more strongly from biological opinion than those of the most parsimonious method. However, multidimensional scaling of a difference matrix of ribonuclease sequences resulted in a representation with all rodent sequences, both from muroids and hystricognaths, grouping together in an extensive area separate from all other mammalian sequences. This indicates that the muroid and hystricognath groups of sequences are not completely unrelated (Westenbrink, K., Beintema, J. J., Seijen, H. G. and Groenier, K, unpublished).

The amino acid sequence of rat ribonuclease differs considerably from those of other muroid rodents studied (mouse, hamster, muskrat) and it is positioned separately from the others when methods employing quantitative measures of (dis)similarity are used. However, be-

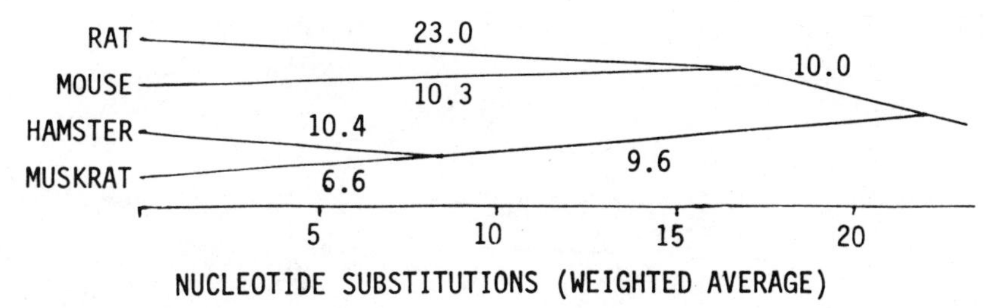

Fig. 3. Most parsimonious tree of muroid rodent ribonucleases. The number on each branch is the minimum number of nucleotide substitutions to account for the descent from the ancestor to its immediate descendant in the tree. Fractions result from averaging over more than one most parsimonious solution. The nodes are placed at a height equal to the weighted average number of nucleotide substitutions between the node and its descendant sequences.

cause rat and mouse ribonucleases share several derived (synapomorphous) substitutions, they are grouped separately from the other two sequences in a most parsimonious tree (Fig. 3). In our previous most parsimonious analysis, the rat ribonuclease sequence was still positioned separately from the other muroid sequences (Beintema and Lenstra, 1982). However, a correction of the rat sequence recently (MacDonald et al., 1982; Beintema, 1983b) has resulted in the topology presented here.

Among the hystricognath rodents, the most interesting feature is the grouping of the African porcupine with the South American caviomorphs. This result was obtained by counting the numbers of differences between ribonucleases of African porcupine, caviomorphs and other eutherian mammals (including muroid rodents), by the maximum parsimony procedure (Fig. 4), and by a separate character-state analysis of a large number of amino acid residue positions, where African porcupine shares derived replacements with the caviomorphs. This analysis has been described more extensively elsewhere (Beintema and Martena, 1982). Because the African porcupine sequence is more similar to the hypothetical sequence of the caviomorph ancestor than those of the investigated extant members of the caviomorphs, joining the African porcupine sequence with either the group consisting of the coypu, casiragua and chinchilla, or with the guinea pig actually requires one substitution less in the most parsimonious tree, than does placing the African porcupine separate from the caviomorphs (Fig. 4).

Coypu, casiragua and chinchilla group together in the most parsimonious tree. It saves one substitution to group casiragua with

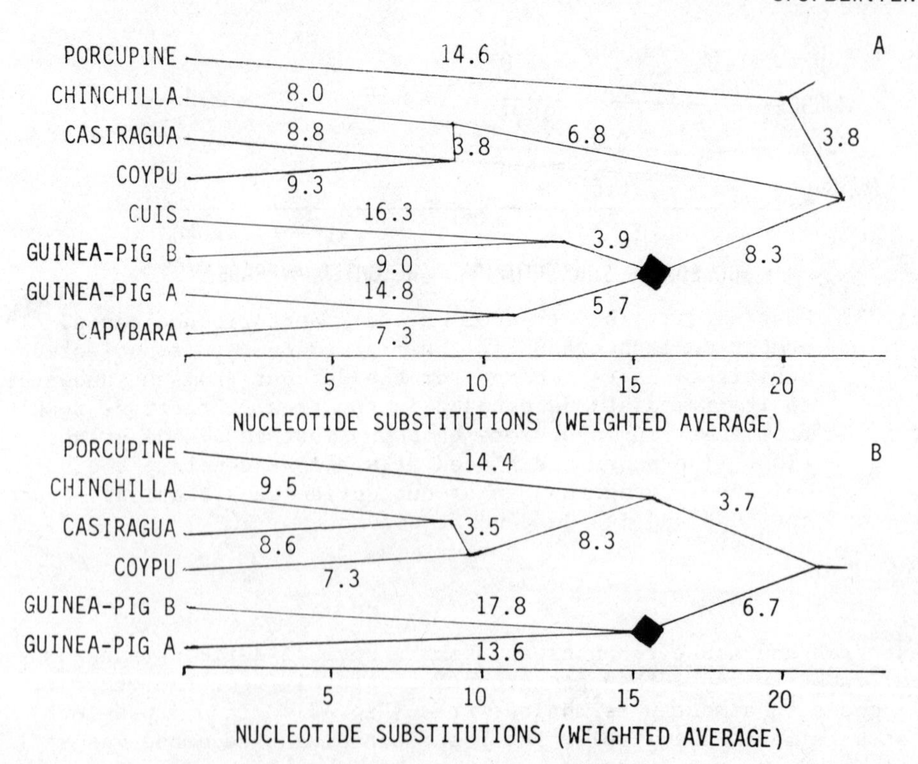

Fig. 4. A. Hystricognath rodent ribonuclease tree which combines current biological opinion with evidence from amino acid sequences studies of this protein. B. Most parsimonious tree of six hystricognath rodent ribonucleases. Black square, gene duplication resulting in two paralogous gene products (Fitch, 1977). For other details, see legend to Fig. 3.

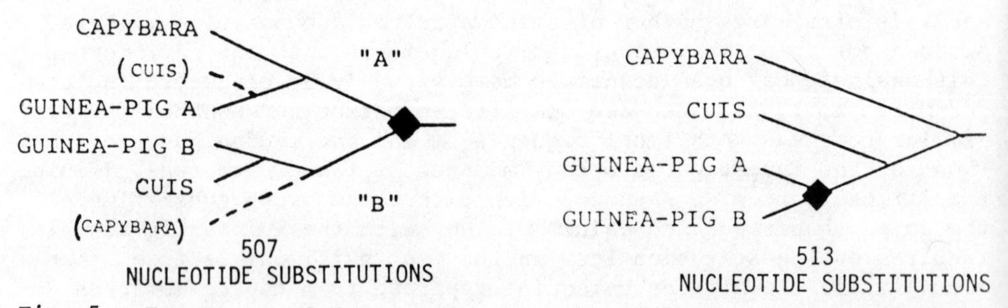

Fig. 5. Two trees of four pancreatic ribonuclease sequences from guinea-pig, capybara and cuis. Black square, gene duplication; ——, gene expressed; – – –, gene silenced or disappeared.

chinchilla, although other biological studies favor grouping casiragua with coypu (see Woods, this volume). However, this particular substitution occurs at a position with many parallel and back substitutions and is of little value for deriving phylogenetic relationships. Other characteristics, such as the presence of four identical additional residues at the C-terminus of the molecule, favor the grouping of casiragua with coypu (Fig. 4).

The ribonucleases of the guinea pig, cuis and capybara share a number of derived substitutions which are in agreement with the grouping together of these species on the basis of other biological criteria. However, here we encounter a phenomenon that one should always keep in mind when comparing homologous sequences of nucleic acids or proteins. Guinea pig is the only species investigated to date with two non-allelic pancreatic ribonucleases resulting from a gene duplication. These two ribonucleases differ at 31 of 124 positions (Van den Berg and Beintema, 1975). In pancreatic ribonucleases of cuis and capybara only one ribonuclease is found. In a most parsimonious tree, the gene duplication leading to both guinea pig ribonucleases occurred before the divergence of the guinea pig, capybara and cuis, but after the divergence of these three from the other caviomorphs investigated. In addition, we found that in the most parsimonious tree one of the guinea pig ribonucleases groups with the capybara enzyme and the other with the cuis enzyme. In Figure 5, two trees of four pancreatic ribonuclease sequences from guinea pig, capybara and cuis are shown. The tree in which the gene duplication leading to both guinea pig ribonucleases occurred after the divergence of guinea pig from the other two species requires six substitutions more than the tree in which this gene duplication occurred before the divergence of the three species. This suggests that in both capybara and cuis one of the ribonuclease genes has been lost or is not expressed, and that the capybara and cuis ribonucleases are not orthologous but paralogous gene products (Fitch, 1977), with an older common ancestry than that of the species themselves.

INSULIN

The therapeutic value of ox and pig insulin for treating human diabetes, usually without severe immunological problems, is due to the great similarity in primary structure of the insulins in the three species concerned. Most mammalian and avian insulins differ at very few positions (Blundell and Wood, 1975; Beintema, 1977). However, only the mature proteins have these very similar structures, while its precursor - proinsulin - contains an additional peptide segment that varies much more strongly among species (Blundell et al., 1982). Insulins with these structural characteristics also occur in muroid rodents (rat, mouse, hamster and spiny mouse). Rat and mouse have an identical pair of non-allelic insulins that differ at two amino acid residue positions (Fig. 6).

```
INSULINS     B-CHAIN  1         2         3  A-CHAIN  1         2
                 5    0    5    0    5    0      5    0    5    0
PIG          FVNQHLCGSHLVEALYLVCGERGFFYTPKA  GIVEQCCTSICSLYQLENYCN-
HAMSTER                                   S     D
SPINY MOUSE                               S     D
RAT/MOUSE 2    K                         MS     D
CHINCHILLA      K        D       D       M      D       T
RAT/MOUSE 1    K     P                    S     D
PORCUPINE                        ND    R        D    GV      Q
GUINEA-PIG     SR     N   T  S  QDD    I  D     D    GT TRH  QS
CUIS          F R     N  D   V  KDK   SR  -     D    R  TS   R
COYPU        Y S R    Q  DT  S  RH   - R ND     D    N   RN  MS   D
CASIRAGUA    Y G R    Q  DT  S  KH   - R SE     D    N   RN  LT
```

Fig. 6. The amino acid sequences of pig (Sus scrofa) and rodent insulins in the IUB one-letter code (Stryer, 1981). Only the differences with the pig insulin sequence are given. Deletions are indicated with (-). References to sequences: Smith, 1966; Markussen, 1971; Bünzli and Humbel, 1972; Neelon et al., 1975; Blundell and Wood, 1975; Horuk et al., 1979, 1980; Bajaj et al., 1984.

The insulins occurring in the hystricognath rodents, however, are quite different. The first indication was the observation that the guinea pig produces antibodies against the insulins of other mammalian species (Moloney and Coval, 1955). Later, it was found that the amino acid sequence of guinea pig insulin deviates strongly from those of other mammalian insulins. Further studies showed that the coypu and cuis as well have insulins with very deviating primary structures, and that there are many differences among these three hystricognath rodent insulin sequences. Other investigated hystricognath insulins are those from casiragua, which differs only slightly from that of the coypu, and from the chinchilla, which is an exception because its structure is similar to those of the muroid insulins. The African porcupine insulin occupies an intermediate position (Fig. 6).

The very well conserved insulin sequences in birds and most mammals can be explained by the requirement to form a very compact zinc containing crystalline hexameric structure in the storage granules o the pancreatic β-cells. However, the insulins from the hystricognat species with deviating amino acid sequences, including that from the

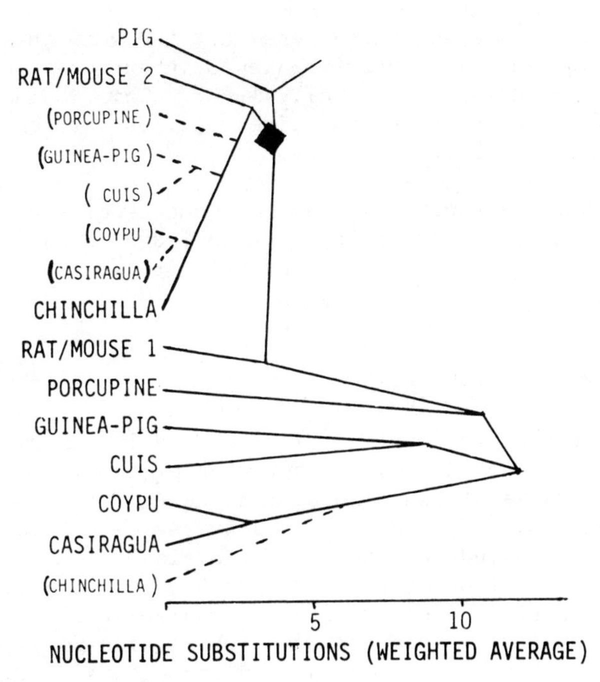

Fig. 7. Insulin tree which combines current biological opinion with evidence from amino acid sequence studies. The sequences in Fig. 6 were used to derive this tree. For other details, see the legends to Figs. 3 and 5.

African porcupine, lost the capability to form zinc-binding hexamers, and this may explain their strongly increased evolutionary rates after divergence from the zinc-binding insulins. Not only has the zinc-binding histidine residue been replaced, but also several apolar residues in the inter-subunit contact area of the hexamer have become changed to polar residues, which stabilize the conformation of the monomeric circulating form of the hormone (Blundell and Wood, 1975).

Yet, this cannot be the complete picture, because the insulin from chinchilla does not have a deviating sequence. Analysis of all rodent insulin sequences with the parsimony method suggests that the insulin gene duplicated before divergence of muroid and hystricognath rodents. Rat and mouse still express both gene products; in other species, only one of the two is expressed. In hystricognaths, one of the genes changed little and is found only in the chinchilla. The second gene underwent much more change, especially after divergence of African porcupine from other hystricognaths, and it is expressed in African porcupine, guinea pig, cuis, coypu and casiragua (Fig. 7).

However, there are also indications that the gene that has changed little is expressed at a very low level in extrapancreatic cells of the guinea pig (Rosenzweig et al., 1983). This would mean that both genes are still present in the guinea pig, but are expressed in different tissues at different rates.

Therefore, although amino acid sequences of rodent insulins contribute little at present to our knowledge of the phylogeny of this mammalian order, they show several irregularities that may be encountered while studying protein evolution.

EVOLUTIONARY RATES

Separate proteins exhibit different evolutionary rates, which are more or less characteristic for each of them. This observation has been the basis of the molecular-clock hypothesis for the evolution of proteins, which is explained by assuming that amino acid replacements observed in homologous proteins are the result of neutral evolution (Wilson et al., 1977). However, many exceptions to this generalization have been found (Goodman et al., 1982). Several examples have already been presented while discussing the molecular evolution of ribonuclease and insulin in rodents.

A number of proteins have been sequenced in both the mouse and rat, and the differences between both species may be compared with the average evolutionary change of these proteins in mammals. Slowl evolving proteins, such as cytochrome *c*, insulin and alpha crystalli A chain, have identical sequences in the rat and mouse. Faster evolving proteins, on the other hand, such as the beta and alpha chains of hemoglobin, ribonuclease and the immunoglobulin kappa chai C region, show increased evolutionary rates in the two murids (Table 2). If this is found to be a general phenomenon, it raises some interesting questions:

(1) The morphology of rodents varies less than that of other mammalian orders. Nevertheless, this order encompasses the largest number of species, living in the most diverse types of habitats. Has the evolution of rodents been influenced more by metabolic adaptations to changing environments than the evolution of other mammals?

(2) In general, the faster evolving proteins have a more direct rela tionship with the environment than the slowly evolving proteins. Ha the recent success of the Muridae had something to do with quick met abolic adaptations facilitated by rapid evolutionary changes of the proteins involved?

(3) Are shorter generation times, as occur in rodents, correlated with faster evolutionary rates? There are conflicting opinions concerning this point (Wilson et al., 1977).

Table 2. Comparison of evolutionary rates of some proteins from rat and mouse and average evolutionary rates of these proteins in mammals.

Protein	Different amino acids in sequences	Evolutionary rates (in PAMs per 100 million years; Dayhoff, 1978)	
	Rat - Mouse	Rat - Mouse[1]	Average of mammals[2]
<u>Slowly evolving proteins</u>			
Cytochrome <u>c</u>	0%	0	2.2
Insulin	0%	0	4.4[3]
Alpha crystallin A chain	0%	0	5.0
<u>Faster evolving proteins</u>			
Hemoglobin beta chain	10%	54	12
Hemoglobin alpha chain	15%	83	12
Ribonuclease	23%	137	21
Immunoglobulin kappa chain C region	26%	159	37

[1] Values of percentage difference were corrected for superimposed replacements using Table 36 in Dayhoff (1978), and rates were calculated using a common ancestry of rat and mouse 10 million years ago.
[2] From Table 1 in Dayhoff (1978).
[3] Not including hystricognath rodent insulins. For the precursor proinsulin, an average evolutionary rate of 12.7 PAMs per 100 millon years was calculated from Matrix 20 in Dayhoff (1976b). The amino acid sequence of mouse proinsulin, however, is not yet known. References to sequences not listed previously: cytochrome <u>c</u>, Carlson et al., 1977; alpha crystallin A chain, de Jong, this volume; hemoglobin beta chain, Garrick et al., 1978; Gilman, 1976; hemoglobin alpha chain, Chua et al., 1975; Popp, 1967; immunoglobulin kappa chain C region, Starace and Querinjean, 1975; Svasti and Milstein, 1972.

ACKNOWLEDGMENTS

Many thanks are due to Drs. W. M. Fitch and R. Niece (Madison, Wisconsin, USA) for their help in deriving ribonuclease and insulin trees, and to Dr. R. N. Campagne for carefully reading the manuscript.

REFERENCES

Bajaj, M., Blundell, T. L. and Wood, S. P. 1984. Evolution in the

insulin family: molecular clocks that tell the wrong time. Biochem. Soc. Trans. 12.
Baldwin, E. 1949. An Introduction to Comparative Biochemistry, 3rd edition, Cambridge University Press, Cambridge.
Barnard, E. A. 1969. Biological function of pancreatic ribonuclease. Nature 221: 340-344.
Beintema, J. J. 1977. Orthologous nature of mammalian insulin genes. J. Mol. Evol. 9: 363-366.
Beintema, J. J. 1983a. Molecular evolution of mammalian pancreatic ribonucleases. In: Numerical Taxonomy, J. Felsenstein, ed., pp. 479-483, Springer-Verlag, Berlin.
Beintema, J. J. 1983b. Rat pancreatic ribonuclease: agreement between the corrected amino acid sequence and the sequence derived from its messenger RNA. FEBS Lett. 159: 191-195.
Beintema, J. J. and Gruber, M. 1973. Rat pancreatic ribonuclease. II. Amino acid sequence. Biochim. Biophys. Acta 310: 161-173.
Beintema, J. J., Knol, G. and Martena, B. 1982. The primary structures of pancreatic ribonucleases from African porcupine and casiragua, two hystricomorph rodent species. Biochim. Biophys. Acta 705: 102-110.
Beintema, J. J. and Lenstra, J. A. 1982. Evolution of mammalian pancreatic ribonucleases. In: Macromolecular Sequences in Systematic and Evolutionary Biology, M. Goodman, ed., pp. 43-73, Plenum Press, New York.
Beintema, J. J. and Martena, B. 1982. Primary structure of porcupine (Hystrix cristata) pancreatic ribonuclease. Close relationship between African porcupine (an Old World hystricomorph) and New World caviomorphs. Mammalia 46: 253-257.
Beintema, J. J. and Neuteboom, B. 1983. Origin of the duplicated ribonuclease gene in guinea-pig: comparison of the amino acid sequences with those of close relatives: capybara and cuis ribonuclease. J. Mol. Evol. 19: 145-152.
Beintema, J. J., Scheffer, A. J., van Dijk, H., Welling, G. W. and Zwiers, H. 1973. Pancreatic ribonuclease: distribution and comparison in mammals. Nature (New Biol.) 241: 76-78.
Blundell, T. L., Pitts, J. E. and Wood, S. P. 1982. The conformation and molecular biology of pancreatic hormones and homologous growth factors. CRC Crit. Rev. Biochem. 13: 141-213.
Blundell, T. L. and Wood, S. P. 1975. Is the evolution of insulin Darwinian or due to selectively neutral mutation? Nature 257: 197-203.
Bünzli, H. F. and Humbel, R. E. 1972. Isolation and partial structural analysis of insulin from mouse (Mus musculus) and spiny mouse (Acomys cahirinus). Z. Physiol. Chem. 353: 444-450.
Carlson, S. S., Mross, G. A., Wilson, A. C., Mead, R. T., Wolin, L. D., Bowers, S. F., Foley, N. T., Muijsers, A. O. and Margoliash, E. 1977. Primary structure of mouse, rat and guinea pig cytochrome c. Biochemistry 16: 1437-1442.
Chua, C. G., Carrell, R. W. and Howard, B. H. 1975. The amino acid sequence of the α chain of the major haemoglobin of the rat

(Rattus norvegicus). Biochem. J. 149: 259-269.
Dayhoff, M. O. 1976a. The origin and evolution of protein superfamilies. Fed. Proc. 35: 2132-2138.
Dayhoff, M. O. 1976b. Atlas of Protein Sequence and Structure, Vol. 5, Suppl. 2. National Biomedical Research Foundation, Washington, D. C.
Dayhoff, M. O. 1978. Atlas of Protein Sequence and Structure, Vol. 5, Suppl. 3. National Biomedical Research Foundation, Washington, D. C.
Fitch, W. M. 1971. Toward defining the course of evolution: minimum change for a specific tree topology. Syst. Zool. 20: 406-416.
Fitch, W. M. 1977. Phylogenies constrained by the crossover process as illustrated by human hemoglobins and a thirteen-cycle, eleven-amino-acid repeat in human apolipoprotein A-I. Genetics 86: 623-644.
Fitch, W. M. and Yasunobu, K. T. 1975. Phylogenies from amino acid sequences aligned with gaps: The problem of gap weighting. J. Mol. Evol. 5: 1-24.
Garrick, L. M., Sloan, R. L., Ryan, T. W., Klonowski, T. J. and Garrick, M. D. 1978. Primary structure of the major β-chain of rat haemoglogins. Biochem. J. 173: 321-330.
Gilman, J. G. 1976. Mouse haemoglobin beta chains. Comparative sequence data on adult major and minor beta chains from two species, Mus musculus and Mus cervicolor. Biochem. J. 159: 43-53.
Goodman, M., Romero-Herrera, A. E., Dene, H., Czelusniak, J. and Tashian, R. E. 1982. Amino acid sequence evidence on the phylogeny of primates and other eutherians. In: Macromolecular Sequences in Systematic and Evolutionary Biology, M. Goodman, ed., pp. 115-191, Plenum Press, New York.
Horuk, R., Blundell, T. L., Lazarus, N. R., Neville, R. W. J., Stone, D. and Wollmer, A. 1980. A monomeric insulin from the porcupine (Hystrix cristata), an Old World hystricomorph. Nature 286: 822-824.
Horuk, R., Goodwin, P., O'Connor, K., Neville, R. W. J., Lazarus, N. R. and Stone, D. 1979. Evolutionary change in the insulin receptors of hystricomorph rodents. Nature 279: 439-440.
Jekel, P. A., Sips, H. J., Lenstra, J. A. and Beintema, J. J. 1979. The amino acid sequence of hamster pancreatic ribonuclease. Biochimie 61: 827-839.
Lenstra, J. A. and Beintema, J. J. 1979. The amino acid sequence of mouse pancreatic ribonuclease. Extremely rapid evolutionary rates of the myomorph rodent ribonuclease. Eur. J. Biochem. 98: 399-408.
Lenstra, J. A., Hofsteenge, J. and Beintema, J. J. 1977. Invariant features of the structure of pancreatic ribonuclease: A test of different predictive models. J. Mol. Biol. 109: 185-193.
MacDonald, R. J., Stary, S. J. and Swift, G. H. 1982. Rat pancreatic ribonuclease messenger RNA. The nucleotide sequence of the entire mRNA and the derived amino acid sequence of the preenzyme. J. Biol. Chem. 257: 14582-14585.

Mahlberg, P. G. and Pleszczynska, J. 1983. Phylogeny of Euphorbia interpreted from sterol composition of latex. In: Numerical Taxonomy, J. Felsenstein, ed., pp. 500-504, Springer-Verlag, Berlin.

Markussen, J. 1971. Mouse insulins - separation and structures. Int. J. Pept. Prot. Res. 3: 149-155.

Moloney, P. J. and Coval, M. 1955. Antigenicity of insulin: Diabetes induced by specific antibodies. Biochem. J. 59: 179-185.

Neelon, F. A., Delcher, H. K., Steinman, H. M. and Lebovitz, H. E. 1975. A comparison of the structure of hamster pancreatic insulin and the insulin extracted from a transplantable hamster islet-cell carcinoma. Biochim. Biophys. Acta 412: 1-12.

Peacock, D. and Boulter, D. 1975. Use of amino acid sequence data in phylogeny and evaluation of methods using computer simulation. J. Mol. Biol. 95: 513-527.

Popp, R. A. 1967. Hemoglobins of mice: Sequence and possible ambiguity at one position of the alpha chain. J. Mol. Biol. 27: 9-16.

Richardson, P. M. 1983. Methods of flavonoid data analysis. In: Numerical Taxonomy, J. Felsenstein, ed., pp. 495-499, Springer-Verlag, Berlin.

Rosenzweig, J. L., Le Roith, D., Lesniak, M. A., MacIntyre, I., Sawyer, W. H. and Roth, J. 1983. Two distinct insulins in the guinea pig: The broad relevance of these findings to evolution of peptide hormones. Fed. Proc. 42: 2608-2614.

Smith, L. F. 1966. Species variation in the amino acid sequence of insulin. Am. J. Med. 40: 662-666.

Starace, V. and Querinjean, P. 1975. The primary structure of a rat κ Bence Jones protein: Phylogenetic relationships of V- and C-region genes. J. Immunol. 75: 59-62.

Stryer, L. 1981. Biochemistry, 2nd ed. W. H. Freeman and Company, San Francisco.

Svasti, J. and Milstein, C. 1972. The complete amino acid sequence of a mouse κ light chain. Biochem. J. 128: 427-444.

Van den Berg, A. and Beintema, J. J. 1975. Non-constant evolution rates in pancreatic ribonucleases from rodents. Amino acid sequences of guinea pig, chinchilla and coypu ribonucleases. Nature 253: 207-210.

Van den Berg, A., van den Hende-Timmer, L. and Beintema, J. J. 1976. Isolation, properties and primary structure of coypu and chinchilla pancreatic ribonuclease. Biochim. Biophys. Acta 453: 400-409.

Van den Berg, A., van den Hende-Timmer, L., Hofsteenge, J., Gaastra, W. and Beintema, J. J. 1977. Guinea-pig pancreatic ribonucleases. Isolation, properties, primary structure and glycosidation. Eur. J. Biochem. 75: 91-100.

Van Dijk, H. , Sloots, B., van den Berg, A., Gaastra, W. and Beintema, J. J. 1976. The primary structure of muskrat pancreatic ribonuclease. Int. J. Pept. Prot. Res. 8: 305-316.

Wilson, A. C., Carlson, S. S. and White, T. J. 1977. Biochemical evolution. Ann. Rev. Biochem. 46: 573-639.

NEW PHIOMORPHA AND ANOMALURIDAE FROM THE LATE EOCENE OF NORTH-WEST AFRICA: PHYLOGENETIC IMPLICATIONS

Jean-Jacques Jaeger[1], Christiane Denys[1], and Brigitte Coiffait[2]

[1]Laboratoire de Paléontologie des Vertébrés, Université P. et M. Curie, UA 720 du C.N.R.S., 4, Place Jussieu F-75230 Paris, France; and [2]Laboratoire de Géologie des Ensembles sédimentaires, Université Nancy I, B. P. 239 54506 Vandoeuvre-Les-Nancy, France

INTRODUCTION

Until recently, the ancient fossil family Phiomyidae (suborder Hystricognathi) was recorded only from the early Oligocene site of the Fayum, Egypt (Andrews, 1906; Osborn, 1908; Schlosser, 1910, 1911; Wood, 1968). As defined by Lavocat (1973), the infraorder Phiomorpha includes many other representatives that occurred in the middle or late Oligocene of the Balearic islands (Adrover and Hugueney, 1975; Adrover et al., 1978), in the Miocene of Africa (Stromer, 1926; Lavocat, 1973), of India and Pakistan (Black, 1972; Flynn and Jacobs, 1982; Flynn et al., 1983; Jaeger et al., 1980), and of the eastern Mediterranean (Chios, Greece) (Tobien, 1968). Phiomorphs are still represented in Africa by two living families: the Thryonomyidae and Petromyidae (or Petromuridae). At the Fayum site, Wood (1968) described five new genera and nine species of phiomyids, showing a great variety of dental patterns (three to six crests in upper and lower molars, for example), but possessing already in the Oligocene all the characteristics of the modern representatives of the thryonomyoids: retention of dP4/4; hystricomorphy; hystricognathy; and multiserial enamel on the incisors.

Wood (1968) concluded that phiomyids had radiated locally from an unknown Eocene immigrant. Based in part on the fact that P/4 of the Fayum species was less molarized than that of theridomyids, an Eocene-Oligocene hystricomorphous but sciurognathous European group of rodents, Wood also suggested that the Fayum phiomyids may have originated directly from Ischyromyidae (= Paramyidae), the only early

Eocene group of rodents recorded in the fossil record at that time.

More recently, Hussain et al. (1978) have described a new middle Eocene family of supposed hystricomorphous rodents from Pakistan - the Chapattimyidae. They suggested affinities of this group to the ancestry of the Phiomorpha, but no precise character analysis demonstrated this relationship. During a revision of the family Chapattimyidae, Hartenberger (1982) included it in the Ctenodactylidae, and he concluded: "Ctenodactylids show clear similarities to ischyromyids and to the *Microparamys* group of rodents, but possible relationships with other more modern groups like Phiomorpha or Caviomorpha are very speculative and no characteristics support this latter hypothesis."

The new fossil rodent fauna discovered in the late Eocene site of the Nementcha Mountains (Eastern Algeria) by Coiffait et al. (1984) provides supplementary information about the earliest stage of phiomyid evolution. These new findings allow us to define more precisely the primitive condition of phiomyid dental structures and to clarify the phylogenetic relationships among the different members of the family Phiomyidae (*sensu stricto* Wood, 1968). The results of this phylogenetic analysis have important paleobiogeographic implications for rodent phylogeny.

The oldest known representatives of the Anomaluridae, an extant family whose phylogenetic relationships are presently unknown, have also been discovered associated with the oldest phiomyids at the Nementcha locality. Their earliest previously known fossil record was from the early Miocene of East Africa (Lavocat, 1973), where they already showed the modern characteristics of this family. The affinities of the anomalurids have been much discussed, and three general tendencies in their classification can be recognized.

First, Simpson (1945) allocated Anomaluridae, Pedetidae, and Theridomyidae, "faute de mieux," to a single superfamily, the Anomaluroidea, but he included them in the suborder Sciuromorpha. Bugge (1971, 1974), on the basis of the similarities in the cephalic arterial pattern, confirmed the relationship of Anomaluridae and Pedetidae, and he proposed a new, independent suborder Anomaluromorpha for the two groups. In 1980, Parent obtained similar conclusions based on the study of middle ear anatomy. Luckett (1971, this volume) has demonstrated the importance of the fetal membrane and placental characters in assessing the relationships among higher taxa of rodents. According to him, the fetal membranes of *Pedetes*, *Anomalurus*, and *Ctenodactylus* are quite similar; however, most of these similarities are believed to be shared primitive characteristics. This prevented Luckett from drawing any conclusions about the origins and possible affinities of anomalurids with other "hystricomorphous" rodents.

In another scheme, based also on similarities in dental pattern

Stehlin and Schaub (1951) placed the Anomaluridae in an isolated family and interpreted their molar type as derived from the Theridomys plan. Wood (1955) described "their cheek tooth pattern as reminiscent of that of the Theridomyoidea from which it is theoretically possible that they might have been derived." Lavocat (1962) also admitted possible relationships between Anomaluridae and Theridomyidae, but after discovery of Miocene representatives of this family, he changed his mind and considered the Anomaluridae more distant (Lavocat, 1973). Chaline and Mein (1979) placed the anomalurids in the superfamily Anomaluroidea, but separated them from the Pedetidae (which they placed in another superfamily). They regrouped the Anomaluroidea with the Theridomyoidea within the infraorder Theridomorpha without further discussion.

A more radical proposal, that of Schaub (1958), placed the Anomaluridae closer to the Eomyidae within the superfamily Eomyoidea. This was founded only on the simple basis of the general resemblance of the molar structures. (For a discussion of the geomyoid affinities of the Eomyidae, see Fahlbusch, this volume.)

These differing speculations concerning the systematic position of the Anomaluridae are related to the fact that they were unknown before Miocene times. The discovery of late Eocene anomalurids described in this paper allows us to contribute to a clarification of their systematic and phylogenetic affinities.

TAXONOMY AND MORPHOLOGY

Order Rodentia Bowdich, 1821

Family Phiomyidae Wood, 1955

Protophiomys, new genus

Type species: *Protophiomys algeriensis*, new species.

Distribution: Late Eocene of Nementcha mountains, Eastern Algeria.

Diagnosis: Primitive phiomyids of the size of *Phiomys andrewsi* Wood 1968, with low cusps and crests on cheek teeth, and without mesoloph or mesolophid. Upper molars with oblique metaloph never connected to posteroloph. Hypocone large with strong anterior arm connected to a median cusp. Lower molars with strong hypoconulid, moderately developed posterior arm of protoconid, and without anterior cingulum. Lower molars with two anterior and one posterior roots.

Protophiomys algeriensis, new species

Type specimen: UON 841, isolated right upper M1 (Plate I, Fig. 7).

PLATE I →

SEM photographs of cheek teeth in the phiomyid Protophiomys and the anomalurid Nementchamys. Scale bar = 0.5mm.

Protophiomys algeriensis

Fig. 1. UON84-2. Right dP/4.

Fig. 2. UON84-3. Left M/1.

Fig. 3. UON84-4. Left M/1.

Fig. 4. UON84-5. Right M/3.

Fig. 5. UON84-6. Left M/3.

Fig. 6. UON84-7. Left M1/ or M2/.

Fig. 7. UON84-1. Right M1/ or M2/. (Type specimen)

Fig. 8. UON84-8. Right M1/ or M2/.

Fig. 9. UON84-10. Right M3/.

Fig. 10. UON84-9. Left dP4/.

Nementchamys lavocati

Fig. 11. UON84-26. Left P/4.

Fig. 12. UON84-25. Right M/1. (Type specimen)

Fig. 13. UON84-27. Right M/3.

Fig. 14. UON84-28. Right P4/.

Fig. 15. UON84-29. Left M3/.

Table 1. Teeth measurements of a new species of phiomyid (Protophiomys algeriensis) from Eastern Algeria. All measurements are given in millimeters.

Tooth	N	Length		Width	
		$\overline{X}$	Min.-Max.	$\overline{X}$	Min.-Max.
dP/4	3	1.60	1.57 - 1.65	1.73	1.64 - 1.80
M/1, M2/	6	1.44	1.37 - 1.56	1.49	1.29 - 1.70
dP/4	2	1.45	1.43 - 1.47	1.25	1.21 - 1.28
M/1	4	1.61	1.56 - 1.68	1.52	1.48 - 1.58
M/2	5	1.77	1.71 - 1.86	1.58	1.51 - 1.65
M/3	3	1.91	1.79 - 1.97	1.72	1.58 - 1.82

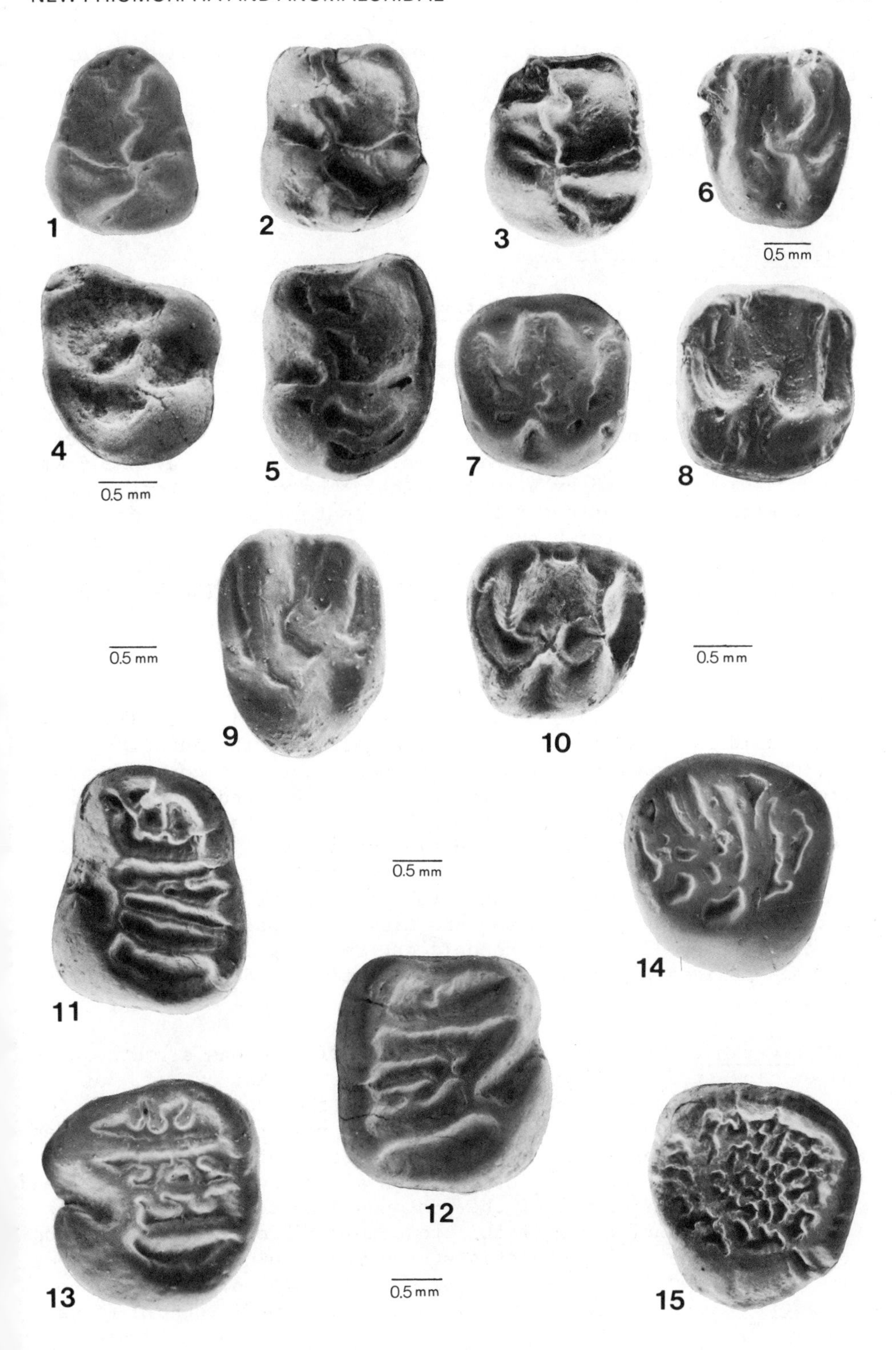
1
2
3
6
0.5 mm
4
5
7
8
0.5 mm
9
10
0.5 mm
0.5 mm
11
0.5 mm
14
12
13
15
0.5 mm

Hypodigm: UON 842 to UON 8422; 5 upper M1 or M2; 3 left upper dP4; 2 lower dP4; 3 lower M3; 9 lower M1 and M2.

Type locality: Nementcha locality, southern slope of Nementcha mountains, Eastern Algeria.

Diagnosis: As for the genus.

Measurements: See Table 1.

Description of Dentition of Protophiomys algeriensis

Lower dP4. This tooth is oval shaped, with a narrow anterior part and a large posterior part (Fig. 1). The main cusps are well individualized, and the crests connecting them are low. Metaconid is situated more anteriorly than the protoconid. An anterior arm limits the anterior part of the crown. This arm is separated from the conical protoconid by a small notch. The short posterior arm of the metaconid is oriented toward the posterolabial end of the tooth. It does not reach the protoconid. An obliquely directed ectolophid connects the posterior wall of the protoconid to the hypolophid, which it meets close to the hypoconid. The hypoconulid is well developed on the posterolophid. A small mesostylid cusp is developed between the metaconid and entoconid.

Lower M1. We identified as M/1 a rectangular tooth on which the cusp arrangement is similar to that of dP/4 (Figs. 2, 3). No anteroconid or anterolophid can be recognized on the anterior wall. The protoconid and metaconid are subequal. The protoconid shows a strong posterior arm which decreases in strength and ends in a large depression, limited posteriorly by the hypolophid and anteriorly by the metaconid. There is no indication of a mesolophid. The hypoconulid is well developed on the posterolophid, which ends on the lingual side at the base of the entoconid. There is only a slight depression separating the hypoconid from the hypoconulid. A sinusid is well developed, and no mesoconid can be recognized on the ectolophid. As in dP/4, cusps are well defined, and connecting crests are low.

Lower M2. This tooth is larger than M/1 and has a rectangular shape. The structure is basically the same as that of M/1, but the crests are better developed, and therefore the tooth has a more lophate pattern. Only one of about nine M/2 referred to this species shows a small anterolabial cingulum indicative of an incipient anteroconid. The anterior crest occupies a very anterior position and connects the anterior arm of the protoconid to the metaconid. There is always a posterior arm of the protoconid, which is sometimes connected to some enamel crenulations. No mesolophid can be observed. In two specimens, the posterior arm of the protoconid is connected anteriorly to the anterior crest, isolating a small rounded fovea.

The hypoconulid is weaker than on M/1 but is still recognizable as an enlargement on the posterolophid. A small crest-like mesostylid is present between the entoconid and metaconid.

Lower M3. This tooth (Figs. 4, 5) appears to be larger than M/2. The posterior part of the tooth is narrower than the anterior part. As in M/1 and M/2, a posterior arm of the protoconid is always present, and there is no mesolophid. The posterolophid usually fuses with the entoconid, isolating a small, deep, oval-shaped and transversely directed depression. One specimen shows a posterior arm of the protoconid running parallel to the longitudinal crest and connected to the hypolophid. There are two anterior roots and one posterior root, as for M/2.

Upper dP4. The proportions of this tooth (length greater than width) and the weakness of the anteroloph, not connected to the protocone, suggest that this tooth may be a dP4/ (Fig. 10). The protocone is strong and is obliquely directed. Its anterior part is connected to a straight protoloph issued from the paracone. The hypocone is as high as the protocone, but it is slightly smaller and prolongated posteriorly into a strong posteroloph. The metaloph is parallel to the protoloph in its more labial part, then it turns forward to join a central cusp that is connected to the protoloph and metaloph, and to the protocone and hypocone, with an X-shaped pattern of crests (Fig. 10). This central cusp, which is closer to the lingual than to the labial cusps, may correspond to either an enlarged and displaced metaconule, or to a new cusp, the mesocone, as interpreted by Wood (1968) for Phiomys. A small mesostyle develops on the labial wall, between the paracone and metacone. Another tooth, also considered to be dP4/, differs from the former by having a more crest-like metaconule connected to the hypocone and to the metaloph only. From the anterior side of the metaconule, a small crest stretches in the direction of the labial side of the tooth; this could represent an incipient mesoloph. However, it does not reach the small mesostyle. The metaloph is curved normally and is oriented toward the protocone.

Upper M1 and M2. Six upper molars can be attributed to M1/ or M2/ (Figs. 6, 7, 8). They have a square or rectangular outline, but they are usually wider than long (Table 1). Cusps and lophs are low. The anteroloph is always connected to the anterior part of the protocone. Thr protoloph is straight and located slightly anterior to the protocone. The metaloph is oblique, directed at its lingual extremity toward the protocone, and it is never connected to the posteroloph, contrary to Oligocene phiomyids. The metaloph ends in the central cusp (metaconule or mesocone - see discussion above), which occupies a more distal position than on dP4/. The position of the central cusp, and the crests connecting to adjacent cusps, are highly variable. In most cases, the central cusp is connected to the metaloph, hypocone, and protocone, the two former connections being always the strongest. The last connection delimits an anteriorly directed, shallow sinus that never extends very far labially. In some specimens this connection is weakly developed or absent. A small crest develops sometimes on the labial side of the metaconule, in the direction of the labial wall (which it never reaches). This can be considered as an incipient mesoloph. The anterior part of the tooth, including the protoloph and anteroloph, is narrow in comparison with the posterior part of the tooth and represents only about one-third of the occlusal surface of the molar. The lingual cusps are more important than the labial ones. A single upper molar shows a cingulum on the lingual wall.

Upper M3. This tooth is characterized by its more or less heart-shaped outline (Fig. 9). The hypocone is located more labially than on the other molars, hence the shortness of the posteroloph. The metacone is always lower than the paracone, and the posterolabial edge of the crown shows a more or less rounded outline. A small mesostyle is sometimes present. The other features of the tooth are the same as those of M1/ and M2/.

Character Analysis, Locality, and Affinities of *Protophiomys*

Protophiomys can be attributed to the family Phiomyidae, mainly on the basis of the structure of the molars and dP4, the lower molars being strikingly similar to those of what can be considered as the most primitive Fayum species *Phiomys paraphiomyoides* (see Wood, 1968). *Protophiomys* differs from the latter species by a more regular development of the posterior arm of the protoconid, by the stronger and more rounded hypoconulid, and by the lesser development of the anterior cingulum. Some lower molars in individuals of *Phiomys paraphiomyoides* have a posteriorly displaced metaconid buccal end, which has not been observed in the Algerian material. This would indicate a derived character state for the Fayum population, an interpretation also strongly supported by comparison of the upper molars. The Algerian molars show a more primitive structure. Crests already exist but are not so well differentiated as in the Oligocene phiomyids. There is no indication of a lingual anterior cingulum. The anteroloph is low on the anterior cheek tooth (dP4/ or P4/), where it is not connected to the protocone in the early stages of wear. No mesoloph is present, but in some individuals a small spur stretches toward the labial border of the tooth, where a tiny mesostyle is sometimes present. More important differences are shown by the central and distal part of the tooth. In both taxa, the hypocone is the strongest of the four main cusps. It possesses a strong anterior arm related to the central cusp, which can be identified as either a mesocone, or as an anteriorly displaced metaconule. This central cusp is always connected to the metaloph, which shows an oblique direction, toward the protocone, such as is found in more primitive rodents like the middle Eocene Chapattimyidae (Hussain et al., 1978). The metaloph is almost never connected to the posteroloph, as is the rule among Oligocene phiomyids. The lingual sinus is shallow in Algerian phiomyids, and the crest connecting the central cusp to the protocone and closing the lingual sinus is highly variable in its development.

In general, resemblances between these two phiomyid species are due more to the common occurrence of shared primitive characters than to any true shared derived traits. Thus, it seems premature to consider these species as closely related phylogenetically. *Protophiomys algeriensis* can be considered as an upper Eocene primitive morphotype in the evolution of African phiomyids. This new phiomyid species discovered in the late Eocene levels of Algeria indicates that this group entered Africa long before the Oligocene, as suggested by Wood (1968). Because *Protophiomys* is older than the Fayum species, it allows us to reevaluate definitions of relatively primitive versus derived conditions among phiomyids, and therefore to assess phylogenetic relationships among phiomyid species. On the other hand, they also allow us to reconstruct with more precision the hypothetical phiomyid ancestor, and therefore to search for affinities with other Eocene rodent families. The new phylogenetic insight they provide into the earliest rodent evolution can then be correlated with paleogeographic events controlling rodent dispersion during Eocene times.

The Nementcha locality in Algeria is considered to be late Eocene in age, because of the geological setting of this continental deposit, just above a marine middle Eocene formation (Coiffait et al., 1984). Several species of phiomyids, differing mostly in size, are recorded in this locality, and they can be related to some of the Oligocene Fayum phiomyids. Only the most abundant new species is described here. The fact that phiomyids are diversified indicates that they entered Africa before the late Eocene. This event could be related to the early immigration of Theridomyidae into western Europe during middle Eocene times, but it may also be completely independent from that event. This depends largely on the possible phylogenetic relationships between the two families. Reconstruction of early Tertiary mammalian evolution in Africa has shown that several faunal exchanges must have occurred between Africa and the surrounding Palearctic unknown land masses, in order to explain the composition of Paleocene and Eocene African faunas (Cappetta et al., 1978; Mahboubi et al., 1983, 1984a, b; Buffetaut, 1982; Bown and Simons, 1984; Jaeger and Martin, 1984). The age of these events and the precise nature of immigrant taxa are largely unknown, due to the scarcity of the fossil record. These immigrations of Palearctic mammals were usually followed by rapid endemic evolution, but, up to now, it has not been possible to localize the area of origin of any immigrant, because too much time separated immigration from fossilization. Therefore, it is expected that reconstruction of the phiomyid ancestor will allow us to locate the precise area of early evolution of this group, because it is generally acknowledged, but not demonstrated, that rodents did not originate in Africa, but in Asia (Li and Ting, this volume; Hartenberger, this volume).

The African Oligocene phiomyids are clearly hystricomorphous and hystricognathous, as demonstrated by Wood (1968, this volume); therefore, they represent, with the earliest South American caviomorphs (Hoffstetter and Lavocat, 1970; Lavocat, 1976; Patterson and Wood, 1982), the earliest known true hystricognathous and hystricomorphous rodents. Unfortunately, the new Algerian fossils do not allow us to recognize these cranial features in *Protophiomys*. All previously studied phiomyids, however, show multiserial incisor enamel, which is not the case for our specimens. The enamel microstructure of *Protophiomys* bears an intermediate pattern between pauciserial and multiserial types; this can be related to a transitional stage of enamel evolution, if our interpretation is correct (Coiffait et al., 1984). However, a detailed investigation is necessary before any conclusions can be drawn relative to the significance of this enamel pattern.

The lower molars of the new Algerian species are strikingly similar to those of some primitive Fayum phiomyids; dP/4 are identical, and M/2 and M/3 have the same size as those of *Phiomys andrewsi*. There is no mesolophid on the molars, and the posterior arm of the protoconid is developed on all teeth. The ectolophid is well developed and does not show any indication of a mesoconid. The hypoconulid is more cusp-like and higher than in the Oligocene taxa. The anterior cingulum is absent or weakly developed (one specimen only), which seems to indicate that the strong development of this crest in Fayum and later phiomyids represents a derived character within the family.

The main differences are shown by the upper molars, which are unlike those of the Fayum species. The main differences result from the orientation of the metaloph, which is directed toward the protocone, at least for its lingual part, where it is connected with a central cusp. The homology of this central cusp represents an important point for assessing the phylogenetic relationships of the earliest phiomyids. It can be interpreted in two different ways. According to Wood (1968), it corresponds to a mesocone in *Phiomys*, which by definition is a new cusp that develops on the ectoloph connecting the protocone and and hypocone. On the basis of the new Algerian material, however, we suggest alternatively that it may correspond to a modified metaconule. Because most middle Eocene rodents show both a protoconule and metaconule, one must propose either early disappearance of both conules among the earliest phiomyids, or else the disappearance of the protoconule and displacement of the metaconule. Only more primitive African fossils will permit us to choose between these two hypotheses. In several individuals, there are one to three crests radiating from this central cusp toward the protoloph, the metaloph, or toward the labial edge of the tooth. This latter short crest corresponds to what Wood (1968) has interpreted as the early development of a mesoloph in the Fayum species. However, in other Eocene rodents, such as European theridomyids, the mesoloph develops from the mesostyle toward the lingual part of the tooth. This incipient mesoloph never reaches the labial wall in *Protophiomys algeriensis*. The central cusp is connected by a strong crest to the hypocone, a character that seems to be one of the few apomorphic characters of phiomyids, in addition to hystricomorphy, hystricognathy, and a tendency to retain dP4. The hypocone is strong, sometimes stronger than the protocone. It develops from the posterior cingulum and is well separated from the protocone by a distinct sinus. This sinus is not as deep as that in the Fayum phiomyids.

In comparison to the generalized pattern of Oligocene phiomyids, the new Algerian species shows upper molars wider than long, an oblique metaloph, a strong central cusp (homologous to either the metaconule or a mesocone), a strong connection between the central cusp and the hypocone, a shallow sinus, an incipient mesoloph, and a small mesostyle. These characters, together with those of the lower dentition and the pauciserial structure of the incisor enamel, can be separated into three categories: derived, primitive, and indeterminate. We consider the strong connection between the central cusp and the hypocone, the disappearance of the protoconule, the appearance of an anterior cingulum on some lower molars, and the molarization of dP4 as shared derived characters with the Oligocene Fayum phiomyids. Therefore, this new genus appears to be a good representative of the upper Eocene ancestral morphotype of the family Phiomyidae.

These conclusions allow us to reassess some of the phylogenetic relationships among the Fayum phiomyids. The lack of a mesoloph (id) is definitely primitive. The five-crested and three-crested patterns of some Egyptian Oligocene phiomyids and even of more recent thryonomyoids must be considered as derived from an ancestral four-crested condition, with development of a strong mesoloph (id) as a derived feature. This leads us to refute some of the relationships among Fayum Oligocene phiomyids, as established by Wood (1968). *Phiomys andrewsi* can not be considered any longer as an ancestor of *Phiomys paraphiomyoides*.

Other characters, such as the strong hypoconulid (at least on the anterior molars), the strong metaconule (if the central cusp is really a metaconule), the regular development of the posterior arm of the protoconid, absence of an anterior cingulum on the lower molars, may have been better developed characters during middle Eocene times. On surrounding land masses at that time, only Chapattimyidae (Hussain et al., 1978) are recorded in Pakistan, and pseudosciurid theridomyoids and ischyromyids are found in western

Europe. The Chapattimyidae are recorded from several middle Eocene localities from the Indian plate (Hartenberger, 1982), at a time when this land mass was already welded to Asia (Sahni et al., 1982; Patriat and Achache, 1984).

A large land communication was open for land mammal dispersion south of the Tethys sea during the Eocene. The Algerian Phiomyidae share several derived characters with the Pakistan Chapattimyidae, such as the molarized dP/4, strong development of the hypocone and hypoconulid, the strong hypolophid and the strong metaconule (if the central cusp is really homologous with the metaconule). They differ from chapattimyids by several more derived features, such as loss of the protoconule and the mesoconid, and development of a central cusp showing a strong connection with the hypocone, and a weaker connection to the metaloph and protocone. The molar pattern is also different, with the transverse crests being better developed than in chapattimyid molars, but less so than in Oligocene phiomyids. P/4 are not yet recorded from the new Algerian locality. The P/4 of Oligocene phiomyids have no hypolophid, a character state interpreted as primitive by Wood (1968). Chapattimyidae have molarized P/4 with a well developed hypolophid. At the present time, however, we can not exclude that this character is derived, and that the reduction of P/4 may be related to the tendency toward persistence of dP/4 among phiomyids. On the other hand, some chapattimyid species have a mesoloph and mesolophid. These species cannot be considered as direct ancestors of *Protophiomys algeriensis*.

Chapattimyidae appear to constitute the sister group of the African Phiomyidae, which seem to be derived from an immigrant group of Asiatic chapattimyids, as suggested previously by Hussain et al. (1978). This interpretation would also allow us to interpret the Asiatic origin of the family Hystricidae. These hystricomorphous and hystricognathous rodents have not been found in the fossil record before middle Miocene times in South Asia. The do not retain dP4 and possess multiserial enamel. In our interpretation, they are the living representatives of Asiatic Chapattimyidae. This group has been recorded recently in late Eocene or Oligocene deposits of Pakistan (Flynn and Jacobs, 1982). This interpretation implies that hystricognathy and multiserial enamel structure developed independently at least twice, in African Phiomyidae and in Asiatic Hystricidae. Such an interpretation (Hypothesis 1) leads us to artificially group the Hystricidae and Chapattimyidae only on the basis of a paleobiogeographical scenario, and not on the existence of shared derived characters. A more anatomically parsimonious hypothesis (Hypothesis 2), in which hystricognathy would appear only once, implies a more complex paleobiogeographical history. However, the transition from sciurognathy to hystricognathy is not yet documented by the fossil record.

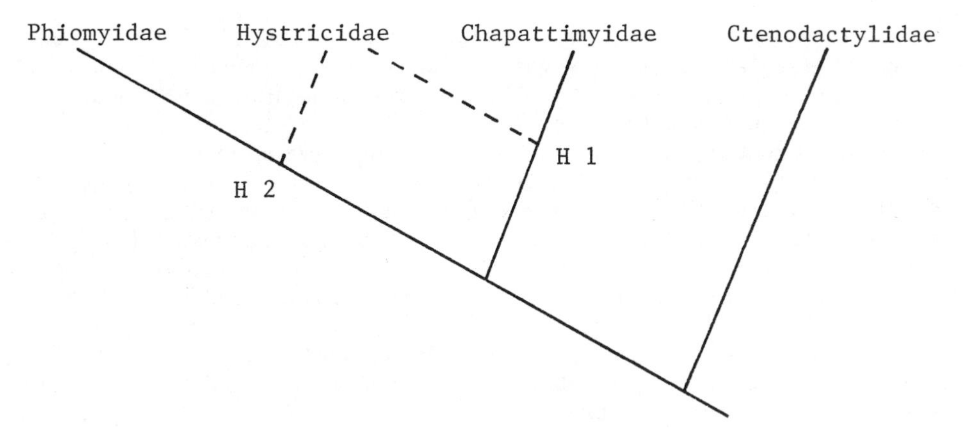

Fig. 16. Cladogram illustrating phiomyid relationships, as deduced from study of Protophiomys teeth. Alternative hypotheses (H 1 and H 2) for the placement of Hystricidae are discussed in the text. Caviomorpha cannot be placed on this cladogram at the present time.

Hartenberger (1982) has suggested that Chapattimyidae should be given the rank of subfamily, and he justifies their inclusion in the family Ctenodactylidae on the basis of several shared characters with primitive Ctenodactylidae described from central Asia (Shevyreva, 1972, 1976). However, he emphasized the differences between the North Tethysian Ctenodactylidae and the South Tethysian Chapattimyinae. It seems to us that if both shared a common ancestor, they evolved independently. On the northern shore of the Tethys sea, Ctenodactylidae can be characterized by the reduction of the posterior part of P/4, strong molar size increase from M/1-M/3, lack of a mesoloph (id), increasing size of the posterolophid and hypoconulid, and anterior shifting of the hypolophid (see Wood, 1977). In contrast, on the southern shore of the Tethys, the Chapattimyidae developed molarized P/4 and a mesoloph (id). The two groups can therefore be considered as vicariant taxa originating from a common ancestor and evolving respectively North and South of the narrow Eocene Tethys sea. These interpretations suggest the cladistic relationships illustrated in Figure 16.

During late Eocene times, Theridomyidae had become the dominant group of rodents on the northwestern shore of the Tethys sea (Hartenberger, 1973). Some of them show general resemblances to the Phiomyidae, and therefore several authors (Wood, 1955, Lavocat, 1962) have speculated about a theridomyid origin for the Phiomyidae. The earliest theridomyids appeared during middle Eocene times (Hartenberger, 1968, 1969) and already possessed several derived features, such as a reduced hypoconulid, strongly developed anteroconid and anterolophid on the lower molars related to the anteriorly displaced metaconid, and a well developed mesoloph (id). The upper molars have a large anterior valley (between the protoloph and anteroloph), in which some additional cusps are sometimes developed. Compared with Algerian phiomyids, theridomyids also show some expected primitive features, such as the presence of a well developed protoconule. It is therefore not possible at this stage of our knowledge to relate them to the Ctenodactyloidea, an interpretation sustained by Hartenberger (1982), who stated that "the gap between European *Microparamys* and theridomyids, for example, is probably less pronounced than that between primitive ctenodactylids and phiomorphs..."

The new Algerian phiomyids provide no additional information about the possible relationship between Phiomyidae and South American Caviomorpha. The molar structure of the Algerian *Protophiomys*, with a four-crested pattern and bunodont cusps, is more primitive than that of any known South American caviomorph. This is contrary to Wood's (1974) opinion, where he considered that a five-crested pattern was the primitive condition for Phiomorpha, whereas the four-crested pattern is the primitive state in Caviomorpha. More paleontological information is therefore necessary to evaluate the origin of the Caviomorpha (see Hoffstetter and Lavocat, 1970; Hussain et al. 1978; Patterson and Wood, 1982; Wood, this volume).

TAXONOMY AND MORPHOLOGY OF A NEW ANOMALURID

Family Anomaluridae Gill, 1872

Nementchamys, new genus

Type species: *Nementchamys lavocati*, new species.

Distribution: Late Eocene of Nementcha mountains, Eastern Algeria.

Diagnosis: Pentalophodont molars with complex mesoloph (id). Strong anterolophid on lower molars. Molarized P4. Occlusal surface complex, with numerous enamel crenulations.

Nementchamys lavocati, new species

Type specimen: UON 84-25, right lower M1 (Plate I, Fig. 12), collections of the University of Oran (Algeria).

Hypodigm: UON 84-26 to 84-30; one P/4 or dP/4; one M/3; two upper molars and one upper molar fragment.

Type locality: Nementcha locality, southern slope of Nementcha mountains, Eastern Algeria.

Diagnosis: As for the genus.

Measurements: See Table 2.

Description of Dentition of Nementchamys lavocati

Lower P4 or dP4. This is a large unworn tooth, only slightly shorter than M/1 (Fig. 11). The posterior part is larger than the anterior part. The cusps are reduced to crests. Eight transverse crests can be recognized from back to front. Identification of these crests is difficult. The most posterior one, which isolates a kidney-shaped depression, is the posterolophid. It issues from the hypoconid, the highest and largest crest-like cusp of the tooth, and it constitutes the posterior limit of the tooth. On its lingual extremity the posterolophid fuses with the entoconid. No hypoconulid can be recognized. The entoconid is related to the hypoconid by a straight transverse crest that joins the anterior arm of the hypoconid. The anterior arm of the hypoconid is connected to the protoconid by a straight ectolophid. A deep retroverse sinusid is developed but not yet expressed, due to the unworn condition of the tooth. The four following crests are difficult to interpret. Two of them can be interpreted as mesolophids, because they originate from the same point. The anterior one is divided at its lingual extremity into two arms. All three crests can be attributed to a complex mesolophid. At their lingual extremity they develop small cusps that close the lingual wall of the tooth. The next crest, in front of the more anterior mesolophid arm, can be interpreted as a metalophulid I, connecting the protoconid and metaconid. The protoconid cannot be recognized as a distinct cusp. An important depression is limited anteriorly, lingually, and labially by a prolongated ectolophid, and posteriorly by the metalophulid. The anterior crest can be interpreted as an anterolophid, a structure that is already present in the lower molars. Several small additional crests oriented anteroposteriorly originate from the anterior wall, but they

Table 2. Teeth measurements of a new species of anomalurid (*Nementchamys lavocati*) from Eastern Algeria. All measurements are given in millimeters.

Tooth	N	Length		Width	
		$\overline{X}$	Min.-Max.	$\overline{X}$	Min.-Max.
P/4 or dP/4	1	2.45		2.00	
M/1	1	2.49		2.21	
M/3	1	2.60		2.45	
M1/ or M2/	2	2.24	2.26 - 2.22	2.16	2.12 - 2.24

do not reach the metalophulid I. The labial wall of the tooth is somewhat inclined toward the lingual part. The occlusal surface is slightly concave and appears to be very similar to the molars.

Lower M1. This molar has a rectangular outline (Fig. 12). The occlusal surface is concave, and, as on the P/4, the cusps are not distinctly recognizable, being fused into crests. Only seven transverse crests can be observed. The anterior one originates in front of the metaconid, from which it is separated by a small notch. It extends transversely to join the protoconid on its anterior wall. It defines a transverse depression partially interrupted by some enamel crenulations, arising from its distal part. The metalophulid I is a straight, transverse, concave crest. From the posterior part of the protoconid there is a well developed ectolophid, which limits a well differentiated oblique sinusid, directed toward the posterior part of the tooth. The mesolophid shows two successive dichotomies that give rise to three crests, all reaching the lingual wall. The hypolophid is slightly oblique and joins the anterior arm of the hypoconid at the level of the deepest point of the sinusid. As on P/4, the posterolophid is united with the entoconid, delimiting a kidney-shaped transverse depression, and no hypoconulid can be recognized. All crests are enlarged on their lingual extremity. These small cusps, sometimes reminiscent of main cusps, are all fused to form an uninterrupted lingual wall.

Lower M3. The M/3 is of rectangular shape, with the trigonid being slightly larger than the talonid. The posterior part of the tooth is rounded. The general structure is similar to that of M/1. It differs from the latter tooth, however, in several features. The hypolophid is transverse, and the mesolophid is doubled. A short anterior one is followed by a more complicated one, composed of a main transverse crest that reaches the labial border of the tooth, from which several short transverse crests originate. The main mesolophid crest only reaches the labial border. The anterior depression is partially interrupted by three posteriorly directed small enamel crests arising from the anterolophid. The occlusal surface is concave, and the lingual wall is continuous and is higher than the labial one (Fig. 13).

Upper P4. The small size of this tooth and the presence of a posterior facet suggest that it is P4/ (Fig. 14). The general outline is rounded, and the occlusal surface is slightly concave. The anteroloph is fused to the paracone. Between it and the protoloph there is a large valley, in which a sinuous supplementary crest is present; its lingual end is fused to the anteroloph. The lingual side of the occlusal surface is difficult to interpret. The hypocone appears to be as high as the protocone and is related to it by two connections, a lingual one and a more labial one separated by a crescent-like isolated depression. A similar situation can be observed on an upper molar fragment. This structure can be interpreted as derived from the closure of an ancestral sinus. The metaloph and posteroloph can scarcely be identified. The metaloph seems to correspond to a large oblique and sinuous crest that ends lingually on the labial edge of the former sinus. If this interpretation is correct, the metaloph is oblique, directed toward the protocone. From its labial part there are two crests running obliquely toward the labial wall. These may correspond to a double mesostyle, because this structure is similarly complex on the lower molars. As on the lingual wall of the lower molars, the labial wall is continuous, and no cusp can be recognized. The posteroloph is fused to the metaloph on its labial extremity, and to the hypocone on its lingual extremity. The small basin isolated between the metaloph and posteroloph is divided into two equal parts by a transverse supernumerary crest.

Upper M3. The upper molars show a rounded outline, and their occlusal surface is covered by numerous small enamel crenulations, which inhibit recognition of the molar crown structures. One very peculiar molar can be recognized as an M3/ (Fig. 15). It has a rounded outline and a concave occlusal surface, as on P4/. The anterior part of the tooth is also similar in structure to that of P4/. The highest recognizable cusp is the paracone. It is related to the protocone by an oblique protoloph. The anteroloph limits a crescent-like anterior depression, which is interrupted by several small transverse enamel crenulations. The labial side of the tooth makes a continuous crest closing the occlusal surface. The posterolabial part of the tooth is rounded, and no metacone can be identified. The occlusal surface is covered by numerous enamel crenulations which mask the structure of the tooth. Posterolingually, a small notch may separate a very posteriorly situated hypocone from a main central protocone. The posteroloph is well developed. The reduction of the posterior part of the molar and the lack of any facet on the posterior wall of the tooth allow us to identify this molar as M3/.

Character Analysis and Suggested Affinities of *Nementchamys*

The molars of *Nementchamys* are very unlike those of any other described extant or fossil rodent. They belong to a rather large animal that possessed, as early as the late Eocene, several derived characters, including molarized P/4, complex mesoloph (id), enamel crenulations, a strong anteroconid, and concave occlusal surfaces. No cranial features are associated with these dental remains; therefore, the search for phylogenetic relationships must be limited to dental characters. This genus differs strongly from the other taxa recorded in the same locality by their large size, complex mesoloph (id), and strong anteroconid. They can scarcely be related to the phiomorphs, and their presence in association with this latter group indicates that phiomorphs were not the only Eocene rodents in Africa.

Although only phiomorphs have been discovered in the lower Oligocene of the Fayum (Wood, 1968), other endemic groups of African rodents, such as Anomaluridae and Pedetidae, are recorded in the lower Miocene of East Africa (MacInnes, 1957; Lavocat, 1973). The newly described molars of Nementchamys share several derived characters in common with some fossil and living Anomaluridae, especially with the genus Anomalurus (Ellerman, 1940; Friant, 1945; Stehlin and Schaub, 1951; Wood, 1962; Lavocat, 1973). These include: (1) the pentalophodont structure of the molars; (2) the strong anterolophid of the lower molars, associated with the development of a valley between the anterolophid and metalophulid I; (3) the absence of a posterior arm of the protoconid (metalophulid II); (4) mesoloph connected to the metaloph; (5) the continuous lingual wall on the lower molars, and the continuous labial wall on the upper molars; and (6) the concave occlusal surfaces.

These shared derived characters, if not the result of independent parallel evolution, are indicative of a relationship between Nementchamys and the Anomaluridae. However, the late Eocene fossils are too specialized in some features to be considered as direct ancestors of Miocene or Recent Anomaluridae. The molar structure of these late Eocene African rodents is also reminiscent of that in the Theridomyidae, an Eocene and Oligocene endemic group of rodents from Western Europe (see Hartenberger, 1973; Vianey-Liaud, 1972, 1976). Shared derived characters between Anomaluridae and Theridomyidae molars are, in addition to the similarity in cranial and mandibular features, the pentalophodont pattern, the strong development of the anteroconid and anterolophid, molarization of the P4, and development of enamel crenulations. Present knowledge and the incomplete fossil record do not allow us to further test this hypothesis, which was proposed earlier by several authors (Simpson, 1945; Wood, 1955; Lavocat, 1962; Chaline and Mein, 1979).

CONCLUSIONS AND SUMMARY

The new discovery and description of the oldest known species of Phiomyidae (Protophiomys algeriensis) and Anomaluridae (Nementchamys lavocati) from the late Eocene of the Nementcha mountains (Algeria) in northwest Africa allowed us to assess the primitive versus derived morphological characters of the teeth in the two groups. This also allowed us to evaluate some ideas about the origins of these families.

Subsequently, the comparison of these new taxa with the other possibly related "hystricomorphous" rodents known from Eocene and Oligocene sites in Africa, Asia, and Europe permitted us to propose some new phylogenetic hypotheses at the family level. Finally, as another result of this study, it appears that the simultaneous occurrence of phiomyids and anomalurids in the late Eocene of Africa can be considered to be due to the same wave of immigration.

Attributed to the family Phiomyidae (suborder Hystricognathi), the new genus Protophiomys shows more primitive characters than do the most primitive members of the family from the Oligocene Fayum. Thus, this new genus could represent an ancestor (or the ancestral morphotype) of the Phiomyidae. Compared to the Asiatic Eocene family Chapattimyidae, the Algerian phiomyids have several derived characters shared in common, which allow us to consider the two families as sister groups. However, the Chapattimyidae can not be considered as the direct ancestors of Protophiomys. This has important implications for the origin of this family, and a paleogeographic hypothesis is proposed to explain its relationships with the other Phiomorpha.

The presence in the same Algerian site of a true anomalurid (Nementchamys) indicates that the Phiomyidae were not the only Eocene rodents present in Africa. The closest affinities that the anomalurids shared were with the European Theridomyidae; this allows us to reformulate previous hypotheses about the systematic position of the Anomaluridae.

Discoveries of additional fossil material, especially of older fossil rodents in Africa, are now indispensable for understanding precisely the respective origins of the Phiomyidae and Anomaluridae.

ACKNOWLEDGMENTS

The authors are indebted to Philippe Blanc (Laboratoire de Micropaléontologie, Univ. Paris VI) and to Claude Abrial (Laboratoire de Paléontologie des Vertébrés, Univ. Paris VI, UA 720 du C.N.R.S.) for the scanning electron microscope (SEM) photographs.

REFERENCES

Adrover, R. and Hugueney, M. 1975. Des rongeurs (Mammalia) africains dans une faune de l'Oligocène élevé de Majorque (Baléares, Espagne). Nouv. Arch. Mus. Hist. nat. Lyon. 13, suppl.: 11-13.

Adrover, R., Hugueney, M., Moya, S. and Pons, J. 1978. Páguera II, nouveau gisement de petits mammifères (Mammalia) dans l'Oligocène de Majorque (Baléares, Espagne). Nouv. Arch. Mus. Hist. nat. Lyon. 16, suppl.: 13-15.

Andrews, C. W. 1906. A Descriptive Catalogue of the Tertiary Vertebrata of the Fayûm, Egypt. Brit. Mus. (Nat. Hist.), London.

Black, C. C. 1972. Review of fossil rodents from the Neogene Siwalik Beds of India and Pakistan. Paleont. 15: 238-266.

Bown, T. M. and Simons, E. L. 1984. Reply to J. J. Jaeger and M. Martin's 1984 paper. Nature 312: 379-380.

Buffetaut, E. 1982. A ziphodont mesosuchian crocodile from the Eocene of Algeria and its implications for vertebrate dispersal. Nature 300: 176-178.

Bugge, J. 1971. The cephalic arterial system in New and Old World hystricomorphs, and in bathyergoids, with special reference to the systematic classification of rodents. Acta Anat. 80: 516-536.

Bugge, J. 1974. The cephalic arterial system in insectivores, primates, rodents and lagomorphs, with special reference to the systematic classification. Acta Anat. 87 (supl. 62): 1-160.

Cappetta, H., Jaeger, J.-J., Sabatier, M., Sigé, B., Sudre, J. and Vianey-Liaud, M. 1978. Découverte dans le Paléocène du Maroc des plus anciens mammifères euthériens d'Afrique. Geobios 11: 257-263.

Chaline, J. and Mein, P. 1979. Les Rongeurs et l'Evolution. Doin Edit., Paris.

Coiffait, P.-E., Coiffait, B., Jaeger, J.-J. and Mahboubi, M. 1984. Un nouveau gisement à Mammifères fossiles d'âge Eocène supérieur sur le versant sud des Nementcha (Algérie orientale): découverte des plus anciens Rongeurs d'Afrique. C. R. Acad. Sci. Paris 299: 893-898.

Ellerman, J. R. 1940. The Families and Genera of Living Rodents. Brit. Mus. Nat. Hist., London.

Flynn, L. J. and Jacobs, L. L. 1982. Effects of changing environments on Siwalik rodent faunas of Northern Pakistan. Palaeogeog., Palaeoclim. Palaeoecol. 38: 129-138.

Flynn, L. J., Jacobs, L. L. and Sen, S. 1983. La diversité de Paraulacodus (Thryonomyidae, Rodentia) et des groupes apparentés pendant le Miocène. Ann. Paléont. 69: 355-367.

Friant, M. 1945. La dentition jugale de l'Anomalurus, écureuil volant d'Afrique. Rev. Zool. Bot. afr. 38: 206-211.

Hartenberger, J.-L. 1968. Les Pseudosciuridae (Rodentia) de l' Eocène moyen et le genre Masillamys Tobien. C. R. Acad. Sci. Paris 267: 1817-1820.

Hartenberger, J.-L. 1969. Les Pseudosciuridae (Mammalia, Rodentia) de l'Eocène moyen de Bouxwiller, Egerkingen et Lissieu. Paleovertebrata 3: 27-61.

Hartenberger, J.-L. 1973. Etude systématique des Theridomyoidea (Rodentia) de l'Eocène supérieur. Mém. Soc. Géol. Fr., N. S. 52, Mém. 117: 1-76.

Hartenberger, J.-L. 1982. A review of the Eocene rodents of Pakistan. Contr. Mus. Pal. Univ. Michigan 26: 19-35.

Hoffstetter, R. and Lavocat, R. 1970. Découverte dans le Déséadien de Bolivie de genres pentalophodontes appuyant les affinités africaines des rongeurs caviomorphes. C. R. Acad. Sci. Paris 271: 172-175.

Hussain, S. T., De Bruijn, H. and Leinders, J. M. 1978. Middle Eocene rodents from the Kala Chitta Range (Punjab, Pakistan). Proc. Kon. Ned. Akad. Wetensch. Amsterdam, Ser. B 81: 74-112.

Jaeger, J.-J. and Martin, M. 1984. African marsupials vicariance or dispersion? Nature 312: 379.

Jaeger, J.-J., Michaux, J. and Sabatier, M. 1980. Premières données sur les rongeurs de la formation de Ch'orora (Ethiopie) d'âge Miocène supérieur. I: Thryonomyidae. Palaeovertebrata, Mém. spéc. Jubil. R. Lavocat: 365-374.

Lavocat, R. 1962. Réflexions sur l'origine et la structure du groupe des rongeurs. Coll. Intern. C. N. R. S. 163: 491-501.

Lavocat, R. 1973. Les rongeurs du Miocène d'Afrique. Mém. Trav. E.P.H.E. Montpellier 1: 1-284.

Lavocat, R. 1976. Rongeurs Caviomorphes de l'Oligocène de Bolivie: II. Rongeurs du bassin Déséadien de Salla-Luribay. Palaeovertebrata 7: 15-90.

Luckett, W. P. 1971. The development of the chorio-allantoic placenta of the African scaly-tailed squirrels (Family Anomaluridae). Am. J. Anat. 130: 159-178.

MacInnes, D. G. 1957. A new Miocene rodent from East Africa. Fossil Mammals of Africa, Brit. Mus. (Nat. Hist.) 12: 1-36.

Mahboubi, M., Ameur, R., Crochet, J.-Y. and Jaeger, J.-J. 1983. Première découverte d'un Marsupial en Afrique. C. R. Acad. Sci. Paris 297: 691-694.

Mahboubi, M., Ameur, R., Crochet, J.-Y. and Jaeger, J.-J. 1984a. Earliest known proboscidean from early Eocene of North-West Africa. Nature 308: 543-544.

Mahboubi, M., Ameur, R., Crochet, J.-Y. and Jaeger, J.-J. 1984b. Implications paléobiogéographiques de la découverte d'une nouvelle localité Eocène à vertébrés continentaux en Afrique nord-occidentale: El Kohol (Sud-Oranais, Algérie). Geobios 17: 625-629.

Osborn, H. F. 1908. New fossil mammals from the Fayûm Oligocene, Egypt. Bull. Am. Mus. Nat. Hist. 24: 265-272.

Parent, J. P. 1980. Recherches sur l'oreille moyenne des rongeurs actuels et fossiles. Anatomie-Valeur systématique. Mém. Trav. E.P.H.E. Inst. Montpellier 11: 1-286.

Patriat, P. and Achache, J. 1984. India-Eurasia collision chronology has implications for crustal shortening and driving mechanism of plates. Nature 311: 615-621.

Patterson, B. and Wood, A. E. 1982. Rodents from the Deseadan Oligocene of Bolivia and the relationships of the Caviomorpha. Bull. Mus. Comp. Zool. 149: 371-543.

Sahni, A., Kumar, K., Hartenberger, J.-L., Jaeger, J.-J., Rage, J. C., Sudre, J. and Vianey-Liaud, M. 1982. Microvertébrés nouveaux des trapps du Deccan (Inde): mise en évidence d'une voie de communication terrestre probable entre la Laurasie et l'Inde à la limite Crétacé-Tertiaire. Bull. Soc. Géol. Fr. (7), XXIV, 5-6: 1093-1099.

Schaub, S. 1958. Simplicidentata. In: Traité de Paléontologie, J. Piveteau, ed., Vol. 6 (Pt. 2), pp. 659-818, Masson et Cie, Paris.

Schlosser, M. 1910. Über einige fossile Säugetiere aus dem Oligocän von Ägypten. Zool. Anz. 35: 500-508.

Schlosser, M. 1911. Beiträge zur Kenntnis der Oligozänen Landsäugetiere aus dem Fayûm (Ägypten). Beit. z. Paläont. u. Geol. Österreich-Ungarns u. d. Orients. 24: 51-167.

Shevyreva, N. S. 1972. New rodents in the Paleogene of Mongolia and Kazakhstan. Pal. J. Moscow 3: 134-145. [in Russian]

Shevyreva, N. S. 1976. Paleogene rodents of Asia. Acad. Nauk. SSSR., Trans. Palaeont. Inst. 158: 1-113. [in Russian]

Simpson, G. G. 1945. The principles of classification and a classification of mammals. Bull. Amer. Mus. Nat. Hist. 85: 1-350.

Stehlin, H. G. and Schaub, S. 1951. Die Trigonodontie der Simplicidentaten Nager. Schweiz. Pal. Abh. 67: 1-385.

Stromer, E. 1926. Reste Land- und Süsswasser-bewohnender Wirbeltiere aus den Diamantfeldern Deutsch-Sudwestafrikas. In: Die Diamantenwüste Südwestafrikas, E. Kaiser, ed., Vol. 2, pp. 107-153, D. Reimer, Berlin.

Tobien, H. 1968. Paläontologische Ausgrabungen nach jungtertiären Wirbeltieren auf des Insel Chios (Griechenland) und bei Maragheh (N. W. Iran). Jahrb. Verein. Freunde des Univ. Mainz 1968: 51-58.

Vianey-Liaud, M. 1972. L'évolution du genre Theridomys à l'Oligocène moyen. Intérêt biostratigraphique. Bull. Mus. Nat. Hist. Nat. (3), 98, Sc. Terre 18: 295-372.

Vianey-Liaud, M. 1976. Les Issiodoromyinae (Rodentia, Theridomyidae) de l'Eocène supérieur en Europe occidentale. Palaeovertebrata 7: 1-115.

Wood, A. E. 1955. A revised classification of the rodents. J. Mammal. 36: 167-187.

Wood, A. E. 1962. The juvenile tooth patterns of certain African rodents. J. Mammal. 43: 310-331.

Wood, A. E. 1968. Early Cenozoic mammalian faunas, Fayum province, Egypt. Part II. The African Oligocene Rodentia. Bull. Peabody Mus. Nat. Hist. 28: 29-105.

Wood, A. E. 1974. The evolution of the Old World and New World hystricomorphs. Symp. Zool. Soc. London 34: 21-60.

Wood, A. E. 1977. The evolution of the rodent family Ctenodactylidae. J. Pal. Soc. India 20: 120-137.

PROBLEMS IN MUROID PHYLOGENY:

RELATIONSHIP TO OTHER RODENTS AND ORIGIN OF MAJOR GROUPS

Lawrence J. Flynn[1], Louis L. Jacobs[2], and Everett H. Lindsay[3]

[1]Dept. Vertebrate Paleontology, American Museum of Natural History, New York, NY 10024; [2]Dept. Geological Sciences Southern Methodist University, Dallas, TX 75275; [3]Dept. Geosciences, University of Arizona, Tucson, AZ 85721

INTRODUCTION

The Muroidea include most of the diverse mouse-like rodents living today. The extant families of muroid rodents recognized by us are Muridae (true rats and mice), Cricetidae (hamsters, diverse hypsodont groups, and many American lineages), Gerbillidae (gerbils, sand rats and jirds) and several smaller groups, most of which have been given familial rank elsewhere. These are Nesomyidae (including Afrocricetodontinae), Rhizomyidae, Dendromuridae, Petromyscidae, Spalacidae, Cricetomyidae, Platacanthomyidae, and Lophiomyidae. Arvicoline (microtine) genera are not considered to constitute a family because they are late derivatives of advanced cricetids and because they form a polyphyletic group (C. A. Repenning, personal communication).

There is little agreement among various students even as to which groups of muroids should be given family status. Some authors include most all muroids in the family Muridae. Others recognize the Cricetidae as a taxon including all Muroidea with the exception of true mice (family Muridae), resulting in a highly paraphyletic family Cricetidae. Chaline et al. (1977) and Chaline and Mein (1979) reduce this problem by recognizing eight muroid families.

Recently the living muroids were reviewed by Carleton and Musser (1984). While they considered fossil data, our task is to examine possible relationships of muroids from a paleontological point of view. This involves less complete material than analysis of extant animals, with samples often comprising isolated teeth or jaws. However, this approach allows inclusion of extinct genera known only from the fossil record and provides a time scale for muroid evolution. Of course, the dental and skeletal morphology of extant rodents is part of the foundation for fossil studies. The systematic distribution of most of the characters discussed below is shown in Figure 1. Results of analyses of organ systems not known from fossils and approaches other than morphological ones, many of which are found in this volume, can be used to test hypotheses based on the fossil record.

Our main conclusions are that the Myomorpha may be the sister group of Ctenodactyloidea plus Hystricognathi, that true myomorphy defines a monophyletic group of muroids (Cricetidae, sensu stricto) exclusive of most Paleogene genera, and that most of the extant muroids are the descendants of a Miocene radiation. Disregarding usage of the familial rank in our analysis, the main departure from other classifications is restriction of nonmyomorphous muroids from the family Cricetidae.

Neogene rodent faunas from Pakistan provide important information on the diversity of fossil ctenodactyloids in south Asia and on the Miocene radiation of muroid rodents. Most of the Pakistani fossil specimens come from Siwalik Group rocks of northern Pakistan, and are roughly 16 to 1.8 million years old. They represent mainly muroids of modern aspect (murids, cricetids, rhizomyids) and less abundantly, glirids, sciurids, ctenodactylids, and thryonomyids. A rodent fauna completely different from that of the Siwaliks is known from the older (early Miocene) Bugti beds of Baluchistan in western Pakistan. The rodent fauna from Bugti is exclusively ctenodactyloid; that from the Siwaliks is predominantly muroid. The details of how the Siwalik fauna replaced the earlier Bugti fauna are currently unknown, but the former documents part of the important Miocene radiation of muroids, and the latter documents a young phase in diversification of ctenodactyloids in south Asia. The Bugti ctenodactyloids are not particularly closely related to Siwalik ctenodactyloids.

The Bugti Rodent Fauna

Rodents from the early Miocene Bugti beds, Baluchistan, Pakistan were first reported by Jacobs et al. (1981). No conclusions were reached concerning the affinities of the Bugti taxa, although some of the Bugti specimens were considered similar in dental pattern to chapattimyids from the Eocene of Pakistan, and perhaps to phiomyids and thryonomyids. Further study by Flynn et al. (ms) indicates the presence of at least six endemic new genera, most of which are apparently more closely related to each other than to any other taxon. The relationship of one of the new genera to the rest is unclear.

All of the new taxa are represented primarily by isolated teeth; only one maxillary fragment with two teeth is known from the entire Bugti rodent fauna. Nevertheless, based on occlusal pattern and other dental traits, the fauna is distinct from that of the Siwaliks but similar to the Eocene chapattimyid fauna of the Indo-Pakistan sub-

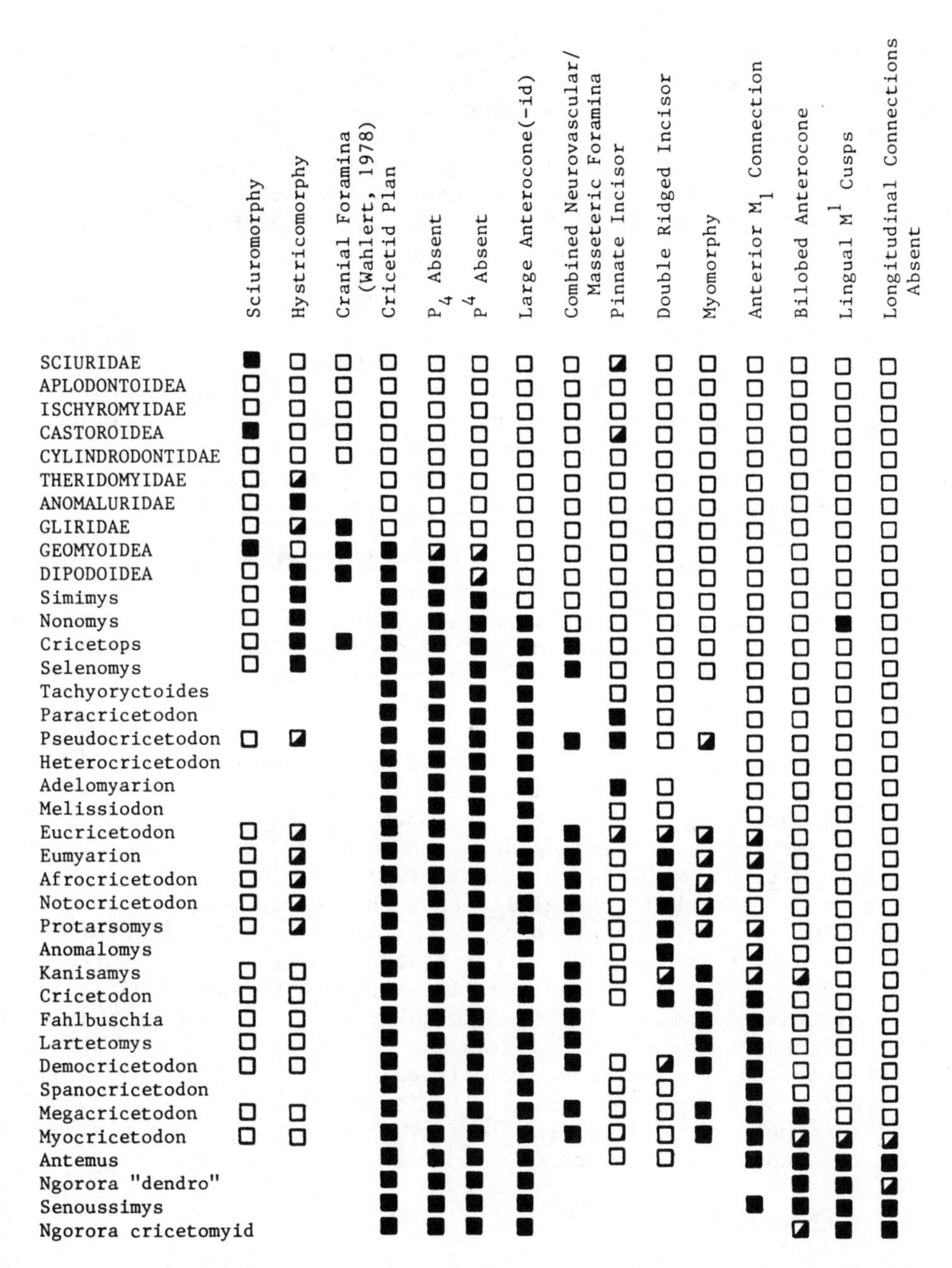

Fig. 1 Distribution of some derived traits of the hard anatomy in major groups of rodents with uniserial enamel and among selected muroids. Filled and empty boxes indicate presence and absence of traits; half filled boxes indicate presence in some members only or incomplete character transformation; no box means that material does not exist or was not surveyed.

continent, even though there are no Oligocene samples from the area that link the two assemblages. Thus there was a radiation of chapat timyids on the Indo-Pakistan subcontinent, distinct from that of other ctenodactyloids in other parts of Asia. This is a significant factor in recognizing Chapattimyidae in that early members of the family resemble early Ctenodactylidae. Regardless of our recogniti of the chapattimyids as a distinct family, the pattern of evolution with respect to Bugti rodents is clear.

The Bugti fauna is too late in time and too fragmentary in nature to provide strong historical evidence for a relationship between chapattimyids and phiomyids, a relationship postulated by Hussain et al. (1978). Probably derived similarities between chapat timyids and early Oligocene *Phiomys* include strong metalophulid II, low anterior cingulum on M_{1-3}, and traits of the P_4 and DP_4 (molariform talonids, anteroconid on P_4 or DP_4 of some chapattimyids and *Phiomys*, possibly mesolophulid on DP_4, presence of mesoloph).

Most chapattimyids and ctenodactylids share multiserial enamel with phiomyids and other hystricognaths. Multiserial enamel can be considered a derived condition because all studied earliest rodents have pauciserial enamel and no eurymyloids have the multiserial mic structure seen in rodents. Hystricognaths could have inherited mult serial enamel from ctenodactyloids with this microstructure.

Phiomyids, like other thryonomyoids, caviomorphs, hystricids a bathyergoids, have the hystricognath angular process. The only kno jaw of a chapattimyid, that of the Eocene *Birbalomys*, is sciurognat ous. Nevertheless, hystricognathy is considered a derived characte state, and *Birbalomys* antedates phiomyids and other unequivocal hystricognaths. Korth (1984) claims that the "incipiently hystricognat ous" franimorphs are sciurognathous. It is possible that hystricognathy and other derived character states evolved later than the shared dental characters and hystricomorphy that unite a clade larg than Hystricognathi, yet smaller than Rodentia. Accepting this argu ment, ctenodactyloids and pedetids may be related to hystricognaths (thryonomyoids, caviomorphs, hystricids and bathyergids). All six are united by hystricomorphy, similarities in cheek teeth, and mult serial enamel. Hystricognathy is a synapomorphy that defines a mor derived clade at a lower taxonomic level.

Muroid Systematics

If ctenodactyloids, pedetids, and hystricognaths comprise one major clade of rodents, to which clade do muroids belong and how do it fit into the pattern of rodent evolution? Muroids and other myomorphs are traditionally thought to derive from Sciuravidae (most recently Korth, 1984), which are generally considered distinct from ctenodactyloids. Hartenberger (1980) is not satisfied with this hypothesis. There is some evidence that early muroids were hystricomorphous (references in Vianey-Liaud, this volume) as are dipodoids and anomalurids. All of these groups are primitive in being sciurognathous. Hystricomorphy as a derived trait could support association of a muroid-dipodoid-anomalurid clade with a ctenodactyloid-pedetid-hystricognath clade, a grouping suggested in part by Wood (1980, p. 57) and Vianey-Liaud (this volume). Of course, hystricomorphy in early muroids is not precisely similar to the very derived hystricomorph conditions seen in the Hystricognathi (see Lavocat, 1973). On the other hand, true myomorphy is quite distinct from the plesiomorphic hystricomorphous condition of early muroids. We agree with Vianey-Liaud's suggestion that primitive hystricomorphy unites muroids with ctenodactyloids and hystricognaths. The clade containing muroids is derived in having uniserial enamel, as opposed to multiserial (primitively [?] pauciserial) enamel in its sister group. If this relationship is rejected on other grounds, one is forced to rationalize that hystricomorphy evolved independently in these two clades. These alternative hypotheses can be tested by developmental studies and by an improved fossil record.

MYOMORPHA AND MUROID MONOPHYLY

Content of the Suborder Myomorpha remains a major problem in rodent systematics. The Muroidea are taken to constitute the heart of the suborder, but phyletic affinity with other groups is difficult to establish on evidence from hard tissues alone. Zygomatic structure does little to unify the Myomorpha, especially as myomorphy exists only in later Muroidea.

Close relationship between Muroidea and Dipodoidea is well founded on myological (Klingener, 1964), cephalic arterial (Bugge, 1971) and fetal membrane (Luckett, this volume) organ systems. Molar morphology is also basically similar and the two groups share loss of the fourth upper premolar. In addition, dipodoids and early muroids share hystricomorphy. Closeness of the Muroidea and Dipodoidea led Schaub (1958) and Emry (1981) to unite them in Infraorder Myodonta. This taxon emphasizes a sister group relationship.

Relationship of Geomyoidea (Eomyidae, Geomyidae, and Heteromyidae) to Myodonta is based on myological and skeletal evidence enumerated in Hill (1937, p. 161), on cranial foramina (Wahlert, 1978 and 1984a) and on fetal membranes (see Luckett, this volume). Furthermore, Eomyidae share with Dipodoidea and Muroidea a similar molar crown plan that is derived with respect to sciuravid teeth. This has been called the cricetid plan and can be defined as follows: presence of four major cusps with narrow, transverse lophs joining them and longitudinal mures joining protocone(-id) to hypocone(-id); a mesoloph(-id), posteroloph(-id) and an anteroloph(-id) are variably developed additional transverse crests. This cricetid plan appears in all three groups at about the same time (late Eocene) and is modified in later members of each superfamily.

However, phyletic affinity of geomyoids and muroids is difficult to demonstrate (Fahlbusch, this volume, and Hill, 1937). Consequently Geomyoidea are placed in Myomorpha but not in Myodonta. Relation to Ctenodactyloidea is more distant than that suggested by Hartenberger (this volume). The sciuromorphous snout of geomyoids apparently evolved independently from that of Sciuridae. The superficially sciuromorphous zygoma of Rhizomys has been shown to be derived from myomorphy (Wahlert, 1978; Flynn, 1982).

The systematic position of the Gliridae is more equivocal. Although some cranial (Wahlert, 1978) and macromolecular (Sarich, this volume) evidence argues for their association with Myomorpha, the evidence of Bugge (1971) and Lavocat and Parent (this volume) does not support this. Alternatively, some middle ear structures (Lavocat and Parent, this volume) suggest that glirids are related to sciurids and aplodontids. Living members of this group are, in fact, poorly surveyed for most other traits. The group is apparently very old, isolated teeth occurring back to at least the middle Eocene in Europe, thus antedating the earliest Myodonta and Geomyoidea. See Hartenberger (1971) for a possible glirid origin. Living glirids are superficially similar to Myodonta because they have hystricomorphous-like infraorbital foramina, but this condition was acquired by fossil glirids in a unique way (Vianey-Liaud, this volume; Klingener, 1984).

Two other problematical groups, Theridomyidae and their possible relatives Anomaluridae, are not generally associated with Myomorpha, but Stehlin and Schaub (1951) noted dental similarities of eomyids and anomalurids. Like Anomaluridae, Theridomyidae are considered hystricomorphous, although large infraorbital foramina are best developed in older, dentally primitive genera, and are not strictly like those of other hystricomorphs (Hartenberger, 1973). Most theridomyids and glirids and all studied anomalurids, geomyoids and Myodonta have uniserial incisor enamel. As presently understood, enlarged infraorbital foramina and uniserial enamel are derived character states that unite all of these groups.

These groups may form a clade separate from ctenodactyloids plus hystricognaths (see above), but if so their common origin must occur early in rodent evolution, i.e. early Eocene. Inflation of the infraorbital foramen could be primitive for Myomorpha, and modified in Geomyoidea. Uniserial enamel occurs in some other rodents, notably sciurids and cylindrodontids, but Boyde (1978) suggests that the enamel microstructure of sciurids and muroids may differ fundamentally. Clearly, resolution of infraorbital, enamel and other character states must be refined before those traits can be used to formally define monophyletic groups. Studies like those of Vianey-Liaud, Hartenberger, and Boyde show that such structural categories may be composites of independently derived traits.

In summary, Myomorpha include at least the closely related Muroidea plus Dipodoidea (Myodonta) and the Geomyoidea (Carleton, 1984). Inclusion of Gliridae or other groups in the suborder requires further character analysis. Within Myomorpha, Muroidea form a monophyletic group by absence of P^4. The derived cheek tooth formula 3/3 occurs convergently in some other rodents (e.g. the geomyoid *Diplolophus* and some dipodoids). Other molar and zygomatic traits characterize Muroidea, but appear after loss of P^4 (see Fig. 1). Muroids are derived in presence of a distinct anterocone on M^1 and anteroconid on M_1 (see Emry, 1981) and in confluence of the masseteric and neurovascular foramina of the anterior zygoma.

EARLY MUROID FOSSILS

The earliest rodent with cheek tooth formula known to be 3/3 is the genus *Simimys* Wilson, 1935, from late Eocene deposits in southern California (Wilson, 1935 and 1949; Lindsay, 1968 and 1977; Lillegraven and Wilson, 1975), and questionably from early Oligocene deposits of Texas (Wood, 1980). These authors and Emry (1981) detail the unsettled taxonomic history of *Simimys*, which continues. Emry (1981) considers *Simimys* a muroid that is not assignable to any living family. Based on molar morphology, Wood (1980, see also Vianey-Liaud, this volume) concludes that *Simimys* is a dipodoid. The anterocone of *Simimys* M^1 is small, low, and lingual; the anteroconid of *Simimys* M_1 is small, slightly labial, and usually isolated, unlike later muroids. However, this cusp morphology is primitive for both muroids and dipodoids, while tooth formula is an apomorphy uniting *Simimys* with Muroidea.

Simimys is hystricomorphous, a primitive trait shared with dipodoids. A small neurovascular foramen next to a large masseteric infraorbital foramen is a common feature in dipodoids, absent in known unequivocal muroids, and present in *Simimys*. This condition may be primitive (Emry, 1981), but if derived would argue for transferal of *Simimys* to Dipodoidea. *Simimys* is peculiar in having a very large neurovascular foramen and a thick bar separating the latter from the infraorbital foramen (Emry, 1981). On the other hand, Lindsay (in Emry, 1981, p. 9) suggested that a third small foramen in *Simimys* is the neurovascular and that Emry's remarkably large neurovascular foramen is really the infraorbital. Furthermore, Emry's masseteric "foramen" in *Simimys* would be the anterior part of the orbit.

Nonomys (including *Subsumus* Wood, 1974) from the early Oligocene of North America, also has cheek tooth formula 3/3. It has a buccal anterocone on M^1 and a prominent anteroconid on M_1 (Emry, 1981). All of these traits support its allocation to Muroidea. Like *Simimys*, *Nonomys* is hystricomorphous and has two large openings that Emry (1981) interprets as neurovascular and infraorbital foramina. Lindsay interprets these openings as the infraorbital foramen and the orbit.

Nonomys is peculiar in lacking well developed transverse lophs and in possessing prominent, cuspidate cingula (labial in lower molars, lingual in upper molars) on its cheek teeth. Nonomys and Simimys are best considered Muroidea, incertae sedis. They are muroids based on cheek tooth formula 3/3, but they retain primitive infraorbital morphology and Simimys retains primitive tooth crown pattern. They show no clear relationship to each other or to living taxa.

Aside from Simimys simplex, the only other Eocene fossils occasionally referred to the Muroidea come from the Chai-li Member, He-ti Formation of Shansi Province, China (Li and Ting, 1983). Hartenberge et al. (1975) argue for their late Eocene age. Zdansky (1930) figure and described two isolated rodent molars, a right M_2 (holotype, M3434 of the Uppsala collection) and a left M_3. Zdansky (1930) named these Cricetodon schaubi, creating a new species for a cricetid genus known primarily from Neogene deposits of Europe. Another specimen from the Uppsala collection that Zdansky (1930, p. 10) thought might be a leptictid M_3, is a rodent left mandibular fragment with incisor and one anterior cheek tooth. Vianey-Liaud (this volume; see her figures) shows that M3434 is not a cricetid of modern grade and thus is not referable to Cricetodon. She refers the species based on M3434 to Eucricetodon, a generic assignment not demonstrable with this single M_2. Vianey-Liaud notes the dipodoid (primitive) affinity of the anterior cheek tooth that Zdansky (1930) considered a "?leptictid" and identifies it "? Parasminthus".

That both teeth could represent the same dipodoid or eomyid species should be considered. The anterior cheek tooth has an outline and minute anterior cusp that resemble P_4 of Namatomys fantasma (a late Eocene eomyid; Lindsay, 1968) and conceivably some morphs of M_1 of Simimys. M3434 agrees closely with M_1 of Namatomys lloydi (early Oligocene; Black, 1965). The Shansi specimens are somewhat larger than the holotype of N. lloydi. Unlike Namatomys, M3434 lacks a labial anterolophid extension and both teeth are longer than wide, which is typical for Myodonta. The zygoma of Namatomys is unknown, but later eomyids are sciuromorphous. Based on the present sample, the Shansi material is referable to Myomorpha, possibly with primiti muroid affinity for M3434, but familial assignments are tentative.

The earliest taxa unquestionably referable to the Muroidea are early Oligocene in age and occur in Chadronian deposits of North America. They are rare and are referred to Eumys elegans or a close ly allied species (Wood, 1980; see Martin, 1980, p. 17 and Korth, 1981, p. 297 for varying opinions). Eumys is part of an Oligocene radiation of primitive muroids in North America, but is derived in having a ventrally constricted infraorbital foramen with strongly inclined anterior plate. Martin (1980) assigned some North America Oligocene muroids that are primitive in having wide infraorbital fo amina to new genera Coloradoeumys, which is derived in its shortene snout, and Eoeumys. The two species Martin (1980) placed in Eoeumy are united by characters that are primitive for eumyids (e.g. buccal position of the anterocone, double saggital crest, pinnately ridged incisor). On the other hand, Martin (1980) found traits that link "Eoeumys" exiguus to Scottimus and "Eoeumys" vetus to Leidymys, where Dawson and Black (1970) and Wood (1980) placed them. Korth (1981) most recently evaluated this problem, removing all but one specimen from the genus Eoeumys. Wilsoneumys, Scottimus, and Leidymys are advanced derivatives of early North American muroids; none belong to Eurasian lineages (contra Martin, 1980). Oligocene and most early Miocene North American muroids comprise a clade distinct from Old World muroids.

Muroids are not certainly known from the early Oligocene of Asia, but they are diverse in middle Oligocene faunas of Mongolia. Specimens referred to Eucricetodon asiaticus (Mellett, 1968; Kowalski, 1974; Lindsay, 1978) include at least three species and perhaps more than one genus. Lindsay (1978) surveyed cranial and dental characters of these samples, showing that E. asiaticus may be a primitive morphotype for Miocene cricetids. Cricetops dormitor is primitive in its hystricomorphous zygomatic structure but advanced in dentition (Wahlert, 1984b). Selenomys mimicus is a muroid; its hystricomorphy is plesiomorphous. Species assigned to Tachyoryctoides (including Aralomys) are primitive muroids unrelated to later Rhizomyidae. Muroidea probably were present and perhaps originated in Asia earlier than the middle Oligocene, with this middle Oligocene radiation resulting from autochthonous evolution.

In contrast, the European fossil record shows that muroids are absent in the early Oligocene, although four genera occur in the middle Oligocene (Pseudocricetodon, Paracricetodon, Eucricetodon, and Heterocricetodon). This apparently sudden immigration is a characteristic of the Grand Coupure.

THE CRICETIDAE: AVOIDING PARAPHYLY

Most classifications of muroids are paraphyletic to some degree; i.e. recognized groups apparently have a single ancestor but these groups do not include all of their descendants. As knowledge increases, systematists reduce paraphyly by splitting major taxa. Hence, Chaline and Mein (1979) relegate nesomyids, dendromurids, and cricetomyids to familial status equal to that of cricetids and murids. They include supposed descendant groups as subfamilies of families.

A major systematic problem in Muroidea is the content of the Cricetidae. The bulk of living taxa assigned to this family probably form a group with relatively close common ancestry. However, certain living taxa (see Carleton and Musser, 1984) and many middle Tertiary genera lie outside this array and in our view should be placed in separate families. Carleton and Musser (1984) recognize many of these groups, but elect not to assign them to families, instead giving them all equal subfamilial status, pending new data. Efforts toward forming an evolutionary classification can be continued by identifying the paleontological appearance of derived character states (see below).

Myomorphy

True myomorphy is probably the most distinctive apomorphy of advanced muroids. Evolution of myomorphy from hystricomorphy is supported by studies of Wilson (1949) and Klingener (1964). Lindsay (1977) suggested three stages in the transformation of the muroid zygoma. These stages are: (1) enlargement of the infraorbital foramen coincident with anterior expansion of the medial masseter onto the rostrum anterior to the infraorbital foramen; (2) expansion of the anterior, ventral surface of the zygoma coincident with expansion of the lateral masseter; and (3) vertical inclination of the anterior plate of the zygoma, with constriction of the lower part of the infraorbital foramen as the lateral masseter expands upward lateral to the infraorbital foramen. The process was virtually completed during the Oligocene. Vianey-Liaud (1974; this volume) documented incomplete myomorphy in European species of *Eucricetodon* and *Pseudocricetodon*. In both genera the infraorbital foramen is wide and the anterior plate of the zygoma is nearly flat. The anterior plate is more inclined with the infraorbital foramen slightly constricted ventrally in *Eucricetodon*, relative to *Pseudocricetodon*.

Lindsay (1977) uses inclination of the anterior plate of the zygoma as a measure of the "degree" of myomorphy (Fig. 2). Increase in the angle of plate inclination indicates dorsal migration of the masseter and stronger myomorphy. Hystricomorphous rodents have negative angles (e.g., *Simimys* = -28°; *Cricetops* = -7°), primitive myomorphous rodents have angles below about 40° (e.g., *Pseudocricetodon* = 10°, *Eucricetodon* = 29-39°, *Eumyarion* = 35°), and advanced myomorphou rodents have angles greater than 40° (e.g., *Eumys* = 53°, *Democricetodon* = 73°, *Copemys* = 76°).

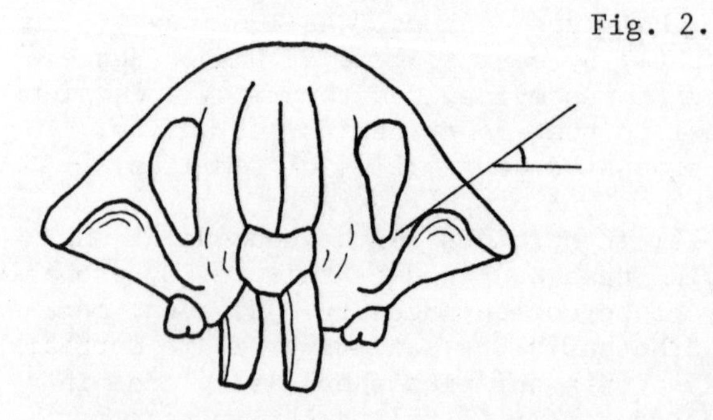

Fig. 2. Angle of inclination of the anterior plate of the muroi zygoma, determined by a horizontal plane and a line passing through the ventral margin of the infraorbital foramen and the dorsal limi of the anterior plate.

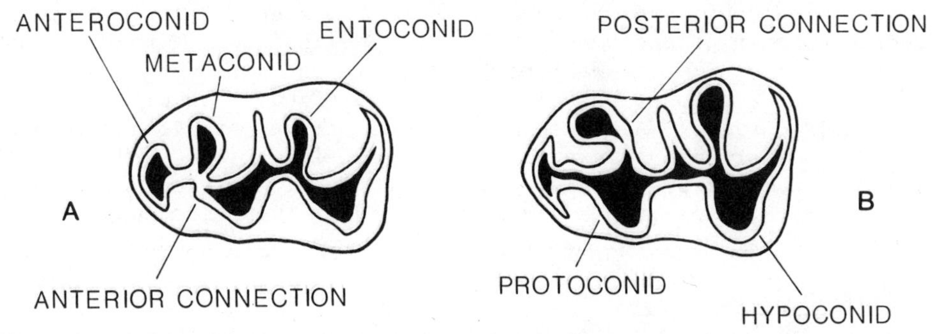

Fig. 3. Lower first molar crown patterns in two idealized muroids, indicating the derived anterior connection (A) and primitive posterior connection (B) between the protoconid and the metaconid.

Lower First Molar

Primitively in the lower first molar, the posterior arm of the protoconid joins the metaconid (Fig. 3; Schaub, 1925). This "alte Vorjochkante" is modified in later forms in which the metaconid joins the anterior arm of the protoconid ("neue Vorjochkante"). Some muroids have both a weak anterior and posterior connection *and* a mesolophid. In other muroids, the posterior connection takes the place of the mesolophid when the anterior connection develops. The apomorphic condition is the anterior protoconid-metaconid connection on M_1. This derived trait occurs in high frequency in and unites *Cricetodon*, *Megacricetodon*, *Democricetodon*, *Fahlbuschia*, *Lartetomys*, *Spanocricetodon*, and most later muroids. Some unworn *Anomalomys* show an anterior connection that may be a homologue. *Adelomyarion* lacks both anterior and posterior connections. M_1 of some variants of *Eumyarion* and of some species of *Eucricetodon* (see Thaler, 1966;, Engesser, 1979; Brunet, 1979) have a weak anterior connection and retain a posterior protoconid arm that joins both metaconid and mesolophid. Perhaps these show incomplete development of the apomorphic condition. A dentary (Fig. 4) from the early Miocene site Legetet, Kenya, (attributed to *Protarsomys macinnesi*) exhibits low anterior junction of the metaconid and protoconid. It is similar to *Lartetomys* in having a small anteroconid on M_1 and a reduced M_3.

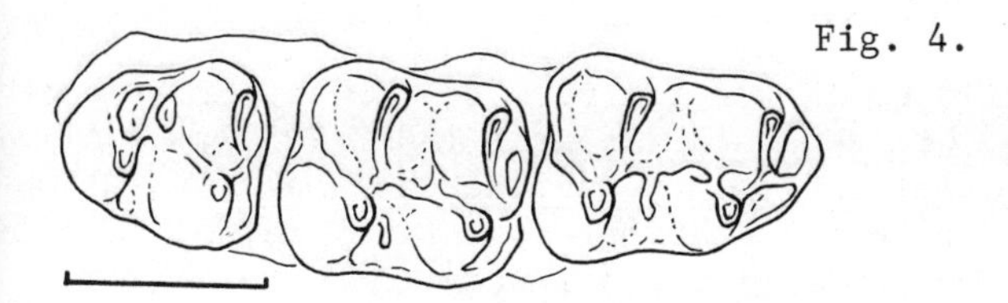

Fig. 4. Crown view of molars of KNM-LG 2188, a dentary in the National Museums of Kenya collections referred to *Protarsomys macinnesi*. Bar = 1 mm.

Excluded from the presumed clade defined by anterior connection of the protoconid and metaconid are most Oligocene Holarctic muroids, including eumyids and some later forms: Melissiodon, afrocricetodontines, and Lophiomys. Wahlert (1984b) argues that Lophiomys may be related to Cricetops, but his association of hamsters with these two genera is dubious. At least some of the derived traits hamsters and Lophiomys share probably evolved in parallel (Wahlert, 1984b, p. 12). Others (anterior protoconid-metaconid connection, strong myomorphy) support derivation from advanced Cricetidae. Opposite cusps, double anterocones and lack of mesoloph(-id) ally Megacricetodon to hamsters.

Incisor Ornamentation

A variety of ridge and groove ornamentation may occur on rodent incisors. Martin (1980) documented an array of ornamentations on lower incisor enamel of Oligocene muroids from North America. Double longitudinal ridges, laterally offset from the midpoint of the lower incisor (Fig. 5), occur in many mid-Tertiary Old World muroids. The antecedent to this condition is unknown, but nonridged eomyids, zapodids, and early muroids suggest that smooth enamel is plesiomorphous. Double ridged enamel then is derived, and is also preceded by pinnate

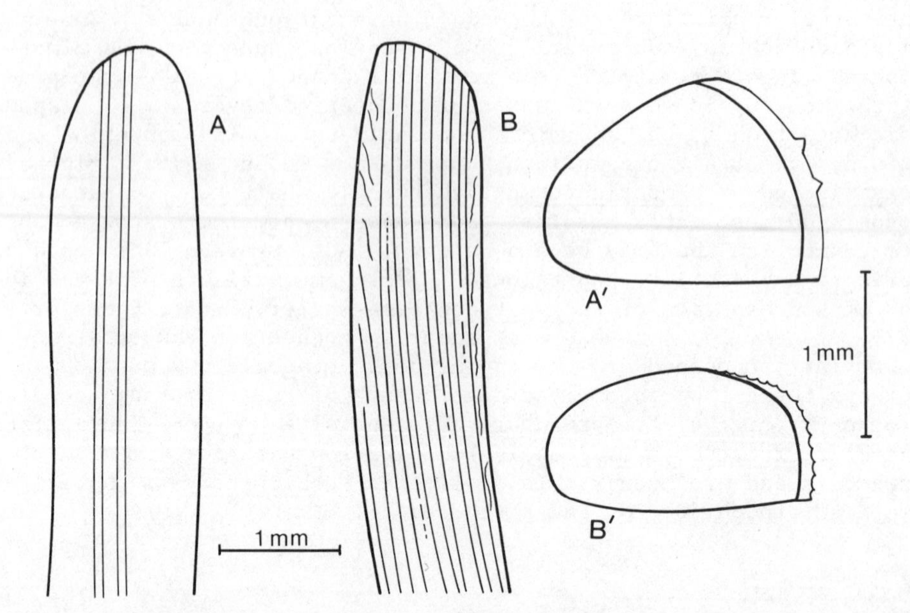

Fig. 5. Lower incisor ridges in ventral view (A, B) and cross section (A', B'). Cricetodon albanensis (AMNH 99557, A, A') has double parallel ridges. Paracricetodon sp. (AMNH 11068 B, B') has pinnate ornamentation, many fine, raised parallel lines, some of which terminate abruptly.

ridging, which occurs in Paracricetodon, Pseudocricetodon, Adelomyarion, some eumyids and some species of Eucricetodon. Other Eucricetodon have fewer ridges. Eucricetodon quercyi has at least three lateral ridges and AMNH 98413, Eucricetodon cf. collatus has two. Eucricetodon asiaticus exhibits variation of two or three ridges. Rodents with double ridged incisors may have evolved from ones with greater numbers of ridges.

As a fixed trait, the presence of two longitudinal ridges displaced laterally from the midline of the lower incisor unites Cricetodon, Eumyarion, Anomalomys, Prospalax (although widely spaced), Democricetodon (from Daud Khel, Siwaliks) and the three early Miocene east African genera, Notocricetodon, Afrocricetodon, and Protarsomys. Two ridges are also present in Kanisamys indicus, an early rhizomyid, but these are reduced to one in tachyoryctines or lost in rhizomyines. There is one in Ruscinomys and none in Copemys, Megacricetodon, Myocricetodon, Spanocricetodon, and most species of Democricetodon. Double ridge incisor ornamentation appeared early as a derived trait in an Oligocene clade of muroids that was developing myomorphy and advanced M_1 morphology (Fig. 1). It helps to identify early members of that clade, but these ridges were lost independently in several subgroups.

Definition of Cricetidae

Derived incisor morphology, zygomatic structure, and molar loph pattern (Fig. 1) collectively define a restricted group of muroids that evolved during the late Oligocene, radiated during the Neogene, and was the source of most of the surviving diverse Muroidea. This clade should be recognized in classification as Cricetidae sensu stricto. Most Oligocene muroids are excluded from the family. Those taxa with full myomorphy and a strong anterior connection between the protoconid and metaconid on M_1 (Fig. 1) should be assigned with certainty to Cricetidae. Some other taxa in Figure 1 with double ridged incisors but less derived zygoma and M_1 (Eumyarion and some species of Eucricetodon) probably can be placed in Cricetidae with justification. Others (Kanisamys and Anomalomys, see below) are assigned to other monophyletic families. North American Oligocene muroids are considered an endemic family Eumyidae; Eumys is enigmatic in its developed myomorphy and deserves further attention as possibly antecedent to Cricetidae sensu stricto or convergent with them. We do not attempt comprehensive allocation of Eurasian Oligocene genera to families.

SPECIAL PROBLEMS IN MUROID RADIATION

Once true myomorphy evolved, further muroid radiation occurred during the Miocene, including the origin of hamsters, murids, gerbils, bamboo rats, dendromurids, cricetomyids, voles, and pack rats. The cladistic relationships of this complex radiation are difficult to sort out because it involves the vicariant diversification of widespread ancestral stocks. In the late Miocene and Pliocene, evolution significant at the subfamily level occurred in several groups, including for example, Otomyinae as an African subfamily of the Muridae. These lower level systematic groups, most of which are extant, possess unique distinctive features (see Carleton and Musser, 1984).

Origin of the Muridae

The evolution of the rodent family Muridae (*sensu stricto*) is of particular significance considering the abundance, diversity, and economic impact of wild species, and the relevance to laboratory research of various strains of *Rattus* and *Mus*. Jacobs (1977, 1978, 1979) considered the most primitive murid genus to be *Antemus* from the Miocene Chinji Formation, Pakistan. This was also the oldest known murid, and indicated that the divergence of *Rattus* and *Mus* occurred later than about 14 million years ago (Jacobs and Pilbeam, 1980). Jacobs did not accept the hypotheses of Petter (1966) and Lavocat (1967), which proposed a seriation of various living or fossil genera, including *Myocricetodon*, *Dendromus*, and *Cricetomys*, as representing the stages through which murids evolved. Rather *Antemus* was thought to derive from a south Asian genus, then undiscovered, "perhaps somewhat similar to *Megacricetodon*" (Jacobs, 1977, p. 9).

Wood (1980, p. 51) considered *Antemus* the "most cricetid-like murid yet known," but because it was "an unquestionable murid" he apparently did not accept *Antemus* as direct evidence of murids being derived from cricetids. Brandy (1981), on the other hand, did not consider *Antemus* to be a murid. He correctly pointed out that inclination of cusps is an evolutionary trend in the Muridae, while *Antemus* has relatively vertical cusps. He considered *Antemus* to be a dendromurid although dendromurids also have inclined cusps.

In 1981 Brandy listed three Miocene localities with undescribed dendromurids: Bou Hanifia, Algeria; Ngorora, Kenya; and Ch'orora, Ethiopia. Recently Ameur (1984) studied the Bou Hanifia dendromurid, *Senuoussimys*, and concluded that Dendromuridae, Muridae, and Cricetomyidae have close common ancestry. Ameur (1984, p. 173) placed *Antemus* near dendromurid ancestry rather than murid ancestry, although she showed that *Antemus* shares the t_1-t_2 (anterostyle-lingual anterocone) connection with Muridae. Our interpretation of the origin of Muridae is influenced by several specimens from Ngorora in the National Museums of Kenya and by larger collections from the lower Siwaliks (Fig. 6). Relationships between major groups are summarized in Figure 7.

An additional argument raised by Brandy (1981) against the murid affinities of *Antemus* was its age relative to that of *Progonomys*, the genus that Brandy considered the oldest "true" murid. *Progonomys*

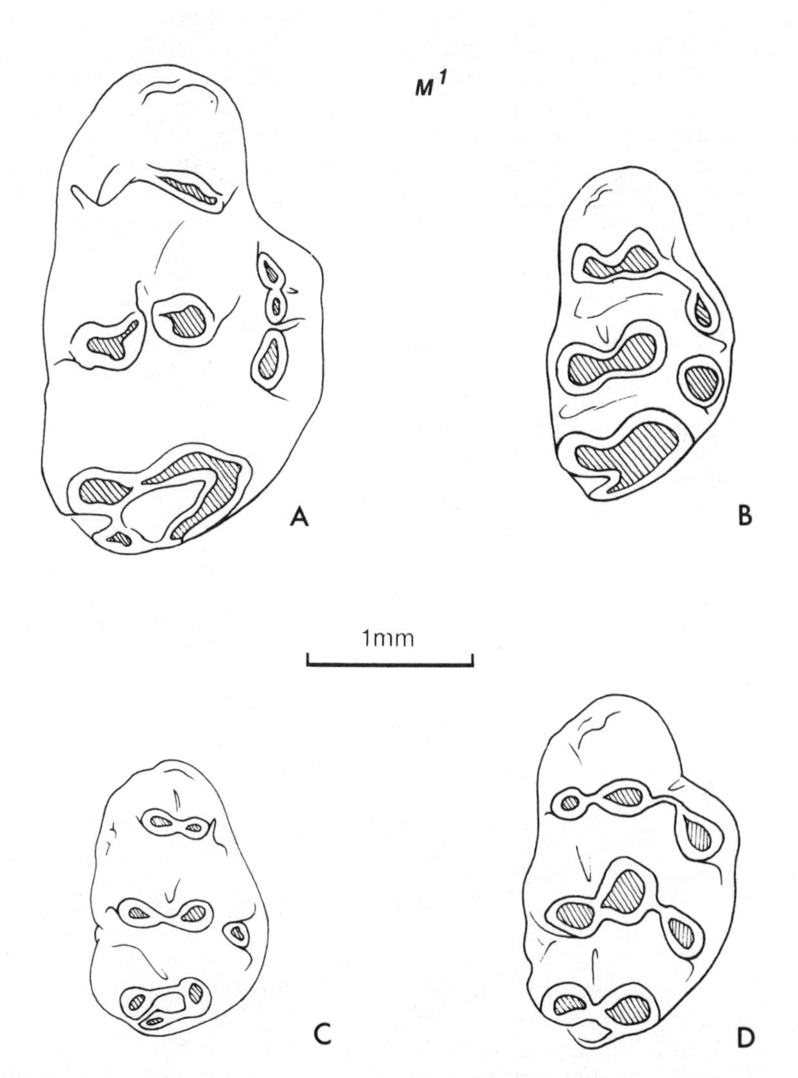

Fig. 6. (A) undescribed cricetomyid from the Ngorora Formation (Fm.), Kenya. Note weakly bilobed anterocone and lack of connection to lingual cusp. (B) Antemus chinjiensis, Chinji Fm., Pakistan, at a relatively greater stage of wear than other teeth illustrated here. Note the unconnected enterostyle and protocone. (C) "Antemus" primitivus, Chinji Fm., Pakistan, after Wessells et al. (1982), considered here a primitive dendromurid rather than a murid. (D) Progonomys sp., upper part of the Chinji Fm., Pakistan, probably 11-12 m.y. old.

occurred about 11-12 million years ago (Brandy, 1981; Ameur et al., 1976) at Bou Hanifia, with dendromurids. Brandy indicated an age of 16-17 m.y. for Antemus and implied (incorrectly) that Jacobs suggested or accepted that age for YGSP 41, the type locality (see Jacobs and Pilbeam, 1980, and references therein). Nevertheless, Brandy rejected a relationship between Antemus and Progonomys because there was inadequate time (roughly four m.y. minimum estimate by his figures) to derive Progonomys from Antemus. ("L'écart chronologique paraît court entre YGSP 41 et Bou Hanifia pour avoir permis l'indivi-

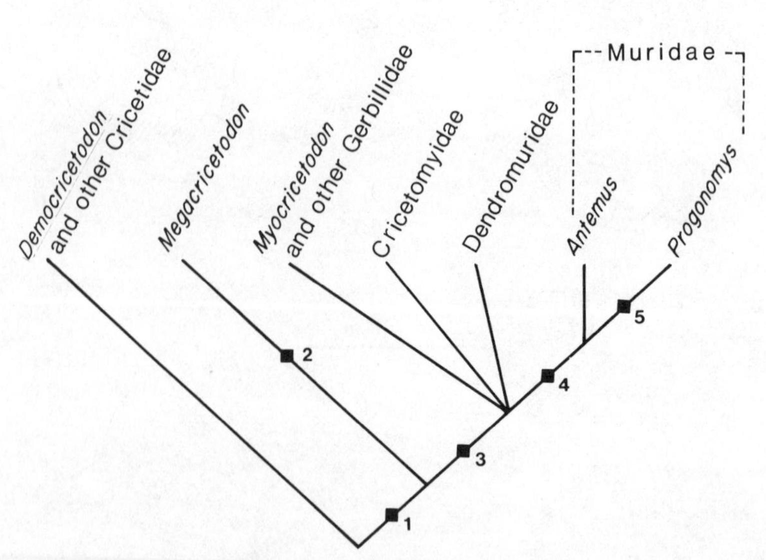

Fig. 7. Proposed relationship of Megacricetodon, gerbils, cricetomyids, dendromurids, and murids. Democricetodon and other cricetids constitute the outgroup. Numbers correspond to acquisition of the following derived characters. (1) bilobed anterocone on M^1 variably developed among members of species. (2) bilobed anterocone stabilized in population; based on this trait Megacricetodon may form a monophyletic group with other muroids (see text). (3) two traits: reduction of longitudinal crests on molars (completely lost in many species retained to varying extent in others, as in Pseudomeriones and Delanymys); presence of a lingual cusp on M^{1-2}. Four groups have these traits, but the tetrachotomy can be reduced to a trichotomy by redefinition of Myocricetodon so that no gerbils have extra lingual cusps (see text). (4) two lingual cusps form two chevrons in M^1: the anterostyle (t_1) is posterolingual to the lingual anterocone (t_2) and joins it in all known species; the enterostyle (t_4) is posterolingual to the protocone (t_5). (5) the enterostyle joins the protocone

dualisation d'un Progonomys typique a partir d'Antemus," Brandy, 1981, p. 154). This argument assumed an undocumented and probably unnecessarily slow and constant rate of evolution. We still consider Antemus to be the oldest known murid and we outline below our hypothesis of its relationship to myocricetodontines, dendromurids and cricetomyids.

Antemus is a murid because it possesses the defining character of the family (4 on Fig. 7): upper first molars with three cusps in the anterior two transverse chevrons. Position of the cusps is important; the anterostyle (t_1) is posterolingual to the lingual anterocone (t_2) and the enterostyle (t_4) is posterolingual to the protocone (t_5). More derived murids may have this pattern modified, but the fundamental arrangement is unambiguous. It is true that Antemus has less inclined cusps than Progonomys or other murids, but the degree of inclination of cusps seen in Antemus is primitive for murids (and dendromurids), whereas the association of an additional cusp in each of the two anterior chevrons of M^1 is derived. Antemus is a murid based on the synapomorphy of cusp arrangement in chevrons; it is not necessarily a dendromurid because of the symplesiomorphy of slight cusp inclination. Dendromuridae have a single lingual cusp; definition of that family would be overextended by inclusion of Antemus. Dendromuridae are one of several diverse middle Miocene clades that exhibit lingual development of cusps on upper cheek teeth.

Jacobs (1977) named Antemus on the basis of four isolated teeth. Subsequently, larger samples have been collected from Siwalik deposits that range from older to slightly younger than the type locality for Antemus. Progonomys is now known from locality 634 and higher in the upper third of the Chinji Formation, directly superposed about 236m over sites yielding Antemus (Fig. 6B, D). Progonomys specimens from locality 634 are similar to Antemus in arrangement of lingual cusps, cingulum on the anterior border of the tooth, and roundness of cusps. These specimens are assigned to Progonomys because the enterostyle is connected to the protocone (5 on Fig. 7). Localities 41 and 634 are considered about 13 and 11-12 m.y. old, respectively (Barry et al., ms.). Thus, Progonomys has an antiquity in the Pakistan Siwaliks comparable to its earliest records elsewhere. Further collecting and study of murids from the Chinji Formation will test the morphological continuity between the dentitions of Antemus and Progonomys.

In addition to murids, living African cricetomyids and dendromurids have upper molars expanded by the addition of lingual cusps. Cricetomys and other cricetomyids are distinguished from dendromurids, among other characters, by having (1) usually two or more lingual cusps situated opposite and posterior to the protocone (derived) without connection to the anterocone, (2) broad anterocones that are usually less strongly bilobed relative to the condition seen in dendromurids and (3) a large posterior cingulum usually present. Modern dendromurids (Dendromus, Megadendromus, Deomys and others) usually have strongly bilobed anterocones and inclined cusps (derived condition). Dendromurids have a single lingual cusp associated with the protocone, derived with respect to gerbils, but primitive with respect to murids. Two undescribed taxa from the Ngorora Formation, Kenya, dated about 12.5 m. y. old (Hill et al., ms.), are considered Cricetomyidae based on morphology of the anterocone, addition of lingual cusps, and size (Fig. 6A). The Miocene record of muroids with strongly bilobed anterocones and lingually expanded upper molars is diverse and is known mainly from Africa and south Asia. This group includes Antemus, dendromurids and myocricetodontines, many of which are undescribed. Some of these are difficult to assign to extant families, but pending systematic revision of those families, may be considered "primitive dendromurids." This revision entails accommodating early taxa that lack the derived molar morphology seen in Dendromus or Deomys and determining how Petromyscus and Delanymys relate to each other and to cricetomyids and dendromurids, questions raised by Carleton and Musser (1984, p. 332).

Myocricetodontinae Lavocat, 1961, were redefined by Jaeger (1977 a) and are characterized by weak longitudinal cusp connections, among other traits. Presence of lingual cusps on M^1 and the bilobed condition of the anterocone are highly variable in Myocricetodontinae, both within and between species (Jaeger, 1977b). Myocricetodon as presently conceived includes species that lack and some that have a lingual cusp on upper molars. Jaeger (1977b) showed that those Myocricetodon lacking lingual cusps lie near gerbil origin (see remarks by Carleton and Musser, 1984, p. 325, 332). Those Myocricetodon with a lingual cusp (and Dakkamys) might better be considered "primitive dendromurids". Other primitive dendromurids include "Antemus" primitivus Wessels et al. (1982), several undescribed lower Siwalik taxa, and perhaps a species from Ngorora. Shamalina from the Miocene of Saudi Arabia (Whybrow et al., 1982) and some undescribed lower Siwalik taxa have bilobed anterocones and lingual cingula, but no lingual cusps (cf. Jacobs, 1977, p. 8). Primitive dendromurids show reduction of longitudinal crests and addition of a lingual cusp on M^{1-2} (3 on Fig. 7), but have no apparent autapomorphies of cricetomyids, dendromurids, or murids. Redefinition of these groups will clarify assignments. One step in this direction would be restriction of Myocricetodontinae to those forms lacking lingual cusps, inclusion of that subfamily in the Gerbillidae, and separation of Cricetomyidae, Dendromuridae, and Muridae from Gerbillidae by acquisition of lingual cusps. Thus, the tetrachotomy is reduced to an unresolved trichotomy.

The bilobed anterocone (1 on Fig. 7) occurs earliest in Megacricetodon and is derived with respect to Democricetodon. However, the bilobed anterocone is a stable structure in all Megacricetodon including the earliest species, which appear during the early Miocene (2 on Fig. 7). A bilobed anterocone appears to be primitive for those muroids that add lingual cusps during the middle Miocene. Cricetomyids have weakly bilobed anterocones, myocricetodontines are variable, and Siwalik muroids with lingual cingula are strongly bilobed. Antemus shares a strongly bilobed anterocone with African and Siwalik primitive dendromurids.

Antemus occurs in the Siwaliks at least 13 m. y. ago. It is stratigraphically succeeded in the upper Chinji Formation by primitive species of Progonomys. It is stratigrapically preceded by muroids, which for want of a better term, are called "primitive dendromurids". Antemus is known in the Siwaliks when cricetomyids and "primitive dendromurids", but not murids, are known from the Ngorora Formation of Kenya. Myocricetodontines and perhaps "primitive dendromurids" are known from Africa (Lavocat, 1961; Jaeger, 1977a) and Asia preceding the first known record of Antemus. Modern murids, cricetomyids, and dendromurids may all be vicariant siblings derived from widely distributed "primitive dendromurids". Murids probably did not pass through a cricetomyid grade of evolution because the isolated anterior lingual cusp of cricetomyids is associated with the protocone rather than the anterocone. Nor did they pass through a modern dendromurid grade.

This outline of murid origin is easily reconciled with the hypotheses of Petter (1966) and Lavocat (1967). Finally, the origin of murids as presented here is consistent with Simpson's Rule (see Simpson, 1971; Jacobs 1979), which states that a taxon cannot occur earlier in time than its ancestor. However, consistency of the fossil record with Simpson's Rule is not necessarily proof of the accuracy of our interpretation of murid origins. We expect that further collecting will provide tests for these ideas.

Independent Origins of Rhizomyidae and Spalacidae

Under ideal conditions, fossils can be a powerful source of evidence bearing on phyletic relationship. Such conditions prevail for the muroid families Rhizomyidae and Spalacidae. Strong similarity of the bamboo rats and blind mole rats results from their common muroid heritage and from convergence to a fossorial mode of life. Consequently they have been classified in the same family, but are known now to be distinct muroid derivatives.

Rhizomyidae originated as part of the early Neogene diversification of myomorphous muroids. Early rhizomyids (Kanisamys and Prokanisamys) exhibit true myomorphy and a double ridge on the lower incisor, but neither the anterior nor the posterior protoconid-metaconid connection on M_1 are well developed (Flynn, 1982). Crown patterns of early rhizomyids are derivable from a primitive cricetid morphotype and resemble Eumyarion in both structure and relative tooth size. Apparently a form like Prokanisamys de Bruijn et al. (1981) split from advanced muroid stock before the anterior connection became a stable structure. Early rhizomyids have a weakly bilobed anterocone on M^1. Siwalik fossil rhizomyids provide an outline for the evolution of Asian Rhizomys. Flynn (1982) shows that rhizomyids began to diversify in Asia before strong burrowing adaptations appeared. They may have radiated initially into a dietary niche and later exploited a burrowing habitus. Asian ancestry for African Tachyoryctes is supported by Siwalik fossils (Flynn, 1982), but that hypothesis is less certain than is the origin of Rhizomys (Flynn and Sabatier, 1984) and should be tested.

Spalacidae lack a well developed metalophid on M_1 and the anterior zygomatic plate is not inclined strongly. These traits indicate that spalacids evolved from hystricomorphous muroids; that is, spalacids evolved from more primitive muroids than did rhizomyids. Reduced mesolophids on M_{1-3} is a spalacid trait. Fejfar (1972) included early Miocene to Pliocene Anomalomys in Spalacidae and postulated that Oligocene Tachyoryctoidinae (Tachyoryctoides, Argyromys, Aralomys) could be ancestral to some later spalacids. Tachyoryctoidines are certainly not rhizomyids; their derived dental traits (short mesolophids and mesolophs, narrow lophs) and the flaring masseteric crest of the mandible preclude that relationship. Pending results of ongoing research by Hans de Bruijn (Utrecht, Netherlands) and Engin Unay (Ankara, Turkey), Tachyoryctoidinae and Anomalomyinae are tentatively considered subfamilies of Spalacidae.

Spalacidae and Rhizomyidae originate separately from different muroids. Rhizomyids certainly are present by about 16 m.y. ago in Pakistan; their split from sister muroids could be as great as 20 m.y. ago. Spalacidae certainly occur in the early Miocene and their origin probably antedates 25 m.y. ago; Oligocene tachyoryctoidines may indicate a much earlier origin for Spalacidae.

South American Cricetids

Living cricetid rodents in South America are diverse and comprise a great number of species. They are generally distinct from their living North American relatives in possessing complex penile morphology and separate suites of parasites. South American cricetids as a whole may not constitute a simple dichotomy with neotomine-peromyscines (Carleton, 1980; Rogers and Heske, 1984). The antiquity and origin of the various South American cricetids have long been hotly debated (see summaries in Simpson, 1980; Slaughter and Ubelaker, 1984; Jacobs and Lindsay, 1984; and also see the references therein), mainly because it is not known how long a time period is needed to achieve such diversity. Sarich and Cronin (1980), by utilizing molecular "clock" techniques, determined that cricetids first entered South America 60 to 40 m.y. ago. This is in direct conflict with the fossil record, which can demonstrate the presence of cricetids in South America no earlier than about 3.5 m.y. ago. The known fossil record shows that prior to 3.5 m.y. ago genera related to South American forms were present in North America, while noncricetid rodents were diverse in South America. The following summary of the fossil record relative to muroid radiation in South America is consistent with data from penile morphology and parasites and can be reconciled with the molecular "clock" (see Sarich and Cronin, 1983; Sarich, this volume).

The oldest known fossil that can be reasonably assigned to the group including South American cricetids is *Calomys* (*Bensonomys*) from White Cone, Arizona (Baskin, 1978, 1979), of Hemphillian age (late Miocene), roughly 6 m.y. old. Somewhat younger representatives of several South American cricetid groups (including phyllotines, akodontines, oryzomyines, and sigmodontines) are known from Kansas, New Mexico, Arizona, and Mexico at Hemphillian or Blancan (latest Miocene and Pliocene) localities. These records support a North American origin and initial diversification of South American cricetids by strict interpretation of the chronological sequence of the fossil record. Staging areas for diversification of cricetids prior to their spread through South America as proposed most recently by McKenna (1980) and Marshall (1979) are unnecessary, or at least constrained by the fossil record in North America.

Calomys (*Bensonomys*) is primitive relative to living *Calomys* (*Calomys*) in being lower crowned, and having M^1 with a strongly bifurcated anterocone, a short mesoloph, and a posteriorly directed lophule on the paracone. In all the characters listed above, *Calomys* (*Bensonomys*) is similar to Old World *Megacricetodon*. *Megacricetodon* occurs (among other places) in China in the Miocene Chetougou Formation (Qiu, Li, and Wang, 1981), older than the earliest known occurrence of *Calomys* (*Bensonomys*).

The cricetid fauna of North America from at least 16 m.y. ago to about 7 m.y. ago is remarkably uniform, comprising only *Copemys*-like forms. At about 7 m.y. the fauna becomes more diverse with the introduction of microtines and perhaps other rodent taxa from Asia. Possibly the ancestry of most South American cricetids can be traced through *Calomys* (*Bensonomys*) back to the Old World Miocene radiation of muroid rodents, an interesting hypothesis that deserves further investigation.

SUMMARY

The Myomorpha are a distinct clade that evolved by the late Eocene. Their relationship to other Eocene rodents may be indicated by their zygomatic structure. Fossils from Pakistan and India support arguments that ctenodactyloids plus hystricognaths form a monophyletic group. This clade in turn shares hystricomorphy with the earliest Myomorpha. Postulating sister group status for Myomorpha and ctenodactyloids plus hystricognaths invites testing by study of all organ systems as well as future fossil discoveries.

Muroidea are the heart of the Myomorpha and are clearly closely related to Dipodoidea. Hystricomorphy, separated neurovascular and infraorbital foramina, and the "cricetid plan" are primitive for these superfamilies. Muroids are distinguished by their derived tooth formula 3/3 and (except in *Simimys*) well developed anterocone and anteroconid on first molars. Geomyoidea are included in the Myomorpha on the strength of dental, skeletal, myological and fetal membrane evidence, but this is questioned by some researchers (e.g., Fahlbusch, this volume). Glirid affinities are less certain.

The internal systematics of the Muroidea are far from settled and no definitive revision of the superfamily is attempted here. Our fundamental proposal is that the appearance of derived features be used to define families. This leads to one major departure from other classifications. In defining Cricetidae by full myomorphy and by presence of an anterior protoconid-metaconid connection on M_1, Eocene and most Oligocene muroids are excluded from the taxon. That these derived traits are useful in systematics has long been recognized (Vianey-Liaud, 1974; Lindsay, 1977; Schaub, 1925; Engesser, 1979); they should be employed in defining higher taxa.

Mein and Freudenthal (1971) began analyzing the distribution of incisor enamel ornamentation. Incisors, like other teeth, display an array of characters, from deep grooves on upper incisors, to fine crenulations, to narrow longitudinal ridges on lower incisors. One such feature, a pair of ridges on lower incisors, is present in many late Oligocene and early to middle Miocene muroids. It is a derived trait that appeared before full myomorphy and the anterior lower molar connection. These and other apomorphies point to monophyletic groups when taken in combination.

As indicated by Figure 1, Cricetidae would include early Miocene *Cricetodon*, *Democricetodon*, *Megacricetodon* and their allies. *Eucricetodon* and *Eumyarion* exhibit partial myomorphy, but might be assigned to the family based on incisor and molar evidence. *Kanisamys* is myomorphous, but excluded from the Cricetidae because it lacks a strong anterior M_1 connection. Afrocricetodontines are excluded because this connection is generally poorly developed in them and because they show incomplete myomorphy. Spalacids lack the connection and are primitively hystricomorphous. They probably evolved during the Oligocene before rhizomyids (*Prokanisamys*), which appeared in the late early Miocene. Hamsters and voles are late cricetid derivatives. Various living American cricetid groups seem closely allied to *Megacricetodon*, to *Democricetodon* and *Copemys*, or perhaps to *Cricetodon*.

Muridae are one of the diverse groups that radiated once full myomorphy appeared. Several of these exhibit bilobed anterocones on upper molars, weakening of longitudinal molar crests, and lingual expansion of upper molars. All appeared by the middle Miocene and their relationships are difficult to sort out. Figure 7 is a working model meant to test relationships. The tetrachotomy can be resolved into dichotomies by reanalysis of the content of the major groups. For example, gerbils should be recognized as the sister group to those muroids with lingual cusps on upper molars. Among the latter group, Muridae M^1 are distinguished by having two lingual cusps that are posterolingual to the lingual anterocone and the protocone. A number of middle Miocene fossils show features intermediate between living gerbils, cricetomyids, dendromurids, and murids. They necessitate the recognition of traits that can be used to redefine higher taxa and call for testing of relationships, such as those hypothesized in Figure 7.

ACKNOWLEDGMENTS

We are grateful to our friend Will Downs, whose energies are largely responsible for the fine collections of Siwalik fossils. Richard E. Leakey and the National Museums of Kenya made several Kenyan fossils available for our study. We thank Guy G. Musser for his comments, Philip D. Gingerich for providing casts of material from the collections of Uppsala, Sweden, and Lorraine Meeker and Lisa Lomauro for preparing four of the figures. This work was partly supported by NSF grants DEB 8120988 to us and BN 8140818 to David Pilbeam.

REFERENCES

Ameur, R. 1984. Découverte de nouveaux rongeurs dans la formation miocène de Bou Hanifia (Algérie Occidentale). Géobios 17(2): 165-175.

Ameur, R., Jaeger, J. J., and Michaux, J. 1976. Radiometric age of early Hipparion fauna in North-west Africa. Nature 261(5555): 38-39.

Barry, J. C., Johnson, N. M., and Raza, S. M., and Jacobs, L. L. Manuscript. Mammalian faunal change in southern Asia: correlation to global patterns.

Baskin, J. A. 1978. Bensonomys, Calomys, and the origin of the phyllotine group of Neotropical cricetines (Rodentia: Cricetidae). J. Mammal. 59: 125-135.

Baskin, J. A. 1979. Small mammals of the Hemphillian age White Cone Local fauna, northeastern Arizona. J. Paleont. 53(3): 695-708.

Black, C. C. 1965. Fossil mammals from Montana. Part 2. Rodents from the early Oligocene Pipestone Springs Local Fauna. Ann. Carnegie Mus. 38: 1-48.

Boyde, A. 1978. Development of the structure of the enamel of the incisor teeth in the three classical subordinal groups of the Rodentia. In: Development, Function and Evolution of Teeth, P. M. Butler and K. A. Joysey, eds., pp. 43-58, Academic Press, London.

Brandy, L. D. 1981. Rongeurs muroïdes du Néogène supérieur d'Afghanistan. Evolution, biogéographie, corrélations. Palaeovertebrata 11(4): 133-179.

de Bruijn, H., Hussain, S. T., and Leinders, J. J. M. 1981. Fossil rodents from the Murree formation near Banda Daud Shah, Kohat, Pakistan. Proc. Konink. Nederl. Akad. Wetensch., Amsterdam, Ser. B 84(1): 71-99.

Brunet, M. 1979. Les Cricetidae (Rodentia, Mammalia) de la Milloque (Bassin d'Aquitaine): Horizon Repère de l'Oligocène supérieur. Géobios 12(5): 653-673.

Bugge, J. 1971. The cephalic arterial system in mole-rats (Spalacidae) bamboo rats (Rhizomyidae), jumping mice and jerboas (Dipodoidea) and dormice (Gliroidea) with special reference to the systematic classification of rodents. Acta Anat. 79: 165-180.

Carleton, M. D. 1980. Phylogenetic relationships in neotomine-peromyscine rodents (Muroidea) and a reappraisal of the dichotomy within New World Cricetinae. Mus. Zool. U. Mich. Misc. Pub. 157: 1-146.

Carleton, M. D. 1984. Introduction to rodents. In: Orders and Families of Recent Mammals of the World, S. Anderson and J. K. Jones, eds., pp. 255-265, John Wiley and Sons, New York.

Carleton, M. D. and Musser, G. G. 1984. Muroid rodents. In: Orders and Families of Recent Mammals of the World, S. Anderson and J. K. Jones, eds., pp. 289-379, John Wiley and Sons, New York.

Chaline, J. and Mein, P. 1979. Les Rongeurs et l'Evolution. Doin, Paris.

Chaline, J., Mein, P. and Petter, F. 1977. Les grandes lignes d'une classification évolutive des Muroïdes. Mammalia 41(3): 245-252.

Dawson, M. R. and Black, C. C. 1970. The North American cricetid rodent "Eumys" exiguus, once more. J. Paleont. 44(3): 524-526.

Emry, R. J. 1981. New material of the Oligocene muroid rodent Nonomys, and its bearing on muroid origins. Novitates 2712: 1-14.

Engesser, B. 1979. Relationships of some insectivores and rodents from the Miocene of North America. Bull. Carnegie Mus. Nat. Hist. 14: 1-68.

Fejfar, O. 1972. Ein neuer Vertreter der Gattung Anomalomys Gaillard, 1900 (Rodentia, Mammalia) aus dem europäischen Miozän (Karpat). N. Jb. Geol. Paläont., Abh. 141(2): 168-193.

Flynn, L. J. 1982. Systematic revision of Siwalik Rhizomyidae (Rodentia). Géobios 15: 327-389.

Flynn, L. J., Jacobs, L. L., and Cheema, I. U. Manuscript. Miocene Rodentia from Baluchistan, with a description of the new subfamily Baluchimyinae (Chapattimyidae).

Flynn, L. J. and Sabatier, M. 1984. A muroid rodent of Asian affinity from the Miocene of Kenya. J. Vert. Paleont. 3(3): 160-165.

Hartenberger, J.-L. 1971. Contribution à l'étude des genres Gliravus et Microparamys (Rodentia de l'Eocène d'Europe. Palaeovertebrata 4(4): 97-135.

Hartenberger, J.-L. 1973. Etude systématique des Theridomyoidea (Rodentia) de l'Eocène supérieur. Mém. Soc. Géol. France, nouv. ser. 52(1-5), Mém. 117: 1-76.

Hartenberger, J.-L. 1980. Données et hypothèses sur la radiation initiale des rongeurs. Palaeovertebrata, Mém. Jubil. R. Lavocat: 285-301.

Hartenberger, J.-L., Sudre, J., Vianey-Liaud, M. 1975. Les mammifères de l'Eocène supérieur de Chine (gisement de River Section) leur place dans l'histoire des faunes eurasiatiques. 3e R.A.S.T., Montpellier: 186.

Hill, J. E. 1937. Morphology of the pocket gopher mammalian genus Thomomys. U. Cal. Pub. Zool. 42(2): 81-172.

Hill, A., Drake, R., Tauxe, L., Monaghan, M., Barry, J. C., Behrensmeyer, A. K., Curtis, G., Jacobs, B. F., Jacobs, L., Johnson, N., and Pilbeam, D. Manuscript. Calibration of the middle and late Miocene of Kenya.

Hussain, S. T., de Bruijin, H., and Leinders, J. M. 1978. Middle Eocene rodents from the Kala Chitta Range (Punjab, Pakistan). Proc. Konink. Nederl. Akad. Wetensch., Amsterdam, Ser. B 81(1): 74-112.

Jacobs, L. L. 1977. A new genus of murid rodent from the Miocene of Pakistan and comments on the origin of Muridae. PaleoBios 25: 1-11.

Jacobs, L. L. 1978. Fossil rodents (Rhizomyidae and Muridae) from Neogene Siwalik deposits, Pakistan. Museum of Northern Arizona Press, Bull. Ser. 52: I-XI, 1-103.

Jacobs, L. L. 1979. Tooth cusp homology of murid rodents based on Miocene fossils from Pakistan. Casopis pro mineralogii a geologii 24(3): 301-304.

Jacobs, L. L., Cheema, I. U., and Shah, S. M. I. 1981. Zoogeographic implications of early Miocene rodents from the Bugti Beds, Baluchistan, Pakistan. Géobios 15(1): 101-103.

Jacobs, L. L., and Lindsay, E. H. 1984. Holarctic radiation of Neogene muroid rodents and the origin of South American cricetids. Jour. Vert. Pal. 4(2): 265-272.

Jacobs, L. L., and Pilbeam, D. 1980. Of mice and men: fossil-based divergence dates and molecular "clocks." J. Human Evolution 9: 551-555.

Jaeger, J. J. 1977a. Rongeurs (Mammalia, Rodentia) du Miocène de Beni-Mellal. Palaeovertebrata 7(4): 91-125.

Jaeger, J. J. 1977b. Les rongeurs du Miocène moyen et supérieur du Maghreb. Palaeovertebrata 8(1): 1-166.

Jaeger, J. J., Michaux, J., and Sabatier, M. 1980. Premières données sur les rongeurs de la formation de Ch'orora (Ethiopie) d'age Miocène supérieur. I: Thryonomyidés. Palaeovertebrata, Mém. Jubil. R. Lavocat: 365-374.

Klingener, D. 1964. The comparative myology of four dipodoid rodents (Genera Zapus, Napeozapus, Sicista, and Jaculus). Mus. Zool. U. Mich. Misc. Pub. 124: 1-100.

Klingener, D. 1984. Gliroid and dipodoid rodents. In: Orders and Families of Recent Mammals of the World, S. Anderson and J. K. Jones, eds., pp. 381-388, John Wiley and Sons, New York.

Korth, W. W. 1981. New Oligocene rodents from western North America. Ann. Carnegie Mus. 50(10): 289-318.

Korth, W. W. 1984. Earliest Tertiary evolution and radiation of rodents in North America. Bull. Carnegie Mus. Nat. Hist. 24: 1-71.

Kowalski, K. 1974. Middle Oligocene rodents from Mongolia. Results Polish-Mongolian Palaeont. Exped. 5, Palaeont. Polonica 30: 147-178.

Lavocat, R. 1961. Le gisement de vertébrés Miocènes de Beni Mellal (Maroc). Etude systématique de la faune de mammifères et conclusions générales. Notes et Mém. Serv. Géol. 155: 1-144.

Lavocat, R. 1967. A propos de la dentition des rongeurs et du problème de l'origine des Muridés. Mammalia 31: 205-216.

Lavocat, R. 1973. Les rongeurs du Miocène d'Afrique orientale: I. Miocène inférieur. Mém. et Trav. Ecole Prat. Haut. Etud., Inst. de Montpellier 1: 1-284.

Li Chuan-Kuei and Ting Su-Yin 1983. The Paleogene mammals of China. Bull. Carnegie Mus. Nat. Hist. 21: 1-93.

Lillegraven, J.A. and Wilson, R.W. 1975. Analysis of _Simimys simplex_, an Eocene rodent (?Zapododae). J. Paleont. 49(5): 856-874.

Lindsay, E. H. 1968. Rodents from the Hartman Ranch Local Fauna, California. PaleoBios 6: 1-22.

Lindsay, E. H. 1977. _Simimys_ and the origin of the Cricetidae (Rodentia, Muroidea). GéoBios 10: 597-623.

Lindsay, E. H. 1978. _Eucricetodon asiaticus_ (Matthew and Granger), an Oligocene rodent (Cricetidae) from Mongolia. J. Paleont. 52(3): 590-595.

Marshall, L. G. 1979. A model for paleobiogeography of South American cricetine rodents. Paleobiology 5(2): 126-132.

Martin, L.D. 1980. The early evolution of the Cricetidae in North America. Univ. Kansas Paleontol. Contrib. 102: 1-42.

McKenna, M. C. 1980. Early history and biogeography of South America's extinct land mammals. In: Evolutionary Biology of the New World Monkeys and Continental Drift, R. L. Ciochon and A. B. Chiarelli, eds., pp. 43-77, Plenum Press, New York.

Mein, P. and Freudenthal, M. 1971. Une nouvelle classification des Cricetidae (Mammalia, Rodentia) du Tertiare de l'Europe. Scripta Geologica 2:1-37.

Mellett, J. S. 1968. The Oligocene Hsanda Gol Formation, Mongolia: A revised faunal list. Novitates 2318: 1-16.

Petter, F. 1966. L'origine des Muridés. Plan cricétin et plans murins. Mammalia 30:205-225.

Qiu Zhuding, Li Chuan-Kuei, and Wang Shijie 1981. Miocene mammalian fossils from Xining Basin, Qinghai. Vert. Palasiatica 19(2): 156-173.

Rogers, D. S. and Heske, E. J. 1984. Chromosomal evolution of the brown mice, genus Scotinomys (Rodentia, Cricetidae). Genetica 63: 221-228.

Sarich, V. M. 1983. Appendix: Retrospective on hominoid macromolecular systematics. In: New Interpretations of Ape and Human Ancestry, R. L. Ciochon and R. S. Corruccini, eds., pp. 137-150, Plenum Press, New York.

Sarich, V. M., and Cronin, J. E. 1980. South American mammal molecular systematics, evolutionary clocks, and continental drift. In: Evolutionary Biology of the New World Monkeys and Continental Drift, R. L. Ciochon and A. B. Chiarelli, eds., pp. 399-421, Plenum Press, New York.

Schaub, S. 1925. Die hamsterartigen Nagetiere des Tertiärs und ihre lebenden Verwandten. Abh. Schweiz. Paläont. Ges. 45: 3-112.

Schaub, S. 1958. Simplicidentata. In: Traité de Paléontologie, 6(2), J. Piveteau, ed., pp. 659-818, Masson et Cie, Paris.

Simpson, G. G. 1971. Status and problems of vertebrate phylogeny. Simposio International de Zoofilogenia, 353-368. Univ. de Salamanca, Facultad de Ciencias.

Simpson, G. G. 1980. Splendid Isolation. The Curious History of South American Mammals. Yale University Press, New Haven.

Slaughter, B. H., and Ubelaker, J.E. 1984. Relationship of South American cricetine rodents to rodents of North America and the Old World. Jour. of Vert. Paleont. 4(2): 255-264.

Stehlin, H. G. and Schaub, S. 1951. Die Trigonodontie der Simplicidentaten Nager. Schweiz. Paläont. Abh. 61: 1-385.

Thaler, L. 1966. Les rongeurs fossiles du Bas-Languedoc dans leurs rapports avec l'histoire des faunes et la stratigraphie du Tertiare d'Europe. Mém. Mus. Nat. Hist. Nat., Sér. C, 27: 1-284.

Vianey-Liaud, M. 1974. L'anatomie crânienne des genres Eucricetodon et Pseudocricetodon (Cricetidae, Rodentia, Mammalia); essai de systématique des Cricetidés oligocènes d'Europe Occidentale. Geol. Mediter. 1(3): 111-132.

Wahlert, J. H. 1978. Cranial foramina and relationships of the Eomyoidea (Rodentia, Geomorpha). Skull and upper teeth of Kansasimys. Novitates 2645: 1-16.

Wahlert, J. H. 1984a. Relationships of the Florentiamyidae (Rodentia, Geomyoidea) based on cranial and dental morphology. Novitates 2769: 1-23.

Wahlert, J. H. 1984b. Relationships of the extinct rodent Cricetops to Lophiomys and the Cricetinae (Rodentia, Cricetidae). Novitates 2784: 1-15.

Wessels, W., de Bruijin, H. Hussain, S. T., and Leinders, J. J. M. 1982. Fossil rodents from the Chinji Formation, Banda Daud Shah, Kohat, Pakistan. Proc. Konink. Nederl. Akad. Wetensch., Ser. B 85(3): 337-364.

Whybrow, P. J., Collinson, M. E., Daams, R., Gentry, A. W., and McClure, H. A. 1982. Geology, fauna (Bovidae, Rodentia) and flora from the early Miocene of Eastern Saudi Arabia. Tertiary Res. 4(3):105-120.

Wilson, R. W. 1935. Cricetine-like rodents from the Sespe Eocene of California. Nat. Acad. Sci., Proc. 21(1): 26-32.

Wilson, R. W. 1949. Additional Eocene rodent material from southern California. Carnegie Inst. Washington, Pub. 584: 1-25.

Wood, A. E. 1974. Early Tertiary vertebrate faunas Vieja Group Trans-Pecos Texas: Rodentia. Texas Memorial Mus. Bull. 21: 1-112.

Wood, A. E. 1980. The Oligocene rodents of North America. Trans. Amer. Phil. Soc. 70(5):1-68.

Zdansky, O. 1930. Die Alttertiären Säugetiere Chinas nebst stratigraphischen Bemerkungen. Palaeont. Sinica, Ser. C. 6(2): 1-87.

ORIGIN AND EVOLUTIONARY RELATIONSHIPS AMONG GEOMYOIDS

Volker Fahlbusch

Institut für Paläontologie und historische Geologie
der Universität München, Richard-Wagner-Str. 10
D-8000 München 2, West Germany

INTRODUCTION

Contents, relationships, and origin of geomyoid rodents have been commented on quite differently during the past, in zoology as well as in paleontology. On the part of zoologists this superfamily contains the well defined families Heteromyidae and Geomyidae, both of them restricted more or less to North America. According to jaw musculature they are mostly assigned to the Sciuromorpha. Following Wilson's (1949) revision of some Oligocene Eomyidae, these also were incorporated into the Geomyoidea by many paleontologists. Moreover, as Wilson (1949) discussed morphological relations to some myomorphs (in which, however, he did not put them), the whole superfamily later became attributed to the Myomorpha in paleontological literature (e.g. Wood, 1955a), a fact, I believe, with which zoologists became scarcely familiar. Some people even considered the whole group to be a separate suborder Geomorpha (Thaler, 1966). Due to the lack of adequate fossils, little is known of the origin of these rodents, except for some necessarily preliminary assumptions dealing with various groups of the Protrogomorpha as possible ancestors.

In the following discussion, I will consider the content of the Geomyoidea, the relationships of the included families to each other, and the questions of descent according to the present stage of knowledge and fossil record.

FAMILY GEOMYIDAE

The family Geomyidae ("pocket gophers") is identified by a series of common features among which the osteological adaptations to fossorial life are most striking, as well as the tendency toward simplified and hypsodont to evergrowing cheek teeth in the living species. Ellerman (1940, p. 469) commented on them: "In some ways the Geomyidae appear to me at their highest development to be among the most highly specialized of all living Rodents". By inclusion of the fossil forms the uniformity of the family remains almost unchanged back to the Lower Miocene. At that time, however, the cheek teeth are still brachydont to slightly hypsodont (mesodont), although hypsodont teeth are developed in some evolutionary lines quite rapidly (Rensberger, 1971).

The oldest and, referring to tooth morphology, most primitive geomyids are *Tenudomys* and the *Florentiamys-Sanctimus* group from the Arikareean (Rensberger, 1973a, b). I follow Rensberger (1973b) in assigning the subfamily Florentiamyinae to the Geomyidae, whereas Wood (1936) originally described them as heteromyids. The bunodont cheek teeth, still more or less low crowned, with beginning lophodonty and large P4/4 prove these genera to be initial but true representatives of the Geomyidae.

A completely different interpretation for this group has been given by Wahlert (1983). On the basis of about a dozen specimens, including three well preserved skulls, he described several new species of *Florentiamys* and *Sanctimus*, and he classified both genera as belonging to a distinct family Florentiamyidae. He defined this family by a few features of the skull and dentition, among which he considers most prominent the "unique participation of the palatine in the edge of the anterior-alar fissure" (cf. Wahlert, 1983, Fig. 6) which is in his interpretation the anterior end of the alisphenoid canal.

It may be dangerous to contradict such a taxonomic interpretation without having seen and compared the specimens to other ones. However, for two reasons I have considerable misgivings. First, I doubt whether we know enough about the taxonomic value of a feature observed in such a few specimens, and for which we have no idea of its variation. Secondly, the absolute security is overwhelming in which Wahlert's (1983) discussion on the anterior-alar fissure in the "Florentiamyidae" ends up in considering it a primitive feature. He has drawn his conclusion after having compared the special situation here to the one seen in *Sciuravus* and *Paramys* on the one hand, and in the eomyid *Viejadjidaumo* (1 skull only) on the other.

Accordingly, Wahlert (1983) considered these so-called "Florentiamyidae" to be a group of geomyoid rodents with "retained primitive characters" that "suggest that the florentiamyids diverged early in geomyoid phylogeny." It should be repeated here that the only two florentiamyid genera *Florentiamys* and *Sanctimus* are stratigraphically restricted to the Lower Miocene or - as far as the Lower Arikareean is part of the Upper Oligocene (Wahlert, 1983) - to late

Upper Oligocene through early Lower Miocene.

The argumentation for Wahlert's (1983) phylogenetic interpretation is based on a set of "Primitive Cranial and Dental Characters Shared by Florentiamyids and other Rodent Groups" listed in his Table 4. Again, I doubt the conclusiveness for all these features to be considered absolutely as primitive ones. I may not be sufficiently knowledgeable about the four groups on the right side of his table (Geomyidae, Heteromyidae, Pleurolicus, Entoptychus). However, I know for the Eomyidae that at the most only one or the other of these features can be checked in one or the other of the skulls so far available; there are about a dozen skulls known for seven eomyid genera, out of about 20-25. To designate that column in Wahlert's table as "Eomyoidea" seems to me to be highly premature, and presented only to "prove" the reliability of Hennigian cladism. In any case, I am unable to agree with Wahlert's phylogenetic interpretation and the following systematics. For the present, the Florentiamys-Sanctimus group, following Rensberger (1973b), is considered best to be a subfamily of the Geomyidae that still exhibits morphological similarities to the stratigraphically older Heteromyidae, but at or near the base of the Geomyidae.

Except for this recent controversy, the taxonomic disposition of Geomyidae below the familial level is handled rather traditionally, although there is not complete agreement on the number and exact content of the subfamilies (e.g., Wood, 1980). These problems will not be discussed here, but it should be noted that there is some good evidence for the monophyletic origin of the Geomyidae (Rensberger, 1973b; Gawne, 1975). According to Gawne, however, "it remains possible" that the individual subfamilies were derived from different lineages within the heteromyids. I am certain that it will take quite a while and will require much additional heteromyid material to understand these evolutionary processes in detail. In spite of a remarkable diversity in, and rapid evolution during, the Lower Miocene, no definite geomyids have been recorded yet from the Oligocene.

In this place, some problematic genera, which have been attributed to different taxa, will be mentioned briefly. Some authors have designated these as possible or questionable geomyids. The genus Diplolophus (Troxell, 1923; Wood, 1980) from the North American Orellan displays a bilophodont tooth pattern and an anterior cheek tooth that are reminiscent of certain geomyids. According to the tooth formula - only three cheek teeth, no matter which they are - I do not consider it possible to be a geomyoid. However, as I am not able to offer any alternative interpretation, I shall leave it as a most questionable genus, only tentatively with geomyoids.

Nonomys from the Lower Oligocene of North America, originally referred to the family Cricetidae by Emry and Dawson (1972, 1973), has been considered by Wood (1980) to be possibly related to the Geomyidae. He stated (p. 51): "The sequence Eomyidae - *Nonomys* - *Diplolophus* seems to me as logical," and therefore "*Nonomys* is tentatively assigned to Geomyoidea, *incertae sedis*, but probably related to *Diplolophus*." New material with additional information has been described by Emry (1981), according to whom this animal, as well as the Upper Eocene *Simimys*, "are best interpreted as members of an early radiation of hystricomorphous muroid rodents, neither being in any presently defined family, and probably neither being in the direct ancestry of any modern rodent." Thus, I think that any relationship of *Nonomys* to geomyoids may be excluded.

Finally, I should mention the relatively large genus *Diatomys* (Li, 1974) from the Middle Miocene of China, with clearly bilophodont, but very low crowned, cheek teeth. The upper and lower dentitions originate from a complete skeleton, but its bad preservation provides little information. Li (1974) referred it questionably to the Geomyoidea, but without any assignment to one of the known families. Without additional material, further interpretation is impossible.

At the moment, these three genera (if they have any relationship to Geomyoidea at all) are of no importance for the problem of evolutionary relationships among geomyoids. Nevertheless, they do demonstrate that there are undoubtedly several evolutionary lines of uncertain affinities. However, these are so poorly documented that any phylogenetic reconstruction is impossible. This, on the other hand, illustrates that we are far from attaining a clear phylogenetic conception of the geomyoids.

FAMILY HETEROMYIDAE

The second well established family of geomyoid rodents, the Heteromyidae, contains mainly scampering and jumping forms, some of them with remarkable saltatorial adaptations and a highly inflated posterior part of the skull. As with geomyids, they are restricted to North (and Middle) America, with only a few Recent forms having invaded northern South America. With their dentition of P4/4 - M3/3 only a few forms show complete hypselodonty. Normally, they preserv a bilophodont structure, with a bunodont basic pattern often recognizable, a lingual cingulum on the upper cheek teeth, and a buccal cingulum on the lower ones. The bunodont tooth pattern prevails in fossil genera, which can be traced back into the Lower Oligocene without widening the diagnosis of morphological features for the family. The only true Oligocene heteromyids are *Heliscomys* (Lower Oligocene) and *Proheteromys* (Upper Oligocene). Only in the Miocene does a greater differentiation take place. An unsolved question is whether the small premolar (especially the missing protoconid of P/4 is a primitive or advanced feature. This is related to the significance of Meliakrouniomys, which will be discussed later.

There was never any doubt about the differences between Geomyidae and Heteromyidae, or their taxonomic rank as independent families, in either zoology or paleontology (except for some special problems, e.g., the assignment of the Florentiamyinae). Nor has there been doubt concerning the close relationship between the two families. In spite of the differences between living representatives of both groups (mainly fossorial Geomyidae, mainly scampering or saltatorial Heteromyidae), with their acquired adaptations of the skull and skeleton, the two families share the same special type of sciuromorphous masseter musculature, combined with a basically bilophodont cheek tooth pattern. The more or less heavy M. masseter lateralis has a broad zygomatic plate at its disposal, as well as a region of the snout, whose upper part - parallel to the posterior portion of the upper incisor - may be reached by this muscle. The infraorbital foramen is fairly small to tiny, situated in the lower part of the M. masseter medialis. This is an important condition. common to both families, to which great importance is attached, and which is found similarly in all fossil genera for which skulls are available. Combined with this type of jaw musculature is a dentition consisting always of P4/4 - M3/3, including a more or less molariform P4/4. Except for highly specialized genera with hypsodont or evergrowing teeth, this dentition exhibits a generally bilophodont pattern in the cheek teeth, either built of two parallel lophs, or, in more primitive species, with each composed of distinct cusps.

In terms of cladism, these two families would be called sister groups. By this terminology, however, the obvious evidence is suppressed that one sister, the geomyid group, is at least 10 million years younger than the other one. It is also worth noting that the younger sister (Geomyidae) evolved much faster than the older one (Heteromyidae).

The origin of Geomyidae in the Lower Miocene from certain species of Heliscomys can be demonstrated as fairly plausible by the examples of Florentiamys/Sanctimus and Tenudomys (Rensberger, 1973b), although this is not completely clear. This close relationship between the two families Heteromyidae and Geomyidae has been documented at all times by uniting them in the superfamily Geomyoidea.

FAMILY EOMYIDAE

Another well defined group is the entirely fossil family Eomyidae (see Fahlbusch, 1979). Their taxonomic rank as a family is supported by the combination of sciuromorphous jaw musculature and a brachydont and primarily quadritubercular to pentalophodont dentition, consisting of M1/1 - M3/3 and more or less molariform P4/4. This group is documented in North America from Upper Eocene through Pliocene times. In addition, eomyids appear in Europe after the "Grande Coupure" (at the boundary Headonian/Suevian, which corresponds approximately to the boundary of Lower/Middle Oligocene). [Editor's note: For some other authors, the Headonian/Suevian boundary corresponds to the Eocene/Oligocene boundary; see Vianey-Liaud, this volume.] Eomyids flourished in Europe during the Middle Tertiary, and they did not become extinct there until the Lower Pleistocene (Fahlbusch, 1979; Fig. 1). Recently, they have been recorded from the Late Miocene/Early Pliocene of China (Fahlbusch et al., 1983). No potential ancestors are known from Europe prior to the "Grande Coupure." The only possible explanation for this is to consider the European eomyids as immigrants from North America; this is supported by the absolute morphological conformity of _Eomys_ (the oldest European genus) and the North American Oligocene genus _Adjidaumo_ (the two names being preserved only for practical reasons; Fahlbusch, 1973).

The sciuromorphous type of zygomasseteric structure is preserved in all eomyids, as far as can be determined from the few forms that are documented by skull fragments (Wilson, 1949; Wahlert, 1978). The dentition only rarely shows a slight tendency toward an increase in crown height (e.g., _Paradjidaumo_). Derived from a cusp-crested crown, a distinct pentalophodont pattern is attained in some evolutionary lines (e.g., _Centimanomys_, _Pseudotheridomys_); in others, a marked bilophodont tooth pattern develops secondarily (e.g., _Ritteneria_), independently and convergent to that observed in Heteromyidae and Geomyidae. Several times, the bunodont ("cricetoid") chewing surface is replaced by a flat one (e.g., _Ligerimys_), which certainly meant a rather drastic change in chewing mechanism.

A separate position within the family Eomyidae is held by _Yoderimys_ from the Lower Oligocene of North America. This is the only eomyid with a small, single rooted P3/ in front of a fairly large P4/ (in crown morphology, the teeth resemble those of Oligocene cricetids, rather than other eomyids). Wood (1955b, 1980) stressed this fact by assigning _Yoderimys_ to a separate subfamily Yoderimyinae opposite to all other eomyids (Eomyinae). This seems reasonable, although it should be emphasized that there are several other genera which, I admit, exhibit the normal tooth formula, but are relatively different in tooth morphology when compared to the central _Adjidaumo-Eomys_ group. Subfamilies could likewise be established for these groups. This is true for some genera from the North American Lower Oligocene (e.g., _Namatomys_, _Centimanomys_) or Upper Miocene (e.g., _Kansasimys_, _Ronquillomys_), as well as for some European forms (e.g., _Ritteneria_, _Ligerimys_, _Apeomys_). However, not even the complete dentition is known for some of these genera. Therefore, it seems premature to attempt a subfamilial subdivision of the Eomyidae. It would only complicate systematics without reflecting our real knowledge. Such a subdivision would require much more complete documentation.

PHYLOGENETIC RELATIONSHIPS AMONG GEOMYOIDS

What do we know about the relationships between Eomyidae on the one hand, and Heteromyidae + Geomyidae on the other? Both groups exhibit the same type of sciuromorphous jaw musculature, with an enlarged zygomatic plate and a small, deep lying infraorbital foramen. This was the main reason that Wilson (1949) assigned the Eomyidae to the superfamily Geomyoidea. This arrangement has been followed by most other paleontologists. The reasons why Rensberger (1973b) and Gawne (1975) excluded eomyids from the Geomyoidea or referred to Geomyoidea _sensu stricto_ were not explained by these authors. As far as one gives systematic value to jaw musculature in rodent classification, this is the prominent morphological feature shared by eomyids and heteromyids + geomyids. Wahlert (1978) has corroborated the close relationship between Eomyidae and other Geomyoidea in his investigation of various cranial foramina. Accord- to the schematizing theory of cladism, he concluded that Eomyidae is the sister group of the remaining Geomyoidea, and he grouped them as a separate superfamily, the Eomyoidea. This taxonomic subdivision is supported by tooth morphology, in so far as geomyids and heteromyids are certainly more closely related to each other than either is to eomyids. I will discuss later whether or not it is necessary to follow Wahlert (1978) in his conclusion to establish the superfamily Eomyoidea, and consequently to put both superfamilies into the next higher taxon (suborder or infraorder), for which Thaler's (1966) suborder Geomorpha was already available.

Because the family Geomyidae apparently forms an Early Miocene (or probably Late Oligocene) offshoot of Heteromyidae, only the latter group will be discussed in the following. _Heliscomys_, in spite of being undoubtedly a primitive heteromyid, is far from being a primitive rodent. It is also distant from all known eomyids, although these seem to be the most comparable rodents concerning tooth morphology. _Heliscomys_ is separated from all known eomyids by its specialized dental pattern, marked by four cusps arranged in two double-cusped rows on the molars, to which is added a broad cingulum and/or with additional cusps (lingually in upper, labially in lower teeth). Whether its small premolars mark a primitive condition or an advanced one is not yet decided definitely, although the latter seems to be more likely, compared with many other sciuromorphous rodents from the Upper Eocene. However, in the Middle Eocene sciuravid _Pauromys_ (to be discussed later), the P4/4 seem to be somewhat reduced already. The relative size of the heteromyid P4/4 is also important in connection with the origin of geomyids, which all exhibit relatively large P4/4. In some fairly complete evolutionary lineages of European eomyids we can observe a reduction in size of those teeth, followed by a later enlargement (e.g., in the lineage _Eomys_ - _Pseudotheridomys_ - _Ligerimys_; Fahlbusch, 1970). Such a process, which is related to largely unknown changes in mastication, should not be excluded in heteromyids also.

The Lower Oligocene genus Meliakrouniomys is, according to Wood (1974, 1980), closely related to Viejadjidaumo, and he has assigned it to the Eomyidae, although considering it (Wood, 1980, p. 43) as "evolving toward the Heteromyidae." In contrast, Emry (1972) considered it to be a typical heteromyid on the basis of its zygomasseteri structure. I am unable to see any important differences between eomyids and heteromyids in this feature. These different opinions of Wood and Emry only prove the close relationship between eomyids and heteromyids during the Lower Oligocene. If considered to be a heteromyid, Meliakrouniomys is quite different from other species of Lower Oligocene age, at least concerning tooth structure, including a relatively large P/4. On the other hand, if this genus is considered to be an eomyid, it is likewise far removed from typical eomyids in having lost (rather than not developing) a mesoconid and/or ectolophid. Unfortunately, upper dentitions are not yet known for this genus. Meliakrouniomys might best be considered to represent an early side brance of eomyids, with somewhat simplified dentition, a condition that appeared later in this family several times.

Even more puzzling is the Upper Eocene genus Griphomys (Wilson, 1940; Lillegraven, 1977), which superficially seems to resemble some later geomyids. Following Wood (1980), I would classify it as an eomyid, but representing another lineage with simplified tooth pattern, rather than anticipating an evolutionary step toward Heteromyidae. Both examples (Meliakrouniomys and Griphomys) are reminiscent of trends in several Late Miocene cricetids, which became very similar to arvicolids, while remaining distinct from them.

Following an examination of lower incisor enamel from a few eomyid genera, Wahlert (1978, p. 2) concluded that the "Eomyidae ... cannot have given rise to the Heteromyidae." However, the data give by Wahlert are, in my opinion, much too poor to justify such a definite conclusion.

In summary, it may be stated that, according to zygomasseteric structure, as well as to tooth morphology, the Heteromyidae and Eomyidae are certainly related. Furthermore, when taking into account the stratigraphic record, the Eomyidae may well be the ancestors of the Heteromyidae (Fig. 1). The present state of knowledge of early eomyid history, however, is not sufficient to identify one of the known genera or genus groups as definite ancestors of the Heteromyidae. Also, at present it can not be excluded absolutely that both families had an older common ancestor.

For the early history of Eomyidae, and that means probably for the Geomyoidea in general, many problems remain unsolved. In the Lower Oligocene of North America, the only continent from which eomyids are known by that time, they exhibit a relatively great diversity in tooth morphology (Fahlbusch, 1973, 1979). On the other hand only the genus Protadjidaumo (one mandible from the Duchesnean of

Utah) has been described from the Upper Eocene. In addition, good material, not yet published, has been found in the somewhat older Uintan of Badwater, Wyoming (Black and Dawson, 1966). To judge only by the lower dentition, *Protadjidaumo* is relatively close to the main group of Oligocene eomyids (*Adjidaumo*, *Paradjidaumo*). Whether or not all the other Lower Oligocene genera are descended from the same or related forms remains highly questionable, especially for *Namatomys* and *Centimanomys*. Certainly, a separate position is held by *Yoderimys* (assuming it really is an eomyid). Thus, the question arises as to whether the family Eomyidae in its present content represents a monophyletic group.

No other eomyids older than the above mentioned ones from Badwater are known as yet. Looking for possible ancestral groups, it seems reasonable to concentrate on North America, which can be regarded as the most probable center of origin, based on the current existing documentation. In my opinion, the only probable stem group is the protrogomorphous family Sciuravidae. The specialized geomyoid type of sciuromorphous jaw structure in eomyids is close enough to the protrogomorphous pattern to be derivable from it directly. In its tooth morphology, the Middle Eocene sciuravid genus *Pauromys* is the one closest to the general eomyid tooth pattern. This is true for the size relations of individual teeth, as well as for arrangement of cusps and their connections. Dawson (1968) described an undetermined species of *Pauromys* from the Middle Eocene of Utah, in which P3/ was possibly already missing. In addition, its zygomasseteric pattern, although still protrogomorphous, exhibited certain progressive features; Dawson (1968) stated that it "is certainly the most progressive zygomasseteric region of the maxilla that has been found in a North American Bridgerian rodent." In many details, the dental pattern of this species is specialized in the direction of eomyids more than that of any other protrogomorphous rodent of the North American Eocene.

All of these features do not mean that this genus, or even this species of *Pauromys*, is the direct ancestor of eomyids. Many details of the genus, including intraspecific variation, are still unknown. However, it indicates that in sciuravids there were several evolutionary trends, out of which the later eomyids may have evolved during the Middle/Upper Eocene.

CONCLUSIONS

In reviewing the evolutionary history of geomyoid rodents, many questions have remained unanswered. Nevertheless, the following conclusions can be drawn.

Eomyidae, Heteromyidae, and Geomyidae each represents a well defined family, and each is probably of monophyletic origin. In addition, the Heteromyidae and Geomyidae are more closely related to each other than either is to the Eomyidae.

Considering the problems in the descent of the heteromyids and in the early history of the eomyids, it seems premature and unnecessary to raise the Eomyidae to a separate superfamily Eomyoidea, opposite to the Geomyoidea (including only Geomyidae and Heteromyidae). Both superfamilies would have to be united in the next higher taxon the Geomorpha (cf. Wahlert, 1978, 1983). Of course, such formalist systematics complies with the rules of Hennigian phylogenetic systematics, but it does not contribute at all to our real knowledge; instead, it pretends detailed information which is not yet available. Being completely conscious of the preliminary status of our knowledge, it seems more appropriate for the moment to retain the three families side by side in the superfamily Geomyoidea. To assign to them the higher rank of a suborder or infraorder would appear to overemphasize the geomyoids, in comparison to other rodent groups, and to be unnecessary as well.

The relationships of the Geomyoidea, which most probably arose from sciuravids, to other Myomorpha are largely unknown. This is true for the Gliroidea, which may have been descended from microparamyines (Hartenberger, 1971), as well as for Dipodoidea and Muroidea. The fossil record is much too poor in this respect for any decision, even for a preliminary comment. Wahlert's (1978) conclusions, based on investigation of some skulls and their foramina (summarized in his cladogram), may be a useful impulse of thought. They are, however, much too uncertain and sparse to be a starting point for phylogenetic reconstructions, or even for systematic arrangements.

To present my view on the relationships among major geomyoid groups (Fig. 1), I feel obliged to take into account our present knowledge of stratigraphic distribution of all known genera. Phylogeny as a historical process should not neglect the time factor. In conclusion, I prefer the old fashioned phylogenetic tree, damned so much by Hennigian disciples. To me, it offers much more information, and it is far from being based on mere fantasy. Furthermore, there is room for question marks to point out unsolved problems, which one never finds in a cladogram.

As long as zygomasseteric structure is not replaced by a more reliable feature in classification, there is scarcely any reason to transfer the Geomyoidea from Sciuromorpha to Myomorpha. This does not exclude a possible closer relationship among certain geomyoids, muroids, and dipodoids. We know that the Sciuromorpha, and the Myomorpha as well, do not form phylogenetic units. However, we know even less about the relationships between geomyoids and myomorph rodents. To replace one uncertainty by another even larger one means very little progress in knowledge. Therefore, the superfamily

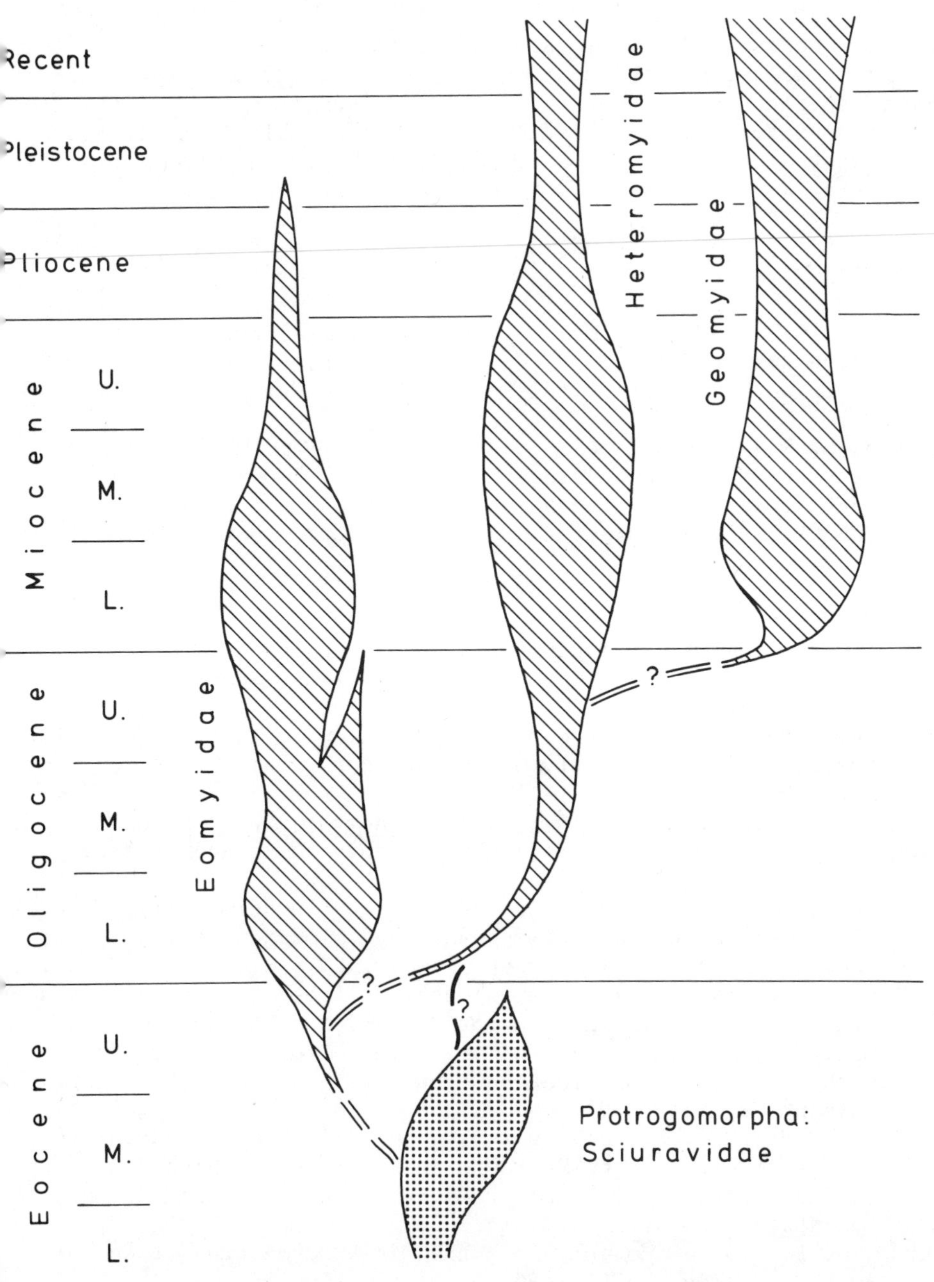

Fig. 1. Phylogenetic relationships among geomyoid rodents.

Geomyoidea, including the three families Eomyidae, Heteromyidae, and Geomyidae, should remain in the Sciuromorpha, at least for the time being.

The unresolved questions on the origin of geomyoids and their phylogenetic relationships are well known. Their answers do not depend on theoretical constructions and reconstructions, but instead on the investigation of material already available in collections still unstudied, and it depends particularly on additional fossils which are, we hope, still to be discovered in the field.

REFERENCES

Black, C. C. and Dawson, M. R. 1966. Paleontology and geology of the Badwater Creek Area, Central Wyoming. Part 1. History of field work and geological setting. Ann. Carneg. Mus. 31: 297-307.

Dawson, M. R. 1968. Middle Eocene rodents (Mammalia) from Northeastern Utah. Ann. Carneg. Mus. 39: 327-370.

Ellerman, J. R. 1940. The Families and Genera of Living Rodents, Vol. 1. Rodents Other than Muridae. British Museum (Nat. Hist.), London.

Emry, R. J. 1972. A new heteromyid rodent from the Early Oligocene of Natrona County, Wyoming. Proc. Biol. Soc. Wash. 85: 179-190.

Emry, R. J. 1981. New material of the Oligocene muroid rodent Nonomys, and its bearing on muroid origins. Amer. Mus. Nov. 2712: 1-14.

Emry, R. J. and Dawson, M. R. 1972. A unique cricetid (Rodentia, Mammalia) from the Early Oligocene of Natrona County, Wyoming. Amer. Mus. Nov. 2508: 1-14.

Emry, R. J. and Dawson, M. R. 1973. Nonomys, new name for the cricetid (Rodentia, Mammalia) genus Nanomys Emry and Dawson. J. Paleont. 47: 1003.

Fahlbusch, V. 1970. Populationsverschiebungen bei tertiären Nagetieren, eine Studie an oligozänen und miozänen Eomyidae Europas. Abh. Bayer. Akad. Wiss. Math.-naturw. Kl., N. F. 145: 1-136.

Fahlbusch, V. 1973. Die stammesgeschichtlichen Beziehungen zwischen den Eomyiden (Mammalia, Rodentia) Nordamerikas und Europas. Mitt. Bayer. Staatsslg. Paläont. hist. Geol. 13: 141-175.

Fahlbusch, V. 1979. Eomyidae - Geschichte einer Säugetierfamilie. Paläont. Z. 53: 88-97.

Fahlbusch, V., Qiu, Z., and Storch, G. 1983. Neogene mammalian faunas of Ertemte and Harr Obo in Nei Mongol, China. -1. Report on field work in 1980 and preliminary results. Scient. Sin. Ser. B. 26: 205-224.

Gawne, C. E. 1975. Rodents from the Zia Sand Miocene of New Mexico. Amer. Mus. Nov. 2586: 1-25.

Harris, J. M. and Wood, A. E. 1969. A new genus of eomyid rodent from the Oligocene Ash Spring Local Fauna of Trans-Pecos Texas. Texas Mem. Mus. Pearce-Sellards Ser. 14: 1-7.

Hartenberger, J.-L. 1971. Contribution à l'étude des genres Gliravus et Microparamys (Rodentia) de l'Eocène d'Europe. Palaeovertebrata 4: 97-135.

Li, C.-K. 1974. A probable geomyoid rodent from Middle Miocene of Linchu, Shantung. Vert. PalAsiat. 12: 43-53.

Lillegraven, J. A. 1977. Small rodents (Mammalia) from Eocene deposits of San Diego County, California. Bull. Amer. Mus. Nat. Hist. 158: 221-261.

Rensberger, J. M. 1971. Entoptychine pocket gophers (Mammalia, Geomyoidea) of the Early Miocene John Day Formation, Oregon. Univ. Calif. Publ. Geol. Sci. 90: 1-209.

Rensberger, J. M. 1973a. Pleurolicine rodents (Geomyoidea) of the John Day Formation, Oregon. Univ. Calif. Publ. Geol. Sci. 102: 1-95.

Rensberger, J. M. 1973b. Sanctimus (Mammalia, Rodentia) and the phyletic relationships of the large Arikareean geomyoids. J. Paleont. 47: 835-853.

Thaler, L. 1966. Les rongeurs fossiles du Bas-Languedoc dans leur rapports avec l'histoire des faunes et la stratigraphie du Tertiaire d'Europe. Mém. Mus. Nat. Hist. Nat., sér. C. 17: 1-295.

Troxell, E. L. 1923. Diplolophus, a new genus of rodents. Amer. J. Sci. 5: 157-159.

Wahlert, J. H. 1978. Cranial foramina and relationships of the Eomyoidea (Rodentia, Geomorpha). Skull and upper teeth of Kansasimys. Amer. Mus. Nov. 2645: 1-16.

Wahlert, J. H. 1983. Relationships of the Florentiamyidae (Rodentia, Geomyoidea) based on cranial and dental morphology. Amer. Mus. Nov. 2769: 1-23.

Wilson, R. W. 1940. Two new Eocene rodents from California. Carneg. Inst. Wash. Publ. 514: 85-95.

Wilson, R. W. 1949. On some White River fossil rodents. Carneg. Inst. Wash. Publ. 584: 27-50.

Wood, A. E. 1936. A new subfamily of heteromyid rodents from the Miocene of Western United States. Amer. J. Sci. 31: 41-49.

Wood, A. E. 1955a. A revised classification of the rodents. J. Mammal. 36: 165-187.

Wood, A. E. 1955b. Rodents from the Lower Oligocene Yoder Formation of Wyoming. J. Paleont. 29: 519-524.

Wood, A. E. 1974. Early Tertiary vertebrate faunas Vieja Group, Trans-Pecos Texas: Rodentia. Texas Mem. Mus. Bull. 21: 1-112.

Wood, A. E. 1980. The Oligocene rodents of North America. Trans. Amer. Phil. Soc. 70: 1-68.

EVOLUTIONARY DATA ON STEPPE LEMMINGS (ARVICOLIDAE, RODENTIA)

Jean Chaline

Centre de Géodynamique sédimentaire et Evolution géobiologique, U. A. CNRS 157, Laboratoire de Préhistoire et Paléoécologie du Quaternaire, Institut des Sciences de la Terre, 21100 Dijon, France

INTRODUCTION

In testing for speciation models, the analysis must be at the species level. The test, in order to be fully satisfactory, must have well preserved data, as well as documented lineages to fully span the geographical and temporal range. Such examples are rare. The steppe lemmings or sagebrush vole lineages of Lagurus (Arvicolidae, Rodentia) are thus of great evolutionary interest in this respect. These lineages are fossilized in large numbers and have a high frequency of occurrence across Eurasia (France to China) throughout the Pleistocene period.

CHARACTER ANALYSIS OF ARVICOLID DENTAL TRAITS

The steppe lemmings have been studied by Kormos (1938), Kretzoi (1956), Janossy (1964), Adamenko and Zazhigin (1965), Schevtschenko (1965), Topachevsky (1965), Konstantinova (1967), Terzea (1968, 1970), Vangengeim and Zazhigin (1969), Hall and Kelson (1969), Zazhigin (1970), Chaline (1972, 1975), and Rabeder (1981). The lineages are characterized paleontologically by their molar structure lacking both roots and cement. Their prismatic teeth are formed by enamel triangles.

Plesiomorph and apomorph conditions of dental characters can be clearly established at the arvicolid family level by their stratigraphic occurrence in the fossil record (Table 1). Despite this clear documentation of character states in the fossil record, there is a problem at the species level analysis. Tooth morphology is

Table 1. Character analysis of selected arvicolid dental traits

Primitive	Derived
1. Lower incisor in diagonal position under molar row	1. Lower incisor completely lingual to molar row
2. Molar roots appear early (= rhizodonty)	2. Molar roots appear later in ontogeny, then progressively disappear (= arhizodonty)
3. Molar growth limited before closing of roots	3. Molars grow continuously
4. Molar crown brachyodont to hypsodont	4. Molar crown hypsodont to hyperhypsodont
5. Slightly sinuous lateral enamel tracks; with wear, no enamel interruption on grinding surface	5. Very strong lateral enamel tracks; with wear, enamel interruptions occur on the grinding surface
6. Relatively low number (3-5) of M_1 triangles	6. Higher number (6-9) of M_1 triangles (= polyisomery)
7. Molar triangles broadly confluent	7. Molars with closed triangles (= deltodonty)
8. M triangle rhombus present	8. M triangle rhombus absent
9. Cement absent in molar re-entrant angles	9. Cement progressively appears, then becomes abundant, in re-entrant angles
10. Enamel uniformly thick	10. Enamel differentiated, with thick and thin areas
11. Enamel Schmelzmuster asymmetrical	11. Enamel Schmelzmuster symmetrical
12. Molar enamel radial	12. Molar enamel tangential, with Hunter-Schreger bands
13. Enamel islet present on lower M1 and upper M3	13. Enamel islet absent on lower M1 and upper M3
14. *Mimomys* fold present (M_1)	14. Mimomys fold absent (M_1)
15. Teeth symmetrical	15. Teeth asymmetrical
16. M^3 lacks additional microangle	16. M^3 with additional microangle (Lagurini)
17. M^3 lacks angle reduction	17. M^3 with angle reduction (Pliomyini)
18. Small size	18. Medium to large size

highly variable within species. For some characters (number of triangles, degree of confluence between triangles and anterior loop), the new apomorph condition appears as a novelty together with the plesiomorph one. Thus, it is impossible to construct cladograms. Moreover, because apomorph characters of Recent voles appear in many parallel and distinct lineages, it is impossible to do a traditional character analysis.

Nevertheless, because of the very fine temporal and geographic fossil record of lagurins, we are able to trace a well documented phylogenetic tree. Most of the species described from Central and Western Europe have been discussed in a typological concept of species (Topachevsky, 1965; Schevtschenko, 1965; Rabeder, 1981). In this splitting approach, nearly all species have received a distinct generic and subgeneric name. A new taxonomy of steppe lemmings is proposed as a consequence of the present evolutionary study.

Taxonomy, Stratigraphical and Geographical Record of Steppe Lemmings

Eight species of lagurins are presently known and distinguished by their structure of lower M1 and upper M3. These distinguishing features include: (1) the number of alternate or confluent enamel triangles, giving a symmetrical or asymmetrical appearance; (2) the structure of the anteroconid complex, with a more or less broad confluence between the anterior loop and triangles components; and (3) size. These species are morphologically different and are recorded in two distinct lineages (*Lagurus* and *Prolagurus*). Reconsidered in a biological populational context, the two lineages can not be considered as distinct genera, but instead are designated here as subgenera. Taking into account the international nomenclature, the following revised lagurin taxa are proposed.

Lagurus (*Lagurus*) *arankae*. Lower M1 has 6 (*arankae* asymmetric type variant) to 7 (less than 5%) triangles (*arankae* symmetric type variant). The anterior loop, fourth and fifth triangles (anteroconid complex) are more or less broadly confluent. The sixth external triangle presents an enamel interruption. The anteroconid complex is more or less asymmetric. Upper M3 has three enamel triangles, and an asymmetric posterior loop. *Lagurus* (*Lagurus*) *arankae* occurs in the middle Lower Pleistocene from Austria, Hungary, Rumania, and the USSR (Ukraine, Moldavia, Western Caucasus, Southwest Siberia).

Lagurus (*Lagurus*) transient *transiens*. Lower M1 anterior loop and upper M3 posterior loop have slightly concave lingual and labial sides, resulting in the appearance of two new enamel triangles (sixth and seventh on lower M1; fourth and fifth on upper M3). All triangles are separated. This transient species occurs in the lower Middle Pleistocene of Hungary and the USSR (Southern part of the Russian plain and Southwestern Siberia).

Lagurus (*Lagurus*) *lagurus*. Lower M1 possesses 7 enamel triangles. The anterior loop shape is variable, broadly open in the sixt and seventh triangles of the anteroconid complex. The sixth externa triangle presents an enamel interruption. Upper M3 has 5 enamel triangles. This species occurred from upper Middle and Upper Pleistocene to the present, from England, France, Germany, Poland, Hungar Rumania, and the USSR (Ukraine, Altaï, Transbaïkal).

Lagurus (*Prolagurus*) *pannonicus*. Lower M1 has 5 to 6 (1%) or 7 (less than 6%) triangles. When the anterior round loop exhibits poorly developed re-entrant angles, it is always narrowly separated from the fourth and fifth broadly confluent triangles of the anteroconid complex. Upper M3 has enamel triangles similar to *L. arankae*. This species is found in the middle Lower Pleistocene from Germany, Austria, Hungary, Czechoslovakia, Yugoslavia, Rumania, and the USSR (Ukraine, Moldavia, Western Caucasus, Southwest Siberia, western part of Transbaïkal area).

Lagurus (*Prolagurus*) transient *argyropuloi*. Lower M1 has 3 closed triangles, and an anteroconid complex composed of a circular anterior loop more or less separated from the fourth and broadly confluent triangles. Upper M3 with 3 enamel triangles, as in *P. pannonicus* and *Lagurus arankae*. This transient species occurs in lower Middle Pleistocene from the USSR (South Russian plain and Southwestern Siberia).

Lagurus (*Prolagurus*) *posterius*. Lower M1 has 5 triangles, and a roundish rectangular anterior loop. All components of the anteroconid complex are separated, particularly the fourth and fifth triangles. Upper M3 is as in *P. pannonicus*. This species is found in the lower Middle Pleistocene from the USSR (Azov coast, Volga-Don interfluve, south of Western Siberia).

Lagurus (*Prolagurus*) *luteus*. Lower M1 has 5 closed enamel triangles and an anterior roundish loop. Molars are of large size. Upper M3 is similar to that of *P. argyropuloi* and *L. arankae*. This species occurs in the lower Middle Pleistocene to Recent from Rumania and the USSR (South Russian Plain, Crimea, Southwestern Siberia to Altaï and Mongolia).

Lagurus (*Lemniscus*) *curtatus*. Lower M1 has 5 closed enamel triangles and an anterior dissymmetric loop. This species lives at the present time in the West of the United States.

PHYLOGENY AND EVOLUTIONARY TRENDS IN STEPPE LEMMINGS

Evolutionary Origin of Lagurins

As suggested by Rabeder (1981), *Mimomys* (*Borsodia*) *petenyi* may be a likely possible ancestor for *Lagurus arankae*, the oldest known steppe lemming (Lagurini) in the fossil record (Fig. 1). They share in common: (1) the same grinding surface of lower M1, with the sixth external triangle featuring an enamel interruption; and (2) the concave border of the posterior loop of lower M1. Zazhigin (1970) suggested that the steppe lemmings are closely related to *Villanya* (*Kolundomys*) ex. gr. *fejervaryi*. However, Rabeder (1981) demonstrated that this form, which he named *Borsodia fejervaryi*, can not be ancestral to the *Lagurus* lineages, because it possesses two apomorphic characters (cement and large dimensions) not found in the earliest *Lagurus* species. Instead, it appears to be a lateral ephemeral species. *Borsodia petenyi* and *Lagurus arankae* differ by the occurrence of roots in *B. petenyi*, and their disappearance in *L. arankae*. Because there is no record of the upper M3 of *B. petenyi*, no information is available about its possible possession of the upper M3 lagurin microangle.

The disappearance of roots, as documented in the arvicolid fossil record, occurs progressively. During the course of time, they appear later and later in the ontogeny of each organism. This phenomenon involves an heterochrony of development, occurring through a neotenic process (Chaline, in prep.).

First Cladogenesis

Lagurus arankae is generally considered to be the oldest lagurin in the fossil record, in comparison to *Prolagurus pannonicus*. Terzea (1970) placed *L. arankae* as the common ancestor of modern steppe lemmings. Zazhigin (1970) considered *L. arankae* as the most primitive, because its dental pattern is conservative, but he also considered this species to be an end branch. He thought that *L. arankae* and *P. pannonicus* had a common ancestor. *L. arankae* looks in fact to be the arhizodont phase of the *Borsodia petenyi* lineage, and a direct ancestor to the *Lagurus transiens*-*L. lagurus* lineage (Fig. 1). The relationships between *L. arankae* and *P. pannonicus* can be deduced by their population variability. *L. arankae*, as documented in the locality of Tsimbal (Western Caucasus), presents 95% of *L. arankae* morphological type variants and 5% of specimens featuring the characters of the *L. transiens* morphological type variants, and some of them have the typical *P. pannonicus* morphology (Zazhigin, 1970).

P. pannonicus populations, in Razdolie (Western Siberia), show great morphological variability. Thus, 1% have the "*arankae*" structure, 13% an "*allophaiomys*" type of structure (a roundish anterior loop confluent with the anteroconid components), 80% with the *pannonicus* structure (the anterior roundish loop being separated from the anteroconid components), and, finally, 6% with the *transiens* morphological structure. It is evident that these populations are closely related. The variability of *L. arankae* yields some variants that will be more abundant in *P. pannonicus*, and, moreover, other new variants appear in P. pannonicus. We should consider here the question of whether or not the variations within populations may have been clinal variations. Because L. arankae and P. pannonicus occupied approximately the same geographic area, two possibilities must be considered. Either L. arankae gave rise to P. pannonicus through a gradual phase of new balancing, or else through a polyphased sequence of speciation occurring by a peripheral bottleneck (Chaline, in press).

Rabeder (1981) suggested that P. pannonicus originated from

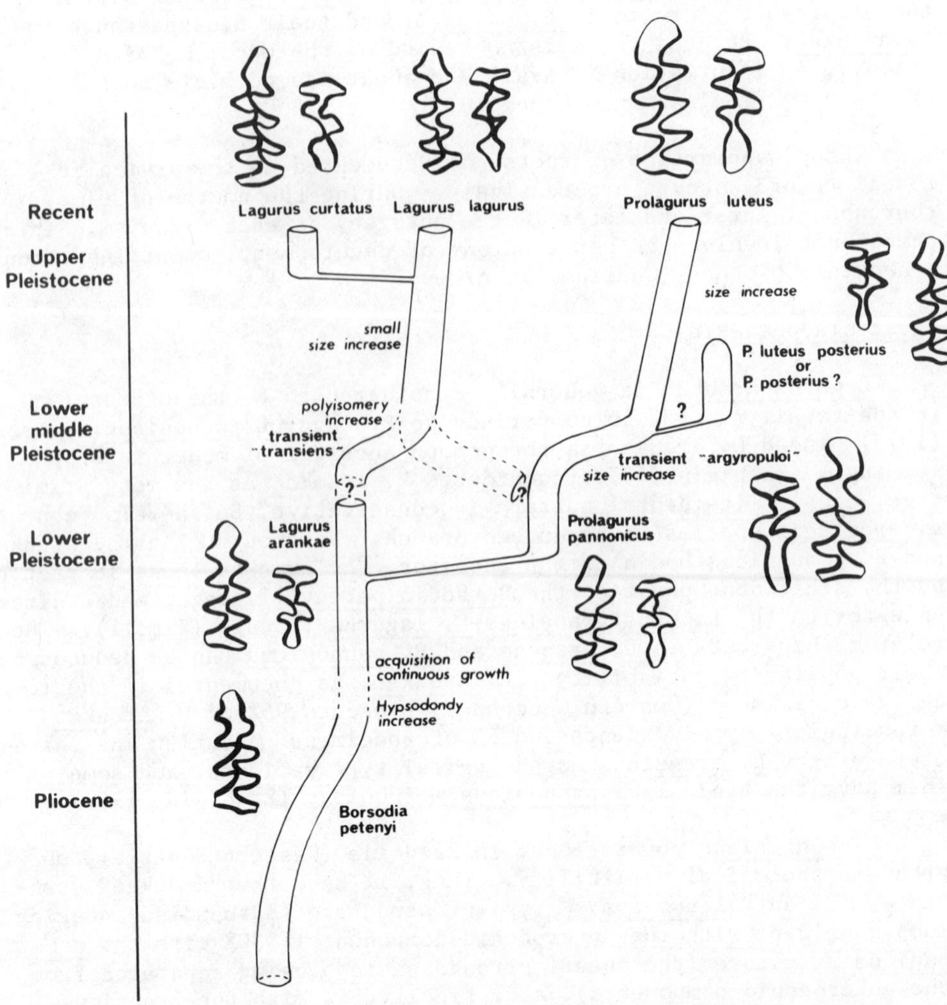

Figure 1. Evolutionary tree of lagurin evolution, showing two distinct morphological trends (polyisomery or size increase of M_1 and M^3) in two parallel lineages.

Borsodia petenyi through an intermediate form - *Borsodia hungarica*. This hypothesis is refuted by the nature of the morphological variability within species, as described above in *L. arankae* and *P. pannonicus*. *P. pannonicus* appears to be derived from *L. arankae*. This speciation event determines cladogenesis. Two distinct lineages occur then - *Lagurus* and *Prolagurus* (Fig. 1). It is not completely clear whether *L. lagurus* originated in *L. arankae*, as developed below, or in *P. pannonicus*.

Evolution of the Lagurus Lineage

The *Lagurus* lineage is characterized by: (1) the persistence of the enamel interruption on the sixth external triangle; and (2) the progressive increase in triangle numbers in lower M1 and upper M3, a polyisomery trend. *Lagurus transiens* acquired slightly concave lingual and labial sides on their lower M1 anterior loop and upper M3 posterior loop. This resulted in the appearance of two new enamel triangles. Because these features first appear as variants in *L. arankae* and are completely realized in *Lagurus lagurus*, through the *L. transiens* intermediate, these three species can be considered as three transients of a single lineage. It seems that the *arankae* stage lasted throughout the Lower Pleistocene. The *transiens* intermediate occurred only during a short period in the lower Middle Pleistocene. This transient probably gave rise to *Lagurus lagurus*, which has stabilized since that period. The *Lagurus arankae-transiens-lagurus* lineage can be considered as a chronospecies, with three respective transient stages. The evolution of the *Lagurus* lineage probably occurred without any drastic population reduction. After the gradual transition from *Borsodia petenyi* to *Lagurus arankae*, the latter species remained in stasis during the Lower Pleistocene (some hundred thousand years). An accelerated phase of gradual evolution (*L. transiens*) occurred then, during a short phase (fifty thousand years?) of the lower Middle Pleistocene. This led to a small size increase in *L. lagurus*, which has conserved its morphological stasis since that period. The *transiens* phase, without population reduction, corresponds to a phyletic gradualism phase.

In the Upper Pleistocene, the *Lagurus lagurus* lineage migrated to North America via the Bering Strait. This migration certainly only involved small peripheral populations of Eastern Siberia. Their settling in the new area led to a present day distinct polytypic species, *L. curtatus*. This splitting corresponds to a late cladogenesis within the *L. lagurus* lineage.

Evolution of the Prolagurus Lineage

Prolagurus pannonicus split from *L. arankae* at the beginning of the Lower Pleistocene (Fig. 1). During the speciation phase, certain characteristics, such as the sixth external triangle with enamel interruption, disappear. In *P. pannonicus* the anterior round loop becomes more or less separated from the fourth and fifth broadly confluent triangles. The evolution of the lineage thus presents two major evolutionary features: (1) a great increase in size, and (2) the progressive closing of the fourth and fifth triangles on lower M1.

P. argyropuloi has been described on anthropogene material from an unknown age and locality. It looks like a large *P. pannonicus*. A specimen found in Chou-Kou-Tien (China), and described under the name of *simplicidens*, seems to be synonymous. In more recent times, the closing of the anteroconid lower M1 triangles leads to *P. luteus*.

In Eastern countries, from the Volga, Azov sea, to Southern Siberia, a new form, morphologically similar, but smaller than *P. luteus*, occurred during the lower Middle Pleistocene. This form has been described as *P. posterius*. It is thus difficult to decide whether this form should be considered as a distinct species, sympatric with *P. luteus*, or only as a small subspecies of it. The greatest size increase occurred in the lower Middle Pleistocene, with morphological changes occurring shortly afterwards. The late Middle Pleistocene *P. luteus* seems to be relatively stable.

The Trends in Steppe Lemmings

The two lagurin lineages, as documented in the fossil record, have had a different evolution. In the *Lagurus arankae-transiens-lagurus* lineage, the major evolutionary feature is an increase of polyisomery and a small increase in size, whereas, in the *Prolagurus pannonicus-argyropuloi-luteus* lineage, the most important evolutionary trend corresponds to a large increase in size, with little morphological change (Fig. 1). These two mechanisms can be interpreted as different adaptations to solve a unique problem - how to increase the grinding surface of molars.

STEPPE LEMMINGS DATA AND SPECIATION

After the splitting of *L. arankae*, the two lineages evolved in parallel. After a stasis of *L. arankae* during the Lower Pleistocene, the *Lagurus arankae-transiens-lagurus* lineage evolved rapidly during the short period of the lower Middle Pleistocene (transient *transiens*), and then *L. lagurus* remained in stasis until the present day. In the *Prolagurus pannonicus-argyropuloi-luteus* lineage, after the *pannonicus* stasis during the Lower Pleistocene, the *P. argyropuloi* transient occurred during the lower Middle Pleistocene, preceding the *P. luteus* stasis leading to the present day species.

The two lineages evolved in parallel yet differ distinctly. A first phase of stasis developed before a short transient gradual phase. This led to a new stasis. Evolutionary changes occurred during the lower Middle Pleistocene. This was also the period when one of the most important cooling trends occurred in Eurasia, the so-called Mindel glacial complex. Gradual morphological changes occurred throughout the geographical area of the two lineages, but without any reduction of population sizes. Data on the evolution of steppe lemmings can be summarized as follows:

(a) Gradual transition phase leading from Borsodia to Lagurus arankae.
(b) Speciation phase giving rise to Prolagurus pannonicus, probably through a peripatric speciation event.
(c) Lagurus arankae stasis phase.
(d) Lagurus transiens gradual transition phase.
(e) Lagurus lagurus stasis phase.
(f) Speciation phase giving rise to Lagurus curtatus, either through a peripatric event (small migrating founder population), or through a dumbbell speciation event.
(g) Prolagurus pannonicus stasis phase.
(h) Prolagurus argyropuloi gradual transition phase.
(i) Prolagurus luteus stasis phase.
(j) Prolagurus posterius speciation or subspeciation phase?

These data support the polyphased model of evolution (Chaline, in press), which postulates that evolution appears as a result of species equilibria, punctuated by environmental or stochastic disequilibria. They induce polyphased sequences of various spatial, temporal, and populational size modes, leading to new species equilibria through anagenetic or cladogenetic patterns.

In steppe lemmings, new species probably arose through a peripatric event (Prolagurus pannonicus, P. posterius?, Lagurus curtatus). In the two lagurin lineages, the gradual transitional morphological phase of L. transiens and P. argyropuloi appears as phyletic gradualism phases, without any size reduction. L. transiens and P. argyropuloi can be considered as transients within a diachron or chronospecies (Chaline, 1983). Most evolutionary changes in steppe lemmings appear to be more or less correlated or directly linked with environmental modifications, very often climatic ones. For instance, the gradual transitional phase between Borsodia petenyi and Lagurus arankae, where continuous growth has been acquired, occurred during the Eburonian cold phase. The same phenomenon developed in the Mimomys-Microtus lineage at the same time and in the same area. The increase in hypsodonty, and the acquisition of continuous growth can be considered as responses to environmental changes through paedomorphic features involving neotenic processes (Chaline, in prep.).

The second gradual phase related to L. transiens and P. argyropuloi is contemporaneous with the so-called Mindel glacial complex. The speciation event of L. curtatus is a direct consequence of climate. The passage of small East Siberian populations to North America via the Bering Strait occurred during the last glacial regression (Wisconsin).

The nature of climatic changes acts as a stimulus initiating a specific process. The morphological responses seem to be adaptive, but they vary in nature (size increase, acquisition of cement, increase of hypsodonty, acquisition of continuous growth, change in the grinding surface, deltodonty, polyisomery, etc.). There exist no rules to respond to environmental changes. Certain climatic changes, such as some cold phases, seem not to have played any role in steppe lemmings, whereas they have acted as determinants in other groups.

CONCLUSIONS

Steppe lemmings are well documented in the fossil record, both in time and space. They demonstrate that evolution occurred either by the peripatric event of speciation (involving size reduction), or by gradual population changes (eliminating size bottlenecks). Climate seems to have played a major role in initiating evolutionary changes. Two steppe lemming lineages evolved in parallel during the Pleistocene, although very different morphological changes occurred. However, these evolutionary changes were responses to the same stimuli. Lagurin data support the polyphased model of evolution.

REFERENCES

Adamenko, O. M. and Zazhigin, V. S. 1965. The fauna of small mammals and the geological age of Kochkovsk suite in southern Kulunda. In: Stratigraphic Importance of Small Mammalian Anthropogen Fauna, K. V. Nikiforova, ed., 7th Cong. INQUA, pp. 162-171, Nauka, Moscow.

Chaline, J. 1972. Les Rongeurs du Pléistocène moyen et supérieur de France. Cahiers de Paléontologie. CNRS, Paris.

Chaline, J. 1975. Taxonomie des Campagnols (Arvicolidae, Rodentia) de la sous-famille des Dolomyinae nov. dans l'hémisphère nord. C. R. Acad. Sci. Paris D 261: 115-118.

Chaline, J. 1983. Les roles respectifs de la spéciation quantique et diachronique dans la radiation des Arvicolidés (Arvicolidae, Rodentia), conséquences au niveau des concepts. In: Modalités, rythmes et mécanismes de l'évolution biologiques: gradualisme phylétique ou équilibres ponctués?, Colloque int. CNRS, No. 330, J. Chaline, ed., pp. 83-90, CNRS, Paris.

Janossy, D. 1964. Evolutionvorgänge bei pleistozänen Kleinsäugern. Z. Säugetierk. 29: 285-289.

Konstantinova, N. A. 1967. The Anthropogen of Southern Moldavia and Southwestern Ukraine. Nauka, Moscow.

Kormos, T. 1938. Mimomys newtoni Major und Lagurus pannonicus Kormos. Zwei gleichzeitige verwandte Wühlmäuse von verschiedener phylogenetischer Entwicklung. Math. Naturw. Anz. Ungar. Akad. Wiss. 57: 356-379.

Kretzoi, M. 1956. Die Altpleistozänen Wirbeltier faunen des Villanyer gebirges. Geol. Hung. Ser. Palaeont. 27: 1-264.

Rabeder, G. 1981. Die Arvicoliden (Rodentia, Mammalia) aus dem Pliozän und dem älteren Pleistozän von Niederösterreich. Beiträge zur Paläontologie von Osterreich, Wien: 1-343.

Schevtschenko, A. I. 1965. Faunistic complexes of small mammals from Upper Cenozoic deposits in the Southwestern part of the Russian plain. In: Stratigraphic Importance of Small Mammalian Anthropogen Fauna, 7th Congr. INQUA, K. V. Nikiforova, ed., pp. 7-59, Nauka, Moscow.

Terzea, E. 1968. Observatii a supra speciilor de Lagurus descoperite in pleistocenul Romaniei. Lucrarile Inst. de Speologie "Emil Racovita" 7: 271-290.

Terzea, E. 1970. Sur l'apparition et l'évolution de quelques genres d'Arvicolidés (Rodentia) pendant le Pleistocène. Livre du centenaire Emil Racovita: 499-511.

Topachevsky, V. A. 1965. Nasekomoiadnye i grizuny Nogaiskoi Pozdne Pliocenovoi Fauny. Naukova Dumka, Kiev: 1-162.

Vangengeim, E. A. and Zazhigin, N. S. 1969. Eopleistocene mammals of Siberia as compared to those of Eastern Europe. In: The Main Problems of Anthropogen Geology in Eurasia, 8th Congr. INQUA, I. M. Gromov and K. V. Nikiforova, eds., pp. 47-59, Nauka, Moscow.

Zazhigin, V. S. 1970. Significance of lagurins (Rodentia, Microtinae, Lagurini) for the stratigraphy and correlation of eopleistocene deposits of Eastern Europe and Western Siberia. Paleogeography, Paleoclimatology, Paleoecology 8: 237-249.

KARYOTYPE VARIABILITY AND CHROMOSOME TRANSILIENCE IN RODENTS: THE CASE OF THE GENUS MUS[1]

Ernesto Capanna

Department of Animal and Human Biology
University of Rome
Rome, Italy

HYPOTHESES AND FACTS CONCERNING CHROMOSOMAL TRANSILIENCE

In the evolutionary processes a basic distinction must be made between adaptive divergence and transilience (Templeton, 1982), because the latter involves a discontinuity in which some sort of reproductive barrier is overcome by different evolutionary forces in which chromosomal rearrangements (chromosomal transilience) have to be considered as preeminent. Oversimplifying this semantic dichotomy, natural selection drives adaptive divergence, while transilience occurs in spite of it. This non-Darwinian formulation of transilience speciation obviously must be accepted with all due caution. As will be shown below, some sort of selective advantage must be involved in the chromosomal rearrangement itself in order to allow the fixation of the new karyotype in the population.

Moreover, the role of chromosomal rearrangement in the evolutionary process is evaluated differently by different authors, as far as its relevance to the process of reproductive isolation and speciation is concerned. However, this problem is easy to state and is thus simple to verify. If there are numerous cases in which closely related species display different karyotypes, an important role must obviously be played by some karyotype transformation mechanism in the speciation process. If, on the other hand, reciprocal

1 This report on chromosomal evolution in rodents was planned by Alfred Gropp and me as a joint effort to summarize the results of a decade of collaborative work on *Mus* cytogenetics. Unfortunately, Alfred Gropp died suddenly in the autumn of 1983. As a token of a profound friendship, I dedicate this paper to his memory.

cases are numerous, i.e., of taxonomically related species and/or genera maintaining the same karyotype inside the families and orders it must be acknowledged to be a comparatively insignificant sort of speciation, at least as far as the evolutionary history of that particular taxon is concerned. However, the whole problem hinges on the exact interpretation of the word numerous. Mayr (1970) was undeniably right when he claimed that "... cases in various groups of organisms are known of well defined and reproductively isolated species that agree completely in their chromosomal structure and differ only in their gene content." However, he was mistaken when he described these cases as numerous.

When referring to rodents, the term "numerous" is certainly muc more appropriate to a high interspecific karyotype variability. One of the many possible examples is that of the genus *Gerbillus*, in which the diploid number varies from 76 to 32 (Wahrman and Zahavi, 1955). If the Gerbillinae are considered as a whole, this range of variability can be further extended to take in *Ammodillus imbellis* (Capanna and Merani, 1981), which shows a diploid number of 18 (Fig. 1).

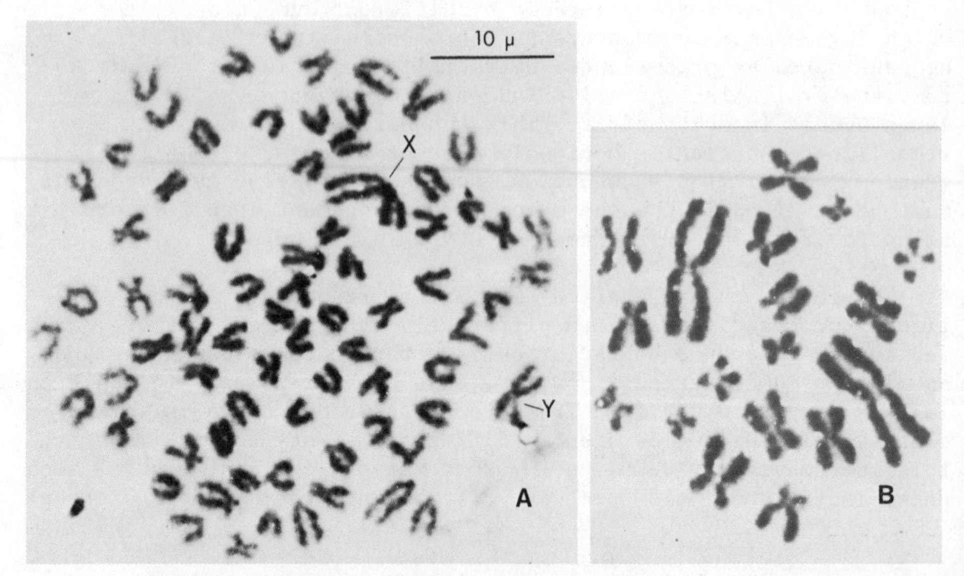

Fig. 1. Extreme karyotype variability characterizes the Gerbillinae Somatic metaphases of: (a) *Gerbillus dunni*, showing a diploid number of 74; and (b) *Ammodillus imbellis*, with a diploid number of 18. Both photographs are at the same magnification. X 2000.

But this is not all. Karyotype variability becomes a frequent phenomenon also within rodent populations/species. Again, with reference to the Gerbillinae, _Gerbillus pyramidum_ displays a karyotype variability, with diploid numbers varying from 66 to 38 (Wahrman and Gourevitz, 1973). However, the greatest variability within the same genus is shown by the burrowing South American rodents of the genus _Ctenomys_, for which Reig and Kiblinski (1969) have indicated a complex system of inter- and intraspecific variability, involving an extremely large number of chromosomal rearrangements, in which the diploid number of the genus ranges from 68 to 22.

It thus seems clear that karyotype rearrangement in rodents is a speciation mechanism that can often be identified _a posteriori_ in the course of phylogenetic and taxonomic comparisons, but it is also present in ongoing processes of chromosomal transilience. This pattern of evolutionary models linked to karyotype rearrangement also includes the stasipatric model proposed by White (1968) for the Orthoptera of the Australian subfamily Morabinae, but it also appears applicable to various cases of chromosomal speciation in rodents (White, 1978a). As will be discussed in greater detail below, White (1978b) postulated a limited vagility of the species as an effective and necessary way of overcoming the negative heterosis of the initial heterozygotes. By means of forced inbreeding, this limited vagility allows the rapid homozygotic fixation of the chromosomal variant. Indeed, the heterozygote for chromosomal rearrangement is affected by an impairment of its relative fertility, on which an effective post-mating reproductive barrier is based when speciation is complete. Nevertheless, such fertility impairment represents a serious handicap for the initial fixation of the new karyotype and its exploitation in the population/species.

Different types of chromosomal rearrangement mean different forms of interference during meiotic chromosomal pairing. Inversion consists of a rearrangement which occurs within the same chromosome. The position of one chromosomal segment is inverted so that, although there is no quantitative or qualitative change in its gene content, the way that the latter is ordered along the chromosome structure does change (Fig. 2). If the centromere is included in the inverted section (pericentric inversion), it can be identified easily, because it converts a metacentric chromosome into an acrocentric one, and vice versa. On the other hand, if the inversion does not include the centromere (paracentric inversion), it is unable to vary its position, and the mutation will be hard to detect among the somatic chromosomes, except by an accurate G- or R-banding analysis, or from evidence indicating the position of constitutive heterochromatin masses.

Systems of karyotype variability through pericentric and paracentric inversions are frequent in rodents such as _Peromyscus_, _Perognathus_, _Thomomys_, _Proechimys_, etc. Structural heterozygosity

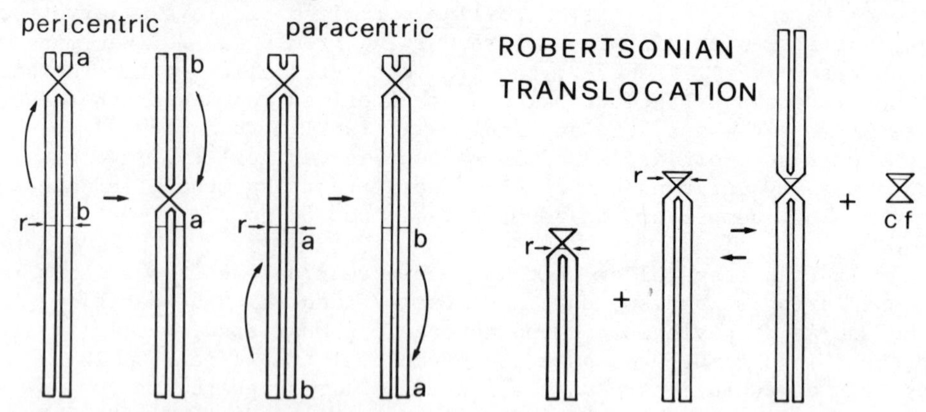

Fig. 2. The main chromosomal rearrangements. Within the inversions, only in the case of pericentric inversion does the inverted fragment comprise the centromere, and the chromosome rearrangement is evidenced by the transformation of an acrocentric into a metacentric chromosome, or vice versa. Robertsonian fusion is interpretable as the translocation of a whole acrocentric on the invisible short arm of another acrocentric chromosome. r, point of the chromosomal break; cf, centric fragment.

through both pericentric and paracentric inversion involves the formation of reversed loops during meiosis. In the case of genetic crossovers set up inside the loops, both duplications and deletions occur in chromatids, which are unevenly shared by the gametes. A certain percentage of gametes will thus be unbalanced, thereby affecting the relative fertility of the hybrids.

More severe fertility impairment occurs in the case of structural heterozygosity due to centric fusion or fission. This kind of chromosomal rearrangement known as the Robertsonian process consists of the fusion of two acrocentric chromosomes at the level of the centromere in order to form a metacentric chromosome, or else the fission of a metacentric and the consequent formation of two acrocentrics. Karyotype transformations of this type are very frequent in rodents, especially in the Myomorpha. Polymorphic Robertsonian systems have been described in many geomyoid and muroid genera, such as *Akodon*, *Thomomys*, *Geomys*, *Gerbillus*, *Ellobius*, *Acomys*, *Rattus*, *Mus*, *Thamnomys*, *Vandeluria*, *Spalax*, etc. However, cases of Robertsonian karyotype variability have been described also in Sciuridae (*Citellus*) and Caviomorpha (*Proechimys*). Very often a polymorphic system of karyotype variability includes both Robertsonian type rearrangements and peri- and paracentric inversions.

In the Robertsonian heterozygote, the pairing of the two acrocentrics with the homologous arms of the metacentrics forms a meiotic trivalent. This trivalent can take up a position on the spindle such that it can segregate either symmetrically, giving rise to balanced gametes, or asymmetrically, in which case it produces hyper- and hypo-haploid gametes. Therefore, not even this type of structural heterozygosity necessarily interferes with the fertility of the hybrids, although they have a lower relative fertility than the homozygotes. This is extremely important from an evolutionary standpoint, because it is compatible with the survival of the hybrid, and, in the early stages of the isolation process, allows the homokaryotype to be fixed. Once the new karyotype variant has been fixed in a population, it provides a means of achieving postmating reproductive isolation.

In addition, the heterochromosomes may be involved in the chromosomal rearrangement processes of both centric fusion and pericentric inversion. Examples of this process are frequent in rodents. Once again, _Gerbillus_ provides examples of heterochromosome rearrangements during speciation processes (Matthey, 1954; Viegas-Péquignot et al., 1982). The most numerous and best studied examples of sex chromosome involvement in karyotype rearrangements have been described by Matthey (1966) and Jotterand (1972) in the African pygmy mice _Nannomys_ (Fig. 3). Clearly, these cases of chromosomal rearrangement involving heterochromosomes have more drastic consequences in heterozygote fertility, in so far as any unbalanced gametes can lead to the formation of zygotes with incorrect sex complement, and consequently fully sterile progeny.

The presence of supernumerary heterochromatic minute chromosomes (B-chromosomes) has been detected in recent years in rodents (Volobuyev, 1980). In 26 murid species karyotype variability has been found to be linked to the presence of B-chromosomes. This number is extremely significant, because it represents two-thirds of the cases described in all mammals. The role played by these chromosomes in evolution remains to be clarified. It is considered by some to be a mere byproduct of processes of more substantial chromosomal rearrangement (White, 1973), and by others as having instead a definite positive significance in evolutionary processes, owing to the direct influence these chromosomes have on increasing chiasma frequency and thus the genetic variability of the populations.

CHROMOSOME VARIABILITY AND EVOLUTION IN THE GENUS MUS

One of the more interesting of the many cases of chromosomal evolution present in rodents is that offered by the species of the genus _Mus_. Not only does the presence of karyotype differences among the species and subgenera allow us to reconstruct the phylogenetic relationships within the genus, but an extensive intraspecific and

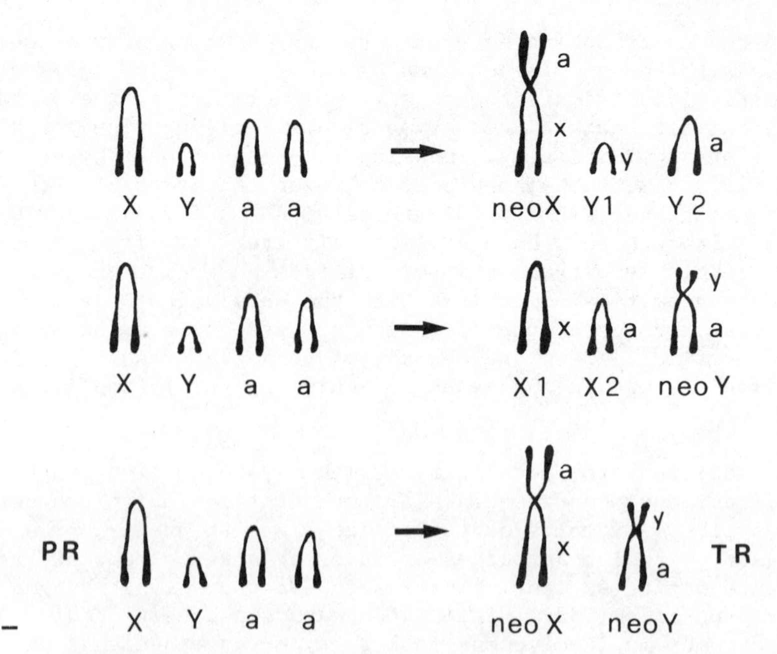

Fig. 3. The chromosomal rearrangements occurring in the heterochromosomes involve changes in the sex complement. Although the translocation X-autosome originates the X Y1 Y2 sex complement, the Y-autosome translocation produces the X1 X2 Y. Peculiar to the chromosomal variability of the African pigmy mice (Nannomys) is the translocation of both X and Y heterochromosomes on the same pair of autosomes. In this case the primitive acrocentric sex complement (PR) changes into the TR sex complement composed by both X and Y metacentric heterochromosomes.

intrapopulation karyotype variability allows us to make a detailed analysis of the role played in evolutionary transilience by karyotype transformation factors, on the one hand, and by the non-chromosomal demographic, ethological and ecological parameters on the other.

The Taxonomic Content

A recent review of the species attributed to the genus Mus (Honacki et al., 1982) lists 41 species. Although the validity of Mus poschiavinus Fatio, which is included in this list, is still debatable, this genus clearly comprises a large number of species. The small sized and undifferentiated murine forms make it difficult to establish a systematic subdivision within the genus. The division

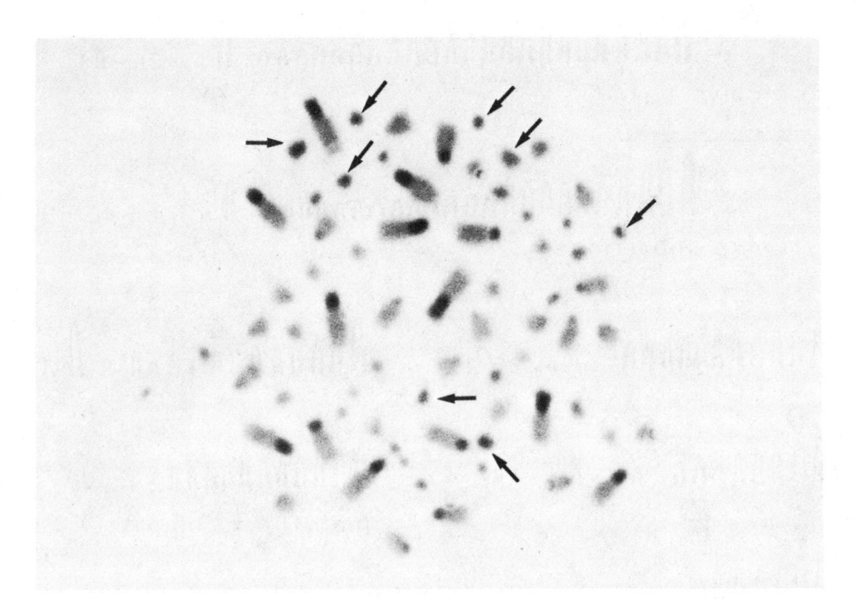

Fig. 4. C-banded somatic metaphases of the African climber rat *Tamnomys* (*Grammomys*) *gazellae* where minute C-positive chromosomes (B-chromosomes) are shown (arrows).

into six species groups proposed by Ellerman and Morrison-Scott (1966), i.e., *M. musculus*, *M. booduga*, *M. platythrix*, *M. bufo-triton*, *M. minutoides* and *M. tenellus* represented a fairly certain taxonomic reference point up until the long series of cytotaxonomic observations carried out by Matthey (1966, 1967, 1973) and continued by his coworkers (Matthey and Petter, 1968; Matthey and Jotterand, 1970; Jotterand, 1970; Petter and Genest, 1970) undermined the foundations of this taxonomic construct, especially as far as the African groups *M. bufo-triton*, *M. minutoides* and *M. tenellus* were concerned.

The Asian Species

In re-examining the systematics of the genus *Mus* in Asia, Marshall (1977) attached great importance to the karyological data. He proposed to subdivide the 16 species held to be valid into three sub-genera: the spiny mice *Pyromys*, with 5 species, *M. shortridgei*, *M. saxicola*, *M. platythrix*, *M. phillipsi* and *M. fernandoni*; the shrew mice *Coelomys* with the species *M. pahari*, *M. crociduroides*, *M. vulcani*, *M. mayori* and *M. famulus*; and lastly the house and rice field mice of the subgenus *Mus* comprising the species *M. caroli*, *M. cervicolor*, *M. cooki*, the *Mus booduga-dunni* complex and the *M. musculus castaneus* complex.

The cytotaxonomic situation regarding these Asian forms is

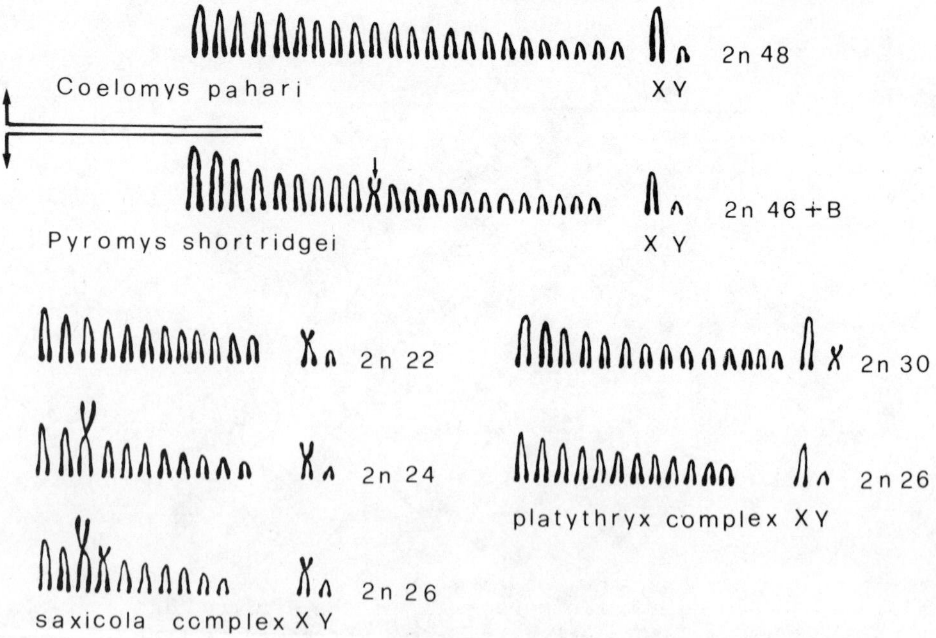

2n 48
Coelomys pahari XY

2n 46+B
Pyromys shortridgei X Y

2n 22
2n 30
2n 24
2n 26
platythryx complex XY
2n 26
saxicola complex XY

Fig. 5. Haploid sets and heterochromosomes of four species of Asian Mus. On the top the sole karyotype known among species of the subgenus Coelomys. On the lower part of the figure, the karyotypes of three species of the subgenus Pyromys, one of which, Mus shortridgei, shows a karyotype variable throughout B-chromosomes; the arrow indicates a variable element of the karyotype. The additional two species, M. saxicola and M. platythryx, show polymorphic chromosomal systems.

extremely interesting and confirms the high degree of karotype variability that exists both between the species and inside the population/species and is typical of murid biology (Fig. 5).

If the three sub-genera are considered separately, it is seen that for Pyromys the karyotypes are known for three out of the five species comprising it. Mus shortridgei displays what may be considered an ancestral karyotype (Gropp et al., 1973) with 46 acrocentric chromosomes, including the heterochromosomes. To this must be added a variable B-chromosome fraction which can increase the diploid number to 49. Two different karyotypes have been attributed to Mus platythrix; one (Dhanda et al., 1973), for a population from central-west India (Poona) with a diploid number of 30, in which the X heterochromosome is still an acrocentric while the Y is a metacentric; a second karyotype (Satya Prakash and Aswathanarayana, 1973) refers to a southern Indian population with diploid number 26. In this system of karyotype variability it is no easy matter to identify simple chromosomal rearrangements to render these karyotypes congruent. Moreover, in view of the great geographic distances separating the two populations, it is unlikely that the two chromosomal forms belong to one and the same polymorphic system. Thus, these Indian spiny mice must be considered as representing a group of sibling species which are identifiable only cytologically, at least for the time being.

Also, in the case of *Mus saxicola* there is a considerable karyotype variability, both between populations of the same species; 2n = 24 in the Poona population (Dhanda et al., 1973) and 2n=22 in that of Nahan (Pathak, 1970); and within the Poona population itself. The latter comprises individuals with 26 chromosomes, all acrocentrics; with 24 chromosomes, including two metacentrics; and hybrid individuals with 25 chromosomes, with only one metacentric. On the whole, therefore, *Mus saxicola* could be postulated as a case in which there is a polymorphic Robertsonian system which is widely distributed in northern India. However, available evidence is still extremely fragmentary.

As far as the subgenus *Coelomys* is concerned, the only known karyotype is that of *Mus pahari*. However, the karyotype of this species (2n=48) closely resembles that of *Mus shortridgei* (Gropp et al., 1973), which we have postulated to be ancestral.

The six species of the subgenus *Mus* have all been found to have an identical 40-chromosome, all acrocentric, karyotype (Matthey and Petter, 1968; Markvong et al., 1973, 1975; Dev et al., 1975; Wurster-Hill et al., 1973). One exception to this standard karyotype is *Mus dunni* (Markvong et al., 1973) which, although still having a diploid number 40, also possesses a certain number of submetacentric autosomes characterized by a short heterochromatic arm. The deviation of *Mus dunni* from the subgenus norm would thus appear to be due to a process of variability in the size of the pericentromeric heterochromatin mass. This hypothesis seems to be confirmed by the fact that the number and size of the short arm vary according to which population the animal belongs to, for instance, that of Madras investigated by Matthey and Petter (1968), and that of Poona analysed by Dhanda et al., (1973). In both populations the X heterochromosome is a large metacentric, which is an unusual occurrence in the genus *Mus* and cannot easily be explained.

Each of the three subgenera could thus have an underlying ancestral type with 46-40 chromosomes, all acrocentrics, including the sex chromosomes. However, a comparison of the G-banding patterns of *Mus musculus*, *Mus pahari*, and *Mus shortridgei* carried out by Hsu et al. (1978) has revealed an entirely different pattern, indicative of phylogenetic relationship that is more remote than the similarities of the diploid number and chromosomal morphology seem to indicate.

The African Pygmy Mice

The situation regarding the African species is much more complex. Matthey (1966, 1967, 1970) and later Jotterand (1972) have investigated more than 400 animals from all over the African continent. The wide distribution of the sample has revealed a high degree of intraspecific karyotype variability between almost all the species of the *M. bufo-triton*, *M. minutoides* and *M. tenellus* groups, a phenomenon that takes on record proportions when the period in which it was described (1953) is considered.

However, even if the diploid number varies considerably within the species and species groups, it is still possible to reconstruct patterns of phylogenetic relationships, particularly by the combined use of diploid numbers and the presence of two quite separate sex complements (Fig. 6).

Two large cytotaxonomic groups may be distinguished. The first is characterized by heterochromosomes, both of which are acrocentric as in the basic *Mus* karyotype, a complement that Matthey (1966) indicates as PR, i.e., primitive. The second is indicated as TR, i.e., translocated, since the pair of heterochromosomes, both of which are metacentrics, are apparently derived from heterochromosomes of the PR complement which have been translocated on to acrocentric autosomes of the same pair. It would thus appear to be legitimate to attribute to the species group having PR chromosomes a more primitive status than the ones with TR sex chromosomes. Moreover, the species having PR sex chromosomes also include *Mus setulosus*, with a 36-chromosome, all acrocentric, karyotype and morphology and diploid number akin to those of the Asian *Mus* of the subgenus *Mus* (the group comprising *Mus musculus*, *M. caroli*, *M. castaneus*, etc.). Furthermore, the G-banding pattern analysis recently performed by Jotterand (1981a, b) shows there are considerable similarities in the banding sequence of *Mus setulosus* with that described by Hsu et al. (1978) for six Asian species of the subgenus *Mus*. Five autosomes of *Mus setulosus* coincide exactly with the banding in *Mus musculus*. A number of pairs display a partial correspondence, thereby revealing deletions or translocations. Only five karyotype elements of *Mus setulosus* display G-banding sequences not found in *Mus musculus*.

On the other hand, an African colonization by *Mus* migrating from Asia over the Miocenic links between the two continents is today considered the most probable hypothesis (Thenius, 1969). It is thus legitimate to postulate that a group of 36-chromosome species with PR heterochromosomes, i.e. *Mus setulosus*, *Mus mattheyi*, *Mus indutus* and *Mus bufo*, split off from 40-chromosome species of the

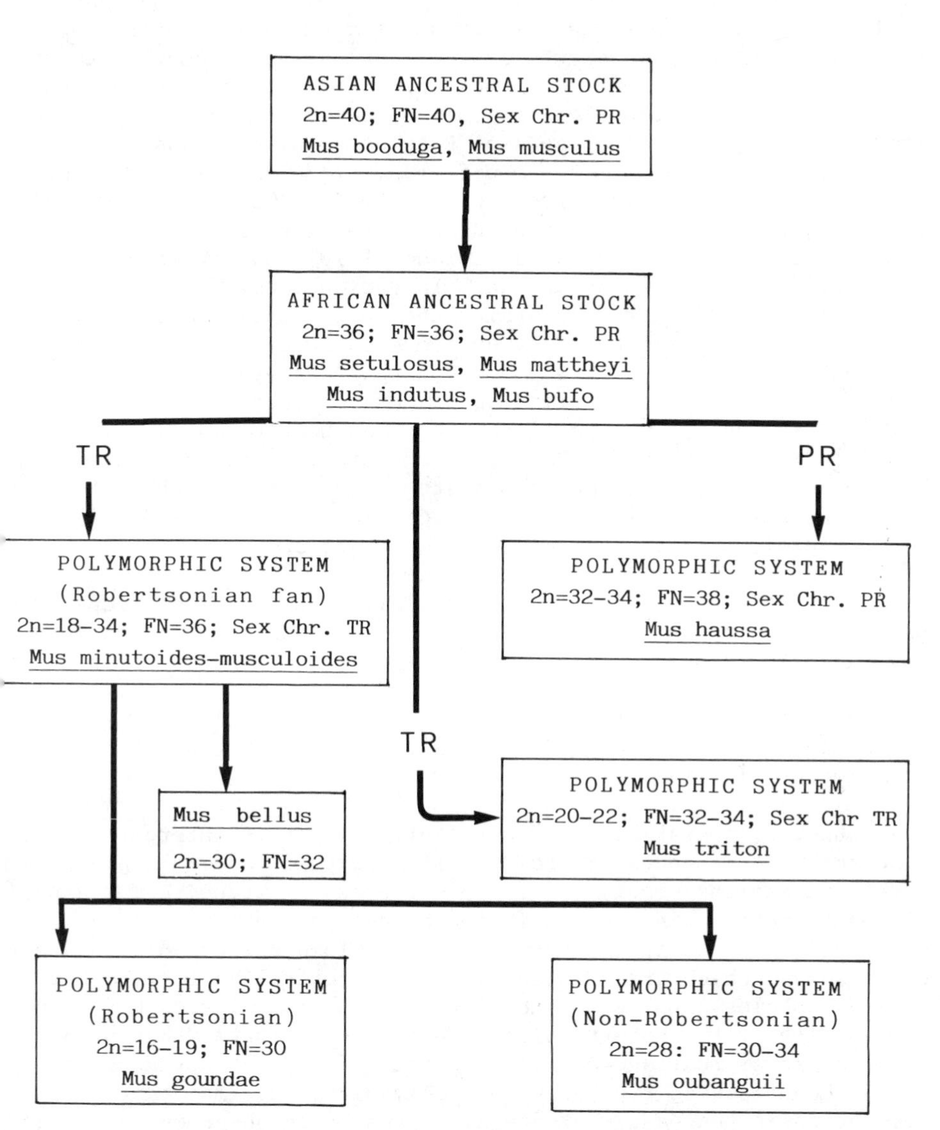

Fig. 6. The phyletic interrelationships of the African pigmy mice, according to the karyological data of Matthey (1966, 1970) and Jotterand (1972). TR indicates the occurrence of a X- and Y-autosome translocation, giving rise to a TR sex complement; PR indicates retention of the ancestral PR acrocentric sex complement.

subgenus *Mus*. At various stages, further evolutionary lines probably split off from this stock. One of these, in which the PR form of the sex chromosomes was preserved, probably gave rise to *Mus haussa*, a form still characterized by diploid numbers close to the original ones, i. e., 32-34. Two other lines then quite independently achieved the translocation of the sex chromosomes onto autosomes, thereby producing the two TR cytotaxonomic groups. On the one hand, there was probably a differentiation of the species of th *M. minutoides-musculoides* group with diploid numbers varying from 18 to 34, but whose fundamental number (FN) was fixed at 36. This resulted in what Matthey and Jotterand (1970) and Jotterand (1975) have called the Robertsonian fan. In various stages, successive processes of pericentric inversion then split off from the polymorphic *M. musculoides-minutoides* system such species as *Mus oubangi*, *Mus goundae*, and *Mus bellus*, in which the fundamental number was decreased, owing to the formation of large acrocentrics through the pericentric inversion of metacentrics. Another evolutionary line leading directly from the *Mus setulosus* group presumably gave rise to the forms of the polymorphic system of *Mus triton*, whose fundamental number is still close to the original ones (FN = 32-34), and whose diploid numbers vary widely from 20 to 32.

As mentioned earlier, the subdivision into species groups proposed by Ellerman and Morrison Scott (1966) is not reflected in the phylogenetic reconstruction based on karyotype. The *Mus bufo-trito* group seems to be unnatural, as do the *M. minutoides* and *M. tenellu* groups. It is better to consider all the African mice as belonging to one and the same group of species (Petter and Matthey, 1975; Bonhomme et al., 1982).

However, as far as the general problem we are interested in is concerned, it is necessary to stress the significant role played by karyotype rearrangement, i. e., by a source of extensive intraspecific polymorphism, in the speciation process of the African *Mus*. With the exception of the forms having diploid number 36 and PR sex chromosomes, which are considered to be ancestral, all the other African species display a high degree of interpopulation karyotype variability of the Robertsonian type. This indicates that processes of chromosomal rearrangement, which are mainly, but not exclusively of the Robertsonian type, have not only played an important role in speciation in the past, but are also present in phenomena of chromosomal transilience currently in progress.

CHROMOSOMAL TRANSILIENCE IN THE EUROPEAN HOUSE MOUSE

The parameters involved in the process of chromosomal transilience may be clarified by considering the case of the European *Mus* species.

The Taxonomic Context

On the basis of the review of Marshall and Sage (1981), which takes into account the biochemical data of Hunt and Selander (1973) and of Bonhomme et al. (1978), five taxa can be distinguished in Europe which can be considered as species or, according to Thaler et al. (1981) as semispecies. Two of these, namely *Mus musculus* and *Mus domesticus*, are more specifically commensal, while *Mus abbotti*, *Mus hortulanus* and *Mus spretus* are more often related to open field, rural, steppe or Mediterranean 'Maquis' environments. Karyologically speaking, all the species have the same 40-chromosome acrocentric karyotype corresponding to that of the Asian species of the subgenus *Mus*, as far as both the G-banding and the distribution of the constitutive heterochromatin are concerned (Dev et al., 1975). However, the genetic isolation of the taxa is betrayed by discriminant alleles that allow a subdivision into four biochemical groups, in good agreement with the morphological identification of the taxa (Table 1) according to Marshall and Sage (1981).

Karyotype Variability in the House Mouse

This karyotypic monotony was broken when Gropp et al. (1970) described a 26 chromosome karyotype for a population of house mice in Val Poschiavo in the Swiss Alps, i.e., with 8 pairs of Robertsonian metacentrics. The most obvious interpretation at the time was

Table 1. Taxonomic context of the European *Mus*.

Species (a)	Biochemical groups (b)	2n	Ecology	Distribution
M. domesticus	Mus 1	40–22	Commensal	Cosmopolitan
M. spretus	Mus 2	40	Semicommensal	Spain and Southern France
M. musculus	Mus 3	40	Commensal	Eastern Europe
M. hortulanus	Mus 4	40	Semicommensal	Southeastern Europe
M. abbotti	(Mus 4?)	40	Semicommensal	From Macedonia to Caspian Sea

(a) Marshall and Sage, 1981; (b) Thaler et al., 1981

to consider the Val Poschiavo population as a separate species (*Mus poschiavinus* Fatio). This conclusion *is no longer tenable* now that an enormous number of populations of feral house mice characterized by Robersonally rearranged karyotypes have been described. This descriptive work has been done both in Italy, mainly through the collaborative efforts of Prof. Gropp and me (Capanna, 1980, 1982; Capanna et al., 1976, 1977; Gropp et al., 1982), but also in more distant geographic areas ranging from the Orkney Islands (Berry et al., 1981) to the Antarctic Convergence (Robinson, 1978), from Spai (Adolph and Klein, 1981) to Greece (Grohe et al., 1981) and India (Chakrabarta and Chakrabarta, 1977).

The most significant fact concerning this extensive system of karyotype variability is that only *Mus domesticus* is involved in this process, in which the diploid number varies from 40 to 22. As in the case of the African *Mus*, and to a lesser extent the Asian *Mus* (*Pyromys*), only a few populations display any karyotype variabil ity, while others show a complete absence of any chromosomal rearrangement.

However, one advantage that our case of karyotype variability in the West European house mouse has over the polymorphic system of the African pygmy mice is the greater ease with which the former animal can be captured, raised and experimentally studied. Over th past decade the combined work of the Lubeck group and the Rome grou has enabled several thousand animals to be studied karyotypically, electrophoretically and biometrically. This combined research has led to the recognition and precise characterization of the Robertsonian fusion pattern of 13 homozygous populations and numerous hybridation areas in the Alps (Capanna and Valle, 1977; Capanna and Riscassi, 1978; Gropp et al., 1982), in the Apennines (Capanna et al., 1977) and the Eolian Islands (Godena et al., 1978) (Fig. 7).

One peculiar characteristic of the karyological polytypism of *Mus domesticus* is the randomness of the Robertsonian fusion pattern which, in contiguous populations, involves different chromosomes of the all-acrocentric standard mouse karyotype. Consequently, we hav different populations, either adjacent or lying far apart, with the same diploid number 22, as in the case of two Apennine populations and one from the Alps, which have completely different karyotypes because the Robertsonian metacentrics they possess are obtained by the fusion of different acrocentrics of the standard 40 chromosome karyotype. For example (Table 2), one metacentric is shared by the CD and the CB Apennine karyotypes, i.e., the fusion Rb (9.16), and one is common to the CB Apennine and to the Orobian Alps karyotypes, i.e., the fusion Rb (5.15). All other metacentrics show different fusion patterns. Consequently, only two appeared more than once out of 27 metacentrics.

The assessment of the fusion pattern was carried out on the

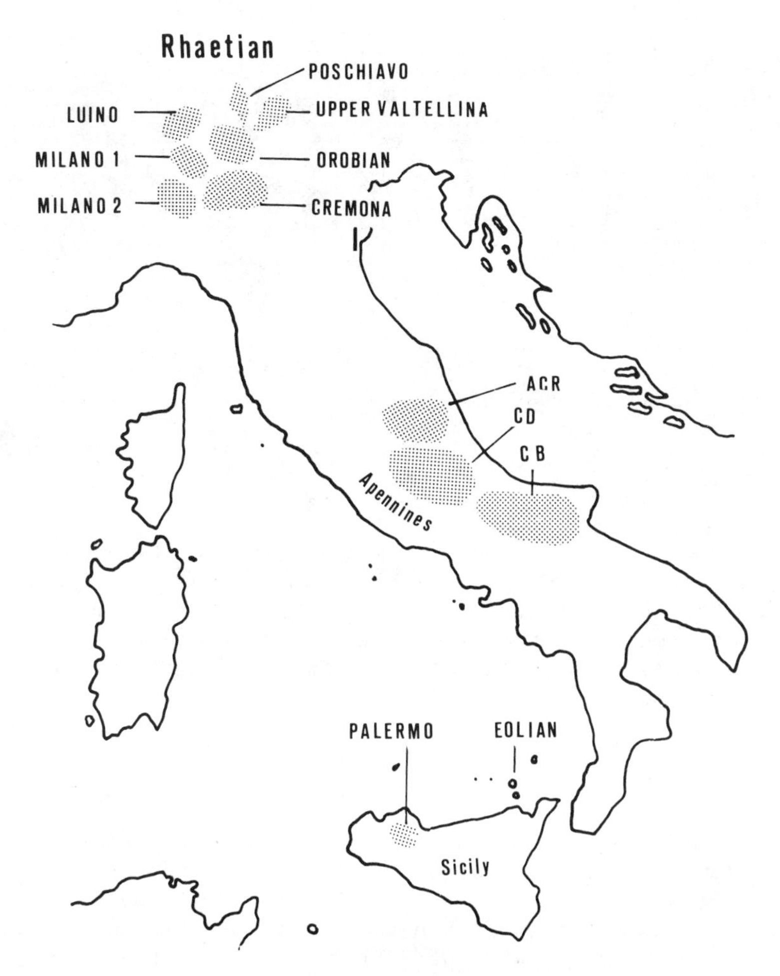

Fig. 7. Distribution of the feral mouse population with a Robertsonally rearranged karyotype in Italy.

basis of the G-banding (Fig. 8) and confirmed in the analysis of the meiotic diakinesis of hybrids. During the meiotic prophase, the Robertsonian metacentrics pair off arm-to-arm, thus producing astonishing multivalent long chain or ring patterns in a cytological domino effect (Fig. 9).

In the present report attention has been focused on the mosaic of Robertsonian populations distributed throughout the Alpine system of the Rhaetian Alps and also covering part of the Po Valley lying between the Ticino and the Mincio Rivers (Fig. 10). This "Rhaetian"

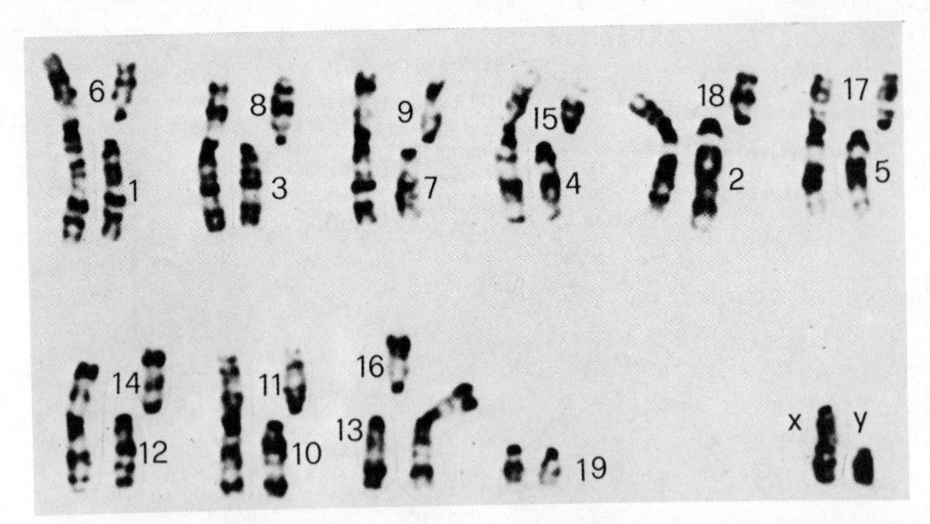

Fig. 8. G-banded karyotype of a hybrid between 22-chromosome Apennine mouse (CD strain) and a 40-chromosome all-acrocentric. The acrocentrics of 40-chromosome origin (numbers according to the standardized genetic nomenclature of mice) are put adjacent to the homologous arm of the metacentrics of 22-chromosome origin.

system constitutes an enthralling puzzle of seven interlocking piece represented by seven populations. The diploid number of the seven populations varies from 26 to 22 (Table 3); two populations have a 22 chromosome karyotype, four have 24 chromosomes and only one has 26 chromosomes. The number of different metacentrics present in the whole system is only 19; there is in fact a high degree of sharing o metacentric chromosomes among the various populations comprising the system. Some Robertsonian metacentrics are common to all the populations, others are present only in the northern populations, while

Table 2. Independent fusion patterns in three 22-chromosome karyotypes in feral mouse populations in Italy. Numbers indicate the acrocentrics of the standard mouse karyotype involved in the formation of the Robertsonian metacentrics.

Apennine CD	7.1	13.6	8.3	15.4	11.10	18.2	17.5	14.12	16.9
Apennine CB	6.7	17.2	18.1	11.4	13.3	15.5	14.8	12.10	16.9
Orobian Alps	3.1	8.2	6.4	11.13	12.10	15.5	14.9	16.17	18.7

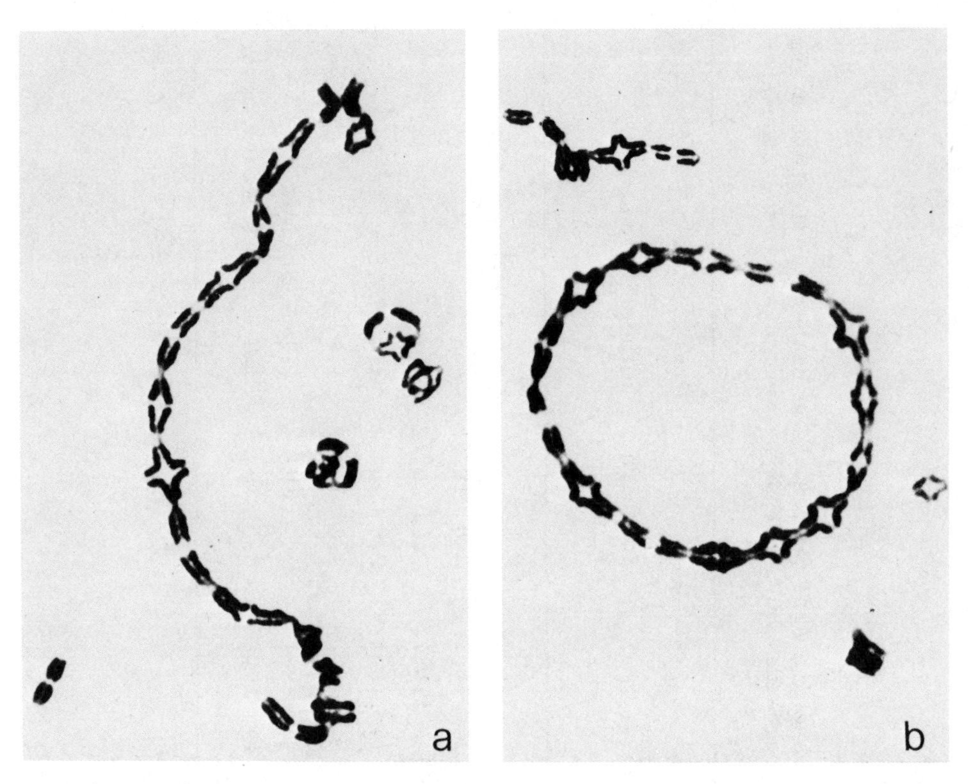

Fig. 9. Diakinesis figures of hybrids between different Apennines Robertsonian races from oocyte meiosis. The arm-to-arm pairing of the monobrachial homologous metacentrics (according to what was foreseen on the basis of the G-banding pattern) builds up such astonishing multivalent superchains or rings. (a) The hybrid CD-22-chromosomes X ACR-24-chromosomes; (b) the hybrid CB-22-chromosomes X ACR-24-chromosomes.

others again have been observed only in the southeastern ones. Only a small number of metacentrics appear in individual populations and become their characteristic karyotypic element. On the basis of this elaborate sharing of Robertsonian metacentrics in the Karyotypes of the Rhaetian populations, it was possible to construct (Capanna, 1982) a cladistic diagram (Fig. 11) for the derivation of different populations in the Lombardy-Rhaetian area, starting from an ancestral stock and going through a process of accumulation by both multiple succeeding mutations (Capanna et al., 1977) and by interdeme migration.

In order to analyse the evolutionary process in *Mus domesticus*, a time sequence derived *a posteriori* was used to shed light on the interrelationships between populations during the phase of fixation

Table 3. Robertsonian metacentrics in the Rhaetian Alps System. The numbers indicate the arm assignment to the acrocentrics of the 40-chromosome mouse karyotype according to the standardized genetic nomenclature.

		ARM COMPOSITION												
Localities	2n	Northern Group		Ancestral Group				Southern Group		Other Fusions				
POSCHIAVO	26	1.3	4.6	5.15	11.13	9.14	16.17				8.12			
U. VALT.	24	1.3	4.6	5.15	11.13	9.14	16.17	2.8	10.12					
OROBIAN	22	1.3	4.6	5.15	11.13	9.14	16.17	2.8	10.12		7.18			
CREMONA	22			5.15	11.13	9.14	16.17	2.8	10.12	3.4	7.18			1.6
MILANO 1	24			5.15	11.13	9.14	16.17	2.8	10.12	3.4				6.7
MILANO 2	24			5.15	11.13	9.14	16.17		10.12	7.8	2.4	3.6		
LUINO	24					9.14	16.17		10.12	3.8	2.4	11.18	5.13	6.7

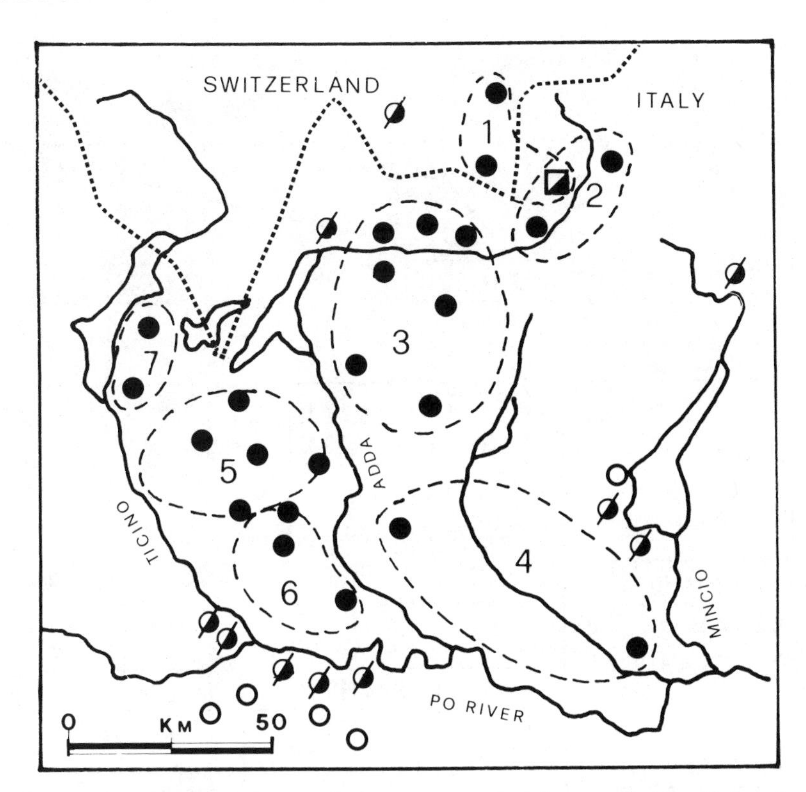

Fig. 10. Distribution map of seven populations of the Robertsonian polymorphic system of mice in the Rhaetian Alps. 1, Poschiavo valley; 2, Upper Valtellina; 3, Orobian Alps (comprising lower Valtellina); 4, Cremona; 5, Milano 1st; 6, Milano 2nd; 7, Luino. Black circles indicate the localities where Robertsonian homozygous populations were found, white circles 40-chromosome populations, and black-white circles hybrid populations. The black and white square locates a site where a sympatric occurrence of two Robertsonian races was found without natural hybridization.

and accumulation of the Robertsonian metacentrics. However, the historical reconstruction of the evolution of the Robertsonian populations in the Rhaetian Alps does not answer the fundamental question raised by this speciation model in *Mus domesticus*, namely; what are the factors allowing the fixation and accumulation of such an unusually large number of centric fusions?

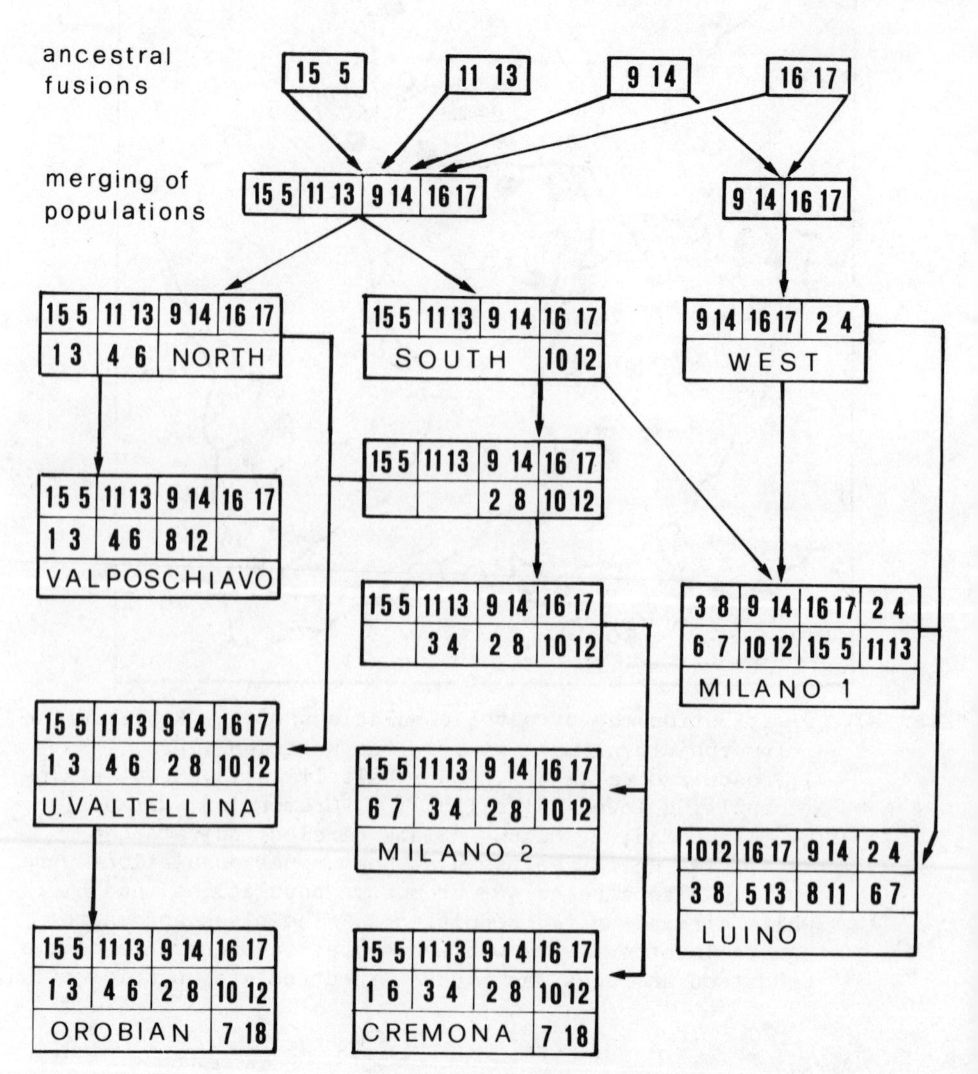

Fig. 11. An entangled network of interrelationships between the populations of the Rhaetian Alps is evidenced by the sharing of their karyotype metacentrics derived from a common ancestor or acquired by interdeme migration. On this basis it is possible to reconstruct the genesis of the seven populations. Numbers indicate the acrocentrics involved in the fusions.

DEMOGRAPHIC FACTORS INVOLVED IN CHROMOSOMAL TRANSILIENCE IN MUS

Attention must thus be focused on factors of paramount importance in the field of population biology, i.e., factors related to the size of the distribution area of the species, behavioral factors, and factors related to the structure of the populations which allow homozygosity to be achieved rapidly for the new metacentrics and their fixation in the population.

Recent algebraic models by Hedrick (1981) and Lande (1979) have been used to evaluate the various factors of particular relevance in this connection. They attach greatest importance to four parameters, i.e., the selective advantage of the new homokaryotype, meiotic drive, genetic drift, and inbreeding consequent on the actual minimal size of the deme.

All these factors may be identified in the biology of the populations of _Mus domesticus_. Only as far as the selective advantage of the new homokaryotype is concerned is it perhaps difficult to imagine that this advantage could be due to the Robertsonian fusion itself. Nevertheless, it can be assumed that some sort of adaptive and advantageous character can appear in the populations on the outskirts of the distribution area of the house mouse and can be positively selected in an environment with high levels of extinction and recolonization. In these demographic conditions, which actually occur in mountain populations of house mice, the appearance of the metacentrics may be assumed to guarantee that the area effect will be maintained and to give the metacentric carrying such a gene system a high degree of adaptiveness.

The other factors, such as inbreeding and genetic drift, may be said to derive from a truly minimal deme size resulting from the strictly commensal condition of _Mus domesticus_ in a primitive rural environment, such as is found in high mountain areas. Under these conditions human dwellings, and thus also the house mouse populations, are broken up into tiny demes. This process is enhanced by the ethology of the mouse itself, which is characterized by territorialism and by a highly restricted vagility. Furthermore, the reproductive behavior of the mouse is characterized also by rigid male and female hierarchies, which keep the number of genes transmitted in a population down to a minimum. As a result of all these conditions, inbreeding becomes unavoidable, genetic drift highly likely, and the homozygosis of the new homokaryotype is rapidly achieved.

From these premises, it may legitimately be surmised that each initial population has achieved homozygosity through single Robertsonian metacentrics, under conditions of allopatric isolation made possible by a commensal context in an archaic system of alpine agriculture.

Table 4. Evolutionary steps in Mus domesticus chromosomal transilience.

1.	Allopatric fixation of new homokaryotypes		
2.	Accumulation stage: (A) Allopatric successive fusions	and/or	(B) Parapatric merging of populations
3.	Parapatric postmating isolation		
4.	Sympatric premating isolation		

The present populations of Mus domesticus with a Robertsonian rearranged karyotype have reached a considerable size, and their karyotypes have a rather large number of metacentric chromosome pair i.e., 7, 8 or 9. This phase of the spread of the populations and the accumulation of metacentrics in them are the product of two mechanisms (Table 4). One is that of successive fusions inside the populations already undergoing homozygosis, and the other the mergin of populations characterized by homozygosis for metacentrics without any arm homology (Capanna, 1982). This gives rise to those processe of meiotic drive whose importance has already been stressed in the algebraic model of Lande (1979). Indeed, a low number of structural heterozygosities gives the hybrids a low degree of fertility impairment. This allows them to reproduce, but drives them to acquire a new homozygosity for both metacentrics involved in the hybridation process of the two populations that are homozygous for a single metacentric.

However, when a large number of metacentrics has accumulated in the population, a postmating isolation occurs to block the gene flow betweeen populations, that is, between the Robertsonian populations themselves, because the gametogenetic process is blocked by multivalent chains, and between Robertsonian populations and the all-acrocentric populations surrounding them, because the large numb of structural heterozygosities (between seven and nine) raises the percentage of aneuploid gametes produced to about 70% (Capanna, 1976

At this stage, premating isolation systems are rapidly selected in cases of sympatric or parapatric contact. In fact, the price to be paid in terms of a decrease in the reproductive potential of the populations is too high if the reproductive isolation is entrusted to the postmating mechanisms alone.

There has been extensive evidence of this reinforcement in the form of a case of sympatric inhabitation of two chromosomal races in the polymorphic system of *Mus domesticus* in the Rhaetian Alps (Capanna and Corti, 1982). The Val Poschiavo population, characterized as it is by a 26 chromosome complement, recently entered Valtellina, where it encountered a different population characterized by a different karyotypic situation - diploid number 24 with two metacentrics also displaying monobrachial homology. Despite the slight karyotype difference, this situation causes full sterility in the male hybrids and serious fertility impairment in the female hybrids. In such cases, behavioral diversification in inter-male aggressive patterns has been observed to play a decisive role in instituting premating isolation (Capanna et al., 1984).

This wide range of actual apparently latent reproductive barriers in the Robertsonian populations of *Mus domesticus* again open the question of the definition of species, race and population, a problem for which a satisfactory solution has been proposed in the form of the Modern Synthesis of Evolution.

It is in fact often difficult to apply the biological concept of species as defined by Dobzhansky (1970) and Mayr (1970) to taxa undergoing rapid speciation processes or such active and often incomplete processes as that of the Robertsonian races of the *Mus domesticus* complex.

A solution to which I am personally opposed is that of giving each isolated population a different taxonomic name. This position reflects my philosophical attitude toward the following question. To what extent is it possible to establish whether these "species in status *nascendi*" have achieved the rank of species or not? Above all, what satisfaction can there be in thinking that a problem can be solved merely by attributing a specific name to it?

ACKNOWLEDGEMENTS

This work has been supported by grants of the Italian Ministry of Education (MPI 40) and of the National Research Council (CNR 82.02722.04).

REFERENCES

Adolph, S. and Klein, J. 1981. Robertsonian variation in *Mus musculus* from Central Europe, Spain, and Scotland. J. Hered. 72: 219-221.

Berry, R. J., Brooker, P. C., Lush, I. E., Nash, H. R., and Newton, M. F. 1981. Robertsonian translocations in wild mice. Mouse News Letter 64: 65-66.

Bonhomme, F., Britton-Davidian, J., Thaler, L., Triantaphyllidis, C.

1978. Sur l'existence en Europe de quatre groupes de souris (genre Mus) du rang espèce et semiespèce demontrée par la génétique biochimique. C. R. Acad. Sci. Paris 287 (D): 631-633.

Bonhomme, F., Catalan, J., Gautun, J. C., Petter, F., and Thaler, L. 1982. Caractérisation biochimique de souris africaines référables au sous-genre Nannomys Peters, 1876. Mammalia 46: 110-113.

Capanna, E. 1976. Gametic aneuploidy in the mouse hybrids. Chromosomes Today 5: 83-89.

Capanna, E. 1980. Chromosomal rearrangement and speciation in progress in Mus musculus. Folia Zool. 29: 43-57.

Capanna, E. 1982. Robertsonian numerical variation in animal speciation. Mus musculus, an emblematic case. In: Mechanisms of Speciation, C. Barigozzi, ed., pp. 155-174, Alan R. Liss, New York.

Capanna, E. Civitelli, M. V. and Cristaldi, M. 1977. Chromosomal rearrangement, reproductive isolation and speciation in mammals The case of Mus musculus. Boll. Zool. 44: 213-246.

Capanna, E. and Corti, M. 1982. Reproductive isolation between two chromosomal races of Mus musculus in the Rhaetian Alps (Northern Italy). Mammalia 46: 107-109.

Capanna, E., Corti, M., Mainardi, D., Parmigiani, S. and Brain, P. 1984. Karyotype and intermale aggression in wild house mice: ecology and speciation. Behavior Genet. 14: 195-208.

Capanna, E., Gropp, A., Winking, H., Noack, G. and Civitelli, M. V. 1976. Robertsonian metacentrics in the mouse. Chromosoma 58: 341-353.

Capanna, E. and Merani, M. S. 1981. Karyotypes of Somalian rodent populations. 2. The chromosomes of Gerbillus dunni (Thomas), Gerbillus pusillus Peters and Ammodillus imbellis (De Winton) (Cricetidae, Gerbillinae). Monit. Zool. Ital. N. S., Suppl. 14: 227-240.

Capanna, E. and Riscassi, E. 1978. Robertsonian karyotype variability in natural Mus musculus populations in the Lombardy area of Po valley. Boll. Zool. 45: 63-71.

Capanna, E. and Valle, M. 1977. A Robertsonian population of Mus musculus in the Orobian Alps. Acc. Naz. Lincei, Rend. Sc. Mat. Fis. Nat. 62: 680-684.

Chakrabarta, S. and Chakrabarta, A. 1977. Spontaneous Robertsonian fusions leading to karyotype variation in the house mouse. Experientia 33: 175-176.

Dev, V. G., Miller, D. A., Tantravah, R., Schrech, P. P., Roderig, T. H., Erlanger, B. P. and Miller, O. J. 1975. Chromosome markers in Mus musculus: differences in M. m. musculus and M. m molossinus. Chromosoma 53: 335-344.

Dhanda, V., Mishza, A. C., Bhat, U. K. M. and Wagh, U. V. 1973. Karyological studies on two sibling species in the spiny mouse Mus saxicola and Mus plathythrix. The Nucleus 16: 56-59.

Dobzhansky, T. 1970. Genetics and the Evolutionary Process. Cambridge Univ. Press, New York.

Ellerman, J. R. and Morrison-Scott, T. C. S. 1966. Checklist of Palaearctic and Indian Mammals, 2nd ed. British Museum (Nat. Hist.), London.

Godena, G., D'Alonzo, F., and Cristaldi, M. 1978. Corrélation entre caryotype et biotype chez le Lérot (Eliomys quercinus) et autres Rongeurs de l'île de Lipari. Mammalia 42: 382-383.

Grohe, G., Gropp, A., Gropp, D., Jüdes, U., Kolbus, U., Noack, G., and Winking, H. 1981. Robertsonian chromosomes in mice from North-eastern Greece. Mouse News Letter 19: 84-85.

Gropp, A., Marshall, J., and Markvong, A. 1973. Chromosomal findings in the spiny mice of Thailand (genus Mus) and occurrence of a complex intraspecific variation in M. shortridgei. Z. Säugetierk. 38: 159-168.

Gropp, A., Tettenborn, U., and Lehmann, E. von 1970. Chromosomenvariation von Robertson'schen typus bei der Tabakmaus, Mus poschiavinus, und ihren hybriden mit der Laboratoriusmaus. Cytogenetics 9: 9-23.

Gropp, A., Winking, H., Redi, C. A., Capanna, E., Britton-Davidian, J., and Noack, G. 1982. Robertsonian karyotype variation in wild house mice from Rhaeto-Lombardia. Cytogenet. Cell Genet. 34: 67-77.

Hedrick, P. W. 1981. The establishment of chromosomal variants. Evolution 35: 322-332.

Honacki, J. H., Kinman, K. E., and Koeppl, J. W. (eds.) 1982. Mammal Species of the World. Allen Press, New York.

Hsu, T. C., Markvong, A., and Marshall, J. T. 1978. G-band patterns of six species of mice belonging to subgenus Mus. Cytogenet. Cell Genet. 20: 304-307.

Hunt, W. G. and Selander, R. K. 1973. Biochemical genetics of hybridization in European house mouse. Heredity 31: 11-33.

Jotterand, M. 1970. Un nouveau système polymorphe Robertsonien chez une nouvelle espèce de Leggada (Mus goundae Petter). Experientia 26: 1360-1361.

Jotterand, M. 1972. Polymorphisme chromosomique de Mus (Leggadas) Africains. Cytogénetique, zoogéographie, evolution. Rev. Suisse Zool. 79: 287-359.

Jotterand, M. 1975. The African Mus (pigmy mice): the role of chromosomal polymorphism in speciation. Caryologia 28: 335-344.

Jotterand, M. 1981a. Le caryotype et la spermatogénèse de Mus setulosus (bandes Q, C, G et coloration argentique). Genetica 56: 217-227.

Jotterand, M. 1981b. La formule chromosomique de Mus setulosus (Leggada). In: Wirbeltierzytogenetick, H.-J. Müller, ed., pp. 44-57, Birkhauser Verlag, Basel.

Lande, R. L. 1979. Effective deme size during long-term evolution estimated from rates of chromosomal rearrangements. Evolution 33: 234-251.

Markvong, A., Marshall, J. T., and Gropp, A. 1973. Chromosomes of rats and mice of Thailand. Nat. Hist. Bull. Siam Soc. 25: 23-32.

Markvong, A., Marshall, J. T., Pathak, S., and Hsu, T. C. 1975. Chromosomes and DNA of Mus: the karyotypes of M. fulvidisventri and M. dunni. Cytogenet. Cell Genet. 14: 116-125.

Marshall, J. T. 1977. A synopsis of Asian species of Mus (Rodentia Muridae). Bull. Amer. Mus. Nat. Hist. 158: 177-220.

Marshall, J. T. and Sage, R. D. 1981. Taxonomy of the house mouse. Symp. Zool. Soc. Lond. 47: 15-25.

Matthey, R. 1954. Un cas nouveaux de chromosomes sexuels multiples dans le genre Gerbillus (Rodentia, Gerbillinae). Experientia 10: 464-466.

Matthey, R. 1966. Le polymorphisme chromosomique des Mus Africains du sous genre Leggada. Révision générale portant sur l'analyse de 213 individus. Rev. Suisse Zool. 73: 585-607.

Matthey, R. 1967. Un nouveau système chromosomique polymorphe chez des Leggadas Africains du groupe tenellus (Rodentia, Muridae). Genetica 38: 211-226.

Matthey, R. 1970. L' "évental Robertsonienne" chez le Mus (Leggada Africains du groupe minutoides-musculoides. Rev. Suisse Zool. 77: 625-629.

Matthey, R. 1973. Leggadas (Mus. sp.) de Moundou (Tchad). Observations d'un caryotype aberrant chez une femelle. Genetica 44: 71-79.

Matthey, R. and Jotterand, M. 1970. Nouveau système polymorphe non-Robertsonien chez de Leggadas (Mus sp.) de République Centro-Africaine. Rev. Suisse Zool. 77: 630-637.

Matthey, R. and Petter, F. 1968. Existence de deux espèces distinctes, l'une chromosomiquement polymorphe chez de Mus indiens du groupe booduga. Etude cytogénétique et taxonomique. Rev. Suisse Zool. 75: 461-498.

Mayr, E. 1970. Population, Species and Evolution. Harvard Univ. Press, Cambridge.

Pathak, S. 1970. The karyotype of Mus platythrix Bennet (1832), a favorable mammal for cytogenetic investigations. Mamm. Chrom. Newsletter 11: 105-106.

Petter, F. and Genest, H. 1970. Liste préliminaire des Rongeurs Myomorphes de République Centro-Africaine. Description de deux espèces nouvelles: Mus oubanguii et Mus goundae. Mammalia 34: 451-458.

Petter, F. and Matthey, R. 1975. Genus Mus. In: The Mammals of Africa: An Identification Manual, Part 6.7, J. Meester and H. W. Setzer, eds., pp. 1-4, Smithsonian Inst. Press, Washington.

Reig, O. A. and Kiblinski, P. 1969. Chromosome multiformity in the genus Ctenomys (Rodentia, Octodontidae). Chromosoma 28: 211-244.

Robinson, T. J. 1978. Preliminary report of a Robertsonian translocation in an isolated feral Mus musculus population. Mamm. Chrom. Newsletter 19: 84.

Satya Prakash, K. L. and Aswathanarayana, N. V. 1973. The chromosomes of the spiny mouse Mus platythrix. Mamm. Chrom. Newsletter 13: 120-121.

Templeton, A. R. 1982. Genetic architecture of speciation. In: Mechanisms of Speciation, C. Barigozzi, ed., pp. 105-121, Alan R. Liss, New York.

Thaler, L., Bonhomme, F., and Britton-Davidian, J. 1981. Processes of speciation and semi-speciation in the house mouse. Symp. Zool. Soc. Lond. 47: 27-41.

Thenius, E. 1969. Phylogenie der Mammalia: Stammesgeschichte der Säugetiere (einschliesslich der Hominiden). W. de Gruyter, Berlin.

Viegas-Péquignot, E., Benazzon, T., Dutrillaux, B., and Petter, F. 1982. Complex evolution of sex chromosomes in Gerbillidae (Rodentia). Cytogenet. Cell Genet. 34: 158-167.

Volobuyev, V. T. 1980. The B-chromosome system of mammals. In: Animal Genetics and Evolution, N. N. Vorontsov and J. M. van Brink, eds., pp. 333-337, W. Junk, The Hague.

Wahrman, J. and Gourevitz, P. 1973. Extreme chromosome variability in a colonizing rodent. Chromosomes Today 4: 399-424.

Wahrman, J. and Zahavi, A. 1955. Cytological contributions to the phylogeny and classification of the rodent genus *Gerbillus*. Nature 175: 600-602.

White, M. J. D. 1968. Models of speciation. Science 159: 1065-1070.

White, M. J. D. 1973. Animal Cytology and Evolution, 3rd ed. Cambridge Univ. Press, London.

White, M. J. D. 1978a. Modes of Speciation. W. H. Freeman, San Francisco.

White, M. J. D. 1978b. Chain process in chromosomal speciation. Syst. Zool. 27: 285-298.

Wurster-Hill, D. H., Hsu, T. C., Gropp, A., Zech, L., and Marshall, J. T. 1973. Q-, G- and benzimidazole banding comparisons in several species of Eurasian *Mus*. Mamm. Chrom. Newsletter 14: 85-86.

ELECTROMORPHS AND PHYLOGENY IN MUROID RODENTS

Francois Bonhomme[1], Djoko Iskandar[1],
Louis Thaler[1], and Francis Petter[2]

[1]Institut des Sciences de l'Evolution, USTL
Place E. Bataillon, 34060 Montpellier Cedex
France, and [2]Muséum National d'Histoire Naturelle
55 rue Buffon, 75005 Paris, France

INTRODUCTION

How do we infer phylogenetic relationships from electrophoretic studies? What is the value of such phylogenies? What can we infer from these phylogenies about the processes of divergence? These are the questions that we have tried to answer empirically by analyzing genetic variability in 76 taxa of rodents, most of them being muroids. Our approach is comparable to that of Patton and Avise (1983), and is aimed at comparing the values of quantitative and qualitative analyses of electrophoretic data. However, their material and ours differ in evolutionary patterns, and the results differ accordingly in some respects.

MATERIALS AND METHODS

Seventy-six taxa at species or semi-species level, most of them being represented by a few individuals, were analyzed by standard starch gel electrophoresis at 24 protein loci. Their systematic positions are as follows:

Suborder Myomorpha: 51 murids (21 genera); 21 non-murid muroids including 1 dendromurid, 6 arvicolids, 12 cricetids, and 2 gerbillids. Suborder Glirimorpha: 2 glirids. Suborder Caviomorpha: 2 echimyids (superfamily Octodontoidea).

The list of genera surveyed appears in Figure 1; the nomenclature is according to Chaline et al. (1977).

However, it is known that routine techniques detect only a fraction of the amino acid substitutions that may affect a protein. Thus, we have addressed ourselves to different electrophoretic conditions to unravel hidden polymorphisms. For this reason, the present study includes the results obtained at 11 protein loci by sequential electrophoresis (implying a series of gels at different pH values) (Aquadro and Avise, 1982; Iskandar and Bonhomme, in press).

ANALYSIS OF DATA

We have undertaken several types of analyses as follows. The quantitative analysis (= phenetic analysis) is based on the estimation of the amount of divergence between taxa. Various indexes of protein similarity (or distance) were used to draw dendrograms (or phenograms) by the mediation of various clustering algorithms. Among the drawbacks of such classical methods, we can note first that as a prerequisite they need that the rates of evolution be constant or at least uniformly varied, so that measures of similarity reflect relationships. To free ourselves of the loss of information and of the eventual distortions brought on by the clustering techniques themselves, we have performed AFC (Analyse Factorielle des Correspondances), following Benzécri (1973), for 388 characters. Each electromorph represents one character, with a simple binary coding. This technique allows a multidimensional view of the divergence between species.

In addition, a qualitative analysis has been based on cladistic principles by trying to detect the synapomorphies (shared derived character states), which are keys for the grouping of species in monophyletic clades.

Preliminary Remarks

When available, data about intrageneric variations showed that this quantity is generally much lower than intergeneric variation. Table 1 shows the overlap of the values obtained. This fact is reinforced by our demonstration that the hidden electrophoretic variability unraveled by sequential electrophoresis is almost entirely restricted to intergeneric comparisons (Iskandar and Bonhomme, in press). This gives support to the morphological definition of most genera commonly used by zoologists, with exceptions detected in the striking cases of the two genera *Apodemus* and *Mus*. *Apodemus* (*Apodemus*) *agrarius* is as distant from the other species of *Apodemus* (*Sylvaemus*) studied here as it is from other murid genera. In the same way, the African pigmy mice *Mus* (*Nannomys*) appear distinct from other members of the genus *Mus*, to which they are not related any closer than to any other murids. This is the reason why we recognize *Sylvaemus* as a genus distinct from *Apodemus*, and *Nannomys* as distinct from *Mus*.

Table 1. Ranges of variation for a few dissimilarity indexes computed for 76 species over 24 protein loci (indexes as described in Jambu and Lebeaux, 1979).

Index	Intrageneric	Intrafamily	Interfamily
Euclidian	1.0 - 4.9	4.8 - 6.5	4.8 - 7.1
X^2-Benzécri	0.6 - 5.0	2.7 - 6.1	2.9 - 7.6
Czekanowsky	0.02 - 0.52	0.45 - 0.87	0.47 - 1.0
Jaccard	0.04 - 0.68	0.58 - 0.93	0.74 - 1.0
Sokal and Sneath	0.08 - 0.81	0.77 - 0.96	0.82 - 1.0

Phenograms

We have used 12 different similarities or distance indexes, and then five different clustering algorithms, for producing trees. In all cases, the phenograms produced show numerous phylogenetic contradictions between themselves and with accepted zoological classifications. The groupings are correct only at the family and superfamily levels (i.e., for the separation of Muroidea, Gliroidea, and Echimyidae). At the generic level, however, not only is there no stable configuration from one dendrogram to another, but there is intermixing of different families within the Muroidea, Thus, the ranges of intergeneric, intrafamilial, and interfamilial distances were largely overlapping (Table 1). The details of these results are presented elsewhere (Iskandar, 1984). An example of the dendrograms obtained is presented in Figure 1.

Hence, it appears, in the family Muridae, for instance, as if we had a set of independent genera, individualized for a long time, and among which our analytic techniques were helpless to show evidence of levels of divergence clearly different from one another.

Multidimensional Analysis

Figure 2 shows unambiguously that, when projected along the two axes having the greatest informational content, the images of the three superfamilies are clearly distinct from one another. If we now restrict our analysis solely to the muroids, we always obtain more scattered pictures (Figs. 3, 4). Nevertheless, the five muroid families studied (Arvicolidae, Cricetidae, Gerbillidae, Muridae, and Dendromuridae) do not appear to be well separated from each other, in keeping with the conclusions of the previous section. Arvicolids and cricetids are always linked, however, and Steatomys (the only

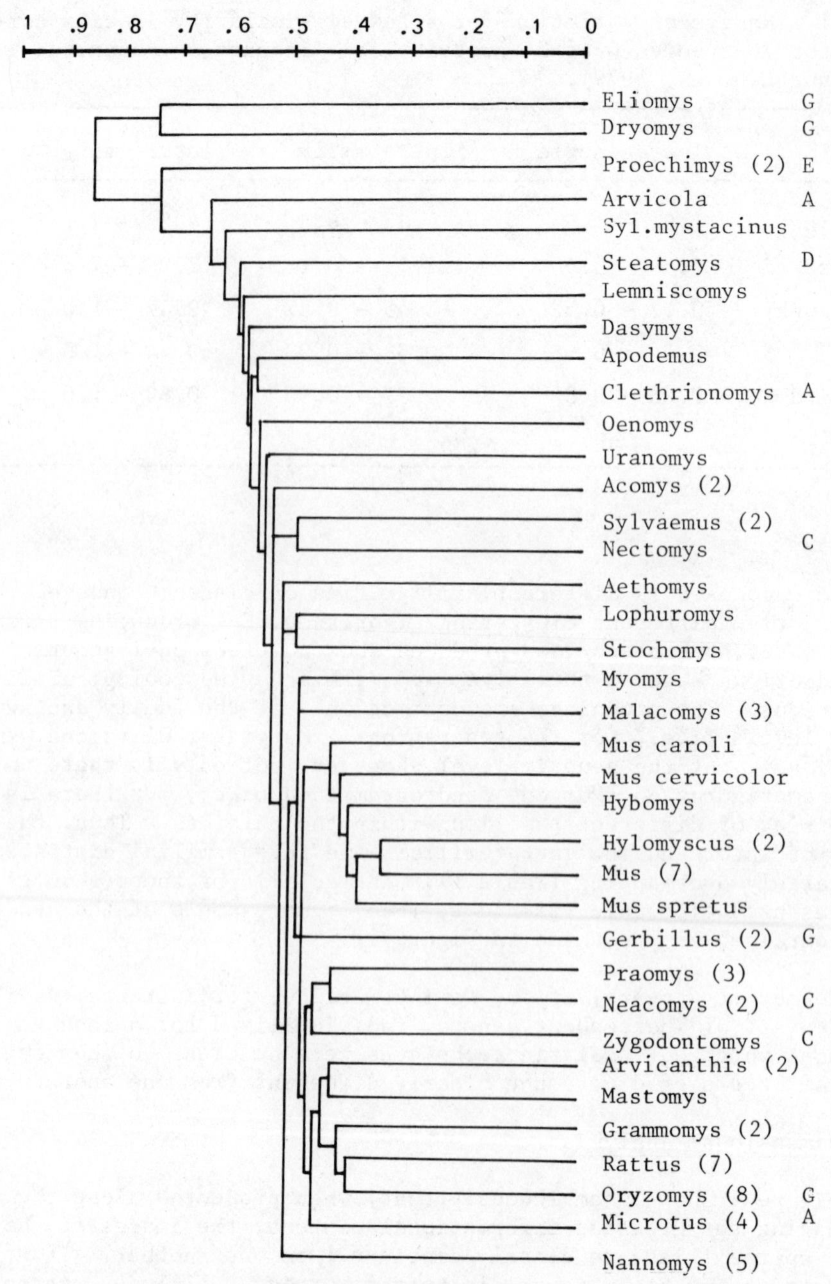

Fig. 1. Dendrogram of 76 rodent species (WPGMA of a Benzécri's X^2 distance matrix computed for 24 protein loci). Number of species studied for each genus indicated in ( ). A, arvicolids; C, cricetids; D, dendromurids; E, echimyids; G, gerbillids; other genera are murids.

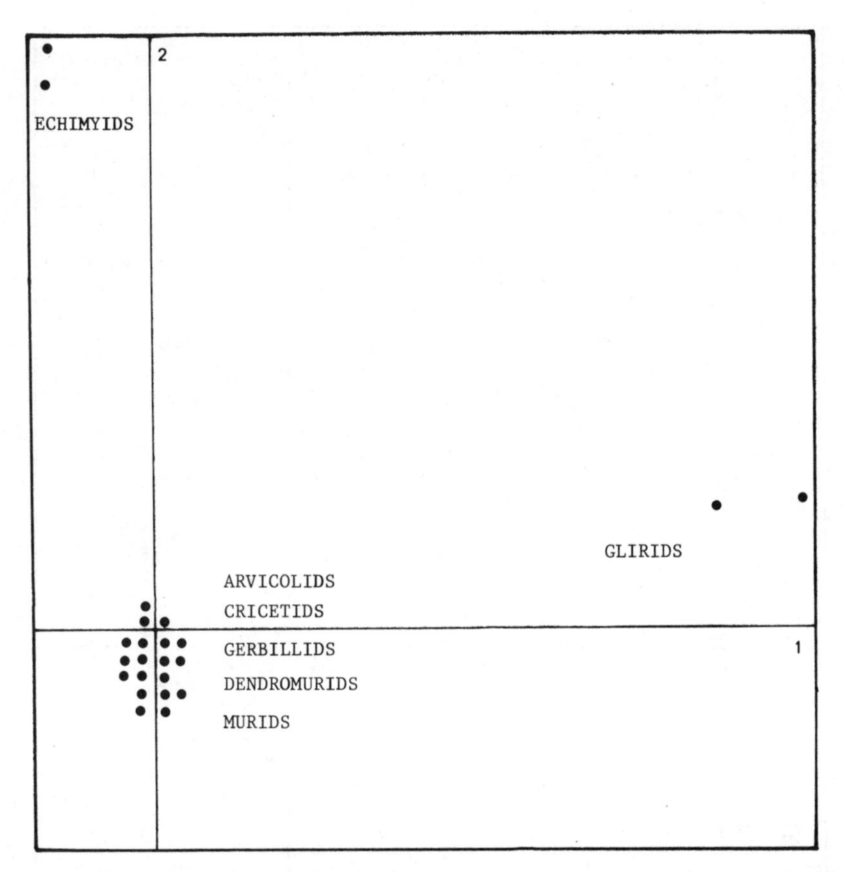

Fig. 2. AFC projection for seven rodent families, based on 388 character states at 24 protein loci. Projection along axes 1 and 2.

dendromurid investigated) does not separate from the African murids. These African murids are grouped relatively close, with the exception of Acomys (and perhaps Malacomys), which show centrifugal tendencies relatively as important as the Eurasian genera Apodemus, Sylvaemus, Mus, and Rattus. Amazingly enough, Rattus appears to be the most divergent genus by this type of analysis, which is very sensitive to the presence of unique characters (autapomorphies).

Cladistic Analysis

We have based our analysis on the data set restricted to the 10 least variable loci, as shown in Table 2. For the other loci studied, each genus displayed its own unique electromorphs (autapomorphic traits), thus allowing no phylogenetic groupings. We analyzed 33 muroid genera, among which the non-murid genera were used

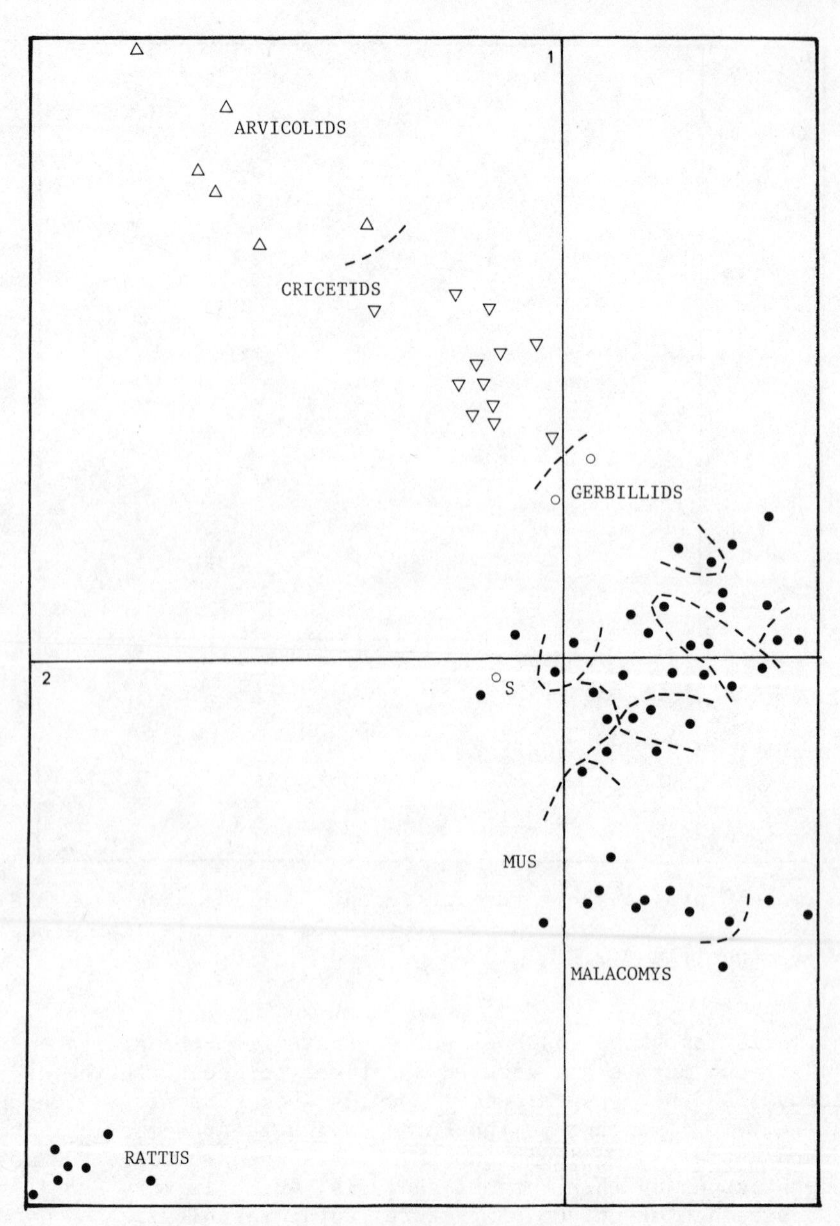

Fig. 3. AFC projection for five muroid families. Projection along axes 1 and 2. S, Steatomys (dendromurid).

as outgroups for defining primitive (plesiomorphous) and derived (apomorphous) states at each locus within the murid family.

Some murid genera are completely divergent; they exhibit primitive muroid states at about one-third of their loci, while the other

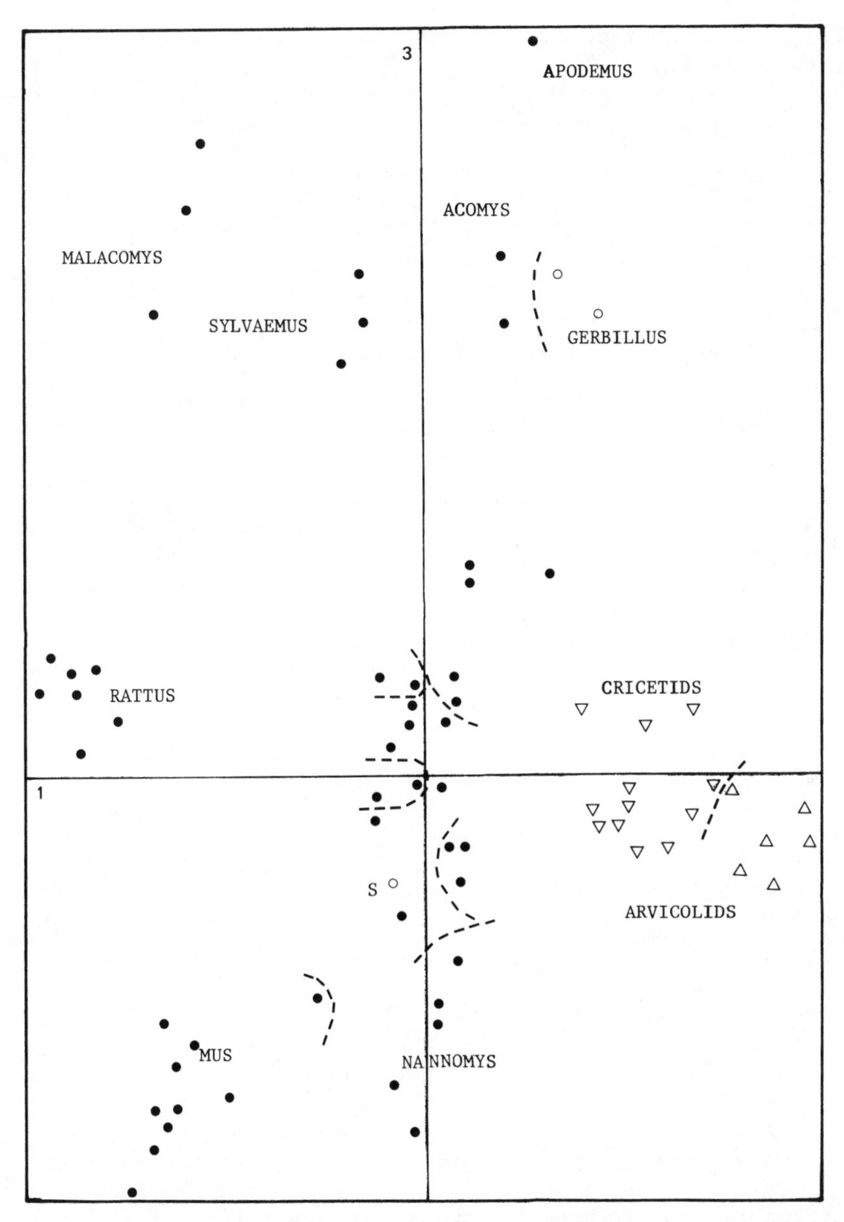

Fig. 4. AFC projection for muroids (same data set as in Fig. 3). Ppojection along axes 1 and 3.

loci are autapomorphic. This is the case for Mus, Apodemus, Acomys, Sylvaemus, and Rattus. These genera appear to be simultaneously divergent from each other, or nearly so, as suggested earlier by the phenetic analysis.

Table 2. Distribution of character states for 10 protein loci in 33 muroid genera.

Group	Genus	Mdh-1	Aat-2	Idh-2	Sod-2	Mdh-2	Pqm-2	Ldh-B	Idh-1	Gpi	Mpi	
M	Apodemus	A	A	A	A	A	A	A	A	A	A	ancestral
M	Mus	B	A	A	B	A	B	B	B	B	B	ancestral
M	Rattus	B	A	A	C	A	B	B	C	C	B	ancestral
M	Sylvaemus	B	A	A	D	A	C	C	D	D	B	ancestral
M	Acomys	C	A	A	B	A	B	B	E	E	C	ancestral
M	Malacomys	B	A	A	B	A	B	B	F	F	D	derived
M	Nannomys	B	A	A	B	A	C	B	G	G	E	derived
M	Lemniscomys	B	A	B	B	A	B	B	H	G	F	derived
M	Uranomys	D	A	A	B	A	B	B	I	H	G	derived
M	Lophuromys	E	B	A	B	A	B	B	J	G	G	derived
M	Dasymys	F	A	A	B	A	B	B	I	G	D	derived
M	Arvicanthis	G	A	A	B	A	B	B	I	G	F	derived
M	Stochomys	B	B	A	B	A	B	B	I	I	E	derived
M	Grammomys	B	A	A	B	A	B	B	I	I	D	derived
M	Hybomys	B	A	B	B	A	B	B	I	I	F	derived
M	Aethomys	B	A	B	B	A	B	B	K	I	E	derived
M	Hylomyscus	B	A	A	B	A	D	B	I	I	D	derived
M	Mastomys	H	A	B	B	A	D	D	I	I	G	derived
M	Myomys	B	A	B	B	A	E	D	I	G	F	derived
M	Praomys	I	A	A	B	A	E	D	I	G	F	derived
M	Oenomys	B	A	B	B	A	B	B	G	I	D	derived
D...	Steatomys	K	C	C	B	A	B	B	L	J	B	ancestral
G...	Gerbillus	L	A	B	B	A	B	B	M	K	A	ancestral
C	Neacomys	M	A	A	B	A	B	B	N	L	A	ancestral
C	Nectomys	M	A	B	B	A	E	B	O	M	A	ancestral
C	Oecomys	M	A	B	B	A	B	B	P	N	A	ancestral
C	Oligoryzomys	M	A	B	B	A	B	B	P	N	A	ancestral
C	Oryzomys	M	A	B	B	A	B	B	P	L	A	ancestral
C	Zygodontomys	M	A	B	B	A	B	B	P	L	A	ancestral
A	Clethrionomys	O	D	B	F	A	B	B	Q	P	C	ancestral
A	Arvicola	M	D	B	E	A	C	B	R	O	C	ancestral
A	Microtus	M	D	B	E	A	B	B	S	Q	C	ancestral
A	Pitymys	M	D	B	E	A	B	B	S	Q	C	ancestral

When looking for further groupings within the remaining murids, the first problem encountered arises at the *Mpi* locus, where we have four character states (electromorphs D, E, F. and G) distributed among them. Do they define different monophyletic clades within the set of 16 African genera that possess them? The following facts lead us to believe that this is not the case. If we now examine the *Gpi* locus (Table 2), we see two electromorphs (I and G) widely distributed among these 16 genera of murids. But what is peculiar here is that each of the two subsets of genera bearing one *Gpi* morph to the exclusion of the other also possesses the four *Mpi* character states at the same time. The same phenomenon happens on a smaller scale at the *Idh-1* locus (G and I forms). At the *Ldh-B* locus, the group defined by the distinct synapomorph D is also heterogeneous for both *Gpi* and two *Mpi* forms, as are also the two genera grouped by the synapomorphic *Aat-2* character states.

These facts can not be accomodated under any simple cladistic rule, and they are strictly contradictory with the existence of independent clades within these African murids. Thus, we consider the four *Mpi* morphs (D, E, F, and G) to be variants of a single basic synapomorph shared by all African murids (*Acomys* excepted), whereas the three electromorphs A, B, and C are considered similarly as several states of the same plesiomorph, found simultaneously in some murids and their outgroups, as exemplified in Table 2. We will discuss the further implications of these unusual assumptions below.

Within the set of 16 genera thus defined as forming the African murid clade, convincing information is available at the *Ldh-B* locus (Table 2). It is clear that the D morph is a synapomorph for the three genera *Mastomys*, *Myomys*, and *Praomys*. This helps us to analyze *Pgm-2*, where we see that the D form is plesiomorphic for the *Praomys-Myomys-Mastomys* group, using the other murids as an outgroup, whereas the E form is the apomorph derived from it. The E-D combination subsequently groups together *Hylomyscus* and the above three genera in a monophyletic group. This group is often referred to by morphologists as the genus *Praomys*, *sensu lato*. Another synapomorphy at the *Aat-2* locus groups together *Lophuromys* and *Stochomys*. All these relationships are tentatively explained in Figure 5.

At this point, we would like to return to the implications of the incongruences observed in the distribution of character states between different loci among different genera. To explain this, we can formulate several hypotheses, and it is possible that each of them may be partially valid at the same time.

(1) Homoplasy hypothesis. Electrophoretic parallelism and convergenc may be frequent enough so that electromorph identity would in fact be misleading. To us, this seems unlikely to happen on a large scale. There are many character states available for a given electromorph, especially when one takes into account part of the

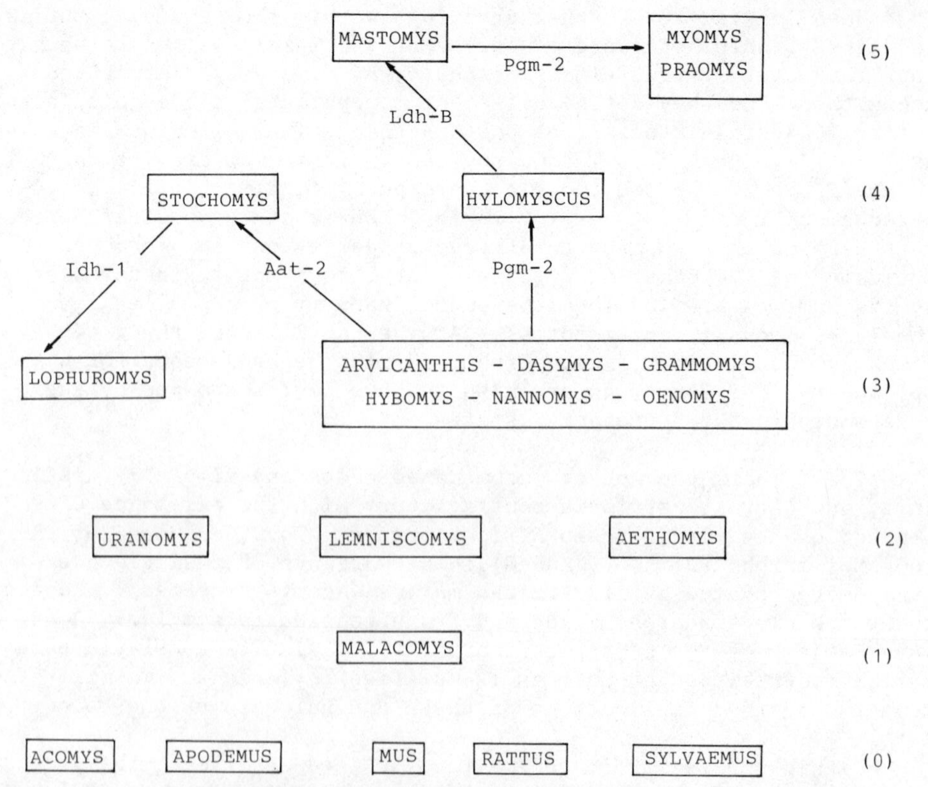

Fig. 5. Cladistic relationships among 21 murid genera. Levels 0 to 5 indicate the number of probable derived character states proper to African murids, which are possessed by each genus. Arrows indicate the probable evolutionary direction (ancestral to derived), when inferable.

hidden variability with techniques such as sequential electrophoresis. Thus, homoplasy may occur occasionally (such as the case of the rare Pgm-2 E electromorph found in Nectomys), but it seems unlikely to us that within each of the "groups" defined by the four Mpi morphs, the Gpi variants arose independently, and the reciprocal would be even more improbable.

(2) Ancestral polymorphism hypothesis. These loci would have expressed polymorphisms in the ancestral species for all the synapomorphic alleles they now display, so that the subsequent distribution would be phylogenetically meaningless. This hypothesis would imply that the ancestral species was more polymorphic than its present day descendants, although this is unknown on biological grounds.

If we believe this unlikely to be true, we may still assume

that alternative sequences of the different electrophoretic forms were already present in the ancestral genome, in the form of non-expressed copies in multigene families. These copies might eventually become evident again after gene conversion or regulatory switch and be differentially expressed in the daughter species, so that the subsequent distribution would once again be cladistically meaningless. Of course, this is purely speculative, but it may explain why two or more character states at a time may be retained in the genome for a long enough period, so that they "pass" through further dichotomies, such as evidenced by the distribution of Mpi and Gpi within the groups defined by restricted synapomorphies at the Aat-2, Pgm-2, and Ldh-B loci.

(3) Reticulate evolution hypothesis. These synapomorphic character states may be rather old, in fact, almost as old as the first African murids, and their present-day distribution may reflect appreciable genetic exchange between taxonomic units at an early stage of differentiation of the entire group. This view is compatible with what we know about gene exchange within extant species complexes (for instance, semi-specific interactions). For examples in the genus Mus, see Boursot et al. (1984, and literature cited within). Thus, the phylogeny of an individual gene is not necessarily congruent with that of the present-day species that carries it (for the genus Mus, see the mtDNA example in Fort et al., in press).

DISCUSSION AND CONCLUSIONS

Electrophoresis has proved to be an exceedingly useful method for biosystematics at the population and species levels. In some instances, it may also help to check the homogeneity of previously defined genera. Nevertheless, when trying to trace phylogenetic relationships, one is faced with several difficulties. In the work reported here, we have applied various techniques to a large data set illustrating biochemical variation in 76 species of rodents, belonging to three superfamilies. Other than the distinction of these three groups, the phylogenetic resolution we have obtained by either the quantitative or qualitative analyses is rather poor. Our clearest result is that the African murids (Acomys excepted) probably represent a monophyletic subgroup within the Muroidea.

To discuss this lack of resolving power, we will not address further questions to the phenetic approach. Historically, it was the first method to be applied to electrophoretic data, and its qualities and drawbacks have been discussed extensively elsewhere (see Farris, 1981; Felsenstein, 1984). Instead, we shall address ourselves more specifically to the cladistic approach, which has been applied only recently to data such as ours.

This matter has already been discussed in detail by Patton et al. (1981) and Patton and Avise (1983). These authors have successfully applied Hennigian cladistics to electrophoretic results in peromyscine rodents and anatine birds. In this regard, we may ask ourselves why we have been much less successful with our own data set?

The first part of the answer is related to the very nature of electrophoretic data. Electromorphs are character states, but they can not be ordered along a chronocline, because one has no a priori knowledge of the sequential order of charge changes that have taken place. Thus, the number of electromorphs useful for Hennigian cladistics is limited by our ability to define, among the different charge states at a given locus, those that are plesiomorphic and those that are relatively apomorphic. For electromorphs, this can only be done by the outgroup method, and to be fully effective this requires: (1) an average duration for a given character state about as long as the life span of the phyletic groups themselves, and (2) a relatively large number of species in the taxon chosen as an outgroup. The determination of a symplesiomorphy requires its detection in the outgroup, which may not be easy if the above conditions are not satisfied. Another difficulty arising from the non-chronocline nature of electromorphs is that there are no ways to distinguish between an autapomorphic trait hiding a synapomorphy, from which it has recently been derived, and an old derived character state arising independently from the basal plesiomorph of the entire group. This type of ambiguity is exactly what has happened with our data.

In practice, all of these difficulties can be overcome best when the loci have mutated only once (i.e., presenting only two character states) within the group of species studied. However, this reduces considerably the amount of useful information for cladistic analysis, especially when the branching events to be detected are ancient.

The second part of the answer may be inherent to the taxonomic structure of the studied group itself, and to the nature of the phyletic relationships to be disclosed. If the branching pattern of the evolutionary lineages occurred as a series of dichotomies separated by long time intervals, it will be possible to distinguish it. If, on the other hand, the time intervals are short, it will be impossible to distinguish and order the dichotomies; eventual synapomorphies, if any, are very likely to be masked by subsequent autapomorphic evolution, if the splitting events are old. It would be virtually impossible to distinguish between a series of dichotomies and a true polychotomy. We believe that this is the chief reason for the relative lack of success for our analysis of muroids, as compared to that of Patton and Avise (1983).

To conclude, we will focus on the African murids, taking into account that: (1) no hierarchical relationships are found among the different genera, which appear to be anciently derived and nearly independent lineages, and (2) there is evidence of contradiction between synapomorphic states at different loci, thereby allowing no precise phylogenetic inference. Therefore we suggest a phylogenetic model with branching events occurring in a relatively short time, but which are not instantaneous and which allow simultaneously for genetic divergence and introgression. This may have been the situation at the time of the initial murid radiation. Semi-specific interactions, as displayed by Recent populations, support this view.

REFERENCES

Aquadro, C. F. and Avise, J. C. 1982. Evolutionary genetics of birds. VI. A reexamination of protein divergence using varied electrophoretic conditions. Evolution 36: 1003-1019.

Benzécri, J. P. 1973. L'analyse des données. II. L'analyse des correspondances. Dunod, Paris.

Boursot, P., Bonhomme, F., Britton-Davidian, J., Catalan, J., Yonekawa, H., Orsini, P., Guerasimov, S. and Thaler, L. 1984. Introgression différentielle des génomes nucléaires et mitochondriaux chez deux semi-espèces européennes de Souris. C. R. Acad. Sci.299: 365-370.

Chaline, J., Mein, P. and Petter, F. 1977. Les grandes lignes d'une classification évolutive des Muroidea. Mammalia 41: 245-252.

Farris, J. S. 1981. Distance data in phylogenetic analysis. In: Advances in Cladistics, V. A. Funk and D. R. Brooks, eds., pp. 3-23, New York Botanical Garden, New York.

Felsenstein, J. 1984. Distance methods for inferring phylogenies: a justification. Evolution 38: 16-25.

Fort, P., Bonhomme, F., Darlu, P., Piechaczyk, M., Jeanteur, P. and Thaler, L. in press. Clonal divergence of mitochondrial DNA versus populational evolution of nuclear genome. Evol. Theory.

Iskandar, D. 1984. Evolution génétique de la superfamille des Muroidés révélée par électrophorèse classique et électrophorèse séquentielle. Thèse de Doct. Spécialité. Université Montpellier, Montpellier.

Iskandar, D. and Bonhomme, F. in press. Variabilité électrophorétique totale à onze locus structuraux chez las Rongeurs muridés (Muridae, Rodentia). Canad. J. Cyt. Genet.

Jambu, M. and Lebeaux, M. O. 1979. Classification automatique pour l'analyse des données, Vol II. Dunod, Paris.

Patton, J. C. and Avise, J. C. 1983. An empirical evaluation of qualitative Hennigian analyses of protein electrophoretic data. J. Mol. Evol. 19: 244-254.

Patton, J. C., Baker, J. R. and Avise, J. C. 1981. Phenetic and cladistic analysis of biochemical evolution in peromyscine rodents. In: Mammalian Population Genetics, M. H. Smith and J. Joule, eds., pp. 288-308, Univ. Georgia Press, Athens.

EVOLUTIONARY RELATIONSHIPS AMONG RODENTS:

COMMENTS AND CONCLUSIONS

W. Patrick Luckett[1] and Jean-Louis Hartenberger[2]

[1]Department of Anatomy, University of Puerto Rico School of Medicine, San Juan, Puerto Rico, USA, and [2]Institut des Sciences de l'Evolution, LA 327, U. S. T. L., Place Eugène Bataillon, Montpellier, France

INTRODUCTION

The Paris Symposium on Rodent Evolution brought together a variety of biologists (paleontologists, comparative anatomists, embryologists, molecular biologists, and geneticists) who shared a common interest in the assessment of phylogenetic relationships among rodents and other mammals. A diversity of methodological strategies for phylogenetic reconstruction was also represented, including traditional paleontological searches for anagenetic or ancestral-descendant relationships; cladistic analyses of dental and cranial features in fossil and extant rodents; phenetic and cladistic analyses of soft anatomical and developmental features; form-functional considerations of selected postcranial, dental, and soft anatomical traits; and maximum parsimony, cladistic, and immunological distance analyses of molecular data. Surprisingly, perhaps, this variety of methodologies and data was accepted with tolerance by most participants, and there was a genuine sense of willingness to listen and to learn on the part of the vast majority of participants. Semantic disagreements over the "best" approaches to phylogenetic analysis were kept to a minimum, perhaps because most participants were anxious to learn about any evidence that might shed light on the uncertainties surrounding the possible evolutionary relationships among the 50 families of extant and fossil rodents (as recognized by Carleton, 1984).

During the past 30 years, greater emphasis has been placed on the classification of rodents than on the reconstruction of their phylogeny. Such classifications have focused almost entirely on the relative merits of different components of the masticatory apparatus (dentition, zygomasseteric complex, and mandibular morphol ogy) as key features for understanding the major pathways of rodent evolution. The conflicting hypotheses based on these data have rarely been tested by the analysis of other types of data, especiall non-osteological features. This is related in part to the fact that the evolutionary history of rodents has been studied mainly by pale-ontologists. In their studies, emphasis is generally placed on the discovery of ancestral-descendant relationships, and less attention has been devoted to the assessment of cladistic or sister-group relationships among rodent taxa. Hypotheses of ancestor-descendant relationships are restricted to studies of the morphology and strati graphic position of incomplete fossil remains, and such hypotheses can be difficult to corroborate. On the other hand, evolutionary analyses of sister-group relationships can utilize observations from both fossil and extant taxa, and the addition of soft anatomical, developmental, molecular, cytogenetic, and functional data to the traditionally-studied dental and cranioskeletal features provides a more extensive assemblage of biological attributes for character analysis and phylogeny reconstruction. Both anagenetic and cladisti hypotheses are important aspects of phylogenetic reconstruction, and character analyses of sister-group relationships can help to focus the search for possible ancestor-descendant relationships (Szalay, 1977).

Given the considerable amount of parallelism and convergence that is believed to have occurred in virtually all components of the rodent masticatory system (Simpson, 1945, 1959; Wood, 1955, 1959, 1965, this volume), as well as the uncertainties that surround the recognition of homology or homoplasy, we believe that those phylogenetic hypotheses which can be corroborated by data from several different (and preferably unrelated) organ systems are more likely to reflect the true phylogeny of a group, than are those hypotheses corroborated by only a single character complex. Therefore, in this brief report we wish to discuss those hypotheses of rodent phylogeny that we believe to be most strongly corroborated by multidisciplinar analyses in this volume.

ORIGIN AND SISTER-GROUP RELATIONSHIPS OF RODENTIA

The relationship of rodents to other eutherian mammals has been a matter for speculation during the last century. Although rodents and lagomorphs were long considered to be the two major subgroups of the order Glires (Tullberg, 1899), most students of mammalian evolution during the 20th century have recognized ordinal distinction for the two groups, following Gidley (1912). Wood (1957) subsequently suggested that rodents and lagomorphs are not closely related, and that their shared similarities are due to the convergent evolution of enlarged, evergrowing incisors, and to the retention of numerous primitive eutherian features. This distinction was further supporte by the hypothesis of a separate origin of rodents from primitive "insectivores" on the one hand, and of lagomorphs from condylarths on the other (Wood, 1957).

Wood (1962) later proposed a possible origin of rodents from Paleocene plesiadapid primates, based primarily on the presence of a pair of enlarged "rodent-like" incisors in these early primates, as well as some generalized similarities of the cheek teeth and skull. Despite the lack of a careful analysis of the homologous or homoplastic nature of these shared similarities, many paleontologists accepted Wood's suggestion of rodent-primate affinities (McKenna, 1969; Van Valen, 1971), without further discussion. However, analyses in the present volume of the astragalocalcaneal complex (Szalay), skull (Novacek), incisor homologies and fetal membrane development (Luckett), and molecular data (Shoshani et al.; Sarich) provide no support for this hypothesis. Butler has discussed the cuspidate and tribosphenic condition of the molars in primitive rodents, and he notes the similarities of cusps, crests, and wear facets between primitive rodents and early primates. These similarities in molar pattern appear to us to be relatively primitive attributes shared by numerous early Tertiary eutherians (e.g., hyopsodontid condylarths). Butler concludes that the molars of early rodents and primates had advanced in the same direction, "without succumbing to [the] temptation" of allying the origin of rodents with primates.

The Glires Concept

The idea of a close rodent-lagomorph relationship was never completely abandoned; this can be attributed in part to Simpson's (1945) comment that "the resemblances formerly used to unite them were not, after all, imaginary." However, Simpson (1945, p. 196) considered union of the two orders in a Cohort Glires as "frankly hypothetical," and he further qualified this hypothesis by the oft-quoted statement that "it is permitted by our ignorance rather than sustained by our knowledge." In truth, evidence for a close relationship between rodents and lagomorphs, within an order Glires, was based on shared similarities of the dentition, skull, skeleton, brain, reproductive system, and placentation (Tullberg, 1899; Weber, 1904; Gregory, 1910). Gregory (1910) acknowledged that many of these shared similarities are primitive eutherian features, but he emphasized that the *combination* of primitive and specialized traits shared by rodents and lagomorphs warranted a close relationship between these taxa, and distinguished them from all other eutherians.

The hypothesis of a higher taxon (superorder or cohort) Glires is supported by multiple character analyses in the present volume of cranial morphology (Novacek; also, Shoshani, pers. comm.), dental development and fetal membrane development (Luckett). Lopez Martinez has evaluated numerous biological features in the order Lagomorpha, and she notes that rodents share a number of derived craniodental traits with lagomorphs. However, she believes that none of the shared similarities of the two orders are uniquely derived, and that therefore these shared, derived features do not corroborate a hypothesis of sister-group relationship between Rodentia and Lagomorpha.

Lopez Martinez notes that some of the dental and cranial similarities shared by rodents and lagomorphs also occur in hyracoids, artiodactyls, and perissodactyls, although she acknowledges that it is uncertain at present whether these similarities occurred in the morphotype of the latter taxa. However, analyses of molecular data in this volume (De Jong; Shoshani et al.), as well as fetal membranes (Luckett, unpublished) and cranial foramina (Shoshani, pers. comm.), support the traditional clustering of hyracoids with other paenungulates or subungulates, and the wide separation of these groups from rodents and lagomorphs. Koenigswald has also emphasized that the incisor enamel of hyracoids lacks the decussatin Hunter-Schreger bands that characterize the enamel of rodents and incisors. These and other data indicate that the masticatory similarities between hyracoids and lagomorphs are due to convergence. We suspect that the same evolutionary processes account for the lagomorph-ungulate similarities, although we have not carried out a complete analysis of this hypothesis.

The available molecular data provide no corroboration as yet for the sister-group relationship of the Glires. It should be noted however, that amino acid sequence data are still lacking for most rodent families, including ctenodactylids, sciurids, geomyoids, and dipodoids. It seems likely that the placement of rodents relative to lagomorphs and other eutherians might change significantly with the addition of such data to maximum parsimony analyses. Moreover, Shoshani et al. note that a parsimony tree that "costs" only three more nucleotide replacements than their "best" hemoglobin parsimony tree would unite Rodentia and Lagomorpha. These authors also admit that their present clustering of Rodentia with other Eutheria by hemoglobin data is "weakly founded," and they conclude that their sequence data are insufficient at present to decisively place the Rodentia within Eutheria.

Rodent-Leptictid Affinities

As an alternative to the hypothesis of rodent-lagomorph affinities, Szalay (this volume) has proposed that the North American fossil family Leptictidae provided the ancestral stock for Rodentia. This hypothesis is based on an evolutionary assessment of the functional morphology of the cruropedal complex, with emphasis on the distal tibia and astragalocalcaneal relationships. On the other hand, Szalay found no postcranial evidence to support a close relationship between rodents and lagomorphs. Szalay has emphasized the value of functional and transformational considerations when assessing the ancestral-descendant relationships of taxa, and his analysis suggests that the early paramyids (and morphotypic rodents) were arboreal and specialized for claw-climbing.

We have several reservations about accepting Szalay's hypothesis of ancestral-descendant relationships between leptictids and rodents. We are able to identify only a single trait from his study - the posterior tibial process - that appears to be uniquely shared by leptictids and rodents (the presence or absence of this feature has not been determined for eurymyloids or zalambdalestids). Furthermore, while emphasizing his functional and transformational approach to phylogenetic reconstruction, Szalay rejects "character distribution analysis" as a biologically important part of character analysis methodology. Consequently, we are unable to determine the distribution of shared cruropedal similarities of leptictids and primitive rodents among most other eutherian orders from his report. Perhaps even more questionable is the restriction of Szalay's analysis to a limited portion of the hind limb, despite the fact that nearly complete skeletons are known for paramyids (Wood, 1962) and leptictids (Novacek, 1977, 1980).

In a preliminary survey, Novacek (1977) acknowledged that leptictids share a number of derived postcranial features with rodents, but he considered these to be "not exclusive to rodents and leptictids, but merely suggest affinities of these taxa with eutherians having more specialized locomotory adaptations." He further emphasized that rodents are morphotypically more primitive than known leptictids in several hind limb and foot structures, and that significant derived similarities of the auditory and basicranial regions are lacking between the two groups. Many of the derived postcranial traits shared by leptictids and rodents also occur in macroscelidids, tupaiids, or lagomorphs (Novacek, 1977, 1980). Character analyses of cranial and dental features of leptictids, rodents, and other eutherians in the present volume (Novacek; Li and Ting; Luckett) also fail to corroborate Szalay's hypothesis of an ancestral-descendant (or even a cladistic) relationship between the former taxa. A more comprehensive analysis of both cranial and postcranial features of leptictids and other eutherians does not corroborate a special leptictid-rodent relationship (Novacek, in prep.).

Rodent-Lagomorph-Eurymyloid Affinities

We believe that the findings of the present volume are most consistent with the hypothesis of close evolutionary relationships among Rodentia, Lagomorpha, and Eurymyloidea, as suggested by Li and Ting. This is based on common possession of numerous shared and derived attributes of the dentition and cranium, whereas only a few available postcranial traits support this hypothesis. We believe this to be the result of two limiting factors. (1) The earliest unquestioned fossil lagomorphs are known from the late Eocene of Mongolia, China, and North America (Li, 1965; Dawson, 1970), but it seems likely that their early history was much more ancient. Thus, many morphotypic features of Lagomorpha, especially cranial and postcranial features, are based on Oligocene-Recent forms, whereas we have little or no knowledge of these traits in Eocene lagomorphs. (2) Postcranial features of eurymyloids are known incompletely from only the middle Eocene Rhombomylus and the ?Oligocene Mimolagus, and therefore it is unclear at present what differences, if any, exist between the morphotypes of Paleocene eurymylids and mimotonids.

Character analysis of the available dental, cranial, and postcranial features of mimotonids, lagomorphs, rodents, and eurymylids does not clearly resolve the cladistic relationships among the four taxa, and we agree with Li and Ting that the best solution at present is to retain Eurymylidae and Mimotonidae as members of the superfamily Eurymyloidea. We propose that Rodentia, Lagomorpha, and Eurymyloidea should be considered as members of the cohort or superorder Glires. Further analysis of the available data may indicate that Eurymylidae and Rodentia are more closely related, and the same may be true for Mimotonidae and Lagomorpha, as suggested previously by Li (1977). Hypotheses of possible affinities among Glires, anagalids, leptictids, zalambdalestids, or pseudictopids require further testing.

INTRAORDINAL RELATIONSHIPS OF RODENTIA

Virtually every aspect of comparative biology examined in this volume, from paleontological to molecular, corroborates monophyly of the order Rodentia, in relation to other Eutheria. Within the Rodentia, the most corroborated hypotheses of monophyly for higher taxa are those for Hystricognathi, Muroidea, and Geomyoidea, whereas there is still considerable uncertainty or controversy about the possible suprafamilial relationships of other rodent taxa.

Monophyly and Content of the Suborder Hystricognathi

The rodent higher taxon most investigated by paleontologists and neontologists is the suborder Hystricognathi. All the investigators in this volume who presented data on hystricognaths have corroborated their monophyly relative to other rodents, although there are two different viewpoints as to the contents of the suborder. This difference relates to whether the North American Eocene franimorphs, as defined by Wood, should or should not be included within the Hystricognathi. This controversy, in turn, is related to the contrasting hypotheses for the origin and sister-group relationship of Hystricognathi.

Studies in this volume on the comparative biology of extant rodents, including cluster and shared-character analyses of blood vascular, reproductive, chromosomal, and skeletal features (George); fetal membranes and placenta (Luckett); carotid arterial pattern (Bugge); and protein immunological and amino acid sequence data (Sarich; Shoshani et al.; Beintema; De Jong) all support the hypothesis of hystricognath monophyly, which was based traditionally on musculoskeletal features of the masticatory apparatus in fossil and living rodents, as discussed by Wood. All the molecular and anatomical studies on extant hystricognaths, including the myological data (Woods) and middle ear features (Lavocat and Parent), provide overwhelming support for hystricognath monophyly, and all indicate that it is virtually impossible to distinguish between Old and New World hystricognaths as separate taxa (Phiomorpha and Caviomorpha). Indeed, in some myological and molecular studies, Old World taxa may cluster more closely with New World taxa than with other Old World groups (Woods; Sarich), and the same is true for a study of cranial foramina (Shoshani, pers. comm.).

In contrast to the evidence for strict monophyly of extant Old and New World hystricognaths, Wood distinguishes five separate groups of fossil and extant Hystricognathi: (1) Hystricidae; (2) Thryonomyoidea; (3) Bathyergoidea; (4) Caviomorpha; and (5) Franimorpha. Furthermore, he suggests that many of the features that characterize Hystricognathi (other than hystricognathy) may have evolved by parallelism. Wood believes that the first four groups above were derived from Eocene franimorphs. Hystricids and thryonomyoids are thought to have originated from unknown Asiatic franimorphs; caviomorphs to be derived from Middle American Eocene franimorphs; and Miocene-Recent bathyergoids of Africa are believed to originate from Asiatic Oligocene tsaganomyids, which in turn were derived from Asiatic cylindrodontids.

Wood further suggests that "incipient" to "fully developed" hystricognathy is the only cranioskeletal or dental feature shared by all hystricognaths, and that this feature probably arose only once during rodent phylogeny. In contrast, he believes that other shared and derived features of hystricognaths, including hystricomorphy, multiserial enamel, and perforation of the pterygoid fossa, may have arisen in parallel more than once after the differentiation of hystricognathy, because these traits are also found in some other taxa that are not members of the Hystricognathi.

We have two major objections to Wood's evolutionary scenario: (1) inclusion of Franimorpha in the suborder Hystricognathi, and (2) reconstruction of the ancestral morphotype for the Bathyergoidea, and their possible affinities with Tsaganomyidae. In turn, these areas of disagreement are related in part to differing approaches to the evolutionary analysis of characters.

Wood's (1975, this volume) definition of the Franimorpha includes mostly primitive rodent features shared also by other Eocene protrogomorphs, the sole exception being the "incipient" development of hystricognathy. However, this incipient hystricognathy has not been recognized generally by other paleontologists who have studied the same specimens (Dawson, 1977; Korth, 1984; Hartenberger, unpublished), and there is little or no evidence of other "typical" hystricognath features in most franimorphs. Indeed, Korth (1984) has specifically denied the occurrence of hystricognathy in any franimorph, including Mysops and Oligocene cylindrodontids; these latter taxa have played a major role in Wood's franimorph hypothesis (also see Wood, 1984). Our character analysis of craniodental features (Fig. 1) in cylindrodontids, hystricognaths, and other hystricomorphous taxa discussed by Wood (this volume) does not corroborate a hypothesis of monophyly between Hystricognathi (sensu Lavocat, 1973) and Eocene-Oligocene cylindrodontids (or other franimorphs). In contrast, a suite of cranial derived features corroborates the traditional hypothesis of Hystricognathi monophyly (excluding franimorphs). The fact that some of these individual traits, such as hystricomorphy or a perforated pterygoid fossa, can occur in other rodent taxa does not diminish the value of the entire suite of features for characterizing Hystricognathi, contrary to the implications of Wood. Nor should this lead to the assumption that perforation of the pterygoid fossa or attainment of hystricomorphy occurred independently within several subgroups of Hystricognathi, in the absence of any evidence to support this belief.

The fossil record clearly demonstrates that some characteristic features of hystricognaths have developed convergently in other taxa. For example, a perforated pterygoid fossa occurs in extant geomyoids and Aplodontia, as emphasized by Wood. However, if we consider the known fossil representatives of both taxa - the eomyids and prosciurines, respectively - it is evident that these early groups lacked a perforated fossa (Wahlert, 1974), and that therefore this feature developed convergently within Geomyoidea and Aplodontidae. In contrast, this character occurs in all known hystricognaths for which this region is available, and it can be assumed to have occurred in their last common ancestor, according to the parsimony principle of hypotheses.

A major argument by Wood against the strict hypothesis of hystricognath monophyly is his belief that protrogomorphy represented the ancestral condition of Miocene-Recent bathyergids, and that, consequently, hystricomorphy must have arisen more than once within Hystricognathi. However, Lavocat (1973) has clearly demonstrated that the three known Miocene genera of bathyergids were hystricomorphous, with a moderately large infraorbital foramen that almost certainly transmitted part of the medial masseter muscle. This foramen is very small (= protrogomorphous) in most extant bathyergids, with the notable exception of Cryptomys, in which there is a moderately large foramen which transmits a part of the medial masseter (see illustrations of muscles in Boller, 1970). Given the relatively large size of the foramen in the earliest known fossils, its intermediate size in Cryptomys, as well as the "reciprocal illumination" provided by a consideration of other biological attributes that cluster bathyergids with other hystricognaths, it is most parsimonious to conclude that hystricomorphy also occurred in the ancestral bathyergids (Table 1, Fig. 1). This is further corroborated by the functional considerations of Boller (1970) and Lavocat (1973), who suggested that reduction in size of the infraorbital foramen is one of several features associated with the fossorial adaptations of bathyergids, possibly correlated with enlargement of the upper incisors. A reduction in size of the infraorbital foramen also takes place in Miocene-Recent rhizomyids, apparently associated with the acquisition of fossorial specializations (Flynn, 1982).

Corroboration for Lavocat's (1973) hypothesis that bathyergids may have originated from Oligocene thryonomyoids is provided by the available albumin immunological data, in which the bathyergid Bathyergus clusters with the thryonomyoids Thryonomys and Petromus, whereas Hystrix and caviomorphs are further removed, although all group together as members of a distinct monophyletic hystricognath clade (Sarich, this volume). Amino acid sequence data are not yet available to test this hypothesis.

Our character analysis, based on a reconstructed morphotype for each major group, does not corroborate Wood's hypothesis that cylindrodontids or other franimorphs should be included in the suborder Hystricognathi (Table 1, Fig. 1), nor his belief that franimorphs are the ancestral group of Hystricognathi. In contrast to Wood's opinion, we do not believe that there was a significant difference in the morphotypic condition of retention of the internal carotid canal (and presumably the artery), or of replacement of dP4 in the ancestral thryonomyoids and caviomorphs. Lavocat and Parent have reviewed the evidence for occurrence of the internal carotid canal in a Miocene thryonomyoid skull (basicrania are not adequately known for any Oligocene genus). Wood's (1968) original description clearly demonstrated that dP/4 was replaced by P/4 in the Oligocene phiomyid Phiomys andrewsi, and that dP4/ was replaced in Gaudeamus. This suggests that dP4 replacement occurred in the phiomyid and thryonomyoid morphotype, in agreement with Wood's (1968, p. 81) original suggestion that retention of dP4 in Fayum phiomyids "may well not have characterized the original immigrants to North Africa."

The most problematic (and least investigated) group of possible hystricognaths is the Asiatic Oligocene family Tsaganomyidae. As emphasized by Wood (1974b), most of the resemblances between Tsaganomyidae and Bathyergidae (excluding hystricognathy), such as the general skull shape, enlarged digging incisors, and forwardly sloping occiput, may be due to convergence for extreme fossorial adap-

Table 1. Character analysis of craniodental features in hystricognathous and hystricomorphous rodents. Int = intermediately derived.

Primitive	Derived
1. Mandible sciurognathous	1. Mandible hystricognathous
2. Infraorbital foramen protrogomorphous	2. Infraorbital foramen hystricomorphous
3. Shallow pterygoid fossa	3. Deepened pterygoid fossa, opens into orbit
4. Pars reflexa of M. masseter superficialis small	4. Pars reflexa of M. masseter superficialis large, attached to medial side of mandible
5. Malleus and incus unfused	5. Malleus and incus closely appressed or fused
6. Internal carotid artery and canal present	6. Internal carotid artery and canal absent
7. Stapedial artery present	7. Stapedial artery absent
8. Molars cuspidate or with early crest formation	8. Molars 4-crested (Int.) or 5-crested
9. dP4 replacement normal	9. dP4 not replaced, but retained

tations. Many of these features, including a perforated pterygoid fossa, are also found in the highly fossorial myomorph Spalax (see Boller, 1970). However, there is no evidence that the hystricognathy of tsaganomyids is related to fossorial activity. Wood has suggested that Asiatic cylindrodontids were ancestral to tsaganomyids, although the craniodental evidence for this has yet to be clearly presented. It appears to be based mainly on similarities in the cheek tooth pattern and general fossorial adaptations of the skull.

Possible Affinities of the Ctenodactyloidea and Hystricognathi

An alternative hypothesis to the franimorph origin of the Hystricognathi is based on character analyses of cranial, soft anatomical, reproductive, developmental, and chromosomal features in living rodents (George). In her cluster and shared-character analyses of a wide range of biological characters, George identified two major groups among extant rodents: (1) a hystricognath-ctenodactylid group, and (2) a myomorph-sciuromorph group. The families Pedetidae and Anomaluridae fall between these two major groups. These results are similar to another cladistic analysis of cranial, developmental, and reproductive traits (Luckett), which shows that ctenodactylids share more derived traits with Hystricognathi than with any other group of Recent rodents. The middle ear features described by Lavocat and Parent are also consistent with the hypo-

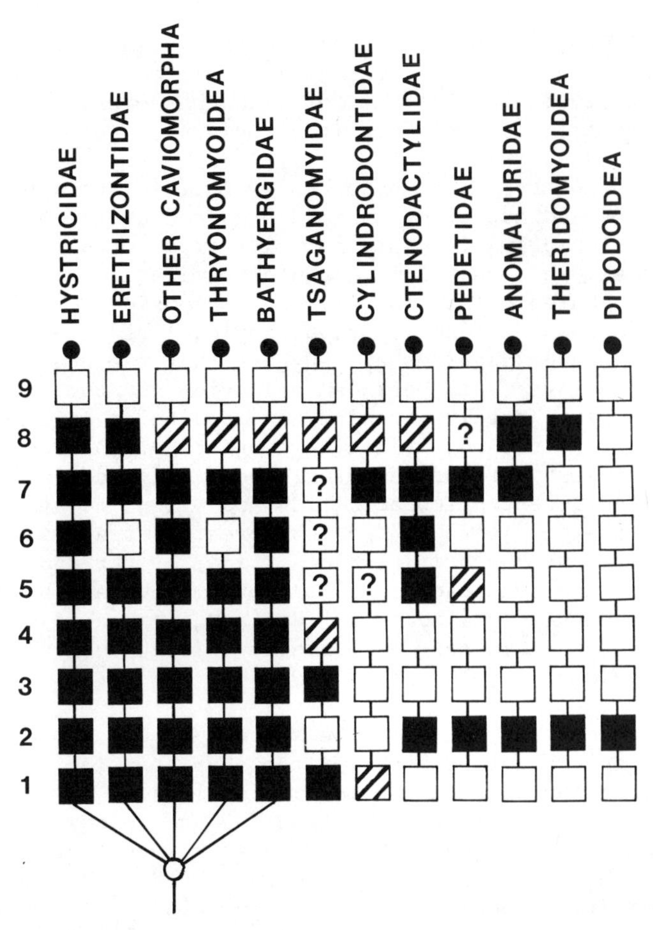

Fig. 1. Character phylogeny of selected craniodental features in hystricognaths and other hystricomorphous rodents. The reconstructed morphotypic condition for each character is shown for each taxon. Derived character states are shown as black squares, open squares are primitive characters, and intermediately derived conditions are shown with diagonal lines. See Table 1 for the characters studied.

thesis of close ctenodactylid-hystricognath affinities. These analyses support an earlier suggestion, based on incomplete fossil evidence, that certain Asiatic Eocene ctenodactyloids might be close to the ancestry of both African and South American hystricognaths (Hussain et al., 1978). To a lesser degree, it is consistent in part with the hypothesis of Hartenberger (1980) for an early dichotomy of Eocene rodents into Asiatic ctenodactyloids and European-North American ischyromyoids, although he was (and remains) uncertain as to which of these groups might be closer to Hystricognathi. To date, however, there are no molecular data available to test the hypothesis of a possible ctenodactyloid-hystricognath relationship. A parsimony analysis of 78 skull foramina characters also clusters extant Ctenodactylidae with Hystricognathi (Shoshani, pers. comm.).

Further support for the hypothesis of close affinities between ctenodactyloids and hystricognaths is provided by the new late Eocene dental remains of primitive thryonomyoids from North Africa described by Jaeger et al. in this volume. Dental similarities of these early phiomyids are shared both with Oligocene phiomyids of Africa and with Eocene ctenodactyloids (family Chapattimyidae) of the Indo-Pakistan region of Asia, and Jaeger et al. suggest that the chapattimyids are the likely ancestors of the African phiomyids and the Afro-Asian hystricids. Unfortunately, cranial and mandibular evidence is lacking to corroborate the phiomyid-hystricognath affinities of these new dental remains, but they do provide dental evidence to bridge the gap between Eocene ctenodactyloids and Oligocene hystricognaths. In addition, they indicate that a four-crested molar pattern was also likely to represent the primitive condition for thryonomyoids (as believed to be the condition also for caviomorphs and bathyergids); this diminishes the significance of another supposed evolutionary distinction between Old and New World hystricognaths (see Fig. 1), discussed previously by Wood (1974a).

Jaeger et al. suggest that the Miocene-Recent Hystricidae are the "living representatives" of Asiatic chapattimyids, and therefore that hystricognathy and multiserial incisor enamel may have evolved independently, in parallel, in Hystricidae and Phiomyidae. The only shared character that they cite to support this chapattimyid-hystricid clade is the normal replacement of dP4 by P4. But certainly this is a primitive rodent (and primitive eutherian) feature, and, as such, it carries no weight in a character analysis, especially when there are numerous shared, derived craniodental features that corroborate the monophyly of Hystricidae with other hystricognaths, to the exclusion of ctenodactyloids (Fig. 1). There are also abundant soft anatomical and developmental data that corroborate this latter hypothesis (see contributions by George and Luckett).

Ctenodactyloidea as the Stem Group of Rodents?

For many years, the available fossil record of early Eocene rodents was limited to the Protrogomorpha of North America and Europe, and, of these, the Paramyidae were generally considered to be the ancestral group of most, if not all, later groups of rodents (see Wood, 1959, 1962, 1965). In recent years, however, the increasing fossil evidence of Asiatic Eocene rodents, as well as different approaches to the problem of evolutionary analysis, have suggested an alternative solution to the question of the rodent stem group. Some authors (Korth, 1984; Dawson and Krishtalka, 1984; Hartenberger, this volume) are inclined to consider that the earliest Asiatic ctenodactyloids (= cocomyids) are the most ancient and earliest differentiated group of rodents. This is based in part on their numerous similarities to the eurymylid Heomys, including the non-molariform nature of the premolars, and the well developed hypocone on the upper molars. Implicit in this proposal is that early ctenodactyloids form the sister group of all other rodents. It is difficult to test this hypothesis until a more complete description of the skull and upper dentition of Cocomys is presented, and until cranial and basicranial features of other Eocene ctenodactyloids are described and illustrated. It is becoming increasingly clear, however, that differences in premolar and molar morphology between early Eocene paramyids and cocomyids (Dawson et al., 1984) support the hypothesis (Hartenberger, 1980) of an early dichotomy during rodent phylogeny between Asiatic cocomyids and North American-European paramyids.

Origin and Composition of the Suborder Myomorpha

There is a general consensus among contributors to the volume, based on cranial morphology, dental morphotypes, and fetal membrane data (Vianey-Liaud; Flynn et al.; Hartenberger; Luckett), and on the available immunological data (Sarich), that Muroidea and Dipodoidea are sister groups within the suborder Myomorpha. The basicranium and associated internal carotid circulation are consistent with this hypothesis, although most shared resemblances of this system are thought to be primitive retentions (Lavocat and Parent; Bugge). Both superfamilies were hystricomorphous primitively, and Vianey-Liaud suggests that they may have been derived from Asiatic Eocene ctenodactyloids, an opinion also endorsed by Flynn et al. Other than hystricomorphy, there is little evidence to support this hypothesis (Fig. 1).

Paleontologists have traditionally considered muroids to be derived from North American sciuravids, based mainly on general dental resemblances (Wood, 1959; Korth, 1984; Flynn et al., this volume). This hypothesis does not appear to have been carefully tested, nor does it appear to be supported by cranial morphology. Although Wahlert (1978, this volume) has not specifically addressed the question of possible sciuravid-muroid or sciuravid-geomyoid affinities, he has not demonstrated any shared derived cranial traits among these taxa. Moreover, the unique pattern of apparent "absence" of the medial internal carotid artery, and the occurrence instead of a promontory artery in sciuravids (Wahlert, 1974), would seem to us to falsify a hypothesis of ancestral-descendant relationship between sciuravids and muroids (or geomyoids), assuming that the Sciuravus pattern is characteristic of other sciuravids.

Relationships within the superfamily Muroidea remain unclear, related in part to the lack of a complete data set for all proposed families. Flynn et al. have proposed a hypothesis of sister-group relationships among Muroidea, based on character analysis of a few selected dental and cranial traits in fossil and extent rodents. Within Muroidea, they suggest that spalacids originated early during muroid evolution, so that they were probably derived from primitive hystricomorphous muroids, whereas later evolving muroids (including rhizomyids, cricetids, arvicolids, gerbillids, and murids) are united by true myomorphy, a condition derived from hystricomorphy during muroid evolution. Rhizomyids appear to fall outside of the murid-cricetid group.

This working hypothesis can be tested by examining the available molecular data in this volume. Both the limited hemoglobin and ribonuclease sequences (Shoshani et al.; Beintema) show that a common arvicolid-cricetid group is the sister taxon of the murids, and, in turn, this combined clade is the sister group of spalacids. These sequence data are consistent with Flynn et al.'s hypothesis, although no information was available for rhizomyids and gerbillids. In contrast, the more slowly evolving α-crystallin sequences can not distinguish between murids, gerbillids, and cricetids (De Jong). More extensive information is available from albumin immunological studies by Sarich. In opposition to the hypothesis of Flynn et al., Spalax clusters closer to other muroids than does Rhizomys. Sarich was not able to resolve a trichotomy among arvicolids, Old World cricetids, and New World cricetids (= Sigmodontinae), whereas Brownell's (1983) DNA hybridization data clustered Arvicolidae and Sigmodontinae together before this branch joined with Old World cricetids. Both the albumin and DNA studies show that Gerbillidae are as distant from the arvicolid-cricetid group as are the murids, corroborating their familial separation. Several derived middle ear features also separate gerbillids from both cricetids and murids (Lavocat and Parent). In addition, Brownell's (1983) DNA data support the hypothesis of Flynn et al. that gerbillids are closer cladistically to murids than to cricetids.

The affinities of Gliridae and Geomyoidea with the dipodoid-muroid clade (as members of the suborder Myomorpha) are more controversial. Glirids have been considered to be related to the muroid-dipodoid clade by many authors (e.g., Simpson, 1945; Wood, 1965; Wahlert, 1978, this volume), based mainly on their possession of myomorphy. However, other authors (Dawson and Krishtalka, 1984; Flynn et al., this volume) believe that glirids acquired their myomorphy independently from that of muroids, and the fossil evidence discussed by Vianey-Liaud in this volume corroborates this suggestion. Fossil skulls indicate that the "pseudomyomorphy" of glirids arose from a primitive protrogomorphous condition, in contrast to an ancestral state of hystricomorphy for muroids. Analysis of middle ear features (Lavocat and Parent) and the associated internal carotid pattern (Bugge) also supports the separation of glirids from muroids. Instead, glirids share more derived basicranial features with sciurids and aplodontids than they do with muroids, although fossil glirid ear regions are unknown. On the other hand, the albumin data presented by Sarich cluster the glirid Eliomys with the muroid clade. Neither amino acid sequences nor fetal membrane features are available to test the alternative hypotheses of glirid affinities with other rodents.

The inclusion of Geomyoidea within a monophyletic suborder Myomorpha has been generally endorsed by paleontologists in recent years (Wood, 1965; Wahlert, 1978, this volume; Dawson and Krishtalka, 1984), although some contributors to this volume have questioned the validity of this proposal (Fahlbusch; Hartenberger; Vianey-Liaud). There is a general consensus that the earliest geomyoids (Eomyidae) were probably derived from Eocene sciuravids, and, as noted by Flynn et al., some early eomyids, dipodoids, and muroids share a common cheek tooth pattern (the "cricetid" plan). However, as discussed above, there is some doubt concerning the origin of dipodoids and muroids from sciuravids, and it is also possible that these groups had an Asiatic origin (Vianey-Liaud). Hartenberger has proposed a possible origin of sciuravids from ctenodactyloids, based on some dental similarities among ctenodactyloids, sciuravids, and geomyoids, and he has tentatively included these taxa in an early ctenodactyloid clade. Character analysis of the available cranial and dental evidence in Eocene-Oligocene sciuravids, eomyids, muroids, and dipodoids is essential for testing these different but interrelated hypotheses of ancestral-descendant and cladistic relationships.

Soft anatomical and molecular data from extant rodents also provide conflicting evidence for geomyoid affinities. Fetal membrane characters of geomyoids are most similar to those of dipodoids, and they appear to represent an intermediate condition in the derivation of the highly advanced muroid pattern (Luckett). Nevertheless, none of the shared derived similarities is unique to a "myomorph' clade; instead, it is the totality of their developmental pattern that is indicative of relationships. In contrast, the albumin data do not appear to support inclusion of Geomyoidea in a "myomorph" clade (Sarich). Their albumins are about as distant from the muroid-dipodoid clade as is the aplodontid-sciurid clade.

Fahlbusch emphasizes that sciuromorphy is shared by geomyoids, sciurids, and castorids, and that, in the absence of convincing evidence for the myomorph affinities of geomyoids, it might be best to retain geomyoids in the suborder Sciuromorpha. It is difficult to refute (or further corroborate) this hypothesis at present.

In contrast to the ambiguities of higher taxon relationships for geomyoids, there is universal agreement, based on both dental and cranial evidence in fossil and extant groups (Fahlbusch; Wahlert;

Vianey-Liaud), and on soft anatomical, molecular, and fetal membrane data in the extant families (George; Sarich; Luckett), that Geomyidae, Heteromyidae, and the fossil family Eomyidae form a monophyletic clade Geomyoidea. The evidence also indicates that Eomyidae is the sister group (and probably includes the ancestral stock) of the heteromyid-geomyid clade.

Affinities of Sciuridae, Aplodontidae, and Castoridae

Vianey-Liaud has reviewed the dental and cranial fossil evidence which suggests that Oligocene-Recent sciurids and aplodontids shared a common ancestor, and that they should be considered as sister groups. This hypothesis is corroborated by analyses of derived traits of the cranial foramina (Wahlert) and middle ear (Lavocat and Parent). Some important differences are detected in the carotid circulation to the middle ear of these families, in particular, the obliteration of the stapedial artery in *Aplodontia*, but its retention in sciurids (Bugge). However, this vessel is apparently also present in primitive aplodontids, such as prosciurines (Wahlert, 1974), and thus it would have been present in the aplodontid morphotype.

The albumin immunological data also support a sister-group relationship between sciurids and aplodontids, but sequence data are not yet available to test this hypothesis. The fetal membrane developmental pattern is virtually identical in the two families, but this is believed to reflect retention of the primitive rodent condition (Luckett). As such, this does not provide special evidence for corroborating a sister-group relationship between the two families, but it is consistent with such a hypothesis.

The affinities of the sciuromorphous and pentalophodont Castoroidea to other rodents remain obscure. Wahlert and earlier authors have considered sciuromorphy to be derived independently in castorids, sciurids, and geomyoids, although the reasons for this have not been adequately discussed. Hartenberger considers the castoroids and sciurids to be a monophyletic assemblage that share the derived feature of sciuromorphy, and aplodontids were also included in this clade, because of the craniodental evidence linking them with sciurids. Bugge emphasizes that castorids differ from sciurids by lacking the stapedial artery; however, this vessel does occur in some fossil castoroids (Wahlert, 1977). Fetal membrane evidence contributes little to understanding castorid affinities; they share numerous primitive retentions with sciurids, and also exhibit some autapomorphous traits (Luckett).

Wilson (1949) suggested that the Aplodontidae, Sciuridae, and Castoroidea were derived directly from Eocene paramyines, and he included the three groups in his suborder Sciuromorpha. If castorids are not related cladistically to the aplodontid-sciurid group, then there is no clear evidence relating them to other rodent higher taxa.

Affinities of the Hystricomorphous Families Anomaluridae and Pedetidae

It is generally agreed that the African families Anomaluridae and Pedetidae are not closely related to the Hystricognathi (Wood), but their affinities with other rodents or with each other remain unclear. As noted by Wood, these two families and Ctenodactylidae have often been considered to be related, in part because of their shared hystricomorphy, sciurognathy, and recent restriction to Africa. Anomalurid teeth have recently been identified in the late Eocene of North Africa by Jaeger et al., and they have discussed some derived similarities of these teeth shared with those of the Eocene-Oligocene European theridomyids. As noted by Hartenberger and by Wood, a theridomyid-anomalurid relationship remains possible, based on shared dental and cranial similarities, but this requires further testing. Middle ear features of theridomyids approximate the primitive rodent condition (Lavocat and Parent), and therefore can not corroborate theridomyid-anomalurid affinities.

Pedetids occurred in Africa, Greece, and Turkey during the Miocene (Wood), but they are presently confined to Africa. The bilobed cheek tooth pattern of pedetids is highly derived and difficult to compare with other rodents. Wood has noted the similarities in the closely appresses malleus and incus, hystricomorphy, and multiserial incisor enamel between pedetids and hystricognaths, but he considers all of these to be parallelisms, rather than indicative of a phyletic relationship between the two taxa. De Jong notes that *Pedetes* groups with *Cavia*, rather than with the muroid clade, on the basis of α-crystallin data. However, this is based on a single amino acid position difference, and De Jong acknowledges that this may not be phylogenetically significant. Sarich found that *Pedetes* clusters with neither the hystricognath nor the sciurognath-myomorph clades, on the basis of albumin data.

A number of derived middle ear features are shared by Recent pedetids and anomalurids (Lavocat and Parent), including a virtually identical pattern of carotid arterial branches and anastomoses (Bugge). These data could support a grouping of the two families in a superfamily Anomaluroidea, as noted by Bugge, although this hypothesis would appear to be in conflict with a hypothesis of theridomyid-anomalurid affinities. George observes that pedetids and anomalurids do not associate readily with either a sciuromorph-myomorph clade or a ctenodactylid-hystricognath clade, based on her analysis of numerous biological traits in Recent rodents. Fetal membrane developmental traits are not very useful for assessing anomalurid-pedetid relationships. Their shared similarities with each other and with other rodents are due primarily to retention of shared primitive features, as well as the occurrence of a few autapomorphies in each family (Luckett).

All of this leaves us with the conclusion that the affinities of the families Anomaluridae and Pedetidae are among the most obscur of all rodents. The time-worn appeal for better and earlier fossil remains, as well as the possibility for obtaining more immunological and amino acid sequence data, offer future hope for the clarificatio of the phylogenetic relationships of these families.

European Theridomyidae and Their Possible Relationships

The origin of the middle Eocene-late Oligocene Theridomyidae remains unclear, as noted by Vianey-Liaud. Although they are usuall thought to be derived from European paramyids (Wood), Vianey-Liaud has suggested an alternative origin from Asiatic (ctenodactyloid) immigrants. This appears to be based on the sharing of hystricomorphy by middle Eocene ctenodactyloids and theridomyids. Analysis of dental and additional cranial features remains to be done to test this hypothesis, although the same can be said for the hypothesis of paramyid ancestry. Despite the fact that the theridomyids were the dominant Eocene rodents of Europe, it is unclear whether any later rodents, with the possible exception of anomalurids, were derived from them.

Middle ear morphology of theridomyids is considered by Lavocat and Parent to reflect the primitive rodent condition, including the probable retention of three major branches of the internal carotid artery. This tripartite arterial supply to the middle ear has long been considered as the primitive eutherian pattern, which was modified to varying degrees during later mammalian evolution. However, careful embryological and comparative anatomical studies indicate that there is no unequivocal evidence for the occurrence of *both* a medial and promontorial branch of the internal carotid artery in any mammal, and embryological studies indicate that the "medial" and "promontory" branches are actually the same vessel, whose relative position may become "shifted" during the fetal period (see the excellent discussion by MacPhee, 1981). Consequently, one of us (WPL) believes that the apparent occurrence of a promontory artery in theridomyids, some paramyids, and *Sciuravus* is a derived trait within Rodentia, probably associated with the "absence" of a medial internal carotid branch. If true, this has important implications when considering hypotheses of affinities between these taxa and later families of rodents.

Relationships among Eocene Rodents

Relationships among Eocene rodents of Europe, North America, and Asia have been discussed by Vianey-Liaud in this volume, and additional comments on North American Eocene rodents and on Theridomyidae were presented by Wood. We wish to add only a few comments here to our previous discussions. The early dichotomy of North

American rodents, supported previously by Wilson (1949) on dental evidence, still appears to be a viable hypothesis, when both dental and cranial evidence is considered. Wilson (1949) suggested that the more primitive paramyids retained numerous primitive features in their dentition, and data on cranial (Wahlert) and middle ear features (Lavocat and Parent) generally support this suggestion. Sciuravids form a more derived group on the basis of their molar morphology (Butler), whereas they still retain many primitive cranial features (Wahlert, 1974, this volume), with the possible exception of the middle ear features noted above.

Considerable difference of opinion exists among North American paleontologists concerning the subordinal and familial groupings of Eocene rodents (see Wood, 1980; Korth, 1984; Black and Sutton, 1984; Dawson and Krishtalka, 1984). In particular, there is disagreement on the distinction between "franimorphs" and other Eocene protrogomorphs, on the familial or subfamilial separation of paramyids and ischyromyids, and on the ancestor-descendant and cladistic relationships among the recognized families and subfamilies. We believe that this is due in great part to the lack of integration of cranial and dental features in assessing evolutionary relationships, to the failure to reconstruct an ancestral morphotypic condition for different dental and cranial features, and to the general lack of character analyses to test either anagenetic or cladistic relationships. As an example, cranial foramina features discussed by Wahlert (1974, this volume) provide better evidence for the familial separation of ischyromyids and paramyids than does dental morphology, in agreement with Wood (1980).

Evolution occurs in a mosaic pattern, and it is incorrect (and counterproductive) to designate particular taxa as "primitive" or "advanced." Instead, any taxon may contain a mosaic of both primitive and derived characters, as evidenced by the dental and middle ear features of sciurids. This is particularly important for the growing debate over whether paramyids or cocomyids from the early Eocene are more "primitive." Both groups are protrogomorphous (surely a primitive condition), but there are differences in the structure of the premolars and molars (Korth, 1984; Dawson et al., 1984; Hartenberger, this volume). As cautioned by Butler, dental traits of early rodents should be compared with those of early mammals when assessing relatively primitive conditions (= morphotypes). Thus, the enlarged molar hypocone and hypoconulid are derived eutherian features that occur in cocomyids, whereas a more primitive eutherian condition for these features is found in most paramyids. If, as seems likely, cocomyids and paramyids had a common (Asiatic?) rodent ancestor, its morphotype may have included some dental and cranial characters of both families (including the premolariform P4 of cocomyids). In any case, hypotheses of ancestral-descendant and cladistic relationships should be based on character analyses of the available data, rather than on personal opinions, and it is important to consider the likely primitive conditions of dental and cranial features in other Paleocene-Eocene eutherians when attempting to reconstruct the ancestral rodent morphotype.

Analysis of Phyletic Relationships at the Species and Generic Level

The focus of this volume (and the symposium) was on analysis of evolutionary relationships above the family level. Nevertheless, we recognize the need to assess both anagenetic and cladistic relationships among rodents at lower levels, and that similar methods of character analysis apply for such analyses. Certain types of studies, such as those on chromosomal and electrophoretic data, may not be very useful for assessing higher taxon phylogeny, because of the degree of variation encountered and the resulting difficulties for determining homology. On the other hand, these data can be very helpful in assessing relationships at the species and generic level.

As emphasized by Capanna, diploid chromosomal numbers can vary considerably within rodent families, genera, and even species, and karyotype rearrangement has been a common phenomenon during rodent evolution. An analysis of the probable evolutionary pathways of these rearrangements provides a valuable tool for assessing cladistic relationships among closely related species and genera, as illustrated by Capanna for the genus *Mus*.

Bonhomme et al. presented phenetic and cladistic analyses of protein electrophoretic characters (= electromorphs) in muroids and other selected rodents. Electromorph analyses consistently clustered the genera correctly at the familial and superfamilial levels, whereas there was a considerable lack of resolving power for distinguishing affinities among closely related species and genera of muroids. One possible explanation offered for this is the probable antiquity of the splitting events and subsequent reticulate evolution of electromorphs. In contrast, Woods found that the more slowly evolving electrophoretic loci were useful for testing and corroborating hypotheses of generic relationships among caviomorphs that were based on myological data. Thus, the choice of the electromorphs to be studied may have a profound effect on their value for evolutionary analyses.

Chaline has discussed the methodology for tracing evolutionary lineages of species in the fossil record, based on the stratigraphic position and morphology of abundant specimens of the Eurasian arvicolid genus *Lagurus* during the Pleistocene. Such well-documented lineages provide an opportunity for following the pattern and rate of morphological changes in the dentition among closely related species, as well as the possible influence of climatic changes on this evolutionary pattern.

SOME GENERAL PROBLEMS OF CHARACTER ANALYSIS

Parallelism and Convergence During Rodent Evolution

The concept of parallelism and convergence as widespread phenomena during rodent evolution is a hypothesis that needs to be tested by biological comparisons of similar structures, rather than an opinion to be accepted without question, no matter how authoritative the source. There has been a tendency to believe that, because characters such as molar lophs and crests have evolved many times in unrelated lineages during rodent evolution, convergence or parallelism is also extensive for other dental, skeletal, and soft anatomical features. It may be true, but, unfortunately, this hypothesis has rarely been tested by a careful analysis of the structural, developmental, or anagenetic evidence for similarities and differences. Geographic separation of taxa is not a valid test of the homology or homoplasy of their shared similarities, although this has been stated or implied as an argument for parallelism between South American and African hystricognaths.

Examples of anagenetic and developmental studies that provide tests for a hypothesis of homology or convergence are presented in this volume, but they remain relatively scarce. Paleontological studies can provide evidence for the modification of a character in a sequence of supposed ancestral-descendant lineages in two or more taxa, and such sequences can be used to test the possible homology or convergence of shared similarities. Thus, Vianey-Liaud was able to demonstrate the probable convergent evolution of myomorphy in fossil glirids and muroids by identifying different ancestral conditions (protrogomorphy and hystricomorphy, respectively) for the two taxa. This hypothesis could be further tested by studying the fetal development of the skull in extant glirids and muroids, in order to determine whether different ontogenetic pathways occur during the development of "myomorphy." Unfortunately, few studies on skull development have been undertaken in rodents. Moreover, these are usually limited to one or a few developmental "stages," and the developmental significance of such phylogenetically important features as hystricomorphy, myomorphy, hystricognathy, or a perforated pterygoid fossa has not been discussed. This is due in great part to the lack of knowledge or interest in evolutionary problems by most developmental biologists.

Developmental studies are particularly valuable for assessing the probable homology or convergence of shared similarities, because different ontogenetic pathways are commonly followed by convergently similar characters or character complexes. On the contrary, the virtually identical ontogenetic pattern of the normal and abnormal deciduous incisors in rodents and lagomorphs suggests their probable homology, as discussed by Luckett. Developmental studies are also useful for assessing the probable homology of major and accessory

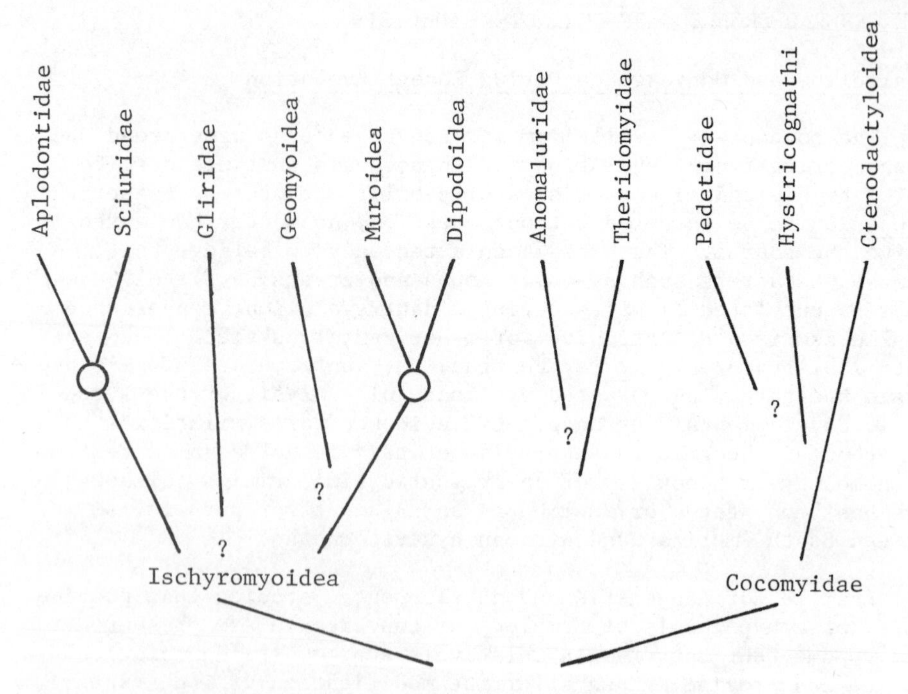

Fig. 2. Working hypothesis of possible evolutionary relationships among major rodent taxa. See the text for discussion of the uncertainties surrounding this hypothesis.

cusps in muroid molars. Unfortunately, paleontologists are often as unaware of such developmental studies and their significance as developmental biologists are of the possible evolutionary significance of their findings!

No single method of character analysis, including comparative anatomy, embryology, stratigraphic position, or functional analysis, is completely reliable or infallible for assessing homology or character state polarity. Consequently, as many different approaches as possible should be used when evaluating evolutionary affinities.

PRESENT AND FUTURE ANALYSES OF RODENT PHYLOGENY

The multidisciplinary analyses of the present volume provide a great deal of comparative data for testing hypotheses of evolutionary relationships among rodents. Our working hypothesis of rodent affinities is summarized in Figure 2. As noted earlier, many of these suggested relationships are weakly founded, and we hope that this working hypothesis will stimulate further analyses of cladistic and ancestral-descendant relationships among rodents.

Suggested Areas of Future Research

We present here our personal views on some important areas for future investigations of rodent phylogeny; obviously, other participants in the volume may not agree with some of our suggestions.

Cranial Morphology and Development. Both comparative and developmental studies of the skull have contributed to an understanding of the form-functional relationships among skeletal elements, nerves, vessels, muscles, and sensory organs in other extant mammals, but such studies are relatively rare for rodents, with the exception of the mouse and rat. These studies are essential as an aid to interpreting the evolutionary significance of skull relationships and foramina in fossil rodents. The study of fossil rodent skulls by Wahlert should be expanded to include fossil and extant ctenodactyloids and hystricognaths, especially for reconstructing the rodent morphotype for skull characters, and for testing hypotheses of relationships between ctenodactyloids and other rodents. Functional and developmental analyses could provide valuable information concerning the possible homology or convergence of hystricomorphy in dipodoids, ctenodactylids, and hystricognaths. An assessment of the adaptive effects of fossorial life on skull morphology could be accomplished by developmental studies on spalacid and bathyergid skulls, in order to supplement the excellent report by Boller (1970) on the functional morphology of the bathyergid skull. Developmental studies might also provide further insight into the interrelationships between bones and muscles of the skull during the differentiation of hystricognathy, myomorphy, and sciuromorphy.

Postcranial Morphology. As demonstrated by Szalay in this volume and elsewhere, postcranial morphology provides an extensive suite of characters that can be used to test hypotheses of mammalian phylogeny. Many fossil postcranial remains of rodents, however, remain undescribed, and efforts are only now beginning for examining these features from a functional perspective in extant rodents. It will be extremely valuable to compare the undescribed postcranial remains of fossil eurymyloids with those of early rodents, to complete the preliminary investigation initiated by Li and Ting.

Incisor Enamel Structure. Based on the papers presented in this volume by Sahni and by Koenigswald, there remains a lack of agreement about the definition of multiserial and pauciserial enamel patterns, and this has a negative effect on attempts to reconstruct the rodent morphotype (see the differing opinions of Sahni, Koenigswald, and Wood). There is agreement that the uniserial enamel pattern is highly derived, but it is unclear whether this derived feature is homologous or convergent in some rodent taxa (see Koenigswald, Flynn et al.). The separation between pauciserial and multiserial enamels may be less distinct than previously believed, and therefore the value of this feature in assessment of higher level rodent systematics may have been overestimated in recent years. This potentially valuable area of research needs careful study and a refinement of the definitions for different types of enamel patterns, before it can be used in a meaningful way to clarify phylogenetic relationships of rodents

Molecular Evolution of Proteins. Collection and analysis of protein molecular data from rodents lag far behind similar studies on primates and artiodactyls. There are extensive gaps in our knowledge of amino acid sequence data from most rodent families for most proteins studied. In particular, sequence data are required for the sciurids, aplodontids, geomyoids, dipodoids, ctenodactylids, and bathyergids. In the absence of such sequence data, it is difficult to interpret the amount of intraordinal variation that occurs in Rodentia, and this may unduly bias the assessment of superordinal affinities of rodents with other eutherians. Immunological studies of serum proteins in rodents are more extensive, thanks to the work of Sarich, but a number of systematically significant families, including Anomaluridae and Ctenodactylidae, remain unstudied. Both immunological and amino acid sequence analysis could provide valuabl tests for the weakly formulated hypotheses of affinities for glirids rhizomyids, spalacids, ctenodactylids, pedetids, and anomalurids.

Fetal Membrane Development. Developmental features of the feta membranes and placenta are very conservative at the family level among rodents and other eutherians, and this makes them valuable for testing hypotheses of suprafamilial and interordinal relationships. Developmental data are lacking for the families Gliridae, Spalacidae and Rhizomyidae, and the earliest stages of anomalurids are unknown. Investigation of each of these groups would supplement other analyse of the higher taxon affinities of these families.

Structural, Functional, and Developmental Studies of Dentition. The dentition will doubtlessly remain as the major source of investigation for studies of rodent phylogeny, because this is the only typ of data that can be compared in all fossil and extant taxa. Nevertheless, overviews of the structure and variability of dental remain from most fossil rodent families are notably lacking, and the morphotypic condition for most dental traits at the family level is poorly known. The extent of data available in Wood's (1962) monograph on Paramyidae can serve as a model for analyses of other families. Lack of dental morphotype reconstruction has hindered the analysis of ancestral-descendant and cladistic relationships among most Eocene-Oligocene fossil families. The addition of functional data that can be gleaned from a study of molar wear facets, as discussed by Butler, can aid in assessing the probability of homology or convergence in molar similarities among fossil and extant rodents. Developmental studies of the molar crown might provide additional insight into possible homologies of primary cusps and crests in lagomorphs and rodents, and studies of premolar development can help clarify the deciduous or successional nature of these teeth in thryonomyoids, echimyids, and pedetids.

Paleontology and the Stratigraphic Framework

The recent discoveries and descriptions of additional fossil rodents from the Eocene of Asia and Africa, and eurymyloids from the Paleocene-Eocene of China (Hussain et al., 1978; Dawson et al., 1984; Li and Ting, this volume; Jaeger et al., this volume), have contributed significantly to a reexamination and restructuring of some of the hypotheses discussed in this volume and symposium. One of the major difficulties encountered when comparing paleontological findings from Asia, North America, Europe, and other continents is the frequent lack of stratigraphic correlations among these continents, as discussed by Vianey-Liaud. For example, different dates are proposed for the Eocene-Oligocene boundary by Vianey-Liaud and Fahlbusch, and their is a considerable difference of opinion between Wood and Woods about the dates of the early Oligocene in South America. Such imprecise dates can effect hypotheses of the transoceanic migration of rodents into South America and Africa, regardless of the source of these immigrants. Paleobiogeographic considerations have also effected some of the hypotheses discussed in this volume. Thus, Jaeger et al. suggested the possibility that Asiatic Chapattimyidae and Hystricidae are sister groups, although they acknowledge that this hypothesis is based only on paleogeographic considerations, and not on shared derived characters. In contrast, the morphological data (see Fig. 1) suggest that Hystricidae are more closely related to other hystricognaths than they are to chapattimyids (and other Ctenodactyloidea). We reaffirm our belief that analysis of biological characters is more meaningful for evaluating phylogenetic relationships than is the paleobiogeographic distribution of taxa.

CONCLUSION

We believe that the multidisciplinary analyses of a wide range of biological attributes in this volume will stimulate a renewed interest in the reconstruction of rodent phylogeny. Clearly, all kinds of biological data, including many not considered in this volume, can be useful in assessing evolutionary relationships at some level. Therefore, we do not accept *a priori* the superiority of one type of data or character complex over the other. There are limitations for the evolutionary analysis of both paleontological and neontological data, and uncertainties always surround the assessment of homology and character state polarity. Most hypotheses of rodent phylogeny are inadequately corroborated, and all of the ones evaluated in this volume can benefit from further testing. We hope that the exchange of ideas between paleontologists and neontologists in this volume and at the symposium will lead to further collaborative efforts among scientists with a common interest in evolutionary analysis of rodents and other mammals.

REFERENCES

Black, C. C. and Sutton, J. F. 1984. Paleocene and Eocene rodents of North America. Carneg. Mus. Nat. Hist. Spec. Publ. 9: 67-84.

Boller, N. 1970. Untersuchungen an Schädel, Kaumuskulatur und äusserer Hirnform von Cryptomys hottentotus (Rodentia, Bathyergidae). Z. wiss. Zool. 181: 7-65.

Brownell, E. 1983. DNA/DNA hybridization studies of muroid rodents: Symmetry and rates of molecular evolution. Evolution 37: 1034-1051.

Carleton, M. D. 1984. Introduction to rodents. In: Orders and Families of Recent Mammals of the World, S. Anderson and J. K. Jones, Jr., ed., pp. 255-265, John Wiley and Sons, N. Y.

Dawson, M. R. 1970. Paleontology and geology of the Badwater Creek area, central Wyoming. Part 6. The leporid Mytonolagus (Mammalia, Lagomorpha). Ann. Carneg. Mus. 41: 215-230.

Dawson, M. R. 1977. Late Eocene rodent radiations: North America, Europe and Asia. Géobios Mém. Spéc. 1: 195-209.

Dawson, M. R. and Krishtalka, L. 1984. Fossil history of the families of Recent mammals. In: Orders and Families of Recent Mammals of the World, S. Anderson and J. K. Jones, Jr., eds., pp. 11-57, John Wiley and Son, New York.

Dawson, M. R., Li, C.-K. and Qi, T. 1984. Eocene ctenodactyloid rodents (Mammalia) of Eastern and Central Asia. Carneg. Mus. Nat. Hist. Spec. Publ. 9: 138-150.

Flynn, L. J. 1982. Systematic revision of Siwalik Rhizomyidae (Rodentia). Geobios 15: 327-389.

Gidley, J. W. 1912. The lagomorphs an independent order. Science 36: 285-286.

Gregory, W. K. 1910. The orders of mammals. Bull. Amer. Mus. Nat. Hist. 27: 1-524.

Hartenberger, J.-L. 1980. Donnees et hypotheses sur la radiation initiale des rongeurs. Palaeovertebrata Mém. Jubil. R. Lavocat: 285-301.

Hussain, S. T., De Bruijn, H., and Leinders, J. M. 1978. Middle Eocene rodents from the Kala Chitta Range (Punjab, Pakistan). Proc. Kon. Ned. Akad. Wetensch., Amsterdam, Ser. B. 81: 74-112.

Korth, W. W. 1984. Early Tertiary evolution and radiation of rodents in North America. Bull. Carneg. Mus. Nat. Hist. 24: 1-71.

Lavocat, R. 1973. Les Rongeurs du Miocène d'Afrique orientale. I. Miocène inférieur. Mém. Trav. E.P.H.E., Inst. Montpellier 1: 1-284.

Li, C.-K. 1965. Eocene leporids of North China. Vert. PalAsiat. 9: 23-36.

Li, C.-K. 1977. Paleocene eurymyloids (Anagalida, Mammalia) of Qianshan, Anhui. Vert. PalAsiat. 15: 103-118.

MacPhee, R. D. E. 1981. Auditory regions of primates and eutherian insectivores. Contrib. Primat. 18: 1-282.

McKenna, M. C. 1969. The origin and early differentiation of therian mammals. Ann. N. Y. Acad. Sci. 167: 217-240.
Novacek, M. J. 1977. Evolution and Relationships of the Leptictidae (Eutheria: Mammalia). Unpublished Ph. D. Thesis, Univ. of California, Berkeley.
Novacek, M. J. 1980. Cranioskeletal features in tupaiids and selected Eutheria as phylogenetic evidence. In: Comparative Biology and Evolutionary Relationships of Tree Shrews, W. P. Luckett, ed., pp. 35-93, Plenum Press, N. Y.
Simpson, G. G. 1945. The principles of classification and a classification of mammals. Bull. Amer. Mus. Nat. Hist. 85: 1-350.
Simpson, G. G. 1959. The nature and origin of supraspecific taxa. Cold Spring Harbor Symp. Quant. Biol. 24: 255-271.
Szalay, F. S. 1977. Ancestors, descendants, sister groups and testing of phylogenetic hypotheses. Syst. Zool. 26: 12-18.
Tullberg, T. 1899. Ueber das System der Nagetiere: eine phylogenetische Studie. Nova Acta Reg. Soc. Scient. Upsala, ser. 3, 18: 1-514.
Van Valen, L. 1971. Adaptive zones and the orders of mammals. Evolution 25: 420-428.
Wahlert, J. H. 1974. The cranial foramina of protrogomorphous rodents; an anatomical and phylogenetic study. Bull. Mus. Comp. Zool. 146: 363-410.
Wahlert, J. H. 1977. Cranial foramina and relationships of _Eutypomys_ (Rodentia, Eutypomyidae). Amer. Mus. Novit. 2626: 1-8.
Wahlert, J. H. 1978. Cranial foramina and relationships of the Eomyoidea (Rodentia, Geomorpha). Skull and upper teeth of _Kansasimys_. Amer. Mus. Novit. 2645: 1-16.
Weber, M. 1904. Die Säugetiere. Gustav Fischer, Jena.
Wilson, R. W. 1949. Early Tertiary rodents of North America. Carneg. Inst. Wash. Publ. 584: 67-164.
Wood, A. E. 1955. A revised classification of the rodents. J. Mammal. 36: 165-187.
Wood, A. E. 1957. What, if anything, is a rabbit? Evolution 11: 417-425.
Wood, A. E. 1959. Eocene radiation and phylogeny of the rodents. Evolution 13: 354-361.
Wood, A. E. 1962. The early Tertiary rodents of the Family Paramyidae. Trans. Amer. Phil. Soc. 52: 1-261.
Wood, A. E. 1965. Grades and clades among rodents. Evolution 19: 115-130.
Wood, A. E. 1968. Early Cenozoic mammalian faunas, Fayum Province, Egypt. Part II. The African Oligocene Rodentia. Bull. Peabody Mus. Nat. Hist. 28: 23-105.
Wood, A. E. 1974a. The evolution of the Old World and New World hystricomorphs. Symp. Zool. Soc. Lond. 34: 21-60.
Wood, A. E. 1974b. Early Tertiary vertebrate faunas, Vieja Group, Trans-Pecos Texas: Rodentia. Tex. Mem. Mus. Bull. 21: 1-112.

Wood, A. E. 1975. The problem of the hystricognathous rodents. Univ. Mich., Papers Paleontol. 12: 75-80.

Wood, A. E. 1980. The Oligocene rodents of North America. Trans. Amer. Phil. Soc. 70: 1-68.

Wood, A. E. 1984. Hystricognathy in the North American Oligocene rodent Cylindrodon and the origin of the Caviomorpha. Carneg. Mus. Nat. Hist. Spec. Publ. 9: 151-160.

CONTRIBUTORS

Jaap J. Beintema
Biochemisch Laboratorium
Rijksuniversiteit Groningen
Nijenborgh 16
9747 AG Groningen
The Netherlands

Francois Bonhomme
Institut des Sciences de l'Evolution, U.S.T.L.
Place Eugène Bataillon
34060 Montpellier
France

Gerhard Braunitzer
Max-Planck-Institut für Biochemie
Abteilung Proteinchemie
Martinsried bei München
West Germany

Jørgen Bugge
Department of Anatomy
Royal Dental College
DK-8000 Aarhus C
Denmark

P. M. Butler
Department of Zoology
Royal Holloway College
Egham, Surrey TW20 0EX
England

Ernesto Capanna
Dipartimento di Biologia Animale e dell'Uomo
Università di Roma
00161 Roma, Italy

Jean Chaline
Centre de Géodynamique sédimentaire et Evolution géobiologique
U. A. CNRS 157
Laboratoire de Préhistoire et Paléoécologie du Quaternaire
Institut des Sciences de la Terre
21100 Dijon, France

Brigitte Coiffait
Laboratoire de Géologie des Ensembles sédimentaires
Université Nancy, I. B. P. 239
54506 Vandoeuvre-Les-Nancy
France

John Czelusniak
Department of Biological Sciences
Wayne State University
Detroit, Michigan 48202, U.S.A.

Wilfried W. de Jong
Laboratorium voor Biochemie
Universiteit van Nijmegen
Geert Grooteplein Noord 21
6525 EZ Nijmegen
The Netherlands

Christiane Denys
Laboratoire de Paléontologie des Vertébrés
Université Pierre et Marie Curie (Paris VI)
U. A. 720 du CNRS
4, Place Jussieu
F-75230 Paris, France

Volker Fahlbusch
Institut für Paläontologie
und historische Geologie
der Universität München
Richard-Wagner-Strasse 10
D-8000 München 2
West Germany

Lawrence J. Flynn
Department of Vertebrate
Paleontology
American Museum of
Natural History
New York, New York 10024
U.S.A.

Wilma George
Department of Zoology
University of Oxford
South Parks Road
Oxford OX1 3PS, England

Morris Goodman
Department of Anatomy
Wayne State University
School of Medicine
Detroit, Michigan 48201
U.S.A.

Jean-Louis Hartenberger
Institut des Sciences
de l'Evolution
LA 327, U.S.T.L.
Place Eugène Bataillon
34060 Montpellier, France

John W. Hermanson
Department of Biology
Emory University
Atlanta, Georgia 30322
U.S.A.

Djoko Iskandar
Institut des Sciences de
l'Evolution, U.S.T.L.
Place Eugène Bataillon
34060 Montpellier, France

Louis L. Jacobs
Department of Geological Sciences
Southern Methodist University
Dallas, Texas 75275
U.S.A.

Jean-Jacques Jaeger
Laboratoire de Paléontologie
des Vertébrés
Université Pierre et Marie
Curie (Paris VI)
U. A. 720 du CNRS, 4 Place Jussieu
F-75230 Paris, France

Wighart von Koenigswald
Hessisches Landesmuseum
Geologisch-Paläontologische
und Mineralogische Abteilung
Friedensplatz 1
D-6100 Darmstadt, West Germany

René Lavocat
Ecole Pratique des Hautes
Etudes, U.S.T.L.
34060 Montpellier, France

Chuan-Kuei Li
Institute of Vertebrate Paleontolog
and Paleoanthropology
Academia Sinica
P. O. Box 643
Beijing 28
People's Republic of China

Everett H. Lindsay
Department of Geosciences
University of Arizona
Tucson, Arizona 85721, U.S.A.

Nieves Lopez Martinez
Departamento de Paleontologia
Facultad de Ciencias Geologicas
Universidad Complutense de Madrid
Madrid 3, Spain

W. Patrick Luckett
Department of Anatomy
University of Puerto Rico
School of Medicine
San Juan, Puerto Rico, 00936, U.S.A

Michael J. Novacek
Department of Vertebrate Paleontology
American Museum of Natural History
New York, New York 10024
U.S.A.

Jean-Pierre Parent
Institut Supérieur d'Agriculture
13, rue de Toul
59046 Lille, France

Francis Petter
Muséum National d'Histoire Naturelle
55, rue Buffon
75005 Paris, France

Ashok Sahni
Centre of Advanced Study in Geology
Panjab University
Chandigarh - 160014, India

Vincent M. Sarich
Departments of Anthropology and Biochemistry
University of California
Berkeley, California 94720
U.S.A.

Jeheskel Shoshani
Department of Biological Sciences
Wayne State University
Detroit, Michigan 48202
U.S.A.

Frederick S. Szalay
Department of Anthropology
Hunter College, City University of New York
New York, New York 10021
U.S.A.

Louis Thaler
Institut des Sciences de l'Evolution, U.S.T.L.
Place Eugène Bataillon
34060 Montpellier, France

Su-Yin Ting
Institute of Vertebrate Paleontology and Paleoanthropology
Academia Sinica
P. O. Box 643
Beijing 28
People's Republic of China

Monique Vianey-Liaud
Institut des Sciences de l'Evolution, U.S.T.L.
Place Eugène Bataillon
34060 Montpellier, France

John H. Wahlert
Department of Vertebrate Paleontology
American Museum of Natural History
New York, New York 10024, U.S.A.

Albert E. Wood
20 Hereford Avenue
Cape May Court House
New Jersey 08210, U.S.A.

Charles A. Woods
Department of Natural Sciences
Florida State Museum
University of Florida
Gainesville, Florida 32611
U.S.A.

INDEX